Atlas of NUCLEAR MEDICINE

Edited by

Douglas Van Nostrand, M.D., F.A.C.P.

Director, Nuclear Medicine Service
Department of Radiology
Walter Reed Army Medical Center;
Associate Professor of Clinical Radiology and Nuclear Medicine
Uniform Services University of Health Sciences
Washington, D.C.

Sheldon Baum, M.D.

Chief, Division of Nuclear Medicine
Lankenau Hospital;
Clinical Professor of Nuclear Medicine
Jefferson Medical College
Philadelphia, Pennsylvania

Atlas of

31 Contributors

J. B. Lippincott Company *Philadelphia*

London Mexico City New York St. Louis São Paulo Sydney

NUCLEAR MEDICINE

Acquisition Editor: Susan Gay
Sponsoring Editor: Delois Patterson
Production Editor: Brenda Lee Reed
Manuscript Editor: Catherine Parry
Indexer: Catherine Battaglia
Design Coordinators: Anita Curry
Susan Hess Blaker
Designer: Tracy Baldwin
Cover Design: Kevin Curry
Production Manager: Kathleen P. Dunn
Production Coordinator: Ken Neimeister
Compositor: Progressive Typographers
Printer/Binder: The Maple Press Company
Color Insert: Imago Sales (USA), Inc.

The opinions or assertions contained herein are the private views of the authors/editors and are not to be construed as official or as reflecting the views of the Uniformed Services University of Health Sciences, United States Army, or the Department of Defense.

6 5 4 3 2 1

Library of Congress Cataloging-in-Publication Data

Atlas of nuclear medicine / edited by Douglas Van Nostrand, Sheldon Baum : 31 contributors.
p. cm.
Includes index.
ISBN 0-397-50789-5
1. Radioisotope scanning--Atlases. I. Van Nostrand, Douglas. II. Baum, Sheldon, 1927–
[DNLM: 1. Nuclear Medicine--atlases. 2. Radionuclide Imaging--atlases. WN 17 A8853]
RC78.7.R4A82 1988 87-24861
616.07′57--dc 19 CIP

The authors, the editors, and publisher have exerted every effort to ensure that drug selection and dosage set forth in this text are in accord with current recommendations and practice at the time of publication. However, in view of ongoing research, changes in government regulations, and the constant flow of information relating to drug therapy and drug reactions, the reader is urged to check the package insert for each drug for any change in indications and dosage and for added warnings and precautions. This is particularly important when the recommended agent is a new or infrequently employed drug.

To Halstead K. and Elizabeth M. Van Nostrand,
for laying the foundation.

D.V.N.

To my wife, Maureen.

S.B.

Contributors

Sue H. Abreu, M.D.
Nuclear Medicine Service
Department of Radiology
Walter Reed Army Medical Center
Washington, D.C.; Assistant Professor of Radiology/Nuclear Medicine
Uniformed Services University of the Health Sciences
Bethesda, Maryland

Frank Atkins, Ph.D.
Nuclear Medicine Service
Department of Radiology
Uniformed Services University of Health Sciences
Walter Reed Army Medical Center
Washington, D.C.

Robert M. Basarab, M.D.
Section Chief
Division of Nuclear Medicine
Lancaster General Hospital
Lancaster, Pennsylvania

Gilles Beauchamp, M.D.
Assistant Professor of Surgery
University of Montreal
Department of Thoracic Surgery
Hôpital Hotel-Dieu de Montreal
Montreal, Canada

Francois Begon, M.D., Ph.D.
Full Professor of Biophysics and Nuclear Medicine
Chief of Nuclear Medicine
Université de Poitiers
C.H.U. de Poitiers
Hôpital Jean Bernard
Poitiers, France

Peter W. Blue, COL, M.D., MC
Chief, Nuclear Medicine Service
Fitzsimmons Army Medical Center
Aurora, Colorado

Tapan K. Chaudhuri, M.D.
Professor of Radiology (Nuclear Medicine) and Chief
Nuclear Medicine Service
Eastern Virginia Medical School and Veterans Administration Medical Center
Hampton, Virginia

B. David Collier, M.D.
Associate Professor of Radiology and Orthopaedics Surgery
Director of Nuclear Medicine
Medical College of Wisconsin
Milwaukee, Wisconsin

Vilma Derbekyan, M.D.
Assistant Professor
Department of Radiology
McGill University;
Assistant Physician
Department of Nuclear Medicine
Royal Victoria Hospital
Montreal, Canada

André Duranceau, M.D.
Professor of Thoracic Surgery
Department of Thoracic Surgery
Hôpital Hotel-Dieu de Montreal
Montreal, Canada

Douglas F. Eggli, M.D.
Division of Nuclear Medicine
Department of Radiology
Georgetown University Hospital;
Nuclear Medicine Service
Department of Radiology
Walter Reed Army Medical Center
Washington, D.C.

Pierre Galle, M.D., Ph.D.
Full Professor of Biophysics and Nuclear Medicine
Chief of Nuclear Medicine
Université Paris XII
C.H.U. de Creteil
Hôpital Henri Mondor
Creteil, France

Julio E. Garcia, M.D.
Nuclear Medicine Service
Department of Radiology
Walter Reed Army Medical Center
Washington, D.C.

Milton D. Gross, M.D.
Associate Professor
Internal Medicine
University of Michigan Medical School;
Chief, Nuclear Medicine Service
Veterans Administration Medical Center
Ann Arbor, Michigan

Michael S. Kipper, M.D.
Assistant Clinical Professor of Radiology
University of California, San Diego;
Director of Nuclear Medicine
North County Medi-Scan, Ltd.
Vista, California

Arthur Z. Krasnow, M.D.
Assistant Professor of Radiology
Medical College of Wisconsin
Milwaukee, Wisconsin

Ralph W. Kyle, CNMT
Nuclear Medicine Service
Department of Radiology
Walter Reed Army Medical Center
Washington, D.C.

Robert Lisbona, M.D.
Professor
Department of Radiology
McGill University;
Director of Department of Nuclear Medicine
Royal Victoria Hospital
Montreal, Canada

S. Bert Litwin, M.D.
Chief
Thoracic and Cardiovascular Surgery
Children's Hospital of Wisconsin
Milwaukee, Wisconsin

Stephen Manier, M.D.
Nuclear Medicine Service
Fitzsimmons Army Medical Center
Aurora, Colorado

Michel A. Meignan, M.D., Ph.D.
Associate Professor of Biophysics and Nuclear Medicine
Director of the Department of Physics in Medicine and Biology
Université Paris XII
Hôpital Henri Mondor
Creteil, France

Javier A. Novales-Diaz, M.D.
Assistant Professor
Department of Nuclear Medicine
McGill University;
Assistant Physician
Department of Nuclear Medicine
Royal Victoria Hospital
Montreal, Canada

Yolanda C. Oertel, M.D.
Professor of Pathology
Director of Cytopathology
George Washington University Medical Center
Washington, D.C.

Brahm Shapiro, M.B., Ch.B., Ph.D.
Professor of Internal Medicine (Nuclear Medicine)
Department of Internal Medicine
University of Michigan
Ann Arbor, Michigan

Walter J. Slizofski, M.D.
Assistant Professor of Radiology
Hahnemann University
Philadelphia, Pennsylvania

James C. Sisson, M.D.
Professor of Internal Medicine (Nuclear Medicine)
Department of Internal Medicine
University of Michigan
Ann Arbor, Michigan

John R. Sty, M.D.
Clinical Professor of Radiology
Medical College of Wisconsin-Milwaukee
University of Wisconsin-Madison;
Radiologist-in-Chief
Children's Hospital of Wisconsin
Milwaukee, Wisconsin

Raymond Taillefer, M.D.
Assistant Professor of Nuclear Medicine
University of Montreal
Department of Nuclear Medicine
Hôtel-Dieu de Montreal
Canada

Leonard Wartofsky, M.D., COL, M.C.
Chief, Endocrinology Service
Department of Internal Medicine
Walter Reed Army Medical Center
Washington, D.C.;
Professor of Medicine
Uniformed Services University of Health Sciences
Bethesda, Maryland

Robert G. Wells, M.D.
Associate Clinical Professor of Radiology
Medical College of Wisconsin-Milwaukee;
Staff Radiologist
Children's Hospital of Wisconsin
Milwaukee, Wisconsin

Sing-Yung Wu, M.D., Ph.D.
Nuclear Medicine Service
Veterans Administration Medical Center
Long Beach, California;
University of California, Irvine
College of Medicine
Irvine, California

Preface

Purpose: *Several excellent atlases of clinical nuclear medicine have been published; however, these suffer from a significant but understandable deficiency. In the attempt to cover a large area of clinical nuclear medicine, the atlases can include only a limited number of images of many procedures. This deficiency has been partially met by atlases that are limited to one procedure or organ; however, many nuclear medicine procedures have not been represented in individual books or do not warrant an entire book. The purpose of* Atlas of Nuclear Medicine *is to present more extensive atlases either of nuclear medicine procedures or of specific aspects of nuclear medicine procedures that have been presented only superficially in previous atlases and/or do not warrant a separate book.*

Readership: *This atlas should benefit the radiologist and nuclear medicine physician, and, because many endocrinologic nuclear medicine procedures are included, the atlas also should benefit the endocrinologist. In addition, residents, teachers, and practitioners should find it valuable, albeit in different ways. For the radiology resident or nuclear medicine fellow, the atlas can be a valuable visual introduction, and because not all training programs perform all procedures extensively, it may complement the resident's training program. For the teacher, the atlas can release him or her from instructing the resident in the ABCs of a procedure, thereby allowing more time to be devoted to teaching the subtleties of imaging. For the practitioner, the atlas should be a valuable review, update, and/or source of new "pearls," and it may also act as part of the foundation for initiating a new procedure.*

Origination of atlas: *The idea for publishing this atlas was given to one of us (DVN) by a friend (although on several occasions during the organization of the book that friendship was questioned) and is based upon the atlas section that frequently appears in the journal,* Clinical Nuclear Medicine. *From approximately twenty-five atlases that had been published previously in* Clinical Nuclear Medicine, *ten were selected, updated, reformatted and/or expanded, and six new atlases were added. We believe the updates and the six new atlases will richly reward the reader.*

Format: *To aid the reader in the use of this atlas, a general structure has been followed in most of the chapters: introduction, technique (imaging procedure, computer acquisition analysis), physiologic mechanism of the radiopharmaceutical, estimated radiation absorbed dose, visual description/interpretation, discussion, atlas (normal, abnormal), and references. Sections such as "imaging technique," "computer technique," and "approach to scintigraphic interpretation" represent the approach used by the authors of the respective chapters.*

Feedback: *Finally, we solicit feedback from the reader regarding any aspect of the atlas such as format, photography, and medical content. Of course, compliments are welcome as we have not yet been successful in eliminating our vanity; however, suggestions for improvements have a greater potential value, and we welcome them at future meetings, by telephone, or by mail.*

Douglas Van Nostrand
Sheldon Baum

Acknowledgments

The editors wish to thank Lizbeth K. Francisco, who assisted in the coordination and preparation of this atlas. In addition, the editors wish to thank the contributors who spent many hours in the preparation of their chapters and gracefully handled the editors' comments, requests, and modifications.

Contents

Raymond Taillefer, Gilles Beauchamp, and André Duranceau
Chapter 1 RADIONUCLIDE ESOPHAGEAL TRANSIT STUDIES *1*

Peter W. Blue
Chapter 2 HEPATOBILIARY IMAGING IN PATIENTS WITHOUT PRIOR SURGERY *41*

Milton D. Gross, and Brahm Shapiro
Chapter 3 SCINTIGRAPHY OF THE ADRENAL CORTEX *52*

Brahm Shapiro, and James C. Sisson
Chapter 4 SYMPATHOADRENAL IMAGING WITH RADIOIODINATED META-IODOBENZYLGUANIDINE *72*

Robert M. Basarab
Chapter 5 PARATHYROID SCINTIGRAPHY *115*

Stephen Manier, Douglas Van Nostrand, Frank Atkins, and Sing-Yung Wu
Chapter 6 IODINE-131 NECK AND CHEST SCINTIGRAPHY *144*

Michel A. Meignan, Francois Begon, and Pierre Galle
Chapter 7 FLUORESCENT THYROID SCANNING *176*

Leonard Wartofsky, and Yolanda C. Oertel
Chapter 8 FINE NEEDLE ASPIRATION OF THYROID NODULES *193*

Stephen M. Manier, Douglas Van Nostrand, Ralph W. Kyle, and Sue H. Abreu
Chapter 9 RENAL TRANSPLANT SCINTIGRAPHY *201*

Robert Lisbona, Vilma Derbekyan, and Javier A. Novales-Diaz
Chapter 10 Tc-99m BLOOD CELL VENOGRAPHY IN DEEP VENOUS THROMBOSIS OF THE LOWER LIMB *239*

Douglas F. Eggli, and Julio E. Garcia
Chapter 11 RADIONUCLIDE IMAGING OF THE ACUTELY PAINFUL SCROTUM *271*

Michael S. Kipper
Chapter 12 INDIUM-111 WHITE BLOOD CELL IMAGING *299*

John R. Sty, Robert G. Wells, and S. Bert Litwin
Chapter 13 PEDIATRIC NUCLEAR CARDIOLOGY *331*

B. David Collier, Walter J. Slizofski, and Arthur Z. Krasnow
Chapter 14 SPECT BONE IMAGING (Lumbar Spine, Hips, Knees, and Temporomandibular Joint) *360*

Tapan K. Chaudhuri
Chapter 15 NUCLEAR DACRYOCYSTOGRAPHY *383*

John R. Sty, Robert G. Wells
Chapter 16 PEDIATRIC RADIONUCLIDE LYMPHOGRAPHY *396*

INDEX *415*

Raymond Taillefer
Gilles Beauchamp
André Duranceau

Chapter 1 RADIONUCLIDE ESOPHAGEAL TRANSIT STUDIES

The relatively recent development of quantitative radionuclide procedures for the assessment of esophagogastric function has significantly advanced the investigation of upper digestive tract diseases. Nuclear medicine offers several advantages over the other standard methods in use in clinical practice in the evaluation of esophageal disorders. The radionuclide esophageal transit study (RETS) is a safe and noninvasive procedure that uses physiologic markers and provides reliable quantitative data on esophageal emptying function.

The atlas in this chapter reviews the technical aspects, scintigraphic patterns, and interpretations of radionuclide esophageal studies. As RETS is primarily a quantitative procedure, data obtained from regions of interest are of prime importance in clinical practice, especially in the follow-up of medical or surgical treatment. However, display of time–activity curves is not particularly well suited to an atlas. Therefore, where appropriate, analog images showing characteristic patterns of different esophageal lesions have been added and related quantitative data are included in the legend.

As with all radionuclide procedures, there are many types of acquisition and data analysis protocols, the extensive discussion of which is beyond the scope of this chapter. There are basically two types of radioactive boluses: liquid and solid. For most of our discussion, RETS performed with liquid bolus will be considered.

TECHNIQUES

Imaging Procedures

Many types of data acquisition protocols have been described in the literature.[1–5] In our laboratory, the type of protocol used depends on the types of symptoms or esophageal lesion to be evaluated.[6,7] The technique is therefore tailored to answer clinical questions. For example, if RETS is performed to evaluate the esophageal emptying in a patient with a known specific esophageal lesion causing either functional or mechanical obstruction, a longer period of observation will be preferable (particularly in the evaluation of achalasia, scleroderma, gastric transposition, or oropharyngeal dysphagia). The standard protocol used in our laboratory is described in the following text.

Patient Preparation

RETS is carried out after a 4-to-6-hour fast. There is no need to discontinue any medication. The most impor-

The authors thank Michèle Mathieu for her secretarial assistance, and the technical staff of the nuclear medicine and audio-visual departments of Hôtel-Dieu de Montrèal Hospital for its collaboration.

tant point in the preparation is to let the patient have several practice swallows before ingestion of the radioactive bolus. It is essential that the patient swallows the bolus in only one "phase." If ingestion is completed in multiple phases, quantitative parameters derived from time–activity curves will be either suboptimal or inadequate. Usually, if patients are allowed one to three practice swallows, a single swallow can then be obtained, unless odynophagia or pharyngeal or upper esophageal sphincter lesions are present.

Positioning of the Patients

RETS can be performed in two positions: upright and supine. The patient is initially evaluated in the upright position (or sitting erect in front of the scintillation camera). This provides a more physiologic evaluation as related to the normal position for swallowing. The upright study is particularly important in the assessment of medical or surgical treatment effectiveness, such as in achalasia, scleroderma, or other motility disorders, because the position reflects usual eating habits. On completion of the upright study, the esophageal function is then evaluated in the supine position where, by partially removing the effects of gravity, esophageal contractions alone become responsible for esophageal emptying. Esophageal motility disorders are easier to demonstrate in this position,[8] particularly in the early stages when a study in erect position can be normal, while distinct abnormalities are detected with the patient supine. When only one technique can be performed, the supine position is definitely the most accurate in demonstrating esophageal involvement. However, when a study is a follow-up evaluation of a specific treatment for relieving pharyngeal or esophageal obstruction (mechanical or functional), it will have more physiologic and clinical relevance if the patient is upright or sitting.

Acquisition Protocol

When both an upright and supine study are required, the upright study is done first. After practice swallows, the patient stands in front of a computer-interfaced, large-field-of-view scintillation camera fitted with a low-energy, all-purpose, parallel-hole collimator. Anterior projection is used on a routine basis. The patient is positioned so that the mouth, hypopharynx, entire esophagus, and ideally the proximal part of the gastric fundus can be clearly visualized in the same field. The patient's head is placed in a slight anterior oblique rotation. A bolus of 0.5 to 1 mCi of technetium-99m (Tc-99m) sulfur colloid or Tc^{99m} albumin colloid diluted in 15 to 20 ml of water is then aspirated in the mouth through a straw and the patient is asked to retain the bolus in the mouth. On completion of the radionuclide oral administration, a 2-minute analog and digital acquisition is started immediately. A few seconds after the beginning of acquisition, the patient is instructed to swallow the entire bolus in only one phase and not to move for 2 minutes (Figs. 1-1 and 1-2). After 30 seconds, the patient is instructed to have a dry swallow, and additional dry swallows are allowed at 15-second intervals for 2 minutes. Analog images are obtained at 2-second intervals for 30 seconds and then at 5-second intervals for 90 seconds. At the end of this dynamic part of the study, a static image (preset time of 60 seconds) is obtained without moving the patient, and a marker (cobalt-^{57} or Tc-99m) is placed over the cricoid cartilage to identify the level of the upper esophageal sphincter. When the upright study is completed, the patient is asked to swallow 100 ml of unmarked water to rinse out residual oropharyngeal and esophageal activity and is then placed supine under the scintillation camera. The same protocol is then repeated for the supine study.

All "dynamic" scintigraphic analog images presented in this chapter were recorded at 2- second intervals in anterior projection. The investigator who would rather use a lower dose of ingested radioactivity (0.2–0.3 mCi) may perform posterior views. They also serve to provide a relatively uniform attenuation of radioactivity throughout the length of the esophagus, thereby avoiding the more important attenuation of the heart.[9] However, with a higher dose (0.5 to 1 mCi), the effect of heterogenous cervical and thoracic attenuation is less important and we have not found significant differences between the two views in the diagnostic accuracy of both qualitative and quantitative data. We use the posterior view when a radionuclide solid bolus study is performed or when the patient cannot be evaluated in the anterior projection because of physical limitations. However, a hypopharyngeal emptying evaluation should be done in the anterior view so that the region of interest is closer to the detector surface.

Computer Acquisition

Hypopharyngeal and esophageal emptying usually occurs in only a few seconds. It is therefore important to study this dynamic event with a relatively fast framing time. In our laboratory the esophageal transit of the radionuclide bolus is recorded in frame mode (64×64 matrix) and stored on a magnetic disk at 0.5-second intervals for a duration of 2 minutes. This acquisition is performed in both positions. Initially we used digital acquisition at 0.1-second intervals but, from a clinical viewpoint, the results were not really different from those obtained at 0.5-second intervals. However, an interval of more than 1 second is not optimal, particularly when the hypopharyngeal emptying function is to be evaluated or when the origin of a specific esophageal regional activity at a given time is to be determined (*i.e.,* of proximal or distal origin).

Computer Analysis

Regions of Interest. Different quantitative analysis approaches can be used, but the overall aim is to provide the clinician with easily comprehensible, clinically relevant data. After data have been recorded, time–activity curves are generated for seven regions of interest: the oral cavity; hypopharynx; proximal, middle, and distal esophageal segments; gastric fundus; and entire esophagus (Fig. 1-3). Choosing the number and localization of regions of interest is essential for an adequate evaluation of esophageal function. Before tracing regions of interest, anatomic structures and morphological lesions must be recognized. This can be done on both the composite image from digital data and the static analog image obtained at the end of each study, using a cricoid marker. The oral activity and hypopharynx are thus easily identified. The presence of a hiatal hernia should be identified on the static or dynamic image and should be excluded from the region of interest of the distal esophageal segment. Regions of interest of the distal esophagus and the gastric fundus should be separated by a few centimeters because respiratory movement can falsely introduce activity from the gastric area into the distal esophageal region.

Monitoring of oral and pharyngeal activity is important in evaluating upper dysphagia or hypopharyngeal emptying disorders. Monitoring is also a way to detect the presence and the effects of an incomplete initial swallow, which can result in a fragmentation of the bolus in the esophagus (multiphasic ingestion). This is particularly true in older patients or in those having odynophagia, who will hesitate on swallowing. In this condition, abnormal parameters are not necessarily secondary to a motility disorder and care must be taken not to mistakenly interpret a technical problem as an esophageal disorder. It is also important to include the proximal gastric area, mainly on the supine study, to check for any gastroesophageal reflux. If there is a sudden rise of activity in the distal esophageal segment, it should be determined whether it is of proximal or gastric origin. Multiple esophageal regions of interest allow characterization of bolus movement, either antegrade or retrograde.

Time–Activity Curves. Figure 1-4 is a schematic representation of a normal RETS and objective parameters derived from time–activity histograms. The quantitative data we have found the most useful for diagnostic and follow-up purposes are the hypopharyngeal, segmental, and global esophageal emptying times. *Emptying time* is defined as the time that it takes for 90% or more of the maximal activity in each region of interest to be eliminated. *Global esophageal emptying time* represents the time from entry of the bolus in the proximal esophagus to the clearance of more than 90% from the entire esophageal region of interest. Data obtained from a multiple ingestion, therefore, are not suited to this form of quantitation and it is usually preferable to repeat the study, depending on the magnitude of the bolus fragmentation.

In many pharyngeal and esophageal disorders, a segmental or global emptying of 90% or more of the maximal ingested activity frequently cannot be achieved at the end of the 2-minute acquisition period. In this case the percentage of residual segmental activity compared to the maximal activity is determined for a fixed time, such as 30, 60, 90, and 120 seconds. Results are expressed in the same manner as those in gastric emptying studies. This form of quantitative data presentation is particularly well suited for the evaluation of treatment effectiveness. Residual activity for each region of interest is also assessed at 2 minutes. It is expressed as the percentage of maximal activity detected for a given region of interest. The value of distal esophageal segment residual activity is somewhat higher than that for proximal and middle segments because of scatter from gastric activity.

Condensed Images. A new, simple, computer-based method of data processing has been recently introduced for facilitating qualitative assessment of radionuclide esophageal transit. With it, dynamic image sequences are reduced to single condensed images. Each original consecutive image frame of the dynamic series is compressed into a single vertical column which shows the vertical radioactivity distribution from mouth to gastric fundus within a short time interval (Fig. 1-5). A condensed image then is reconstructed by assembling these columns side by side according to their frame number. In the condensed image, therefore, the vertical axis shows the spatial distribution of the radioactivity, while the horizontal axis is temporal. As pointed out by Ham and coworkers,[10] this method of data processing is well suited to radionuclide esophageal studies because the principal region to be evaluated is a long vertical rectangle and there is no radioactive contamination outside this region of interest. In daily practice, having a single picture to present to the clinician is certainly a convenience. Furthermore, this type of display is useful to detect patient motion or the presence of gastroesophageal reflux during the test. Other types of computer analysis for radionuclide esophageal studies (such as centroid analysis), are beyond the scope of this chapter and are discussed elsewhere.[11]

Normal Values. As for any other radionuclide quantitative procedure, a range of normal values must be established. Normal data can be obtained from healthy subjects free from any esophageal symptoms,

Table 1-1
Normal Values for Emptying Time

	Upright $T^{9}/_{10}$*(sec) ± 1 SD†	*Supine* $T^{9}/_{10}$*(sec) ± 1 SD†
Hypopharynx	1 ± 0.5	1.5 ± 0.5
Proximal Esophagus	1.5 ± 0.5	1.5 ± 0.5
Middle Esophagus	3 ± 0.5	3.5 ± 1
Distal Esophagus	3.5 ± 1	6 ± 2.5
Entire Esophagus	5 ± 2	9 ± 2.5

* $T^{9}/_{10}$, *emptying time, or the time necessary to clear 90% or more of the maximal activity recorded for a given region of interest. The exception to this rule is when the entire esophagus is under study. In this case, the global emptying time represents the time from entry of the bolus in the proximal esophagus to the clearance of more than 90% of the maximal activity recorded in the entire esophageal region of interest.*

† SD, *standard deviation*

but ideally they should be from asymptomatic healthy subjects with normal esophageal motility and endoscopic studies. Since the composition of the radionuclide bolus, acquisition protocol, instrumentation, and positioning of the patients, to name only a few factors, can differ slightly from one institution to another, each laboratory must evaluate normal values that are adapted to the patient population that will be investigated. These should be established for each quantitative parameter to be evaluated and should differ for children, adults, or elderly patients. Tables 1-1 and 1-2 summarize the average values and upper limits (obtained in healthy adult subjects) for some of the parameters that we use in our laboratory.

PHYSIOLOGIC MECHANISM OF THE RADIOPHARMACEUTICAL

The radiopharmaceutical used for the evaluation of esophageal emptying ideally should be inexpensive, readily available, and easy to prepare — in addition to having the optimal physical properties related to the radionuclide marker. Furthermore, as the goal of the test is to assess the transit of the ingested radionuclide through the esophageal lumen, it should not be absorbed or secreted by the esophageal mucosa. Tc-99m sulfur colloid fulfills these criteria and is, at the present time, the most frequently used radiopharmaceutical for the study of upper digestive tract function. Tc-99m sulfur colloid is not absorbed by or adsorbed to a significant level on a normal esophageal mucosa. However, when the integrity of the mucosa is altered, as in peptic esophagitis or in columnar-lined (Barrett's) esophagus, Tc-99m sulfur colloid may adhere to the mucosal surface.[12,13]

There are some limitations to the use of Tc-99m labeled radiopharmaceuticals for quantitation of esophageal emptying. Considerations of radiation exposure limit the amount of Tc-99m activity that can be ingested and, consequently, lower the count rate acquired during the procedure, thereby limiting the quality of the data. Krypton-81m has been used for the quantitation of esophageal transit.[5] With its ultra-short half-life, Kr-81m has a negligible radiation burden that allows ingestion of several millicuries, and this amount of radioactivity provides better counting statistics. Furthermore, the study can be repeated as often as necessary. This feature can be particularly useful with children or patients who are unable to ingest the liquid bolus in a single swallow. The disadvantages of Kr-81m is that its ultra-short half-life does not allow an adequate study of patients with significantly prolonged esophageal emptying, as in achalasia or scleroderma, and clinically it is expensive and not readily available.

Table 1-2
Normal Values for Residual Radioactivity at 2 Minutes

	Upright %* ± 1 SD†	*Supine* %* ± 1 SD†
Oral Cavity	2 ± 1	3 ± 1
Hypopharynx	3 ± 2	3 ± 2
Proximal Esophagus	2 ± 1	2 ± 1
Middle Esophagus	2 ± 1	3 ± 2
Distal Esophagus	3 ± 1	5 ± 3
Entire Esophagus	4 ± 2	6 ± 3

* %, *residual activity at 2 minutes, expressed as the percentage of the maximal activity recorded for a given region of interest*

† SD, *standard deviation*

ESTIMATED RADIATION ABSORBED DOSES

The estimated radiation absorbed dose following an oral administration of a liquid bolus containing Tc-99m sulfur colloid has been previously described.[14] Data were based on the ingestion of 300 μCi of Tc-99m sulfur colloid diluted in water. The target organ is the proximal large bowel, which receives 0.52 rads/mCi; other organs for which dosimetry was estimated (rads/mCi) are the distal large bowel (0.33), small bowel (0.28), stomach (0.09), ovaries (0.10), and testes (0.007). Ingestion of Tc-99m sulfur colloid yields a whole-body radiation absorbed dose of 0.02 rads/mCi.

VISUAL DESCRIPTION AND INTERPRETATION

Esophageal emptying usually occurs in a relatively short period of time, sometimes even in the presence of an underlying motility disorder. It is therefore useful to evaluate together as many data as possible to study this rapid event. Three types of display are available with the standard RETS: analog, computerized (time–activity curves and condensed images), and cine-format. The range of quantitative measurements is sometimes so wide that the results cannot be interpreted without analog images, so it is essential to analyze both qualitative and quantitative data together. One isolated abnormal parameter is far from being diagnostic of any specific esophageal lesion. Cine-format display is useful to follow the bolus transit through the hypopharynx and esophageal lumen. Although it is possible to detect esophagopharyngeal regurgitation and retrograde esophageal movements of the radioactive bolus on analog images and on time–activity curves, it is easier with cine-format display recorded at 0.25 to 0.5 seconds or with a condensed image. Before interpreting the RETS, it is recommended that all three of these displays be evaluated together for a better overview.

As with other procedures and because the RETS is a complementary test primarily used for evaluating emptying function, knowledge of results of previous investigations or previous upper digestive tract surgery is helpful. Transitory radionuclide esophageal stasis does not have the same meaning in patients with esophageal diverticulae as in patients who were previously submitted to a fundoplication for a significant gastroesophageal reflux disease.

The first step in evaluation of analog images and cine-format display is to verify that the bolus ingestion was completed in a single or multiple swallows. Hypopharyngeal transit of the radionuclide bolus is very short, usually less than 0.5 to 1 second, unless there is underlying pharyngeal or upper esophageal sphincter involvement. As the pharynx contracts and upper esophageal sphincter opens, the bolus rapidly goes through the upper esophagus; two or more swallows are usually easily detected in this first part of the study. At midesophagus, from the level of the aortic arch down to the lower esophageal sphincter and gastric lumen, the radiocolloid moves less rapidly, particularly in the supine position. This slowing of the bolus movement at the middle esophagus might be secondary to external pressure from the aortic arch or tracheal bifurcation, or it may represent the level of transition from striated to smooth muscle.[15]

On completion of the bolus passage through the esophageal lumen, residual radioactivity is assessed for each region of interest (*i.e.,* mouth, hypopharynx, and the three esophageal segments). Normally there is only a slight residual activity in the mouth and at midesophagus, although it will be more pronounced in the supine position. Several lesions may explain an abnormal segmental stasis at 1 or 2 minutes after the radionuclide ingestion. Besides significant esophageal motility disorders, morphological stenotic lesion, diverticulum, or peptic esophagitis should be included in the differential diagnosis. The static image obtained at the end of the study is well suited to evaluating regions of abnormal stasis. However, it cannot provide adequate quantitative data. Although RETS does not have a high anatomic resolution and is not dedicated to the study of anatomic details, the static image is often useful for detecting morphological lesions that may manifest themselves by radiocolloid stasis, such as diverticulae or a hiatal hernia. It can also enhance the definition of extraluminal activity, as encountered in tracheal aspiration or tracheoesophageal fistula.

Once analog and cine-format evaluations have been completed, computerized data are analyzed. Hypopharyngeal, global, and segmental esophageal time–activity curves are assessed, first separately and then together. Radionuclide emptying time (as defined previously) is calculated for each region of interest. When emptying is not achieved at the end of the 2-minute study, the percentage of activity remaining is determined at 1 or 2 minutes. Another way of assessing the esophageal clearance is to determine the time necessary to empty 25%, 50%, 60%, and so on, of the ingested activity.

The shape of the time–activity histograms is also qualitatively analyzed. In the normal patient, the peak of the curve is rapidly attained in a few seconds and there is only one peak for each segment. The 90% emptying is also obtained a few seconds after the curve peak. Multiple peaks can be seen in different lesions. Multiple swallows are the most frequent cause for several peaks. This pitfall can be ruled out by assessing the curves of the hypopharyngeal and proximal esophageal segments. If these time–activity curves are fragmented and there is no residual activity in the hypopharynx, an inadequate ingestion is probable. In mechanical (as seen in peptic esophagitis or esophageal cancer) or functional (as in achalasia) obstructive lesions, the first part of the curve generally has a plateau shape which is followed by a more or less significant emptying depending on the severity of the obstruction. Esophageal retrograde movements of the radionuclide bolus and gastroesophageal reflux can be recognized by analyzing the temporal relationship between curve peaks of adjacent regions of interest in the same manner as is done in the study of gastroesophageal reflux with radionuclides.

The standard liquid radionuclide esophagogram study report used in our laboratory includes the following:

Quality of ingestion (single or multiple swallows)

Hypopharyngeal and esophageal emptying and stasis

Assessment of abnormal movement of the bolus (pharyngonasal, pharyngo-oral, esophagopharyngeal regurgitations, and esophagoesophageal and gastroesophageal reflux)
Description of abnormal extraluminal focus of activity
Assessment of "anatomic" abnormalities

These parameters are done for each study in both positions. When the RETS is performed to evaluate a specific disorder with delayed emptying (including achalasia, scleroderma, upper esophageal sphincter, and pharyngeal lesions), global esophageal or hypopharyngeal time–activity curves are reproduced in the report so the clinician can easily assess and compare the effectiveness of treatment on the radionuclide emptying in the follow-up.

DISCUSSION

Many modalities are currently available for the clinical evaluation of esophageal pathophysiology: motility study, *p*H-metry, endoscopy, and barium contrast esophagogram with or without cinefluoroscopy. Although all of these nonscintigraphic techniques provide useful information about the anatomy and function of the esophagus, they have significant limitations in the study of esophageal motor function. They are all either semiquantitative or nonquantitative and they alter the normal physiology of the esophagus. The barium esophagogram is relatively insensitive to subtle motor disturbances, its interpretation is subjective, and, because of radiation considerations, the function of the esophagus is studied for only a short period of time. Furthermore, barium is not a physiologic marker. Although considered the "gold standard" in the study of esophageal motility disorders, manometry requires intubation and that can possibly induce a bias in the evaluation of esophageal physiology. It is also sometimes poorly accepted by the patient because of the discomfort caused by the introduction of an endoluminal catheter. The resultant lack of patient compliance may be a significant concern, especially in the follow-up of either medical or surgical treatment.

Nuclear medicine with computer data processing offers advantages over other standard clinical methods used in the investigation of esophageal motility disorders. First introduced by Kazem in 1972[16] and subsequently modified to improve its accuracy, RETS fulfills the major criteria for evaluating esophageal emptying function: it is safe, noninvasive, very well accepted by patients, easy to perform, readily available to an average nuclear medicine laboratory, and has a low radiation burden. Most importantly, it uses physiologic markers and provides quantitative data on hypopharyngeal and esophageal emptying functions. When compared to other tests, RETS is probably the most accurate technique in reflecting the physiology of swallowing and esophageal clearance. However, it is essential to emphasize that RETS is not designed to replace the standard radiologic, endoscopic, or manometric evaluation used in clinical esophagology. Rather, it is a noninvasive complementary procedure that gives further functional parameters. It is or should be used as part of the global work-up of esophageal motility disorders. Time–activity curves derived from the digitalized data provide information to the clinician that is unavailable by other means, and RETS can quantitatively assess the esophageal emptying function.

Clinical application of radionuclides in the study of esophageal motility disorders is a relatively new field of interest. It is probably too soon to establish its definitive role. A more frequent use, more extensive correlation data, and longer follow-up period will be necessary before RETS is fully appreciated in clinical practice. We have performed radionuclide esophageal studies in many hundreds of patients and in different types of esophageal diseases and, at the present time, we use this procedure for the purposes summarized here. Specific applications of RETS are discussed in more detail in the atlas section.

Clinical Indications for RETS

Objective evaluation of medical or surgical treatment effectiveness:
- Esophageal motor disorders (*e.g.*, achalasia, scleroderma, diffuse esophageal spasm)
- Pharyngeal and upper esophageal sphincter lesions

Evaluation of dysphagia or atypical chest pain, especially when
- Barium esophagogram is normal
- A complete cardiac work-up is normal or inconclusive

Concomitant with a gastroesophageal reflux study with radionuclides, particularly to detect esophageal motility disorders associated with gastroesophageal reflux

Miscellaneous
- Detection of pharyngeal regurgitation or tracheal aspiration
- Detection of tracheoesophageal fistula

One of the most frequent and interesting indications for performing RETS is the objective evaluation of medical or surgical treatment effectiveness. For example, in esophageal achalasia, the primary goal of treatment is to relieve the obstruction by weakening the lower esophageal sphincter[17]; doing so can improve the esophageal emptying and decrease the complications of achalasia. There are two ways to assess the efficacy of medical or surgical treatment of pneumatic dilation: esophageal manometric study, which assesses the resting esophageal pressure and the lower esophageal sphincter pressure; and radionuclide esophageal study, which provides a physiologic quantitative evaluation of the global esophageal emptying. Patients with achalasia or with other lesions causing either functional or mechanical obstruction are studied in both upright and supine positions with liquid and solid boluses for at least a 15-minute period (instead of the standard 2-minute acquisition). This technique provides a complete physiologic evaluation of esophageal emptying. A variety of solids can be labeled for this study; we usually perform it with a piece of scrambled egg labeled with 0.5 mCi of Tc-99m sulfur colloid. The use of a solid meal with the patient in the upright position exactly reflects the clinical situation, while the supine study, particularly with the liquid bolus, is useful in evaluating the presence or absence of esophagopharyngeal regurgitations and tracheal aspiration.

RETS has also been proven to be very useful in the objective assessment of the results of cricopharyngeal myotomy in patients with upper esophageal sphincter achalasia, muscular dystrophy with impaired pharyngeal muscles, or other causes of oropharyngeal dysphagia. The goal of cricopharyngeal myotomy is to facilitate the passage of food from the pharynx through the upper esophageal sphincter, which is assessed by measuring hypopharyngeal emptying time, derived from the time–activity curve.(This parameter is used in a way similar to the use of global esophageal emptying time to evaluate esophageal clearance in achalasia.) It provides reproducible quantitation of bolus emptying with an intact upper esophageal sphincter or following removal of this sphincter by a cricopharyngeal myotomy. Furthermore, RETS can demonstrate pharyngonasal or pharyngo-oral regurgitations and is very sensitive in detecting tracheal aspiration.

A modification of the standard protocol using a longer period of acquisition allows the study of the emptying function of a gastric transposition. It can assess transit in the esophageal remnant as well as emptying of the intrathoracic stomach. Data obtained provide useful documentation on the function of the transposed organ. Gastroesophageal reflux and tracheal aspiration are quantified and detected more easily with radionuclide techniques than with radiologic procedures. The same type of study can be applied in the functional evaluation of colonic interposition.

The second general indication for performing a RETS is in the investigation of dysphagia or atypical chest pain in specific conditions. Kjellen and colleagues[18] and Russell and associates[19,20] reported that aproximately half the patients investigated for dysphagia with a normal manometry, normal gastroesophageal reflux tests, and radiologic study has scintigraphic abnormalities, suggesting that the radionuclide study is more sensitive than any other test in detecting esophageal motility disorder. We have had similar results with patients suffering from dysphagia secondary to gastroesophageal reflux disease or after an antireflux procedure,[7] possibly because manometry separately measures the duration, velocity, and pressure of the lower esophageal sphincter and peristaltic waves, while RETS evaluates the combined effects of these factors on the segmental and global esophageal emptying; it also takes into consideration bolus retrograde movement. However, more data are required to appreciate the clinical significance of the esophageal motor dysfunction detected by radionuclide study with a solid or liquid bolus. The true role of RETS remains to be established by good prospective correlations between clinical, radiologic, manometric, and radionuclide studies.

Atypical chest pain is a relatively frequent diagnostic problem, mainly when the patient has a normal cardiac work-up, including thallium-201 myocardial imaging and coronary angiogram. Cardiologists at our institution have found RETS useful in this particular context. If the barium esophagogram is normal, a radionuclide study is performed in order to rule out an esophageal motor dysfunction. Any abnormality detected by RETS should be ideally correlated with manometry. In our experience the radionuclide study is significantly more sensitive than barium studies in detecting motility disorder.

The third general clinical indication for RETS is the concomitant study of the esophageal emptying when performing a gastroesophageal reflux study with radionuclides. Because the same radionuclide marker is used in both types of study, the esophageal clearance and the presence of gastroesophageal reflux can be evaluated simultaneously without increasing the radiation burden. Although the radionuclide study does not have a high anatomic resolution, it can demonstrate morphological lesions that may accompany gastroesophageal reflux, such as sliding hiatal hernia, paraesophageal hernia, or esophageal diverticulum.[6] Furthermore, in some cases, peptic esophagitis and columnar-lined esophagus may show an abnormal radionuclide stasis which would correspond to a radiocolloid retention on a damaged mucosa.[12,13] The main reason to perform a RETS in this situation, however, is to detect a possible esophageal motor dysfunction accompanying a gastroesophageal reflux disease.

Pharmacologic intervention is increasingly used in different nuclear medicine procedures. No extensive study has been published yet of RETS used in this application. It would be interesting to perform a radionuclide esophageal study in patients with suspected primary motor abnormality following stimulation with a cholinergic agonist (*e.g.*, bethanechol) or with an acetylcholinesterase inhibitor (*e.g.*, edrophonium) as already studied with manometry.[21] A diagnosis of chest pain of esophageal origin can be made if the pain is reproduced and if an esophageal emptying abnormality is detected on RETS.

A variety of protocols have been used for the radionuclide esophageal study. The atlas that follows predominantly illustrates cases evaluated with our standard study, that is, with a liquid bolus in both positions during a period of 2 minutes. Initially, all patients were studied during at least 10 minutes in each position, but the 2-minute acquisition was found to be more convenient without really modifying the diagnostic accuracy of the test. As the goal of the RETS is not to detect gastroesophageal reflux but to quantitate esophageal emptying (which is usually rapidly completed), the 2-minute study is satisfactory in clinical practice. However, if there is a significant esophageal stasis at the end of the test or if the presence of a specific lesion causing obstruction to the hypopharyngeal or esophageal transit is known before performing the study, then a 10- to 15-minute acquisition becomes necessary. The use of a solid bolus is particularly interesting in this situation and in the evaluation of patients suffering from dysphagia to solids. However, it is important to note that a solid bolus gives a wider normal range of esophageal emptying than a liquid bolus, and the result strongly depends on the composition of the marker. Nevertheless, we think that a study with a solid bolus should be performed in specific conditions because it provides a more complete physiologic evaluation.

The atlas shows a variety of esophageal lesions diagnosed by other modalities; in many of these the radionuclide study was primarily used to evaluate the emptying function. Other lesions are presented as abnormalities serendipitously discovered when performing RETS. Further investigation is still necessary, but it is hoped that radionuclide esophageal study will continue to improve as a useful modality in clinical practice.

REFERENCES

1. Tolin RD, Malmud LS, Reilley J et al: Esophageal scintigraphy to quantitate esophageal transit. Gastroenterology 76:1402, 1979
2. Russell COH, Hill LD, Holmes ER et al: Radionuclide transit: A sensitive screening test for esophageal dysfunction. Gastroenterology 80:887, 1981
3. Fisher RS, Malmud LS: Functional scintigraphy: Diagnostic applications in gastroenterology. Developments in Digestive Diseases, p. 139, 1980
4. Blackwell JN, Hannan WJ, Adam RD et al: Radionuclide transit studies in the detection of esophageal dysmotility. Gut 24:421, 1983
5. Ham HR, Piepsz A, Georges B et al: Quantitation of esophageal transit by means of ^{81m}Kr. Eur J Nucl Med 9:362, 1984
6. Taillefer R, Beauchamp G: Radionuclide esophagogram. Clin Nucl Med 9:465, 1984
7. Taillefer R, Beauchamp G, Duranceau A et al: Nuclear medicine and esophageal surgery. Clin Nucl Med 11:445, 1986
8. Lamki L: Radionuclide esophageal transit study: The effect of body posture. Clin Nucl Med 10:108, 1985
9. Ferguson MK, Ryan JW, Little AG et al: Esophageal emptying and acid neutralization in patients with symptoms of esophageal reflux. Ann Surg 201:728, 1985
10. Ham HR, Georges B, Froideville JL et al: A simple data display for qualitative assessment of esophageal transit. Clin Nucl Med 10:483, 1985
11. Klein HA, Wald A: Computer analysis of radionuclide esophageal transit studies. J Nucl Med 25:957, 1984
12. Beauchamp G, Taillefer R, Devito M et al: Radionuclide esophagogram in the evaluation of experimental esophagitis: Manometric and histopathologic correlations. In Demeester TR, Skinner DB (eds): Esophageal Disorders: Pathophysiology and Therapy, pp 77–81. New York, Raven Press, 1985
13. Taillefer R, Beauchamp G, Devito M et al: Oesophagite expérimentale: évaluation par oesophagogramme radioisotopique (^{99m}Tc-soufre colloïdal) corrélations manométrique et histopathologique. J Biophys Med Nucl 7:131, 1983
14. Wu RK, Malmud LC, Knight LL et al: Radiation dose calculations for orally administered radiopharmaceuticals in upper gastrointestinal disease. In Raynard C (ed): Proceedings of the Third World Congress of Nuclear Medicine and Biology, vol IV, p. 2961. London, Pergamon Press, 1982
15. Rozen P, Gelfond M, Zaltzman S et al: Dynamic, diagnostic, and pharmalogical radionuclide studies of the esophagus in achalasia. Radiology 144:587, 1982
16. Kazem I: A new scintigraphic technique for the study of the esophagus. Am J Roentgenol Rad Ther Nucl Med 115:681, 1972
17. Payne WJ, King RM: Treatment of achalasia of the esophagus. Surg Clin North Am 63:963, 1983
18. Kjellen G, Svedberg JB, Tibbling L: Solid bolus transit by esophageal scintigraphy in patients with dysphagia and normal manometry and radiography. Dig Dis Sci 29:1, 1984
19. Russell COH, Pope CE, Gannan RM et al: Does surgery correct esophageal motor dysfunction in gastroesophageal reflux? Ann Surg 194:290, 1981
20. Russell COH, Hill LD, Holmes ER et al: Radionuclide transit: A sensitive screening test for esophageal dysfunction. Gastroenterology 80:887, 1981
21. Benjamin SB, Richter JE, Cordova CM et al: Prospective manometric evaluation with pharmacologic provocation of patients with suspected esophageal motility dysfunction. Gastroenterology 84:893, 1983
22. Duranceau AC, Beauchamp G, Jamieson GG et al: Oropharyngeal dysphagia and oculopharyngeal muscular dystrophy. Surg Clin North Am 63:825, 1983
23. Duranceau AC, Lafontaine ER, Taillefer R et al: Oropharyngeal Dysphagia and surgery on the upper esophageal

sphincter. In Nyhus L (ed): Surgery Annual 1987, 317–362. New York, Appleton Century Crofts, 1987

24. Rozen P, Gelfond M, Zaltzman S et al: Radionuclide confirmation of the therapeutic value of isosorbide dinitrate in relieving the dysphagia in achalasia. J Clin Gastroenterol 4:17, 1982

25. McKinney MK, Brady CB, Weiland FL: Radionuclide esophageal transit: An evaluation of therapy in achalasia. South Med J 76:1136, 1983

26. Thomas E, Lebow RA, Gubler RJ et al: Nifedipine for the poor-risk elderly patient with achalasia: Objective response demonstrated by solid meal study. South Med J 77:394, 1984

27. Gross R, Johnson LP, Kaminski RJ: Esophageal emptying in achalasia quantitated by a radioisotope technique. Dig Dis Sci 24:945, 1979

28. Henderson RD: Diffuse esophageal spasm. Surg Clin North Am 63:951, 1983

29. Benjamin SB, Gerhardt DC, Castell DO: High amplitude, peristaltic esophageal contractions associated with chest pain and/or dysphagia. Gastroenterology 77:478, 1979

30. Davidson A, Russell C, Littlejohn GO: Assessment of esophageal abnormalities in progressive systemic sclerosis using radionuclide transit. J Rheumato 12:472, 1985

31. Carette S, Lacoursière Y, Lavoie S et al: Radionuclide esophageal transit in progressive systemic sclerosis. J Rheumatol 12:478, 1985

32. Bivins BA, Reed MF, Belin RP et al: Diagnosis of tracheoesophageal fistula by radioscanning. J Surg Res 23:384, 1977

33. Dunn EK, Man AC, Lin KJ et al: Scintigraphic demonstration of tracheoesophageal fistula. J Nucl Med 24:1151, 1983

Atlas Section

Figure 1-1

Normal Analog Study (supine position)

Analog images were obtained in the supine position at 2-second intervals (beginning with the upper left image down to the bottom right image) following ingestion of 0.7 mCi of Tc^{99m} sulfur colloid diluted in 10 ml of water. As pharyngeal contraction and upper esophageal sphincter relaxation occur, the bolus travels rapidly through the upper esophagus. At midesophagus, from the level of the aortic arch down to the lower esophageal sphincter and gastric lumen, the radiocolloid moves less rapidly, although its clearance is completed in a few seconds. There is a normal slight delay in the distal esophageal lumen *(closed arrow)*. Gastric activity seen on the first frame represents residual activity originating from a previous study performed in the upright position. At 18 seconds, there is a second swallow *(open arrow)*, which is less than 10% of the initial activity. Because of the significant delay between the two swallows and the low degree of activity of the second swallow, quantitative parameters can still be adequately evaluated. (Taillefer R, Beauchamp G, Duranceau A, Lafontaine E: Nuclear medicine and esophageal surgery. Clin Nucl Med 11:445–459, 1986)

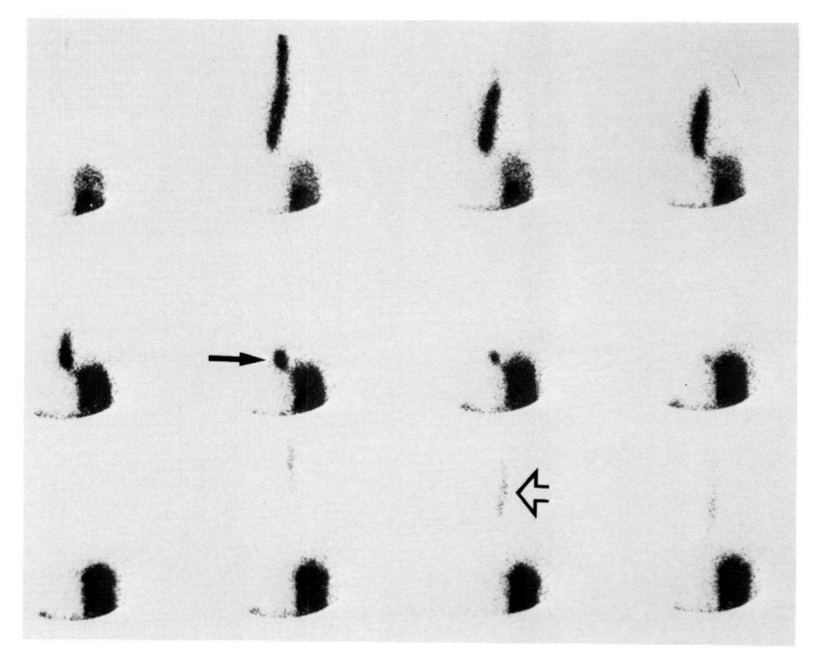

Figure 1-2 ***Normal Analog Study (upright position)***

This RETS was performed at 2-second intervals in a normal patient standing upright in front of the scintillation camera. The liquid radionuclide bolus is aspirated through a straw and kept in the mouth *(arrow)* for at least 2 seconds. Then the patient is asked to ingest the bolus in only one swallow. The bolus moves rapidly through the esophageal lumen, and the global esophageal clearance is completed within 8 seconds after the ingestion. (Gastric activity is not seen because of the patient's height, which did not allow the inclusion of the stomach in the field of view.)

Figure 1-3 ***Regions of Interest and Time–Activity Curves***

(A) After data have been recorded, one region of interest each is placed over the mouth, the hypopharynx, the proximal, middle, and distal esophageal segments, and the entire esophagus. A seventh region of interest should include the gastric fundus if the patient's height allows it. ***(B)*** Corresponding time–activity curves are expressed as absolute activity or as a percentage of the maximal activity as a function of time (0.5 sec/frame) for each region of interest in a normal subject in the upright position. Maximal activity is attained after a few frames, and all activity rapidly clears so that in less than 7 seconds 90% or more of the initially ingested activity has disappeared from the entire esophagus.

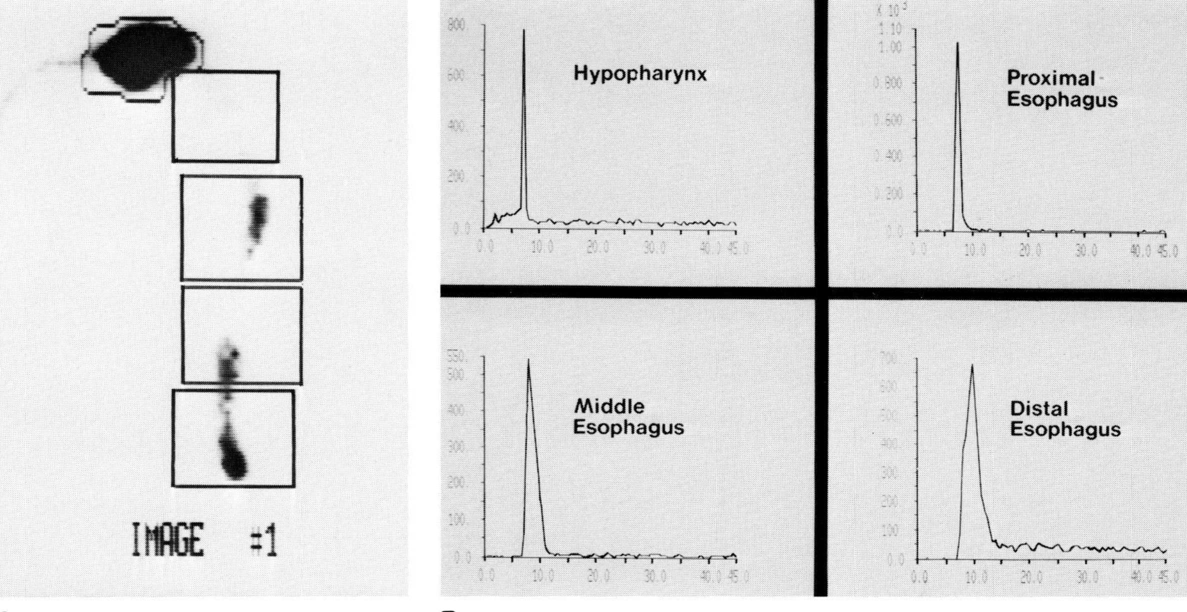

Figure 1-4 *Schematic Representation of a Normal RETS*

Segmental esophageal emptying time *(1)* is defined as the time it takes to eliminate 90% or more of the maximal activity from each segment. The global esophageal emptying time *(2)* represents the time from entry of the bolus in the proximal esophagus to the clearance of 90% or more of the entire esophageal activity.

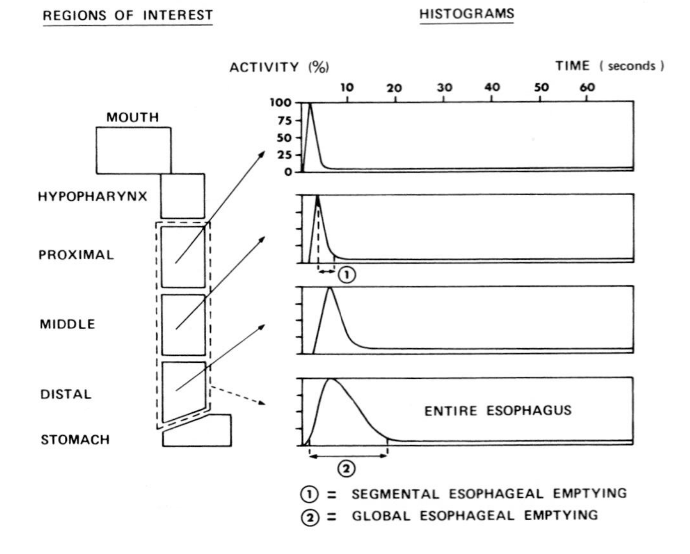

Figure 1-5 *Condensed Images*

In a condensed image, each original consecutive image frame of the dynamic series is compressed into a single vertical column that shows the vertical radioactivity distribution from mouth *(M)* to stomach *(S)* within a short time interval. A condensed image is reconstructed by assembling these colums side by side according to frame number. The vertical axis shows the spatial distribution of the radioactivity while horizontal dimensions are temporal. *(A)*. Condensed image obtained from a normal esophageal transit. There is a slight and normal residual mouth activity. *(B)*. Condensed image in a patient with myotonic dystrophy. There is significant radionuclide hypopharyngeal stasis *(arrow)* during the study.

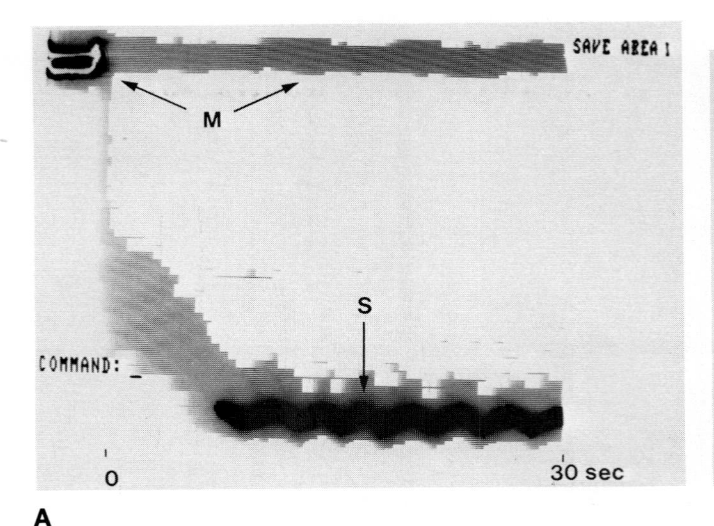

A

B

Figure 1-6

Normal Static Image (upright position)

A 1-minute acquisition static image was obtained in the upright position 3 minutes after the radionuclide bolus ingestion. The oral cavity *(large arrow)* and gastric fundus *(GF)* are clearly visualized. The radiocolloid may adhere to the tongue, and up to 4% of the initially ingested activity can remain in the oral cavity at the end of the 2-minute acquisition. It should be remembered that this radioactivity stasis is "artificially" enhanced on the static image in contrast to the 2-second image of the dynamic study. Persistent midesophageal activity on the static image *(at the level of the small arrow)* may be a relatively frequent finding, particularly in the supine position and in elderly patients. However, it should be less than 5% of the maximal activity recorded at this level. (Taillefer R, Beauchamp G: Radionuclide esophagogram. Clin Nucl Med 9;465-483, 1984)

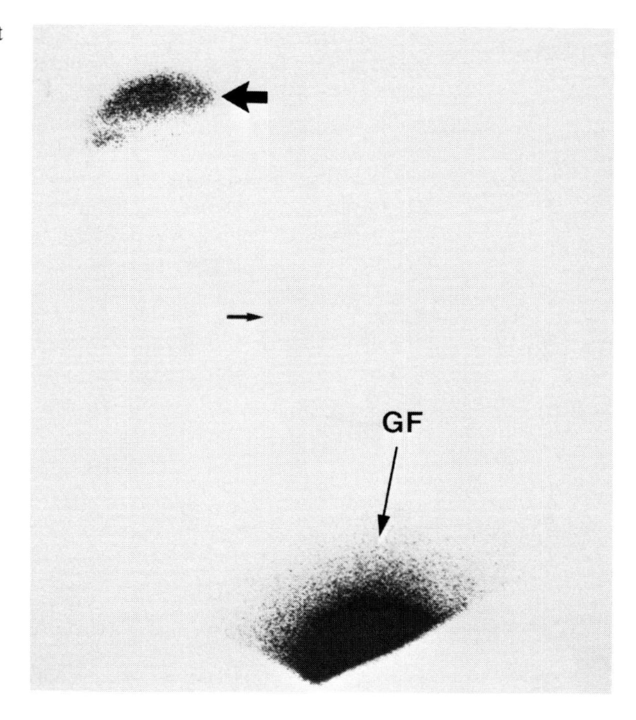

Figure 1-7

Normal "Transient Delayed Emptying" of the Middle Esophagus

The RETS was performed in the supine position in a 68-year-old healthy subject free of significant esophageal symptomatology. Beginning on the fourth frame, there is a transient radionuclide stasis of 15% involving the middle esophageal segment *(arrows)* which disappeared within 7 seconds after a second swallow, with a global esophageal emptying time of 18 seconds. ***Comment:*** This RETS illustrates a normal variant that is usually seen in the elderly patient, particularly in the supine position with a solid bolus. This can be caused by a decreased amplitude in the esophageal body propulsion. It is therefore important to establish a normal range of values in that specific population.

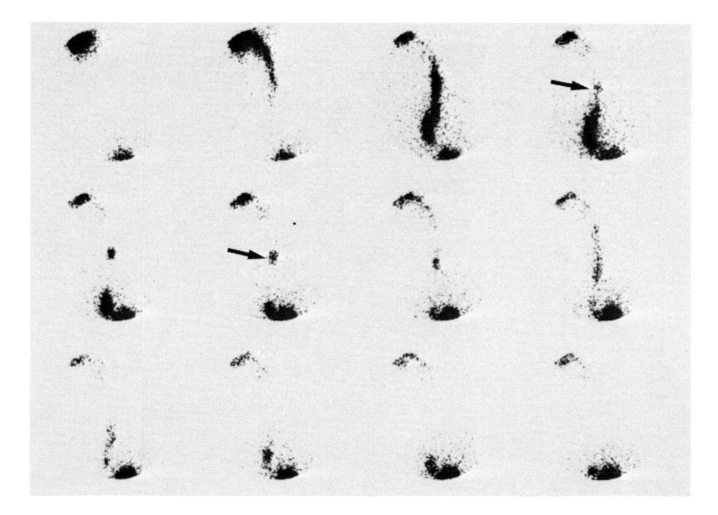

Figure 1-8 ***Diagnosis of Cricopharyngeal Achalasia***

(A) Radionuclide study (upright position, liquid bolus) in a patient suffering from a significant upper dysphagia, in whom manometry found a cricopharyngeal achalasia. There is a marked hypopharyngeal *(H)* radionuclide stasis, which persists throughout the study. This stasis was evaluated to be 85% at 15 minutes on the hypopharyngeal time-activity curve. Esophageal activity was barely seen during the dynamic study. ***(B)*** The same study performed one week after a cricopharyngeal myotomy shows a better pharyngeal clearance, although there still is moderate (25%) radionuclide retention, reached a few seconds after the maximal peak activity and persisting for 15 minutes in valleculae and pyriform sinuses *(H)*. ***Comment:*** RETS provides unique quantitative information to the clinician on the hypopharyngeal emptying function. It is particularly useful in the follow-up of treatment effectiveness, as with cricopharyngeal myotomy. (Taillefer R, Beauchamp G, Duranceau A, Lafontaine E: Nuclear medicine and esophageal surgery. Clin Nucl Med 11:445-459, 1986)

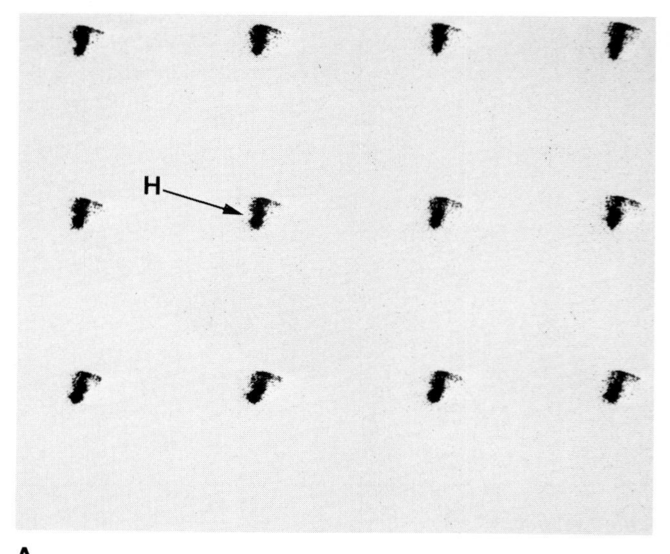

A

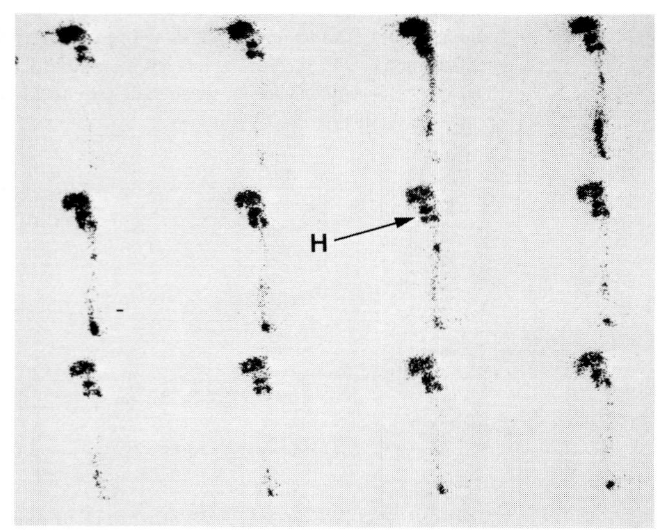

B

Figure 1-9 ***Diagnosis of Oculopharyngeal Muscular Dystrophy (OMD)***

This patient had the characteristic symptoms of OMD: bilateral symmetrical ptosis and progressive oropharyngeal dysphagia occurring late in life. This RETS was performed upright with a solid bolus (a piece of scrambled egg labeled with Tc^{99m} sulfur colloid). There is a significant hypopharyngeal *(H)* stasis and also some degree of pharyngonasal regurgitation *(closed arrow).* Furthermore, the esophageal emptying is very delayed *(open arrows).* ***Comment:*** OMD is a disorder of genetic origin with an autosomal dominant inheritance. Dysphagia usually is accompanied by pharyngo-oral or pharyngonoasal regurgitations and by frequent tracheal aspirations. Motility studies in patients with this lesion have revealed an impaired pharyngeal function characterizied by weak and slow contractions.[22] Manometrically, the upper esophageal sphincter must be considered to be normal, but radiologically, the unrelaxed or incompletely relaxed sphincter suggests that it probably acts as a functional obstruction against the powerless pharynx. Peristaltic activity is reduced or absent in the esophageal body, as seen in this case. There is no curative treatment for this disorder. The goal of treatment is to facilitate the passage of food from the pharynx through the upper esophageal sphincter. Cricopharyngeal myotomy remains the best approach to improve the symptomatology in these patients. Abnormal hypopharyngeal radionuclide stasis is not specific to OMD. It can be found in many muscular or neurological diseases affecting the degree of pharyngeal contraction or in upper esophageal sphincter lesions.[23] Since clinical presentation of OMD is usually very characteristic, RETS is mostly useful to quantitatively evaluate the treatment effectiveness.

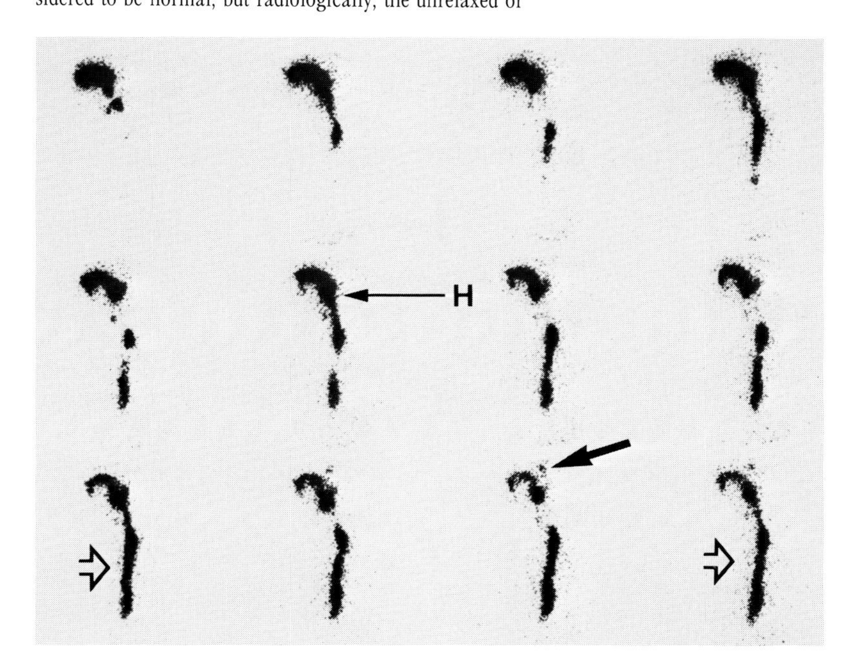

Figure 1-10 ***Diagnosis of OMD, Before and After Surgery***

(A) RETS was performed upright in this patient with OMD to evaluate the hypopharyngeal emptying. There is significant hypopharyngeal radiocolloid stasis *(arrows)* cephalad to the upper esophageal sphincter, including valleculae and pyriform sinuses. Quantification of the hypopharyngeal emptying showed a 35% stasis at 15 minutes. The esophageal transit is normal with no significant residual activity. ***(B)*** The radionuclide study performed in the same patient six months following a cricopharyngeal myotomy shows a significant improvement of the hypopharyngeal *(H)* clearance with only a slight stasis of less than 5% at 10 minutes. There also is a delayed emptying of the distal esophageal segment *(arrows)* which is frequently seen in these patients and attributed to a muscular involvement. Since this abnormality was not present in the preoperative study, the main diagnostic hypothesis is a secondary esophageal motility disorder. In this case, the impaired esophageal clearance was related to a concomitant peptic esophagitis detected on endoscopy. (Taillefer R, Beauchamp G, Duranceau A, Lafontaine E: Nuclear medicine and esophageal surgery. Clin Nucl Med 11:445–459, 1986)

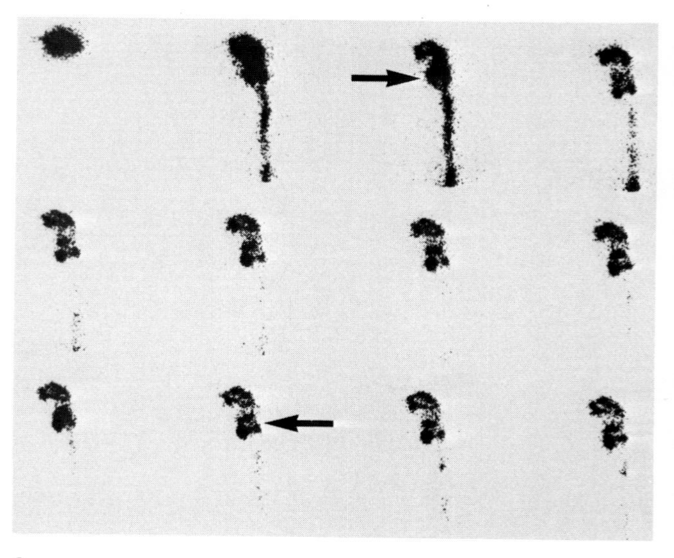

A

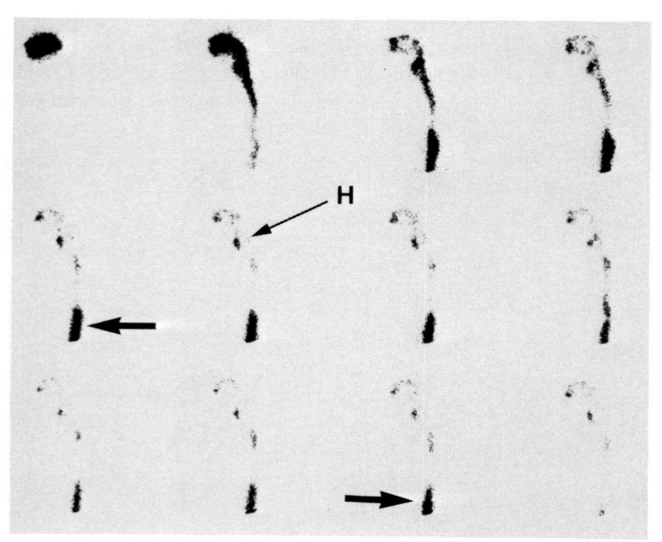

B

Figure 1-11 ***Diagnosis of Right Bronchial Aspiration and Pharyngonasal Regurgitation in OMD.***

(A) This patient with OMD had a preoperative evaluation of the hypopharyngeal emptying function with RETS performed in the uprigt position. Significant delay is noted in the mouth *(M),* hypopharyngeal *(H),* and esophageal *(E)* emptying. At 15 minutes following the ingestion, there still was a hypopharyngeal stasis of 25%. Furthermore, a few seconds after the beginning of the acquisition there is a slight paraesophageal linear focus of activity *(arrow)* corresponding to a right bronchial aspiration of the radioactive bolus. ***Comment:*** RETS is a very sensitive procedure in detecting tracheobronchial aspiration in patients with OMD. When present, this sign reflects the relative severity of the lesion. ***(B)*** Pharyngonasal regurgitation *(arrows),* seen in another patient with OMD, is also indicative of the severity of the lesion. (Taillefer R, Beauchamp G, Duranceau A, Lafontaine E: Nuclear medicine and esophageal surgery. Clin Nucl Med 11:445-459, 1986)

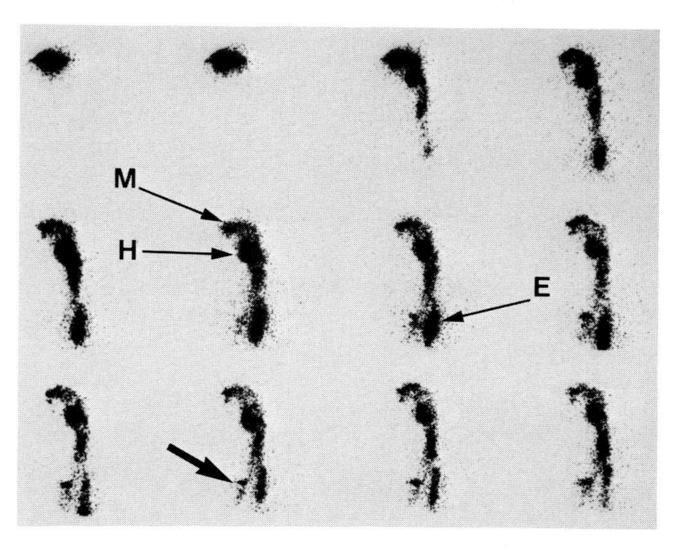

A

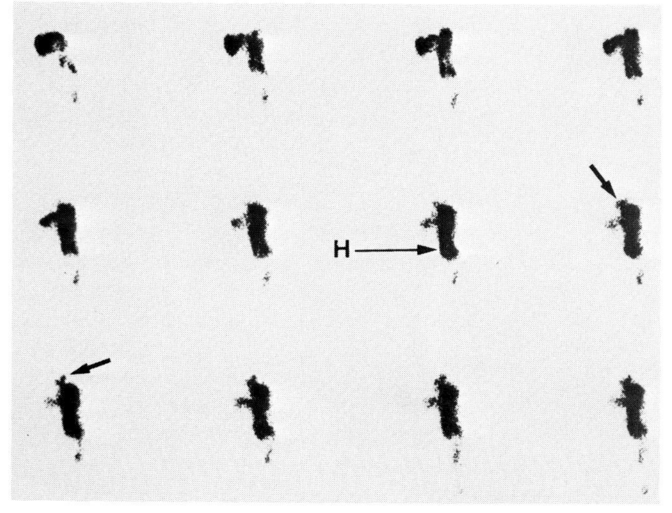

B

Figure 1-12 ***Diagnosis of Severe Pulmonary Aspiration***

Static RETS (upright position, liquid bolus) in a patient with OMD who had transient coughing during the radionuclide ingestion shows significant hypopharyngeal *(H)* stasis, with visualization of the esophagus *(arrowheads)* and the stomach *(S).* There are several sites of abnormal extraesophageal activity *(arrows)* corresponding to a severe pulmonary aspiration with diffuse bilateral bronchial involvement. ***Comment:*** Tracheal aspiration is a relatively frequent complication in OMD as in patients with other causes of oropharyngeal dysphagia. RETS is sensitive in detecting this complication, even if the patient remains asymptomatic during the ingestion. (Taillefer R, Beauchamp G, Duranceau A, Lafontaine E: Nuclear medicine and esophageal surgery. Clin Nucl Med 11:445-459, 1986)

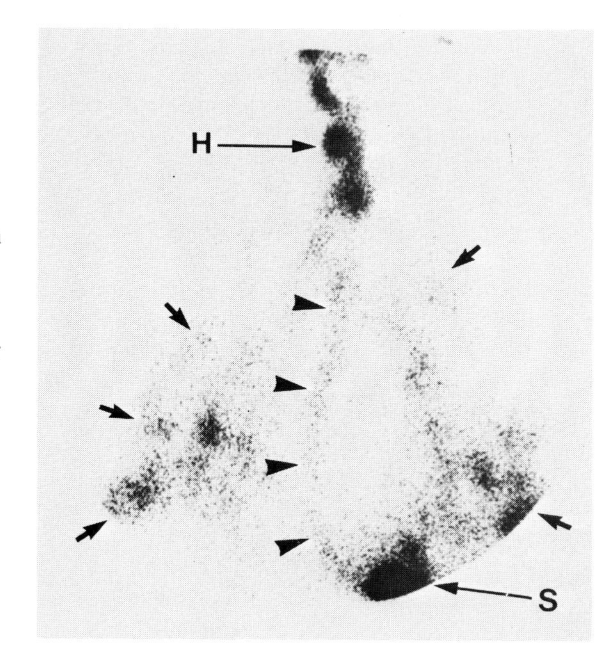

Figure 1-13 **Diagnosis of Polymyositis**

In a patient with polymyositis, the RETS shows a significant hypopharyngeal *(H)* stasis with delayed clearance from each esophageal segment, more pronounced at the distal part *(arrows)*. In contrast to other connective tissue disorders like scleroderma, systemic lupus erythematosus, or Raynaud's disease, dermatomyositis or polymyositis may cause both impaired pharyngeal function and esophageal motility disorder involving proximal (striated muscle) and distal (smooth muscle) esophageal segments.

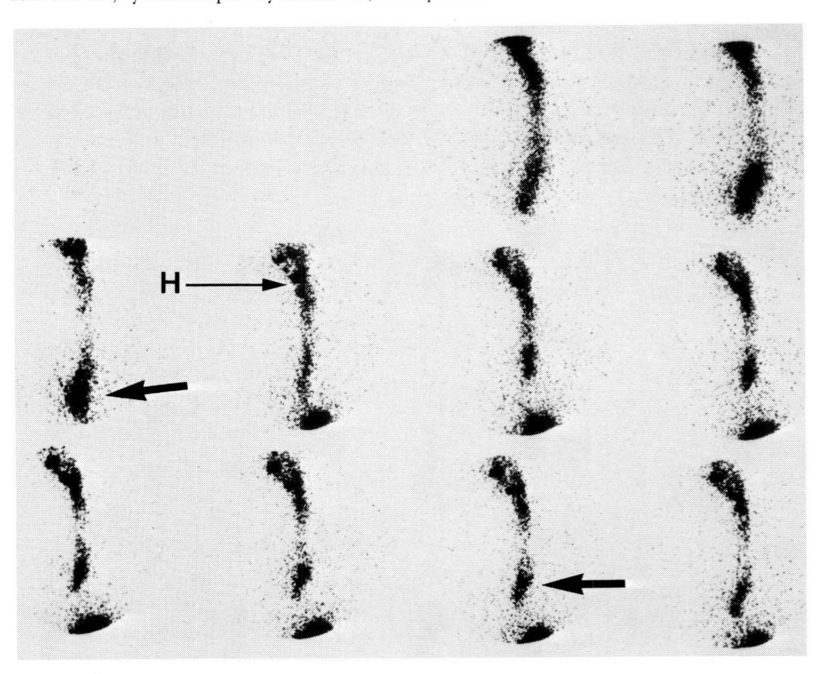

Figure 1-14 **Diagnosis of Myotonic Dystrophy (Steinert's disease)**

RETS was performed to evaluate esophageal involvement. There is a marked radionuclide hypopharyngeal *(H)* stasis during the entire study, which is accompanied by pharyngonasal regurgitation *(closed arrows)*. There is also delayed distal esophageal clearance *(open arrows)* with an esophageal emptying of more than 3 minutes in the upright position.
Comment: Myotonic dystrophy is a disorder of genetic origin which has an autosomal dominant inheritance. Myotonia is characterized by abnormality of muscle relaxation. Esophageal motility dysfunction is a common feature with both smooth and striated muscle involvement, as in this case. There is a decrease in the amplitude of esophageal muscle contraction and a decrease in the number of peristaltic contractions following deglutition.

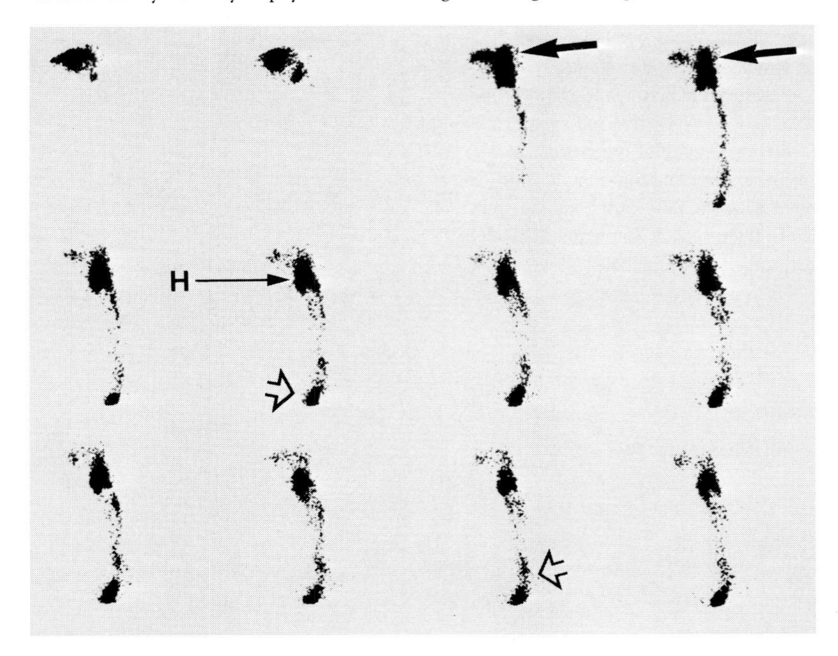

Figure 1-15 ***Diagnosis of Achalasia***

This patient complained of progressive dysphagia over the previous few years. ***(A)*** The RETS (upright position) shows a dilation and complete radionuclide stasis at the level of the upper esophageal lumen *(arrow)*. Until the first minute, the distal esophageal lumen was not visualized. Gastric activity appeared only at 4 minutes following the ingestion. ***(B)*** In supine position, more than 85% of the ingested activity remained in the proximal and distal esophageal segments at 15 minutes *(arrows)*. ***Comment:*** This pattern is suggestive of achalasia, an esophageal motility disorder of unknown etiology characterized by absence of peristalsis in the body of the esophagus and incomplete or absent relaxation of the high pressure zone (lower esophageal sphincter) upon deglutition. Stenotic lesions as seen with esophageal cancer or in complications of peptic esophagitis may also show significant stasis. Complementary procedures are essential to determine the final diagnosis.

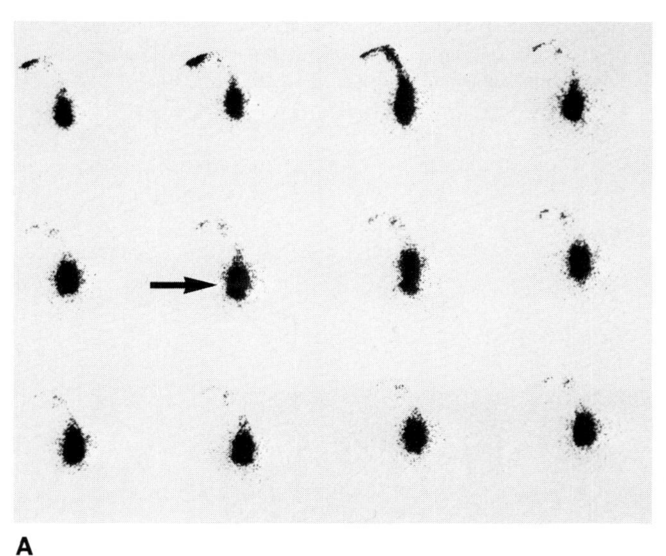

A

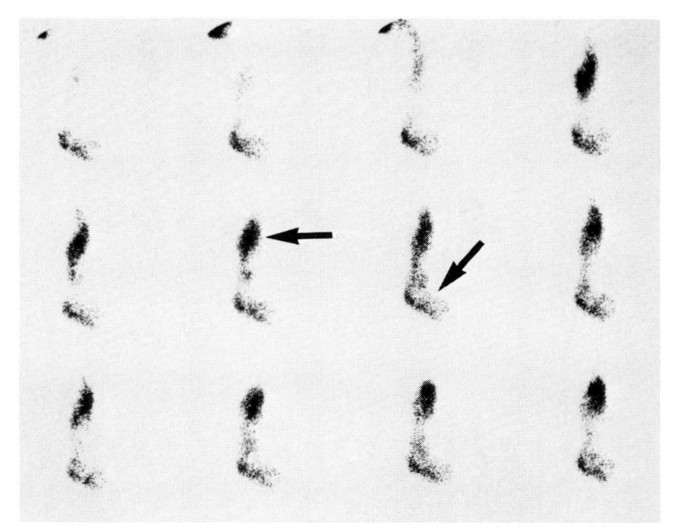

B

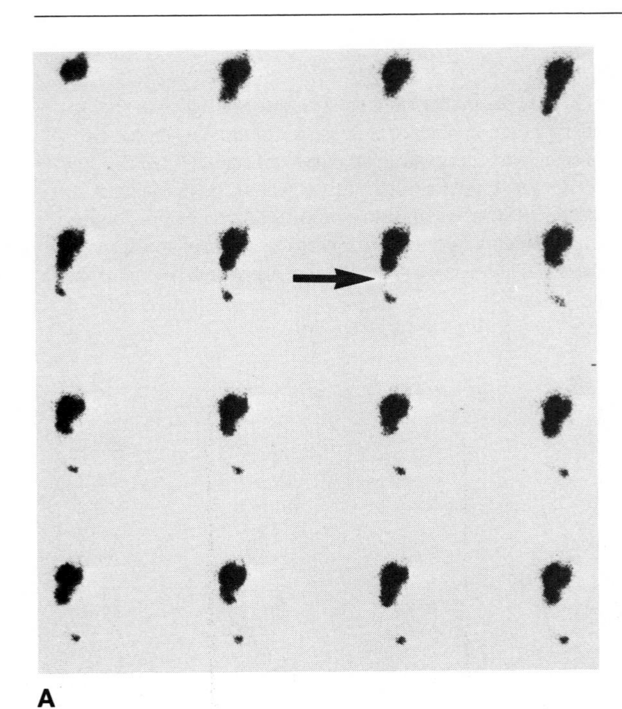

A

Figure 1-16 ***Diagnosis of Severe Achalasia***

A few years following a Heller myotomy for achalasia, this patient was investigated for recurrent progressive dysphagia. ***(A)*** This study (supine position) shows a significant proximal esophageal dilation with subtotal obstruction at the level of the middle segment *(arrow).* Gastric activity was not seen and there was no esophageal emptying at 15 minutes. ***(B)*** The static image obtained at the end of the study in upright position and *(C)* corresponding barium swallow demonstrate a dilated esophagus *(open arrow)* with no passage of the bolus through the esophagogastric junction. ***Comment:*** This is a pattern of esophageal aperistalsis with abnormal lower esophageal sphincter opening. (Taillefer R, Beauchamp G: Radionuclide esophagogram. Clin Nucl Med 9:465 – 483, 1984)

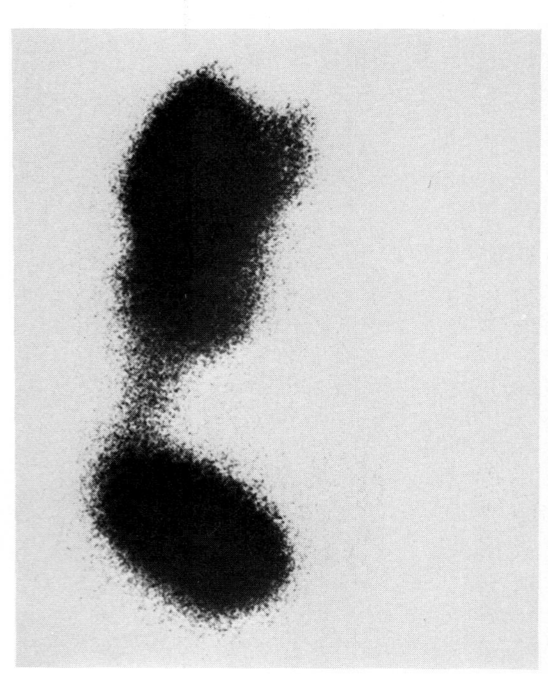

B

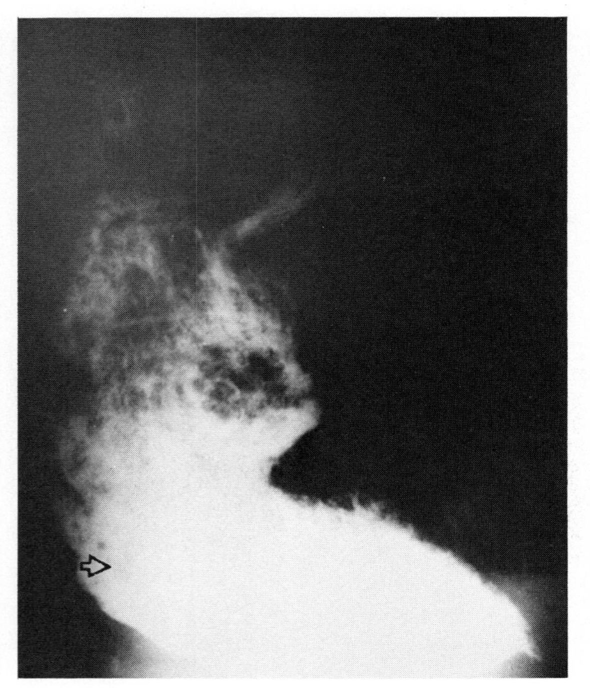

C

Figure 1-17 ***Diagnosis of Achalasia, Before and After Surgery***

(A) Preoperative RETS (upright position) in a patient investigated for dysphagia. There is a very slow progression of the bolus from the proximal to the distal esophageal lumen (20 seconds) with no visualization of gastric activity during the first minute, suggesting the diagnosis of an esophageal motility disorder such as achalasia. *(B)* On the corresponding time-activity curve, the global esophageal radionuclide stasis is 80% at 15 minutes. *(C)* RETS performed in the same patient nine months after a Heller myotomy shows a significant improvement in esophageal emptying. Gastric activity is seen as soon as 8 seconds *(arrow)* after radionuclide ingestion. *(D)* This is quantitatively evaluated on the time-activity curve. ***Comment:*** The principal role of RETS in achalasia is to assess the effectiveness of treatment, either medical (mechanical and pneumatic dilators, drug therapy) or surgical (Heller's myotomy).[15,24-27]

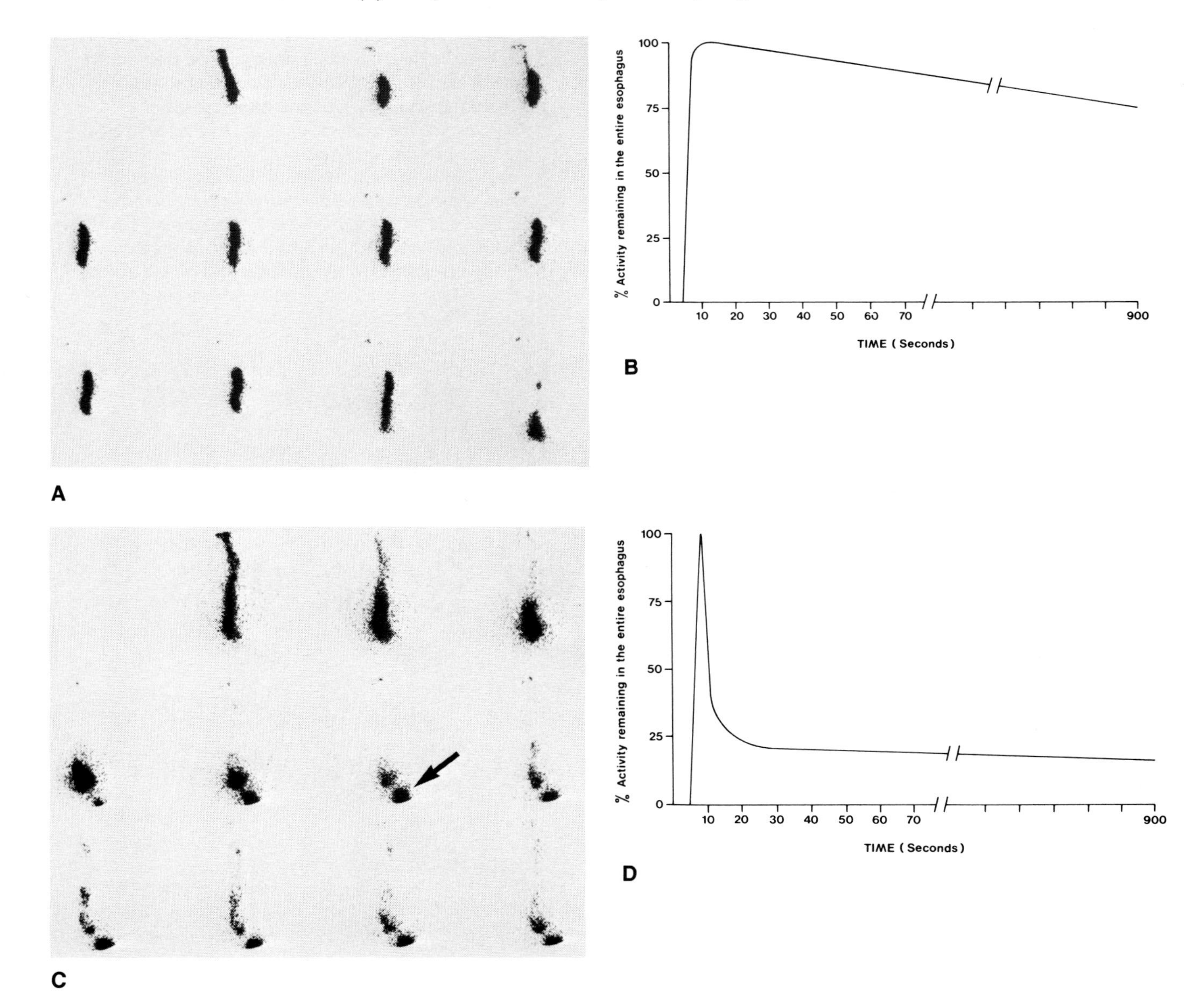

Figure 1-18 ***Diagnosis of Diffuse Esophageal Spasm***

This patient had complained of atypical chest pain for a few years. The results of a complete cardiac work-up were normal. The development of progressive dysphagia raised the possibility that the pain was of esophageal origin. Results of barium swallow, endoscopy, and acid reflux test were also in the normal range. The RETS performed in the upright position showed decreased entire and segmental esophageal emptying with incoordinate passage of the bolus through the esophagel lumen, accompanied by fragmentation and retrograde movements of the bolus *(arrows)*. These findings suggested the diagnosis of diffuse esophageal spasm, which was confirmed at manometry. ***Comment:*** Diffuse esophageal spasm is an uncommon motility disorder characterized by spastic activity of the lower two thirds of the esophagus. In its severe form, it produces chest pain or dysphagia.[28] These symptoms are associated with multiple, synchronous, and repetitive contractions of increased amplitude and duration. The distal segment contracts synchronously with the others, rather than in the normal sequential pattern. Esophageal spasm may appear as a manifestation of esophageal inflammation, or as a consequence of a mechanical or functional obstruction. It may also be an isolated motility disorder increasing in frequency with age. The precise cause of diffuse spasm is not established, but there is some evidence that diffuse esophageal spasm and achalasia may be related diseases, both showing vagal degeneration. RETS can be used as a complementary diagnostic procedure; the majority of patients with diffuse esophageal spasm will have abnormal findings. Although nonpropulsive contractions may be associated with a large number of esophageal motility disorders, when the findings described above are seen in the upright position in a patient with "classic" symptomatology, they are likely to be secondary to a diffuse esophageal spasm with multiple peaks on time-activity curves of esophageal segments (with only one single peak on hypopharyngeal and proximal esophageal curves, thus eliminating multiple swallows). However, results of RETS are often at the normal upper limit in this position. A supine study is much more sensitive but less specific for diffuse esophageal spasm. Retrograde movements or esophagoesophageal reflux can be seen in patients with nonspecific motor disorders. Achalasia or gastroesophageal reflux may also produce this pattern, although these conditions can be differentiated by other findings. (Taillefer R, Beauchamp G: Radionuclide esophagogram. Clin Nucl Med 9:465-483, 1984)

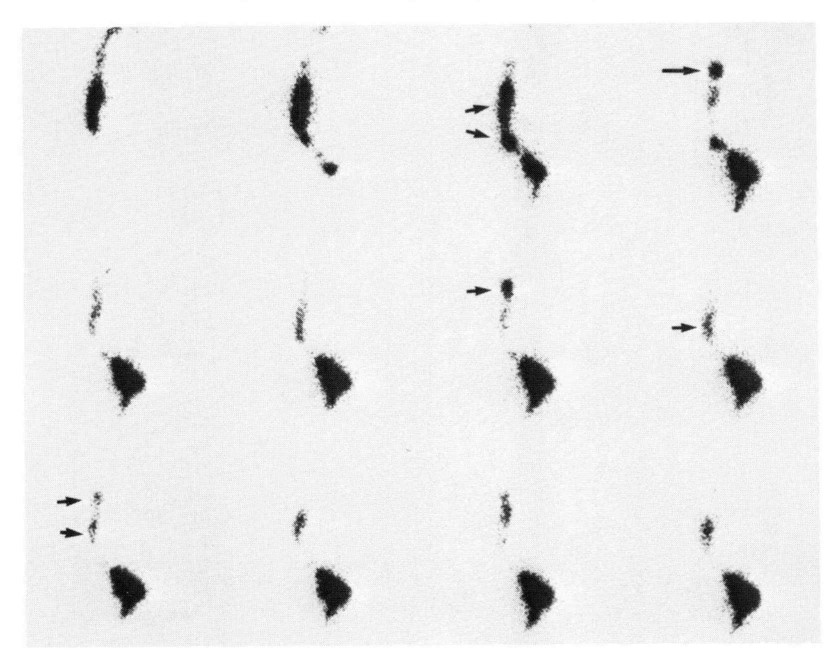

Figure 1-19 ***Diagnosis of Diffuse Esophageal Spasm***

Intermittent dysphagia and substernal discomfort prompted a RETS evaluation in this patient. *(A)* On the upright study, there are retrograde movements of the bolus *(arrows)* in the middle and distal esophagus, suggesting nonperistaltic contractions in the smooth muscle segment of the esophagus, causing proximal and distal displacement from multiple contraction sites. Furthermore, early normal visualization of the gastric activity probably reflects an adequate lower esophageal sphincter relaxation. These findings and the clinical background are compatible with diffuse esophageal spasm. *(B)* Supine study in the same patient showing a more significantly impaired esophageal emptying pattern.

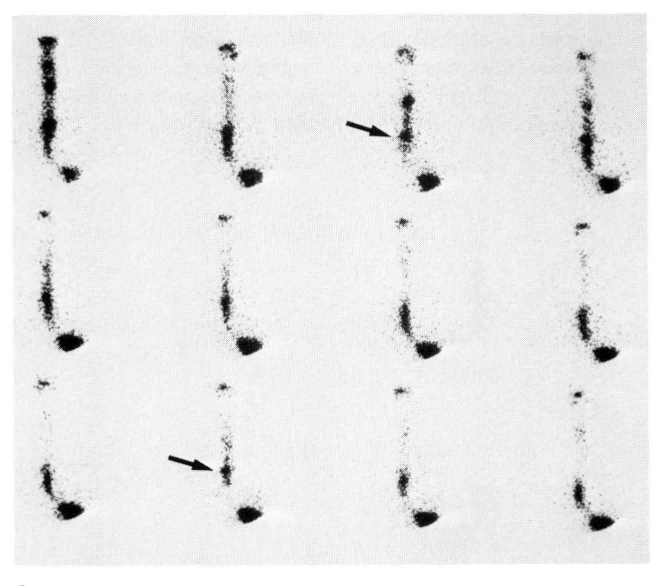

A

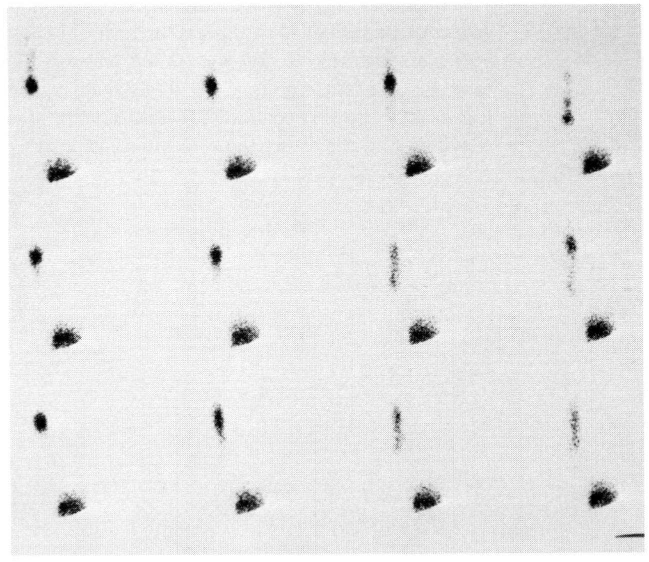

B

Figure 1-20 ***Diagnosis of "Nutcracker" Esophagus***

RETS was performed in this patient to evaluate an atypical chest pain of possible esophageal origin. The esophageal transit is normal in the proximal segment. In the distal segment there is a delay in the emptying and fragmentation of the bolus *(arrows)* with retrograde movements. The middle and distal esophageal emptying time was 18 seconds, which is definitely abnormal. In this clinical context, these findings should raise the possibility of diffuse esophageal spasm or an entity that Benjamin and co-workers[29] have described as the "nutcracker" esophagus. ***Comment:*** The "Nutcracker" esophagus is an esophageal motor disorder characterized by hypertensive peristaltic contractions localized to the distal esophagus. It is one of the most common manometric abnormalities (increased amplitude of the esophageal peristaltic waves) found in patients investigated for chest pain of esophageal etiology. Scintigraphically, the "nutcracker" esophagus is characterized by delayed esophageal emptying and a chaotic aspect of the esophageal bolus transit. Unless there is a diffuse esophageal involvement, it is impossible to distinguish, on a scintigraphic basis alone, "nutcracker" esophagus from diffuse esophageal spasm, because the latter also usually involves the distal esophagus. Furthermore, at the present time it is not known if "nutcracker" esophagus is a true specific primary esophageal motility disorder or if it represents an intermittent diffuse esophageal spasm.

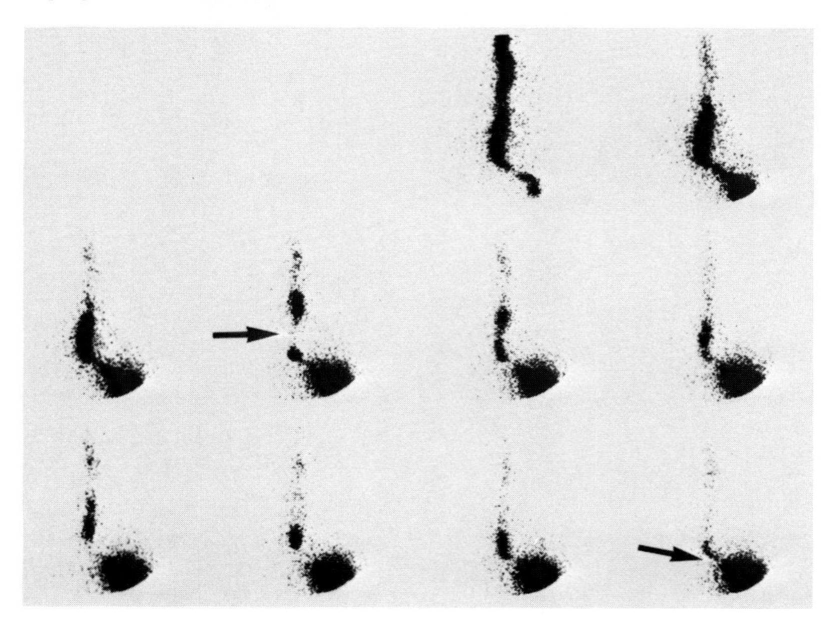

Figure 1-21 ***Diagnosis of Esophageal Motility Disorder Associated with Gastroesophageal Reflux***

This patient complained of dysphagia, pyrosis, and regurgitation. The barium swallow showed mucosal irregularity at the distal third, which was confirmed on endoscopy to be uncomplicated peptic esophagitis. RETS (upright position) was performed to evaluate the dysphagia. The bolus transit is normal in the proximal esophagus, but there is a delay in the clearance of the distal part *(arrow)*, which suggests an esophageal dysmotility. In this case, this is probably related to a gastroesophageal reflux disease. Manometry confirmed the esophageal motility disorder by showing an increased incidence of nonperistaltic contraction waves of low amplitude in the lower esophagus. (Taillefer R, Beauchamp G, Duranceau A, Lafontaine E: Nuclear medicine and esophageal surgery. Clin Nucl Med 11:445-459, 1986)

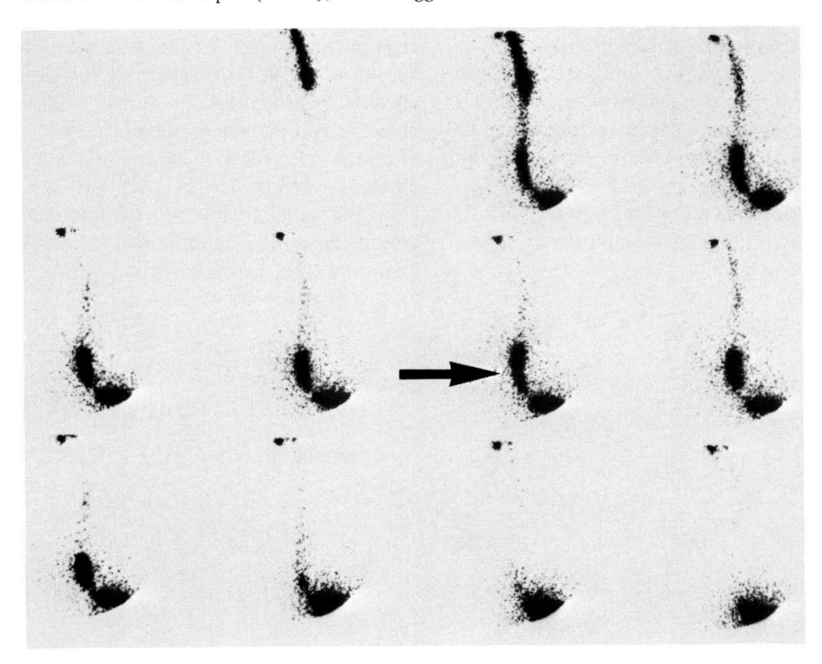

Figure 1-22 ***Diagnosis of Esophageal Motility Disorder and Antireflux Procedure***

Ten months after an antireflux procedure (fundoplication), this patient complained of persistent dysphagia. A barium esophagogram did not show mechanical obstruction, but the RETS showed an abnormal global and segmental esophageal emptying *(arrows)* mainly of the proximal and middle segments with retrograde bolus movement *(large arrow)*. These findings are compatible with an esophageal motility dysfunction which was confirmed at manometry. ***Comment:*** RETS allows sensitive detection of esophageal emptying disorders in the evaluation of dysphagia in a patient submitted to a fundoplication. During the first months after an antireflux procedure, patients may experience dysphagia; if it persists, however, and if a mechanical obstruction is eliminated by a normal barium swallow, then an esophageal motility disturbance should be suspected. Interpretation of the RETS results should ideally be correlated to a preoperative RETS study. The presence of a motor abnormality postoperatively may be due to a persistent preoperative defect or may be secondary to an induced distal obstruction.

Three different RETS patterns of esophageal emptying are seen in patients with a total fundoplication. The first is of a normal radionuclide transit and esophageal emptying, and usually is seen in asymptomatic patients. The second pattern is of a transiently delayed clearance of the radioactive bolus localized to the distal third of the esophageal body, and is seen mostly with the study performed in supine position with a solid bolus. When correlated to a barium swallow, this delayed emptying is often related to the anatomic narrowing created by the fundoplication. If the barium swallow is normal, then a motility dysfunction may be the cause. The third pattern is characterized by a delayed emptying of the bolus in the distal third segment as well as in the middle and sometimes in the proximal esophagus. This has often been seen in patients with esophageal motility dysfunction, but stricture or tight fundoplication has to be ruled out. (Taillefer R, Beauchamp G, Duranceau A. Lafontaine E: Nuclear medicine and esophageal surgery. Clin Nucl Med 11:445–459, 1986)

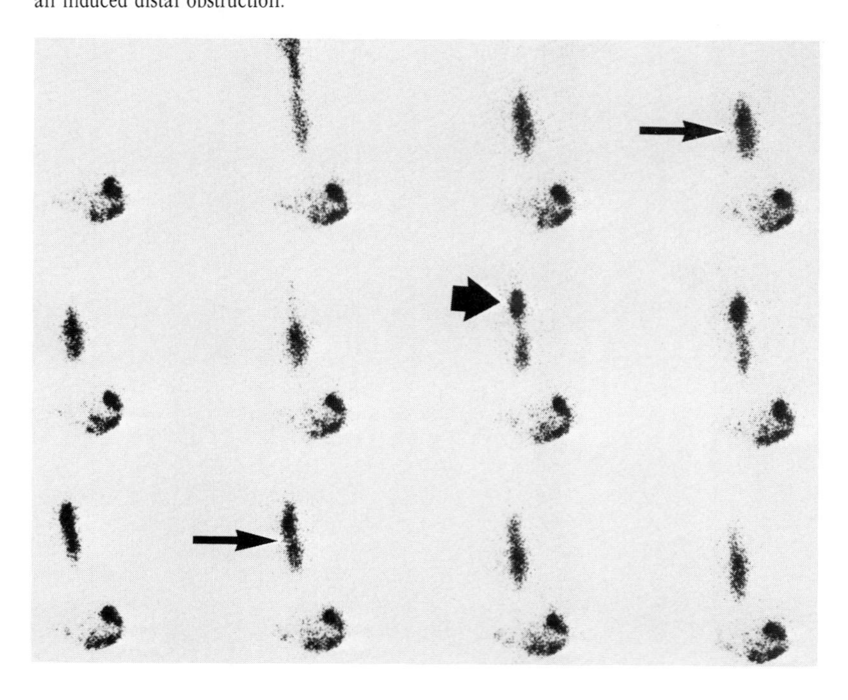

Figure 1-23 ***Diagnosis of Myasthenia Gravis***

With complaints of a progressive oropharyngeal dysphagia for the last three years, this patient was found to suffer from myasthenia gravis. RETS (upright position) shows a moderate transient hypopharyngeal *(H)* stasis with an emptying time of 5 seconds instead of 1 second, which is clearly abnormal. There is a decreased proximal esophageal emptying with retrograde movement *(arrow)*. ***Comment:*** In myasthenia gravis, the peristaltic waves show a decrease in amplitude and, on repetitive swallowing, they disappear in the lower esophagus.

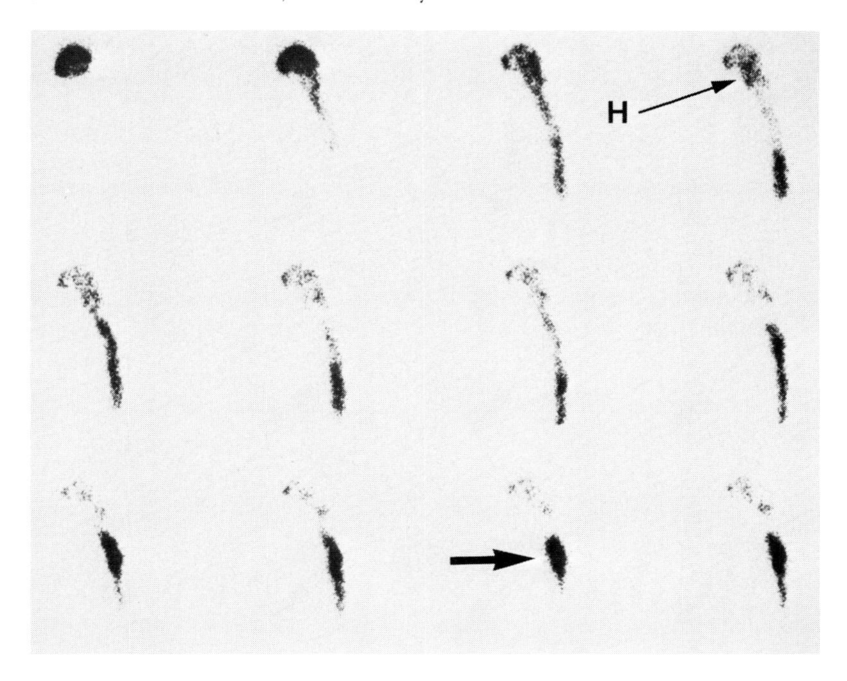

Figure 1-24 ***Diagnosis of Progressive Systemic Sclerosis (upright study)***

This is a RETS in a patient with progressive systemic sclerosis (PSS). The emptying of the proximal esophageal segment is normal while there is significant delay in the clearance of both middle and distal segments *(arrows)* with a stasis of 20% at 2 minutes. ***Comment:*** These findings are not specific and scintigraphically, the main differential diagnosis is of an esophageal motility disorder, as seen in achalasia. Findings have to be interpreted in the clinical context. In the PSS patients we have studied, the esophageal emptying function involvement is usually less severe than in patients with achalasia, depending on the stage of the lesion. RETS is a sensitive method of detecting esophageal involvement in PSS, which is seen in more than 90% of patients with this condition.[30,31] It is important to recognize the impaired esophageal emptying because with as many as 50% of patients asymptomatic, the loss of both esophageal peristalsis and lower esophageal sphincter tone can lead to reflux esophagitis, which can be refractory to treatment.

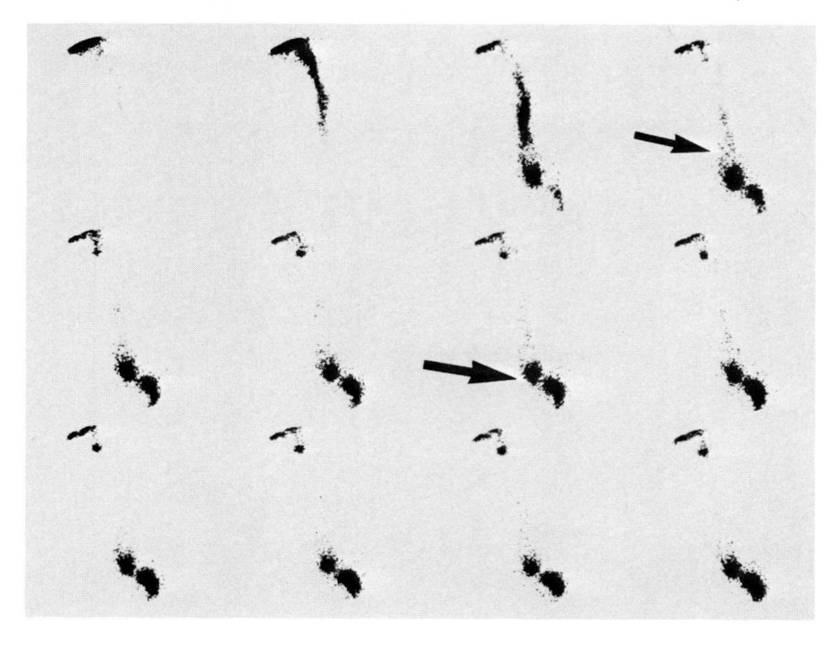

Figure 1-25 ***Diagnosis of Progressive Systemic Sclerosis (supine study)***

This study performed in a patient with PSS shows significant delay in emptying along the entire esophagus *(arrows)* corresponding to nonpropulsive esophageal waves (aperistalsis). This is a severe esophageal involvement. Manometry showed aperistalsis and absent lower esophageal sphincter pressure.

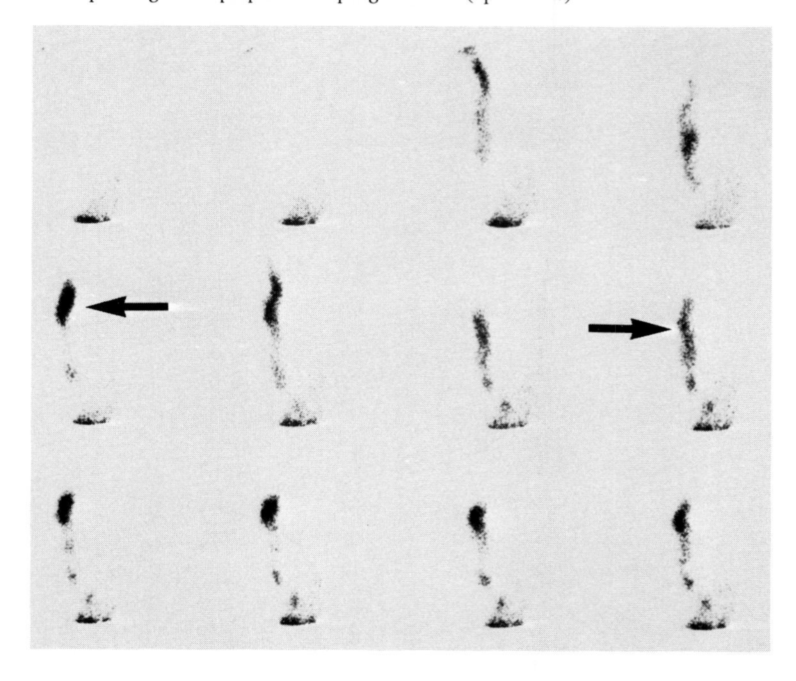

Figure 1-26 ***Diagnosis of Pharyngoesophageal Diverticulum with Regurgitation***

RETS was performed to evaluate an upper dysphagia. There is an abnormal spherical radionuclide retention *(arrow)* at the level of the upper esophageal sphincter, suggesting a pharyngoesophageal (Zenker's) diverticulum. Furthermore, this study shows two distinct episodes of pharyngeal reflux *(arrowheads)*. ***Comment:*** A scintigraphic image of Zenker's diverticulum is different from the one described in upper esophageal sphincter lesions or in powerless pharyngeal contractions. In these latter lesions the radionuclide retention is cephalad to the upper esophageal sphincter, that is, in the valleculae and pyriform sinuses. (Taillefer R, Beauchamp G, Duranceau A, Lafontaine E: Nuclear medicine and esophageal surgery. Clin Nucl Med 11:445-459, 1986)

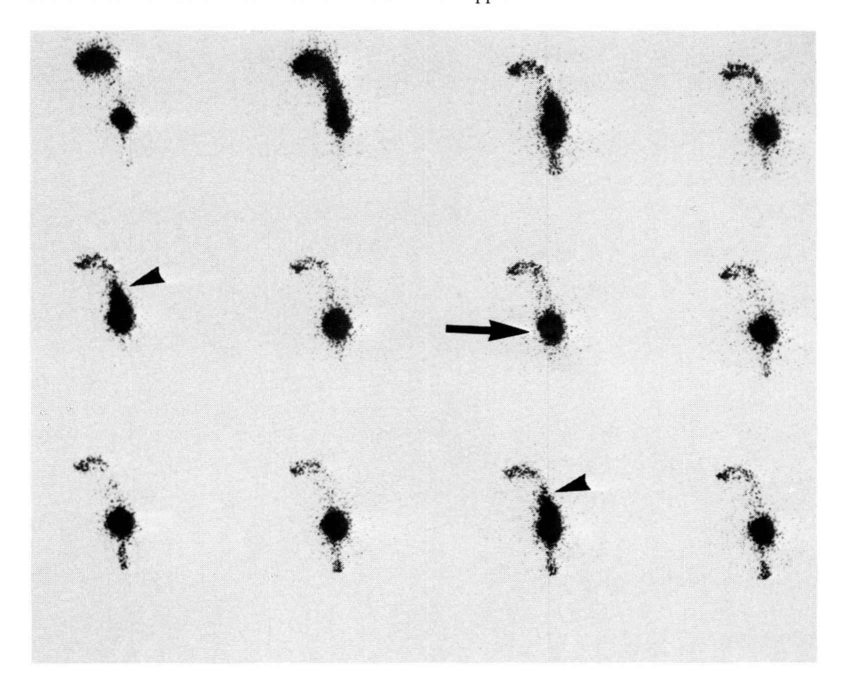

Figure 1-27 ***Diagnosis of Pharyngoesophageal Diverticulum with Esophageal Motility Disorder***

This patient was investigated for a symptomatology suggesting the presence of a Zenker's diverticulum. He also suffered from a lower dysphagia, for which RETS was performed. ***(A)*** The ovoid focus of radionuclide retention *(closed arrow)* at the level of the upper esophageal sphincter corresponds to a pharyngoesophageal diverticulum. Global esophageal emptying is also moderately prolonged *(open arrow)*, indicating an underlying alteration of esophageal motility. A nonspecific esophageal motor disorder was found at manometry. ***(B)*** Corresponding barium esophagogram with the Zenker's diverticulum. (Taillefer R, Beauchamp G: Radionuclide esophagogram. Clin Nucl Med 9:465-483, 1984)

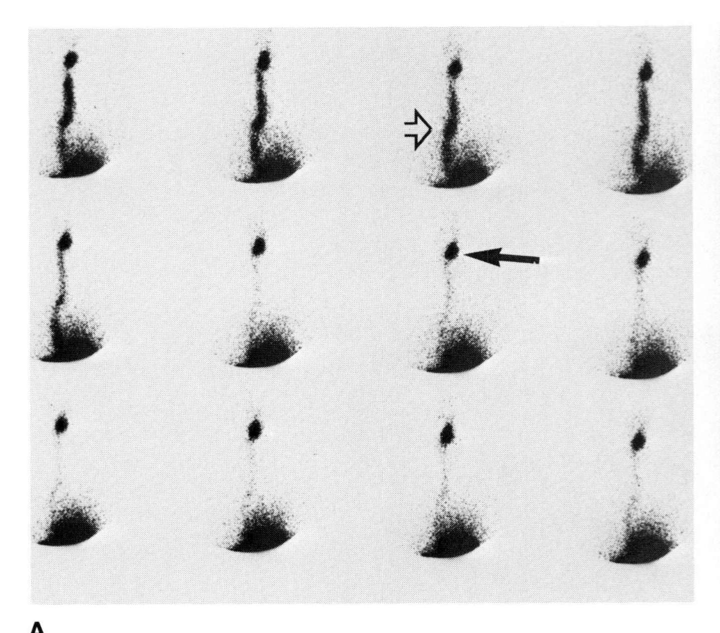

A

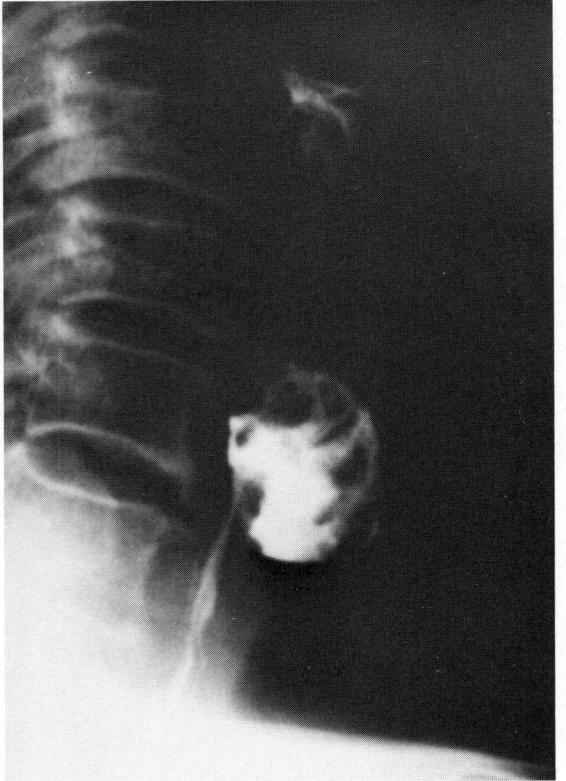

B

Figure 1-28 ***Diagnosis of Sliding Hiatal Hernia***

A static image was obtained in supine position in a patient investigated for heartburn and regurgitations. There is an abnormal focus of radionuclide retention *(arrow)* just above the gastroesophageal junction. ***Comment:*** This focus corresponds to a hiatal hernia, which is frequently seen in patients referred for a gastroesophageal reflux study, particularly in the elderly. Sometimes, peptic esophagitis or a columnar-lined esophagus involving the distal esophageal segment may cause a radiocolloid adsorption to the damaged mucosa. However, these conditions can be differentiated from hiatal hernia by their morphology. They usually have a linear morphology instead of the ovoid or hemispherical shape of the hiatal hernia. Diverticulae may also retain the radionuclide, and oblique anterior or posterior views are helpful to differentiate them from a hiatal hernia. (Taillefer R, Beauchamp G: Radionuclide esophagogram. Clin Nucl Med 9:465-483, 1984)

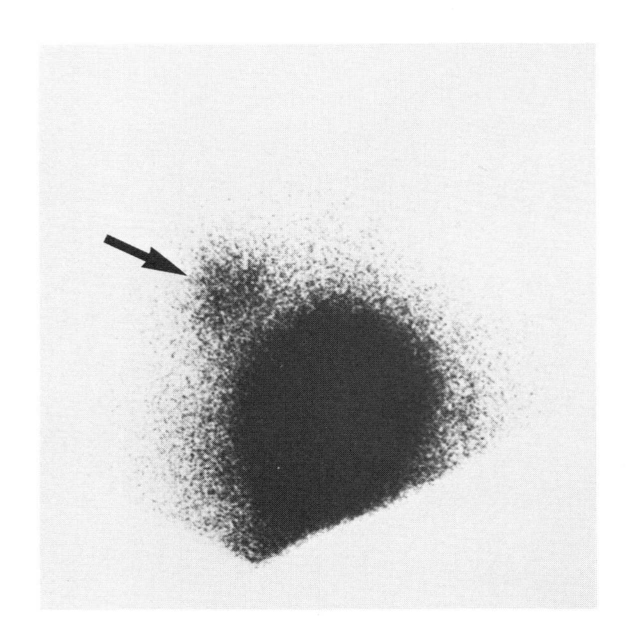

Figure 1-29 ***Diagnosis of Hiatal Hernia with Associated Esophageal Motility Disorder***

Before tracing a region of interest over different esophageal segments, it is important to identify the presence of a hiatal hernia to avoid including it with the distal esophageal segment. Inclusion of a hiatal hernia will falsely give a delay in the emptying of the distal esophageal segment. ***(A)*** In addition to an abnormal transient delayed clearance at the junction of the proximal and middle esophagus, there is a persistent focus of radionuclide retention in projection of the distal esophagus *(arrow)*. ***(B)*** The static image shows that this retention *(arrow)* corresponds to a hiatal hernia.

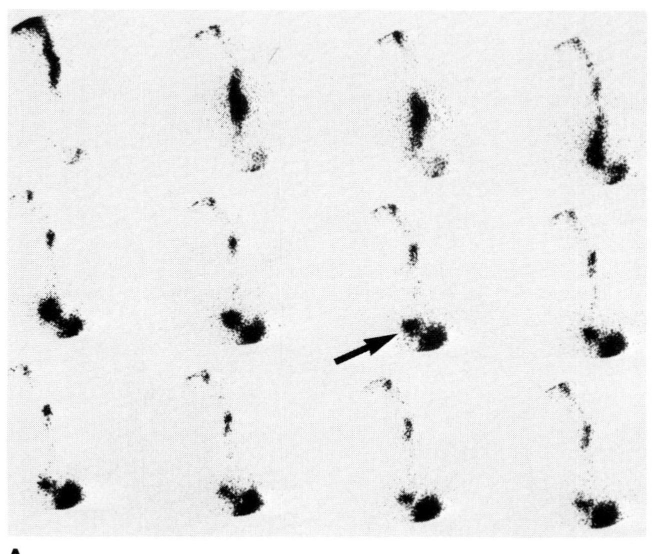

A

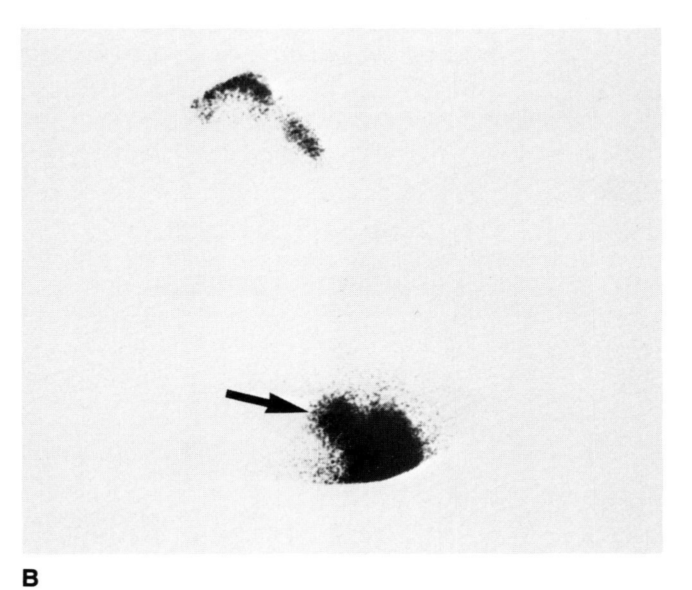

B

Figure 1-30 ***Diagnosis of Paraesophageal Hernia***

RETS was performed to investigate a problem of lower dysphagia. ***(A)*** Dynamic study parameters are normal for the proximal and middle esophageal segments. However, in the projection of the distal segment *(arrow)*, there is an abnormal stasis of radionuclide which slowly empties distally. ***(B)*** The static image at the end of the study was particularly useful to clarify this finding. There is no esophageal stasis, but the focus of marked retention *(arrow)* corresponds to a paraesophageal hernia, which was confirmed on barium swallow. ***(C)*** ***Comment:*** Contrary to the axial hiatus hernia, the cardia remains normally located and does not herniate through the hiatus in the paraesophageal hiatal hernia. This type of hernia occurs much less frequently than the sliding hiatal hernia. (Taillefer R, Beauchamp G: Radionuclide esophagogram. Clin Nucl Med 9:465–483, 1984)

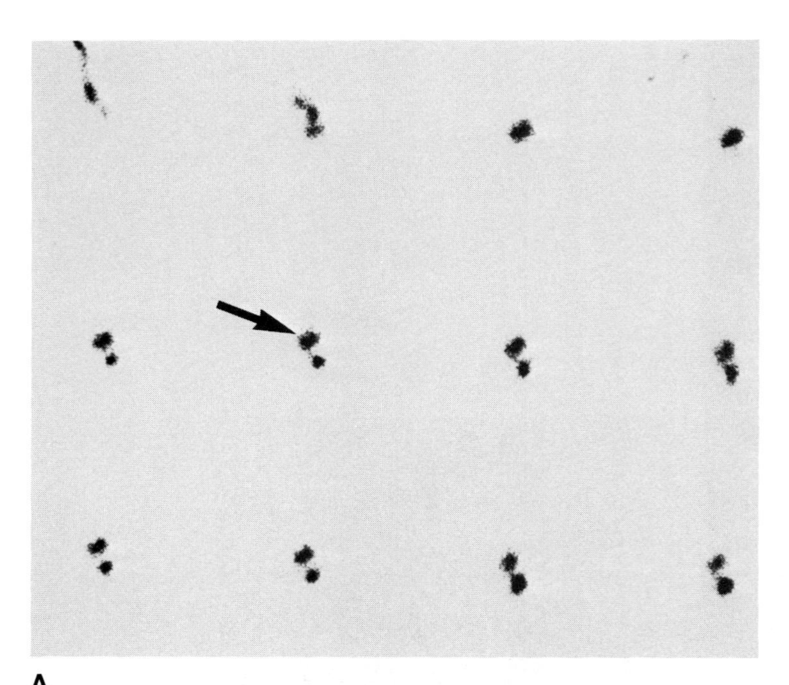

A

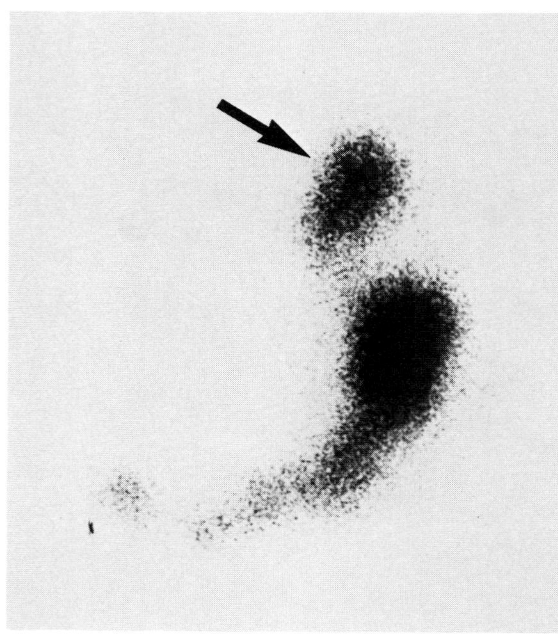

B

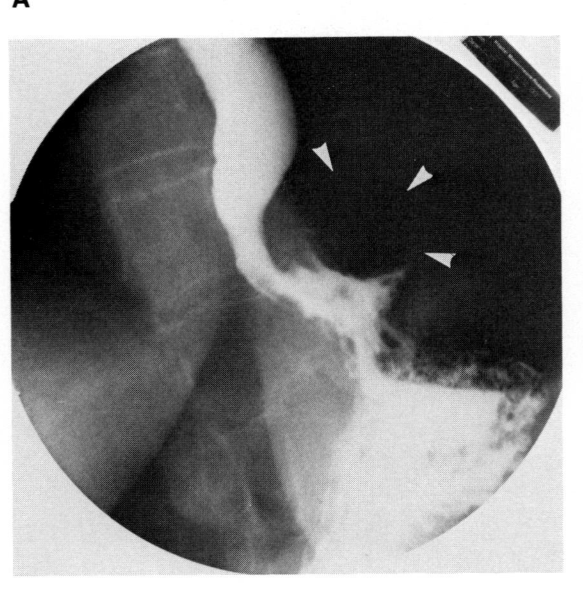

C

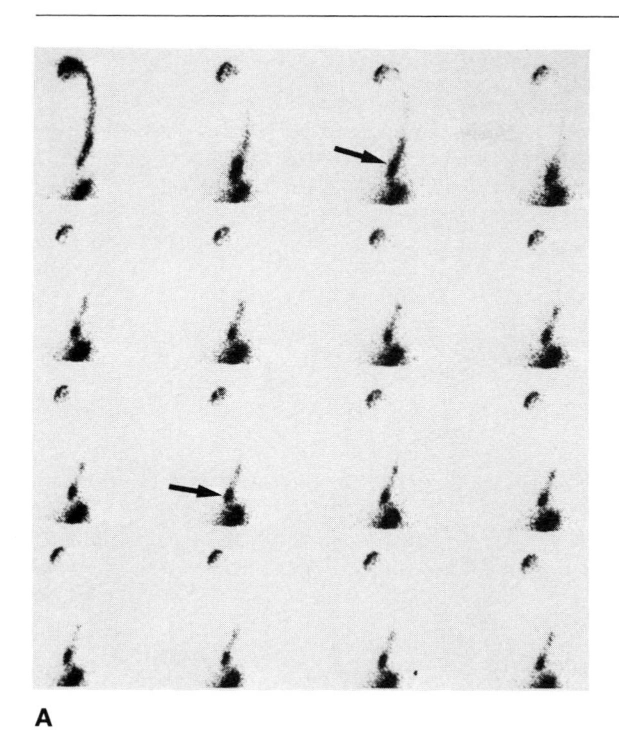

A

Figure 1-31 ***Diagnosis of Esophageal Diverticulae***

This patient was referred for RETS (supine position) for the evaluation of dysphagia and regurgitations. ***(A)*** There is a significant delay in emptying of the distal part of the esophagus *(arrows)* persisting throughout the study. Proximal esophageal emptying is normal. Without static imaging, it is very difficult to determine the exact origin of this emptying delay from any number of causes. ***(B)*** A left anterior oblique static image obtained at the end of the study shows two abnormal ovoid foci of radionuclide retention *(arrows)*. The distal focus has an extraluminal projection. This finding suggests the diagnosis of esophageal diverticulae. ***(C)*** This is confirmed with a barium study. ***Comment:*** RETS has a limited anatomic resolution, but the static image can enhance indirect evidence of a "morphological" lesion. (Taillefer R, Beauchamp G, Duranceau A, Lafontaine E: Nuclear medicine and esophageal surgery. Clin Nucl Med 11:445–459, 1986)

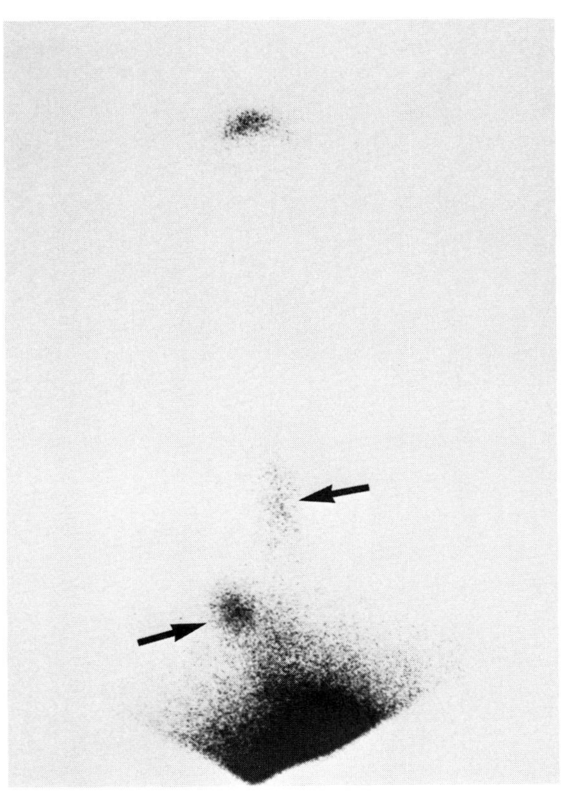

B

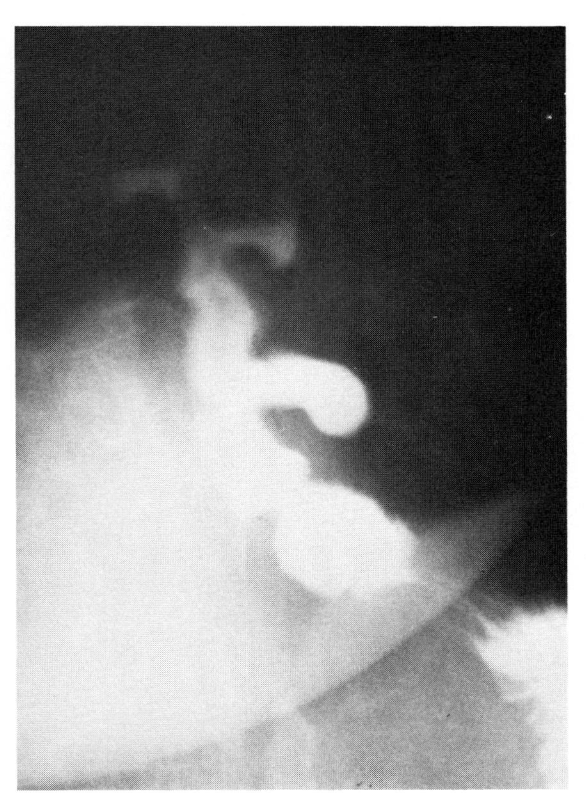

C

Figure 1-32 ***Diagnosis of Epiphrenic Diverticulum***

In this study in the supine position, the proximal and middle esophageal emptying is normal, although there is a biphasic radionuclide ingestion *(arrowhead)*. The distal esophagus shows a significant extraluminal radionuclide stasis *(arrow)* which persisted for 2 minutes, indicating an epiphrenic diverticulum. ***Comment:*** Large diverticulae usually empty only by gravity. Epiphrenic diverticulae located just above the lower esophageal sphincter, as in this case, generally project to the right. (Taillefer R, Beauchamp G: Nuclear medicine and esophageal surgery. Clin Nucl Med 11:445-459, 1986)

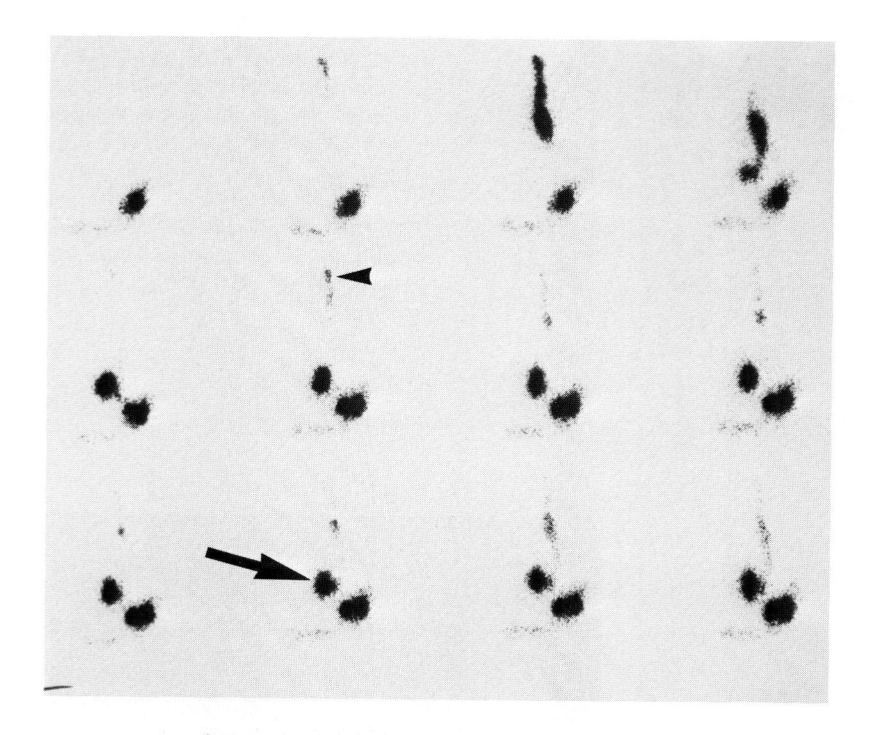

Figure 1-33 ***Diagnosis of Bronchoesophageal Fistula***

This patient, operated on 15 years earlier for a benign tracheoesophageal fistula, complained of dysphagia and transient coughing when swallowing liquids. ***(A)*** The RETS (supine position, liquid bolus) shows an abnormal esophageal emptying of the proximal and middle segments *(arrowheads)*. This was accompanied by an extraluminal focus of activity *(arrow)* on the right, which is more clearly seen on posterior static images ***(B,** arrows)*. This abnormal retention was proven on simultaneous bronchoscopy and esophagoscopy to be a recurrent localized bronchoesophageal fistula involving the right main bronchus. ***Comment:*** Tracheal aspiration of the radionuclide bolus may give a similar pattern; if this is the case, oblique or lateral views and static images can help identify the site of aspiration.

The diagnosis of a tracheoesophageal fistula is sometimes difficult. The symptomatology suggesting its presence (persistent coughing, recurrent pulmonary infection, or cyanosis associated with swallowing) can be confusing. The barium esophagogram is not always diagnostic and simultaneous bronchoscopy and esophagoscopy for dye injection usually necessitate general anesthesia. Pediatric and geriatric patients would benefit from less invasive diagnostic procedures, such as RETS.[32,33] (Taillefer R, Beauchamp G, Duranceau A, Lafontaine E: Nuclear medicine and esophageal surgery. Clin Nucl Med 11:445–459, 1986)

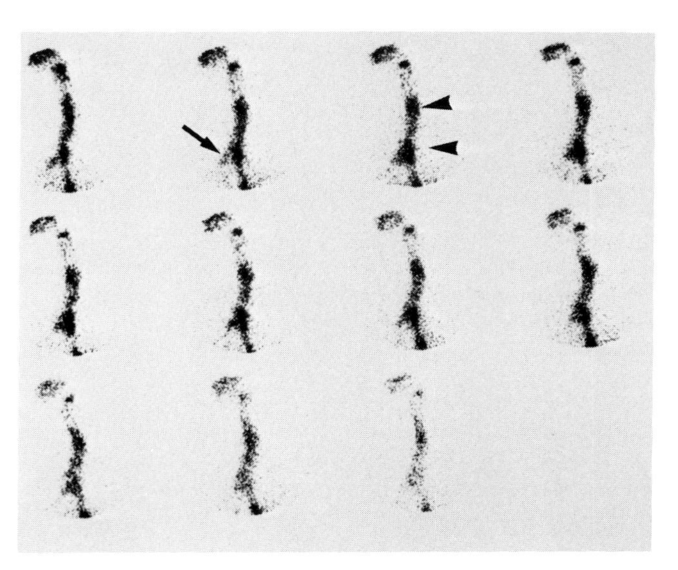

A

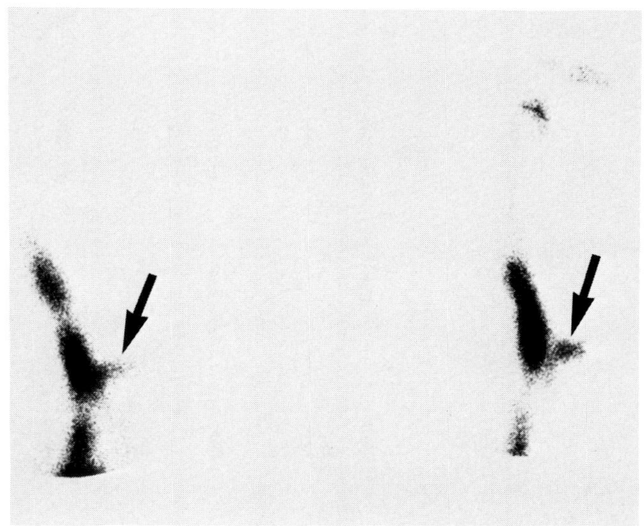

B

Figure 1-34 ***Diagnosis of Columnar-Lined Esophagus with Esophageal Stenosis***

(A) RETS was performed in supine position to evaluate a progressive dysphagia in a patient who was known to have peptic esophagitis secondary to a gastroesophageal reflux of many years' duration. RETS shows an abnormal esophageal emptying *(arrowheads)*, particularly involving the distal and middle segments. Furthermore, there is an area of decreased activity *(arrow)* and partial obstruction to the normal progression of the bolus just above a hiatal hernia *(HH)*, which was clearly seen on the static image. ***(B)*** Tc-^{99m} pertechnetate imaging (2 hours after the intravenous injection of the radionuclide) was done two days later. It shows the hiatal hernia *(arrowheads)* and linear Tc-^{99m} pertechnetate uptake *(arrows)* corresponding to a columnar-lined esophagus (Barrett's). Endoscopy and biopsy results confirmed the presence of peptic esophagitis with partial stenosis accompanying a columnar-lined esophagus involving the distal and a portion of the middle segment. Manometry demonstrated a hypotensive lower esophageal sphincter with esophageal motor dysfunction. (Taillefer R, Beauchamp G: Radionuclide esophagogram. Clin Nucl Med 9:465–483, 1984)

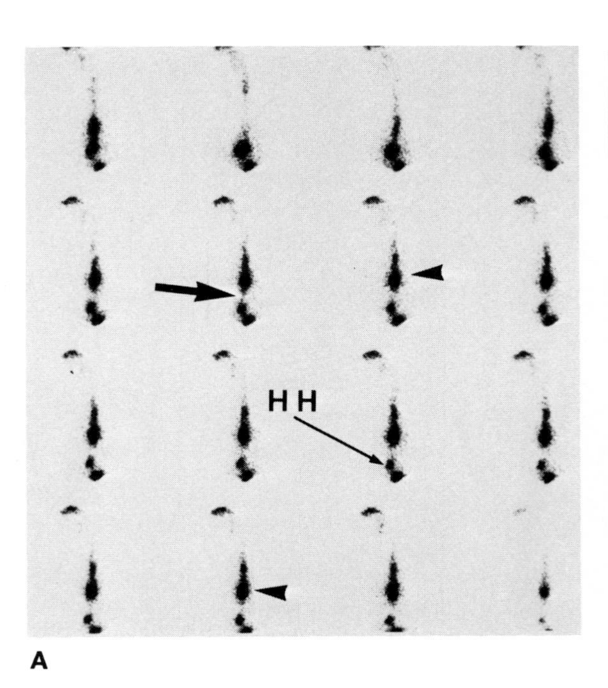

A

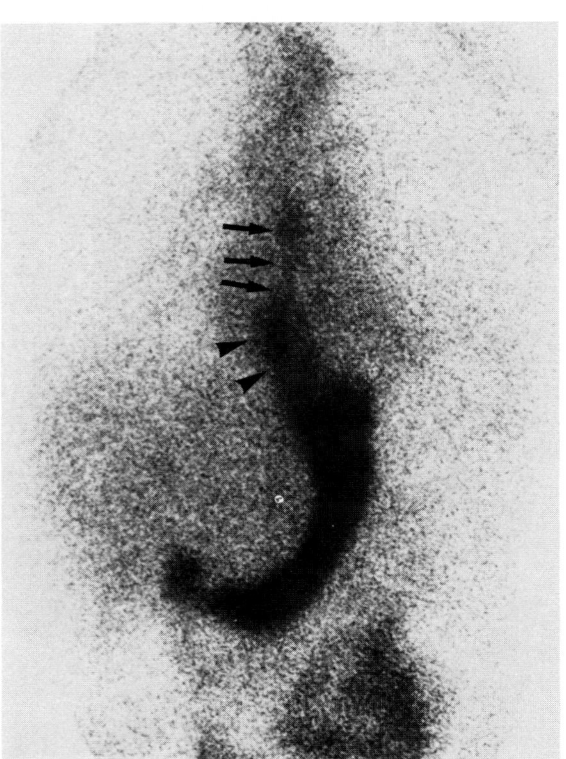

B

Figure 1-35 ***Diagnosis of Gastroesophageal Reflux***

Operated on 4 years earlier for an adenocarcinoma of the cardia, this patient complained of symptoms suggestive of gastroesophageal reflux. During the first minute of the dynamic study (supine position), there were a few episodes of significant gastroesophageal reflux *(arrows)*. Simultaneous analysis of both the distal esophageal segment and gastric time-activity curves allows the detection of a small gastroesophageal reflux. However, the relatively short duration of the standard 2-minute acquisition is not ideal for detecting gastroesophagel reflux and RETS cannot replace the usual radionuclide gastroesophageal reflux study for this specific purpose. (Taillefer E, Beauchamp G: Radionuclide esophagogram. Clin Nucl Med 9:465-483, 1984)

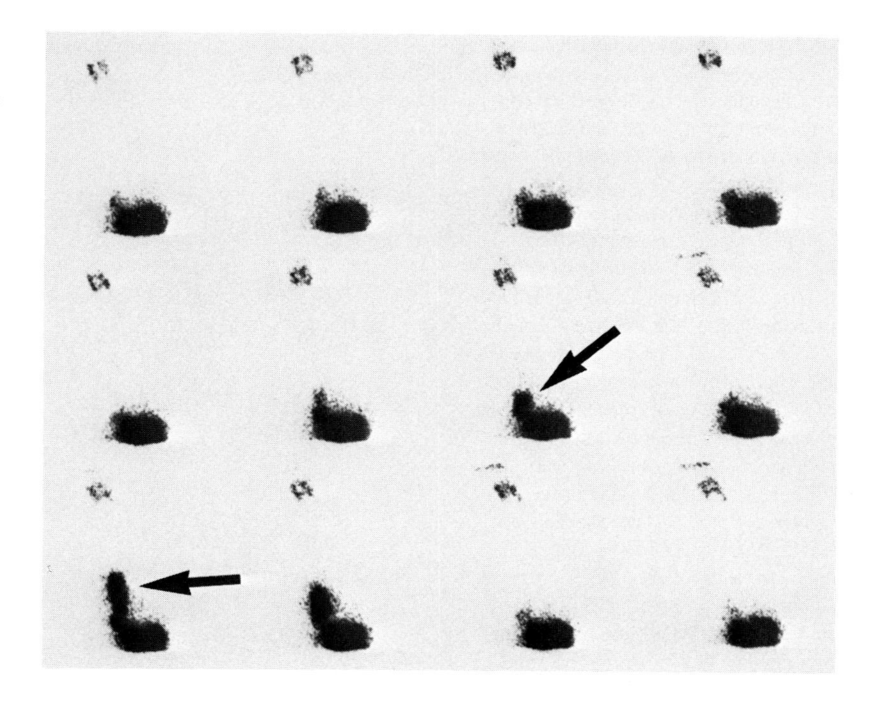

Figure 1-36 ***Diagnosis of Peptic Esophagitis***

This patient underwent an esophagectomy with a gastric pull-up for an epidermoid carcinoma three years previously. RETS was performed because of a recent complaint of heartburn and regurgitation. The intrathoracic stomach *(large arrow)* is well-visualized on the static image. Proximal to the stomach is a persistent and diffuse retention of the radiocolloid for a few centimeters *(small arrows).* Endoscopy revealed grade II esophagitis, corresponding to the radionuclide retention. ***Comment:*** An experimental animal model, based on induction of chemical esophagitis by an esophageal infusion of hydrochloric acid in cats, showed that the site and degree of radiocolloid retention corresponded to the location and severity of esophagitis.[12,13] Unfortunately, experimental data were derived from idealized conditions and do not necessarily reflect the clinical situation. There is a good correlation between esophageal radiocolloid retention and grade III or extensive grade II esophagitis. However, RETS cannot identify grade I and localized grade II esophagitis. Detection of this type of lesion by radionuclide retention index seems dependent on three main factors: degree of macroscopic involvement, extension of inflamed and ulcerated lesions, and site of esophagitis. Patients with localized grade I or grade II esophagitis situated near the gastroesophageal junction do not usually show sufficient radionuclide adsorption to be detected with static images. Activity in the gastric lumen can mask a slight retention near the lower esophageal sphincter. In spite of the lack of accuracy, this observation can be useful when RETS is performed in cases of possible gastroesophageal reflux. Abnormal esophageal radiocolloid retention in the absence of any signs of stricture or significant motility disorder can be explained by mucosal abnormalities, and this possibility should be included in the differential diagnosis of abnormal esophageal radionuclide stasis. (Taillefer E, Beauchamp G: Radionuclide esophagogram. Clin Nucl Med 9:465–483, 1984)

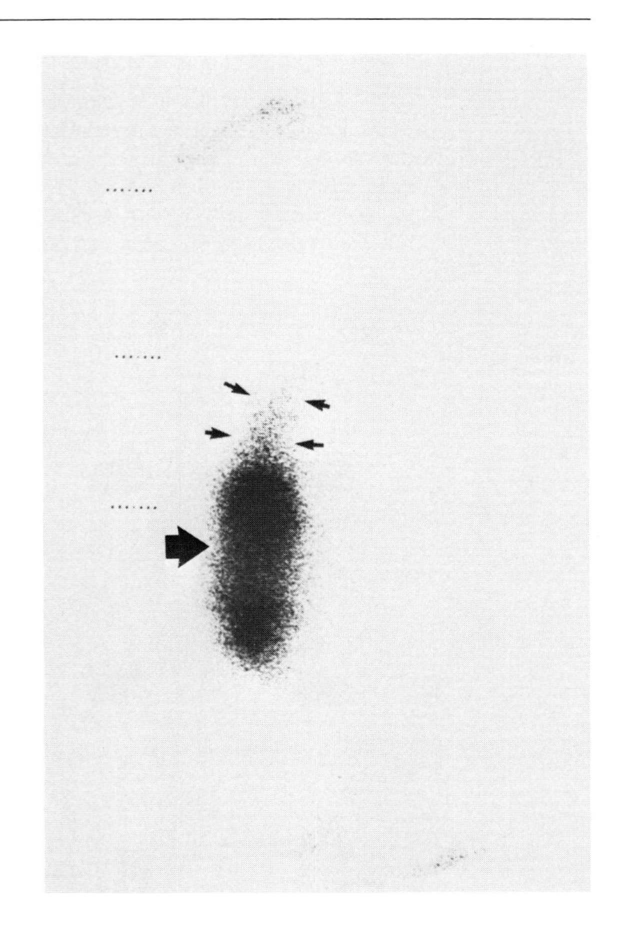

Figure 1-37 ***Diagnosis of Esophagectomy and Gastric Transposition with Bronchial Aspiration*** ▶

This patient had undergone an esophagectomy and gastric transposition for an esophageal carcinoma. The RETS performed in both upright *(A)* and supine *(B)* positions show a radionuclide stasis in projection of the hypopharynx and proximal esophagus. The intrathoracic stomach *(arrowheads)* is better visualized in the supine position because gravity plays an important role in food progression through this partially denervated tube. On the upright study, as soon as 6 seconds after the radionuclide ingestion, there is a left paraesophageal linear focus of activity *(A, arrows).* This corresponds to a left bronchial aspiration, which is well-visualized on the static image *(C, arrows)* done in upright position. *(D)* This patient had a gastric transposition for an esophageal chemical stenosis. RETS performed at 5-second intervals in the supine position with a liquid bolus shows the intrathoracic stomach with a gastroesophageal reflux *(arrow)* localized to the distal part of the esophageal remnant. ***Comment:*** We use a modification of the standard RETS to evaluate patients with esophagectomy and gastric transposition. Patients are studied upright for a 15-minute acquisition and supine for a 1-hour or ideally a 2-hour acquisition. This protocol combines the assessment of transit in the esophageal remnant with the assessment of the emptying of the intrathoracic stomach. The data obtained provide useful documentation on the transposed organ function. In addition, gastroesophageal reflux and tracheal aspiration are quantified and more readily detected when radionuclide techniques are used, as compared to radiologic procedures. (Taillefer R, Beauchamp G, Duranceau A, Lafontaine E: Nuclear medicine and esophageal surgery. Clin Nucl Med 11:445–459, 1986)

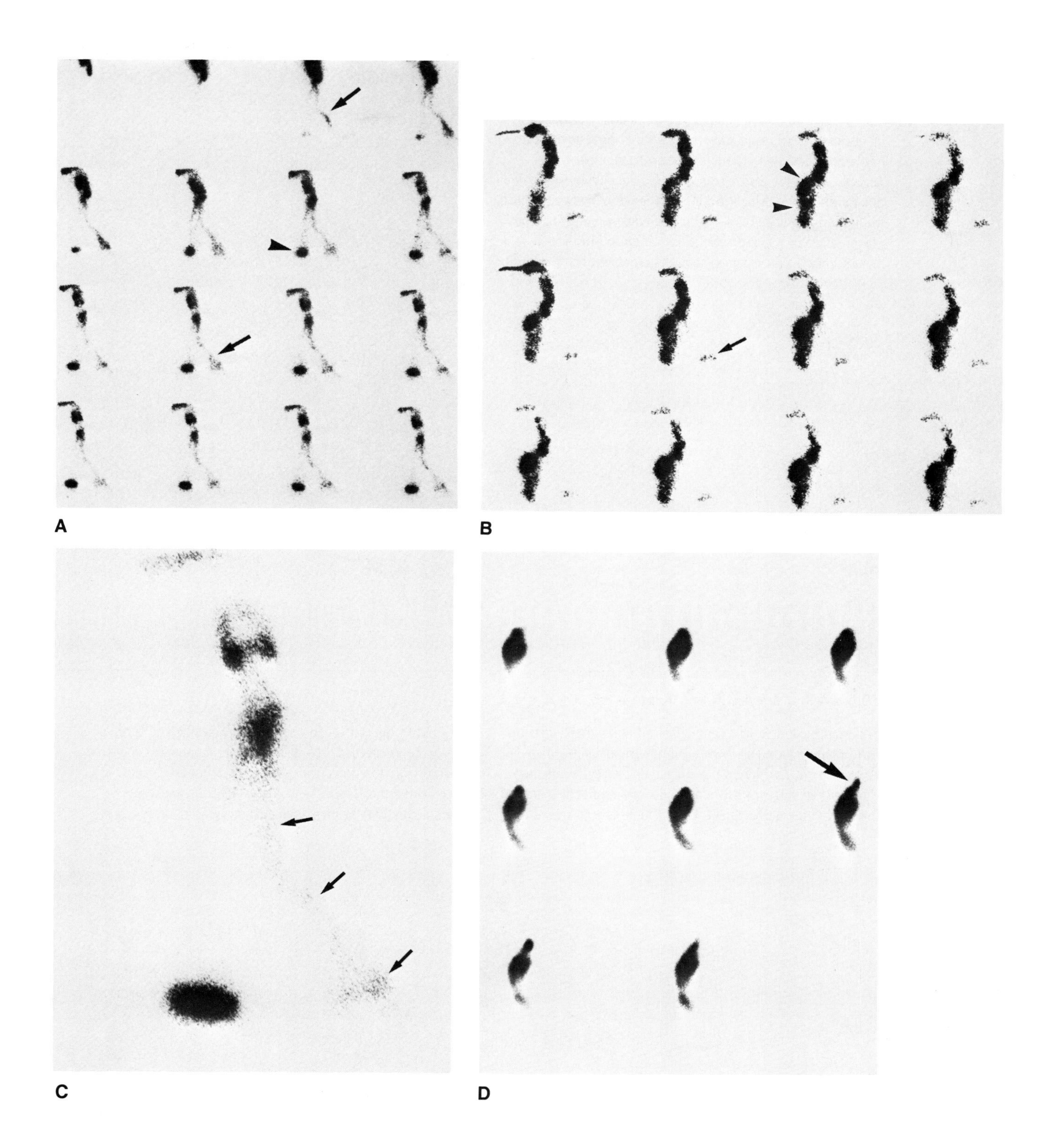
A
B
C
D

Figure 1-38 ***Diagnosis of Esophagectomy and Colonic Interposition***

This patient underwent an esophagectomy with colonic interposition for a severe chemical esophagitis. RETS shows many sites *(arrows)* of radionuclide stasis corresponding to intraluminal activity in the interposed intrathoracic colon. ***Comment:*** As for gastric transposition, the patient with colonic interposition may benefit from a radionuclide study performed with a long observation period to evaluate the emptying and the presence of reflux or tracheal aspiration. (Taillefer R, Beauchamp G: Nuclear medicine and esophageal surgery. Clin Nucl Med 11:445-459, 1984)

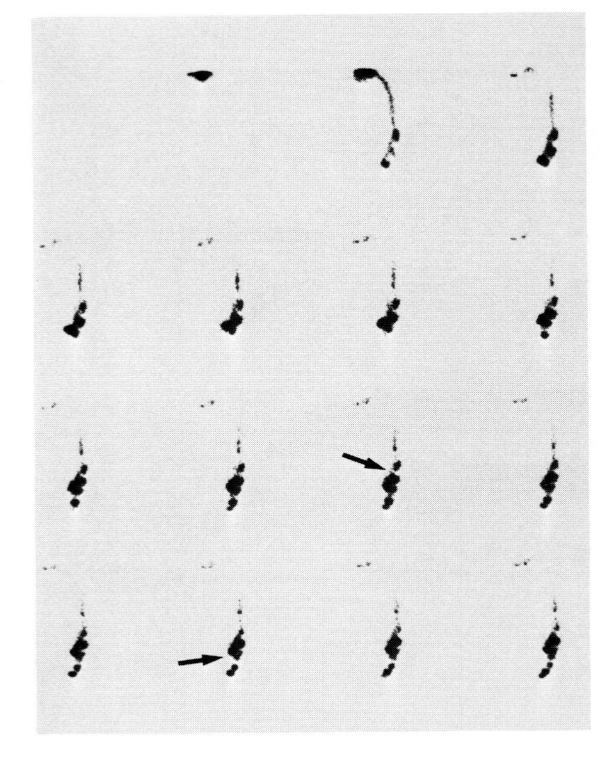

Figure 1-39 ***Diagnosis of Dysphagia Lusoria***

This young adult patient complained of upper dysphagia. On RETS (supine position) there is an abnormal radionuclide stasis *(arrows)* at the level of upper esophageal segment. A transient esophageal stasis may sometimes represent a normal finding in an elderly patient, but it should be considered abnormal in the young. This stasis is not specific and further investigation is needed. The delayed upper esophageal emptying was secondary to an aberrant right subclavian artery that passes behind the esophagus, compressing the esophageal lumen, a condition known as dysphagia lusoria.

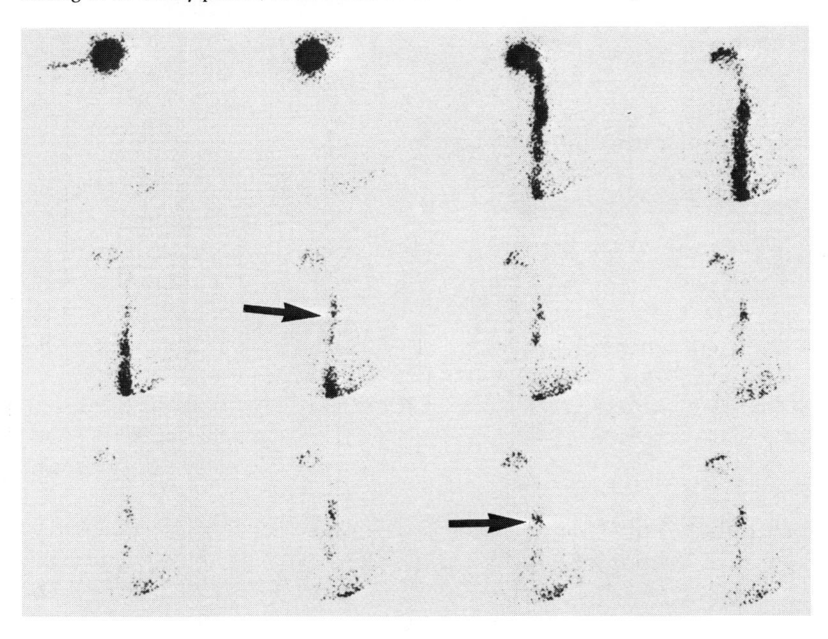

Peter W. Blue

Chapter 2 HEPATOBILIARY IMAGING IN PATIENTS WITHOUT PRIOR SURGERY

The interpretative scheme, flow sheet, and case studies presented in this chapter demonstrate normal and abnormal hepatobiliary imaging using technetium-99m diisopropyl iminodiacetic acid (DISIDA; Disofenin). This chapter applies only to the diagnosis of biliary disease in patients who have had no prior biliary surgical intervention.

TECHNIQUES

The following procedure and agent are recommended for hepatobiliary imaging.

1. Patient Preparation
   If the gallbladder is to be imaged, the patient is kept NPO > 4 hours and < 24 hours. If NPO > 24 hours, imaging is done 2 hours after sincalide (Kinevac) injection (0.04 mcg/kg IV).
   If the gallbladder is not to be imaged, no special patient preparation is necessary.
2. Radiopharmaceutical Administration
   Agent: DISIDA, Disofenin
   Dosage: if bilirubin <2 mg/dl, then 5 mCi (185 mBq); if bilirubin ≥ 2 mg/dl, then 8 mCi (296 mBq)
   Route: intravenous
3. Image Acquisition
   Camera: large field of view
   Collimator: low energy, high resolution (preferred) or low energy, all purpose; parallel-hole.
   Energy window: 20% centered on 140 keV.
   Positioning: all images acquired supine.
   Anterior: upper hepatic margin at top of field and small portion of base of heart visible. Right lateral: upper hepatic margin at top of field.

Views:

Anterior views every 5 minutes from 0 to 60 minutes. All images are for the same amount of time and camera settings; for example, 1500 counts/cm² information density (ID) over liver on first image, rest of views same time and intensity settings.

Right lateral view at 30 to 60 minutes is taken to separate gallbladder activity (anterior) from duodenum (posterior). Oral administration of water to wash out duodenal activity and review of the dynamic study acquired on computer may be helpful.

Overexposed anterior view at 60 minutes

is taken to visualize gastrointestinal (GI) tract if not seen at normal intensity.

At 60 minutes, if gallbladder activity is seen and GI activity is not seen, sincalide (Kinevac) 0.04 mcg/kg may be given to rule out common duct obstruction.

Delayed views: if the gallbladder is not yet seen, obtain a 2-hour image, and if liver or biliary duct activity is still present, obtain 4-hour images. If GI tract is not yet seen, obtain 2-, 4-, 6-, and 24-hour images.

PHYSIOLOGIC MECHANISM OF THE RADIOPHARMACEUTICAL

Tc-99m iminodiacetic acid (IDA) and its analogs share a common anionic transport mechanism with bilirubin and certain bile salts (*e.g.,* taurocholate). Uptake, in particular of the IDA and bilirubin, is competitive and high levels of bilirubin will inhibit IDA uptake even in the presence of normal hepatocyte function. The characteristics that make a good hepatobiliary agent include rapid hepatocyte clearance from plasma, rapid excretion from hepatocyte to bile, resistance to competition from high bilirubin levels, and low renal clearance. Currently DISIDA best meets these criteria and is commercially available.[1] The approximate radiation absorbed doses of Tc-99m DISIDA for various organs are listed in Table 2-1.

VISUAL DESCRIPTION AND INTERPRETATION

Hepatobiliary scintigraphy dynamically follows the course of tracer movement from hepatic clearance

Table 2-1
Radiation Dosimetry

Organ	*Radiation Absorbed Dose (rads/mCi)*
Total body	0.016
Liver	0.039
Gallbladder wall	0.12
Small intestine	0.21
Upper large intestine	0.39
Lower large intestine	0.27
Urinary bladder wall	0.093
Ovaries	0.081
Testes	0.0065
Red marrow	0.028

(Hepatolite Tc-99m-Disofenin. Product information. Boston, New England Nuclear, 1985)

Table 2-2
Normal Results of Hepatobiliary Studies

Finding	*Usual Normal*	*Upper Normal*
Cardiac blood pool activity	Cleared by 5 min	Cleared by 10 min
Time of peak hepatic activity	<10 min	15 min
Leading-edge transit time	<10 min	15 min
Time of peak activity in HD* and CD*	<30 min	45 min
Time of GB* visualization	<30 min	60 min
Time of GI* visualization	<60 min	Nonvisualization until meal or sincalide injection

* HD, *hepatic ducts;* CD, *common ducts;* GB, *gallbladder;* GI, *gastrointestinal tract*

(Kuni CC, Klingensmith WC: Atlas of Radionuclide Hepatobiliary Imaging. Boston, GK Hall Medical Publishers, 1983)

through the biliary system into the gastrointestinal tract. In order to evaluate the study, each component of this course may be individually and sequentially evaluated and combined into a unified explanation and interpretation of the events seen (Table 2-2).

Clearance. Clearance represents the ability of the liver to remove tracer from the blood. The rate of disappearance of activity in the cardiac blood pool is used to estimate clearance.

Hepatic Excretion. Hepatic excretion is the ability of the liver to transfer tracer from the hepatocyte to the most proximal biliary radicles. The time of first appearance of tracer in the biliary system (leading-edge transit time) best reflects this function. The intensity of hepatic uptake (target–background ratio or hepatic uptake curves) results from the combined effect of hepatic clearance and excretion. Cardiac blood pool activity, intensity of hepatic uptake, and leading-edge transit time are thus used in combination to evaluate hepatocyte function.

Biliary Patency and Function. Biliary patency and function are evaluated by the presence, intensity, and timing of appearance of tracer in the hepatic ducts, the common duct, the gallbladder, and the gastrointestinal tract.

Table 2-3 presents a diagnostic interpretive scheme of the usual patterns seen in a variety of normal and abnormal hepatobiliary scintograms. Examples from the atlas section of the chapter are also cited.

Table 2-3
Interpretation of Hepatobiliary Studies

Observations	*Diagnosis*	*Figure (example)*
Rapid HU; rapid vis of HD, CD (peak activity $\leq$ 45 min), GB, and GI	Normal	2-2
Rapid HU, rapid vis of HD, CD (peak activity $\leq$ 45 min), and GB; nonvis of GI	Normal variant; if cholecystokin in peptide (Kinevac) administered, proceed with GI vis	2-3
Rapid HU; rapid vis of HD, CD, and GI; nonvis of GB through 4 hours	Cystic duct obstruction	2-4
Plus acute RUQ pain	Acute cholecystitis	2-4
Plus GB hydrops	Cystic duct Ca, GB ca, or acute cholecystitis (rare)	2-5
Plus liver defects (metastasis)	GB Ca, cystic duct Ca, or acute cholecystitis plus another Ca	2-6
Rapid HU; rapid vis of HD, CD, and GI; delayed ($>$ 1 hr) or decreased GB vis	Chronic cholecystitis	2-7
Rapid HU; nonvis of HD, CD, GB, and GI through 24 hours	Recent (24 - 48 hr) complete CD obstruction	2-8, 2-10
	Intrahepatic cholestasis (rare pattern)	2-12
	Biliary atresia (unusual pattern)	2-9
Mild/moderate decreased HU; nonvis of HD, CD, and GI through 24 hours	Chronic complete CD obstruction (usual pattern)	2-8
Rapid HU; vis of HD, CD, and GB; nonvis of GI through 24 hours	Acute complete CD obstruction (very rare pattern)	2-10
Rapid HU; vis and persistence of tracer in prominent HD and CD (peak activity $>$ 45 min); vis of GB; delayed vis of GI	Partial CD obstruction	2-11
Rapid HU; delayed or decreased vis of HD, CD, GB, and GI	Intrahepatic cholestasis	2-12
Severely decreased HU and poor or nonvis of HD, CD, GB, and GI	Hepatocellular dysfunction	2-13
Bilirubin $>$ 20	Interference by elevated bilirubin (*i.e.*, nondiagnostic study)	2-13

(HU, *hepatic uptake;* HD, *hepatic ducts;* CD, *common ducts;* GB, *gallbladder;* GI, *gastrointestinal tract;* Ca, *cancer;* vis, *visualization;* RUQ, *right upper quadrant)*

REFERENCES

1. Chervu LR, Nunn AD, Loberg MD: Radiopharmaceuticals for hepatobiliary imaging. Sem Nucl Med 12:5, 1982
2. Hepatolite Tc-99m-Disofenin. Product information. Boston, New England Nuclear, 1985
3. Kuni CC, Klingensmith WC: Atlas of Radionuclide Hepatobiliary Imaging. Boston, GK Hall Medical Publishers, 1983
4. Weissman HS, Sugarman LA, Freeman LM: Atlas of Tc-99m iminodiacetic acid (IDA) cholescintigraphy. Clin Nucl Med 7:231, 1982
5. Freeman LM, Blaufox MD (eds): Gastrointestinal disease update. 1. Liver and biliary disease. Semin Nucl Med 12:2, 1982
6. Klingensmith WC, Whitney WP, Spitzer VM et al: Effect of complete biliary-tract obstruction on serial hepatobiliary imaging in an experimental model: Concise communication. J Nucl Med 22:866, 1981
7. Mavro MA, McCartney WH, Melmed JR: Hepatobiliary scanning with 99m-Tc PIPIDA in acute cholecystitis. Radiology 142:193, 1982
8. Weissman HS, Badio J, Sugarman LA et al: Spectrum of 99m-Tc-IDA cholescintigraphic patterns in acute cholecystitis. Radiology 138:167, 1981
9. Szlabick RE, Catto JA, Fink Bennett D et al: Hepato-

biliary scanning in the diagnosis of acute cholecystitis. Arch Surg 115:540, 1980

10. Weissman HS, Frank MS, Bernstein LH et al: Rapid and accurate diagnosis of acute cholecystitis with 99mTc-HIDA cholescintigraphy. AJR 132:253, 1979

11. Saurez CA, Block F, Bernstein D et al: The role of HIDA/PIPIDA scanning in diagnosing cystic duct obstruction. Ann Surg 191:391, 1980

12. Freitas JE: Cholescintigraphy in acute and chronic cholecystitis. Sem Nucl Med 12:18, 1982

13. Ramanna L, Brachman MB, Tanasescu DE et al: Cholescintigraphy in acute acalculous cholecystitis. Am J Gastroent 79:650, 1984

14. Weissmann HS, Berkowitz D, Fox MS et al: The role of technetium-99m iminodiacetic acid (IDA) cholescintigraphy in acute acalculous cholecystitis. Radiology 146:177, 1983

15. Shuman WP, Rogers JV, Rudd TG et al: Low sensitivity of sonography and cholescintigraphy in acalculous cholecystitis. AJR 142:531, 1984

16. Mindelzun RE, McCort JJ: Acute abdomen. In Margulis AR, Burhenne HF (eds): Alimentary Tract Radiology, p. 403. St. Louis, CV Mosby, 1983

17. Floyd JL, Collins TL: Discordance of sonography and cholescintigraphy in acute biliary obstruction. AJR 140:501, 1983

18. Blue PW: Hyperacute complete common bile duct obstruction demonstrated with Tc-99m-IDA cholescintigraphy. Nucl Med Comm 6:275, 1985

19. Majd M: Personal communication.

20. Majd M, Reba RC, Altman RP: Hepatobiliary scintigraphy with 99mTc-PIPIDA in the evaluation of neonatal jaundice. Pediatrics 67:140, 1981

21. Majd M, Reba RC, Altman RP: Effect of phenobarbital on 99m-Tc-IDA scintigraphy in the evaluation of neonatal jaundice. Sem Nucl Med 11:194, 1981

22. Kuni CC, Klingensmith WC, Fritzberg AR: Evaluation of intrahepatic cholestasis with radionuclide hepatobiliary imaging. Gastrointest Radiol 9:163, 1984

23. Rosenthall L: Cholescintigraphy in the presence of jaundice utilizing Tc-IDA. Sem Nucl Med 12:53, 1982

Atlas Section

Figure 2-1 *Biliary Diagnostic Flow Chart*

(***Bold print,*** diagnosis; *italic print,* findings; roman print, choices of findings; →, proceed; — — →, proceed if any tracer is seen in biliary system, otherwise study is indeterminate; ||, diagnosis follows) (Blue P: Biliary Scanning Interpretations Using Technitium-99m DISIDA. Clin Nuc Med 10:742 - 751, 1985)

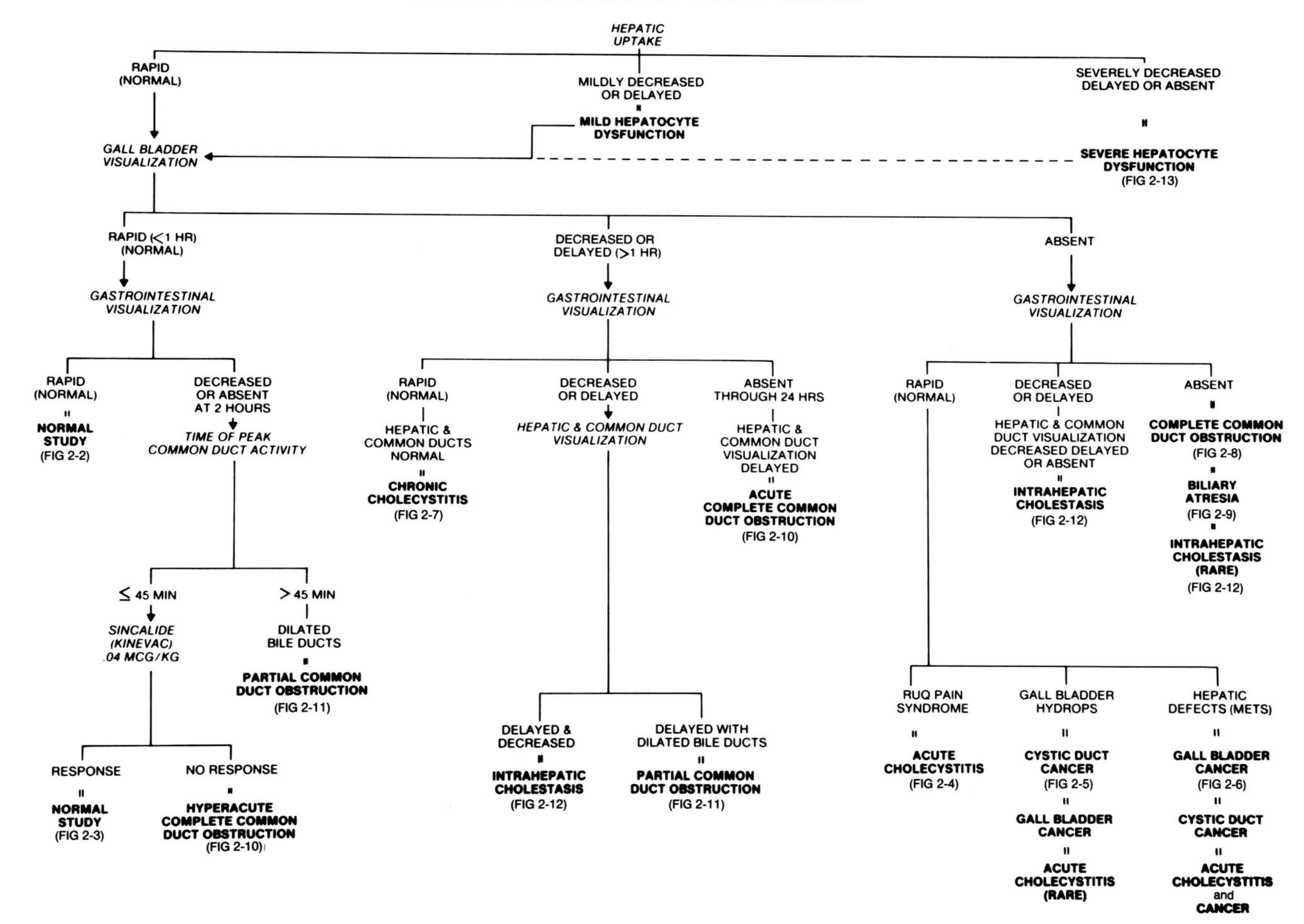

Figure 2-2

Normal Cholescintigram

Normal Tc-99m DISIDA cholescintigram shows normal, rapid hepatic clearance (cardiac blood pool activity cleared by 10 minutes), with rapid sequential visualization of the common duct and gallbladder (20 minutes) and gastrointestinal tract (30 minutes). Renal activity normally may be seen within 10 minutes *(arrows)*.[3] (***GB***, gallbladder; ***CD***, common duct; ***GI***, gastrointestinal tract; ***R LAT***, right lateral view) (Blue P: Biliary Scanning Interpretations Using Technitium-99m DISIDA. Clin Nuc Med 10:742-751, 1985)

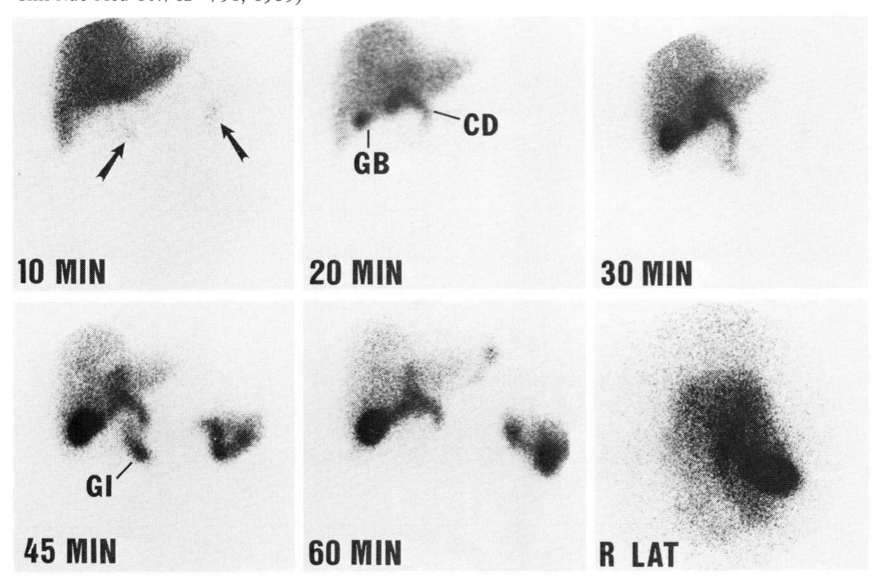

Figure 2-3

Normal Cholescintigram

Normal cholescintigram shows normal hepatic clearance, prompt visualization of the hepatic and common ducts, and prompt gallbladder visualization. Comment: Against a closed Oddi's sphincter and with good water reabsorption by the gallbladder wall, net bile flow is toward the gallbladder.[3-6] Therefore, the GI tract is not seen by 2 hours (even in intentionally overexposed views). Five minutes after sincalide injection (C-terminal octapeptide of cholecystokinin, or CCK), the gallbladder has contracted and GI tracer is well seen. Also, in contrast to patients with partial common duct obstruction (Fig. 2-11) who have persistent tracer in prominent biliary ducts, these normal patients have peak ductal activity early (before 45 minutes) as bile moves to the gallbladder.[3] Ultrasonography may also be used to differentiate these patterns. (Blue P: Biliary Scanning Interpretations Using Technitium-99m DISIDA. Clin Nuc Med 10:742-751, 1985)

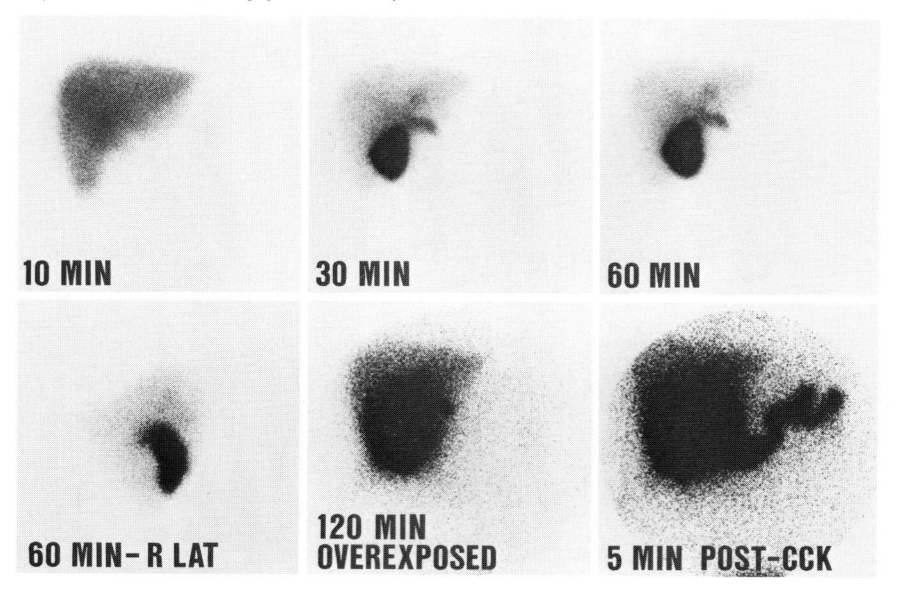

Figure 2-4 ***Acute Cholecystitis***

Acute cholecystitis is evidenced by normal hepatic clearance, excretion into the ductal system, and gastrointestinal visualization. The gallbladder is not seen during the first 2 hours, by which time almost all of the hepatic tracer has been excreted; further images would be nonproductive. Comment: The use of cholescintigraphy in patients suspected of having acute cholecystitis produces a sensitivity, specificity, and predictive accuracy of greater than 95%.[3,4,7-12] Acute acalculous cholecystitis occurs in 2% to 10% of all cholecystectomies,[13] with nonvisualization of the gallbladder occurring in 92% of the cases reported.[13-15] (Blue P: Biliary Scanning Interpretations Using Technitium-99m DISIDA. Clin Nuc Med 10:742-751, 1985)

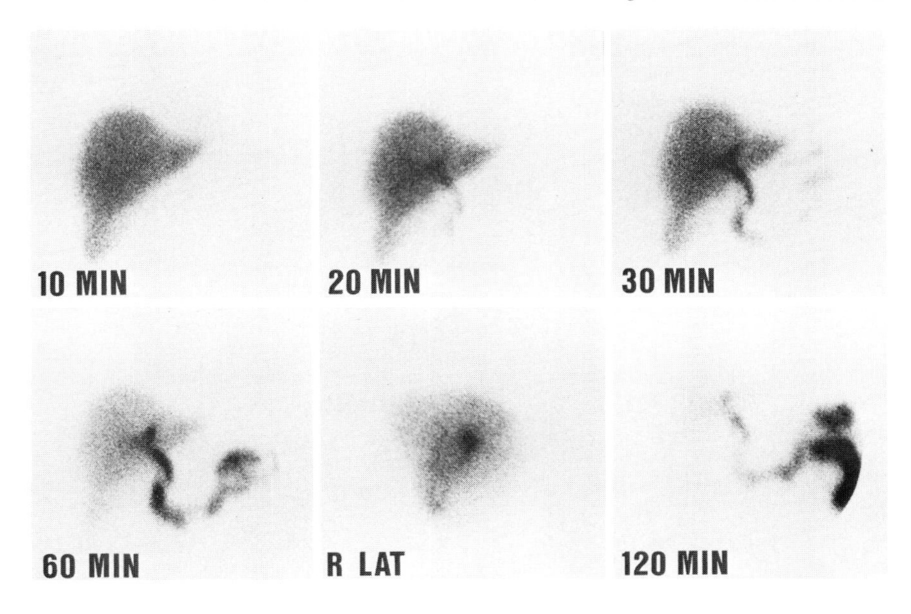

Figure 2-5 ***Hydrops of the Gallbladder in a Patient with Cystic Duct Carcinoma***

Hepatic clearance is moderately diminished (the cardiac blood pool *(CBP)* is still visible at 10 minutes). By 30 minutes, the GI tract is seen. Tracer in the gallbladder is not seen in these images, nor on delayed images obtained through 3 hours. In addition, a massive photon-deficient area is seen and was found ultrasonographically to represent gallbladder hydrops. Comment: Gallbladder hydrops occurs most commonly in obstructive disorders of the biliary system associated with a normal (expandable) gallbladder wall. Pancreatic carcinoma causing gallbladder hydrops is ruled out in this patient by the rapid appearance of GI tracer. Cystic duct obstruction due to carcinoma of the cystic duct or proximal gallbladder are the most likely causes. Although acute cholecystitis may result in hydrops,[4,16] the typical disease-thickened wall generally precludes dilatation and hydrops (Courvoisier's law). (Blue P: Biliary Scanning Interpretations Using Technitium-99m DISIDA. Clin Nuc Med 10:742-751, 1985)

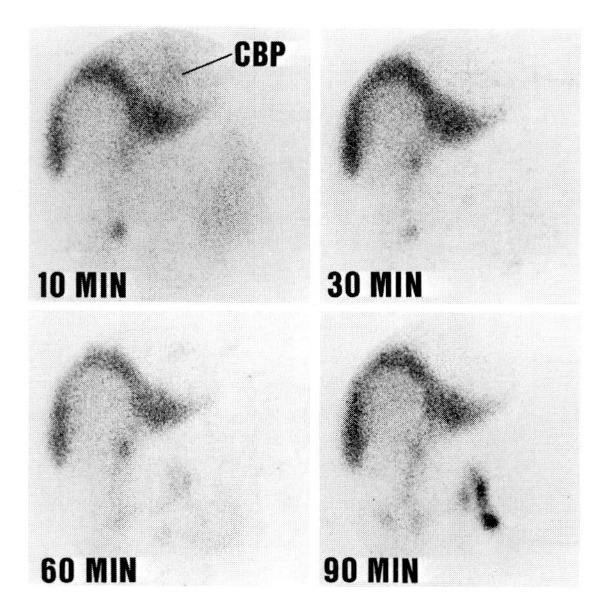

Figure 2-6

Metastatic Gallbladder Carcinoma

Normal hepatic uptake, excretion, and gastrointestinal tract visualization is seen. The gallbladder fossa is well seen *(closed arrow),* but without tracer accumulation, suggesting cystic duct obstruction. In addition, a metastic lesion is seen in the lower right lobe of the liver *(open arrow).* Cystic duct or gallbladder carcinoma with hepatic metastases would account for these findings. A study performed 6 months after the carcinomatous gallbladder was removed demonstrates progression of the metastatic disease. (Blue P: Biliary Scanning Interpretations Using Technitium-99m DISIDA. Clin Nuc Med 10:742 - 751, 1985)

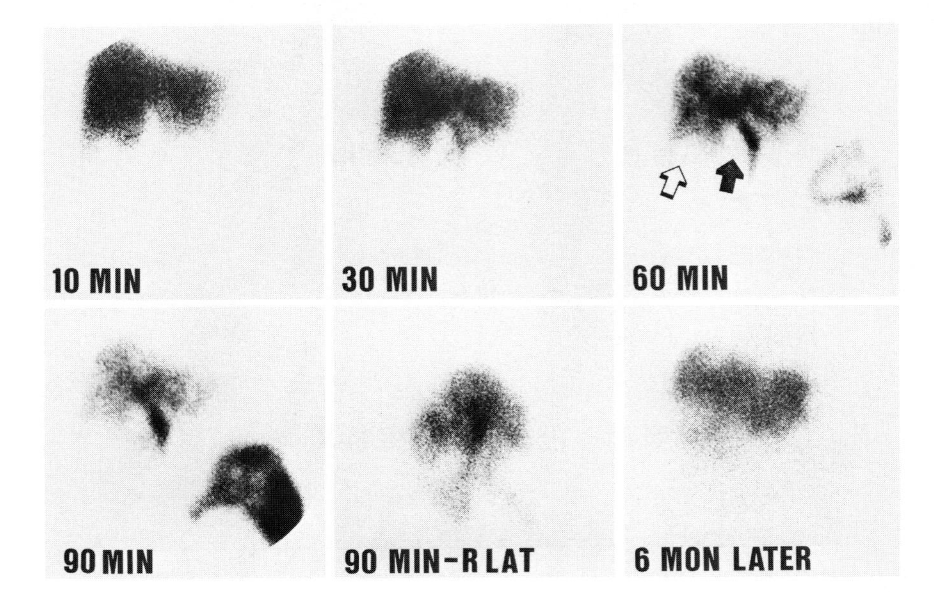

Figure 2-7

Chronic Cholecystitis

Prompt hepatic uptake, excretion of tracer, and gastrointestinal tract visualization is seen. The gallbladder is faintly seen at 30 minutes *(arrow)* and persists faintly throughout the study. Delayed appearance and/or decreased concentration of tracer supports the diagnosis. Comment: This pattern results from a diseased gallbladder wall that cannot concentrate bile effectively. The study must be continued for at least 2 hours, preferably 4 hours, to elicit this pattern and thus exclude patients from the group with acute cholecystitis (gallbladder nonvisualization). (Blue P: Biliary Scanning Interpretations Using Technitium-99m DISIDA. Clin Nuc Med 10:742 - 751, 1985)

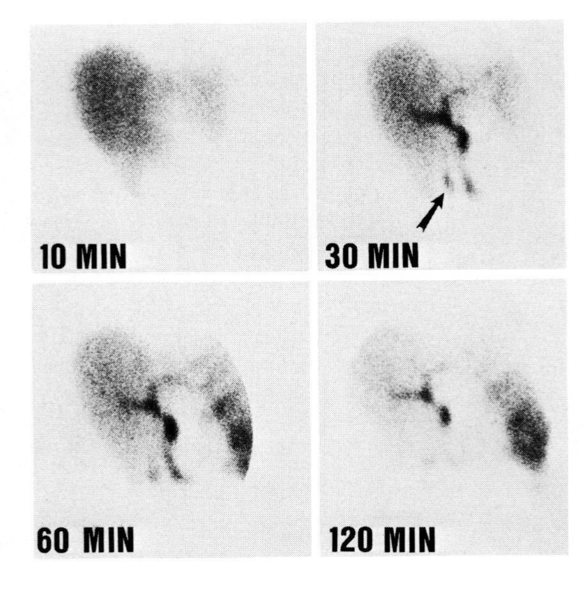

Figure 2-8

Complete Common Duct Obstruction

There is fair hepatic tracer uptake with no visualization of hepatic ducts, the common duct, gallbladder, or gastrointestinal tract through 24 hours of study. Comment: The only route of excretion is through the kidneys (ureter seen in 10-minute image, kidneys seen through 24 hours). Immediately following complete obstruction of the common bile duct, the gallbladder will normally continue to reabsorb water and decompress the biliary system until the bile is maximally concentrated (Fig. 2-10).[2,6,17,18] At this point, net bile flow ceases and is followed by the pattern of nonvisualization of the biliary system. (Blue P: Biliary Scanning Interpretations Using Technitium-99m DISIDA. Clin Nuc Med 10:742–751, 1985)

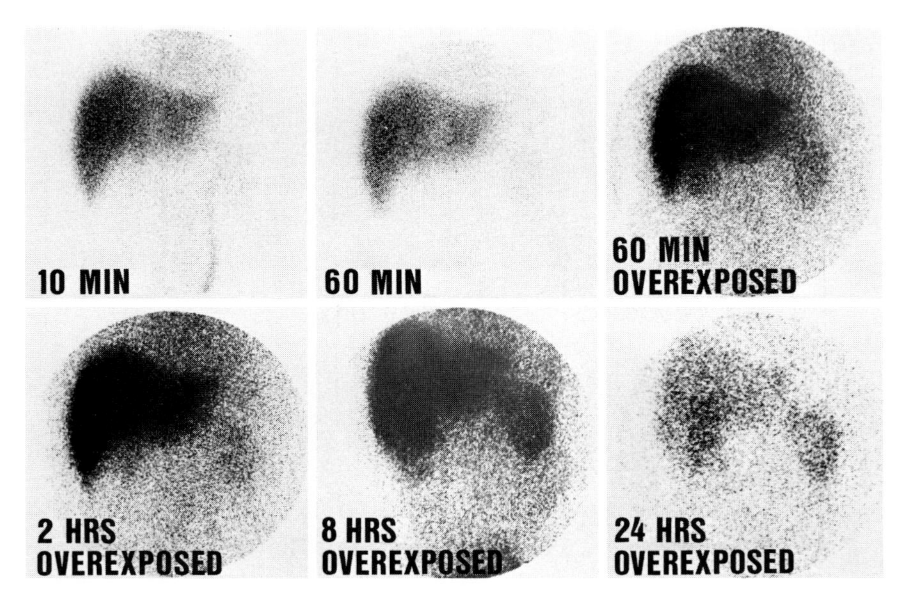

Figure 2-9

Primary Biliary Atresia

Images were performed following phenobarbital pretreatment (2.5 mg/kg b.i.d. for 4 days to maintain a serum phenobarbital level of 14 mcg/dl.[19] Diminished hepatic clearance is noted with no gastrointestinal excretion seen through 24 hours. Comment: In patients pretreated with phenobarbital, this pattern is highly specific for primary biliary atresia and separates this group of patients from those with neonatal hepatitis, in which gastrointestinal excretion is seen.[20,21] (*CBP*, cardiac blood pool; *UB*, urinary bladder) (Blue P: Biliary Scanning Interpretations Using Technitium-99m DISIDA. Clin Nuc Med 10:742–751, 1985)

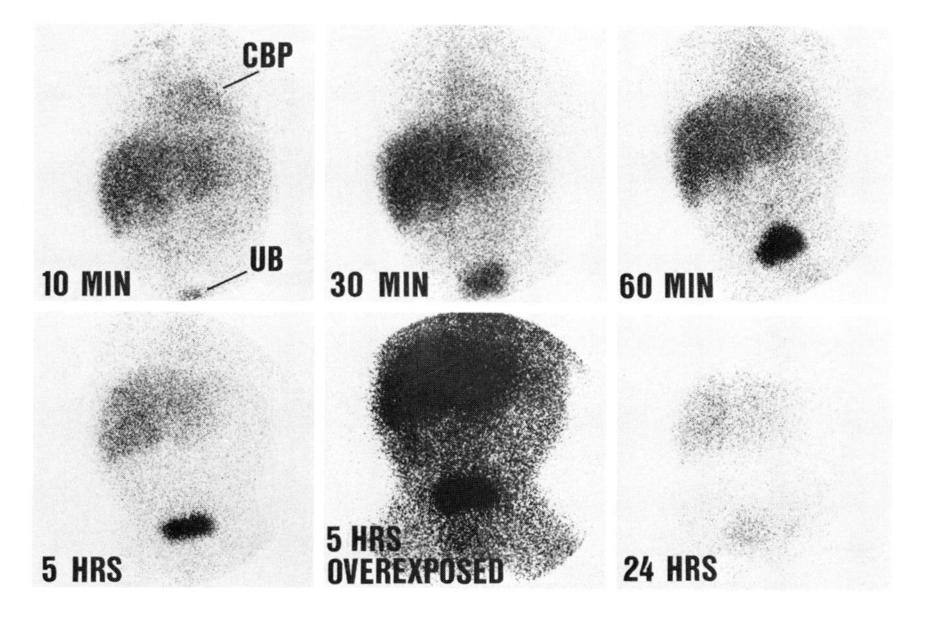

Figure 2-10 ***Acute Complete Common Duct Obstruction***

There is good hepatic clearance, but the hepatic and common ducts are not seen until 3 hours postinjection of sincalide (CCK). At 6 hours, the gastrointestinal tract is not seen and there is no response to sincalide. No tracer is seen in the gastrointestinal tract through 24 hours. ***Comment:*** If a common duct obstruction occurs and the gallbladder bile is not maximally concentrated, net bile flow will occur and the biliary system will be visualized. If the patient is imaged immediately following the obstruction, the study may appear normal, except for GI nonvisualization. Eventually, net bile flow will cease, and the pattern seen in Figure 2-8 will result. No gastrointestinal activity will be seen in any case.[3,6,17,18] (Blue P: Biliary Scanning Interpretations Using Technitium-99m DISIDA. Clin Nuc Med 10:742-751, 1985)

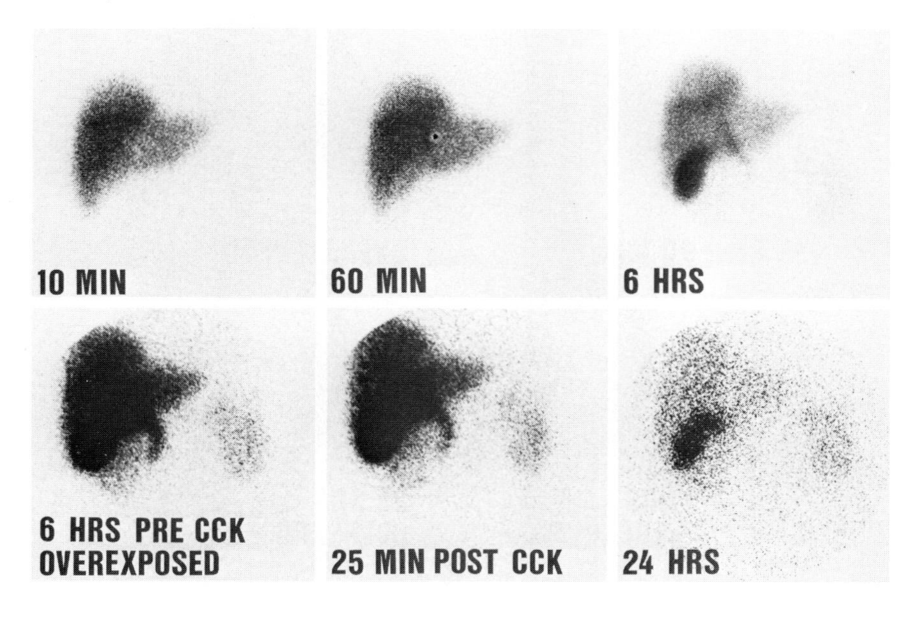

Figure 2-11 ***Partial Common Duct Obstruction***

There is good hepatic clearance of tracer with prompt excretion into the hepatic and common ducts. The gallbladder *(GB)* is rapidly seen. The anterior location of tracer on the right lateral view confirms the collection to be the gallbladder rather than the posterior-lying duodenum. There is persistent tracer seen in prominent biliary ducts. Gastrointestinal tracer is not seen until 3 hours into the study. Comment: This pattern is differentiated from the normal pattern of Figure 2-3 by the persistence of tracer in prominent hepatic and common ducts. In the normal variant (tight Oddi's sphincter) the normal movement of bile into the gallbladder will cause the ducts' activity to peak by 45 minutes and then decrease.[3] Ultrasonography may also be used to differentiate these two patterns. (Blue P: Biliary Scanning Interpretations Using Technitium-99m DISIDA. Clin Nuc Med 10:742-751, 1985)

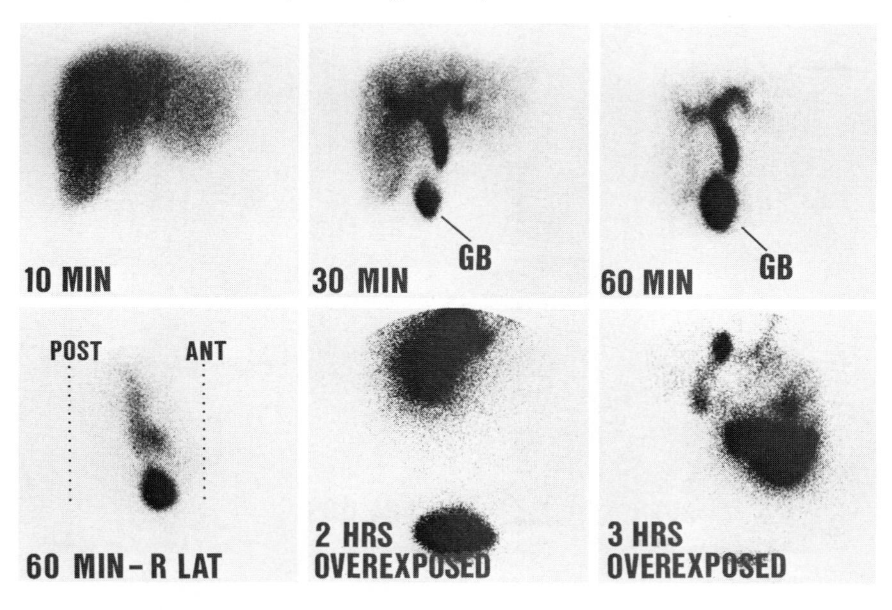

Figure 2-12 ***Intrahepatic Cholestasis***

Good hepatic clearance is seen at 10 minutes, but no tracer is seen in the biliary ducts or gallbladder 60 minutes after injection. The gastrointestinal tract and possibly the gallbladder are seen late in the study (11 hours). ***Comment:*** The appearance of gastrointestinal activity precludes a complete common duct obstruction. Lack of persistent tracer in dilated ducts militates against partial common duct obstruction. Normal or near-normal clearance excludes hepatocyte dysfunction as the cause of this pattern. Therefore, this scintigraphic picture is due to partial obstruction of the hepatocyte excretory mechanism (*e.g.*, drug-induced cholestasis) or of the smallest biliary ducts (ascending cholangitis in this patient).[3,22] Rarely in intrahepatic cholestasis is the obstruction so complete that no gastrointestinal tracer will be seen by 24 hours. Also rarely, gastrointestinal activity may be seen in an almost-total, common duct obstruction in the absence of biliary tract visualization. (Blue P: Biliary Scanning Interpretations Using Technitium-99m DISIDA. Clin Nuc Med 10:742-751, 1985)

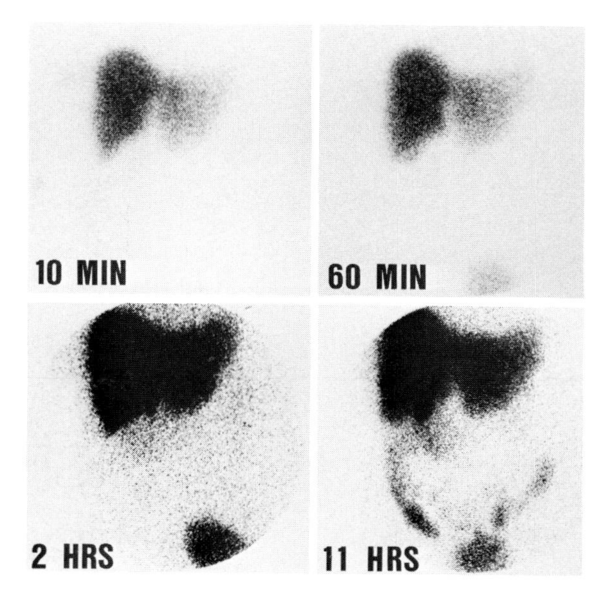

Figure 2-13 ***Hepatocyte Failure***

There is almost no hepatic clearance, as seen by the lack of change in the hepatic-to-cardiac ratio throughout the study. The only route of excretion is renal. Comment: In patients whose bilirubin is great (usually >20 mg/dl for DISIDA[1,3]), no tracer would be expected to reach the biliary ducts, and the scintigram would be indeterminate for biliary pathology. In patients with lesser hepatocyte dysfunction, the time course to biliary tract visualization would be prolonged in direct relation to the height of the bilirubin. If the study in these patients is prolonged sufficiently, biliary diagnosis may be determined. Nonvisualization of the biliary system in patients with lower bilirubin levels suggests complete common duct obstruction. In patients with very high bilirubin levels, visualization of GI tract activity rules out complete common duct obstruction. However, if there is nonvisualization of the biliary tract and GI tract, complete common duct obstruction cannot be separated from hepatocyte failure.[3,23] (***CBP***, cardiac blood pool; ***K***, kidneys) (Blue P: Biliary Scanning Interpretations Using Technitium-99m DISIDA. Clin Nuc Med 10:742-751, 1985)

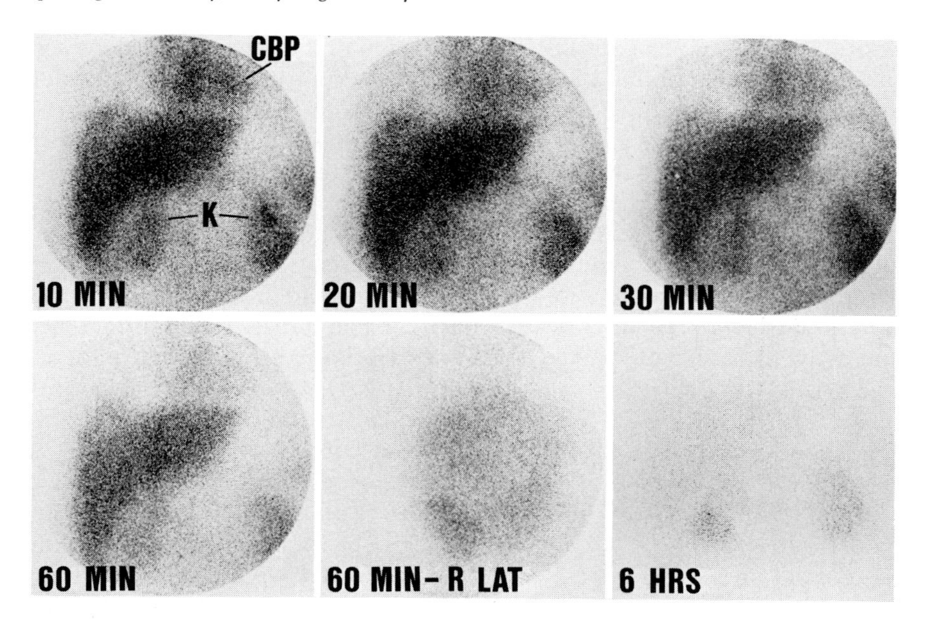

Milton D. Gross
Brahm Shapiro

Chapter 3 SCINTIGRAPHY OF THE ADRENAL CORTEX

Adrenal cortical scintigraphy encompasses the evaluation of dysfunction of the three concentric zones of the adrenal glands. Each zone is responsible for the synthesis and secretion of one or more characteristic hormones necessary for endocrine homeostasis (Table 3-1). The outer cortex or zona glomerulosa (zG) secretes aldosterone, the principal hormone of salt and water balance which promotes sodium/potassium exchange in the distal nephron.[1] As a result of a complicated biochemical scheme, the enzyme renin is secreted from a specialized portion of the kidney, the juxtaglomerular apparatus (JGA).[2] The JGA is a "sensor" comprised of a portion of the afferent renal arteriole and proximal renal tubule. Variation in renal perfusion pressure, sodium, and potassium results in augmentation of renin release, which catalyzes in the lung the degradation of angiotensinogen (renin substrate) to angiotensin I and then to angiotensin II by the enzymatic action of the converting enzyme.[2,3] Angiotensin II interacts with a glomerulosa cell receptor and stimulates the release of aldosterone.[4]

The abnormal aldosterone secretion of primary aldosteronism (PA) can occur as a result of a unilateral adrenal adenoma, adrenal carcinoma, or bilateral adrenal hyperplasia.[5] The major biochemical manifestations of aldosterone excess are hypokalemia and metabolic alkalosis, while hypertension is an important clinical sign of the syndrome.[6] Many screening tests of the zG function have been used to identify PA and to distinguish adenoma from hyperplasia (Table 3-2).[5–8] These studies should be performed in a logical sequence prior to any attempts at adrenal localization.

Inner adrenocortical function comprises hormones produced by the zona fasciculata (zF) and the zona reticularis (zR) (Table 3-1). Cortisol is the principal glucocorticoid produced by the zF and is responsible for the maintenance of intermediary metabolism.[1] It is secreted in response to adrenocorticotropic hormone (ACTH), which in turn is modulated by hypothalamic corticotropin-releasing hormone (CRH).[9] ACTH and cortisol exhibit a diurnal variation; peak levels are observed in the early AM, while a nadir is seen in the late PM.[10] Hypercortisolism manifests as a syndrome of a loss of or accentuated diurnal variation of cortisol and ACTH, along with other well-described signs and symptoms (Cushing's syndrome).[11] The syndrome can be a result of hypothalamic dysfunction or

We wish to thank Drs. T. Mangner and H. Anderson-Davis for the synthesis of NP-59, the Phoenix Memorial Laboratories for the use of their radiochemical facilities, and Ms. Leann C. Beird for the preparation of the manuscript.

Supported by grants from NCI (CA-09015), MIAMDD RO1-AM-21477 RAD, GCRC HEW 3M01-RR00042-21 CLR, NHLB1HL-18575, the Nuclear Medicine Research Fund, and the Veterans Administration Research Service.

Table 3-1
Adrenal Cortex Anatomy and Function

Anatomic Zone	*Principal Hormone(s) Produced*	*Principal Trophic Factors*
Outer Cortex		
Zona glomerulosa	Aldosterone	Renin/Angiotensin
Inner Cortex		
Zona fasciculata	Cortisol	Adrenocorticotropin/CRH
Zona reticularis	Androstenedione Dehydroepiandrosterone	Adrenocorticotropin (luteinizing hormone)

Table 3-2
Biochemical Studies Used to Distinguish Adenoma From Hyperplasia in Aldosteronism

	Adenoma	*Hyperplasia*
Serum aldosterone level	Marked increase	Increase
Serum potassium	Marked decrease	Decrease
Plasma renin activity	Marked suppression	Suppressed
Postural response of aldosterone	NR* or decrease	Increase
Aldosterone response to ACTH	Marked increase	NR or increase
Aldosterone response to dexamethasone	NR	Decrease

* NR, *no response*

Table 3-3
Clinical and Laboratory Findings in Different Forms of Cushing's Syndrome

	Pituitary ACTH-Secreting Tumor	*Hypothalamic ACTH-Dependent*	*Ectopic ACTH*	*Adrenal Adenoma*	*Adrenal Carcinoma*	*Nodular Hyperplasia*
Obesity	+	+	−	+	±	+
Pigmentation	±	+	+	−	−	−
Alkalosis	−	−	+	−	−	−
Urinary 17-ketosteroids	↑	↑	↑↑	↑	↑↑	↑
Urinary 17-OHCS	↑↑	↑↑	↑↑	↑	↑	↑
Hypokalemia	−	−	−	−	−	−
Hyperkalemia	−	+	−	−	−	−
ACTH response	↑	↑	0	0	0	±
Metyrapone response	±	+	0	0	0	±

+, *present;* −, *absent;* ±, *present or absent;* ↑, *increased;* ↑↑, *markedly increased; 0, no change.*

pituitary ACTH hypersecretion (Cushing's disease); or the presence of an ectopic ACTH-secreting neoplasm (lung, pancreas, or carcinoid), an adrenal adenoma, or adrenal cortical carcinoma.[11–13] Numerous biochemical tests of zF function have been proposed to not only diagnose hypercortisolism, but to aid in distinguishing the various etiologies of the syndrome (Table 3-3).

The zR produces circulating androgens, dehydroepiandrosterone, androstenedione, and, to a much lesser degree, testosterone (Table 3-1).[14] These hormones are necessary in women for secondary sex characteristics[14]; their precise function in men is less well understood. In disease, hyperandrogenism in women produces a characteristic constellation of findings that range from hirsutism to virilization, with this progression dependent upon the level of androgens secreted and the duration of exposure.[15] Although other endocrine and nonendocrine tissues secrete androgens (testes, ovary, and peripheral tissues), the adrenal glands may play a contributing role either as a result of congenital adrenal hyperplasia or as a component of other recognized endocrine syndromes (*i.e.*, polycystic ovary syndrome).[15] Furthermore, hyperandrogenism can be the result of benign and malignant adrenal or ovarian tumors.[15]

Etiologies of Androgen Excess

Dysfunctional States
- Adrenals (hyperplasia)
- Ovary (stromal hyperplasia)
- Peripheral tissues (*e.g.*, skin conversion steroid precursors to androgens)

Neoplasms:
- Adrenals: adenoma, carcinoma
- Ovary: arrhenoblastoma

Congenital:
- Enzyme deficiencies (*e.g.*, 17-hydroxylase and 21-hydroxylase deficiencies)

TECHNIQUES

Imaging Procedures

Adrenal gland imaging is optimally performed using a gamma camera interfaced to a digital minicomputer.[16] As a result of the metabolism of the present generation of adrenal cortical imaging agents, scanning procedures are conducted over protracted intervals after radiotracer injection.[16] The imaging intervals are dependent upon the presumed pathophysiology under examination. The usual dose of the radiopharmaceutical, iodine 131-6*β*-iodomethyl-19-norcholesterol (NP-59), is 1 mCi/1.73 m^2.[16] Posterior, anterior, and lateral images are obtained at each scan session. Images of 50,000 to 100,000 counts provide sufficient data in most instances for the first sets of scans; subsequent images are then acquired for an equivalent interval of time.[16] The lateral image is important, not only to discriminate gallbladder uptake of iodocholesterol from the right adrenal, but also to estimate adrenal depth (attenuation correction) for the calculation of iodocholesterol uptake (see Estimation of Radiation Absorbed Doses, below).[17]

Studies of Cushing's syndrome, adrenal enlargement, and anatomic localization are performed after stopping medications that would interfere with adrenal function or alter hypothalamic-pituitary-adrenal axis function.[17] Therefore, in these investigations, usually a 5- to 7-day interval after radiotracer injection is allowed to afford sufficient time for the clearance of background radioactivity from nontarget tissues. In studies to localize and characterize abnormal outer cortical (zG) function (*i.e.*, aldosteronism or the evaluation of hyperandrogenism), dexamethasone is recommended to enhance the functional difference between the zones of the cortex that are not solely dependent upon the tropic influence of ACTH.[18] Furthermore, as a result of de-iodination of the iodocholesterol derivatives, stable iodide, either in the form of Lugol's solution or super-saturated potassium iodide (SSKI), is given prior to iodocholesterol injection. This serves to decrease thyroidal accumulation of free radioactive iodine.[17]

Empiric studies demonstrate that variation of the dose and duration of dexamethasone suppression administered prior to iodocholesterol used in the evaluation of hyperandrogenism and aldosteronism has profound effects upon adrenal gland uptake and imaging.[18] The present dexamethasone suppression regimen is devised to distinguish normal from abnormal adrenal function. Administering 1 mg of dexamethasone, 4 times daily for 7 days before iodocholesterol injection and continued thereafter, the normal adrenal cortex will image as early as the fifth day after injection.[19] Adrenal glands that visualize *prior* to the fifth day after radiotracer injection on this dexamethasone suppression regimen are abnormal. The pattern of visualization, unilateral or bilateral, can thus be used to distinguish adenoma from hyperplasia.[18]

Computer Acquisition

The acquisition of adrenal cortical images on a dedicated nuclear medicine computer system allows for image enhancement and the estimation of iodocholesterol uptake.[17] The level of accumulation achieved can be used much in the same manner as that of 24-hour radioiodine uptake measurements of thyroid function.[18] A semi–operator-independent computer algorithm has been developed for this purpose.[20] The "center" point of each adrenal gland is selected by the operator and the algorithm then searches contiguous points for a maximum count drop. At the point of maximum count drop, a 2-point pixel displacement edge is created surrounding each adrenal gland to allow for an estimate of local background radioactivity. Once the estimated adrenal depth is known, an attenuation correction can be made and is used to correct for adrenal asymmetry (see General Concerns in Adrenocortical Image Interpretation, below).[20] The normal range of iodocholesterol uptake estimated by this approach in studies not performed on dexamethasone suppression in $0.16 \pm 0.05\%$ administered dose/gland.[18]

PHYSIOLOGIC MECHANISMS OF RADIOPHARMACEUTICAL UPTAKE

The iodocholesterol adrenal cortical imaging agents share mechanisms of uptake that are identical to those of native cholesterol.[21] Cholesterol accumulation by the adrenal cortex is mediated by a receptor-dependent process by which cholesterol is conveyed to membrane-bound receptors by specific carriers, the lipoproteins.[22] In humans, low-density lipoproteins (LDL) are recognized by the adrenal receptor and this allows the cholesterol contained within these particles to enter the cortical cells. Control of LDL ingress is modulated by the intracellular cholesterol content, circulating LDL cholesterol levels, and adrenocorticotropic hormone levels.[22] Receptor affinity and number are decreased in states of hypercholesterolemia and in rare syndromes of congenitally absent or abnormal receptor function.[22] Within the adrenal cortical cell, cholesterol (and iodocholesterol) are esterified for storage. However, little if any iodocholesterol is liberated from cholesterol ester storage, and thus circulating radioiodinated steroid hormones or urinary metabolites are not detected after administration of iodocholesterol.[21]

The uptake of iodocholesterol can be manipulated for diagnostic advantage with dexamethasone, as noted above.[18] Enterohepatic circulation of radioiodinated bile salts does occur and will occasionally obscure the region of the adrenals, thus making image interpretation difficult. As a result, mild laxatives can be used to speed the elimination of background radioactivity from the bowel.[23] Additionally, because a large number of drugs will affect adrenal hormone secretion and iodocholesterol accumulation by the adrenals, a complete medication history is mandatory for accurate scan interpretation (Table 3-4).[18]

ESTIMATION OF RADIATION ABSORBED DOSES

The dosimetry of the adrenal cortical imaging agents is listed in Table 3-5.[24] Previous estimates of the radiation dose from the iodocholesterols were exaggerated; the initial studies were performed using tissues from pa-

Table 3-4
Factors that Alter Adrenal Function and Adrenal Scintigraphic Imaging

	Effect	*Mechanism*	*Scan Appearance*
Adrenal Cortex			
Zona fasciculata/reticularis			
Hormones			
Adrenocorticoids:	↓ Cortisol	↓ CRF	↓ Uptake
Dexamethasone	↓ Androgens	↓ ACTH	
Metabolic inhibitors:			
Aminoglutethimide	↓ Cortisol synthesis		
Metyrapone	↑ Adrenal cholesterol uptake	↑ ACTH	↑ Uptake
Mitotane	↓ Cortisol-ACTH	Adrenal pituitary suppression	↓ Uptake
Exogenous ACTH	Direct adrenal stimulation		↑ Uptake
Zona glomerulosa			
Antihypertensives:			
Propranolol	↓ Plasma renin activity	β-receptor blockade	↓ Uptake
Antagonists:			
Spironolactone	↓ Aldosterone	Adrenal suppression	↓ Uptake(?)
Diuretics (all)	↓ Serum sodium ↓ Plasma volume	↑ Plasma renin activity	↑ Uptake
Oral contraceptives (all)	↑ Plasma renin activity	↑ Cortisol secretion rate	↑ Uptake
Excessive salt intake	↓ Aldosterone	↓ Plasma renin activity	↓ Uptake
General			
Cholesterol-lowering agents	↓ Serum cholesterol	(?) Cholesterol pool effect	↑ Uptake
4-aminopyrazolopyrimidine	↓ Serum cholesterol	↑ LDL-receptor activity	↑ Uptake
Hypercholesterolemia	↑ Serum cholesterol	(?) Cholesterol pool effect	↓ Uptake

(Modified from Gross MD et al: The role of pharmacologic manipulation in adrenocortical scintigraphy Sem Nucl Med 11:121-149, 1981)

Table 3-5
Radiation Dosimetry of I-131-6β-iodomethyl-19-norcholesterol (NP-59)

Organ	*Rad/mCi*
Total Body	1.2
Adrenals	26.0*
Ovary	8.0*
Testes	2.3*
Liver	2.4

** A 50% decrease is seen on dexamethasone suppression.*

(Carey et al: Absorbed dose to the human adrenal asymmetry: Explanation and interpretation. J Nucl Med 20:60–61, 1979)

tients with Cushing's syndrome.[24] The administration of dexamethasone results in an approximate 50% decrement of iodocholesterol adrenal uptake.[17]

DESCRIPTION AND INTERPRETATION OF THE ADRENAL CORTICAL SCAN

General Concerns in Adrenocortical Image Interpretation

The patterns of adrenal cortical imaging are sufficiently restricted so that image interpretation alone is not sufficient for a diagnostic evaluation in most patients. Interpretation is in large part dependent upon the results of screening biochemical studies that assist in the diagnostic evaluation of these patients.

Adrenal cortical scintigraphy depicts the adrenals and their anatomic and positional asymmetry in most patients. The right adrenal lies more cephalad and posterior in the abdomen than the left adrenal and, as a result, in the posterior projection the right adrenal appears to contain more activity than the left.[25] Although detailed adrenal anatomy is not a primary concern in adrenal scintigraphy, the right adrenal in many studies has an ovoid appearance and the left adrenal is elongated or crescentic.[25] The lateral scintiscan is important, because gallbladder radioactivity can obscure or mimic the right adrenal in the posterior projection.[17,18] Therefore, the gallbladder (an anterior structure) can be distinguished from the right adrenal (a posterior structure) in the right lateral projection.[18]

Dexamethasone suppression imaging presents some difficulties in scan interpretation, because background radioactivity remains high and only faint adrenal cortical activity is observed in some patients with adrenal dysfunction in the early (prior to Day 5) images.[17] Studies obtained after the fifth day after discontinuation of dexamethasone suppression are useful, as "retrospective" interpretation of the earlier scans can, in some instances, identify the collections of radioactivity seen on images prior to Day 5 as "adrenal" and allow for the correct assignment of adenoma, hyperplasia, or normal (uptake on or after Day 5) adrenal function.[17] Dexamethasone should be discontinued on the fifth day after NP-59 because scans obtained thereafter are only useful for localization purposes; differential adrenal function analysis is dependent upon finding discernible adrenal cortical uptake of the radiotracer prior to Day 5 on the 4 mg/day × 7 day dexamethasone suppression regimen.[18,19]

The diagnostic use of iodocholesterol adrenocortical uptake measurements has limited applicability.[26] Studies suggest that the level of uptake achieved can be useful in documenting function or dysfunction, but is not useful in assigning definitive diagnostic impressions.[18] This is similar to the use of radioiodine uptake in the evaluation of patients with thyroid dysfunction.[27]

Interpretation of Scans in Cushing's Syndrome

The appearance of the adrenals in ACTH-dependent Cushing's syndrome is bilateral and includes excessive iodocholesterol accumulation (Table 3-6). The scintiscan is a direct representation of the pathophysiologic process. In ACTH-independent disease, the adrenal adenoma is depicted as a unilateral accumulation of iodocholesterol; the contralateral adrenal cortex does not visualize as ACTH levels and iodocholesterol uptake are suppressed by the autonomous secretion of glucocorticoids by the adenoma.[28]

Bilateral, autonomous adrenal (cortical nodular) disease is usually anatomically asymmetric and a pattern of marked asymmetric imaging in a patient with suppressed (or paradoxically normal) ACTH levels in

Table 3-6
Patterns of Imaging in Cushing's Syndrome

*Scintigraphic Pattern**	*Form of Cushing's Syndrome*
Bilateral symmetric	ACTH-dependent Hypothalamic Pituitary Ectopic ACTH syndrome
Bilateral asymmetric	ACTH-Independent Nodular hyperplasia
Unilateral	Adrenal adenoma Adrenal remnant (ectopic adrenocortical tissue)
Bilateral nonvisualization	Adrenal carcinoma (functioning tumor suppresses contralateral gland)

** Images obtained 5 to 7 days after iodocholesterol injection*

the presence of signs and symptoms of Cushing's syndrome suggests this variant of glucocorticoid hypersecretion.[27,28] Bilateral nonvisualization is characteristic of a functioning adrenocortical carcinoma; the neoplasm produces glucocorticoids and their precursors, suppresses ACTH and contralateral adrenal iodocholesterol uptake, but does not itself image because the uptake of iodocholesterol is too low for its visualization.[29] Thus, the presence of a mass lesion of the adrenal seen on computed tomography or ultrasound and the bilateral adrenal nonvisualization on the scintiscan strongly suggest the diagnosis of a functioning adrenocortical carcinoma.[27]

Interpretation of Scans in the "Incidentally" Discovered Nonfunctioning Adrenal Mass

In the absence of biochemical dysfunction, the presence of iodocholesterol accumulation within an anatomically abnormal adrenal ("incidentally" discovered adrenal mass) suggests a functioning but not hyperfunctioning adrenal.[30] The absence of discernible or decreased adrenal iodocholesterol uptake in an anatomically abnormal gland as compared to the normal contralateral adrenal is compatible with a space-occupying lesion.[30] The differential diagnosis of these lesions is extensive; the scintiscan may be useful in their functional evaluation.[31]

Incidental Adrenal Masses
Adrenal cyst
Adrenal lipoma/myelolipoma
Metastases to adrenal (*i.e.*, lung, bowel neoplasms)
Adrenal adenoma (nonfunctional, euadrenal, or hyperfunctioning mass without obvious clinical signs or symptoms)
Adrenal lymphoma
Adrenal carcinoma
Adrenal pseudotumor

Interpretation of Scans in Primary Aldosteronism

The patterns of adrenocortical imaging in aldosteronism reflect the abnormal secretion of aldosterone from unilateral (adenoma) or bilateral (hyperplasia) sources (Table 3-7).[32] As unusual variant of aldosteronism, the familial syndrome of dexamethasone-suppressible aldosteronism, is depicted on scintiscans as bilateral nonvisualization, while some mineralocorticoid-secreting adrenal cortical carcinomas and their metastases have been reported to image with NP-59.[27,33]

Table 3-7
Patterns of Imaging in Primary Aldosteronism

Scintigraphic Pattern	*Form of Primary Aldosteronism*
Unilateral early ($<$ 5 days)*	Adenoma
Bilateral early ($<$ 5 days)	Bilateral hyperplasia
Bilateral late ($>$ 5 days)	Dexamethasone-suppressible hyperplasia (nondiagnostic)

** Adrenal imaging occurring before the fifth day after NP-69 injection is abnormal.*

Interpretation of Scans in Hyperandrogenism

The adrenal contribution to abnormal androgen levels can be imaged with dexamethasone suppression (DS)–iodocholesterol scintigraphy (Table 3-8).[34] Bilateral (prior to Day 5 post iodocholesterol on DS) and unilateral early visualization are compatible with adrenal hyperplasia and adenoma, respectively.[34] Alternatively, a normal DS adrenal scintiscan suggests the presence of ovarian or peripheral source(s) of androgen production. Functioning ovarian tumors (arrhenoblastomas) have been imaged with iodocholesterol, and in the absence of abnormal adrenal uptake, scans of the lower pelvis (after laxative administration) may identify these rare neoplasms.[35]

DISCUSSION: PITFALLS IN ADRENOCORTICAL IMAGING

Primary emphasis must be directed toward a biochemical evaluation sufficient for diagnosis prior to any attempt at localization. Adrenal gland dysfunctional states are problematic in their diagnosis, because screening studies and suppression/stimulation tests of hormone secretion are critical, time-consuming, ini-

Table 3-8
Patterns of Imaging in Hyperandrogenism

Scintigraphic Pattern	*Form of Hyperandrogenism*
Unilateral (early) imaging*	Adrenal adenoma
Bilateral (early) imaging	Adrenal hyperplasia
Bilateral increased uptake†	Polycystic ovary disease Congenital adrenal hyperplasia (CAH)
Early nonvisualization Bilateral late visualization (normal adrenal uptake)	Ovarian or peripheral
Focal pelvic uptake	Ovarian dysfunction

** Early is defined as adrenal activity seen before the fifth day after NP-59 injection on dexamethasone suppression.*

† Accentuated (increased) bilateral adrenal NP-59 accumulation seen on non–dexamethasone-suppressed scans.

tial steps in the proper evaluation of patients with suspected adrenal disease. Scintiscan interpretation is not only directly dependent upon the results of biochemical studies, but also is affected by intradiagnostic medications and therapeutic maneuvers.[18] For example, diuretics or spironolactone given during dexamethasone suppression adrenal scintiscanning in primary aldosteronism may result in early (prior to Day 5) imaging of the "normal" contralateral adrenal in patients with adenoma.[18] Hypercholesterolemia, either by virtue of LDL receptor down-regulation or by an increased cholesterol pool effect, will decrease iodocholesterol adrenal uptake. Evaluation of cholesterol levels must be considered when the scintiscans show discordant images of patients in whom elevated iodocholesterol uptake would be anticipated.[35] In Cushing's syndrome due to pituitary or hypothalamic dysfunction, high-dose dexamethasone or antimetabolite administration (such as metyrapone or mitotane) will interfere with the accumulation of iodocholesterol in the adrenal cortex and will result in scans with either normal, low, or no discernible adrenocortical uptake (Table 3-4).[18]

Thus, in the evaluation of adrenal cortical dysfunction, abnormal hormone secretion from all of the functional zones has been shown to be related to the uptake of iodocholesterol in Cushing's syndrome, aldosteronism, and adrenal hyperandrogenism. The scintiscan is thus a sensitive map of adrenocortical function and the pattern and the degree of radiotracer accumulation are useful in the localization of adrenal disease.[18,27,28]

REFERENCES

1. Neville AM, O'Hare MJ: Aspects of structure, function and pathology. In James VHT (ed): The Adrenal Gland, pp 1–15. New York, Raven Press, 1979

2. Nelson DH: The importance of aldosterone in clinical medicine. Med Clin North Am 42:1195, 1958

3. Laragh JT, Angers M, Kelley WG: The effect of epinephrine, norepinephrine, angiotensin II and others on the secretory rate of aldosterone in man. JAMA 174:234, 1960

4. Skeggs LT, Dorer PE, Kahn JR: The biochemistry of the renin angiotensin system and its role in hypertension. Am J Med 60:737, 1976

5. Feriss JB, Bevers DG, Brown JJ et al: Clinical, biochemical and pathological features of low-renin (primary) hyperaldosteronism. Am Heart J 95:375, 1978

6. Feriss JB, Bevers DG, Boddy K et al: The treatment of low renin (primary) hyperaldosteronism. Am Heart J 96:97, 1978

7. Stockigt JR: Mineralocorticoid excess. In James VHT (ed): The Adrenal Gland, pp 197–242. New York, Raven Press, 1979

8. Weinberger MH, Grim CE, Holfield JW: Primary aldosteronism: Diagnosis, localization and treatment. Ann Intern Med 90:386, 1979

9. Nelson DH: The Adrenal Cortex: Physiological Function and Disease, pp 65–88. Philadelphia, WB Saunders, 1980

10. Shima S, Mitsonaga M, Nakao T: Effects of ACTH on cholesterol dynamics in rat adrenal tissue. Endocrinology 90:808, 1972

11. Kreiger DT, Allen W: Relationship of bioassayable and immunoassayable plasma ACTH and cortisol concentrations in normal subjects and in patients with Cushing's disease. J Clin Endocrinol Metab 40:675, 1975

12. West CD, Dolman LI: Plasma ACTH radioimmunoassays in the diagnosis of pituitary-adrenal dysfunction. Ann NY Acad Sci 297:205, 1977

13. O'Neal LW: Correlation between clinical pattern and pathological findings in Cushing's syndrome. Med Clin North Am 52:313, 1968

14. Vermuelen A, Rubens R: Adrenal virilism. In James VHT (ed): The Adrenal Gland, pp 259–282. New York, Raven Press, 1979

15. Givens JR: Hirsutism and hyperandrogenism. Adv Int Med 1:221, 1976

16. Thrall JF, Freitas JE, Beierwaltes WH: Adrenal scintigraphy. J Nucl Med 8:23, 1978

17. Gross MD, Thrall JH, Beierwaltes WH: The adrenal scan: A current status report on radiotracers, dosimetry and clinical utility. In Weissmann H, Freeman L (eds): Nuclear Medicine Annual, pp 127–175. New York, Raven Press, 1980

18. Gross MD, Valk TW, Swanson DP et al: The role of pharmacologic manipulation in adrenal cortical scintigraphy. Sem Nucl Med 9:128, 1981

19. Gross MD, Freitas JE, Swanson DP: The normal dexamethasone suppression adrenal scintiscan. J Nucl Med 20:1131, 1979

20. Koral KF, Sarkar SD: An operator-independent method for background subtraction in adrenal uptake measurements. Concise communication. J Nucl Med 18:925, 1977

21. Gross MD, Swanson DP, Wieland D, Beierwaltes WH: Mechanisms of localization, specificity and metabolism of adrenal gland imaging agents. In Billinghurst MW, Colombetti G (eds): Mechanisms of Localization of Radiotracers, Part B: Study of cellular function, vol. 12, pp 189–223. Boca Raton, CRC Press, 1982

22. Brown M, Goldstein J: Receptor–mediated control of cholesterol metabolism. Science 91:150, 1976

23. Shapiro B, Nakajo M, Gross MD et al: Bowel preparation in adrenocortical scintigraphy with NP–59. J Nucl Med 24:732, 1983

24. Carey JE, Thrall JH, Freitas JE, Beierwaltes WH: Absorbed dose to the human adrenal from iodomethyl-norcholesterol (I-131) "NP-59". J Nucl Med 20:60, 1979

25. Freitas JE, Thrall JH, Swanson DP et al: Normal adrenal asymmetry: Explanation and interpretation. J Nucl Med 19:149, 1978

26. Gross MD, Shapiro B, Freitas JE: Limited significance of asymmetric adrenal visualization on dexamethasone suppression scintigraphy. J Nucl Med 26:43, 1985

27. Gross MD, Shapiro B, Thrall JH et al: The scintigraphic imaging of endocrine organs. Endocr Rev 5:221, 1984

28. Gross MD, Shapiro B, Beierwaltes WH: The functional

localization of the adrenal cortex by quantitative scintigraphy. In Lawrence JH, Winchell HS (eds): Advances in Nuclear Medicine, vol. 5, pp 83–115. New York, Grune and Stratton, 1983

29. Schteingart DE, Seabold JE, Gross MD, Swanson DP: Iodocholesterol adrenal tissue uptake and imaging in adrenal neoplasms. J Clin Endocrinol Metab 52:1156, 1981

30. Gross MD, Smid A, Bouffard JA: Adrenal scintigraphy in the evaluation of the clinical silent adrenal mass lesion (abstr). J Nucl Med 27:908, 1986

31. Gross MD, Wilton GP, Shapiro B et al: Normal adrenal function and asymmetric anatomy: The adrenal goiter (abstr). Clin Res 33:308A, 1985

32. Gross MD, Shapiro B, Grekin RJ et al: Scintigraphic localization of adrenal lesions in primary aldosteronism. Am J Med 77:839, 1984

33. Shenker Y, Gross MD, Grekin RJ et al: The scintigraphic localization of mineralocorticoid-producing adrenocortical carcinoma. J Endocrinol Invest 9:115, 1986

34. Gross MD, Freitas JE, Swanson DP, Beierwaltes WH: Dexamethasone suppression (DS) adrenal scintigraphy in hyperandrogenism. J Nucl Med 20:1131, 1979

35. Taylor L, Ayers JWT, Gross MD et al: Diagnostic considerations in virilization iodomethylnorcholesterol scintigraphy in the localization of androgen-secreting tumors. Surg Gynecol Obstet (in press)

36. Valk TW, Gross MD, Freitas JE et al: The relationship of serum lipids to adrenal gland uptake of 6β-(I-131) iodomethyl-19-norcholesterol in Cushing's syndrome. J Nucl Med 21:1069, 1980

Atlas Section

Figure 3-1

Normal Adrenal Scintiscan (baseline study)

Normal posterior *(A)* and anterior *(B)* NP-59 adrenal scans. In the posterior view, the right adrenal appears to accumulate more radioactivity because it is more posteriorly situated in the abdomen than the left adrenal. In the anterior view, more radioactivity is seen in the left adrenal. (*Arrows,* adrenals; *L,* liver; *B,* bowel) (Freitas JE et al: Normal adrenal asymmetry: Explanation and interpretation. J Nucl Med 19:149-154, 1978)

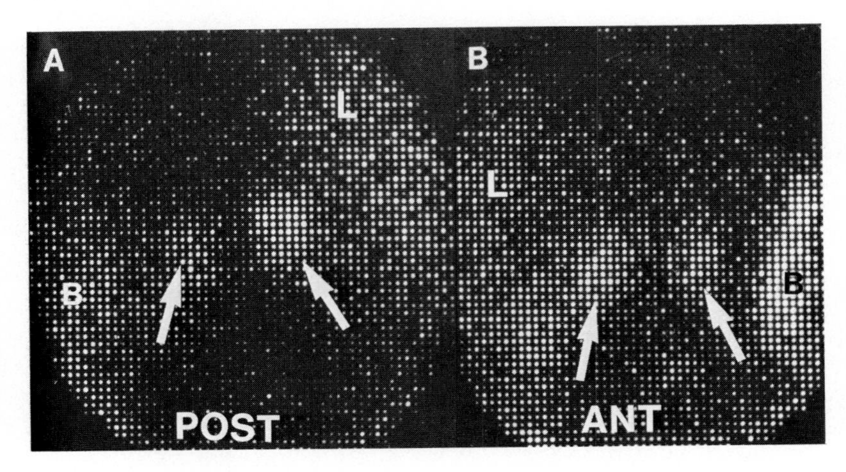

Figure 3-2 ***Normal Adrenal Scintiscan (dexamethasone suppression)***

Normal posterior adrenal scintiscans obtained on constant dexamethasone suppression (4 mg × 7 days prior to and 5 days following NP-59 injection). ***(A)*** On Day 4 after injection, there is no discernible adrenal radioactivity, while on Day 5 bilateral adrenal visualization is noted *(arrows)* on the analog ***(B)*** and digitally enhanced ***(C)*** images. (Gross MD et al: The role of pharmacologic manipulation in adrenal cortical scintigraphy. Sem Nucl Med 11:137, 1981)

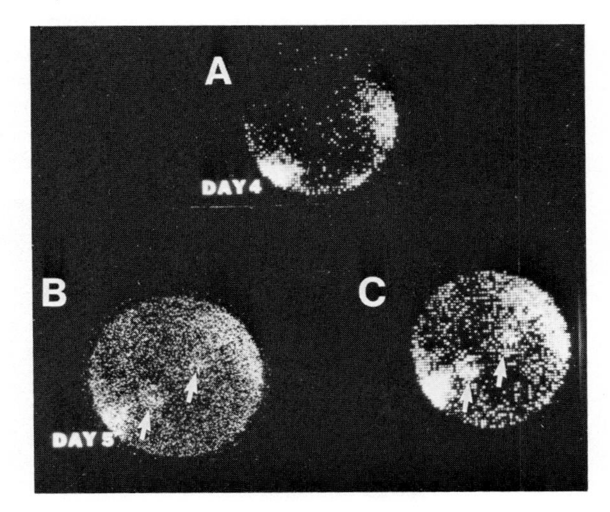

Figure 3-3 ***The Effect of Background Radioactivity on Adrenal Visualization***

Posterior adrenal scintiscans demonstrate the effect of increasing bowel (background) radioactivity upon adrenal visualization. The administration of laxatives ***(A)*** has resulted in rapid clearance of NP-59 from bowel on the Day 3 study without dexamethasone, as contrasted with studies in two other patients (***B*** and ***C***) made Day 3 post-NP-59 injection where bowel preparation was omitted. The adrenals are obscured by background radioactivity in ***(C)*** (Shapiro B et al: Bowel preparation in adrenocortical scintigraphy with NP-59. J Nucl Med 24:732-734, 1983)

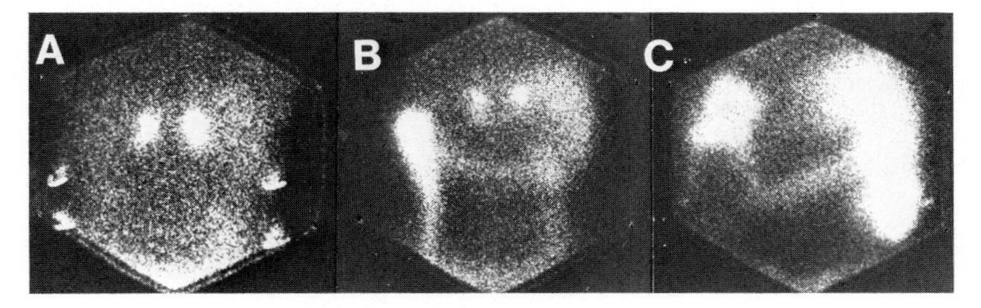

Figure 3-4 ***ACTH-Dependent Cushing's Syndrome — Bilateral Hyperplasia***

Posterior *(A)* and lateral *(B)* adrenal scintiscans in a patient with elevated urinary hydroxycorticosteroids, accentuated diurnal variation of plasma cortisol, and elevated ACTH levels. The bilateral, symmetrical adrenal accumulation of NP-59 is characteristic of ACTH-dependent Cushing's syndrome, which in this example was due to a pituitary adenoma. The arrows identify a linear Ba-133 marker placed along the lumbar spine. (Gross MD et al: The relationship of adrenal iodomethylnor-cholesterol uptake to indices of adrenal cortical function in Cushings syndrome. J Clin Endocrinol Metab 52:1062 - 1066, 1981)

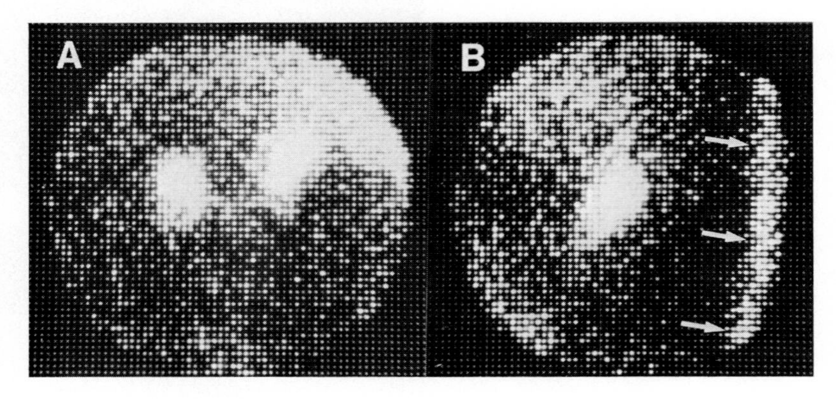

Figure 3-5 ***ACTH-Dependent Cushing's Syndrome — Adenoma***

Posterior analog *(A)* and digitally enhanced *(B)* NP-59 scintiscans of a patient with marked elevation of 17-hydroxy- and 17-ketosteroids, absent diurnal variation of cortisol, and suppressed ACTH levels. The scans depict a large collection of NP-59 activity localized to the right adrenal, a finding characteristic of an autonomous adrenal adenoma. The contra-lateral left adrenal is not seen because pituitary ACTH is suppressed by the abnormal glucocorticoid secretion by the abnormal adrenal. (Gross MD: The scintigraphic localization of adrenal tumors. Urol Rad 3:241 - 244, 1982)

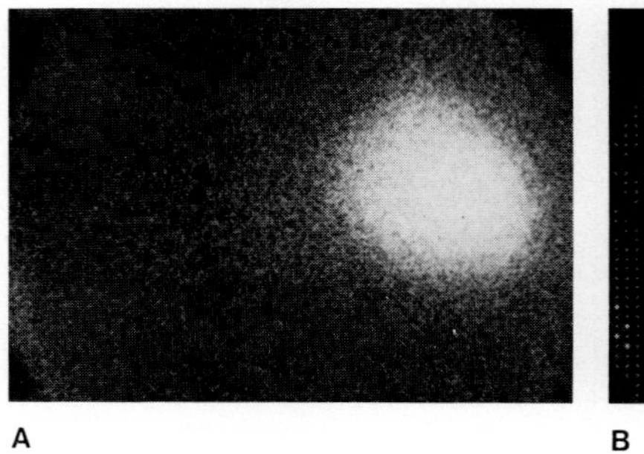
A

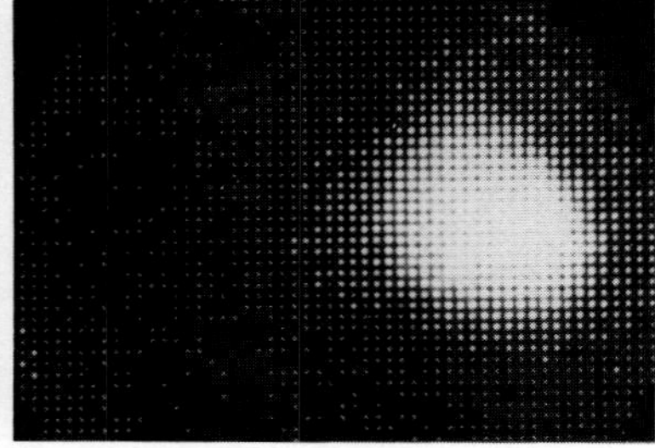
B

Figure 3-6 ***ACTH-Dependent Cushing's Syndrome— Bilateral Nodular Hyperplasia***

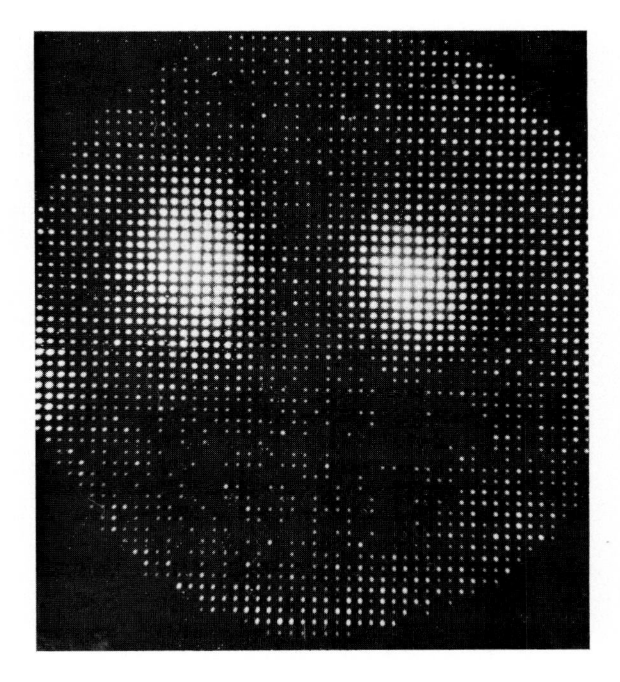

A posterior NP-59 adrenal scintiscan in a patient with elevated urinary and plasma corticosteroids and normal ACTH levels. The scan is characteristic of the bilateral and marked asymmetric nature of the unusual variant of ACTH-dependent Cushing's syndrome, bilateral nodular hyperplasia. In this example the left adrenal has accumulated over twice the NP-59 radioactivity of the right adrenal. (Gross MD et al: The adrenal scan: A current status report on radiotracers, dosimetry and clinical utility. In Freeman L, Weissman H (eds): Nuclear Medicine Annual 1980, p 137. New York, Raven Press, 1980)

Figure 3-7 ***ACTH-Dependent Cushing's Syndrome— Carcinoma***

Anterior chest *(A)* and posterior abdominal *(B)* NP-59 scans in a patient with a large right adrenal mass on computed tomography and biochemical evidence of Cushing's syndrome. The absence of NP-59 accumulation by the adrenal mass, when other reasons for adrenal nonvisualization can be excluded (*i.e.,* hyperlipidemia or subcutaneous injection of NP-59), is characteristic of a functioning adrenocortical carcinoma. (*m,* markers)

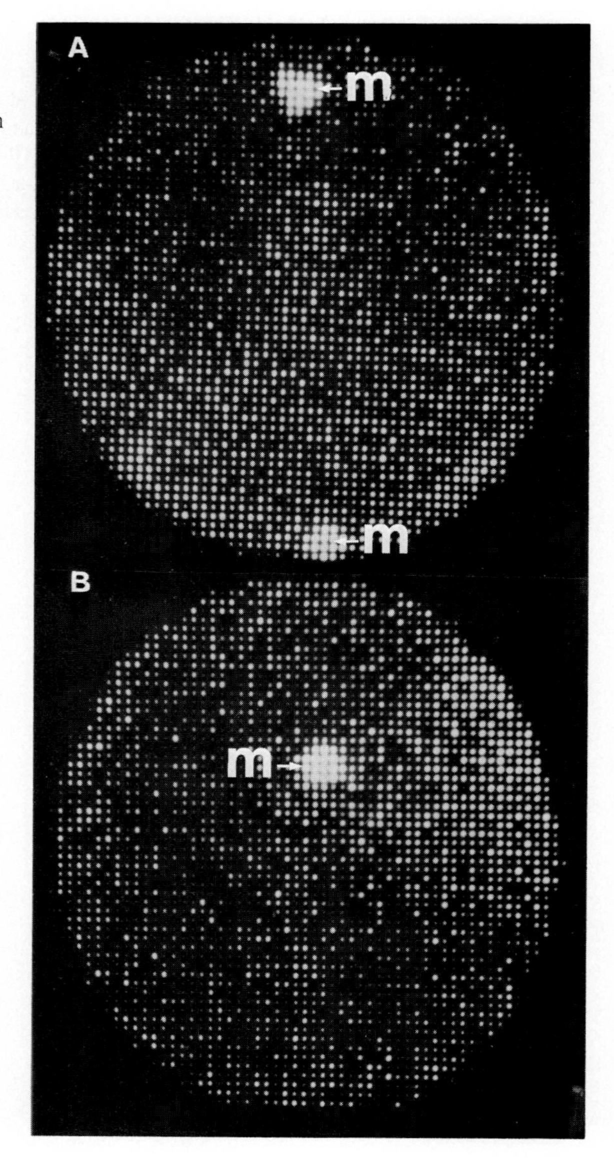

Figure 3-8 **_Primary Aldosteronism — Adenoma_**

Posterior adrenal scintiscans obtained on Day 3 *(A)* and Day 5 *(B)* after NP-59 injection while on constant dexamethasone suppression in a patient with hypertension, hypokalemia, elevated urinary aldosterone excretion, and suppressed plasma renin activity. Left-sided adrenal activity is seen on Day 3 *(arrow)*, whereas on Day 5 bilateral NP-59 uptake is noted *(arrows)*. The faint right adrenal uptake is normal on or after the fifth day on this dexamethasone regimen. A left adrenal adenoma was subsequently resected.

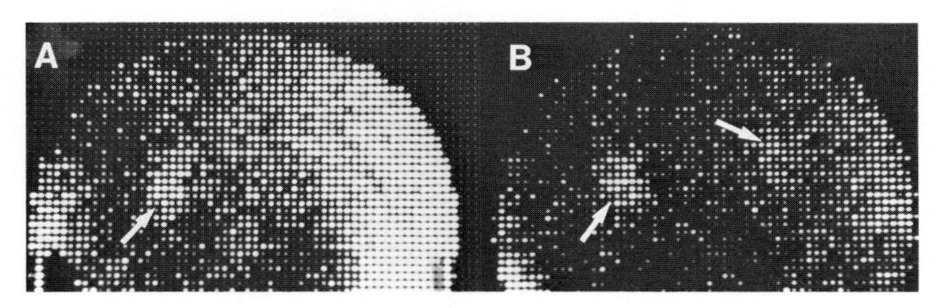

Figure 3-9 **_Primary Aldosteronism — Bilateral Hyperplasia_**

Posterior adrenal scintiscan obtained on Day 3 post-NP-59 injection in a patient with hypertension, hypokalemia, elevated urinary aldosterone excretion, and suppressed plasma renin activity. Bilateral adrenal NP-59 uptake seen at this time interval *(arrows)* is compatible with adrenal hyperplasia. Bilateral zona glomerulosa hyperfunction was proven at adrenal vein hormone sampling. (Gross MD et al: The scintigraphic imaging of endocrine organs. Endo Rev 5:258, 1984)

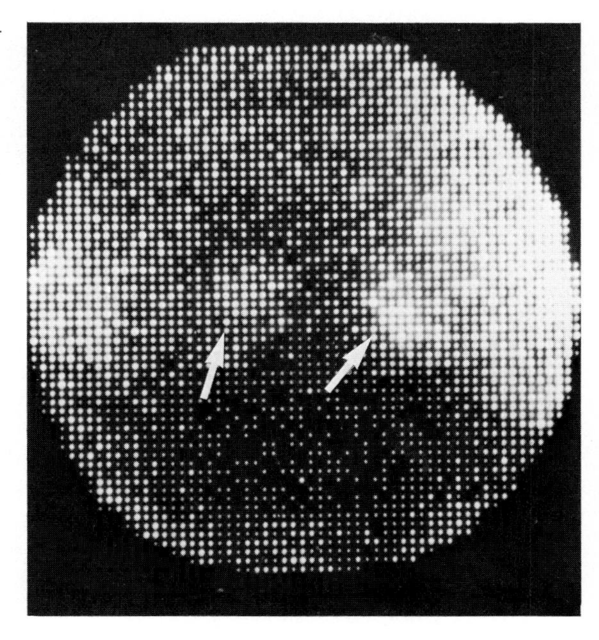

Figure 3-10 *Hyperandrogenism — Adrenal Adenoma*

Posterior adrenal scintiscan obtained on Day 5 post-NP-59 injection on constant dexamethasone suppression (DS) in a woman with elevated urinary 17-ketosteroids, serum testosterone, and findings of amenorrhea and virilization. The marked accumulation of NP-59 in the left adrenal is a result of an androgen-secreting adrenal adenoma. Faint right-sided adrenal uptake is seen on the scan and is normal for DS images at this interval after NP-59 injection. (Juni J et al: Bilateral visualization of adrenal cortical scintigraphy. Sem Nucl Med 13:168-170, 1983)

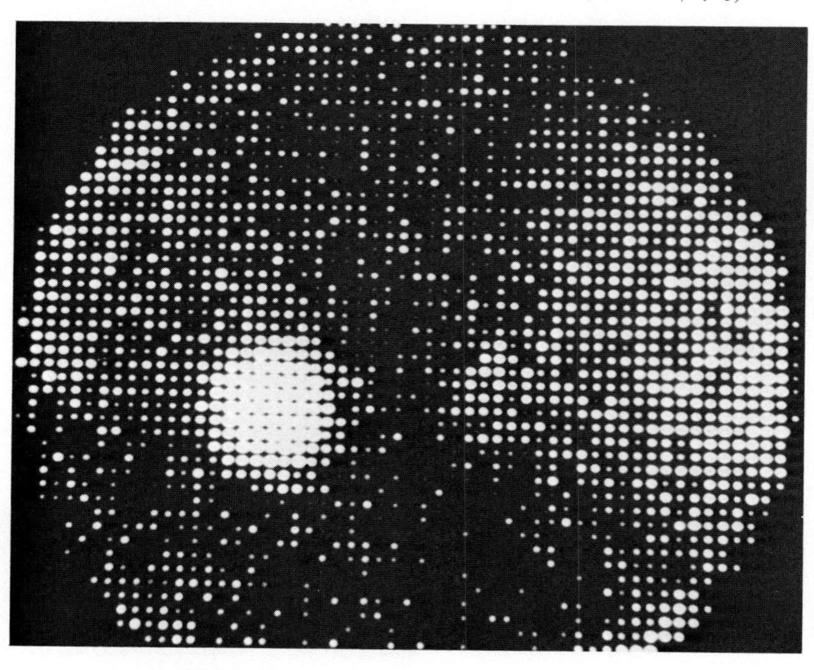

Figure 3-11 *Hyperandrogenism — Adrenal Hyperplasia*

Posterior analog *(left)* and digital *(right)* NP-59 images obtained on Day 3 *(A)* and Day 5 *(B)* on constant dexamethasone suppression in a woman with menstrual irregularities, hirsutism, and hyperandrogenism. The bilateral, early NP-59 uptake *(arrows)* is compatible with adrenal hyperfunction and was confirmed at adrenal vein hormone sampling where bilateral adrenal vein testosterone gradients were documented. (Gross MD, et al: The adrenal scan: A current status report on radiotracers, dosimetry and clinical utility. In Freeman L, Weissman H (eds): Nuclear Medicine Annual 1980, p 152. New York, Raven Press, 1980)

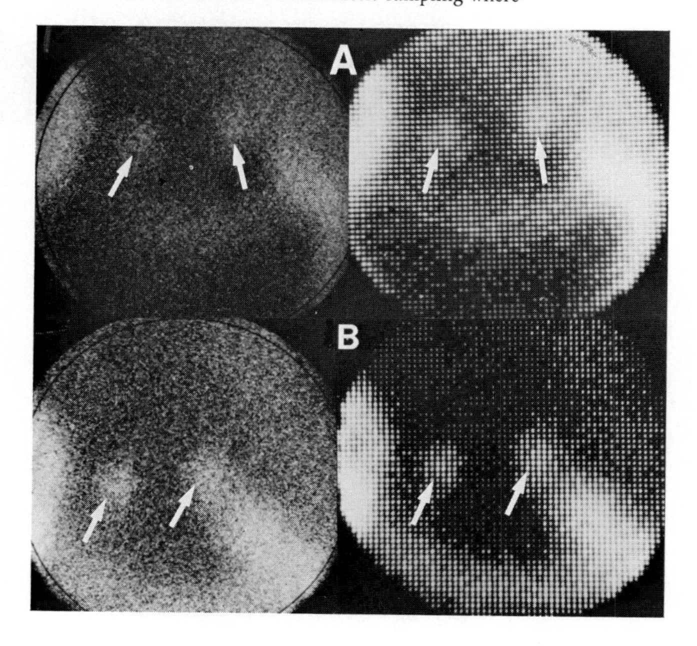

Figure 3-12 ***Hyperandrogenism — Ovarian Androgen-Secreting Tumor***

Anterior pelvic scintiscans obtained on Day 3 *(A)*, Day 5 *(B)*, and Day 7 *(C)* after NP-59 injection in a woman with virilization, amenorrhea, and hyperandrogenism. Dexamethasone was stopped after the Day 5 image. The persistent focus of NP-59 uptake in the right pelvis *(arrows)* was later shown to be an ovarian arrhenoblastoma.

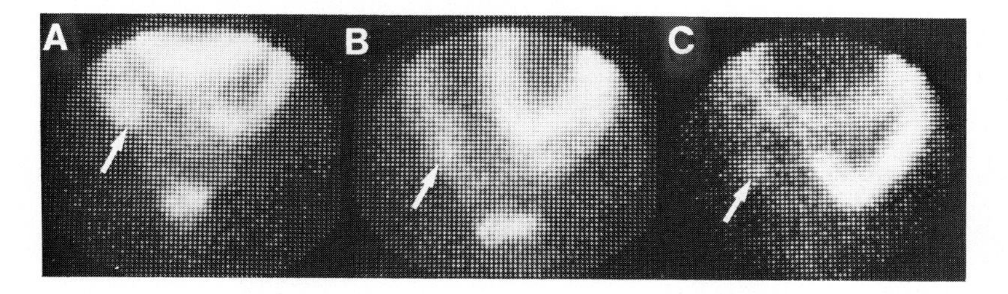

Figure 3-13 ***Nonfunctioning Adrenal Mass — Adenoma***

(A) Posterior NP-59 adrenal scan obtained on Day 5 post-NP-59 was without dexamethasone suppression. *(B)* CT scan is of a patient with a "silent" adrenal mass that was noted on CT exam for staging of prostate carcinoma. All biochemical tests of adrenal function were normal. A 2-cm left adrenal mass *(arrow)* is seen on the CT scan and is depicted on the scintiscan as an asymmetric and focal area of NP-59 uptake. This patient has been followed for an interval of 2 years without change in adrenal function or size.

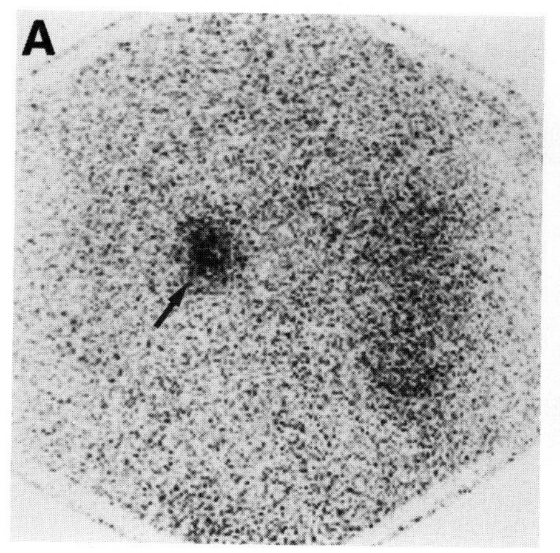

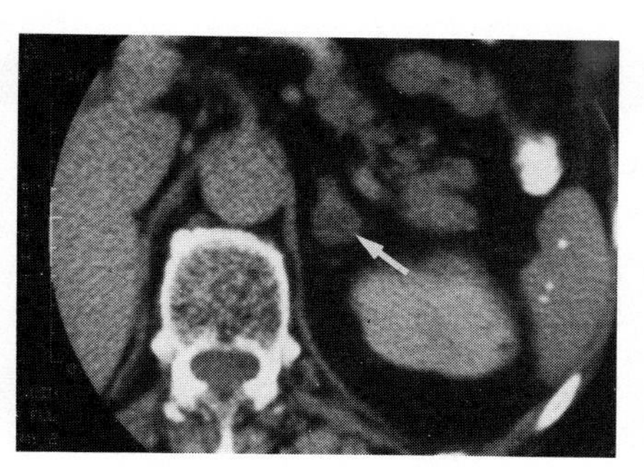

Figure 3-14 ***Nonfunctioning Adrenal Mass— Adrenal Cyst***

Posterior adrenal scan obtained on Day 5 post-NP-59 ***(A)*** injection with no dexamethasome suppression shows decreased accumulation of NP-59 in the superior portion of the left adrenal *(arrow)* in a patient without adrenal dysfunction. This abnormal region of absent NP-59 uptake was later shown to be an adrenal cyst on plain roentgenography. ***(B),*** as confirmed on abdominal ultrasound ***(C)***. Contrast injected percutaneously ***(D)*** further localized and characterized the cyst. (*C,* cyst; *KID,* kidney) (Gross MD, et al: Documentation of adrenal cyst by adrenal scanning techniques. J Nucl Med 19:1092, 1978)

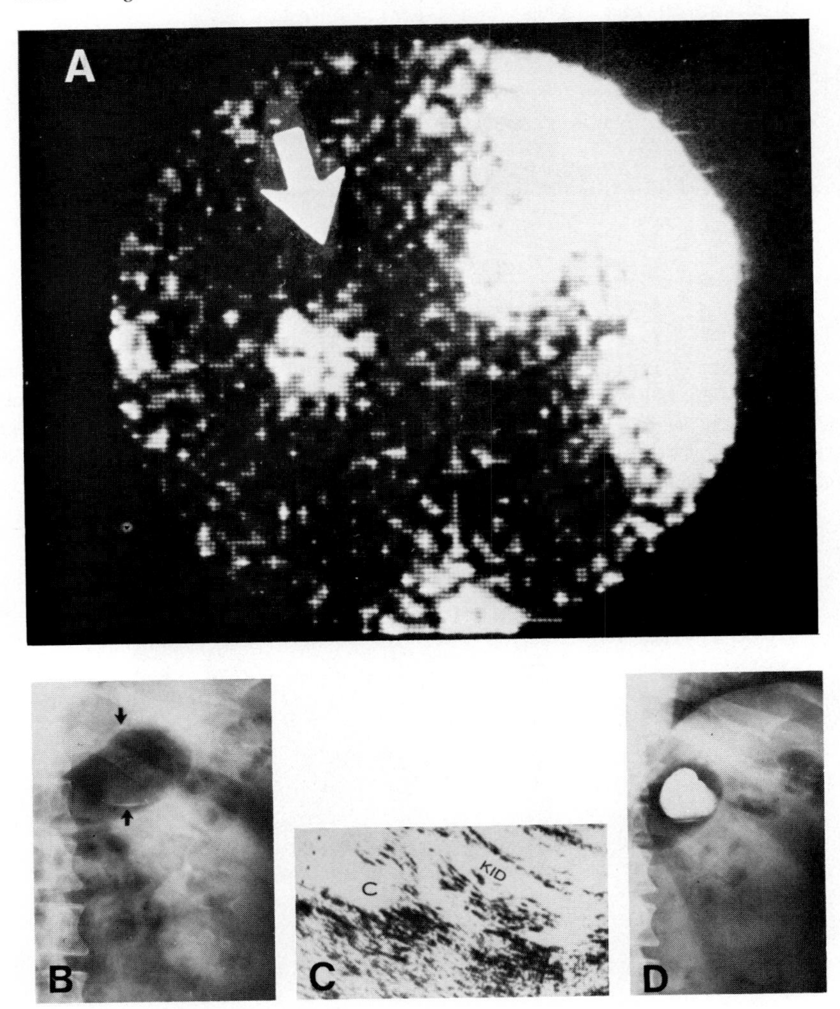

Figure 3-15 *Pitfalls—Antimetabolites in Cushing's Syndrome*

A depressed level of NP-59 adrenal uptake is seen in a patient with ACTH-dependent Cushing's syndrome scanned on Op'DDD therapy (0.18% administered dose). (Gross MD, et al: The role of pharmacologic manipulation in adrenal cortical scintigraphy. Sem Nucl Med 11:135, 1981)

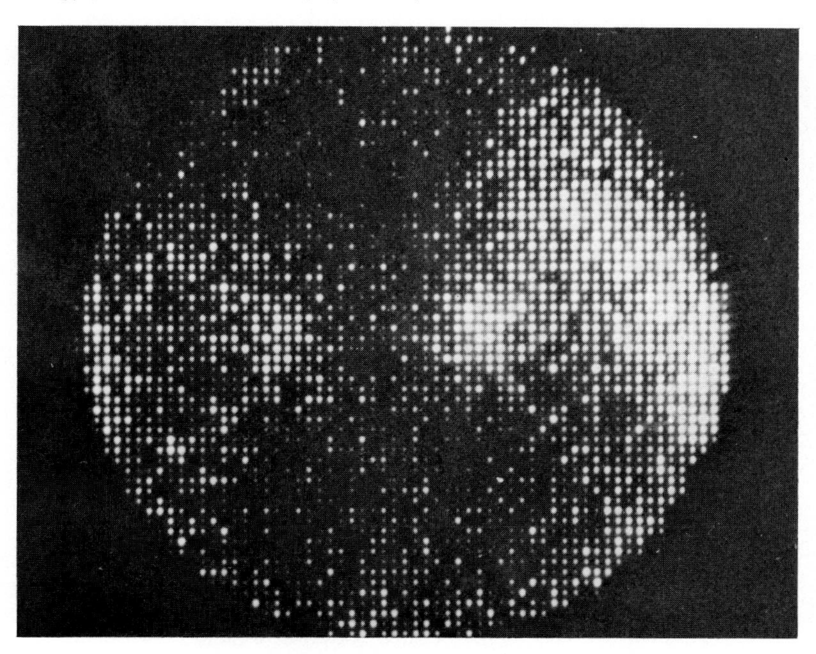

Figure 3-16 *Pitfalls—Effect of Hypercholesterolemia*

A depressed level of NP-59 adrenal uptake is seen in a patient with ACTH-dependent Cushing's syndrome scanned while hypercholesterolemic (390 mg/dl, 0.21% administered dose). (Gross MD, et al: The role of pharmacologic manipulation in adrenal cortical scintigraphy. Sem Nucl Med 11:133, 1981)

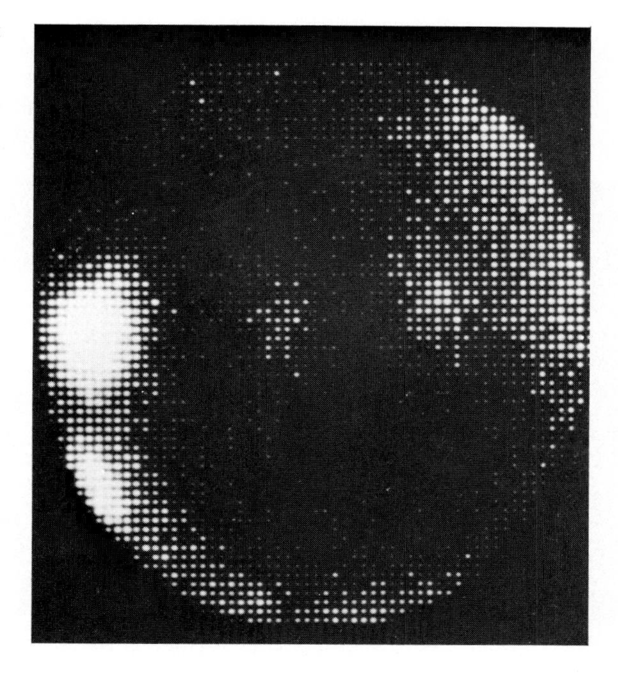

Figure 3-17 ***Pitfalls—Effect of Diuretics***

Bilateral, early (Day 3 with dexamethasone suppression) adrenal gland NP-59 accumulation *(arrow)* in a patient with a right adrenal aldosterone-secreting adenoma. Diuretics were not discontinued prior to the imaging sequence and account for the faint, but discernible left adrenal activity. (Gross MD, et al: The role of pharmacologic manipulation in adrenal cortical scintigraphy. Sem Nucl Med 11:142, 1981)

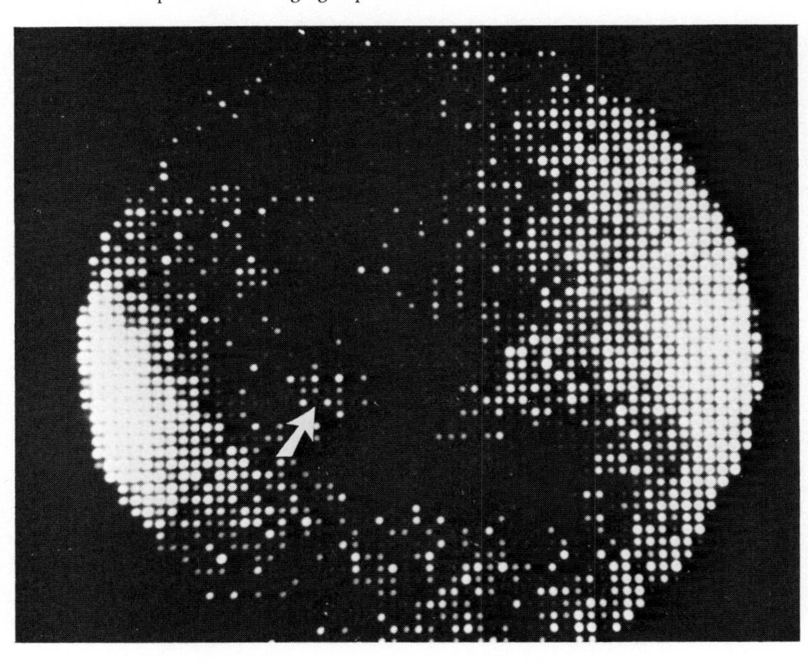

Figure 3-18 ***Pitfalls—Gallbladder Imaging***

Posterior *(A)* and right lateral *(B)* adrenal scintiscans on Day 3 on dexamethasone suppression in a woman with a suspected adrenal androgen-secreting tumor. The bilateral NP-59 uptake seen in the posterior view includes accumulation of NP-59 in the gallbladder as confirmed on the lateral scan. The small arrows identify a Ba-133 marker placed along the lumbar spine. The right adrenal was normal and is shown on day 5 (Fig. 3-10) after the gallbladder had emptied. *GB*, gallbladder. (Juni J et al: Bilateral visualization on adrenal cortical scintigraphy. Sem Nucl Med 13:168 - 170, 1983)

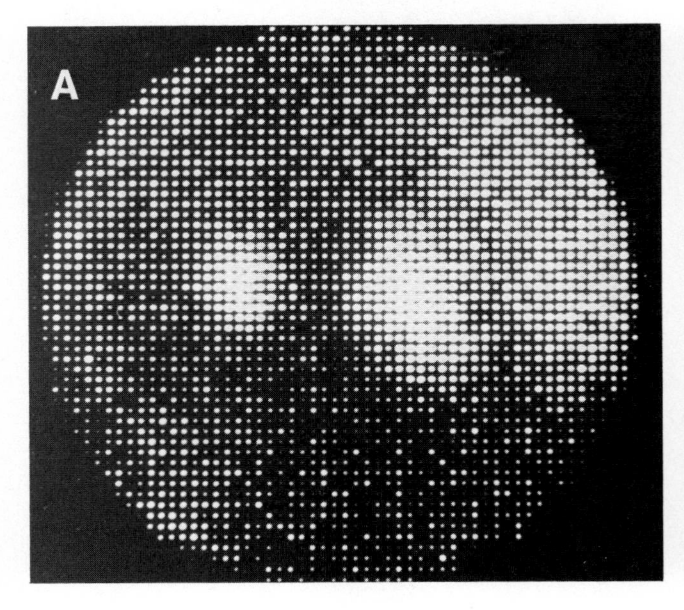

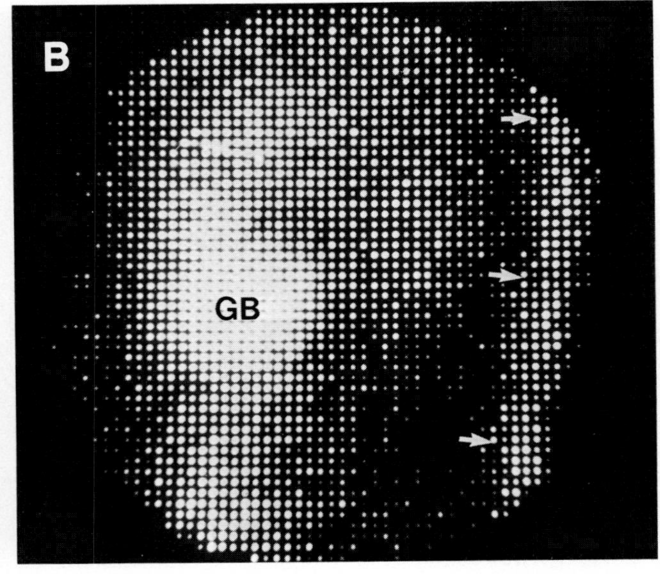

Figure 3-19 ***Pitfalls — Background Radioactivity***

Posterior adrenal dexamethasone suppression scintiscan in a patient with aldosteronism. Intense bowel and liver activity are observed. *(A)* An area of increased uptake in the left abdomen is seen but obscured by bowel *(arrow)*. *(B)* The left posterior oblique image defines the area as outside the bowel and corresponding to that of the left adrenal adenoma *(arrow)*. (Gross MD, et al: The role of pharmacologic manipulation is adrenal cortical scintigraphy. Sem Nucl Med 11:141, 1981)

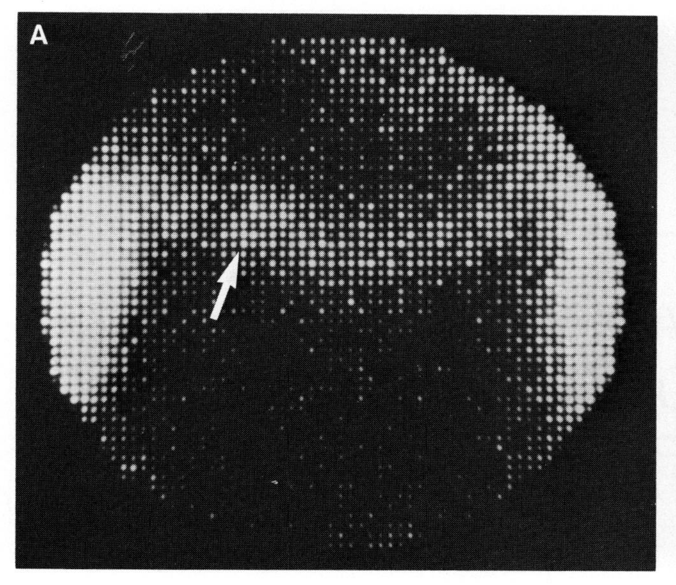

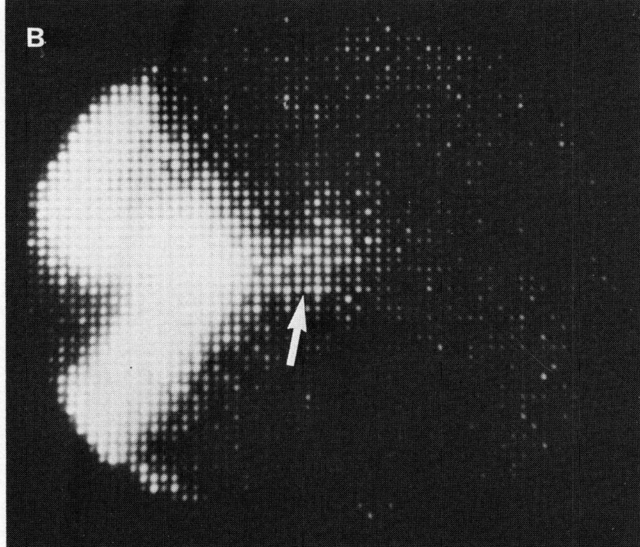

Brahm Shapiro
James C. Sisson

Chapter 4 SYMPATHOADRENAL IMAGING WITH RADIOIODINATED META-IODOBENZYLGUANIDINE

Since its introduction in 1980, iodine-131 meta-iodobenzylguanidine (MIBG) and its analog I-123 MIBG have gained an accepted role in the location of pheochromocytomas and related sympathoadrenal lesions such as neuroblastomas and paragangliomas.

Pheochromocytomas

Pheochromocytomas are somewhat uncommon, but by no means rare tumors. The majority are benign and intra-adrenal. However, multiple primary lesions, usually bilateral intra-adrenal, may occur and are often associated with familial syndromes predisposing to this tumor.[1–4] About 10% arise in extra-adrenal sites. In addition, 10% of pheochromocytomas are malignant tumors[3,4], and in the hypertensive population, the incidence may be as high as 0.1%.[4] With no obvious predominance in either sex, the lesion most frequently is encountered in the fourth and fifth decades, although up to 10% occur during childhood.[5]

Embryologically pheochromocytomas are derived from chromaffin cells of the sympathoadrenal portion of the autonomic nervous system, which extends from the base of the skull to the pelvic floor. The adrenal medullary component is the largest in the system; however, the para-aortic organ of Zuckerkandl below the origin of the inferior mesenteric artery contains the largest extra-adrenal component. The extra-adrenal portion of the system is abundant in infancy, but regresses considerably after age three.[1] There are two nomenclatures used for these lesions. The adrenal medullary tumors are termed *pheochromocytomas,* but the extra-adrenal lesions may either be termed *extra-adrenal functioning paragangliomas* (if they secrete catecholamines) or *extra-adrenal pheochromocytomas.*[1–4] We have chosen to use the latter terminology.

The clinical manifestations of pheochromocytoma are protean and may vary from trivial to disabling. Asymptomatic lesions are not uncommon. The clinical features result primarily from the excessive secretion of catecholamines by the tumors and commonly include hypertension. The hypertension is often labile, severe, and resistant to conventional therapy. Headache, palpitations, and sweating are the most common symptoms; they are characteristically paroxysmal, occurring spontaneously or in response to exercise, bending, eating, or micturition.[1–4] The range of symptoms observed is presented in Table 4-1.

The diagnosis of pheochromocytoma depends on demonstration of excessive catecholamine hypersecretion.[1–4,6] This is now commonly accomplished by an assay of plasma catecholamines using the radioenzymatic or high-pressure liquid chromatography (HPLC) technique. Plasma samples should be obtained under carefully controlled circumstances (with

Table 4-1
Symptomatology in Pheochromocytoma

Symptoms Presumably Caused by Excess Catecholamines or Hypertension	Approximate Percent Paroxysmal (37 patients)	Approximate Percent Persistent (39 patients)
Headaches (severe)	92	72
Excess sweating (generalized)	65	69
Palpitations ± tachycardia	73	51
Anxiety or nervousness (± fear of impending death; panic)	60	28
Tremulousness	51	26
Pain in chest, abdomen (usually epigastric), lumbar regions, lower abdomen, or groin	48	28
Nausea ± vomiting	43	26
Weakness, fatigue, and prostration	38	15
Weight loss (severe)	14	15
Dyspnea	11	18
Warmth ± heat intolerance	13	15
Visual disturbances	3	21
"Dizziness" or faintness	11	3
Constipation	0	13
Paresthesia or pain in arms	11	0
Bradycardia (noted by patient)	8	3
Grand mal seizures	5	3
Miscellaneous		

(Manger WM, Gifford RW: Pheochromocytoma, p. 89. New York, Springer-Verlag, 1972)

the patient fasting and supine for at least 30 minutes with an indwelling needle *in situ* to negate the stress of venipuncture). An alternative approach is to measure catecholamine or catecholamine metabolite excretion rates in urine.[6] In this instance, care must be taken that dietary or drug components do not interfere with the assay of these substances.

Equivocal studies may be further investigated using pharmacological interventions. Clonidine, a centrally acting alpha adrenergic agonist, reduces central stimulation to the sympathoadrenal system and thus suppresses catecholamine levels in normal subjects and in essential hypertension, but not in patients with pheochromocytomas.[7]

Once the diagnosis of pheochromocytoma is strongly suspected on the basis of clinical features and biochemical findings, the site(s) of the tumor must be found. The only curative therapy for pheochromocytoma is complete surgical extirpation of all tumor foci, and therefore preoperative location of all tumor deposits is essential if surgery is to be successful.[2,3,8] To date, conventional radiographic techniques have not been consistently effective in the location of pheochromocytomas.

Large intra-adrenal and pararenal tumors may be depicted by intravenous urography and nephrotomography, but smaller lesions elude detection.[4] Flush aortography has been replaced by selective angiography as a means of depicting the highly vascular lesions.[9] Venography with venous sampling for catecholamine concentrations may point to regions of hypersecretion of catecholamines.[10,11] Ultrasound is helpful in certain instances,[12] particularly for tumors in the pelvis. Computed tomography (CT) was the noninvasive radiological technique with the greatest accuracy until the introduction of I-131 MIBG.[13–15] MIBG scintigraphy has been found to be especially suited to screening the entire body for known or suspected occult primary or secondary pheochromocytoma deposits. Anatomical radiographic techniques are best suited to establishing the anatomical relationships of known lesions.[16] The role of nuclear magnetic resonance imaging is being investigated at present, but high T_2 weighted signal strength seems to be characteristic and may help to define the nature of the tumor.[17,18]

Although surgical resection is the only curative therapy, it requires careful pharmacological preparation with alpha adrenergic blockade to control hypertension and other effects of hypercatecholaminemia.[1–4,8] Once alpha adrenergic blockade has been achieved, beta adrenergic blockade may be introduced if palpitations or tachycardia persist. In patients with inoperable disease, prolonged adrenergic blockade is required. Phenoxybenzamine is the alpha blocker of choice, but the combined alpha and beta blocker, prazosin, is also highly effective. In emergency situations, the short-acting alpha blocker, Regitine, or alternatively, nitroprusside, are the best drugs. Another approach for chronic long-term management is the administration of alpha-methyl-paratyrosine, which acts as a competitive inhibitor of tyrosine hydroxylase, the rate-limiting enzyme in the biosynthesis of catecholamine. Hypersecretion is often reduced to 50% or less by alpha-methyl-paratyrosine therapy, and this effect may remove the clinical manifestations or at least permit the reduction of the doses of other drugs.[19]

Metastatic and unresectable pheochromocytomas are resistant to external beam radiotherapy.[20] The exception to this is that some palliation of painful skeletal metastases is possible. Responses to a wide variety of chemotherapeutic regimens have been inconsistent.[21] Recently some encouraging results have been observed with a combination of vincristine, cyclophosphamide, and DTIC, but this regimen remains experimental.[21] The uptake of I-131 MIBG by many metastatic or irresectable pheochromocytoma deposits raises the possibility of delivering therapeutic doses of radiation through this radiopharmaceutical.[22,23] To be effective, the radiation absorbed doses would have to exceed the maximum doses of external beam radiation. (In selecting patients for such therapy, a tumor's absorbed dose estimate of 2000 rads/100 mCi would be the minimum from which any response might be anticipated.)

Neuroblastomas

These tumors are highly malignant tumors of sympathoadrenal origin, usually occurring in childhood, which are responsible for more cancer deaths than any other solid extra-adrenal tumors of childhood.[1,24,25] They are presumed to be derived from primitive sympathoblasts. The tumor has a great propensity to metastasize to bone marrow and bone, with more than half of cases having such metastases even at the time of presentation. Although responses to chemotherapy and radiotherapy are commonly achieved at the outset, the long-term outlook remains poor for children with metastatic disease. Occasionally these highly malignant lesions will undergo complete or partial maturation toward the more benign ganglioneuroblastoma or ganglioneuroma. The choice of therapy is mandated by the extent and degree of spread, and a wide variety of procedures are used in combination to define this. Ultrasound, CT, technetium-99m MDP bone scan, Tc-99m colloid bone marrow scans, skeletal radiography, and bone marrow biopsy (often multiple) are considered standard but may fail to accurately define the extent of the lesion, especially in the face of residual or recurrent disease.[24,25]

Although the primitive neuroblastoma cell does not efficiently synthesize the catecholamines norepinephrine and epinephrine, many do synthesize the precursor, dopamine, and also excrete the dopamine metabolite, homovanillic acid, in the urine. Despite this fact, the majority of neuroblastomas retain the catecholamine uptake mechanism and, as a consequence, concentrate MIBG, a feature which enables scintigraphic imaging.[24,26]

This same amine uptake pathway is present in a wide range of neuroendocrine systems. Since many of these tumors and tissues share the common property of amine-precursor uptake and decarboxylation (APUD), a number of them have been depicted by MIBG scintigraphy.[27]

TECHNIQUES

Imaging Procedures

I-131 MIBG Scintigraphy

The standard imaging protocol employed at the University of Michigan involves the slow intravenous injection (over 20–30 seconds) of 0.5 mCi I-131 MIBG/1.7m^2 body surface area to a maximum dose of 0.5 mCi. More recently, doses of 1.0 mCi/1.7m^2 have been administered in cases of known or strongly suspected malignant disease, obesity, or unsatisfactory result with 0.5 mCi. If the larger dose is used, more counts than usual (*i.e.,* at least double) should be collected to increase the information content of the images.[28–30]

Thyroidal uptake of free radioiodine present in the preparation and, more importantly, that released by *in vivo* de-iodination is blocked by the administration of stable iodide. We prefer the administration of saturated solution of potassium iodide, one drop three times a day (120 mg of stable iodide/day) begun 24 hours before tracer administration and continued for 7 days thereafter. Alternative regimens include Lugol's solution or potassium perchlorate and thyroxine.[28–31]

Imaging is performed using a large-field-of-view gamma camera equipped with a high-energy collimator and interfaced to a dedicated nuclear medicine computer. Imaging is usually performed 1, 2, and 3 days after radiopharmaceutical administration. A series of overlapping images is obtained to include the entire region from the base of the skull to the pelvic floor, because primary pheochromocytomas may occur anywhere in this region. At a minimum, this requires three views: (1) anterior head, neck, and upper chest; (2) posterior chest and upper abdomen; and (3) anterior abdomen and pelvis. In the past, the first view was posterior rather than anterior. Additional views, including lateral and obliques, may be obtained to better define abnormal or suspicious foci of tracer uptake. When there is a possibility of metastatic disease, it is important to include the whole head and the proximal femurs because metastatic pheochromocytomas have a predilection for these sites. The regions of the thigh and knee (femur and proximal tibia) are especially important in suspected metastatic neuroblastomas.[28-31]

The normal I-131 MIBG scan has little anatomical information, and thus liberal use should be made of surface radioactive markers on anatomical landmarks such as the acromial processes, costal margins, pelvic brim, and greater trochanters.[29]

Abnormal foci of I-131 MIBG uptake may be better localized by the simultaneous imaging of various organs with other radiopharmaceuticals: Tc-99m DTPA for the kidneys and bladder; Tc-99m DMSA for the kidneys; Tc-99m MDP for the skeleton; Tc-99m sulfur colloid for the liver and spleen; Tc-99m–labeled red blood cells for the cardiac and great vessel blood pool; and thallium-201 for the left ventricular myocardial mass. The studies may be stored digitally and thus superimposed to better locate abnormal foci of I-131 MIBG uptake.[28,31,32]

In the usual case using 0.5 mCi I-131 MIBG, images are obtained for at least 100,000 counts/image or 20 minutes imaging time/image. If larger doses than 0.5 mCi are used, images should still be obtained for 20 minutes/image to yield a higher number of counts/image.

The images may be displayed as analog scintigrams on Polaroid or clear nuclear medicine film, or digitized images may be displayed (see Computer Acquisition and Analysis, below).

I-123 MIBG Scintigraphy

I-123 has a number of potential advantages over I-131 as a radiolabel for MIBG: it has a low photon energy that is better suited to modern gamma cameras and results in a far higher detection efficiency; it is more readily collimated; and its decay is by electron capture and the particulate emissions are minimal and thus the radiation dosimetry is far more favorable.[33] Thus 10 mCi of I-123 MIBG delivers approximately the same radiation absorbed dose as 0.5 mCi of I-131 MIBG.[34] The shorter half-life of I-123, while contributing to the low radiation absorbed dose, means that the advantage of high useful photon flux diminishes with the passage of time.

The majority of studies we have performed with I-123 MIBG have utilized 3 to 10 mCi (usually 10 mCi) and the same camera, collimator, and computer system as for I-131 MIBG.[35,36] The use of the high-energy collimator was primarily because we were comparing the effectiveness of I-123 MIBG and I-131 MIBG simultaneously. I-123 does have a number of low-abundance, high-energy photons that degrade the performance of low-energy collimators. We have, however, achieved highly satisfactory images using low-energy, general-purpose collimators.

A further advantage of I-123 MIBG's low photon energy and high photon flux is the ability to perform single photon emission computed tomography (SPECT) using rotating gamma camera systems. This technique permits the distribution of I-123 MIBG to be examined in axial, parasagittal, and coronal sections, which is helpful in studying the uptake in the heart, normal adrenal glands, and abnormal foci.[36]

Because of the shorter half-life, thyroid blockade is continued for 5 rather than 7 days.

Computer Acquisition and Analysis

Our approach has been to acquire all images as analog images on film and simultaneously in digital form as a 64 × 64 word matrix. Unprocessed digitized images with optimal intensity are then presented on film.[28-31] Some use has been made of background subtraction (usually up to 30%) images viewed on video display or on film.

When dual tracer studies are performed, both are acquired in analog and digital form as described above and the I-131 MIBG image may then be added to the second tracer image (Fig. 4-9). Alternatively the second tracer image can be subtracted from the I-131 MIBG image. These images tend to be rather difficult to interpret and we have found it better to place a region of interest around the abnormal focus of I-131 MIBG uptake and transfer this outline to the second radiopharmaceutical image (Figs. 4-10 and 4-16).[32,37]

MIBG studies are particularly well suited to screening the entire body for suspected occult lesions. When such a lesion is disclosed, MIBG-directed studies such as CT will give useful anatomical orientation, especially of primary extra-adrenal and metastatic pheochromocytomas (Figs. 4-12, 4-16, and 4-17).[16,31,37]

The *in vivo* quantification of I-131 MIBG uptake by pheochromocytoma deposits becomes important in the case of metastatic or unresectable lesions in which treatment with therapeutic doses of I-131 MIBG is contemplated. Serial measurements of tumor uptake permits the calculation of the radiation absorbed dose.[22,23,38]

The approach we have taken to this difficult dosimetric problem has been as follows:

1. Volume of the lesion is determined by CT or ultrasound.
2. Tumor uptake is determined daily for 5 to 7 days using the conjugate view–geometrical mean technique. (Recently we have refined the technique to include standard I-131 sources taped to the surface of the patient in proximity to the underlying tumor.)

3. Images are then obtained with and without the standards in place. This technique has been useful in correcting for attenuation, variable counting efficiency, and septal penetration in the collimator.
4. Tumor volume and serial uptake measurements are used with standard medical internal radiation dose (MIRD) formalism to calculate the tumor radiation absorbed dose.[39]

The whole body radiation absorbed dose was calculated from the cumulative urinary excretion determined from serial urine collections (the renal excretion of MIBG being the major route) or whole-body retention measurements using various external counting techniques. The blood radiation absorbed dose was calculated from serial measurements of radioactivity in blood.[22,23,39]

We have not used computerized techniques to make quantitative estimates of adrenal medullary MIBG uptake, but have used a semi-quantitative visual grading scale instead (Fig. 4-3).[29] Quantification of adrenal tracer uptake is a potentially useful approach that has been used with I-131–labeled iodocholesterol derivatives to separate normal from abnormal adrenocortical tracer uptake. Recently I-123 MIBG adrenomedullary uptake has been quantified by the group from St. Bartholomew's Hospital, London, and this may prove useful in distinguishing normal from hyperplastic medullae and in revealing small adrenal pheochromocytomas.[40]

PHYSIOLOGICAL MECHANISMS OF THE RADIOPHARMACEUTICAL

Radioiodo–MIBG structurally resembles the endogenous neurotransmitter hormone, norepinephrine, and the ganglion blocking drug, guanethidine (Fig. 4-1).[41,42] The exact mechanisms for the tissue localization of the agent have not been elucidated in detail but the following data are available:

Tissue Distribution Studies. MIBG uptake occurs in many tissues but the highest uptake and most prolonged retention occur in the adrenal medulla. Organs with extensive sympathetic innervation, such as the heart, spleen, and salivary glands, also show readily discernible uptake. The agent is not taken up by the neurons of the central nervous system because it does not cross the blood–brain barrier.[41,42]

Within the adrenal medulla, MIBG uptake is concentrated in the intracellular storage granules as demonstrated by subcellular tissue fractionation.[41–44]

HPLC of extracts of human pheochromocytomas labeled preoperatively with I-131 MIBG shows that the radiopharmaceutical is taken up and stored as unaltered I-131 MIBG.[43]

Pharmacological Intervention Studies. These studies in animals have shown that reserpine alkaloids, tricyclic antidepressants, cocaine, insulin-induced hypoglycemia, and 6-hydroxydopamine all strikingly reduce adrenomedullary and cardiac MIBG uptake.[41–45]

In humans, tricyclic antidepressants, phenylpropanolamine, and labetalol have all been shown to reduce or abolish MIBG uptake by the adrenal medulla, heart, and salivary glands. These drugs and any others which act by inhibiting tissue catecholamine uptake should be avoided in any patients undergoing scintigraphy.[41,42,44,45]

Experiments of Nature in Humans. Cervical sympathectomy has been observed in experiments of nature in humans to substantially reduce ipsilateral salivary gland MIBG uptake. Also, generalized autonomic neuropathy has shown to be associated with loss of cardiac and salivary gland uptake of MIBG.[46,47]

In vitro *Incubation Studies.* Using bovine adrenal medullary cells or human pheochromocytoma cells in culture, *in vitro* incubation studies have shown that I-125 MIBG and H-3 norepinephrine share similar specific, active, energy, and sodium-dependent uptake mechanisms, and that MIBG and norepinephrine compete with each other for this pathway, although not in a simple competitive fashion.[48,49]

These data have been interpreted as showing that MIBG uptake by the adrenal medulla and sympathetic autonomic neurons occurs by two processes: a specific active process known as type I uptake and a nonspecific, concentration-dependent, diffusional uptake known as type II uptake. Probably there is some nonspecific uptake into nonneuronal tissues.[44,48–50]

Once taken up, the MIBG is stored in the intracellular hormone/neurotransmitter storage granules of the sympathomedullary tissues. Thus the scintigraphic distribution of MIBG reflects the specific and nonspecific uptake and storage capacity.

Metabolism and Excretion. The majority of the MIBG administered intravenously is excreted unchanged by the kidneys, 40% to 60% within 4 days. Excretion is reduced in patients with impaired renal function and this reduction is inversely proportional to the creatinine clearance.[43] Minor degrees of metabolism and conversion to I-131 meta-iodohippuric acid, I-131 meta-iodobenzoic acid, and I-131 4-hydroxy-3-MIBG with urinary excretion have been observed.[43]

Fecal excretion is minor and represents no more than a few percent of the total excretion. Minute quantities of activity have been detected in the saliva, sweat, and exhaled breath of patients receiving therapeutic doses of I-131 MIBG.

Table 4-2
Estimated Radiation Absorbed Doses from I-131 and I-123 MIBG

Organ	Maximal Uptake (% kg dose/g)	Time of Maximal Uptake (hours)	Estimated Human Dosimetry (rads/mCi)		
			I-131 MIBG	"Pure" I-123 MIBG	Contaminated I-123 MIBG
Thyroid	3.40	24.0	35	2.20	2.55
Adrenal medulla	13.60	48.0	100	0.80	2.76
Heart wall	0.50	0.5	0.7	0.03	0.04
Liver	0.36	0.5	0.4	0.05	0.05
Spleen	0.30	0.5	1.6	0.14	0.15
Ovaries	0.14	2.0	1.0	0.06	0.07
Total body			0.1	0.02	0.02

(Swanson DP et al: Human absorbed dose calculations for iodine 131 and iodine 123 labelled MIBG: A potential myocardial and adrenal medulla imaging agent. Health and Human Services publication FDA 81-8166, pp 213–224. Rockville, Maryland, 1981)

ESTIMATED RADIATION ABSORBED DOSES

The estimated radiation absorbed doses for I-131 and I-123 MIBG have been derived from tissue distribution data in rats and dogs and whole-body retention data in man. Standard MIRD formalism and calculations were used (Table 4-2).[34] Subsequent dosimetric data derived from limited tissue sampling at the time of surgical exploration and from measurements related to the therapeutic administration of I-131 MIBG for malignant pheochromocytomas tend to confirm these radiation absorbed doses.[22–24,34,38,39]

The administration of iodides or other thyroid blocking regimens can reduce the thyroidal radiation absorbed dose by a factor of approximately 100.[34,38]

Adequate hydration and frequent bladder voiding will reduce the bladder, gonadal, and whole-body radiation absorbed doses. The whole-body radiation absorbed dose will be increased in cases of renal insufficiency, since the major route of excretion of radioiodinated MIBG is through the kidney.[43,51]

The ability of reserpine alkaloids and tricyclic antidepressants to discharge MIBG from those tissues in which it has been taken up by the specific type I mechanism might provide a means to discharge MIBG after imaging has been completed, but this approach has not been attempted to date.[44]

VISUAL DESCRIPTION AND VISUAL INTERPRETATION

Normal Scintigraphic Appearance of I-131 MIBG Distribution

The normal distribution of I-131 MIBG as depicted by scintigraphy has been described in detail by Nakajo and co-workers (Fig. 4-2).[29]

The normal adrenal medulla is not usually visualized by I-131 MIBG by standard diagnostic doses (0.5 mCi/1.7m^2). Faint uptake may be observed 48 to 72 hours postinjection in up to 16% of cases. Visible uptake at 24 hours is less frequent and probably does not occur in more than 10% of cases. The intensity of adrenal medullary uptake can be graded on a semi-quantitative five-level scale (Fig. 4-3), in which an uptake of grades 3 or 4 is definitely abnormal.[28,29] The administration of larger diagnostic doses of I-131 MIBG (*e.g.*, 2 mCi administered by Brown and colleagues[52]) or therapeutic doses of 100 to 200 mCi regularly and clearly depict the adrenal medulla.[22,23]

The parotid and submandibular salivary glands are normally visualized. This is believed to be largely due to the rich sympathetic innervation of these organs. The visualization of the nasopharynx is believed to result from a similar mechanism. The radioactivity observed is not that of free I-131 and may be much reduced by surgical sympathectomy (Horner's syndrome) or by drugs that block catecholamine uptake.[46]

The extensive sympathetic innervation of the heart results in the visualization of this organ. The uptake is variable and is inversely related to the plasma catecholamine levels. Clear visualization of the heart at 24 hours post–I-131-MIBG injection makes the diagnosis of a secreting pheochromocytoma unlikely.[53]

The single organ with the greatest uptake is the liver. This is maximal at 24 hours and declines rapidly, more so than most pheochromocytomas. Adrenal medulla–liver ratio of as high as 680 : 1 have been demonstrated in dogs.[29,31]

Splenic uptake is seen in all individuals and tends to increase between 24 and 48 hours, which is the reverse of that observed in the liver. The uptake is again believed to be by the sympathetic nerves.[29]

The excretion of MIBG is through the urinary tract with about 60% of the administered dose being excreted in the first 24 hours. Occasionally the kidneys are visualized at 24 hours, but the presence of I-131 MIBG in the urine results in the regular visualization of the urinary bladder. The intensity of bladder visualization declines with time. This activity may obscure pelvic pheochromocytomas, and thus the anterior pelvic view should be acquired immediately following voiding; it may be necessary to empty the bladder more completely by catheterization or even by saline lavage. Alternatively, subtraction of the urinary tract by means of simultaneously acquired Tc-99m DTPA images may be helpful in this setting.[29]

Radioactivity may be seen in the large bowel in 15% to 20% of cases and occasionally this may be sufficient to be confused with or to obscure tumor uptake. The mechanism of I-131 MIBG concentration in the bowel is uncertain and it is not known if the uptake is in the bowel wall (possibly in autonomic neuronal elements) or is caused by secretion into the lumen. In general, bowel uptake is least obtrusive on the 24-hour images.[29]

The lung shows high uptake immediately after and for about 4 hours following injection. This concentration of I-131 MIBG is believed to be nonspecific diffusional uptake by endothelial cells, as a result of the high initial concentration of I-131 MIBG presented to the lung. The rapid fall in lung activity means it is unlikely to interfere with interpretation. The activity is usually most pronounced in the lung bases, probably because of the greater perfusion of these regions.[29]

Faint thyroidal uptake of free I-131, primarily derived from *in vivo* de-iodination, may be observed despite various "blocking" regimens. Generally, the thyroidal uptake is far less than that observed in the salivary glands.[29,43,46]

Normal Scintigraphic Appearance of I-123 MIBG Distribution

The scintigraphic depiction of the distribution of I-123 MIBG is similar to that of I-131 MIBG, but it reveals greater resolution and detail due to the higher useful photon flux. The important points of difference are described below (Fig. 4-5).[35,36,54]

The adrenal medulla is regularly depicted when doses of 3 to 10 mCi are used. Recent work has shown that the uptake may be quantified *in vivo* and represents 0.01% to 0.3% of the administered dose at 22 hours.[40] This technique may prove useful in distinguishing normal from hyperplastic glands and may be superior to visual impression alone.

The salivary glands are particularly well delineated (Fig. 4-6), and in many cases there is faint but definite uptake in the lacrimal glands, which is not seen in I-131 MIBG studies.[35,36]

I-123 MIBG is especially efficient in depicting the sympathetic innervation of the heart and may provide an *in vivo* probe for imaging this system. The percent uptake is not dissimilar to that of Tl-201.[55] This uptake is reduced by high catecholamines, tricyclic antidepressants, and phenylpropanolamine, all of which strongly support a sympathetic neuronal uptake mechanism.[44] Cardiac uptake is also reduced in autonomic neuropathy and, acutely, following myocardial infarction.[47]

The steady increase in splenic uptake is particularly well demonstrated, going from none at 4 hours and rising from 24 to 48 hours using I-123 MIBG.

Early images (less than 18 hours postinjection) show loops of small bowel in many cases, but this pattern tends to diminish by 24 hours and later, to be replaced by large bowel uptake. The latter organ is also sometimes seen on the earliest images.

The ability to perform SPECT with I-123 MIBG provides for the tomographic depiction of the heart in which I-123 MIBG uptake can be presented in a fashion analogous to that for Tl-201 tomography, with short axis and vertical and horizontal long axis sections. The normal adrenal medulla is also well demonstrated on axial and coronal tomographic sections of the abdomen.[35,36,47,55]

Abnormal Distributions of Radioiodinated MIBG

Tumorous Uptake of MIBG

Any focus of activity not conforming to the normal scintigraphic distribution of I-131 or I-123 MIBG uptake (*e.g.,* salivary glands, heart, liver, spleen, bladder, and bowel) must be considered abnormal and due to uptake by pheochromocytoma or another neuroendocrine tumor.[28–31,56] The clinical setting and biochemical findings are essential in determining the type of lesion.

Most pheochromocytomas (more than 80%) are clearly demonstrated on the 24-hour images. Their depiction becomes clearer with the passage of time in most cases because uptake declines more slowly in tumors than in normal organs. Thus some lesions only become evident at 48 or 72 hours after the injection of the radiopharmaceutical.[28–31,56] Indeed there is a report of visualization occurring only as late as 7 days postinjection.[57] A relatively long half-life radionuclide such as I-131 is essential to produce these late images. However, the far higher useful photon flux that can be achieved with I-123 MIBG may permit early imaging of lesions that may only be depicted late, if at all, by I-131 MIBG (Fig. 4-14).[35,36,47] The optimal choice of label remains controversial,[58,59] but a potentially useful algorithm is put forth in Figure 4-34.[59]

In the case of known or suspected metastatic pheochromocytoma or neuroblastoma, both of which have a predilection to spread to bone and bone marrow, careful attention must be paid to the axial skeleton. In the case of neuroblastoma, attention should be directed to the regions of the thigh, knee, calf, and ankle. Normally there is only a diffuse faint tracer uptake in the muscle masses of the limbs; any focal uptake or uptake confined to the bony portion of the limb (especially the knee or ankle) is indicative of tumor.[24,26]

Intra-adrenal pheochromocytomas are usually obvious from intense tracer uptake (Fig. 4-7) and a typical location, which can be confirmed by producing concurrent kidney images with Tc-99m DTPA or Tc-99m DMSA.[28–31,56] It may be difficult to distinguish a faintly visualized adrenal pheochromocytoma from prominent uptake in a normal adrenal medulla. Furthermore, it may be difficult to separate normal adrenal medullary uptake from the increased uptake observed in the adrenal medullary of the MEN-2 syndromes.[29] Because of an evolution towards pheochromocytomas, adrenal medullae of patients with MEN-2 syndromes are visualized more frequently than in normal persons, but in individual cases differentiation of normal from abnormal uptake may not be possible (Figs. 4-3 and 4-4).[29,60,61] Objective quantification of uptake using I-123 MIBG may be helpful in this regard.

Examples of the scintigraphic appearances of a wide range of intra-adrenal, extra-adrenal, and metastatic pheochromocytomas, neuroblastomas, and other neuroendocrine tumors are depicted in the atlas section of this chapter.

Other Abnormalities of MIBG Uptake

Other abnormalities of MIBG distribution have already been alluded to and include the decreased uptake in sympathetically innervated organs, such as the heart, as a consequence of autonomic neuropathy, and salivary glands after surgical denervation. The administration of certain drugs will interfere with uptake of MIBG in all tissues. These phenomena indicate an intraneuronal residence for MIBG and may make I-123 MIBG a useful *in vivo* physiological probe of the sympathetic nervous system.[44,46,47,50,53,62]

The urinary excretion of MIBG means that artifactual abnormalities of distribution may result from urinary contamination (important in children studied for neuroblastoma). Furthermore, urinary tract abnormalities in which MIBG collects or is trapped may be confused with tumorous foci of MIBG uptake. Chief among these problems would be a hydronephrotic or dilated renal pelvis or bladder diverticulum. If artifact due to urinary tract abnormality is suspected, the performance of a simultaneous Tc-99m DTPA study may be helpful. MIBG directed intravenous urography or ultrasonography would be an alternative approach.[28–31]

DISCUSSION

Results of MIBG Scintigraphy in Pheochromocytoma

I-131 Imaging

I-131 MIBG scintigraphy has been shown to be effective in locating a majority of pheochromocytomas of all types.[28–32] These include primary intra-adrenal,[28–31] primary extra-adrenal (both abdominal[28–31] and thoracic[32], recurrent,[63,64] residual,[63,64] and metastatic tumors.[37] In addition, I-131 MIBG scintigraphy has been effective in locating tumors in patients with syndromes predisposing to pheochromocytoma, including the multiple endocrine neoplasia type 2 syndromes (MEN-2a and MEN-2b),[60,61,65] von Hippel–Lindau syndrome, neurofibromatosis,[65,66] Carney's triad (gastric leiomyosarcoma, pulmonary hamartoma, and multifocal extra-adrenal pheochromocytoma), and single familial pheochromocytoma.[67] The overall University of Michigan experience is presented in Table 4-3.

The results from the University of Michigan resemble those from other institutions. A lower sensitivity of 67% has been reported in one series.[68] Table 4-4 presents a summary of the experience from multiple series drawn from the literature.

The reasons for false-negative studies remain obscure; there is no obvious relationship to the site of the lesion, the age or sex of the patient, the type and severity of the biochemical abnormalities, or the histological features of the lesion. Low, rather than completely absent, tracer uptake may be remedied by the administration of greater radioactivity. The constraints of radiation dosimetry limit doing this with I-131 MIBG, but with I-123 MIBG it remains a practical option.[35,36,54]

I-123 Imaging

Results with I-123 MIBG have been most encouraging.[35,36,54] All 21 cases of known or suspected pheochromocytoma studied revealed one or more tumor deposits. In two instances, primary extra-adrenal lesions (one pararenal and one paracardiac) were located, which were not revealed by CT or I-131 MIBG scintigraphy.[35,36,69]

In ten cases of metastatic malignant pheochromocytoma, I-123 MIBG scintigraphy revealed additional metastatic deposits not seen using I-131 MIBG.

The use of SPECT and the dynamic rotating display was found to provide additional information as to the location and relations of abnormal foci of I-123 MIBG uptake in six cases.[36]

These findings are very much in keeping with the experience of other clinicians using I-123 MIBG.[54] This radiopharmaceutical would be the agent of choice if it were more readily available.[35,36,69]

Table 4-3
I-131 MIBG Scintigraphy for Pheochromocytoma (University of Michigan Experience 1980–1985)

Patient Category	Number of Patients Studied	Results			
		True Positive	True Negative	False Positive	False Negative
Sporadic intra-adrenal pheochromocytoma	34	31	0	0	3
Sporadic abdominal extra-adrenal pheochromocytoma	11	9	0	0	2
Sporadic thoracic extra-adrenal pheochromocytoma	11 (1 malig)	11	0	0	0
Unknown site	5	0	0	0	5
MEN-2a and MEN-2b	37 (3 malig)	22	13	0	2
Neurofibromatosis	15 (2 malig)	5	10	0	0
von Hippel–Lindau disease	3 (1 malig)	3	0	0	0
Simple familial	3 (2 malig)	3	0	0	0
Sporadic malignant	65	57	0	0	8
"False positive"	3	0	0	3*	0
Pheochromocytoma excluded (entirely normal biochemistry)	165	0	165	0	0
Pheochromocytoma probably excluded (nondiagnostic elevations of catecholamines, but radiology negative and follow-up negative)	210	0	210	0	0
TOTALS	562	141	398	3	20

* *Three patients who were evaluated for pheochromocytoma had abnormal scans that were not caused by pheochromacytoma. These scans included one case each of retroperitoneal neuroendocrine tumor (noncalecholamine-secreting atypical schwannoma), metastatic choriocarcinoma, and dilated renal pelvis.*

Table 4-4
Summary of I-131 MIBG Scintigraphy in Pheochromocytoma

Study Site	Number of Patients Studied	Results*				Sens* (%)	Spec* (%)	−PDA* (%)	+PDA* (%)	Prevalence (%)
		TP	FP	TN	FN					
University of Michigan (current experience)	562	141	3	398	20	88	99	98	95	26
University of Michigan (previously published experience)[87]	475	124	3	329	19	87	99	95	97	30
Combined German series[88]	191	56	1	126	8	88	99	94	98	34
Southampton, England[89]	46	21	1	21	3	88	95	88	95	52
Mayo Clinic[90]	42	15	1	22	4	79	96	85	94	45
Combined French series[85] †	99	42	2	51	4	91	96	95	92	46
Tours, France[75]	27	8	1	17	1	89	94	94	89	33

* TP, *true positive;* TN, *true negative;* FP, *false positive;* FN, *false negative;* Sens, *sensitivity;* Spec, *specificity;* +PDA, *positive predictive accuracy;* −PDA, *negative predictive accuracy*

† *Equivocal negative added to true negative, equivocal positive added to true positive*

(Modified from McEwan AJ et al: Radioiodobenzylguanidine for the scintigraphic location and therapy of adrenergic tumors. Sem Nucl Med 15:132–153, 1985)

Table 4-5
Summary of MIBG Scintigraphy in Neuroblastoma

Study Site	Number of Patients Studied	Results*				Sens* (%)	Spec* (%)	−PDA* (%)	+PDA* (%)	Prevalence (%)
		TP	FP	TN	FN					
University of Michigan (current experience)	36	28	0	3	5	85	100	38	100	92
University of Michigan (previously published experience)[26]	10	9	0	0	1	90			100	100
Villejuif, France[91] †	24	14	0	4	6	70	100	40	100	83
Amsterdam, Holland[79]	26	21	0	4	1	95	100	80	100	85
Copenhagen, Denmark[24]	16	12	0	0	4	75			100	100
Philadelphia[92] ‡	19	8	2	3	6§	57	60	60	80	74
Tubingen, West Germany[93]	5	4	0	1	0	100	100	100	100	80

* TP, *true positive;* TN, *true negative;* FP, *false positive;* FN, *false negative;* Sens, *sensitivity;* Spec, *specificity;* +PDA, *positive predictive accuracy;* −PDA, *negative predictive accuracy*

† *Majority of cases studied with I-123 MIBG*

‡ *Listed sites of disease in 13 patients.*

§ *In four cases, tumor was present, but was ganglioneuroma or ganglioneuroblastoma*

Results of MIBG Scintigraphy in Neuroblastoma

I-131 Imaging

I-131 MIBG scintigraphy has been shown to be efficacious in depicting the location of primary and metastatic deposits of neuroblastoma and useful in the diagnosis, staging, and follow-up of patients with this disease.[24,26]

Since the initial reports of successful imaging of neuroblastoma,[70–72] extensive experience has been acquired at the University of Michigan and other institutions,[73,74] a summary of which is presented in Table 4-5.

I-123 Imaging

We have not as yet used I-123 MIBG to study neuroblastoma, but the experience of those who have is that, as is the case with pheochromocytoma, it provides greater clarity of image and higher sensitivity than I-131 in the detection of subtle tumor deposits.[75,76]

Results of MIBG Scintigraphy in Other Neuroendocrine Tumors

To date we have studied 57 patients with known or suspected neuroendocrine tumors other than pheochromocytoma and neuroblastoma. These were all performed with 0.5 mCi I-131 MIBG and imaged 1, 2, and 3 days postinjection (same as for studies in pheochromocytomas).[27,77]

Plasma epinephrine and norepinephrine were measured by radioenzymatic assay and the urinary excretion rates of catecholamines, metanephrines, 3-methoxy-4-hydroxymandelic acid, and 5-hydroxyindoleacetic acid by fluorometric and colorimetric assays. Plasma concentrations of cortisol and relevant peptide hormones (gastrin, calcitonin, insulin, and ACTH) were measured by radioimmunoassay. The measurements made in any individual varied with the symptomatology and clinical setting.[27,77]

All resected tumors and biopsies were examined by conventional hematoxylin and eosin histology and by various combinations of silver and immunohistochemical stains to determine the neuroendocrine nature of the lesion and identify its hormonal products. In selected cases, electron microscopy was used to demonstrate the presence of and categorize the morphology of neurosecretory granules.[27,77]

Our results of I-131 MIBG scintigraphy were correlated with all available data, including other imaging techniques (various combinations of plain radiographs, radionuclide liver scans, ultrasound, CT, and angiography), and with surgical findings[27,77] and are presented in Table 4-6. Our rate of detection of carcinoid tumors is similar to that of Fisher and coworkers,[78] Hoefnagel and colleagues,[79] and Blinder and associates.[80] The latter group used a Tc-99m sulfur colloid liver scan and I-131 MIBG subtraction to visualize the lesions in five of nine cases.

Tumors of extra-adrenal sympathetic tissue which secrete catecholamines are termed extra-adrenal pheochromocytomas or secretory (functioning) paragangliomas, while those without catecholamine secretory capacity are termed nonsecretory or nonfunctioning paragangliomas. We visualized I-131 MIBG concentration in all three such cases studied, which is in keeping with the results of Smit and colleagues.[81] These cases demonstrate that MIBG uptake may be independent of whether or not a lesion actively

Table 4-6
Results of I-131 MIBG Scintigraphy in Neuroendocrine Tumors (APUDomas) Other Than Pheochromocytoma and Neuroblastoma

Tumor Type	*Total Number*	*I-131 MIBG Positive*	*Percent*
Carcinoid tumors	10	4	40
Nonsecretory paragangliomas	3	3	100
Chemodectomas	5	2	40
Sporadic medullary carcinoma of thyroid	5	1	20
Medullary carcinoma of thyroid associated with the MEN-2a and MEN-2b syndromes			
Cases with elevated calcitonin	12	1	8
Cases with normal calcitonin	8	0	0
Cases with unavailable calcitonin	6	0	0
Oat cell carcinoma of lung	4	0	0
Metastatic choriocarcinoma	1	1	100
Atypical schwannoma (with neurosecretory granules)	1	1	100
Merkel cell carcinoma of skin	1	1	100
Islet cell carcinoma of pancreas	1	0	0
Total	57	14	25

secretes catecholamines. Carotid body tumors (chemodectomas) are a specialized form of paraganglioma in which we observed I-131 MIBG uptake in two of five cases studied. Like Hoefnagel and co-workers,[79] we were only able to demonstrate I-131 MIBG uptake in a small minority of medullary carcinomas of the thyroid (both sporadic and associated with the MEN-2a and MEN-2b syndromes). This is at variance with the higher incidence of positive scans for this tumor reported by others.[82,83]

Positive imaging of single cases of atypical schwannoma, Merkel cell carcinoma of skin,[84] and metastatic choriocarcinoma indicate that a wide range of neuroendocrine tumors may take up I-131 MIBG. The overall frequency with which lesions are depicted appears to be lower than is the case for pheochromocytomas and neuroblastomas. It is necessary to study larger numbers and a wider range of neuroendocrine tumors to accurately establish the sensitivity and specificity of MIBG scintigraphy in each tumor type. Such a study would determine the usefulness of MIBG scintigraphy in searching for these tumors. At present it appears that I-131 MIBG scintigraphy may be consistently helpful only in paragangliomas and carcinoids. Additional data available include a report of no uptake in any of five melanomas.[79]

The specificity of I-131 MIBG scintigraphy for pheochromocytoma suggested by earlier studies must be modified in the light of the experience with neuroblastomas and other neuroendocrine tumors. Interpretation of positive scans requires that the clinical and biochemical features of the case be taken into account.

Optimizing Strategies using MIBG Scintigraphy

Suggested Diagnostic Algorithms

Much depends on the availability of radioiodinated MIBG. In the United States this remains an investigational new drug and thus is not generally available. The nuclear pharmacy at the University of Michigan Medical Center does supply I-131 MIBG to qualified investigators. I-131– and I-123–labeled MIBG are widely available commercially throughout western Europe and are produced either commercially or in-house in Canada, Australia, South Africa, and Japan.

In the case of suspected pheochromocytoma, attempts to locate a suspected tumor follow the clinical and biochemical establishment of a diagnosis.[28,31] I-131 MIBG scintigraphy has the advantage of permitting the entire body to be screened in a single procedure. I-123 MIBG may serve the same function, but we have reserved this for cases where I-131 MIBG appears to be falsely negative or provides insufficient information. (In this setting a false-negative I-131 MIBG scan is one in which no abnormal focus of uptake is seen and in which the plasma norepinephrine is greater than 2000 pg/ml or urine catecholamine or catecholamine metabolite excretion rates are greater than 5 standard deviations above the mean.)[35,36] The main reason for this policy is the expense of I-123 and the need to freshly synthesize the radiopharmaceutical (in contrast to the 2-week shelf-life of I-131 MIBG). An argument can be made that over 80% of pheochromocytomas are intra-adrenal and the first locating technique should be CT of the adrenal region, which is a method with very high sensitivity for this region. Indeed CT in certain series may be marginally more sensitive than I-131 MIBG scintigraphy.[85] The major advantage of I-131 MIBG lies in its ability to depict extra-adrenal primary lesions and metastatic deposits. Even if an anatomical locating technique such as CT discloses a pheochromocytoma, a strong case can be made for the performance of MIBG scintigraphy to search for otherwise occult additional primary lesions or metastatic deposits. Figure 4-34 depicts an algorithm incorporating I-131 MIBG and I-123 MIBG based on the approach we have taken at the University of Michigan.

In the case of neuroblastoma, MIBG scintigraphy using either I-131 MIBG or I-123 MIBG is emerging as an important tool in a number of situations,[70–76] including these:

In demonstrating the neuroendocrine origin of a tumor mass not readily accessible to biopsy or in which the histology is that of an equivocal small round cell tumor. In the latter case, scintigraphy may provide the answer before immunohistochemical or electron microscopy results become available.

In staging (and thus planning the therapy of) known neuroblastoma, MIBG scintigraphy may in many instances provide as much or more information than all other imaging techniques combined.

In following the response to therapy of patients with known neuroblastoma. Abnormal foci of tracer uptake in patients apparently in complete remission by all other criteria are unequivocal evidence of tumor. The choice between I-131 or I-123 as a radiolabel for studies in neuroblastoma patients remains controversial, but if available, I-123 MIBG with its high useful photon flux is probably preferred.

MIBG Scintigraphy to Directed Anatomical Imaging Procedures

MIBG scintigraphy is a functional imaging modality suited to screening the entire body for pheochromocytoma deposits, which usually are depicted as high-contrast foci of abnormal tracer uptake. The exact anatomical location of these foci may not be immediately obvious; the uptake of MIBG by normal organs provides only a rough guide; liberal use of surface radioactive markers and the simultaneous scintigraphic visualization of other tissues or organs may be required.[28,31,32]

Curative surgical resection requires the clearest possible preoperative depiction of the anatomical relationships of the tumor to adjacent structures. This is especially true of extra-adrenal lesions. We have found CT studies directed by the preceding MIBG scintigraphy to be especially helpful in this regard. Dynamic CT during the injection of a contrast bolus may disclose lesions not observed during routine CT examinations.[16,32] Depending on the location of the lesion, similarly directed venographic,[67] arteriographic, and magnetic resonance imaging studies have proved helpful in certain cases.

In the case of neuroblastoma, bone or bone marrow biopsy directed toward sites of abnormal tracer uptake may demonstrate the presence of tumor in patients otherwise believed to be in complete remission.

The Potential of I-131 MIBG for Therapy

Rationale

If tracer studies reveal that a primary or metastatic tumor has high MIBG uptake and prolonged retention, the potential exists that large doses of I-131 MIBG may be able to deliver therapeutic doses of radiation to the tumor with acceptable whole-body and other critical organ radiation absorbed doses.[22,23,38,86] The analogy here is to that of the use of I-131 iodide in well-differentiated thyroid cancer. This approach appears to be justified in the case of malignant pheochromocytoma, which tends to be resistant to chemotherapy and external beam radiotherapy (the palliation of painful skeletal metastases with external beams being one exception).

Patients with malignant or irresectable pheochromocytomas have been selected on the following basis:

Tumor deposits are not amenable to treatment by any other means.

The patient is not moribund and might be expected to survive a year even without treatment (allowing time for a response to I-131 MIBG therapy to manifest).

Dosimetric tracer studies on at least one representative tumor deposit yield dosimetric estimates of at least 2000 rads/100mCi of I-131 MIBG with the whole-body radiation absorbed dose less than 150 rads and that to the blood less than 50 rads.

The patient must be willing and able to return for evaluation and retreatment at three-to-six-month intervals for at least 2 years.

In neuroblastoma, for which there are accepted standard radiation and chemotherapeutic therapies, I-131 MIBG therapy has been reserved for patients with Stage IV disease which has relapsed despite conventional chemotherapy and, in some instances, also after bone marrow transplantation and experimental chemotherapy.[24] Many of these patients are desperately ill with rapidly progressive tumors. The dosimetric considerations are as for pheochromocytoma patients, but in the case of patients with diffuse bone marrow tumor infiltration, it may be difficult or impossible to calculate the tumor absorbed dose because the tumor infiltrates and is intimately mixed with normal marrow, making tumor volume indefinable. The presence of tumor in the bone marrow may increase the bone marrow radiation absorbed dose by acting as a local radiation source within the marrow space.

Any neuroendocrine tumor with good tracer uptake and retention of MIBG that is not responsive to conventional therapy might also be treated with large doses of I-131 MIBG. Candidate tumors in which isolated attempts have been made include carcinoids and medullary carcinoma of the thyroid.[79]

Technique

Doses of up to 230 mCi of high specific activity I-131 MIBG (containing not more than 6mg MIBG) are infused over 90 minutes using a shielded infusion pump. Remote monitoring of pulse, blood pressure, and ECG are performed every 5 minutes during the infusion and hourly for the next 24 hours. Thyroid blockade with iodides or perchlorate and triiodothyronine are continued for at least four weeks following therapy.[22,23,87]

Results

No acute hemodynamic or ECG changes have been observed, nor have abnormalities of thyroid function, hepatic enzymes, adrenal glucocorticoid secretion, or autonomic nervous system function been detected. Reductions in total white blood counts to as low as 2000/mm^3 and of platelets to as low as 80,000/mm^3 have been observed in adults. These nadirs are reached between 4 and 6 weeks, and recovery to or close to pretherapy levels have followed. In the case of children, the bone marrow suppression has been more marked, especially the platelets, which in some cases have fallen to as low as 20,000/mm^3 and required platelet transfusion. Furthermore, recovery to the pretherapy levels did not always occur. Factors accounting for this may include previous hematosuppressive chemotherapy, bone marrow transplant, and the presence of neuroblastoma in the marrow which may replace normal hematopoietic elements or act like radiation sources within the marrow.

Partial responses, defined as reduction of tumor volume by more than 50% or reduction of catecholamine secretion by more than 50%, have been achieved in a number of patients with malignant pheochromocytomas.[22,23,87]

Preliminary data would suggest that the tumors which respond are those receiving the greatest radiation absorbed dose (mean dose of responders 17,500 rads vs 7500 rads to nonresponders), although other as yet undefined factors may be involved. A number of patients who demonstrated initial partial responses, some of which lasted several years, have demonstrated relapses; not all of these were responsive to subsequent retreatment with I-131 MIBG. This raised the possibility that such therapy may select out I-131 MIBG–resistant clones of tumors.[22,23,87]

In the case of patients with neuroblastoma, I-131 MIBG therapy has been recorded as resulting in subjective reductions in pain and other tumor symptoms, reductions of tumor fever and narcotic doses, as well as objective reductions in tumor volume and urinary HVA excretion. In most cases these responses have been short-lived with eventual disease progression and death of the patient. At least one case of apparent complete remission has been recorded. Patients with very widespread marrow involvement seem to do worse than those with more focal disease.[73] These very preliminary results suggest that I-131 MIBG therapy may have a role in the management of neuroblastoma, perhaps combined with chemotherapy earlier in the course of the disease. This is the focus of active research at present.

With carcinoids, medullary carcinoma of the thyroid, and other neuroendocrine tumors, too few cases have as yet been treated and the period of follow-up is as yet too short to know the place, if any, of I-131 MIBG therapy in these lesions.[79,86]

The Potential of MIBG as an *In Vivo* Probe of the Sympathetic Autonomic Nervous System

As a major fraction of MIBG uptake into tissues such as the salivary glands, heart, spleen, and adrenal medulla is by the type I uptake mechanism, this tracer has the potential of mapping *in vivo* the distribution of catecholamine uptake and storage function.[44,46,48,49] This, especially when combined with quantitation, may provide a unique tool for examining this function noninvasively in humans. The effects of drugs acting on sympathetic autonomic neurons, disease of the autonomic nervous system, and, in the case of the heart, acute ischemia, heart failure, and other primary heart diseases that directly or indirectly influence the cardiac autonomic innervation all need to be studied.[44,46–49,62] This area of MIBG scintigraphy, while experimental, opens exciting new frontiers.

REFERENCES

1. Cryer PE: Physiology and pathophysiology of the human sympathoadrenal neuroendocrine system. N Engl J Med 303:436, 1980

2. Bravo EL, Gifford RW: Pheochromocytoma: Diagnosis, localization and management. N Engl J Med 311:1298, 1984

3. Levine SN, McDonald JC: The evaluation and management of pheochromocytomas. In Shire JT (ed): Advances in Surgery, vol. 17, pp 281–313. New York, Year Book Medical Publishers, 1981

4. Manger WM, Gifford RW: Hypertension secondary to pheochromocytoma. Bull NY Acad Med 58:139, 1982

5. Stackpole RH, Melicow MN, Uson AC: Pheochromocytoma in children. Report of 9 cases and a review of the first 100 published cases with follow-up studies. J Pediatr 63:315, 1964

6. Freier DT, Harrison TS, Donahue SM et al: Rigorous biochemical criteria for the diagnosis of pheochromocytoma. J Surg Res 14:177, 1973

7. Bravo EL, Tarazi RC, Fouad FM et al: Clonidine-suppression test: A useful aid in the diagnosis of pheochromocytoma. N Engl J Med 305:623, 1981

8. Freier DT, Eckhauser FE, Harrison TS: Pheochromocytoma: A persistently problematic and still potentially lethal disease. Arch Surg 115:388, 1980

9. Rossi P, Young IS, Panke WF: Techniques, usefulness and hazards of arteriography of pheochromocytoma: Review of 99 cases. JAMA 205:547, 1968

10. Dunnick NR, Doppman JL, Gill JR et al: Localization of functional adrenal tumors by computed tomography and venous sampling. Radiology 142:429, 1982

11. Jones DH, Allison CA, Hamilton CA et al: Selective venous sampling in the diagnosis and localization of pheochromocytoma. Clin Endocrinol 10:179, 1979

12. Bowerman RA, Silver TM, Jaffe MJ et al: Sonography of adrenal pheochromocytomas. Am J Roentgenology 137:1227, 1981

13. Stewart BH, Bravo EL, Haaga J et al: Localization of pheochromocytoma by computed tomography. N Engl J Med 299:460, 1978

14. Laursen K, Damgaard-Pedersen K: CT for pheochromocytoma diagnosis. AJR 134:277, 1980

15. Adams JE, Johnson RJ, Rickards D et al: Computed tomography in adrenal disease. Clin Radiol 34:39, 1983

16. Francis IR, Glazer GM, Shapiro B et al: Complementary roles of CT scanning and 131-I-MIBG scintigraphy in the diagnosis of pheochromocytoma. Am J Roentgenol 141:719, 1983

17. Schultz CL, Haaga JR, Fletcher BD et al: Magnetic resonance imaging of the adrenal glands: A comparison with computed tomography. AJR 143:1235, 1984

18. Fink IJ, Reinig JW, Dwyer AJ et al: MR imaging of pheochromocytoma. J Comp Assist Tomog 9:454, 1985

19. Hengstmann JH, Gugler R, Dengler JH: Malignant pheochromocytoma. Effect of oral alpha-methyl p-tyrosine upon catecholamine metabolism. Klin Wochenschr 58:351, 1979

20. Drasin H: Treatment of malignant pheochromocytoma. Western J Med 128:106, 1978

21. Keiser HR, Goldstein DS, Wade JL et al: Treatment of malignant pheochromocytoma with combination chemotherapy. Hypertension 7:1–18, 1985

22. Sisson JC, Shapiro B, Beierwaltes WH et al: Treatment of malignant pheochromocytoma with a new radiopharmaceutical. Trans Assoc Amer Phys 96:209, 1983

23. Sisson JC, Shapiro B, Beierwaltes WH et al: Radiopharmaceutical treatment of malignant pheochromocytoma. J Nucl Med 25:197, 1984

24. Munkner T: ^{131}I-metaiodobenzylguanidine scintigraphy of neuroblastomas. Sem Nucl Med 15:154, 1985

25. Jaffe N: Neuroblastoma: Review of the literature and an examination of factors contributing to its enigmatic character. Cancer Treat Rev 3:61, 1976

26. Geatti O, Shapiro B, Sisson JC et al: ^{131}I-metaiodobenzylguanidine (^{131}I-MIBG) scintigraphy for the localization of neuroblastoma: Preliminary experience in 10 cases. J Nucl Med 26:736, 1985

27. Shapiro B, Von Moll L, McEwan AJ et al: 131-I-MIBG scintigraphy of neuroendocrine tumors other than pheochromocytoma and neuroblastoma. J Nucl Med 27:908, 1986

28. Shapiro B, Sisson JC, Beierwaltes WH: Experience with the use of 131-I-metaiodobenzylguanidine for locating pheochromocytomas. In Raynaud C (ed): Nuclear Medicine and Biology (Proceedings of the Third World Congress of Nuclear Medicine and Biology), vol II pp 1265–1268. Paris, Pergamon Press, 1982

29. Nakajo M, Shapiro B, Copp J et al: The normal and abnormal distribution of the adrenomedullary imaging agent m[I-131]iodobenzylguanidine (131-I-MIBG) in man: Evaluation by scintigraphy. J Nucl Med 24:672, 1983

30. Shapiro B, Copp JE, Sisson JC et al: 131-I-metaiodobenzylguanidine for the locating of suspected pheochromocytoma: Experience in 400 cases (441 studies). J Nucl Med 26:576, 1985

31. McEwan AJ, Shapiro B, Sisson JC et al: Radioiodobenzylguanidine for the scintigraphic location and therapy of adrenergic tumors. Sem Nucl Med 15:132, 1985

32. Shapiro B, Sisson JC, Kalff V et al: The location of middle mediastinal pheochromocytomas. J Thorac Cardiovasc Surg 87:816, 1984

33. Myers WG: Radioiodine-123 for medical research and diagnosis. In Lawrence JH (ed): Recent Advances in Nuclear Medicine–Progress in Atomic Medicine, vol. 4, pp. 133–160. New York, Grune and Stratton, 1974

34. Swanson DP, Carey JE, Brown LE et al: Human absorbed dose calculations for iodine 131 and iodine 123 labelled MIBG: A potential myocardial and adrenal medulla imaging agent. Proceedings of the Third International Radiopharmaceutical Dosimetry Symposium, Health and Human Services publication FDA 81-8166, pp 213-224. Rockville, Maryland, 1981

35. Lynn MD, Shapiro B, Sisson JC et al: Portrayal of pheochromocytoma and normal human adrenal medulla by 123-I-metaiodobenzylguanidine (123-I-MIBG). J Nucl Med 25:436, 1984

36. Lynn MD, Shapiro B, Sisson JC et al: Pheochromocytomas and normal adrenal medulla: Improved visualization with ^{123}I-MIBG scintigraphy. Radiology 156:789, 1985

37. Shapiro B, Sisson JC, Lloyd R et al: Malignant pheochromocytoma: Clinical, biochemical and scintigraphic characterization. Clin Endocrinol 20:189, 1984

38. Kimmig B, Bubeck B, Eisenhut M et al: Radiation burden using 131-I-metabenzylguanidine (MIBG) for diagnosis and therapy. Nuc Compact 14:353, 1983

39. Sipila LN, Carey JE, Shapiro B et al: Health physics aspects of I-131-metaiodobenzylguanidine therapies–A prototype for I-131-radiolabeled antibody therapies (abstr). Med Physics 12:520, 1985

40. Bomanji J, Flatman WD, Horne T et al: Quantitation of

123-I-metaiodobenzylguanidine (MIBG) uptake by normal adrenal medulla (abstr). Nucl Med Communications 7:296, 1986

41. Wieland DM, Wu JL, Brown LE et al: Radiolabeled adrenergic neuron blocking agents: Adrenomedullary imaging with [^{131}I]iodobenzylguanidine. J Nucl Med 21:349, 1980

42. Wieland DM, Brown LE Tobes MC et al: Imaging the primate adrenal medulla with ^{123}I and ^{131}I metaiodobenzylguanidine. Concise communication. J Nucl Med 22:358, 1981

43. Mangner TJ, Tobes MC, Wieland DM et al: Metabolism of meta-I-131-iodobenzylguanidine in patients with metastatic pheochromocytoma: Concise Communication. J Nucl Med 27:37, 1986

44. Shapiro B, Wieland DM, Brown LE et al: 131-I-meta-iodobenzylguanidine (MIBG) adrenal medullary scintigraphy: Interventional studies. In Spencer RP (ed): Interventional Nuclear Medicine, chapter 19 pp 451–481. New York, Grune and Stratton, 1984

45. Guilloteau D, Baulieu JL, Hugnet F et al: Metaiodobenzylguanidine adrenal medulla localization: Autoradiographic and pharmaceutical studies. Eur J Nucl Med 9:278, 1984

46. Makajo M, Shapiro B, Sisson JC et al: Salivary gland accumulation of meta [131-I] iodobenzylguanidine. J Nucl Med 25:2, 1984

47. Shen SW, Sisson JC, Shulkin B et al: I-123-metaiodobenzylguanidine (I-123-MIBG) as an index of neuron integrity following myocardial infarction (abstr). J Nucl Med 27:949, 1986

48. Jacques S, Tobes MC, Sisson JC et al: Comparison of the sodium dependency of uptake of meta-iodobenzylguanidine and norepinephrine into cultured bovine adrenomedullary cells. Molec Pharmacol 26:539, 1984

49. Tobes MC, Jacques S, Wieland DM et al: Effect of uptake-one inhibitors on the uptake of norepinephrine and metaiodobenzylguanidine. J Nucl Med 26:897, 1985

50. Nakajo M, Shimabakuro K, Yoshimura H et al: Iodine 131 metaiodobenzylguanidine intra and extravesicular accumulation in the rat heart. J Nucl Med 27:84, 1986

51. Tobes MC, Shapiro B, Meyers L et al: Pharmacokinetics of I-131-metaiodobenzylguanidine (MIBG) in an anephric patient. European Nuclear Medicine Congress, London, England, September 3–6, 1985

52. Brown MJ, Fuller RW, Lavender JP: False positive diagnosis of bilateral pheochromocytoma by iodine-131-labelled meta-iodobenzylguanidine. Lancet i:56, 1984

53. Nakajo M, Shapiro B, Glowniak J, et al: Inverse relationship between cardiac accumulation of ^{131}I-MIBG and circulating catecholamines in suspected pheochromocytoma. J Nucl Med 24:1127, 1984

54. Horne T, Hawkins LA, Britton KE et al: Imaging of pheochromocytoma and adrenal medulla with 123-I-metaiodobenzylguanidine. Nucl Med Communications 5:763, 1984

55. Kline RC, Swanson DP, Wieland DM et al: Myocardial imaging in man with ^{123}I-meta-iodobenzylguanidine. J Nucl Med 22:129, 1981

56. Sisson JC, Frager MS, Valk TW et al: Scintigraphic localization of pheochromocytoma. N Engl J Med 305:12, 1981

57. Lindberg S, Ernest I, Fjalling M et al: The value of late images in 131-I-metaiodobenzylguanidine (131-I-MIBG) scanning of the adrenal medulla (abstr 142). Europ J Nucl Med II: A28, 1985

58. Shapiro B, Fischer M: Summary of the proceedings of a workshop on 131-I-metaiodobenzylguanidine held at Schloss Wilkinghege, Munster, W. Germany, September 27, 1984. Nucl Med Communications 6:179, 1985

59. McEwan AJ, Shapiro B: 131-I-metaiodobenzylguanidine-clinical strategy (abstr). Nucl Med Communications 7:296, 1986

60. Valk TW, Frager MS, Gross MD et al: Spectrum of pheochromocytoma in multiple endocrine neoplasia: A scintigraphic protrayal using ^{131}I-metaiodobenzylguanidine. Ann Intern Med 94:762, 1981

61. Sisson JC, Shapiro B, Beierwaltes WH: Scintigraphy with I-131 MIBG as an aid to the treatment of pheochromocytomas in patients with the MEN-2 syndromes. Henry Ford Med J 32:254, 1984

62. Rabinovitch MA, Rose CP, Chartrand G et al: Imparied myocardial retention of I-131-metaiodobenzylguanidine (MIBG) in chronic volume-overload heart failure (abstr). J Nucl Med 27:1068, 1986

63. Sisson JC, Shapiro B, Beierwaltes WH et al: Locating pheochromocytomas by scintigraphy using ^{131}I-metaiodobenzylguanidine. Ca—A Cancer Journal for Clinicians 34:86, 1984

64. Kalff V, Sisson JC, Beierwaltes WH: Adrenal gland identification: Pre-operative assessment. Surgery 91:374, 1982

65. Kalff V, Shapiro B, Lloyd RV et al: Bilateral pheochromocytomas. J Endo Invest 7:387, 1984

66. Kalff V, Shapiro B, Lloyd R et al: The spectrum of pheochromocytoma in hypertensive patients with neurofibromatosis. Arch Int Med 142:2092, 1982

67. Glowniak JV, Shapiro B, Sisson JC et al: Familial extra-adrenal pheochromocytoma: A new syndrome. Arch Int Med 145, 257, 1985

68. Hattner RS, Huberty JP, Engelstead BL et al: Sensitivity and specificity of MIBG scintigraphy for the detection of pheochromocytoma (abstr). J Nucl Med 24:P54, 1983

69. Shulkin BL, Shapiro B, Francis IR et al: Primary extra-adrenal pheochromocytoma: A case of positive 123-I-MIBG scintigraphy with negative 131-I-MIBG scintigraphy. Clin Nuc Med 11:851, 1986

70. Hattner R, Huberty JP, Engelstad BL et al: Location of M iodo (^{131}I) benzylguanidine in neuroblastoma. AJR 143:373, 1984

71. Treuner J, Feine U, Niethammer D et al: Scintigraphic imaging of neuroblastoma with ^{131}I metaiodobenzylguanidine. Lancet i:333, 1984

72. Kimmig B, Brandeis WE, Eisenhut M et al: Scintigraphy of neuroblastoma with ^{131}I-MIBG. J Nucl Med 25:773, 1984

73. Shapiro B, Geatti O, Sisson JC et al: Diagnosis and therapy of neuroblastoma (neuro) 131-I-meta-iodobenzylguanidine (131-I-MIBG) (abstr C3:23). European Nuclear Medicine Congress, London, England, September 3–6, 1985. Nucl Med Communications 6:555, 1985

74. Feine U, Treuner J, Muller-Schaunburg W et al: Scintigraphic imaging of neuroblastoma with I-131-meta-iodobenzylguanidine (abstr). J Nucl Med 25:P75, 1985

75. Baulieu JL, Guilloteau C, Chambon C et al: Meta-iodo-

benzylguanidine (MIBG) scintigraphy: A one-year experience (abstr). J Nucl Med 25: Plll, 1984

76. Lumbroso J, Hartmann O, Guermazi F et al: The clinical contribution of I-123 and I-131 meta-iodo-benzylguanidine (MIBG) scans in neuroblastoma (abstr). J Nucl Med 27:947, 1986

77. Von Moll L, McEwan AJ, Shapiro B et al: 131-I-MIBG scintigraphy of neuroendocrine tumors other than pheochromocytoma and neuroblastoma. J Nucl Med, March 1986

78. Fischer M, Kamanabroo D, Sanderkamp H et al: Scintigraphic imaging of carcinoid tumors with ^{131}I MIBG. Lancet ii:165, 1984

79. Hoefnagel CA, DeKraker J, Marcuse HR et al: Detection and treatment of neural crest tumors using I-131-metaiodobenzylguanidine (abstr A73). Europ J Nucl Med 11:A17, 1985

80. Blinder RA, Feldman JM, Coleman RE: ^{131}I-MIBG imaging of carcinoid tumors (abstr 56). 32nd Annual Meeting of the Society of Nuclear Medicine, Houston, Texas, June 2–5, 1985. J Nucl Med 26:P17, 1985

81. Smit AJ, Van Essen LH, Hollenca H et al: ^{131}I MBIG uptake in a non-secreting paraganglioma. J Nucl Med 25:984, 1984

82. Sone T, Fukunaga M, Otsuka N et al: Metastatic medullary thyroid cancer: Localization with iodine-131-metaiodobenzylguanidine. J Nucl Med 26:604, 1985

83. Endo K, Shiomi K, Kasagi K et al: Imaging of medullary carcinoma of the thyroid with ^{131}I-MIBG. Lancet ii:233, 1984

84. Sibley RK, Rosai J, Foucar E et al: Neuroendocrine (Merkel cell) carcinoma of the skin: A histological and ultrastructural study of 2 cases. Am J Surg Path 4:211, 1980

85. Chatal JF, Charbonnel B: Comparison of iodobenzylguanidine imaging with computed tomography in locating pheochromocytoma. J Clin Endocrinol Metab 61:769, 1985

86. Hoefnagel CA, Den Hartog Jager FCA, Van Gennip AM et al: Diagnosis and treatment of a carcinoid tumor using iodine 131 metaiodobenzylguanidine. Clin Nucl Med 3:150, 1986

87. Shapiro B, Sisson JC, Eyre P et al: ^{131}I-MIBG, a new agent in diagnosis and treatment of pheochromocytoma. Cardiology (Basel) 72:137, 1985

88. Editorial: Clinical value of adrenomedullary scintigraphy with ^{131}I-MIBG. Nucl Compact 14:318, 1983

89. Ackery DM, Tippett P, Condon B et al: New approach to the localization of phaeochromocytoma: imaging with ^{131}I-MIBG. Brit Med J 288:1587, 1984

90. Swensen SJ, Brown ML, Sheps SG et al: Use of 131-I-MIBG scintigraphy in the evolution of suspected pheochromocytoma. Mayo Clin Proc 60:299, 1985

91. Lumbroso J, Hartmann O, Lemerle J et al: Scintigraphic detection of neuroblastoma using 131-I and 123-I labelled metaiodobenzylguanidine (abstr 71). Europ J Nucl Med 11:A16, 1985

92. Heyman S, Evans AE: I-131-metaiodobenzylguanidine (I-131-MIBG) in the diagnosis of neuroblastoma (abstr). J Nucl Med 27:931, 1986

93. Feine U, Treuner J, Niethammer D et al: Erste untenshungen zur scintigraphischen darstellung von neuroblastomen mit 131-J-meta-benzylguanidin. Nucl Compact 15:23, 1984

Atlas Section

Figure 4-1

Comparison of the structural formulae of the endogenous neurotransmitter hormone, norepinephrine; the hypotensive drug, guanethidine; and meta-iodobenzylguanidine (MIBG).

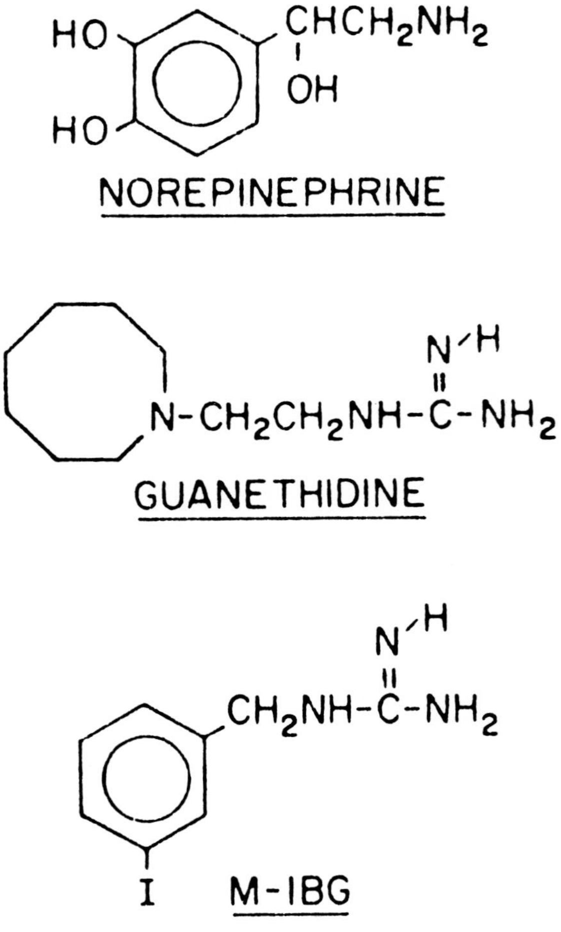

NOREPINEPHRINE

GUANETHIDINE

M-IBG

Figure 4-2 ***Normal I-131 MIBG Scintigram***

Overlapping I-131 MIBG images 48 hours postinjection of 0.5 mCi in a typical, normal subject: ***(A)*** posterior head and chest; ***(B)*** posterior midabdomen and lower chest; and ***(C)*** anterior lower abdomen and pelvis. Comment: Note normal I-131 MIBG accumulation in parotid salivary glands *(S),* heart *(H),* liver *(L),* Spleen *(SP),* and urinary bladder *(B).* Radioactive markers *(M)* have been placed on the axillae, costal margins, and pelvic brim. (Nakajo M et al: The normal and abnormal distribution of the adrenomedullary imaging agent m[I-131] iodobenzyl-guanidine (131-I-MIBG) in man: Evaluation by scintigraphy. J Nucl Med 24:672 - 682, 1983)

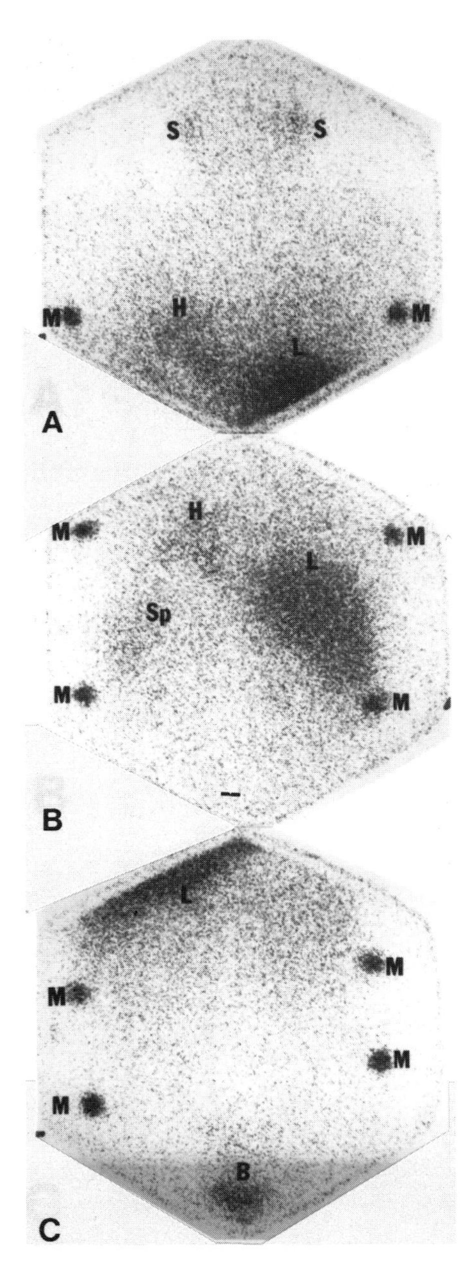

Figure 4-3 ***The Spectrum of Intensity of Adrenal Uptake of I-131 MIBG in Multiple Endocrine Neoplasia Type 2 (MEN-2)***

All images are posterior abdominal views obtained 48 hours following the injection of 0.5 mCi I-131 MIBG. ***(A)*** Patient with MEN-2a; no visible adrenal I-131 MIBG (grade 0). ***(B)*** Patient with MEN-2b; minimal bilateral adrenal I-131 MIBG uptake indicated by arrows (grade 1 on left, grade 2 on right). ***(C)*** Patient with MEN-2a; prominent bilateral adrenal I-131 MIBG uptake indicated by arrows (grade 3). The grading scale is: Grade 0, no visible uptake; grade 1, uptake just visible; grade 2, uptake clearly visible; grade 3, prominent uptake; grade 4, uptake yielding maximum film density. (Nakajo M et al: The normal and abnormal distribution of the adrenomedullary imaging agent m[I-131]iodobenzylguanidine (131-I-MIBG) in man: Evaluation by scintigraphy. J Nucl Med 24:672–682, 1983)

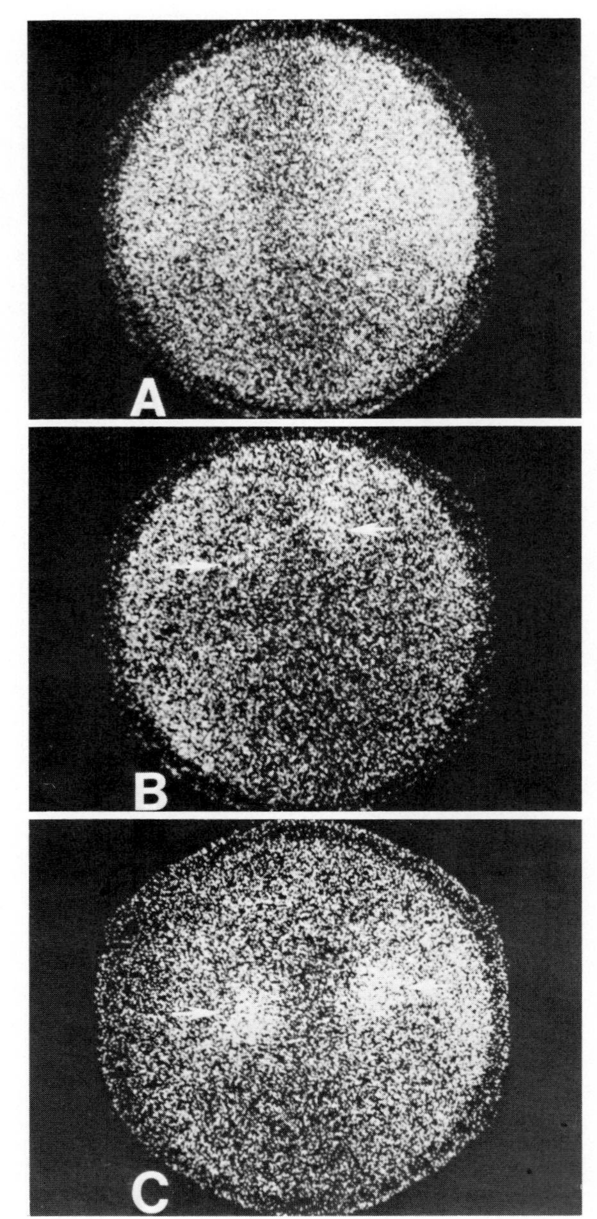

Figure 4-4 ***Frequency and Intensity of I-131 MIBG Uptake by the Adrenal Medulla in Different Groups Following Administration of 0.5 mCi***

Data are expressed as percentage of observations made at 24, 48, and 72 hours following I-131 MIBG administration. The findings are as follows: group I, normal; group II, probably normal; group III, MEN-2 patients; group IV, adrenal pheochromocytomas (*T*, tumor; *C*, contralateral normal gland); group V, indeterminate group (incomplete biochemical data but probably tumor-free). Comment: Note the low incidence of faint adrenal uptake in normal glands and the far greater incidence of uptake in MEN-2 patients (both faint and intense). Note also that the intensity of adrenal I-131 MIBG uptake increases between 24 and 72 hours. (Nakajo M et al: The normal and abnormal distribution of the adrenomedullary imaging agent m[I-131]iodobenzylguanidine (131-I MIBG) in man: Evaluation by scintigraphy. J Nucl Med 24:672–682, 1983)

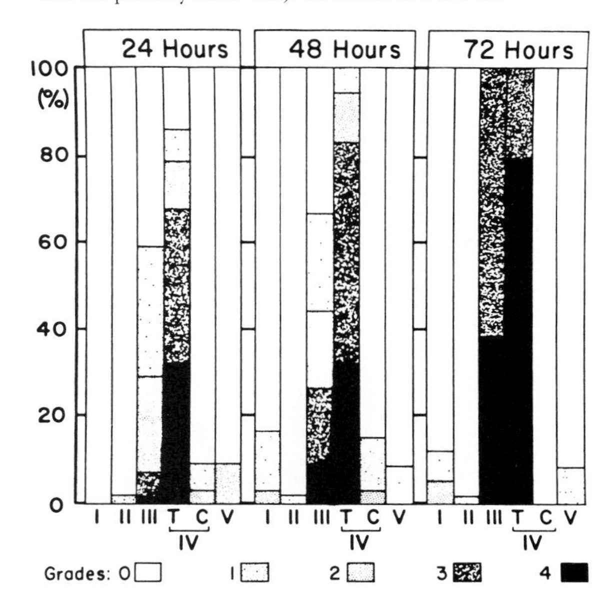

Figure 4-5 ***Comparison of I-123 and I-131 MIBG Scintigraphy***

Thirty-three-year-old male patient with metastatic pheochromocytoma arising from an abdominal, extra-adrenal primary tumor (resected). All images are of the posterior abdomen. ***(A)*** Image was obtained 24 hours following administration of 10.0 mCi I-123 MIBG (484,000 counts in 3.5 minutes). ***(B)*** Image was obtained 48 hours following administration of 10.0 mCi I-123 MIBG (404,000 counts in 15 minutes). ***(C)*** Image was obtained 24 hours following administration of 0.5 mCi I-131 MIBG (100,000 counts in 12 minutes). ***(D)*** Image was obtained 48 hours following the administration of 0.5 mCi I-131 MIBG (100,000 counts in 20 minutes). ***(E)*** Image was obtained 6 days following the administration of a 179-mCi therapeutic dose of I-131 MIBG (200,000 counts in 1.5 minutes). ***Arrowheads*** indicate MIBG uptake in the normal adrenal medullae. *Open arrows* indicate metastatic deposits in bone. *L* normaL MIBG uptake in liver; and in *S*, spleen. (Lynn MD et al: Portrayal of pheochromocytoma and normal human adrenal medulla by 123-I-metaiodobenzylguanidine (123-I-MIBG). J Nucl Med 25:436-440, 1984)

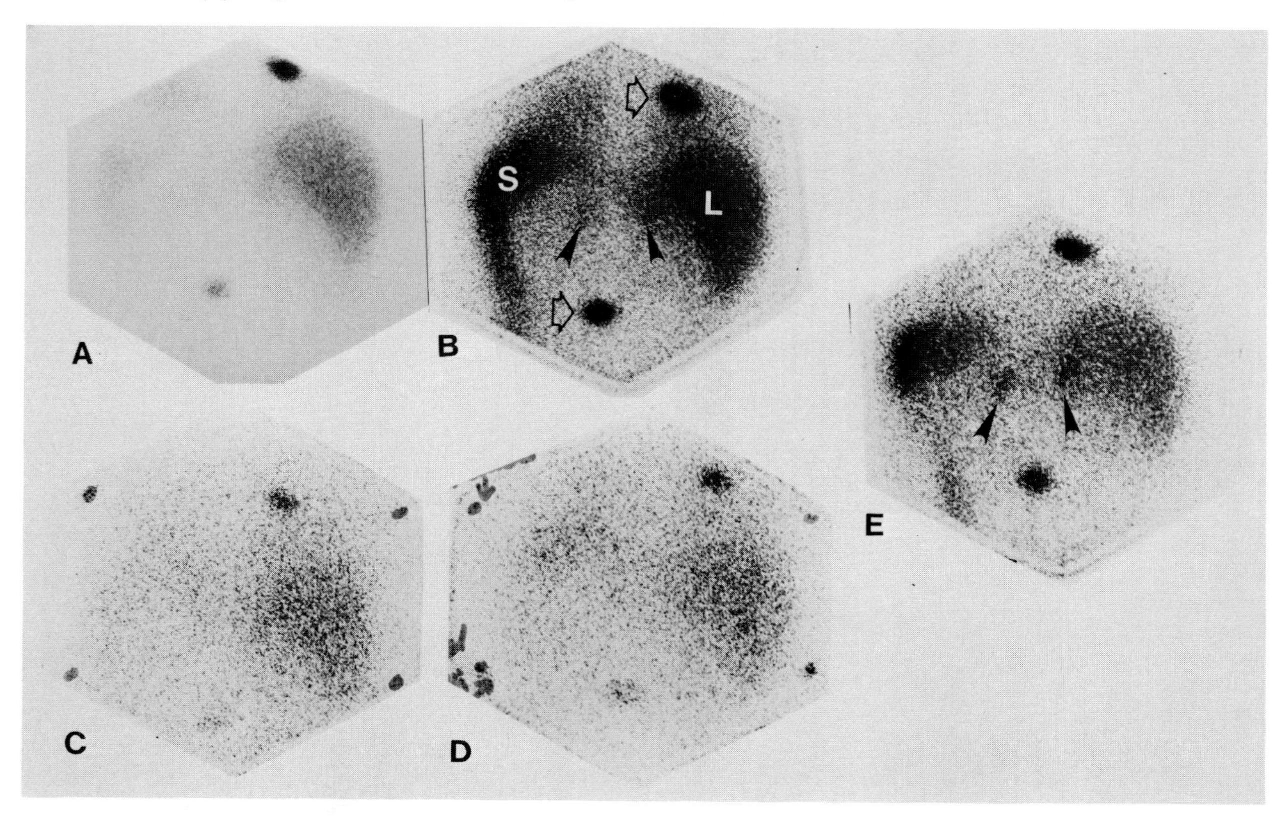

Figure 4-6 ***The Same Patient As in Figure 4-5 Imaged 24 Hours Following 10.0 mCi I-123 MIBG***

(A) Anterior head and neck image, showing intense uptake and excellent resolution of the normal uptake with parotid glands *(P)*, submandibular glands *(S)*, and nasopharynx *(N)*. Skull metastases are indicated by arrows. Faint thyroidal visualization occurs despite thyroidal blockade with stable iodide *(T)*. ***(B)*** Anterior abdomen with normal uptake in liver *(L)* and gut *(G)*. Metastatic foci are indicated by arrows. ***(C)*** Anterior abdomen-pelvis with normal uptake in gut *(G)* and bladder ***(B)***. Metastatic foci are indicated by arrows. Comment: Comparison of I-131 and I-123 MIBG images shows the latter to have far higher photon flux and resolution. Most normal adrenal medullae can be depicted with radio-MIBG if sufficient activity (3 to 10 mCi) is administered. However, with 0.5 mCi I-131 MIBG, such depiction is uncommon and, if present, is faint.

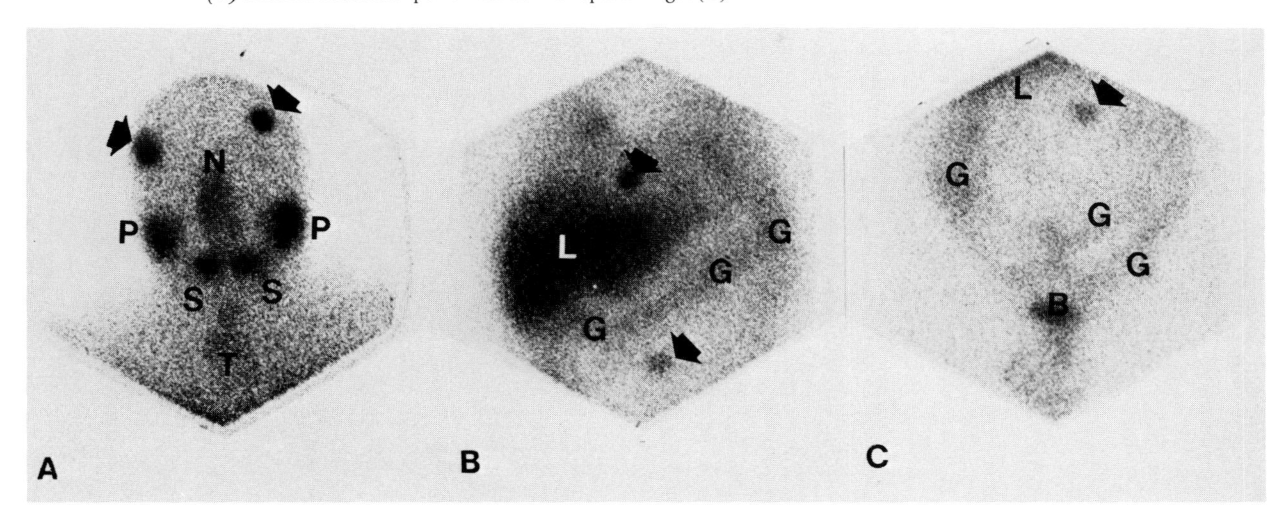

Figure 4-7 ***Intra-Adrenal Pheochromocytoma***

Forty-year-old male with classical symptomatology, 45-fold elevation of plasma epinephrine, and tenfold elevation of plasma norepinephrine. ***(A)*** Posterior image obtained 24 hours following the administration of 0.5 mCi I-131 MIBG. There is abnormal focus of tracer uptake in a left adrenal pheochromocytoma. *L*, normal liver uptake. ***(B)*** Left lateral image of the same patient. Arrows indicate a radioactive marker along the patient's spine. Comment: This intense degree of uptake is observed in the majority of cases of pheochromocytoma. The patient underwent surgery and a 24-g left-adrenal pheochromocytoma was resected. This was followed by complete resolution of the patient's symptoms and biochemical abnormalities. (Sisson JC et al: Scintigraphic localization of pheochromocytoma. New Engl J Med 305:12-17, 1981, Reprinted by permission of the New England Journal of Medicine)

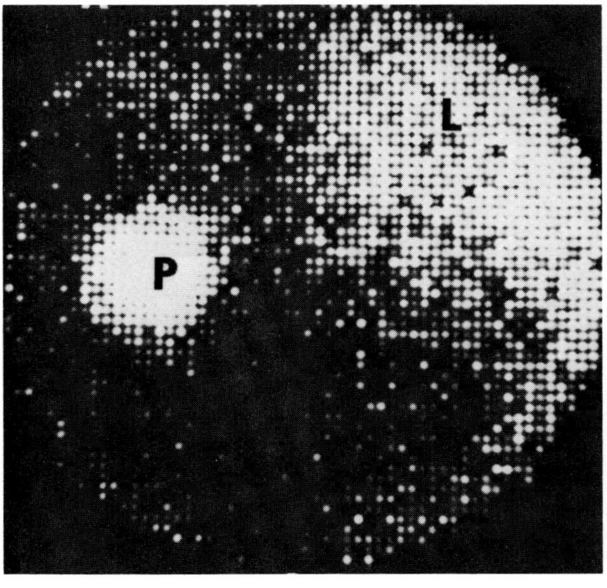

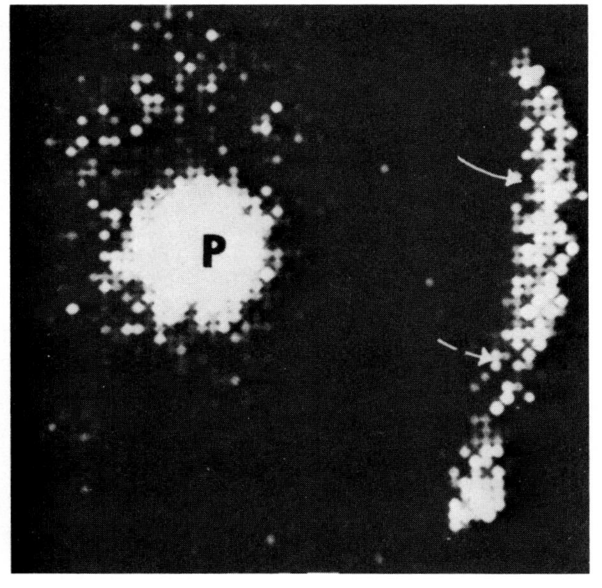

Figure 4-8 ***Spectrum of I-131 MIBG Uptake Encountered in Sporadic Intra-Adrenal Pheochromocytoma***

Posterior abdominal views obtained 48 hours following the administration of 0.5 mCi I-131 MIBG. ***(A)*** Typical intense (grade 4) uptake in a left-sided lesion *(arrow)*. ***(B)***. Less intense uptake (grade 3) in left-sided lesion *(large arrow)*. Note faint (grade 2) uptake in normal right adrenal *(small arrow)*. The normality of the right gland was confirmed at surgery. Comment: In both instances the left-sided adrenal tumor was 3 cm in diameter. ***B*** shows how the intensity of I-131 MIBG uptake in an unusually faint tumor may approach that of an unusually intense normal gland. Ten percent to 15% of lesions may show no I-131 MIBG uptake at all. (Nakajo M et al: The normal and abnormal distribution of the adrenomedullary imaging agent m[I-131] iodobenzylguanidine (131-I-MIBG) in man: Evaluation by scintigraphy. J Nucl Med 24:672 - 682, 1983)

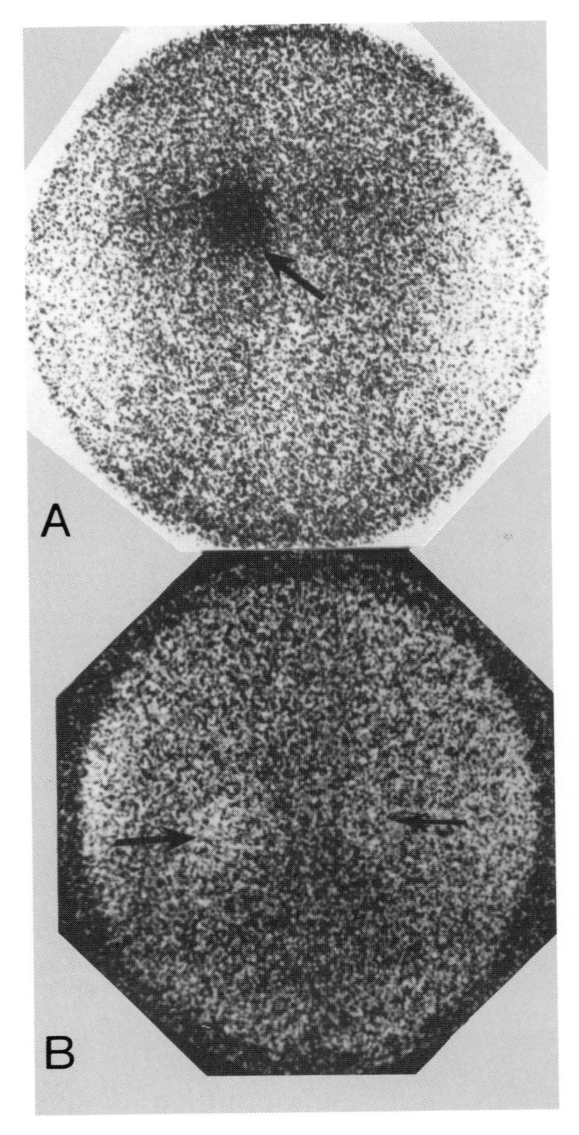

Figure 4-9 ***Usefulness of I-131 MIBG Scintigraphy for Intra-Adrenal Pheochromocytoma in the Case of Extensive Previous Surgery***

An 11-year-old boy with hypertension underwent resection of an atrophic left kidney, after which blood pressure normalized for 2 years. Recurrent hypertension caused by renal artery stenosis prompted an autotransplantation of the right kidney to the right iliac fossa. At this operation an incidentally discovered para-aortic nodule was resected and proved to be a pheochromocytoma. The patient remained well until the age of 15 when hypertension, tachycardia, and sweating occurred. Plasma catecholamines were ten times the upper limit of normal. CT scans were uninterpretable due to the interference of surgical clips. A vena caval blood sampling study showed the highest levels of catecholamines in the superior vena cava. An I-131 MIBG study was performed. *(A)* Posterior abdominal image 24 hours following the administration of 0.5 mCi I-131 MIBG shows an abnormal focus of uptake in the left adrenal region *(arrow)*. *L*, normal liver uptake of I-131 MIBG. *(B)* Posterior abdominal I-131 MIBG image, as in *A*, with superimposition of a Tc-99m DTPA study, demonstrates the autotransplanted kidney *(K)* in the right iliac fossa. Comment: The adrenal pheochromocytoma was resected and the hypertension resolved. The case demonstrates the utility of I-131 MIBG scintigraphy in situations where previous surgery disrupts tissue planes and introduces metallic clips which vitiate CT, and where disruption of normal venous drainage may lead to misleading venous sampling studies. In this case the adrenal pheochromocytoma was believed to be drained by the azygous system into the superior vena cava as a consequence of the previous nephrectomy. (Sisson JC et al: Locating pheochromocytomas by scintigraphy using I-131 metaiodobenzylguanidine. Ca—A Cancer Journal for Clinicians 34:86-92, 1984)

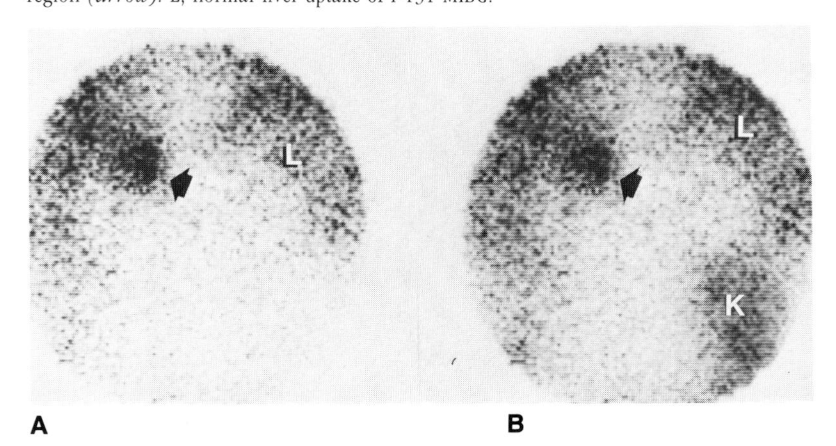

Figure 4-10 ***Usefulness of I-131 MIBG Scintigraphy in a Case of Locally Recurrent Adrenal Pheochromocytoma***

A 12-year-old boy underwent left total and right subtotal adrenalectomies for bilateral pheochromocytoma. The subtotal right adrenalectomy was designed to remove the pheochromocytoma and preserve sufficient adrenocortical tissue to make glucocorticoid replacement therapy unnecessary. The patient did well for 4 years, after which hypertension, headache, and sweating recurred. Biochemical studies were unequivocally diagnostic of pheochromocytoma. CT scan of the abdomen was less than optimal due to the presence of surgical clips, but no tumor was demonstrated.

Posterior abdominal image demonstrates abnormal adrenal focus of activity *(arrow)* 48 hours following the administration of 0.5 mCi I-131 MIBG. Renal outlines are derived from a simultaneously acquired Tc-99m DTPA study. *M,* surface radioactive markers. Comment: This demonstrates the utility of I-131 MIBG scintigraphy in patients with a history of previous pheochromocytoma resection in whom there are recurrent symptoms or biochemical abnormalities. In this case, the cause was a local recurrence probably due to residual tissue left at the time of initial resection. In the majority of cases of recurrent disease we have found the cause to be due to metastatic disease for which I-131 MIBG scintigraphy is often diagnostic (Fig. 4-20).

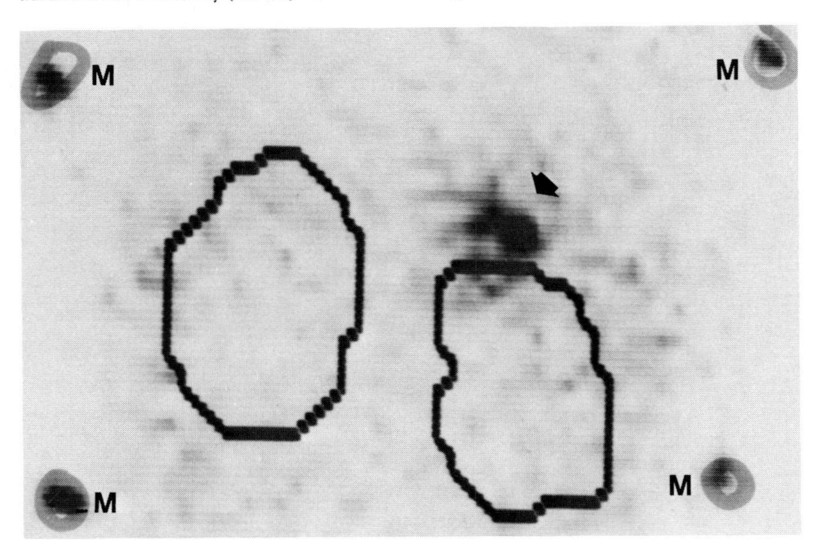

Figure 4-11 ***Bilateral Adrenal Pheochromocytomas Associated with Neurofibromatosis***

A 28-year-old female with a 5-year history of hypertension, headache, and sweating had multiple cutaneous neurofibromas and café au lait spots and was a member of a well-documented neurofibromatosis kindred.

On posterior abdominal image 48 hours following the administration of 0.5 mCi I-131 MIBG, bilateral adrenal pheochromocytomas show increased uptake *(closed arrows).* Radioactive markers on the iliac crests are indicated by open arrows and radioactivity in the bladder by *B.* Both tumors were successfully resected.

Comment: Pheochromocytoma occurs in 1% to 2% of patients with neurofibromatosis, which is at least ten times as frequently as in the general population. Thus the combination of neurofibromatosis and hypertension should strongly raise the suspicion of pheochromocytoma. Abdominal CT scan may be confusing in these cases as there may be multiple intra-abdominal neurofibromas. These, however, do not take up I-131 MIBG, whereas pheochromocytomas do.

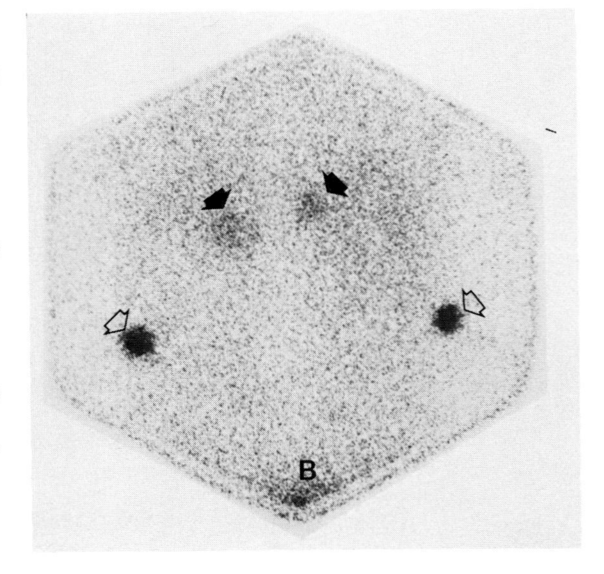

Figure 4-12 ***Extra-Adrenal Abdominal Pheochromocytoma***

A 28-year-old man with an 11-year history of hypertension poorly controlled with conventional therapy had a 10-year-old daughter with a right-sided, pararenal pheochromocytoma (Fig. 4-13). Biochemical studies of the father showed elevation of plasma norepinephrine, urinary norepinephrine, metanephrine, and vanillylmandelic acid. ***(A)*** Anterior abdominal I-131 MIBG image obtained 24 hours following administration of 0.5 mCi shows abnormal focus of I-131 MIBG *(large arrows)*; surface markers are indicated by small arrows. *L*, normal I-131 MIBG uptake in liver; and *S*, in spleen. ***(B)***. Abdominal CT scan shows tumor arising from the right renal hilar region *(arrow)*. ***(C)***. Inferior venacavogram showing indentation of the vena cava at the level of the renal veins. (Glowniak JV et al: Familial extra-adrenal pheochromocytoma: A new syndrome. Arch Int Med 145:257-261, 1985; copyright 1985, American Medical Association)

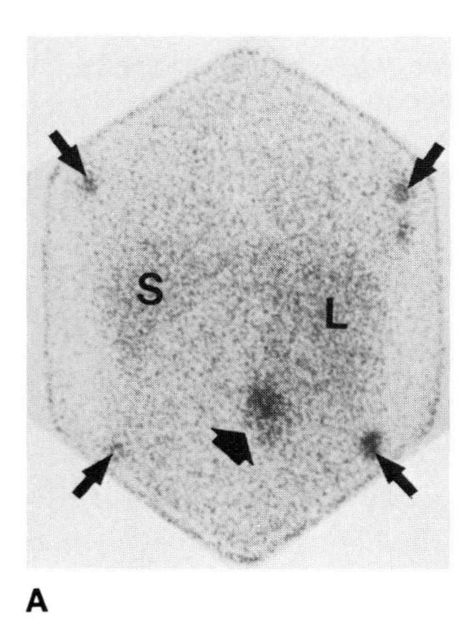

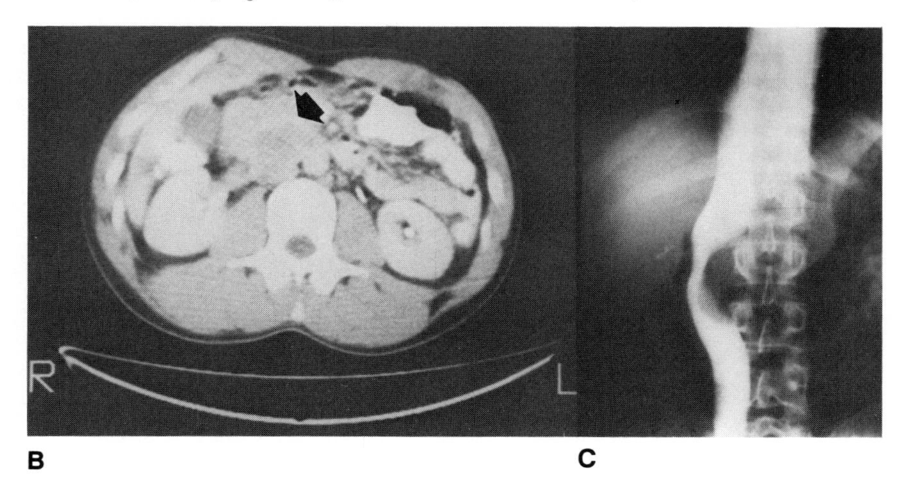

A **B** **C**

Figure 4-13 ***Familial Extra-Adrenal Pheochromocytoma***

In this family tree, the patient described in Figure 4-12 is V-2 and his daughter is VI-5. Note that proven or suspected pheochromocytomas have occurred in five generations and that certain individuals are apparently unaffected obligate carriers (*e.g.*, IV-1). Comment: This case demonstrates that familial pheochromocytoma may occur in the absence of the neuroectodermal syndromes such as neurofibromatosis, MEN-2a and MEN-2b, or von Hippel-Lindau disease.

The patient was operated on and the primary tumor completely removed. Histologically it demonstrated aggressive but no frankly malignant features. Three years later, an I-131 MIBG study performed for back pain demonstrated skeletal metastases. This is in keeping with our finding that extra-adrenal lesions have a greater propensity to metastasize than intra-adrenal lesions and confirms the usefulness of I-131 MIBG scintigraphy in follow-up. (Glowniak JV et al: Familial extra-adrenal pheochromocytoma: A new syndrome. Arch Int Med 145:257-261, 1985; copyright 1985, American Medical Association)

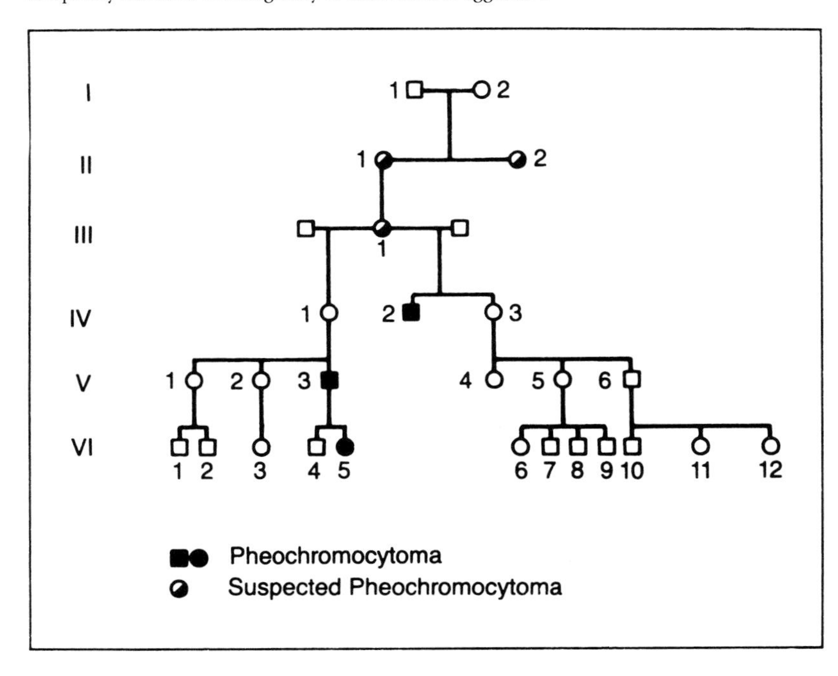

Figure 4-14 ***Extra-Adrenal, Intra-Abdominal Pheochromocytoma Not Demonstrated by I-131 MIBG but Successfully Depicted in I-123 MIBG***

The patient was a 27-year-old woman with hypertension, headache, and sweating. Plasma and urinary catecholamines were unequivocally diagnostic of pheochromocytoma. Intravenous urography with nephrotomography and abdominal CT scan (which was technically suboptimal) were negative. The I-131 MIBG scan was also normal, but I-123 MIBG scintigraphy revealed an abnormal focus of uptake. ***(A)*** Posterior abdominal image obtained 48 hours following the administration of 0.5 mCi I-131 MIBG shows only normal tracer uptake in the liver *(L)*, bladder *(B)*, and both adrenal medullae *(arrows)*. The image of 24 hours also failed to reveal the tumor. ***(B)*** Posterior abdominal image obtained 24 hours following the administration of 10.0 mCi I-123 MIBG shows normal uptake in the liver *(L)*, spleen *(S)*, bladder *(B)*, and adrenal medullae *(small arrows)*. An abnormal focus of activity is present in the left midabdomen.

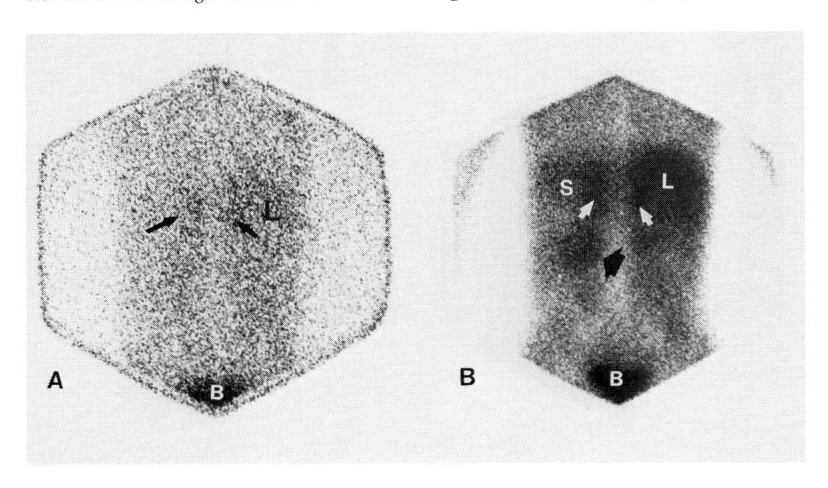

Figure 4-15 ***Single Photon Emission Computed Tomography (SPECT) Using I-123 MIBG***

In the same patient presented in Figure 4-14, SPECT was performed 24 hours following administration of 10.0 mCi I-123 MIBG. ***(A)*** Transaxial section is through center of abnormal focus *(arrow)* of I-123 MIBG uptake at the level of the left renal lower pole. ***(B)*** Coronal section is through abnormal focus of I-123 MIBG uptake *(large arrow)*. Also seen is normal adrenal medullary uptake *(small arrows)*. *L*, normal liver; *S*, normal spleen, *R*, right side. ***(C)*** Parasagittal section is through the abnormal focus of I-123 MIBG uptake *(large arrow)*, normal left adrenal *(small arrow)*, and left lobe of liver *L*. *A*, anterior. A benign extra-adrenal pheochromocytoma was resected from the site indicated by scintigraphy. Comment: This case demonstrates that in cases where MIBG uptake by the tumor is poor, I-123 MIBG may offer the advantages of high photon flux, suitable photon energy, and high camera efficiency, all of which combine to yield a positive study in circumstances where the disadvantages of I-131 MIBG may lead to failure. SPECT permits the abnormal focus to be displayed in multiple anatomic planes that clearly delineate its relationships.

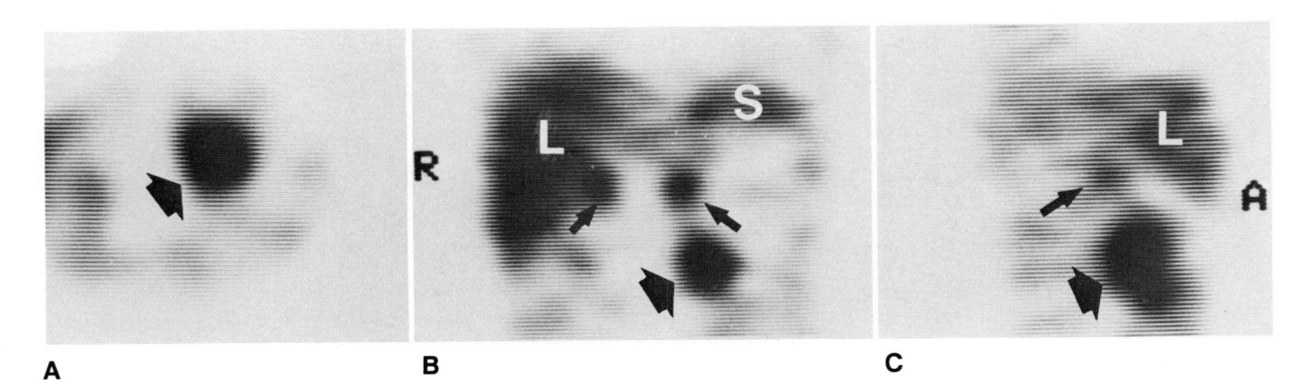

Figure 4-16 ***Primary Thoracic Pheochromocytoma Arising from the Left Cardiac Atrium***

A 26-year-old female with a history of classical pheochromocytoma symptomatology and diagnostic (tenfold) elevations of plasma, urine catecholamines, and urine catecholamine metabolites was referred for I-131 MIBG after extensive negative investigations, including intravenous urography with nephrotomography; abdominal (two), thoracic, cranial, and mastoid CT scans; thoracic laminography; angiography of the abdominal, pelvic, and pulmonary vasculature; three vena caval blood sampling procedures, and a negative laparotomy.

The I-131 MIBG scintigraphy revealed an abnormal focus of uptake in the midchest, but a thoracotomy performed at another institution failed to find the tumor. Repeat I-131 MIBG scan performed 8 months later revealed the same focus of abnormal uptake, the exact location of which was established by multiple radiopharmaceutical scintigraphy and I-131 MIBG-directed dynamic CT scanning. ***(A)*** I-131 MIBG scintigraphy 24 hours following administration of 0.5 mCi: *left,* posterior image of chest; *right,* right lateral image of chest. The arrow points to the abnormal focus of I-131 MIBG uptake *(arrow). M,* external radioactive marker on the spine; *L,* normal liver uptake; *SP,* normal splenic uptake. ***(B)*** Tc-99m-labeled red blood cell images with the location of the abnormal focus of I-131 MIBG uptake superimposed by computer: *left,* anterior chest image; *right,* right lateral image. *K,* kidney; *H,* cardiac blood pool; *A,* aortic arch. ***(C)*** Tc-99m MPD bone scan images with the location of the abnormal focus of I-131 MIBG uptake superimposed by computer: *left,* posterior image of chest; *right,* right posterior oblique image of chest. *K,* kidney; *S,* spine; *ST,* sternum. (Shapiro B et al: The location of middle mediastinal pheochromocytomas. J Thorac Cardiovasc Surg 87:814-820, 1984)

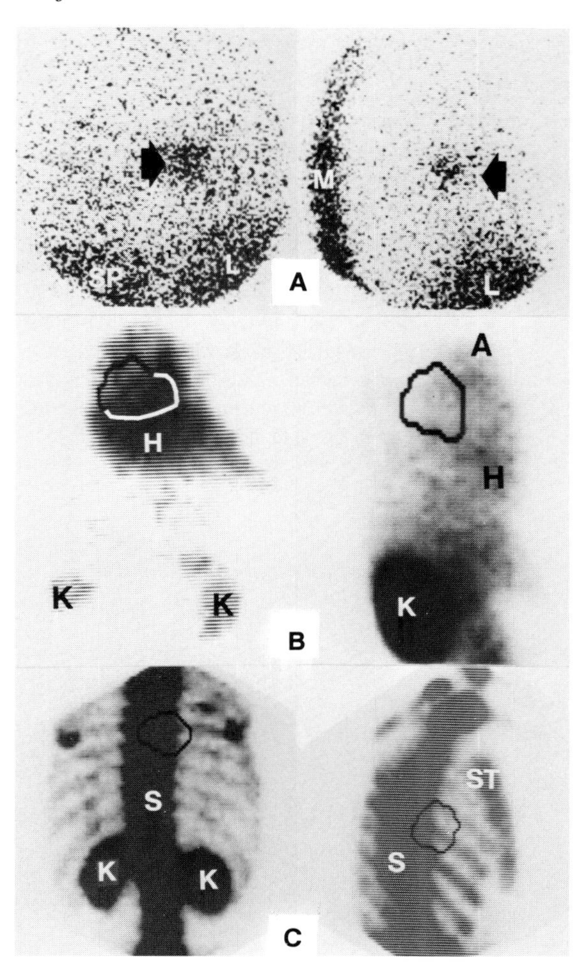

Figure 4-17 ***I-131 MIBG-Directed Dynamic CT Scan Demonstrating a Primary Left Atrial Pheochromocytoma***

Dynamic computed tomographic scans of the chest at the level of the I-131 MIBG abnormality were made of the same patient as in Figure 4-16. ***(A)*** Enhancement of the cardiac chambers 15 seconds following the bolus injection of contrast shows outlines of the nonenhanced pheochromocytoma arising from the epicardial surface of the left atrium *(arrow).* ***(B)*** Following the equilibration of contrast medium, the atrial chambers and tumors are enhanced to the same degree. This explains why the tumor was not appreciated during conventional chest CT scan with drip infusion of contrast.

The patient underwent a second thoracotomy and a 7-cm tumor was resected from the left atrium. The lesion could not be seen or palpated until the pericardium was opened. The patient continues in good health 4 years later. Comment: This case demonstrates the utility of combining I-131 MIBG scintigraphy with simultaneous scintigraphy of other organ systems and computer superimposition of both images. It also demonstrates the complementary roles of I-131 MIBG scintigraphy and CT scan and the utility of dynamic CT scan for these highly vascular lesions. (Shapiro B et al: The location of middle mediastinal pheochromocytomas. J Thorac Cardiovasc Surg 87:814-820, 1984)

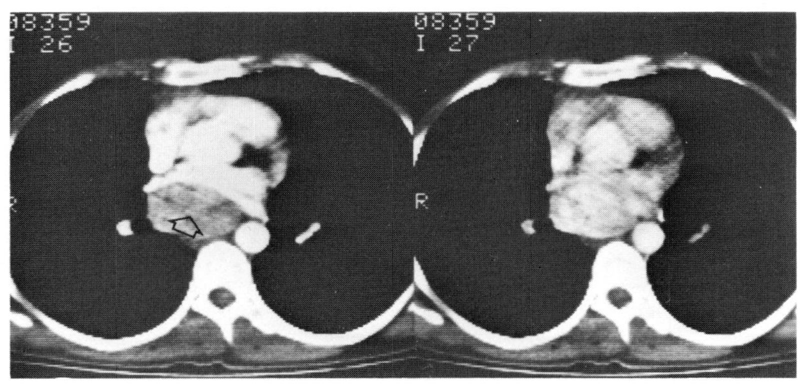

Figure 4-18 ***I-131 MIBG Scintigraphy Demonstrating a Cervical Paraganglioma (Nonfunctioning)***

A 66-year-old woman with an anterior neck mass initially thought to be of thyroid origin underwent an attempted thyroidectomy at another institution, which revealed a highly vascular nonresectable mass. A biopsy subsequently was interpreted as a neuroendocrine tumor. Plasma catecholamines, urine catecholamines, and catecholamine metabolites were normal. Blood and pentagastric-stimulated calcitonin values were normal. ***(A)*** Anterior head and neck image 24 hours following the administration of 0.5 mCi I-131 MIBG reveals normal uptake in the salivary glands and nasopharynx and an abnormal focus of activity *(arrow)* in the right neck corresponding to the palpable lesion. (Note that this differs from the faint thyroidal imaging sometimes seen despite iodide administration; see Fig. 4-6). ***(B)*** Identical study performed four weeks following complete removal of the tumor.

The lesion was successfully resected and was found to be a paraganglioma arising from the carotid sheath. The thyroid was entirely normal both microscopically and histologically. ***Comment:*** This lesion represents a nonfunctioning (in that urine and plasma catecholamine values were normal) paraganglioma and closely resembles the case reported by Smit and colleagues).[81] It also demonstrates that I-131 MIBG uptake and catecholamine secretion may not always be linked. (Von Moll et al: I-131 MIBG scintigraphy of neuroendocrine tumors other than pheochromocytoma and neuroblastoma. J Nucl Med 28:979–988, 1987)

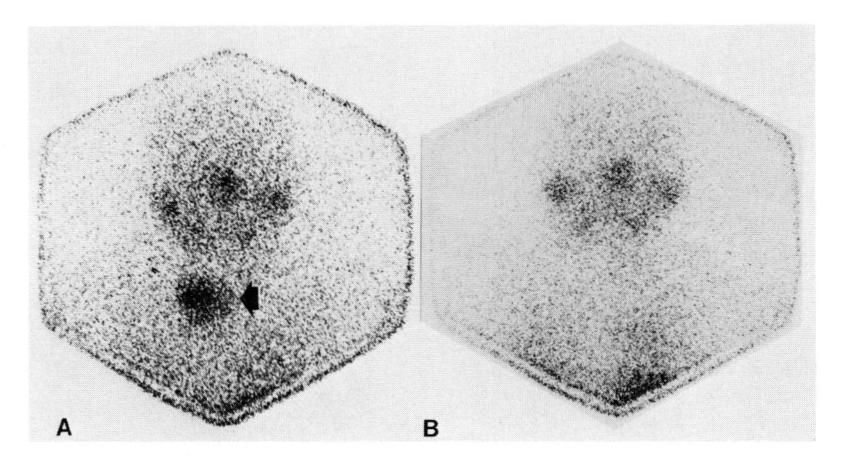

Figure 4-19 ***Malignant Pheochromocytoma — Skeletal Metastases***

A 13-year-old boy presented with back pain and an abnormal gait. He had modest hypertension (140/90 mm Hg) and tachycardia (100 beats/min). A CT scan of the abdomen and Tc-99m MDP bone scan revealed a 13-cm right abdominal (adrenal) mass with central necrosis and metastatic deposits in the T7, T9, and L1 vertebrae. I-131 MIBG scintigraphy revealed widespread metastases throughout the axial skeleton and uptake in the primary tumor. Central necrosis gives rise to the "doughnut sign" *(arrows)*. Images were obtained 24 hours following the administration of 0.5 mCi I-131 MIBG. (***A***, anterior head; ***B***, posterior pelvis; ***C***, posterior abdomen; ***D***, posterior chest; ***E***, anterior pelvis; ***F***, anterior abdomen; ***G***, anterior chest) ***Comment:*** This is an example of a case in which I-131 MIBG scintigraphy was more sensitive than all conventional imaging modalities combined in delineating the extent of disease. (Shapiro B et al: Malignant pheochromocytoma: Clinical, biochemical and scintigraphic characterization. Clin Endocrinol 20:189-203, 1984)

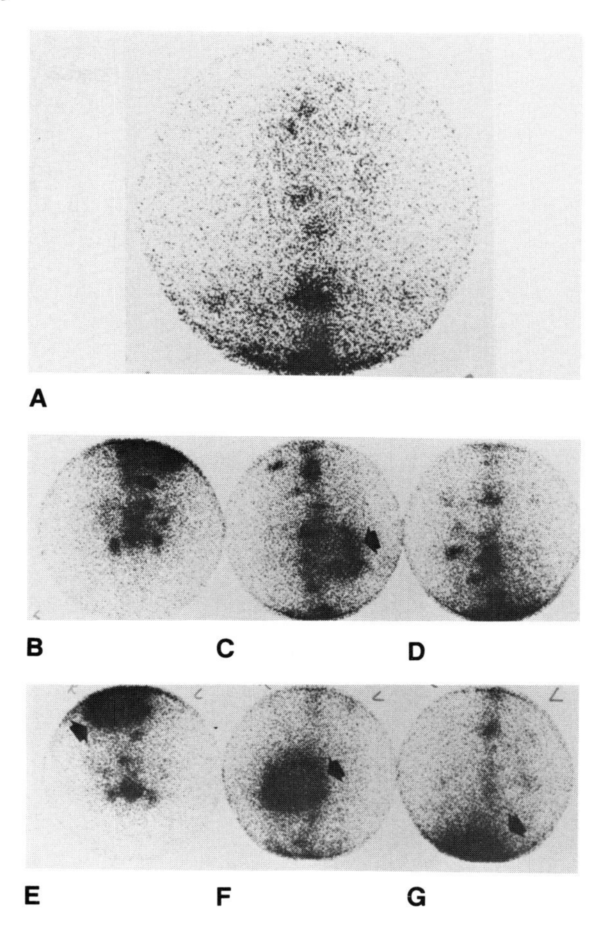

Figure 4-20 ***Malignant Pheochromocytoma — Skeletal Metastases***

A 35-year-old male who had undergone an apparently successful adrenalectomy for pheochromocytoma 6 years earlier represented with mildly abnormal biochemical findings and mild symptoms of catecholamine excess. A CT scan of the abdomen revealed a questionable abnormality at the site of the previous adrenalectomy, but this was difficult to interpret because of surgical disruption of tissue planes and the presence of metallic clips. The patient was scheduled for surgical exploration as a case of suspected local pheochromocytoma recurrence. I-131 MIBG scintigraphy images performed 24 hours following the administration of 0.5 mCi ***(A–E)*** are compared with conventional Tc-99m MDP bone scans ***(F–J)***. Sites of metastatic deposits are indicated by solid arrows and surface markers by *m*. ***(A)*** and ***(F)***. Anterior head: normal I-131 MIBG uptake is present in the nasopharynx *(n)*, and salivary glands *(sl)*. A metastasis is present in the region of the left brow ridge. ***(B)*** and ***(G)***. Posterior thorax: abnormal I-131 MIBG uptake is shown in midthoracic spine *(arrow)*. It is not revealed by Tc-99m MDP bone scan. ***(B) (C)*** and ***(H)*** Posterior abdomen: the vertebral lesion is demonstrated by I-131 MIBG but not by Tc-99m MDP; normal I-131 MIBG uptake is present in liver *(L)* and spleen *(S)*. ***(D)*** and ***(I)*** Anterior pelvis: a lesion in the superior pubic ramus is clearly demonstrated by I-131 MIBG and faintly shown by Tc-99m MDP; normal I-131 MIBG activity is seen in the bladder *(B)*. ***(E)*** and ***(J)***. Anterior leg region with tibial metastasis well-demonstrated by both radiopharmaceuticals. The uptake on Tc-99m MDP bone scan had initially been ascribed to trauma.

Comment: In this case the CT scan proved misleading and almost led to unnecessary and noncurative surgery. The bone scan was initially interpreted as not showing any metastases, but reinterpretation in the light of the I-131 MIBG scan revealed subtle abnormalities at some sites; other lesions were only demonstrated by the I-131 MIBG scan. (Shapiro B et al: Malignant pheochromocytoma: Clinical, biochemical and scintigraphic characterization. Clin Endocrinol 20:189–203, 1984)

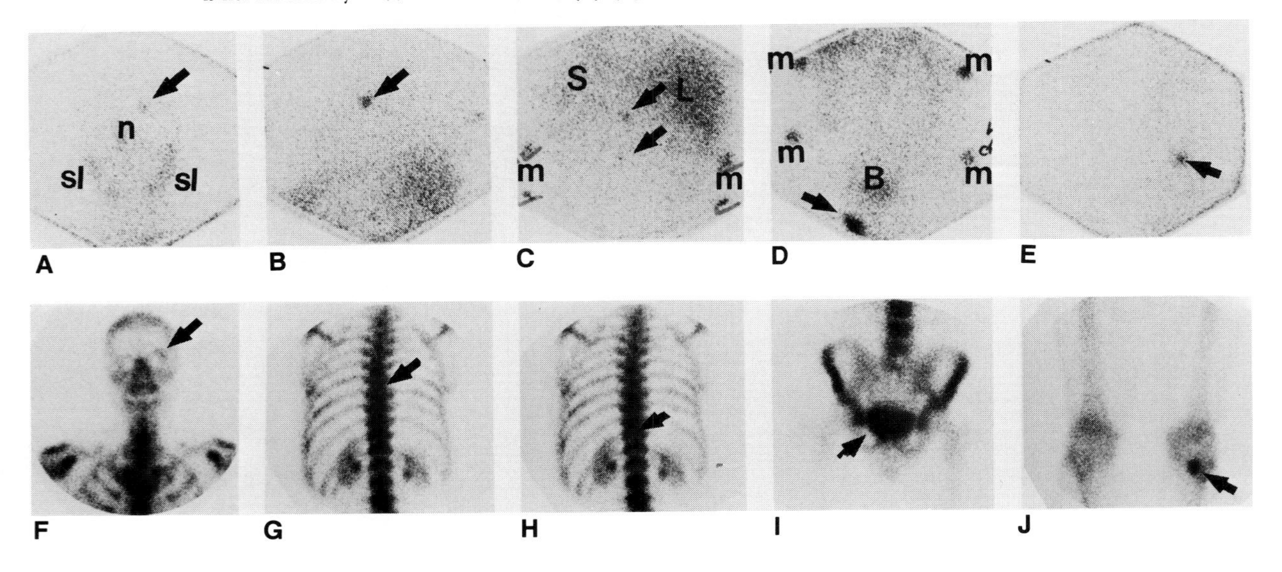

Figure 4-21 ***Malignant Pheochromocytoma — Hepatic Metastases***

The patient was a 69-year-old woman with malignant pheochromocytoma metastatic to liver with a solitary vertebral metastasis. All I-131 MIBG images were obtained 24 hours following the administration of a 0.5 mCi injection. ***(A)*** and ***(B)***. Anterior and posterior abdominal I-131 MIBG scintigrams. Note normal hepatic uptake with focal areas of supernormal concentration at the sites of metastases *(small arrows)*. There is a single midthoracic abnormal focus *(large arrow)*. ***(C)*** Right lateral Tc-99m sulfur colloid liver scan shows a defect due to metastases *(arrow)*. Other views were normal. ***(D)*** Right lateral I-131 MIBG scan shows focal increased uptake *(arrow)* corresponding to the defect seen on the liver scan. ***(E)*** Posterior chest image shows I-131 MIBG hepatic metastases *(small arrows)* and vertebral lesion *(large arrow)*. ***(F)*** Posterior chest image with Tc-99m MDP shows no obvious skeletal abnormality, once again demonstrating that I-131 MIBG scintigraphy may be more sensitive than any other modality. Comment: The liver is normally the organ in which most I-131 MIBG concentration occurs. This may make the detection of liver metastases difficult. Fortunately, in most cases the retention of I-131 MIBG is greater in the tumor than in normal liver. Thus as liver background diminishes over time, the metastases become more apparent. In this setting images at 48 or 72 hours or even later may be most valuable. (Shapiro B et al: Malignant pheochromocytoma: Clinical, biochemical and scintigraphic characterization. Clin Endocrinol 20:189 - 203, 1984)

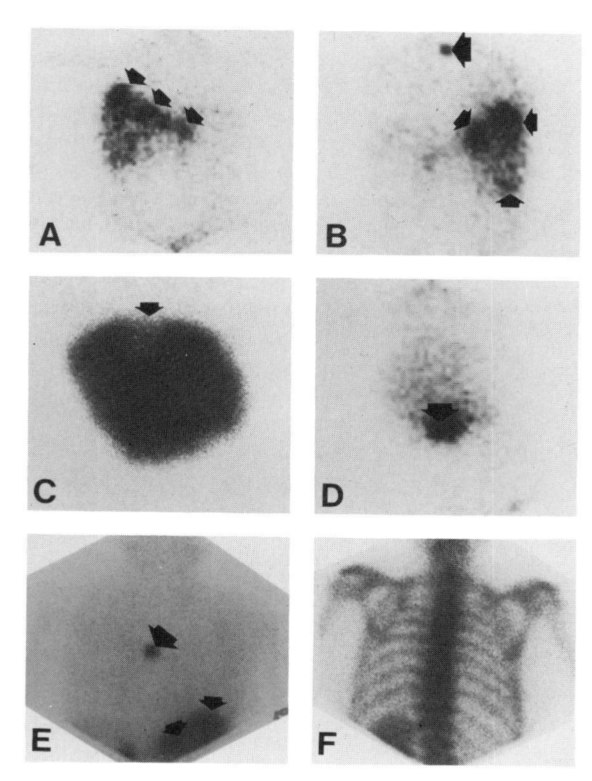

Figure 4-22 ***Malignant Pheochromocytoma — Pulmonary Metastases***

A 58-year-old male presented with a large right-sided, upper abdominal mass and pulmonary nodules and hypertension. The histology of a small biopsy obtained by a percutaneous needle was difficult to interpret; the initial diagnosis was a urothelial tumor. The possibility of pheochromocytoma was raised and an I-131 MIBG scan, special immunohistochemical stains, and plasma catecholamine assay confirmed this diagnosis. I-131 MIBG images were made 24 hours following administration of a 0.5 mCi injection. ***(A)*** Posterior abdominal view demonstrates a large right adrenal primary tumor *(P)* and bilateral pulmonary uptake *(L)*. The location of the kidneys is demonstrated from the computer-superimposed outlines taken from a simultaneous Tc-99m DTPA scan. ***(B)*** and ***(C)*** Posterior thoracic I-131 MIBG scan and the corresponding chest x-ray demonstrate extensive bilateral pulmonary I-131 MIBG uptake *(L)* and uptake in the primary tumor *(P)*. The region of the heart *(H)* shows relatively little uptake. The open arrow indicates the region of the suprasternal notch. ***(D)*** Anterior chest and right shoulder image showing abnormal I-131 MIBG uptake in the humeral head *(solid arrow)*, lung *(L)*, and right adrenal *(P)*. An open arrow indicates a radioactive marker on the costal margin. ***(E)*** Anterior left thigh image with abnormal I-131 MIBG uptake in the site of a pathological fracture *(solid arrows)*. An open arrow indicates radioactive marker on the greater trochanter (Shapiro B et al: Malignant pheochromocytoma: Clinical, biological, and scintigraphic characterization. Clin Endocrinol 20:189 - 203, 1984)

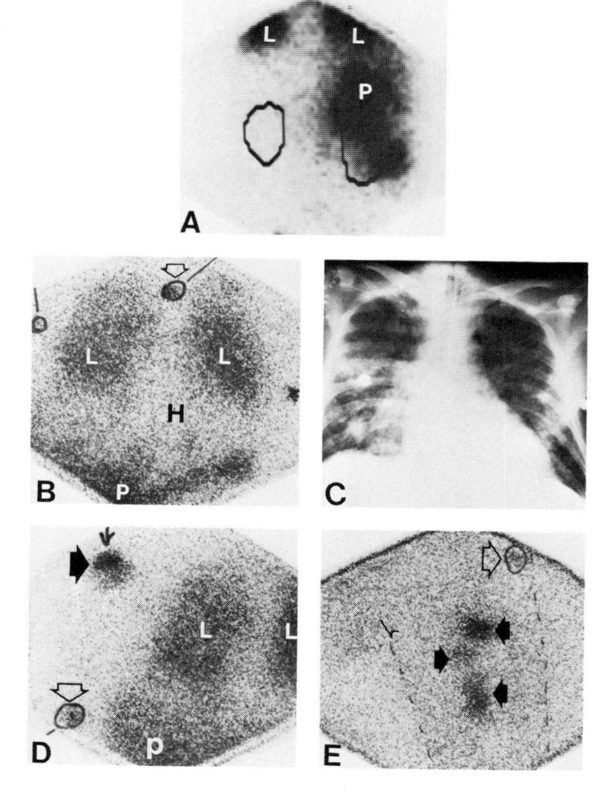

Figure 4-23 *Malignant Pheochromocytoma — Retroperitoneal Lymph Node Metastases*

The patient was a 35-year-old man with malignant right adrenal primary pheochromocytoma and lymph node metastases to adjacent right retroperitoneal nodes. Three attempts had failed to surgically extirpate the lesion, which had recurred on each occasion. ***(A)*** and ***(C)*** Anterior and posterior Tc-99m sulfur colloid liver spleen scans reveal no abnormality. ***(B)*** and ***(D)*** Anterior and posterior images acquired 24 hours following administration of 0.5 mCi I-131 MIBG reveals a large abnormal focus of tracer uptake in the retroperitoneal metastases *(large arrow)*. In addition, a small liver metastasis *(small arrow)* not demonstrable in *A* or *C* was revealed. The open arrows are radioactive markers on the costal margins and axillae. ***(E)*** and ***(F)*** CT scans demonstrate the retrohepatic tumor mass *(large arrow)* and peripheral liver metastasis *(small arrow)*.

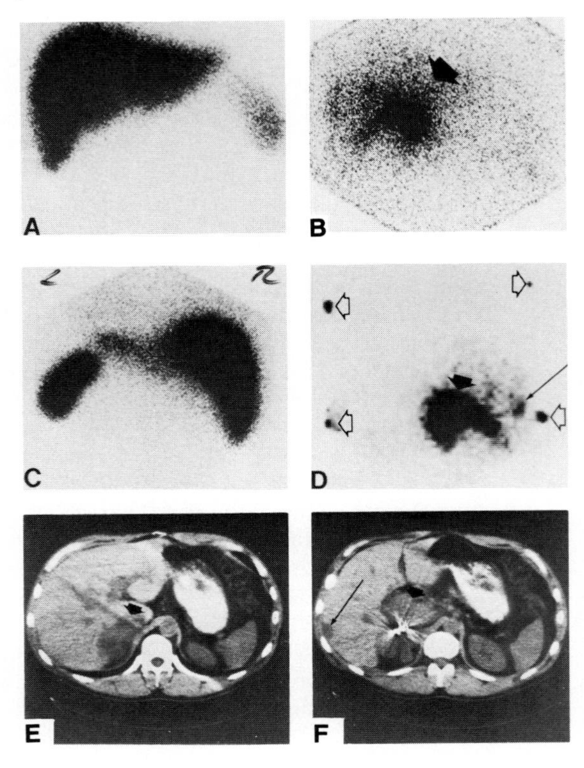

Figure 4-24 *The Spectrum of I-131 MIBG Uptake by the Adrenal Medullae of Patients with the MEN-2 Syndrome*

Many patients with the MEN-2 syndrome have a high propensity to develop pheochromocytoma, usually adrenal and bilateral although the lesions may be asymetrical or asynchronous in their development. The development of rank pheochromocytomas is preceded by adrenomedullary hyperplasia. At this time symptoms may be slight and biochemical abnormalities minimal or intermittent.

Top: A 32-year-old man with MEN-2a had had a previous thyroidectomy for medullary carcinoma of the thyroid but had no hypertension- or pheochromocytoma-related symptoms. Mild and only intermittent elevations of plasma and urinary catecholamines and metabolites were present. A CT scan ***(a)*** shows a normal left and questionably enlarged right adrenal *(arrows)*. Posterior ***(b)*** and left lateral ***(c)*** I-131 MIBG images acquired 48 hours following the administration of 0.5 mCi reveals faint bilateral tracer uptake in the regions of both adrenal glands.

Center: A 14-year-old girl with MEN-2b had had a previous thyroidectomy for medullary carcinoma of the thyroid and had a 10-month history of weakness, headache, and tremor, but was never demonstrated to be hypertensive. Definite but intermittent abnormalities of plasma and urinary catecholamines and metabolites were documented. Posterior abdominal images ***(a)*** obtained 48 hours following the administration of 0.5 mCi/1.7 m^2 I-131 MIBG revealed bilateral symmetrical abnormally increased tracer uptake *(arrows)*. A CT scan ***(b)*** revealed a morphologically normal right adrenal *(arrow)* and left adrenal (not seen at the level of this section). Three months after the CT scan the patient underwent bilateral adrenalectomy. Each gland weighed 6 g and was macroscopically hyperplastic. Histology revealed hyperplasia. Postoperatively all biochemical studies were repeatedly shown to be normal.

Bottom: A 32-year-old man with MEN-2a had had a previous thyroidectomy for medullary carcinoma of the thyroid and had had a number of thyroid carcinoma metastases resected from the neck. He had persistently elevated plasma calcitonin values, a history of mild hypertension, and persistently elevated plasma and urinary catecholamines and metabolites. A CT scan ***(a)*** shows an enlarged right adrenal *(arrow)*. The left adrenal, not seen at this level, was also enlarged. Posterior abdominal image ***(b)*** 48 hours following the administration of 0.5 mCi I-131 MIBG shows markedly increased symmetrical uptake into the tumorous glands. Posterior abdominal image ***(c)*** as in ***b*** is superimposed with a Tc-99m DTPA renal scan to show the relationships of the abnormal foci of I-131 MIBG uptake to the kidneys. The left adrenal gland is displaced medial to the kidney by the patient's scoliosis *(large arrow)*, whereas the right adrenal *(small arrow)* is in the normal location. ***Comment:*** The intensity of adrenal medullary I-131 MIBG uptake appears to mirror the evolution of medullary hyperplasia to pheochromocytoma in MEN syndrome. The earlier stages of hyperplasia may overlap with the extremes of normal in some instances (Fig. 4-4). (Valk, TW et al: Spectrum pf pheochromocytoma in multiple endocrine neoplasia: A scintigraphic portrayal using I-131-metaiodobenzylguanidine. Ann Intern Med 94:762–767, 1981)

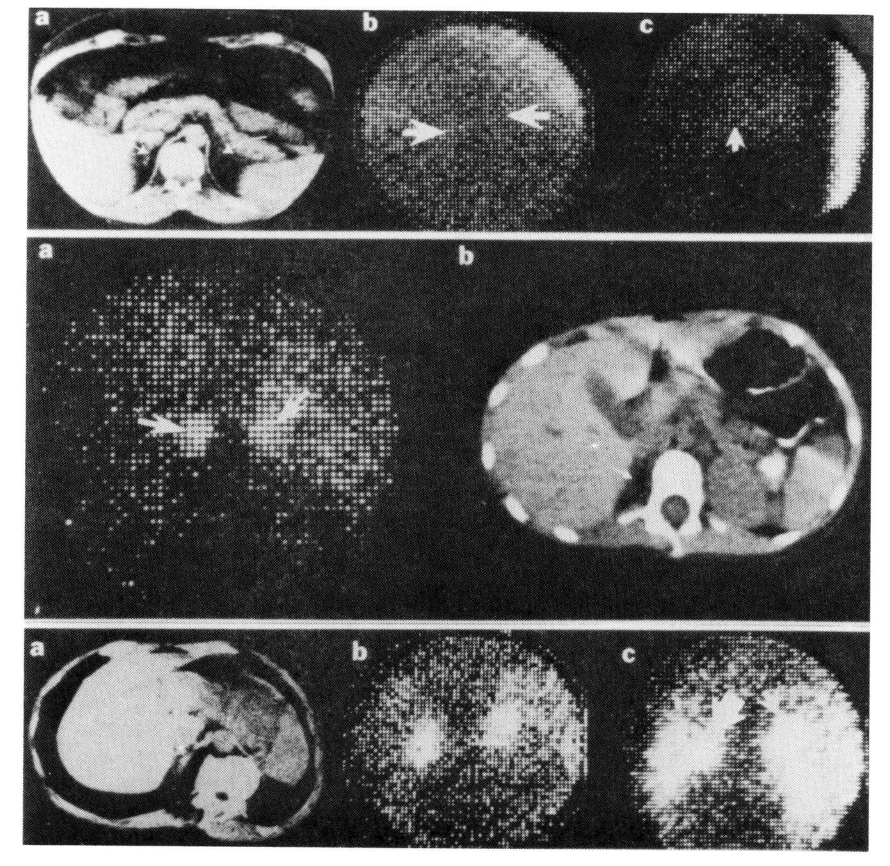

Figure 4-25 ***Malignant Pheochromocytoma with Hepatic Metastases in a Patient with the MEN-2a Syndrome***

A 26-year-old woman with MEN-2a presented with a malignant pheochromocytoma. Both adrenals contained pheohromocytomas and were resected, but spread to the liver had already occurred. *(A)* Posterior abdominal image obtained 72 hours following the administration of 1.0 mCi I-131 MIBG shows abnormal foci of tracer uptake *(arrows)*. *(B)* Posterior abdominal image obtained following Tc-99m sulfur colloid administration shows areas of reduced colloid accumulation *(arrows)* which correspond to the foci of abnormal I-131 MIBG uptake. (Sisson JC et al: Scintigraphy with I-131 MIBG as an aid to the treatment of pheochromocytomas in patients with the MEN-2 syndromes. Henry Ford Hosp Med J 32:254–261, 1984; copyright 1984 by Henry Ford Hospital)

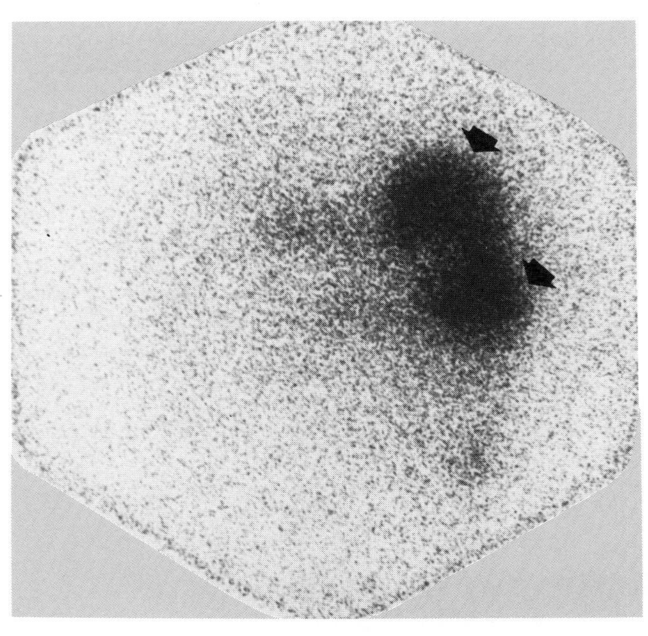

A

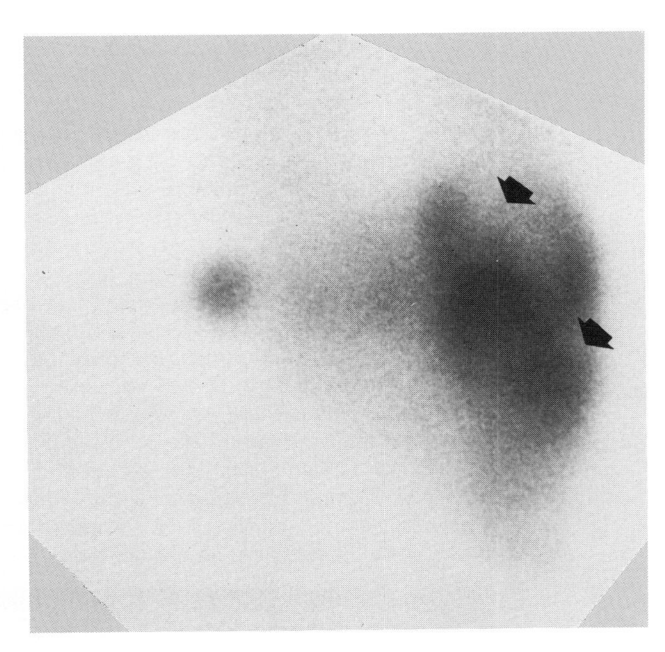

B

Figure 4-26 ***Malignant Pheochromocytoma with Hepatic Metastases— Comparison of I-131 MIBG with I-123 MIBG and SPECT Imaging with I-123 MIBG Studies in the Same Patient as in Figure 4-25***

(A) anterior abdominal image 24 hours following the administration of 1.0 mCi I-131 MIBG. *(B)* anterior abdominal image 24 hours following the administration of 10.0 mCi I-123 MIBG. Note the clear depiction of the multiple liver metastases achieved with the I-123 MIBG. The high photon flux and suitable photon energy permit the performance of rotating gamma camera SPECT. *(C)* Multiple transaxial cuts were obtained through the liver using SPECT 24 hours following the administration of 10.0 mCi I-123 MIBG. Note the central necrosis with reduced tracer uptake in a large lesion in the right lobe *(arrow)*. ***Comment:*** Although unusual, malignant metastatic pheochromocytoma does occur in the MEN-2 syndromes, ten cases having been reported to date.

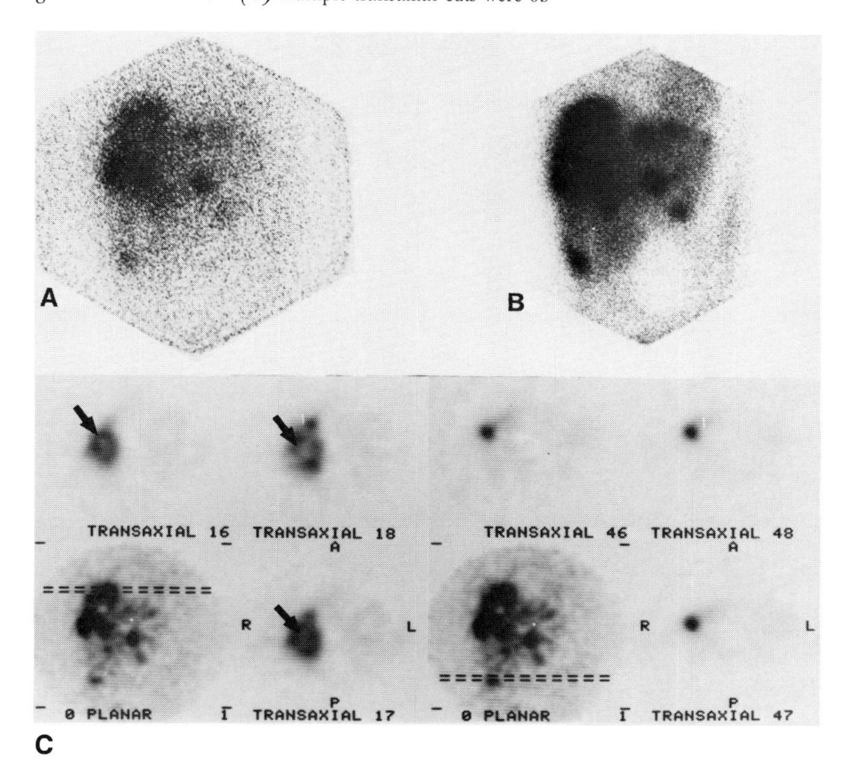

Figure 4-27 ***I-131 MIBG Scintigraphy in Neuroblastoma with Uptake in Primary and Metastatic Lesions***

An 18-month-old female with Stage IV neuroblastoma with midabdominal primary, left orbital, and bone marrow metastases. ***(A)*** Tc-99m MDP bone scan reveals abnormal uptake in the region of the left orbital metastasis *(closed arrow)* and primary tumor in the abdomen *(open arrow).* ***(B)*** I-131 MIBG images 48 hours following 0.5 mCi/1.7 m^2 injection show abnormal uptake in the midabdominal primary tumor *(open arrow),* orbital metastasis *(large closed arrow),* and at multiple sites in the skull, spine, hips, and femurs *(small arrows).* (Geatti, O et al: I-131-metaiodobenzylguanidine (I-131-MIBG) scintigraphy for the localization of neuroblastoma: Preliminary experience in 10 cases. J Nucl Med 26:736-742, 1985)

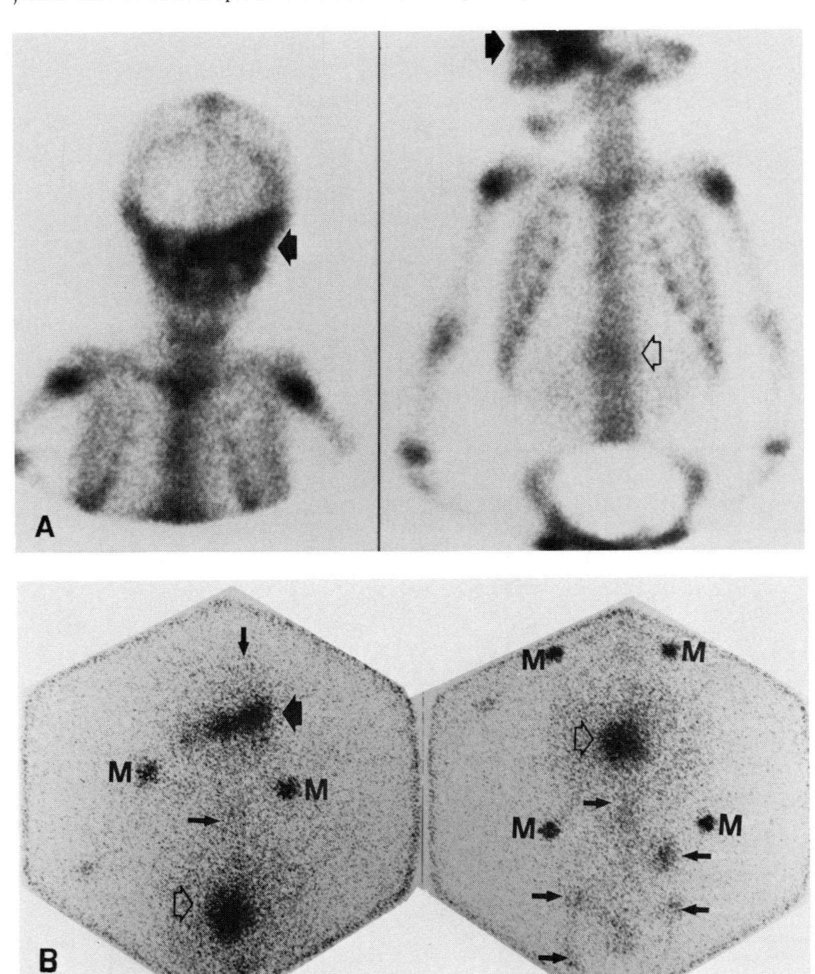

Figure 4-28 ***I-131 MIBG Scintigraphy in Stage IV Neuroblastoma***

Images were taken of a nine-year-old male with Stage IV neuroblastoma: ***(A)*** anterior whole body Tc-99m MDP bone scan revealing multiple skeletal metastases and ***(B)*** anterior I-131 MIBG images obtained 24 hours following administration of 0.5 mCi/1.7 m^2 injection. Note that the skeletal and bone marrow deposits of the neuroblastoma are more clearly delineated than by conventional bone scan. ***Comment:*** These are two examples of the usefulness of I-131 MIBG in neuroblastoma. In other cases a positive scan may occasionally be the only evidence for residual or recurrent disease. (Geatti, O et al: I-131-metaiodobenzylguanidine (I-131-MIBG) scintigraphy for the localization of neuroblastoma: Preliminary experience in 10 cases. J Nucl Med 26:736-742, 1985)

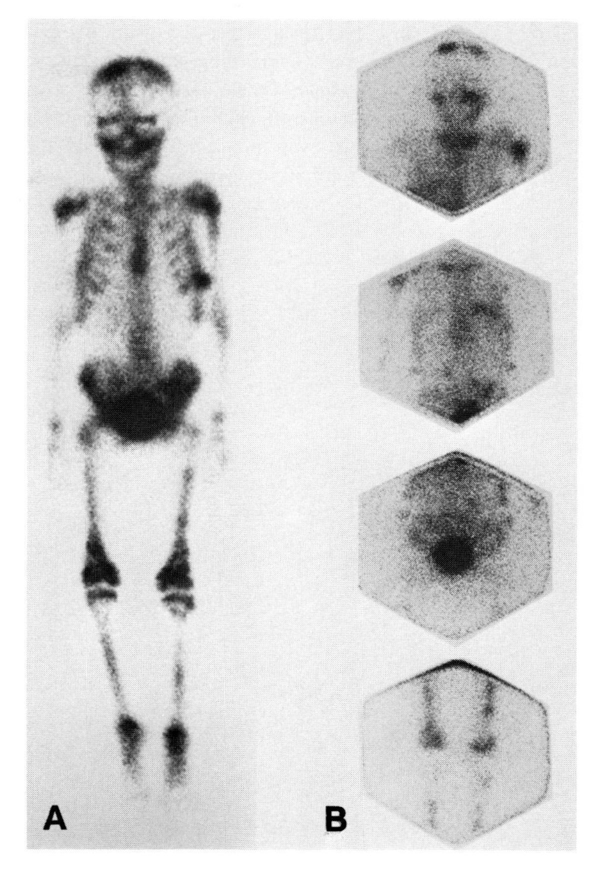

Figure 4-29 *I-131 MIBG Scintigraphy of Carcinoid Tumor Metastatic to Mesenteric Lymph Nodes*

The patient was a 68-year-old man with a history of primary small bowel carcinoid which was resected 2 years previously. Urinary 5HIAA excretion was normal, but a CT scan showed a mass lesion in the root of the mesentery which was believed to be a lymph node metastasis. The image obtained 72 hours following the administration of 0.5 mCi I-131 MIBG reveals an abnormal focus of uptake *(large arrow)* at a site corresponding to the CT abnormality. Normal uptake is seen in the liver *(L)* and bladder *(B)*. Radioactive markers on the costal margin and pelvic brim are indicated by open arrows.
Comment: I-131 MIBG uptake occurs in a majority of sympathoadrenal tumors such as pheochromocytomas, paragangliomas, and neuroblastomas. In addition, uptake has been observed in some carcinoids, medullary carcinoma of the thyroid, chemodectomas, Merkel cell tumors, oat cell bronchial carcinomas, and other neuroendocrine lesions of the APUD series. The frequency of significant uptake appears to be less than for sympathoadrenal lesions, and the clinical utility of I-131 MIBG scintigraphy in these lesions is as yet unclear. Von Moll, L et al (Von Moll et al: I-131 MIBG scintigraphy of neuroendocrine tumors other than pheochromocytoma and neuroblastoma. J Nucl Med 28:979 - 988, 1987)

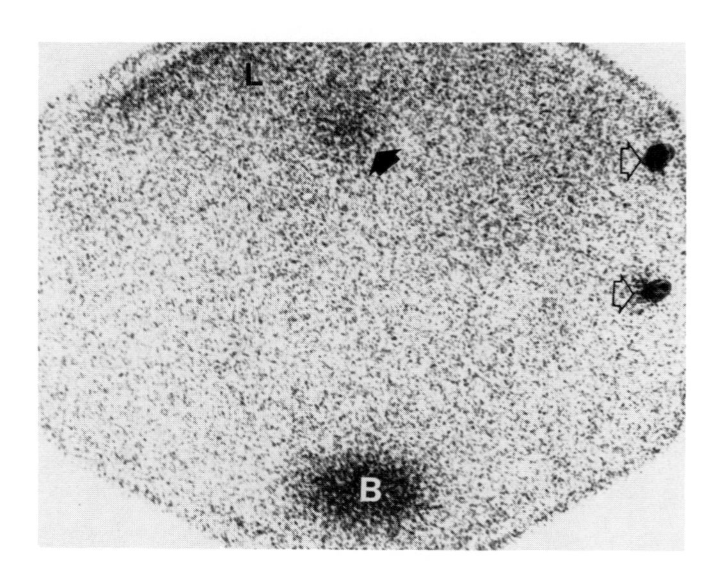

Figure 4-30 ***Effect of Cervical Sympathectomy on Salivary I-131 MIBG Uptake***

A 72-year-old female presented with a right-sided, complete Horner's syndrome following previous thyroidectomy for goiter at age 40. The patient was referred for possible pheochromocytoma which was excluded. ***(A)*** Anterior head view obtained 24 hours following the administration of 0.5 mCi I-131 MIBG shows absence of uptake in the right parotid and submandibular glands. ***(B)*** Anterior head view obtained 20 minutes following the administration of 5 mCi Tc-99m TcO_4 shows normal symmetrical uptake in the salivary glands and nasopharynx. ***Comment:*** This case demonstrates that the uptake of I-131 MIBG by the salivary glands is dependent on the intact sympathetic autonomic innervation of the organs. The glandular secretory component was normal, as demonstrated by the normal symmetrical pertechnetate accumulation. (Nakajo, M et al: Salivary gland accumulation of meta[131-I] iodobenzylguanidine.

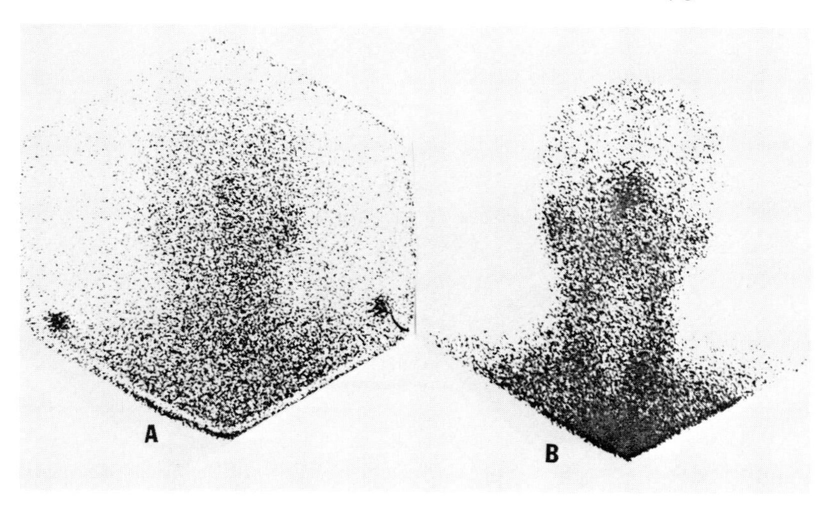

Figure 4-31 ***Influence of Tricyclic Antidepressants on I-131 MIBG Salivary Gland Uptake***

Both studies are posterior head-chest views performed 48 hours following the administration of 0.5 mCi I-131 MIBG. ***(A)*** Study performed 3 weeks following the withdrawal of imipramine HCl administration. ***(B)*** Study performed during the administration of imipramine HCl 75 mg/day. (Note absence of I-131 MIBG uptake by the salivary gland.) Parotid salivary glands are indicated by arrows. *m*, radioactive surface markers; *L*, liver uptake. ***Comment:*** Because tricyclic antidepressants block the uptake of catecholamines and MIBG by sympathetic nerve endings, such drugs must be withdrawn for at least 3 weeks prior to I-131 MIBG scintigraphy. Other potentially interfering drugs include cocaine, reserpine alkaloids, phenylpropanolamine, and labetalol. (Nakajo, M et al: Salivary glands accumulation of meta[131-I] iodobenzylguanidine. J Nucl Med 25:2-6, 1984)

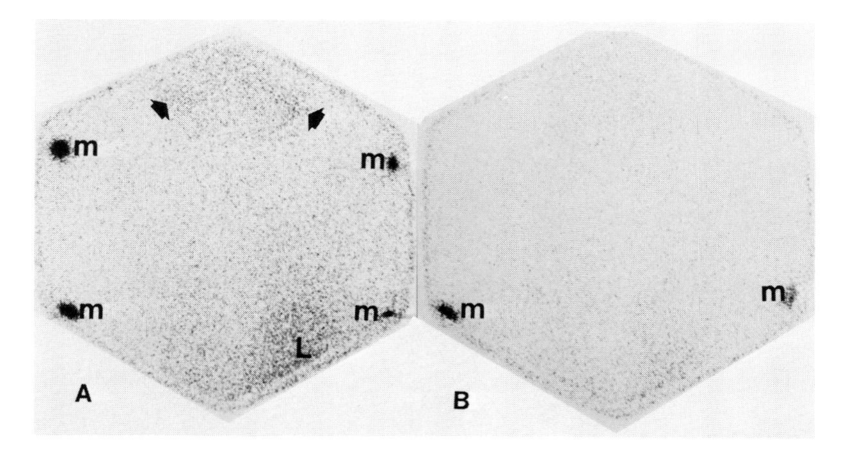

Figure 4-32 ***Cardiac Uptake of I-131 MIBG— Grading of Intensity***

Posterior chest-abdomen views obtained 24 hours following administration of 0.5 mCi I-131 MIBG are graded for intensity of cardiac uptake: ***(A)*** grade 0 (no visible uptake); ***(B)*** grade 1 (visible uptake less than left lobe of liver); ***(C)*** grade 2 (visible uptake equal to left lobe of liver); and ***(D)*** grade 3 (visible uptake equal to right lobe of liver). The region of the heart is indicated by an arrow. (*L*, normal liver uptake; *S*, normal splenic uptake, *m*, surface radioactive markers) (Nakajo, M et al: Inverse relationship between cardiac accumulation of 131-I-MIBG and circulating catecholamines in suspected pheochromocytoma. J Nucl Med 24, 1127-1134, 1984)

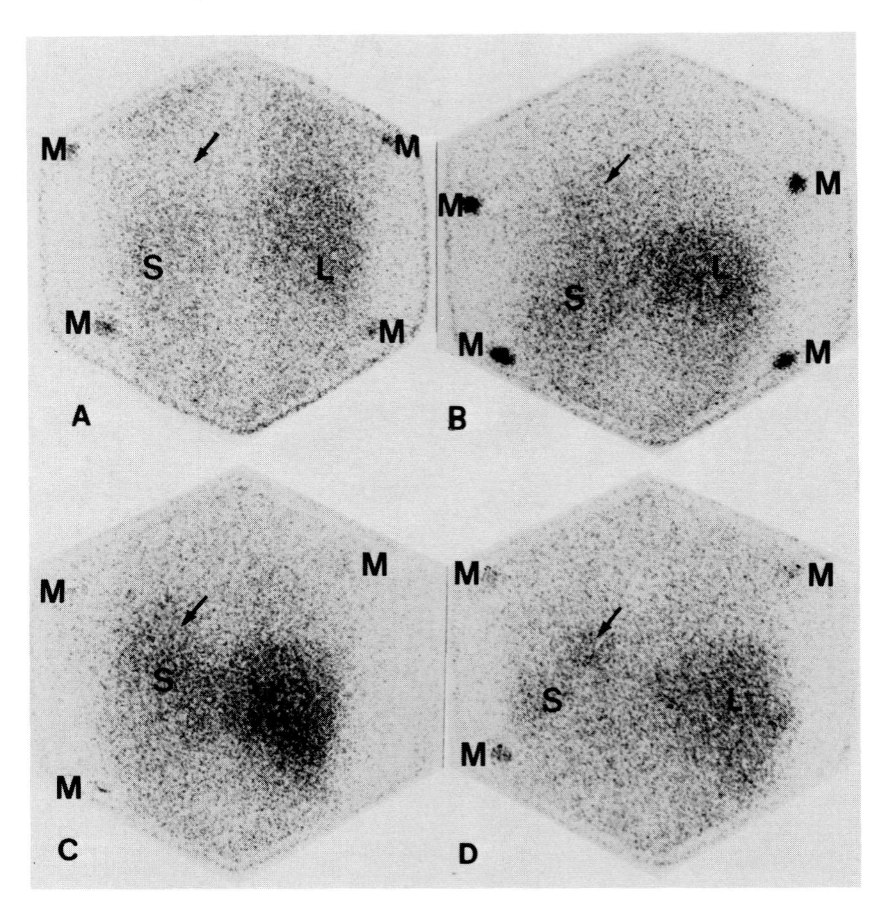

Figure 4-33 ***Cardiac Uptake of I-131 MIBG: The Effect of Hypercatecholaminemia***

Alteration in the intensity of cardiac I-131 MIBG uptake in a patient with neurofibromatosis and bilateral pheochromocytomas is demonstrated in these studies. ***(A)*** Preoperative study 24 hours after the administration of 0.5 mCi I-131 MIBG. Posterior chest - abdomen view shows no cardiac uptake (grade 0) visible *(closed arrow)*. Open arrows indicate the adrenal pheochromocytomas. ***(B)*** Postoperative posterior abdomen view 24 hours after the administration of 0.5 mCi I-131 MIBG. Obvious cardiac uptake (grade 2) is now evident *(closed arrow) L,* normal liver uptake; *S,* splenic uptake; *m,* surface radioactive markers. ***Comment:*** Cardiac (and splenic) uptake of I-131 MIBG are reduced at the time of marked hypercatecholaminemia but are restored 6 weeks after surgical removal of the pheochromocytomas had returned plasma catecholamines to normal. The intensity of cardiac MIBG uptake is inversely proportional to the circulating catecholamines. Clear depiction of the heart very strongly predicts that measurements of plasma and urinary catecholamines will be normal. (Nakajo, M et al: Inverse relationship between cardiac accumulation of 131-I-MIBG and circulating catecholamines in suspected pheochromocytoma. J Nucl Med 24:1127 - 1134, 1984)

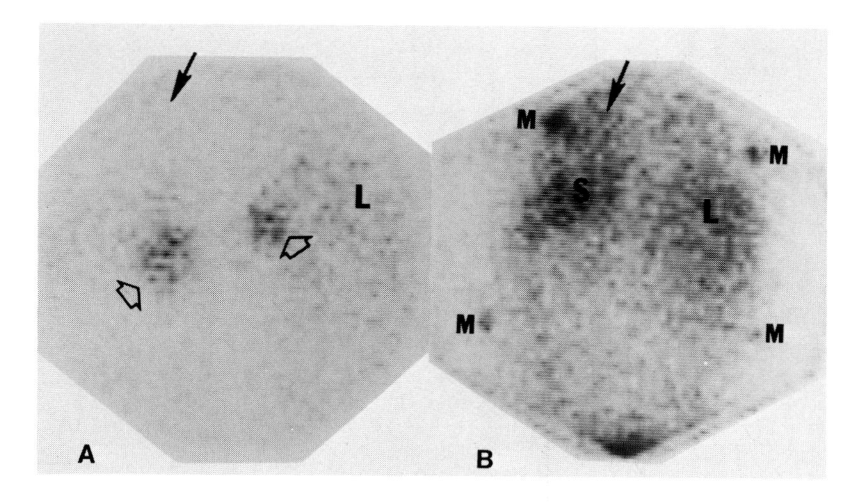

Figure 4-34 *A suggested algorithm for the use of radioiodinated MIBG in the evaluation of patients suspected of having pheochromocytoma.*

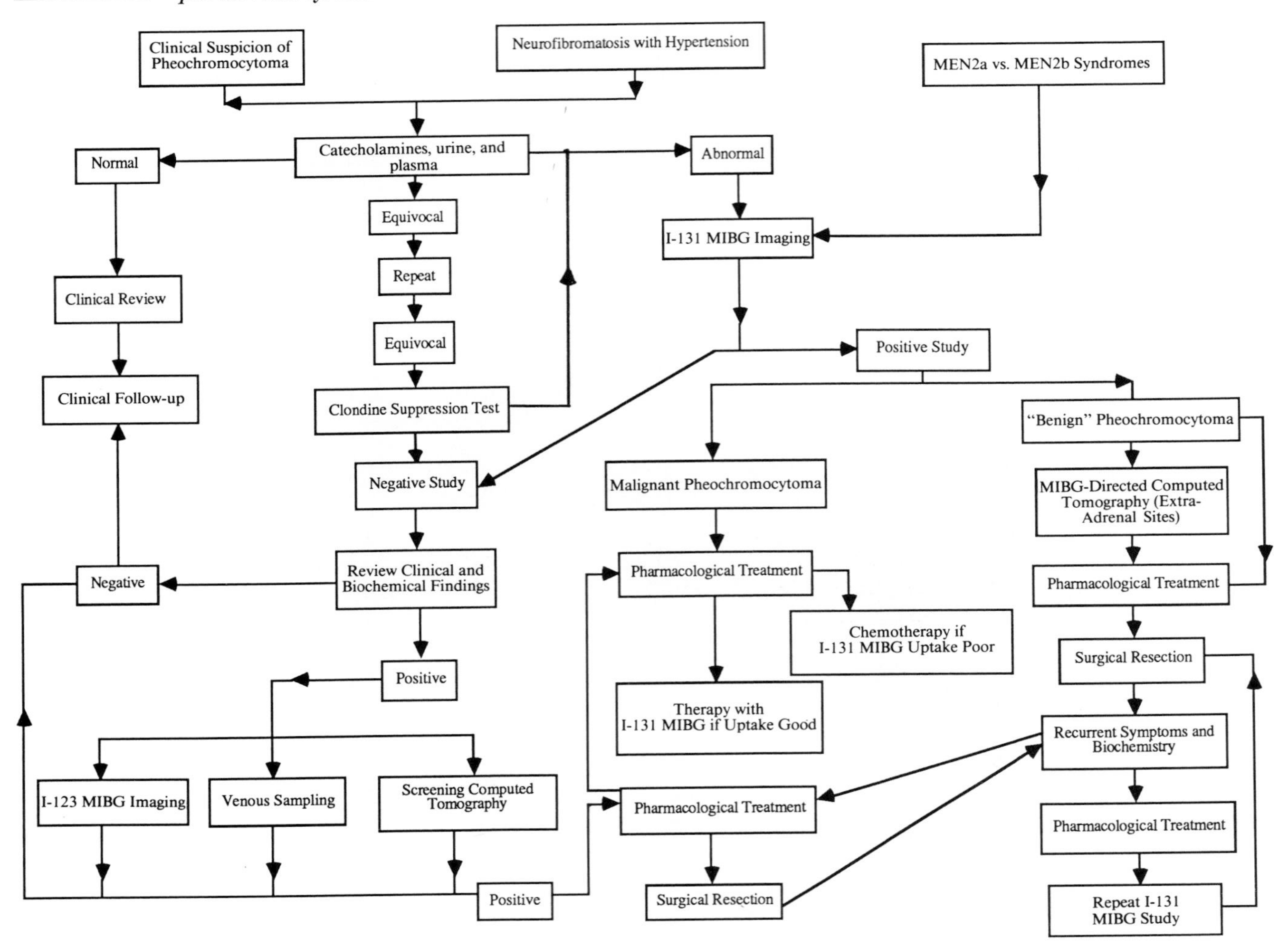

Robert M. Basarab

Chapter 5 PARATHYROID SCINTIGRAPHY

Ninety-five percent of primary hyperparathyroidism is due to excessive secretion of parathyroid hormone (PTH) by a single autonomous adenoma or primary chief cell hyperplasia. In general, approximately 80% of cases are due to an adenoma and 15% to chief cell hyperplasia. The pathologic distinction between these two entities is debated because it cannot be made on the basis of histology of a single gland. The high frequency of microscopic hyperplasia has also blurred the distinction between adenoma and hyperplasia. The diagnosis of an adenoma is usually reserved for those cases in which there is a single enlarged gland with normal or suppressed remaining glands. Parathyroid carcinoma (4%) and water clear cell hyperplasia (1%) account for most of the remaining causes.[1] Rarely, tumors of nonparathyroid origin secrete PTH or a PTH-like substance mimicking the clinical and biochemical manifestations of primary hyperparathyroidism.[2]

Primary hyperparathyroidism is usually an isolated endocrinopathy; however, it can also occur in conjunction with multiple endocrine neoplasia type I (MEN-1) syndrome in association with the Zollinger-Ellison syndrome and pituitary chromophobe adenomas. Familial hyperparathyroidism, an uncommon autosomal dominant hereditary version, also exists.[3]

The most frequent biochemical manifestation of primary hyperparathyroidism is hypercalcemia. In the past, nephrolithiasis or renal colic caused by renal calculi was the most frequent clinical manifestation. With the advent of routine screening of serum calcium levels and sensitive PTH radioimmunoassays, the diagnosis of primary hyperparathyroidism is now frequently made in asymptomatic patients.[4] An estimated 2.5% of the cases of asymptomatic hypercalcemia are due to primary hyperparathyroidism with a prevalence of approximately 0.1% in the general population.[5]

The parathyroid glands originate from the paired third and fourth pharyngeal pouches, designated parathyroid glands III and IV, respectively. Parathyroid glands III migrate caudally in association with the thymus, becoming the inferior parathyroid glands. The parathyroid glands IV migrate a lesser distance, becoming the superior parathyroid glands. There is a great deal of normal variation in the location of parathyroid glands. In his autopsy series, Wang[6] noted that 39% of the inferior parathyroid glands (III) were located in the superior pole of the thymus, 2% in the mediastinum, and 2% in other ectopic sites. There was less variation in the position of the superior parathyroid glands (IV), with 99% located either behind the upper poles of the thyroid lobes or adjacent to the cricoid cartilage. More than four glands were identified in 2.5% of patients. Diseased parathyroid glands are also frequently aberrantly located. In the surgical series of Goldman,[7] 13% of the parathyroid adenomas were found outside the thyroid bed, including 6% in the posterior mediastinum, 4% in the anterior mediastinum, 1.6% along the great vessels, and 1% adjacent to but detached from the thyroid; 1% were retroesophageal and 0.3% retropharyngeal.

The treatment of primary hyperparathyroidism is usually surgical.[8] Some authors believe that subtotal parathyroidectomy with removal of three or three and one-half glands, including any obviously enlarged glands, is necessary to minimize the rate of recurrence.[9,10] However, most now believe that removal of the adenoma with biopsy or inspection of the remaining parathyroid glands is sufficient if the remaining glands are not hyperplastic.[11-13] Harrison and colleagues[14] reported no recurrences in 16 patients undergoing conservative surgery despite microscopic evidence of hyperplasia in the remaining normal-sized glands. The more aggressive approach is now usually reserved for patients with MEN-1 syndrome or familial or secondary hyperparathyroidism. In experienced hands, both approaches lead to a 90% to 95% cure rate without any preoperative localizing procedure. However, even the best technique results in a 5% to 10% recurrence rate, usually related to aberrant or ectopically located glands or recurrent hyperplasia.[15,16] In Wang's surgical reexploration series,[16] almost 72% of the glands not identified at initial exploration were found outside the thyroid bed with 33% in the superior posterior mediastinum, 20% in the anterior mediastinum, 13% in the superior pole of the thymus, 5% in the retropharyngeal area, and 1% at the angle of the jaw. Reexploration is technically more difficult and has a higher morbidity than the initial operation.[17]

It would be great assistance to the surgeon to have an accurate, noninvasive test for the preoperative localization of diseased parathyroid glands. The goal of such a test should be to reduce the incidence of reexploration or simplify the surgical procedure, thus reducing the operative time. In addition, in order to be useful, a preoperative diagnostic test should not only accurately localize juxtathyroidal adenomas but also localize ectopic adenomas. A radiopharmaceutical which selectively localizes in parathyroid tissue would be ideally suited for this task. Currently, no single radiopharmaceutical is adequate.

The first radiopharmaceutical to gain reasonable acceptance for the imaging of diseased parathyroids was selenium-75 (Se-75) selenomethionine. While initial results were mixed,[18,19] later investigators reported better results using dual isotope techniques detecting adenomas as small as 350 mg.[20,21] In view of its poor imaging characteristics, poor sensitivity and specificity, and relatively high radiation dose, the use of Se-75 selenomethionine for parathyroid disorders has not achieved widespread acceptance. With more accurate alternative imaging techniques available, there is little utility for selenomethionine in parathyroid disorders.[22]

Since the initial reports by Ferlin and associates,[23] encouraging results have been reported by several investigators using thallium–technetium or thallium–iodine dual-isotope subtraction. Reported results indicate a sensitivity of 88% to 95% in the detection of parathyroid adenomas and a lesser sensitivity of 56% to 80% in the detection of secondary hyperplasia. Adenomas as small as 60 mg have been reported as detectable by these techniques, although the reliable lower limits appear to be 300 mg to 400 mg. Neither normal parathyroid glands nor normal-sized hyperplastic glands are reliably visualized.

The technique described in this chapter is a variation of that described by Young and co-workers.[24] When used with a converging collimator this technique can be optimized for the detection of ectopic adenomas. A normalized subtraction procedure maintains reasonably high sensitivity for the detection of juxtathyroidal adenomas. As the surgeon is likely to find most juxtathyroidal adenomas without any preoperative localization procedure, the technique's greater utility is in the detection of ectopic adenomas. This approach should result in a decreased need for reexploration and should simplify the operative procedure. Mattar and colleagues[25] demonstrated a 50% reduction in surgical time and a decrease in the number of nonparathyroid tissue biopsies in those patients having undergone a preoperative dual-isotope scan compared to similar patients who did not undergo a preoperative localization scan. Additionally, the technique is effective in detecting adenomas in the postoperative patient with recurrent or persistent hyperparathyroidism.

TECHNIQUE

Imaging Procedure

Patient Preparation

Before scanning any patient, there should be biochemically proven hyperparathyroidism with elevated serum calcium and PTH levels. The patient is screened much as if preparing for a thyroid scan. Patients on interfering medication or thyroid hormone or those who have had recent radiographic iodinated contrast studies should have their studies delayed for an appropriate period of time, usually 6 weeks.

Hyperthyroidism can mimic the clinical and biochemical presentation of hyperparathyroidism. Therefore, if suspected, thyroid function tests are recommended to rule out hyperthyroidism. A routine questionnaire is completed by each patient prior to scanning in order to avoid potential pitfalls. Questions asked include those related to symptoms, medical history, surgical history, medications, and previous studies including contrast studies. No other special patient preparation is necessary.

Equipment

Imaging is described as performed at the Milton S. Hershey Medical Center of the Pennsylvania State University and DePaul Hospital/Eastern Virginia Medical School. At both institutions, imaging is performed on a 10-inch–field-of-view gamma camera interfaced to a digital computer. Converging collimation is preferred to pinhole collimation due to the wider image field and better counting statistics of the former (Fig. 5-9). The thallium and technetium peaks are checked on the gamma camera and recorded prior to the administration of the radiopharmaceuticals to the patient. Twenty percent energy windows are used for both the 140 KeV technetium and 80KeV thallium peaks.

Imaging Sequence

A secure venous access in the arm is started with a butterfly needle and maintained with heparinized saline flush. The patient is injected with 1 to 2 mCi (37–74 MBq) Tc-99m pertechnetate intravenously through the butterfly. The patient is permitted to relax or move during the 15-minute waiting period needed to allow accumulation of the pertechnetate in the thyroid. The head is then immobilized in slight extension, using sandbags or a head holder, allowing a comfortable position with the closest possible approach of the camera to the neck. The patient is positioned accurately so that the thyroid gland is in the center of the imaging field (Fig. 5-2). Paper tape is placed across the brow and chin as a reminder to the patient to remain still.

Simultaneous computer acquisition and analog film images are then obtained in the following sequence (Fig. 5-3):

1. Obtain one 5-minute image on 140 KeV technetium window.
2. Change the window to 80 KeV thallium window.
3. Start five-image (5 minutes each) dynamic study.
4. Allow the first 5-minute image to be acquired *prior* to any thallium. This will be used for background-correcting the subsequent four images.
5. After the first 5-minute image has been acquired, inject 2 mCi (74 MBq) thallium-201 chloride IV via the butterfly without disturbing or distracting the patient.
6. Acquire the four remaining 5-minute thallium images, making sure the patient does not move.
7. After the dynamic acquisition is completed, obtain a 100,000 count thallium image of the mediastinum between the thyroid gland and the heart.

Computer Acquisition

All images are also acquired on a digital computer. At Hershey Medical Center, a 64 × 64 word matrix was used with a magnification factor of 1.48 acquired on a Medical Data Systems (MDS) A² computer. At DePaul Hospital, a 128 × 128 word matrix was used and acquired without magnification on a Technicare 550/560 computer system. The difference in matrix sizes is due to the different statistics (100,000–150,000/image at the former vs. 200,000–250,000/image at the latter institution), the limitation to a maximum 10-bit word depth in Technicare systems (vs. 16 in MDS), and the lack of an intermediate expansion factor between 1 and 2 in the Technicare system. This emphasizes the need to tailor and optimize the examination according to available equipment and software at each institution.

Computer Analysis

The computer processing is straightforward, but does require pixel-by-pixel multiplication, division, and subtraction capability by the software. The following subtraction sequence is the one currently in use (Fig. 5-4):

1. Generate four scatter-corrected thallium images by subtracting the first prethallium injection image of the 80 KeV dynamic sequence from the remaining postinjection images.
2. Check for patient motion by creating an outline region of interest (ROI) around each thyroid lobe and the salivary glands on the pertechnetate image (Fig. 5-5A). Then superimpose this ROI on each of the corrected thallium images. If any motion is detected, the images must be realigned prior to subtraction.
3. Create an irregular ROI composed of central thyroid pixels flagged on the pertechnetate image (Fig. 5-5B). Avoid any cold areas or obvious adenomas (hot areas on thallium). This ROI is used to sample the counts on the pertechnetate and thallium images so that a normalized subtraction can be performed. Record the counts obtained from this ROI for the pertechnetate image and each of the scatter-corrected thallium images.
4. Determine the normalization factors needed to create normalized technetium images. Normalization factors are obtained by dividing the counts obtained (from the ROI created in step 3) from the original pertechnetate image by each scatter-corrected thallium image. This results in four normalization factors which will

be used to create four corresponding normalized pertechnetate images.

5. Create the four normalized pertechnetate images by performing a pixel-by-pixel multiplication of the *original* pertechnetate image by the thallium/pertechnetate ratios (normalization factors) calculated in step 4. This results in four pertechnetate images in which the counts in the central thyroid region are approximately equal to their corresponding thallium images. Save these images.
6. From each scatter-corrected thallium image, subtract its corresponding normalized pertechnetate image.
7. Display the subtracted images with and without the thyroid outline superimposed. Usually the second subtraction image is chosen for photographic hard copy.
8. If desired, the subtraction can be performed after smoothing the images prior to normalization and subtraction.

PHYSIOLOGIC MECHANISM OF THE RADIOPHARMACEUTICALS

Thallium accumulates nonspecifically in both parathyroid and thyroid tissue presumably in relation to the cellularity or vascularity.[26] Time–activity curves demonstrate the peak thyroid–background ratio occurring in the second thallium image, approximately 5 to 10 minutes after injection (Fig. 5-7B). For this reason, the second image is usually the best one for subtraction purposes. Time–activity curves generated from regions placed on parathyroid tissue can result in a later, broader peak than those from thyroid tissue. However, this finding is not consistent and many parathyroid adenomas have behaved identically to thyroid tissue with an early peak and prompt clearance, making it impossible to distinguish reliably between parathyroid and thyroid tissue on the basis of thallium kinetics.

Pertechnetate is trapped in the thyroid but not organified. Pertechnetate has been reported to accumulate in a parathyroid adenoma,[27] which may account for some false-negative subtraction scans. However, the incidence of this phenomenon appears to be sufficiently low to allow a relatively high sensitivity for the detection of parathyroid adenomas.

ESTIMATED RADIATION ABSORBED DOSE

The estimated maximum radiation dose from the administered amounts of Tl-201 chloride (2 mCi) and Tc-99m pertechnetate (1–2 mCi) are shown (in rads) in Table 5-1.

From the above data, it is apparent that the majority of the total radiation dose is delivered by Tl-201. Varying the amount of pertechnetate given will have little effect on the total dosimetry. The kidneys are the critical organs, receiving 3 rads. These exposures are within the generally accepted limits (3 rads to the whole body, active blood-forming organs, lenses of the eyes, and gonads; 5 rads to the other organs). These radiopharmaceuticals have been approved by the Food and Drug Administration in these dose ranges and by the intravenous route of administration for other indications, although currently they are not specifically approved for parathyroid imaging. Their use, therefore, represents an alternative indication not listed in the package insert. As the radiopharmaceuticals are in the same physical and chemical form, are administered by the same route, and are in the same dosage range as usual for approved indications, an Investigational New Drug (IND) application is not necessary.[30]

VISUAL DESCRIPTION AND INTERPRETATION

When the described normalized subtraction is used, there should be no central activity visible in the thyroid bed. There is normally a rim of minimal activity along the outer margins of the thyroid lobes. This rim is usually of equal or less intensity than the adjacent neck uptake and is usually only visible after superimposing the thyroid outline ROI on the subtracted image (Fig. 5-10). Of the four subtraction images obtained, the second is usually the best and is the most reliable for the detection or exclusion of adenomas, provided the patient has not moved. While all four images are rou-

Table 5-1
Estimated Maximum Radiation Dose from Tl-201 Chloride and Tc-99m Pertechnetate

Radiopharmaceutical	*Ovaries*	*Testes*	*Bone Marrow*	*Thyroid*	*Kidney*	*Total Body*
2 mCi Tl-201 chloride[28]	1.13	1.08	1.68	1.49	2.93	0.48
2 mCi Tc-99m pertechnetate[29]	0.018	0.06	0.034	0.26	N/A	0.022
Total	1.148	1.14	1.714	1.75	3	0.502

(Basarab RM, Manni A, Harrison TS: Dual isotope subtraction parathyroid scintigraphy in the preoperative evaluation of suspected hyperparathyroidism. Clin Nuc Med 10(4):300–314, 1985)

tinely processed and subtracted, only one (usually the second) is selected for photography and final data presentation.

For the purpose of analysis, scan patterns can be classified into three broad groups: negative, positive single dominant focus, and positive multiple foci. Occasionally, a small foci of minimal thallium activity can be seen near the margins of the gland; these are usually misregistration or subtraction artifacts and are described in the report so that the surgeon may double-check the area at surgery. However, in the presence of a dominant, well-defined focus elsewhere, these minimal foci are usually not of significance. The dominant focus is usually the adenoma. If there are multiple dominant foci, then most frequently one of the foci is the parathyroid adenoma while the others are false-positive sites, often thyroid adenomas. Primary hyperparathyroidism due to multiple parathyroid adenomas is rare. If one of the multiple dominant foci is outside the thryoid bed, it should be reported because it may well be a parathyroid adenoma (Fig. 5-23). If a patient has known renal failure, then multiple foci usually represent secondary hyperplastic parathyroid glands. In general, the intensity of secondary hyperplastic glands is less than that of adenomas.

Finally, correlation with palpation and the pertechnetate scan is of value in interpreting the results of subtraction. In general, parathyroid adenomas are not easily palpable; palpable thyroid nodules are usually thyroid adenomas. A "cold" nodule or defect on the pertechnetate scan that fills in on the thallium image is likely to be a thyroid adenoma. Most parathyroid adenomas are not detectable as discrete defects on the pertechnetate image, but rather are seen on the thallium image as a focus of increased intensity compared to thallium elsewhere in the thyroid, or will add activity to fill out the contour of one of the poles (usually lower) of the thyroid gland.

These general guidelines are useful in interpreting the subtraction images; however, as in all image interpretation, they are not infallible.

DISCUSSION

Parathyroid Subtraction Scintigraphy

The general principle of dual-isotope subtraction for parathyroid disorders centers on the use of thallium, a radiopharmaceutical that accumulates both in thyroid and parathyroid tissue. In order to identify parathyroid tissue, a second isotope is also given which delineates the thyroid gland. Subtraction of these two images results in an image of the parathyroid tissue (Fig. 5-1). The principle is analogous to digital subtraction angiography, in which a background "mask" image is subtracted from postcontrast injection images, resulting in images containing only contrast.

Dual isotope subtraction has been useful for other nuclear medicine procedures besides parathyroid imaging. For example, it has been used for gallium-67 citrate or indium-111 WBC scans for which a second radiopharmaceutical, such as Tc-99m sulfur colloid (for liver/spleen lesions),[31] Tc-99m methylene diphosphonate (for bone lesions), or Tc-99m glucoheptonate (for renal lesions) is given to diagnose or localize pathology in the corresponding organ system.

There are four general approaches to performing dual-isotope parathyroid computer subtraction scans. Aside from earlier methods using Se-75 selenomethionine, all use thallium for the primary isotope. The secondary isotope that delineates the thyroid gland used is either Tc-99m pertechnetate or I-123 sodium iodide. The four general methods are (1) pertechnetate given first followed by thallium,[24,32,33] (2) thallium given first followed by pertechnetate,[34,35] (3) pertechnetate and thallium given simultaneously,[36] and (4) I-123 given first followed by thallium.[37] There are many different protocols available within these four general methods. We have selected the first method, but corrected the thallium images for the downscatter in the 80KeV thallium window which is present due to the technetium on board. Using a phantom (Fig. 5-6A), the downscatter averages about 16% of the counts in the 140KeV window of technetium (Fig. 5-6B). However, in patients, soft tissue scatter and variable uptake of technetium versus thallium in the thyroid results in the downscatter being as much as 65% of the total thallium counts. For optimal results, thallium images should be corrected for technetium downscatter. The cross-interference of radiopharmaceuticals is not solved by giving thallium first (method 2). Approximately 14% of the thallium 80KeV counts can be measured in the technetium 140 KeV window (Fig. 5-6B), due to thallium's own gamma photons of 135 KeV (3%) and 167 (10%). These photons contribute enough counts to the technetium image that correction by obtaining a background image prior to technetium injection should be performed even when thallium is given first.

Technetium uptake plateaus in the thyroid 15 to 20 minutes after injection (Fig. 5-7C). Therefore, the time between optimum uptake of thallium and technetium requires the patient to be still for a minimum of 40 to 45 minutes when thallium is given first, as opposed to 15 to 20 minutes when technetium is given first. This additional time period increases the likelihood of motion and resulting misregistration artifacts. When thallium and technetium are given simultaneously (method 3), there is no way to correct for interference in either energy window. Additionally, technetium counts would be rising at the same time thallium has peaked and is washing out. Nor is the fourth method of giving I-123 first free of downscatter problems, as approximately 50% of the I-123 counts from the 159 KeV window can be measured within thallium's 80 KeV window (Fig. 5-6C). If I-123 is used, scatter correction should also be performed.

When normalized by the method described, the salivary glands nearly always subtract out completely due to the higher relative uptake of pertechnetate than thallium. The salivary glands can also serve as valuable markers for both adenoma localization and reregistering the images if there has been any patient motion. The salivary glands are not visualized routinely on I-123 thyroid scans. Therefore, when I-123 is used, one would expect the salivary glands to subtract out positive because of their normal thallium accumulation. This would make it difficult to identify parathyroid adenomas in the region of the salivary glands. In our experience two such adenomas have been identified that may have been difficult to distinguish from the submandibular glands if I-123 had been used (Fig. 5-17).

Pertechnetate has been reported to accumulate in a parathyroid adenoma, whereas iodine has not. This should result in fewer false negatives with radioiodine. On the other hand, discordant thyroid nodules that accumulate pertechnetate (but not iodine) would be a more frequent source of false positives when radioiodine is used. In addition, the higher 159 KeV photon of I-123 may increase the subtraction error due to differential soft-tissue attenuation or collimator septal penetration.

Despite the theoretical and practical problems with all the methods, empirically investigators have reported good results with all four basic approaches, often without the scatter correction or the normalized subtraction that I believe is necessary to optimize the subtraction procedure. McFarlane and colleagues[38] demonstrated a 72% sensitivity with dual-isotope imaging in which no computer subtraction was performed, but instead the analog images were inspected for differences. While this approach is not recommended unless computer acquisition is unavailable, it does demonstrate that the analog images are useful and should be inspected in conjunction with the digital images. Occasionally, an ectopic adenoma will be better perceived on the analog images than the subtraction images. Using the described method, the author has found a sensitivity of 92% (23 of 25) for the detection of parathyroid adenomas when analyzed according to site (Table 5-2). False-negative results are relatively uncommon; however, false-positive sites are common. When analyzed according to patients rather than site, Manni and associates[39] reported a sensitivity of 82% (23 of 28) in patients with surgically proven hyperthyroidism due to a single adenoma. Because the number of patients in our experience with negative scans who were subsequently operated upon is small, it is difficult to determine a reliable specificity. The relatively high number of false-positive sites encountered, however, suggest that the specificity is lower than the 85% to 95% generally reported in the literature. Patients with palpable thyroid nodules are not excluded as they were in studies by Okerlund and coworkers.[34] The scan results will still be valid if a thyroid nodule accumulates both thallium and pertechnetate to the same degree, or if it accumulates neither radiopharmaceutical. More importantly, evaluation for potential ectopic parathyroid adenomas can be helpful in these patients. One ectopic parathyroid adenoma has been detected in the presence of a thallium-accumulating thyroid adenoma (Fig. 5-23). However, one must accept an appreciably higher incidence of false-positive sites within the thyroid

Table 5-2
Summary of Confirmed Scan Results for Each Site (not Each Patient)

Parameter	*Number of Patients*	*True-Positive Sites*	*Sensitivity (%)*	*False-Positive Sites*	*False-Negative Sites*
Single dominant focus	19	18	95	1	1
Multiple foci	4	5	100	4	0
Normal exam	2				1
Ectopic foci	7	7	100	0	0
Juxtathyroidal foci	18	16	89	5	2
Prior neck surgery*	7	8	100	0	0
Initial evaluation	18	15	88	5	2
All patients	25	23	92	5	2

** Patients included four with persistent hyperparathyroidism after previous negative exploration, two with recurrent hyperparathyroidism, and one with surgery for previous unrelated thyroid disease.*

bed when imaging patients with nodular thyroid disease.

The technique is extremely useful in the evaluation of the patient who has had previous neck surgery either for previous parathyroid disease or an unrelated disorder such as previous thyroid disease. In the series of Skibber and associates, the scans detected 52% of the causes of recurrent hyperparathyroidism in patients with previous exploration.[40] In our experience, the results have been even higher, with adenomas or hyperplasia correctly identified in all six patients who had previous negative neck exploration or recurrent hyperparathyroidism. The scan should be the first, but not necessarily the only, modality used in these patients. The smallest adenoma detected to date weighed 300 mg and the largest 10 g (Fig. 5-13). Some of the false negatives have been large, however, with at least one 1-g (Fig. 5-20) and one 5-g (Fig. 5-21) adenoma found at surgery that were not visualized prospectively on the scan. One possible explanation may be pertechnetate accumulation in the parathyroid adenomas. If the pertechnetate uptake equals that of thallium, then the adenoma would subtract away completely. Fortunately, this appears to be rare.

False positives have been largely due to thyroid adenomas or hyperplasia. Thallium has been documented to accumulate in multinodular goiters.[41] One of our false positives was due to a thyroid carcinoma (Fig. 5-24). Thallium has been well documented to accumulate in thyroid carcinomas.[42,43] As carcinomas are usually cold on pertechnetate images, this source of false positivity is expected, although again rare. Lymph nodes involved with sarcoidosis[24] and metastatic ovarian carcinoma[44] have been additional reported causes for false-positive subtraction scans.

Normalization or misregistration artifacts can occur, although these are minimized by carefully checking for motion and by avoiding sampling potential adenomas when the ROIs are created. In addition to pitfalls in processing and interpretation, the surgeon must be aware of the limitations of radioisotope techniques. The size and intensity of activity visualized are not necessarily proportional to the actual size of the lesion. Distortion at the edges of a pinhole or converging collimator may incorrectly project the relationship of the abnormal focus to adjacent structures, such as markers placed superficially, or the thyroid or the salivary glands. A significant limitation is the inability to determine depth of a lesion in the anteroposterior plane. This has resulted in missing one lesion at surgery which was located deeper than the surgeon expected. Oblique views have been suggested[45]; however, these are impractical with converging collimation. In addition, the ability to correct for technetium downscatter is lost on the obliques.

One additional pitfall to be aware of is that of suppressed thyroid tissue. Three scans have been performed in patients with previously unknown hyperfunctioning thyroid nodules, with suppression of the remaining portions of the thyroid (Fig. 5-26), and on one patient on thyroid suppression therapy (Fig. 5-27). Unfortunately, while pertechnetate uptake is suppressed, the thallium uptake is not proportionately suppressed. This results in diffuse thallium accumulation after subtraction in the suppressed portions of the thyroid, precluding evaluation for a parathyroid adenoma. In an animal model, Oster and colleagues[26] demonstrated that the differential suppression of thallium and iodine (following triiodothyronine administration) results in a significant increase in the thallium/iodine uptake rates from 2.5% to 3.6%. A similar pitfall is expected when scanning patients following high iodine loads after radiographic contrast administration. We now screen patients as if they were having a thyroid scan and avoid scanning any patient with hyperthyroidism, or we delay scanning for 6 weeks for any patient on thyroid replacement therapy or following radiographic contrast studies.

Theoretically, exogenous thyroid suppression could enhance the detection of parathyroid adenomas, because parathyroid tissue is not TSH-dependent. However, full suppression does not allow adequate visualization of the thyroid gland with pertechnetate (Fig. 5-27). Therefore, positioning the patient and defining the thyroid for purposes of normalization correction and motion correction would be difficult. In addition, if autonomous or nonsuppressible areas existed with the thyroid, there would be normalization difficulties. Partial thyroid suppression in patients with no palpable thyroid nodules may be a feasible method to enhance the relative visualization of parathyroid tissue.

Other Methods

Ultrasound is attractive in view of the ease in performance of the examination, the low cost, the high resolution, and the lack of ionizing radiation. Reported sensitivities for the detection of abnormal parathyroid glands range from 44% to 88%.[46–48] Ectopic (especially substernal) adenomas or minimally enlarged hyperplastic glands are not reliably identified. A positive examination is helpful. A negative examination does not rule out parathyroid disease. Ultrasound and radionuclide subtraction imaging are complementary, each identifying pathology that is missed by the other, allowing for an overall increased accuracy if both modalities are used. Ultrasound excels at finding the juxtathyroidal lesions; however, most of these are found by the surgeon on routine neck exploration. Ultrasound has not been demonstrated to significantly alter the already high cure rate and cannot be justified in terms of cost (despite being relatively inexpensive) in the routine preoperative evaluation of hyperparathyroidism.

The sensitivity of computed tomography (CT) varies from 29% to 88%, with improvement noted with newer generation scanners and special positioning maneuvers.[49–52] Contrast morbidity and mortality are potential drawbacks, especially when screening asymptomatic patients. CT excels in the detection of mediastinal adenomas. Its role is therefore best reserved for the patient with previous negative neck exploration, whose postoperative dual-isotope substraction scan is also negative or suggests a mediastinal focus. When a mediastinal adenoma is identified or confirmed on CT, it may preclude difficult and time-consuming angiographic searches.

Both CT and ultrasound have a higher reported accuracy[53] in the detection of four-gland hyperplasia than does radioisotope techniques, a reflection of superior spatial resolution. When positive, both CT and ultrasound better demonstrate the location, depth, and relationship of a lesion to other structures than does radioisotope imaging. Thyroid adenomas, hyperplasia, and carcinomas that plague radioisotope methods are also causes for false-positive CT and ultrasound scans. In postoperative patients, CT and ultrasound examinations are hampered by residual thyroid tissue, scarring, metal clips, and distortion to a greater degree than radioisotope methods. One patient referred for dual-isotope scintigraphy had a preoperative CT scan, which incorrectly suggested residual thyroid tissue to be the parathyroid adenoma (Fig. 5-16). The use of either CT or ultrasound in the routine preoperative screening of patients has not been demonstrated to improve significantly the already high cure rate.

Angiography, including venous sampling of PTH, has been used to locate adenomas; however, because of its invasiveness, it is usually reserved for finding recurrent or ectopic adenomas following failed initial surgery.[54] Considerable skill and experience are necessary for best results. Digital intravenous angiography[55] is promising; however, currently, the more invasive intra-arterial digital studies are superior.[56]

Magnetic resonance imaging (MRI) holds promise for the identification of many tissue types. Stark and colleagues[57] reported identifying nine parathyroid tumors (three adenomas, five hyperplastic glands, and one adenoma) in six patients. All were greater than 1 cm in size. While the lesions could be separated from the adjacent tissues and scar, the authors were unable to distinguish reliably between the various types of parathyroid tumors, thyroid lesions, and abnormal lymph nodes. Neither normal parathyroid glands nor pathologic glands of less than 1 cm were visualized. At this juncture, MRI has not been used for routine screening of parathyroid disease, but rather to further characterize and delineate lesions usually detected by other modalities. As the technology improves with increasing resolution and tissue characterization capabilities, a time can be envisioned when MRI will become the most accurate modality. Whether it will be sufficiently sensitive, accurate, cost-effective, or practical for the routine preoperative evaluation of all patients with suspected hyperparathyroidism has yet to be proved.

Future Considerations

The subtraction technique can be performed after spatially and temporally smoothing the data (Fig. 5-8B). While some of the images have been more appealing, to date there has been no change in the overall interpretation due to these manipulations. In addition, interpolative background subtraction has been performed on the thallium and technetium images by processing the images through available cardiac analysis software prior to normalized subtraction (Fig. 5-8C, 5-8D). Some software introduces additional image misregistration that must be corrected. In addition, there are artifacts introduced where the target–background ratio is poor, as well as artifacts between the thyroid lobes due to background that is not perceived as background by the software. Unfortunately, the interpolative background subtraction is needed most in cases of poor or ill-defined thallium uptake. Therefore, this maneuver has been abandoned as a routine.

Okerlund and co-workers[34] described a parametric image technique that is claimed to improve the interpretation of results. In general, parametric image displays do not provide information that is not available on routine images. It may, however, present data in a more discernible fashion. Divisor images (rather than subtraction) have been created, but offer no advantage for this parametric technique over subtraction. Perhaps other computer enhancements will be of value.

Administration of oral phosphate has been suggested as a method to enhance visualization of the parathyroids.[58] Improved instrumentation and software may allow single photon emission computed tomography (SPECT) image to be of help and provide some of the needed three-dimensional localization capability. The necessary resolution and time restraints of current systems make SPECT imaging impractical, and currently it is not being used. There is a need for a better radiopharmaceutical with better imaging characteristics, higher parathyroid tissue specificity, and better retention than thallium. Although thallium represents a significant step forward compared to Se-75 selenomethionine, it is far from ideal. Perhaps the greatest help to the surgeon and patient is yet to come with the development of a specific parathyroid-localizing radiopharmaceutical. If effective, the high incidence of hyperparathyroidism and the potential usefulness to the surgeon would guarantee widespread use of such a radiopharmaceutical.

REFERENCES

1. Golden A, Kerwin DM: The parathyroid glands. In Bloodworth JMB Jr (ed): Endocrine Pathology—General and Surgical, pp 205–220. Baltimore, Williams & Wilkins, 1982
2. Sommers SC, Gould VE: Endocrine activities of tumors (ectopic hormones). In Bloodworth JMB Jr (ed): Endocrine Pathology—General and Surgical, pp 221–243. Baltimore, Williams & Wilkins, 1982
3. Aurbach GD, Marx SJ, Spiegel AM: Parathyroid hormone, calcitonin and the calciferols. In Williams RH (ed): Textbook of Endocrinology, pp 922–1031. Philadelphia, WB Saunders, 1981
4. Mundy GR, Cove DH, Frisken R: Primary hyperparathyroidism: Changes in the pattern of clinical presentation. Lancet i:1317, 1980
5. Boonstra CE, Jackson CE: Serum calcium survey for hyperparathyroidism: Results in 50,000 clinic patients. Am J Clin Pathol 5:523, 1971
6. Wang CA: The anatomic basis of parathyroid surgery. Ann Surg 183:271, 1976
7. Goldman L, Gordan GS, Roof BS: The parathyroids: Progress, problems and practice. Curr Probl Surg 8:1, 1971
8. Purnell DC, Scholz DA, Smith LH et al: Treatment of primary hyperparathyroidism. Am J Med 56:800, 1974
9. Paloyan E, Lawrence AM, Baker WH et al: Near-total parathyroidectomy. Surg Clin North Am 49:43, 1969
10. Haff RC, Ballinger WF: Causes of recurrent hypercalcemia after parathyroidectomy for primary hyperparathyroidism. Ann Surg 173:884, 1971
11. Edis AJ, Beahrs OH, Van Heerden JA et al: "Conservative" vs. "liberal" approach to parathyroid neck exploration. Surgery 82:466, 1977
12. Esselstyn CB Jr: Parathyroid surgery. How many glands should be excised? Is there still a controversy? Surg Clin North Am 59:77, 1979
13. Coffey RJ, Lee TC, Canary JJ: The surgical treatment of primary hyperparathyroidism: A 20-year experience. Ann Surg 185:518, 1981
14. Harrison TS, Duarte B, Reitz RE et al: Primary hyperparathyroidism: Four-to-eight year postoperative follow-up demonstrating persistent functional insignificance of microscopic parathyroid hyperplasia and decreased autonomy of parathyroid hormone release. Ann Surg 194:429, 1981
15. Clark OH, Way LW, Hunt TK: Recurrent hyperparathyroidism. Ann Surg 184:391, 1976
16. Wang CA: Parathyroid re-exploration: A clinical and pathological study of 112 cases. Ann Surg 186:140, 1977
17. Brennan MF, Marx SJ, Doppman J et al: Results of reoperation for persistent and recurrent hyperparathyroidism. Ann Surg 194:671, 1981
18. Potchen EJ, Watts HG, Awwad HK: Parathyroid scintiscanning. Radiol Clin North Am 5:267, 1967
19. Colella AC, Pigorini F: Experience with parathyroid scintigraphy. Am J Roentgenol 109:714, 1970
20. Crocker EF, Jellins J, Freund J: Parathyroid lesions localized by radionuclide subtraction and ultrasound. Radiology 130:215, 1979
21. Robinson PJ: Parathyroid scintigraphy revisited. Clin Radiol 33:37, 1982
22. Waldord JC, van Heerden JA, Gorman CA et al: Se-75 selenomethionine scanning for parathyroid localization should be abandoned. Mayo Clin Proc 59:534, 1984
23. Ferlin G, Borsato N, Camerani M et al: Parathyroid scintigraphy with a new double-tracer (99mTc-201Tl) technique. J Endocrinol Invest (suppl 1) 5:101, 1982
24. Young AE, Gaunt JI, Croft DN et al: Location of parathyroid adenomas by thallium-201 and technetium-99m subtraction scanning. Br Med J 286:1384, 1983
25. Mattar AG, Wright ES, Chittal SM et al: Impact on surgery of preoperative localization of parathyroid lesions with dual radionuclide subtraction scanning. Can J Surg 29:57, 1986
26. Oster ZH, Strauss HW, Harrison K et al: Thallium-201 distribution in the thyroid: Relationship to thyroidal trapping function. Radiology 126:733, 1978
27. Alagumalai K, Avramides A, Carter AC et al: Uptake of technetium pertechnetate in a parathyroid adenoma presenting as an iodine-131 "cold" nodule. Ann Intern Med 90:204, 1979
28. Package insert, thallous chloride Tl-201. Product 2095. Medi-Physics, Emeryville, California, February, 1983
29. MIRD dose estimate report no. 8. Summary of current radiation dose estimates to humans from Tc-99m as sodium pertechnetate. Active population. J Nucl Med 17:74, 1976
30. Swanson DP, Lieto RP: The submission of IND applications for a radiopharmaceutical research: When and why. J Nucl Med 25:714, 1984
31. Rovekamp MH, van Royen EA, Folmer SCCR et al: Diagnosis of upper abdominal infections by IN-111 labeled leukocytes with Tc-99m colloid subtraction technique. J Nucl Med 24:212, 1983
32. Ferlin G, Borsato N, Camerani M et al: New perspectives in localizing enlarged parathyroids by technetium-thallium subtraction scan. J Nucl Med 24:438, 1983
33. Basarab RM, Manni A, Harrison TS: Dual isotope subtraction parathyroid scintigraphy in the preoperative evaluation of suspected hyperparathyroidism. Clin Nucl Med 10(4):300, 1985
34. Okerlund MD, Sheldon K, Corpuz S et al: A new method with high sensitivity and specificity of localization of abnormal parathyroid glands. Ann Surg 200:381, 1984
35. Winzelberg GG, Hydovitz JD, O'Hara KR et al: Parathyroid adenomas evaluated by Tl-201/Tc-99m pertechnetate subtraction scintigraphy and high-resolution ultrasonography. Radiology 155:231, 1985
36. Percival RC, Blake GM, Urwin GH et al: Assessment of thallium pertechnetate subtraction scintigraphy in hyperparathyroidism. Br J Rad 58:131, 1985

37. McKusick KA, Palmer EL, Hergenrother J et al: Is there a role for dual tracer imaging in detection of parathyroid disease? Abstract presented at the 31st Annual Society of Nuclear Medicine Meeting. J Nucl Med 25:P19, 1984

38. McFarlane SD, Hanelin LG, Taft DA et al: Localization of abnormal parathyroid glands using thallium-201. Am J Surg 148:7, 1984

39. Manni A, Basarab RM, Pluorde PV et al: Thallium-technetium parathyroid scan: A useful noninvasive test for localization of abnormal parathyroid tissue. Arch Intern Med 146:1077, 1986

40. Skibber JM, Reynolds JC, Spiegel AM et al: Computerized technetium/thallium scans and parathyroid reoperation. Surgery 98:1077, 1985

41. Fukuchi M, Hyodo K, Tachibana K et al: Marked thyroid uptake of thallium-201 in patients with goiter: Case report. J Nucl Med 18:1199, 1977

42. Tonami N, Bunko H, Michigishi T et al: Clinical application of 201-Tl scintigraphy in patients with cold thyroid nodules. Clin Nucl Med 3:217, 1978

43. Hisada K, Tonami N, Miyamae T et al: Clinical evaluation of tumor imaging with 201-Tl chloride. Radiology 129:497, 1978

44. Punt CJA, DeHooge P, Hoekstra BL: False positive subtraction scintigram of parathyroid glands due to metastatic tumor. J Nucl Med 26:155, 1985

45. Urgancioğlu I, Hatemi H, Seyahi V et al: Letter to the editor. J Nucl Med 26:99, 1985

46. Parr JH, Tarkunde I, Ramsay I: The use of ultrasound in the localisation of parathyroid glands in parathyroid disorders. Clin Radiol 34:395, 1983

47. Reading CC, Charboneau JW, James EM et al: High-resolution parathyroid sonography. Am J Roentgenol 139:539, 1982

48. Simeone JF, Mueller PR, Ferrucci JT et al: High-resolution real-time sonography of the parathyroid. Radiology 141:745, 1981

49. Krudy AG, Doppman JL, Brennan MF et al: The detection of mediastinal parathyroid glands by computed tomography, selective arteriography and venous sampling. Radiology 140:739, 1981

50. Stark DD, Gooding GAW, Moss AA et al: Parathyroid imaging: Comparison of high-resolution CT and high-resolution sonography. Am J Roentgenol 141:633, 1983

51. Friedman M, Mafee MF, Shelton VK et al: Parathyroid localization by computed tomographic scanning. Arch Otolaryngol 109:95, 1983

52. Stark DD, Moss AA, Gooding GAW et al: Parathyroid scanning by computed tomography. Radiology 148:297, 1982

53. Takagi H, Tominaga Y, Uchida K et al: Image diagnosis of parathyroid glands in chronic renal failure. Ann Surg 198:74, 1983

54. Doppman JL: Parathyroid localization: Arteriography and venous sampling. Radiol Clin North Am 14:163, 1976

55. Levy JM, Hessel SJ, Dippe SE et al: Digital subtraction angiography for localization of parathyroid lesions. Ann Intern Med 97:710, 1982

56. Krudy AG, Doppman JL, Miller DL et al: Work in progress: Abnormal parathyroid glands. Comparison of nonselective digital arteriography, selective parathyroid arteriography and venous digital arteriography as methods of detection. Radiology 148:23, 1983

57. Stark DD, Moss AA, Gamsu G et al: Magnetic resonance imaging of the neck. Part II: Pathologic findings. Radiology 150:455, 1984

58. Gupta SM, Belsky JL: Letter to the editor. J Nucl Med 26:100, 1985

59. Watson DD, Cambell NP, Read EK et al: Spatial and temporal quantitation of plane thallium myocardial images. J Nucl Med 22:577, 1981

Atlas Section

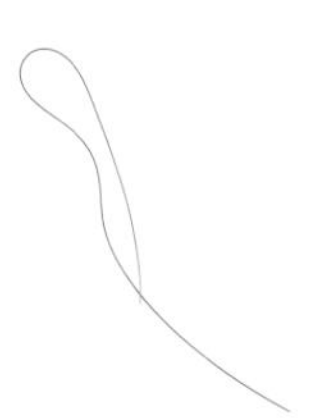

Figure 5-1 ***Principles of Dual Isotope Subtraction***

Thallium accumulates in both thyroid and parathyroid tissue. Pertechnetate accumulation delineates the thyroid. After subtraction, the parathyroid adenoma is revealed.

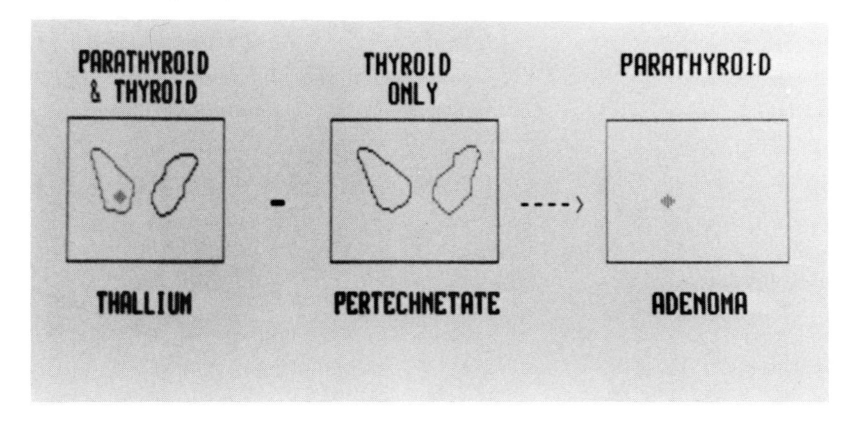

Figure 5-2 ***Patient Positioning***

Accurate positioning is achieved with the thyroid centered in the imaging field. A head holder, sandbags, or tape can be used to minimize patient motion.

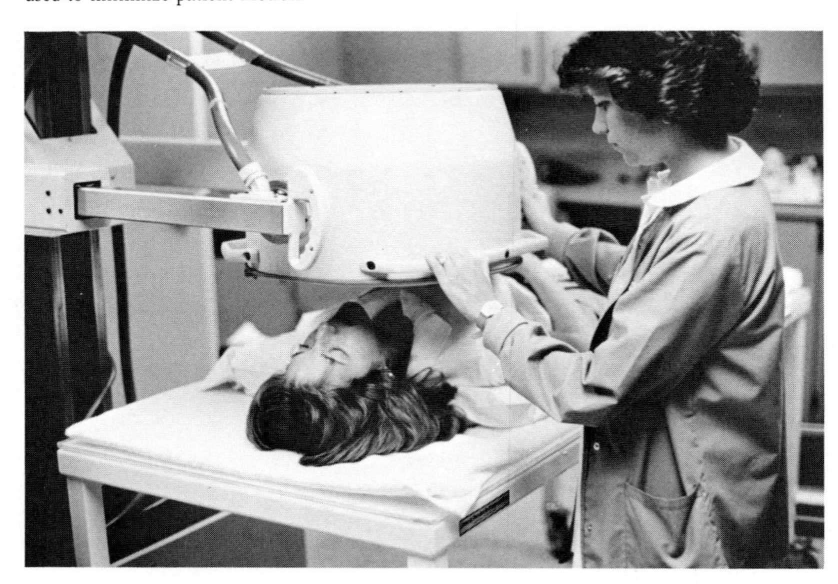

Figure 5-3 ***Acquisition Sequence***

(A) Pertechnetate image. *(B)* Thallium dynamic sequence. Note that thallium is not injected until after the first 5-minute image has been acquired. *(C)* Mediastinal view.

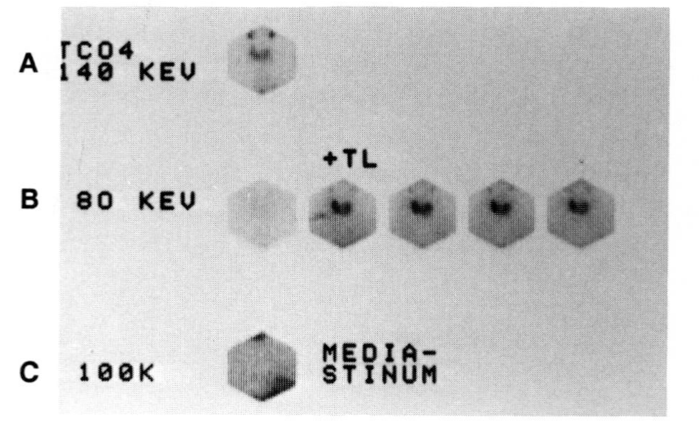

Figure 5-4 *Subtraction Sequence*

(A) Uncorrected thallium dynamic study. *(B)* Scatter-corrected thallium images obtained by subtracting the first image (pre-thallium injection) from each subsequent thallium image. *(C)* Normalized pertechnetate images. One is generated for each post-thallium injection image. *(D)* Subtraction images.

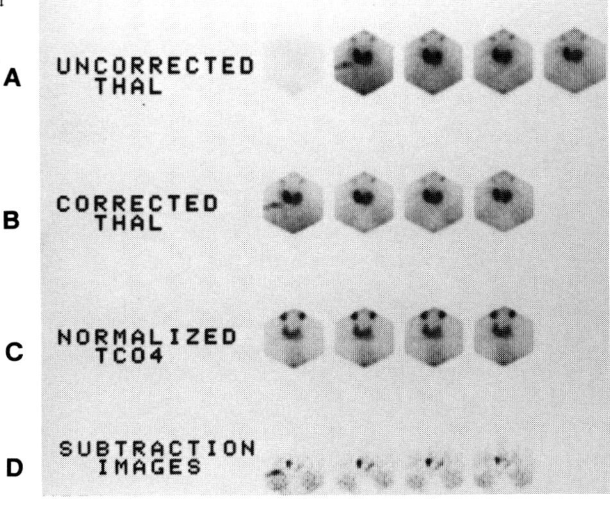

Figure 5-5

(A) Motion Check. A region of interest (ROI) is created that outlines the margins of the thyroid gland and salivary glands (if visible). Using the computer, the same ROI is then superimposed on the thallium image(s). If there is any motion, it will be detected by a shift of the thallium activity to one side, up or down, relative to the ROI. A consistent shift of one or more margins indicates that the patient moved or that the camera electronics are not in registration for different energy peaks (a pitfall that should be tested by bar phantom studies prior to use). A focal bulge outside the ROI margin may be a parathyroid adenoma; therefore, motion correction is only performed if there is consistent malalignment of the two images. *(B)* Normalization factor generation. An irregular ROI composed of central pixels is created from the pertechnetate image. The normalization factor is calculated by dividing the thallium counts by the pertechnetate counts flagged by this ROI. In this example, the normalization factor is 1.5. Therefore, the original technetium image is multiplied by 1.5 prior to subtraction from the thallium image.

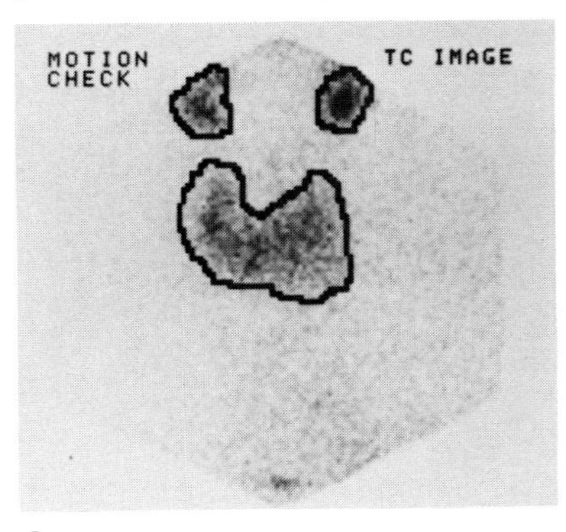

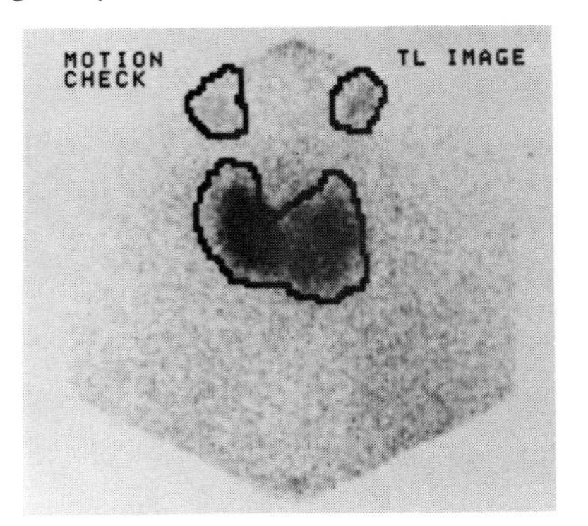

A

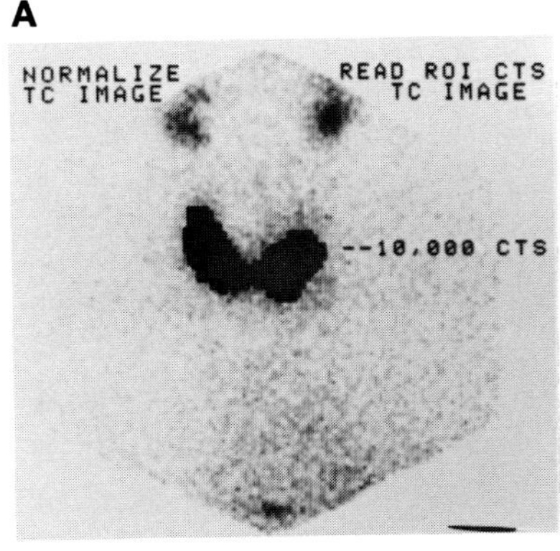

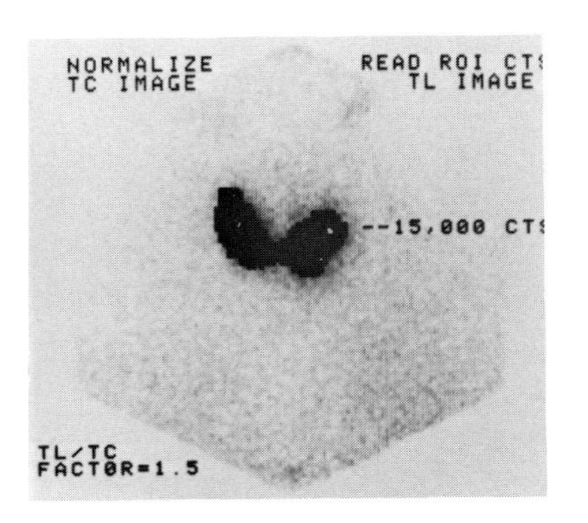

B

Figure 5-6 ***Cross-Interference of Radiopharmaceuticals***

(A) IV-bag phantoms containing thallium *(left)*, pertechnetate *(middle)*, and D5W *(right)* are used to determine relative contribution of counts from each isotope in the other's energy window. *(B)* Images obtained from the phantoms demonstrate approximately equal cross-interference of each isotope. Therefore, whichever isotope is given first, a background correction is suggested prior to subtraction. *(C)* I-123 thyroid image and counts in thallium's 80-KeV energy window obtained in a patient referred for thyroid imaging. Downscatter correction is also suggested if I-123 is used first. (Basarab RM, Manni A, Harrison TS: Dual isotope subtraction parathyroid scintigraphy in the preoperative evaluation of suspected hyperparathyroidism. Clin Nuc Med 10(4):300–314, 1985)

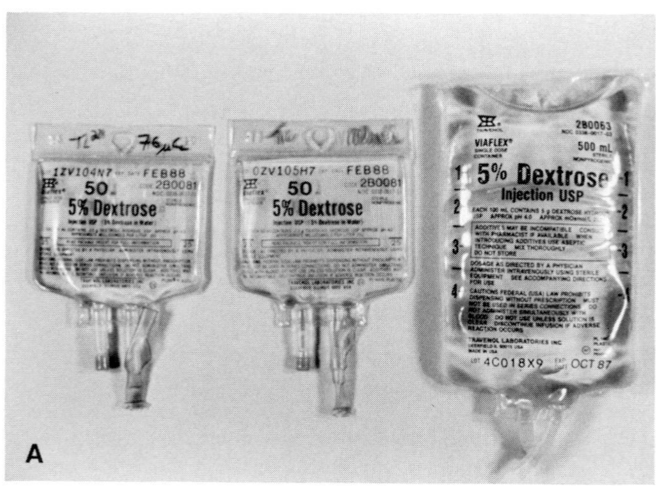

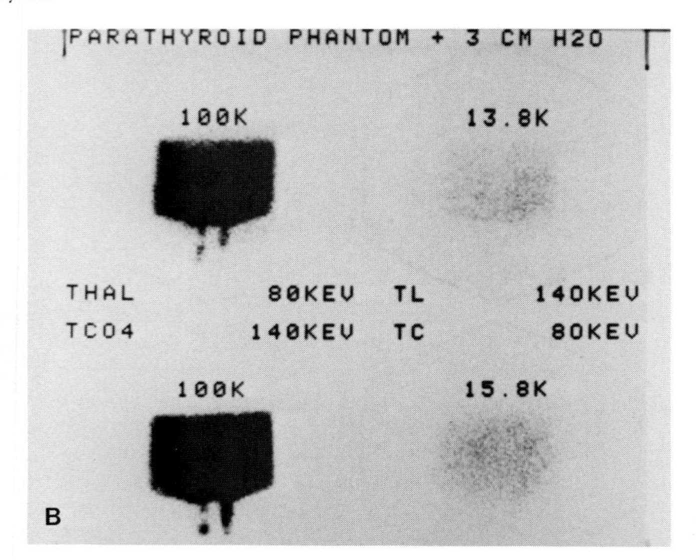

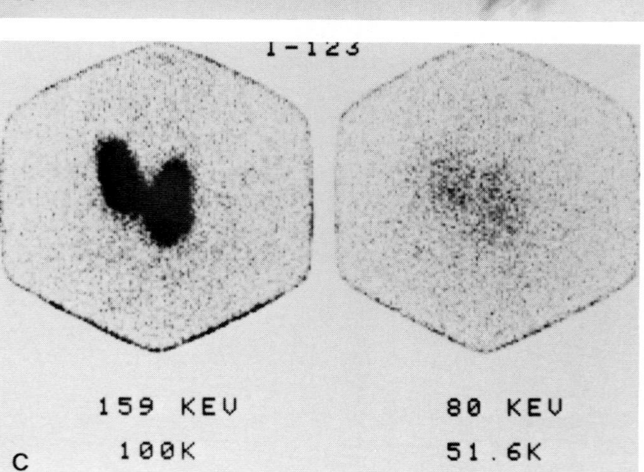

Figure 5-7 ***Thyroid and Parathyroid Time–Activity Curves***

(A) ROIs are used to obtain time–activity curves after thallium injection. A parathyroid adenoma can be seen on this nonsubtracted image just below the left thyroid. *(B)* Background-corrected time–activity curves generated from the ROIs show the prompt uptake and clearance of thallium from the thyroid tissue. The left lower parathyroid adenoma shows slower uptake and greater retention of thallium than the thyroid tissue. *(C)* Background-corrected time–actvity curves generated in a similar fashion in a patient undergoing a pertechnetate thyroid scan show a plateau of uptake of 15 to 20 minutes postinjection. For this reason, imaging of the pertechnetate is best performed 15 to 20 minutes after injection.

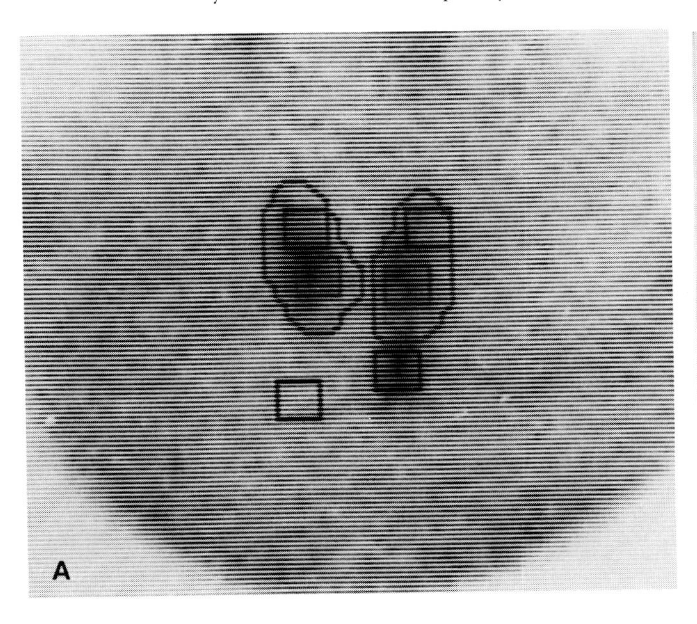

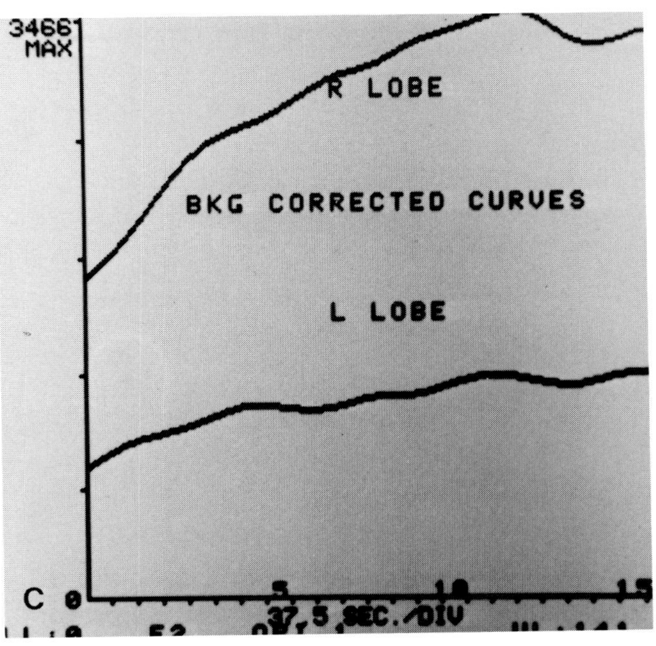

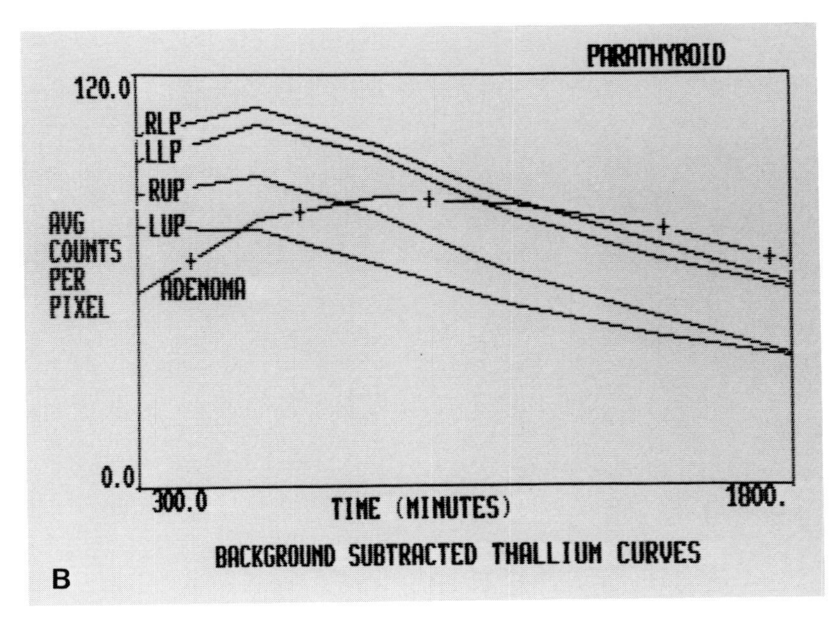

Figure 5-8 ***Various Subtraction Techniques***

Right lower pole parathyroid adenoma is demonstrated using four different subtraction methods. Note the additional artifacts seen in the latter two methods. ***(A)*** Standard method. ***(B)*** Following spatial and temporal smoothing. ***(C)*** Following the interpolative background correction of Watson and colleagues.[57] ***(D)*** Following Medical Data Systems'® interpolative background correction. (Basarab RM, Manni A, Harrison TS: Dual isotope subtraction parathyroid scintigraphy in the preoperative evaluation of suspected hyperparathyroidism. Clin Nucl Med 10(4):300 - 314, 1985)

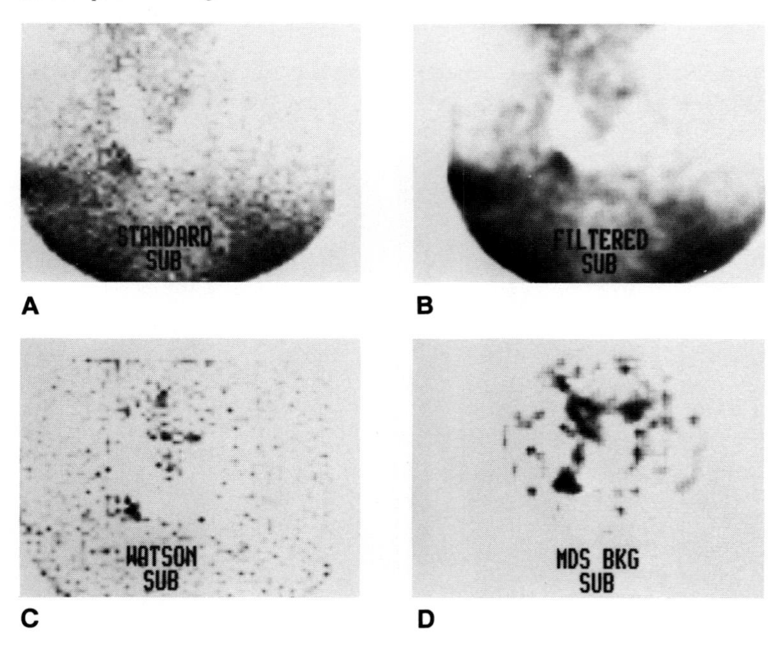

Figure 5-9 ***Pinhole vs. Converging Collimator***

(A) Pinhole examination demonstrates a single dominant focus projected from the left lower pole of the pinhole field. A 2.5-g parathyroid adenoma was confirmed at surgery. ***(B)*** Converging collimator examination, now the preferred method, is performed in a patient with chronic renal failure and either tertiary hyperparathyroidism or a coexisting primary adenoma. A 3.0-g left lower pole adenoma, similar in location to the previous case, is visualized. Note the larger field of view afforded by converging collimation. (Fig. 9A: Basarab RM, Manni A, Harrison TS: Dual isotope subtraction parathyroid scintigraphy in the preoperative evaluation of suspected hyperparathyroidism. Clin Nucl Med 10(4):300 - 314, 1985)

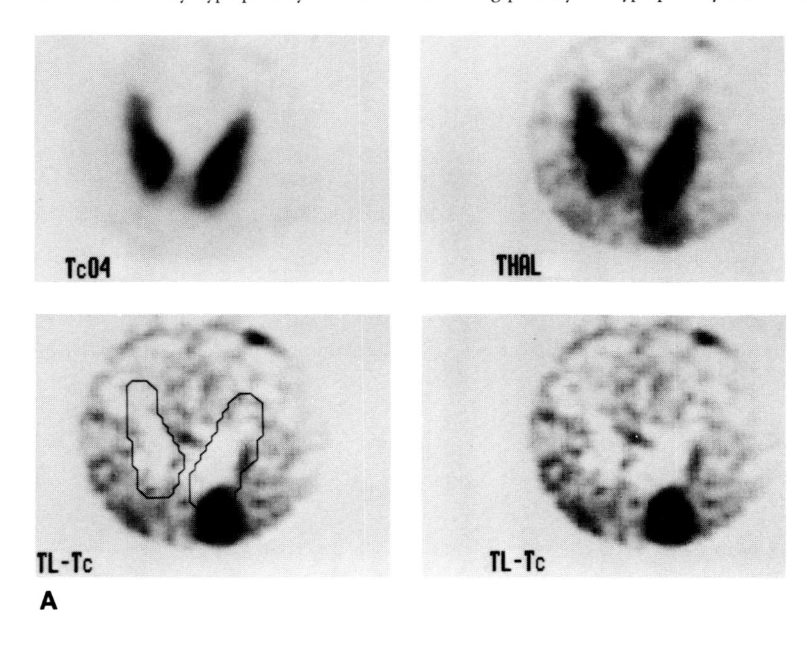

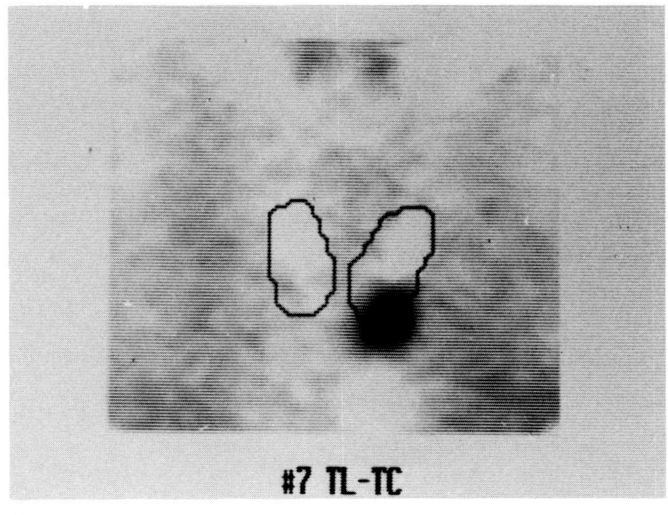

Figure 5-10

Normal Examination

Normal converging collimator examination in a patient whose initial elevated calcium and PTH levels were due to laboratory error (Basarab RM, Manni A, Harrison TS: Dual isotope subtraction parathyroid scintigraphy in the preoperative evaluation of suspected hyperparathyroidism. Clin Nucl Med 10(4): 300 - 314, 1985)

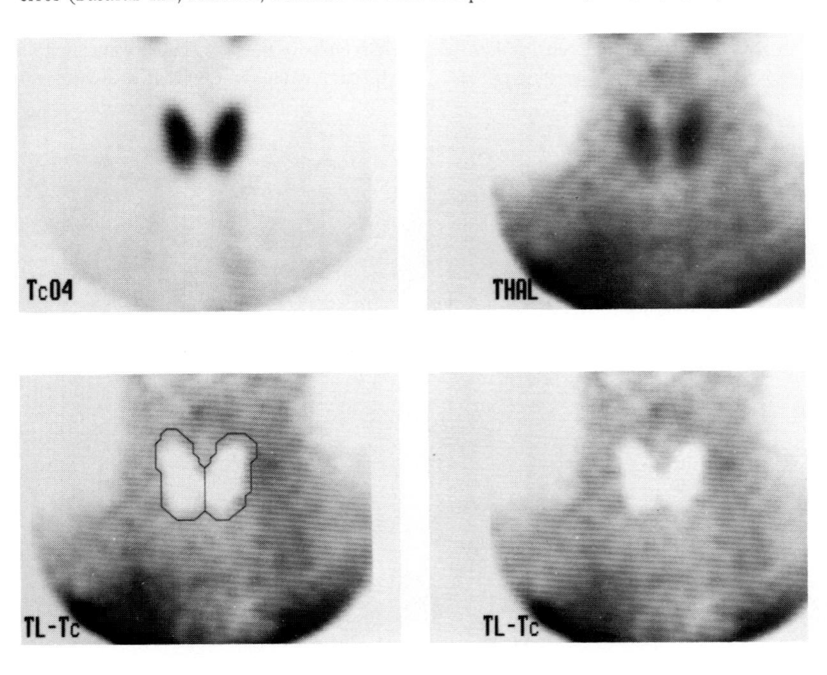

Figure 5-11

Distribution of Sites of Lesions

Composite drawing demonstrates the approximate distribution of 25 abnormal lesions (20 adenomas, 5 hyperplasias) found at surgery in relation to their scan localization. False-negative scan sites are indicated by an asterisk. (Basarab RM, Manni A, Harrison TS: Dual isotope subtraction parathyroid scintigraphy in the preoperative evaluation of suspected hyperparathyroidism. Clin Nucl Med 10(4):300 - 314, 1985)

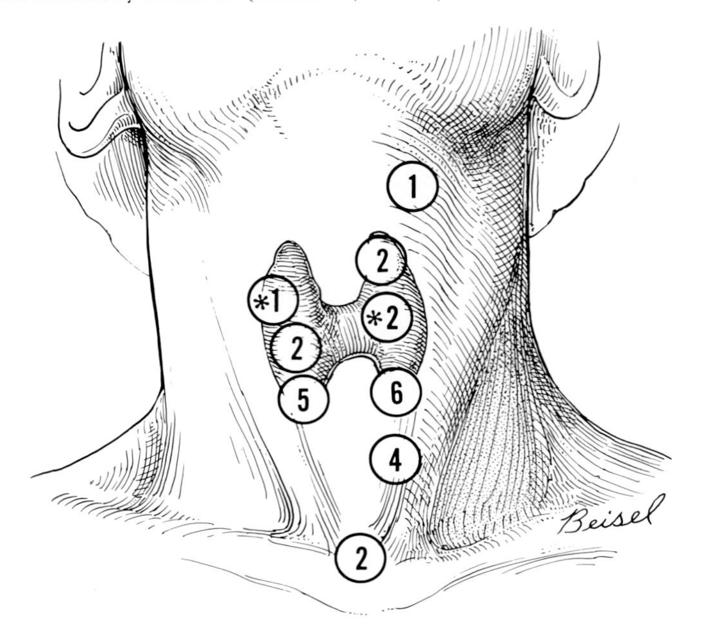

Figure 5-12 ***Parathyroid Adenoma***

(A) Subtraction image demonstrates a focus of thallium accumulation just below the left lower thyroid pole. ***(B)*** A 1.3-g parathyroid adenoma was confirmed at surgery. (Fig. 12A: Basarab RM, Manni A, Harrison TS: Dual isotope subtraction parathyroid scintigraphy in the preoperative evaluation of suspected hyperparathyroidism. Clin Nucl Med 10(4):300-314, 1985)

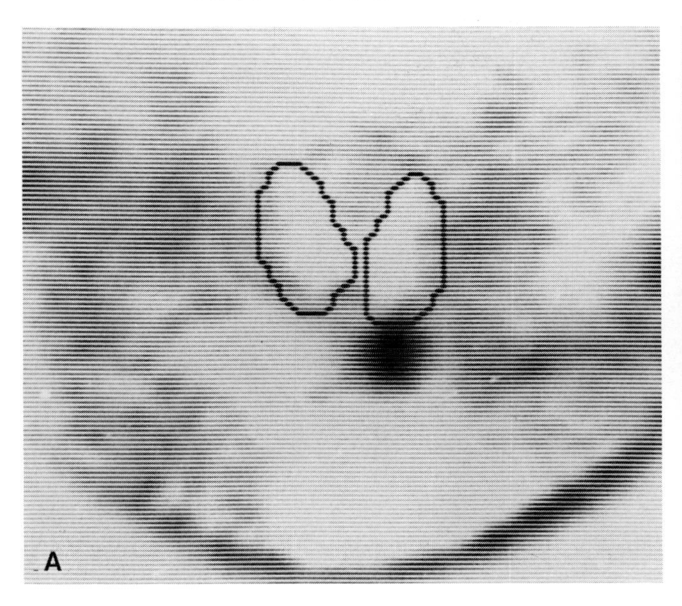

Figure 5-13 ***Ectopic Parathyroid Adenoma***

A 10-g ectopic adenoma can be easily identified without subtraction near the level of the suprasternal notch *(open arrow)*. The patient had a prior negative neck exploration, at which time a right hemithyroidectomy was performed. The remaining left lobe is visualized *(closed arrow)*. (Basarab RM, Manni A, Harrison TS: Dual isotope subtraction parathyroid scintigraphy in the preoperative evaluation of suspected hyperparathyroidism. Clin Nucl Med 10(4):300 - 314, 1985)

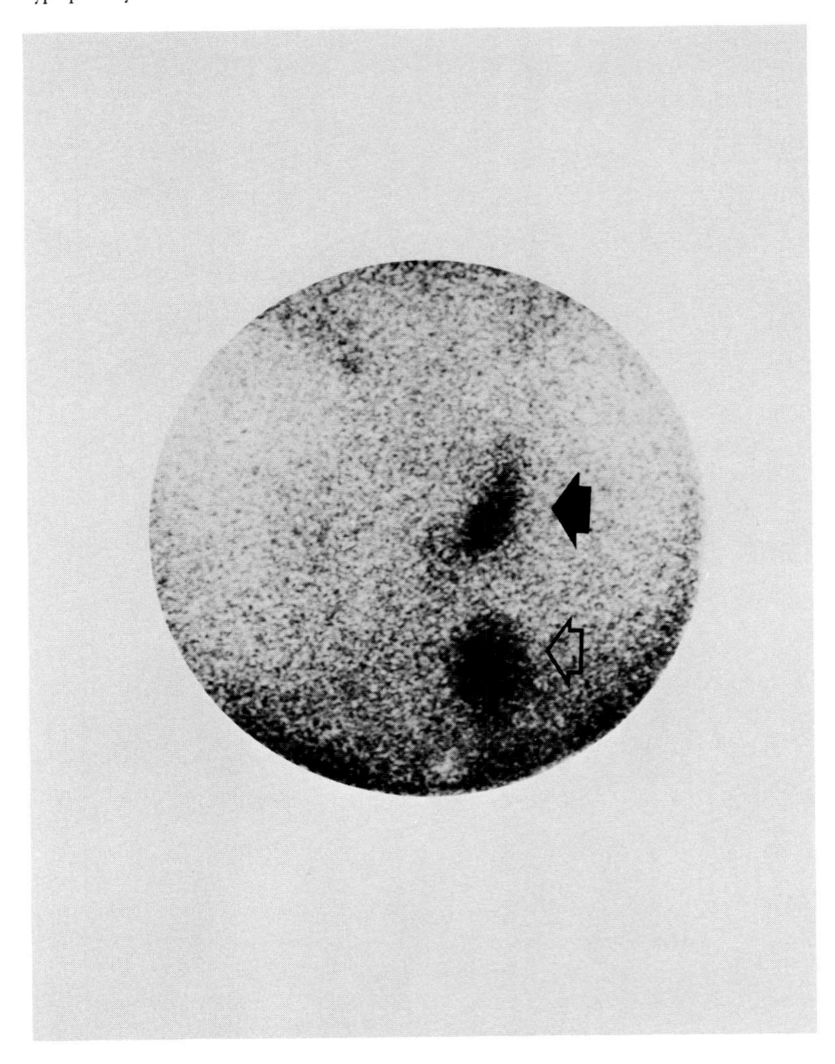

Figure 5-14 ***Parathyroid Adenoma***

(A) A relatively subtle focus is identified on both the analog *(left)* and subtraction image *(right)* below the left thyroid lobe *(arrows)*. ***(B)*** Digital subtraction angiogram, AP view, left thyrocervical trunk injection. The 1.3-g adenoma found at surgery was posteriorly located in the left tracheoesophageal groove. (Basarab RM, Manni A, Harrison TS: Dual isotope subtraction parathyroid scintigraphy in the preoperative evaluation of suspected hyperparathyroidism. Clin Nucl Med 10(4):300–314, 1985)

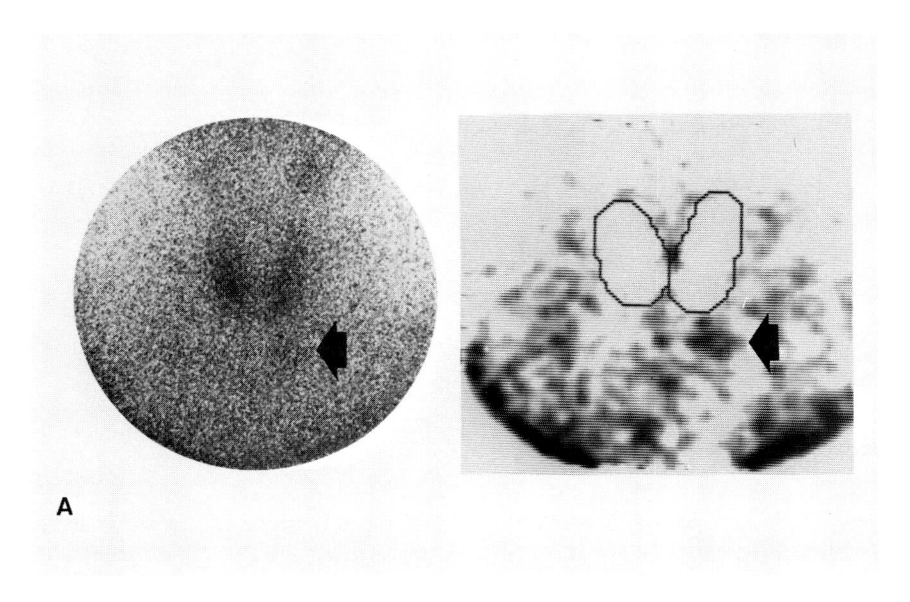

A

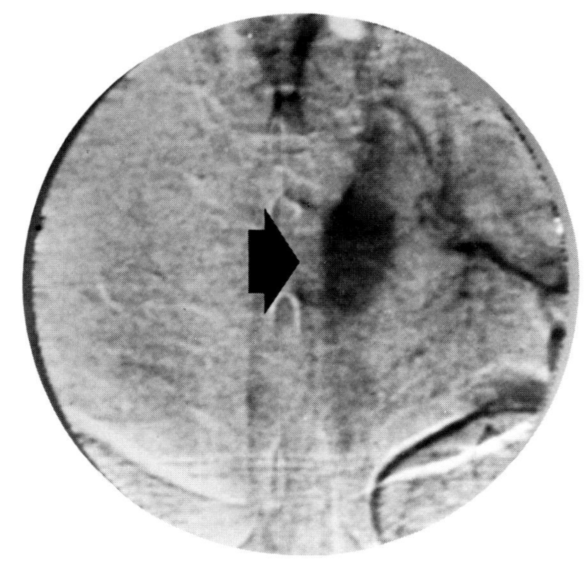

B

Figure 5-15 ***Parathyroid Adenoma***

A 0.9-g right inferior parathyroid adenoma is well-visualized overlying the right lower thyroid pole. (Basarab RM, Manni A, Harrison TS: Dual isotope subtraction parathyroid scintigraphy in the preoperative evaluation of suspected hyperparathyroidism. Clin Nucl Med 10(4):300-314, 1985)

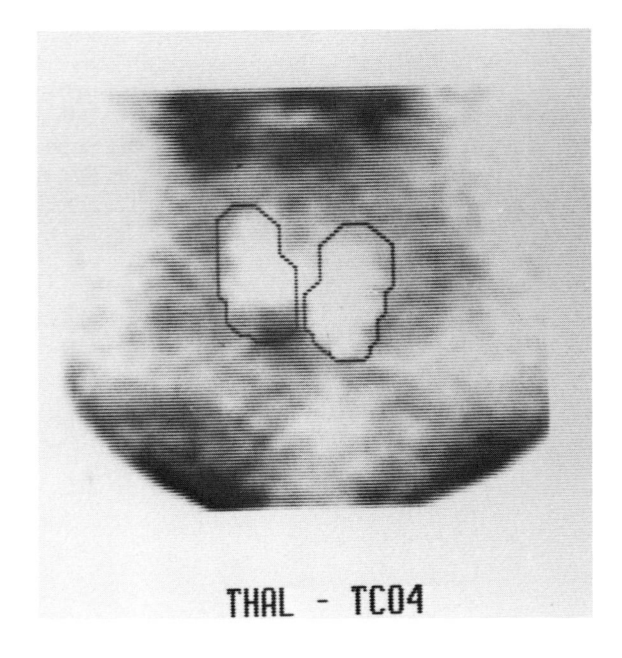

Figure 5-16 ***Ectopic Parathyroid Adenoma***

Positive ectopic site is in a patient who had a partial thyroidectomy performed 30 years previously for unrelated thyroid disease. An adenoma (3.2 cm by 1.5 cm by 1.0 cm) was identified near the sternal notch *(arrow)*. A preoperative CT scan was performed, in which the residual left thyroid lobe was incorrectly identified as the possible adenoma. (Basarab RM, Manni A, Harrison TS: Dual isotope subtraction parathyroid scintigraphy in the preoperative evaluation of suspected hyperparathyroidism. Clin Nucl Med 10(4):300-314, 1985)

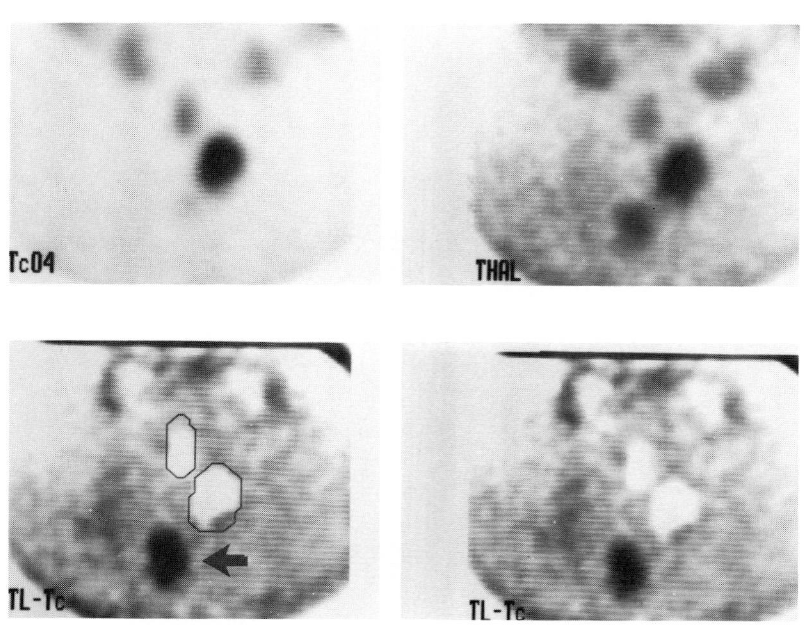

Figure 5-17 ***Ectopic Parathyroid Adenoma***

A positive ectopic site is present in a patient who had a previous negative neck exploration including a right hemithyroidectomy. On reoperation, a left superior parathyroid adenoma (2.6 cm by 1.2 cm by 0.6 cm) was confirmed near the left submandibular gland. (Basarab RM, Manni A, Harrison TS: Dual isotope subtraction parathyroid scintigraphy in the preoperative evaluation of suspected hyperparathyroidism. Clin Nucl Med 10(4):300 - 314, 1985)

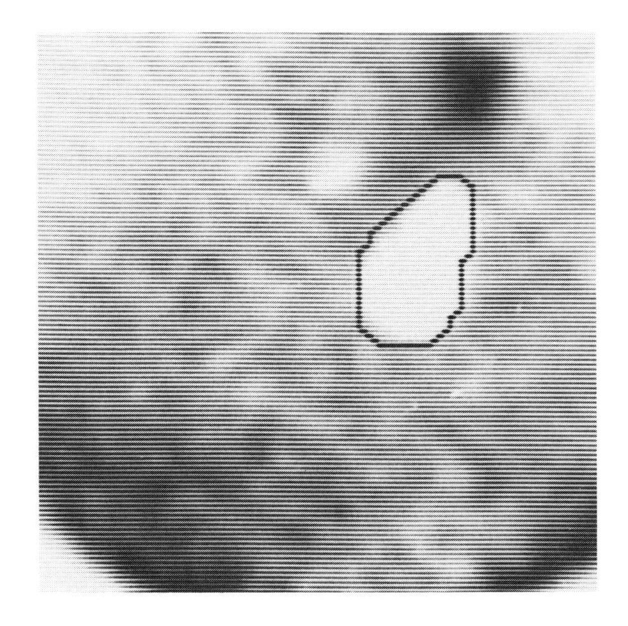

Figure 5-18 ***Parathyroid Hyperplasia Recurrent***

Hyperplastic parathyroid tissue was confirmed in two sites in a patient with MEN-1 syndrome who developed recurrent hyperparathyroidism 5 years after subtotal parathyroidectomy. (Basarab RM, Manni A, Harrison TS: Dual isotope subtraction parathyroid scintigraphy in the preoperative evaluation of suspected hyperparathyroidism. Clin Nucl Med 10(4):300 - 314, 1985)

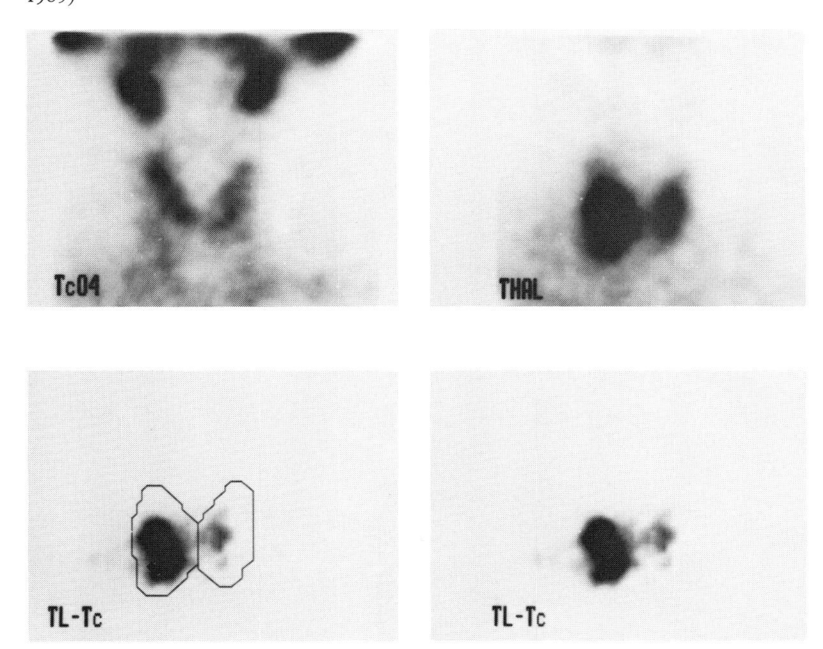

Figure 5-19 ***Parathyroid Hyperplasia — Four-Gland Hyperplasia***

The subtraction image demonstrates four hyperplastic glands in this patient with chronic renal failure and secondary hyperparathyroidism.

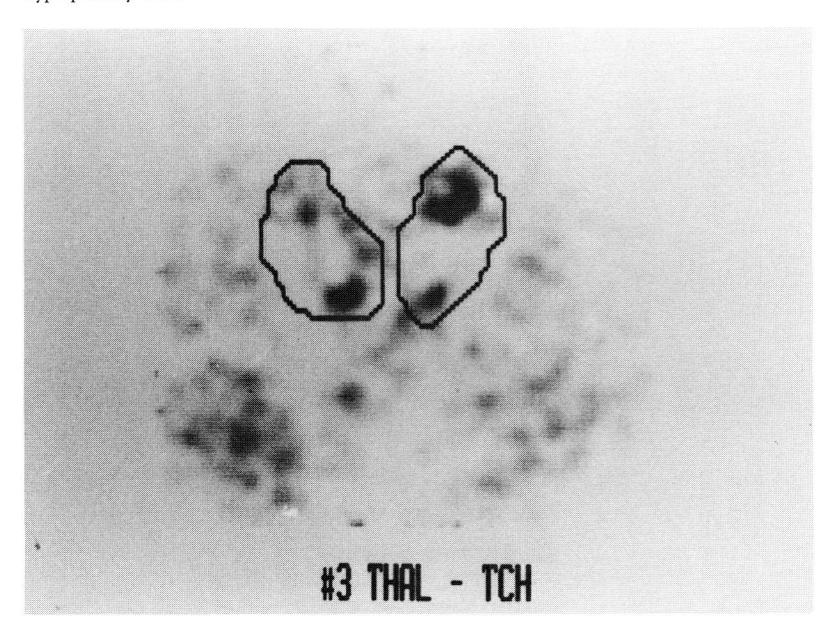

Figure 5-20 ***Parathyroid Adenoma***

Prospectively, this examination was interpreted as negative. In retrospect, a subtle focus can be identified along the right lateral margin of the thyroid. A right superior parathyroid adenoma (2.5 cm by 0.9 cm by 0.4 cm) was identified at surgery.

(Basarab RM, Manni A, Harrison TS: Dual isotope subtraction parathyroid scintigraphy in the preoperative evaluation of suspected hyperparathyroidism. Clin Nucl Med 10(4):300 - 314, 1985)

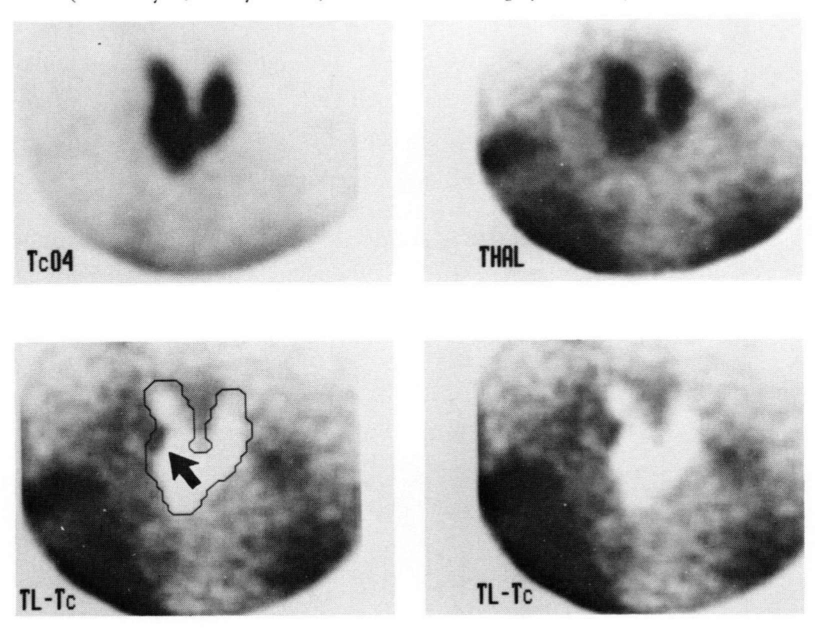

Figure 5-21 *Thyroid Adenoma*

A large area of thallium uptake in the right lobe proved to be a benign thyroid adenoma. Even in retrospect, the 5-g parathyroid adenoma found at surgery on the left side cannot be identified. *(A)* Pertechnetate image. *(B)* Thallium image. *(C)* and *(D)*. Thallium-technetium subtraction images. (Basarab RM, Manni A, Harrison TS: Dual isotope subtraction parathyroid scintigraphy in the preoperative evaluation of suspected hyperparathyroidism. Clin Nucl Med 10(4):300 - 314, 1985)

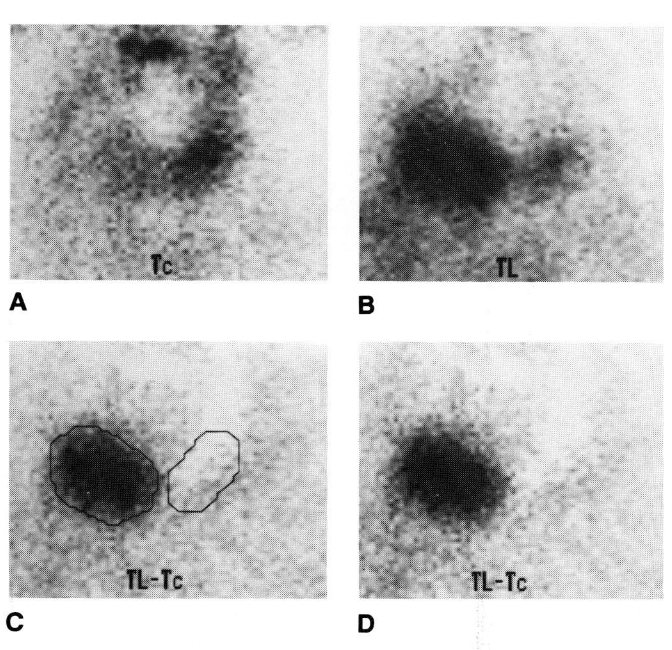

Figure 5-22 *False-Positive (Nodular Thyroid Hyperplasia) and Parathyroid Adenoma*

A 500-mg parathyroid adenoma was found in the upper left site and nodular thyroid hyperplasia in the lower right. (Basarab RM, Manni A, Harrison TS: Dual isotope subtraction parathyroid scintigraphy in the preoperative evaluation of suspected hyperparathyroidism. Clin Nucl Med 10(4):300 - 314, 1985)

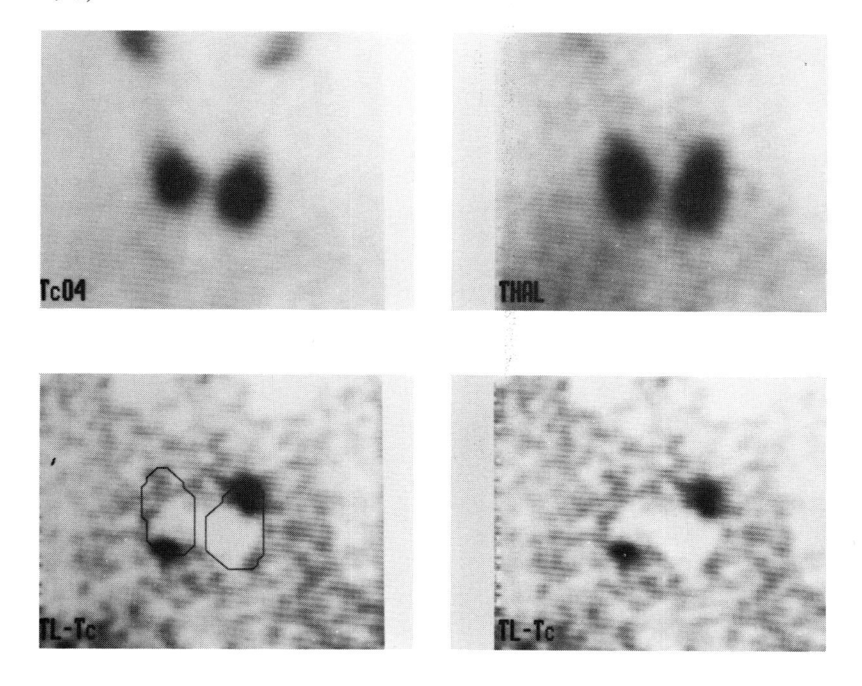

Figure 5-23 ***False-Positive (Thyroid Adenoma) and Ectopic Parathyroid Adenoma — Subtraction Image***

A 400-mg parathyroid adenoma was confirmed at the ectopic site just below the left lower pole. A thyroid adenoma accounted for the uptake on the right. (Basarab RM, Manni A, Harrison TS: Dual isotope subtraction parathyroid scintigraphy in the preoperative evaluation of suspected hyperparathyroidism. Clin Nucl Med 10(4):300 - 314, 1985)

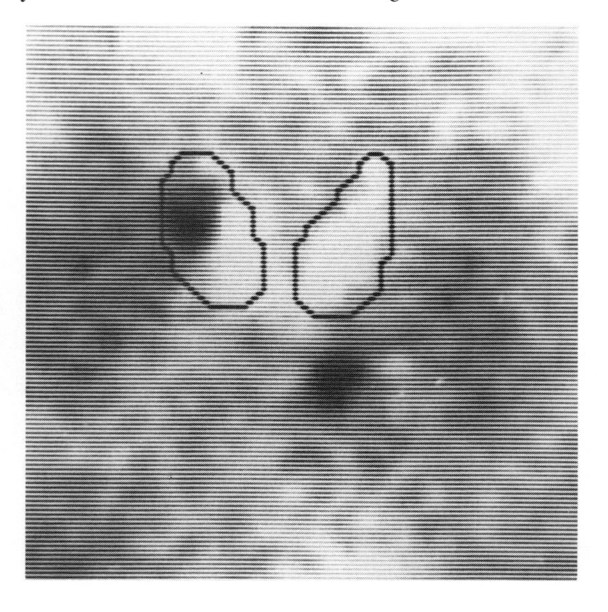

Figure 5-24 ***False-Positive (Thyroid Carcinoma)***

(A) A large focus of thallium accumulation is evident after subtraction in the left middle to lower pole. *(B)* Surgical exploration revealed a 2.4-cm thyroid carcinoma *(arrowheads)*, with both papillary and follicular elements evident microscopically.

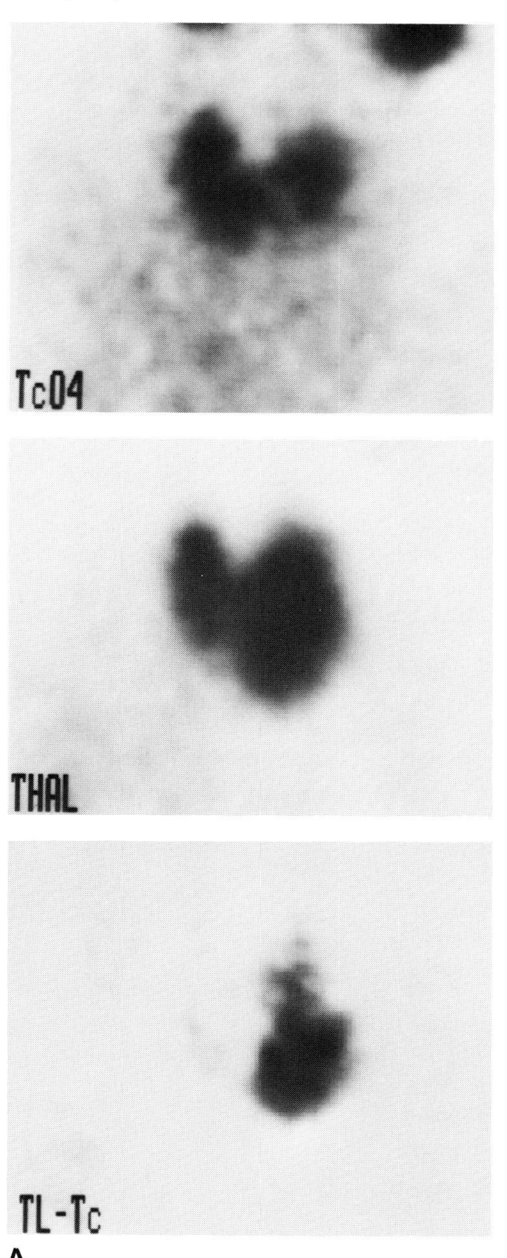

A

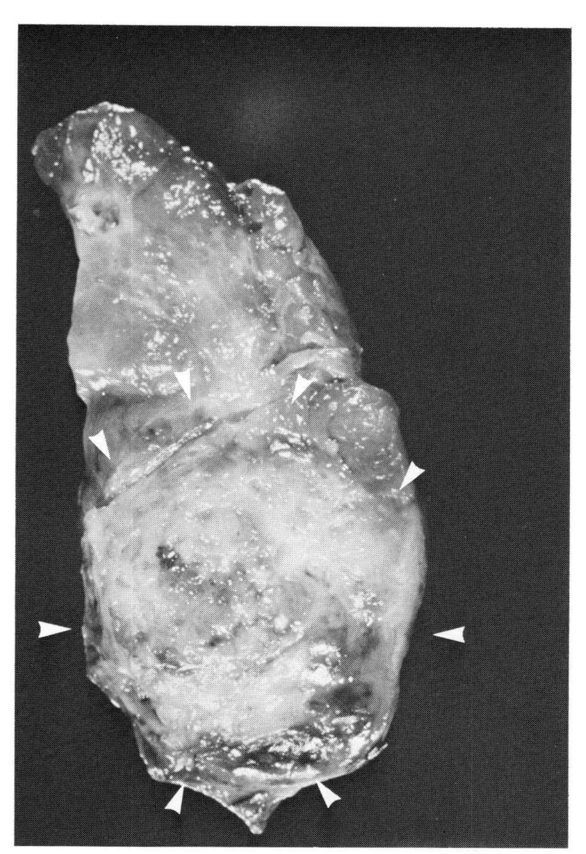

B

Figure 5-25 ***Multiple False-Positive (Thyroid Adenomas) and Parathyroid Adenoma***

This patient had a multinodular goiter. The arrow indicates where a 1-g parathyroid adenoma was found. The remaining foci represent thyroid adenomas. (Basarab RM, Manni A, Harrison TS: Dual isotope subtraction parathyroid scintigraphy in the preoperative evaluation of suspected hyperparathyroidism. Clin Nucl Med 10(4):300-314, 1985)

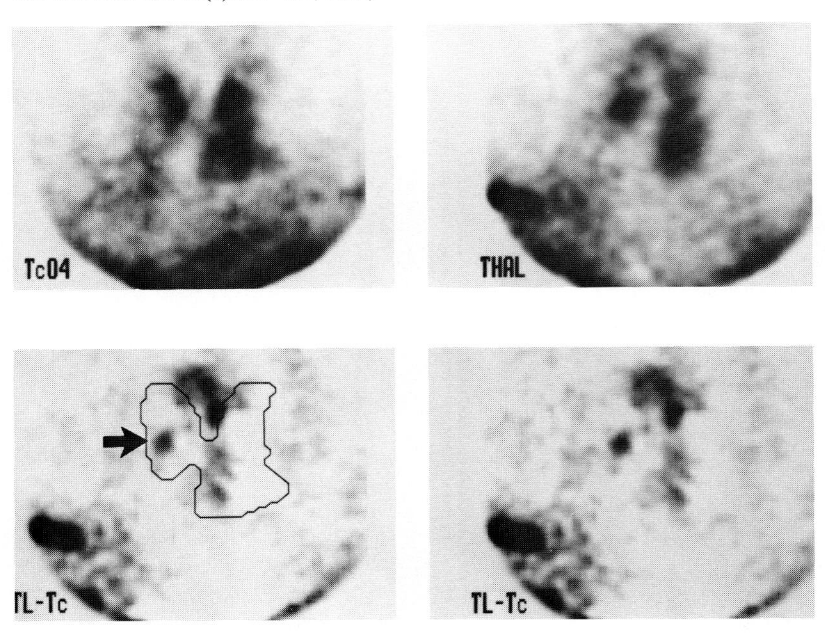

Figure 5-26 ***False-Positive (Hyperfunctioning Thyroid Adenoma)***

This patient was found to be hyperthyroid, although not suspected clinically prior to the examination. ***(A)*** Hyperfunctioning right thyroid adenoma with suppression of the remaining gland. ***(B)*** Thallium uptake can be seen in the suppressed portions of the gland. ***(C)*** and ***(D)***. Subtraction images. (Basarab RM, Manni A, Harrison TS: Dual isotope subtraction parathyroid scintigraphy in the preoperative evaluation of suspected hyperparathyroidism. Clin Nucl Med 10(4):300-314, 1985)

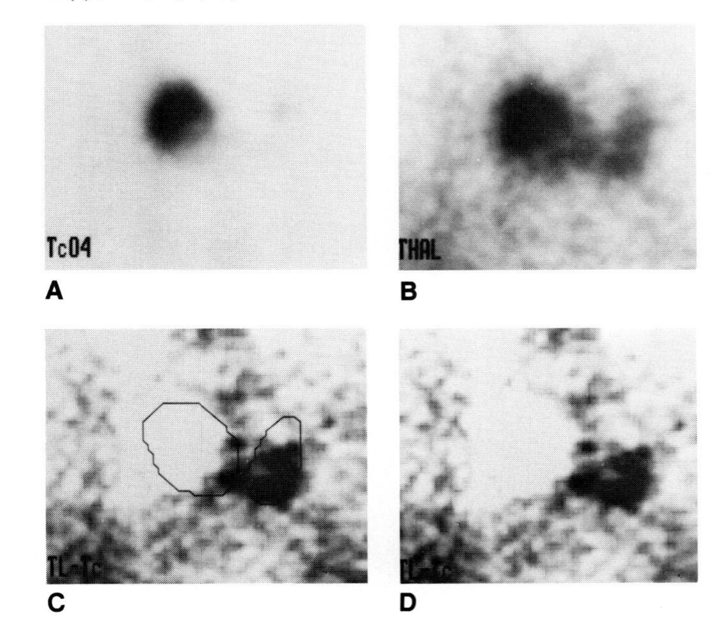

Figure 5-27 ***Thyroid Suppression with Analog Images***

(A) This examination was performed on a patient who was taking levothyroxine. Note the near complete suppression of pertechnetate *(upper left)* but the presence of thallium uptake *(lower left)*. ***(B)*** Repeat examination of the same patient 6 weeks after discontinuation of levothyroxine. Pertechnetate uptake in the thyroid is now seen *(upper right)*. In addition, there is more thallium uptake evident, indicating previous partial thallium suppression due to the levothyroxine *(lower left)*.

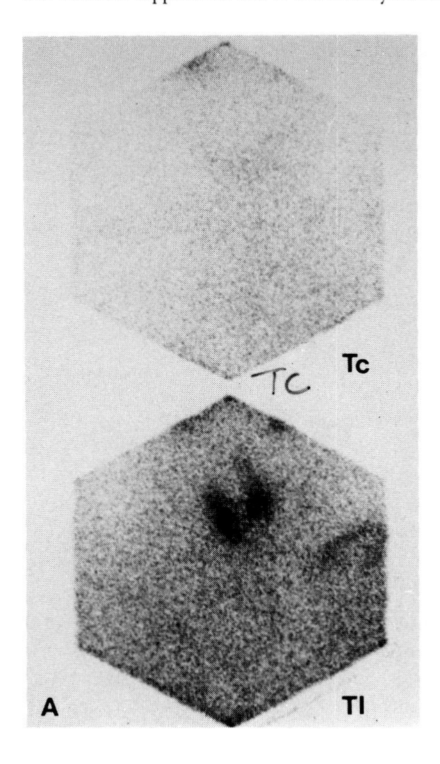

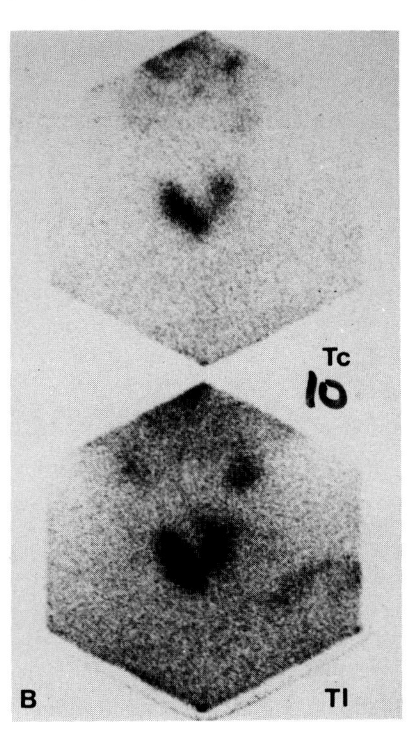

Stephen Manier
Douglas Van Nostrand
Frank Atkins
Sing-Yung Wu

Chapter 6 IODINE-131 NECK AND CHEST SCINTIGRAPHY

This chapter is designed to aid the physician who is either learning to perform and interpret Iodine-131 neck and chest scintigraphy or desires a review of the subject. The text is divided into discussions of patient preparation, imaging technique, quantitative estimation of I-131 clearance, and observations and interpretations of scintigraphs. The atlas section presents representative examples from the spectrum of normal, normal variants, artifacts, and various disease entities.

Indications for I-131 neck-bed ablation or I-131 therapy are not discussed in this chapter, but may be found elsewhere.[1–3]

PATIENT PREPARATION

Thyroid Hormone Withdrawal

Thyroxine is discontinued approximately 5 weeks prior to scintigraphy. Upon discontinuation of thyroxine or immediately post-thyroidectomy, the patient can be started on triiodothyronine (typically 25 μg orally b.i.d. or t.i.d.), to be continued for a minimum of 3 weeks, and then discontinued 2 weeks prior to scintigraphy.

The authors wish to thank Ms. Lorraine Woodruff for her word processing and secretarial support.

Discussion

Because the uptake of I-131 in normal tissue and functioning metastatic thyroid tissue is controlled by thyroid-stimulating hormone (TSH), it is important to increase TSH blood levels in order to maximize the likelihood of detecting any residual thyroid tissue and/or thyroid metastasis.

Although we follow the above procedure, other approaches to maximizing uptake of I-131 have been used, including administration of oral thyrotropin-releasing hormone (TRH), oral antithyroid drugs, or intravenous bovine TSH.[1,4,5]

Low-Iodine Diet

The daily intake of iodine is reduced to less than 20 mg to 30 mg for the 5 days prior to the study. The patient should not eat or drink for 2 hours prior to the administration of I-131.

Discussion

A low-iodine diet depletes the stores of body iodine, thereby theoretically increasing the uptake and possibly the residence time of the I-131 dosage in normal residual thyroid tissue or functioning thyroid carcinoma. This may result in better detection of metastasis on I-131 neck and chest scintigraphy and higher radiation doses delivered to the tumor from I-131 ablations and therapies. When a patient has known metastatic disease, the low-iodine diet is initiated 3 weeks prior to the scintigraphy. Other approaches to facilitate iodine depletion (*i.e.,* diuretics) have been advocated.[6,7] However, the value of iodine depletion has not been confirmed, and iodine depletion may increase total body irradiation for a given therapeutic dosage.[8]

Pregnancy

The physician must interview or review the clinical charts of all female patients to ensure a low likelihood of pregnancy (*e.g.,* patients should be premenarchal, postmenopausal, within 7 days of menses, or have a very reliable history of abstinence or birth control).

Discussion

No consensus exists regarding guidelines for this area, and each institution must establish its own policy. The above is one such guideline, and if the patient can not satisfy any of the established criteria, then the counseling physician, patient, and referring physician must decide whether or not to delay the study.

Blood Tests

Complete blood count (CBC) and blood thyroglobulin samples are obtained prior to dosing.

Discussion

Bone marrow suppression is an infrequent complication of ablations and therapies using dosages of 29 to 200 mCi (1.07 to 7.40 GBq).[1] However, a pre-ablation or pretherapy CBC provides a valuable baseline should subsequent therapies result in significant accumulative amounts of administered radioiodine. Thyroglobulin levels may be of value in following these patients as markers of recurrent metastatic disease.[1]

Thyroid-Stimulating Hormone (TSH) Blood Levels

TSH blood levels are obtained prior to dosing and the blood levels are simultaneously reviewed with the scintigraphic images. If the TSH blood levels are greater than 40 mIU/ml, then the thyroid withdrawal of hormone is arbitrarily considered adequate. Patients who have TSH blood levels of less than 40 mIU/ml, are managed on an individual basis.

Discussion

Until the best approach for the use of TSH levels is established, each institution must develop its own policy. A proposed use of TSH blood levels to modify the scintigraphic and therapeutic approach is presented in Table 6-1 (see last page of the atlas section for this chapter). The chart uses an arbitrary TSH level of 40 mIU/ml and an I-131 uptake of 5%, and the proposed approach is dependent on timely TSH assay results.

IMAGING TECHNIQUE

Numerous imaging techniques for I-131 neck and chest exams are used with variations of such factors as dosage, time to imaging, imaging device, collimator, and duration of imaging. We believe that the methodology described in this section is fundamentally sound and will yield a diagnostic study. However, we do not propose that this is the only or even the best aproach, and obviously other factors such as goals, cost/benefit ratio, and availability of equipment will influence the protocol design for each institution.

Dose

A dose of 5 mCi (185 MBq) I-131 sodium iodide liquid or capsule is given orally.

Discussion

A broad range of diagnostic doses, from 0.5 to 10 mCi (18.5–370 MBq), have been reported. The detection rate of normal residual thyroid tissue and metastasis increases with increasing dosages.[9]

Time to Imaging

Imaging is performed 72 hours or later after dosing.

Discussion

If imaging is performed earlier than 72 hours, the images may be less sensitive to metastasis because significant blood pool radioactivity may obscure I-131 uptake by metastatic foci (Fig. 6-1). However, blood pool radioactivity may persist at 72 hours, which would necessitate 96- or 120-hour delayed images.[10]

Evaluation of the presence of cardiac radioactivity may aid in determining if additional delayed images are required. The use of a 5-mCi (185 MBq) dose or greater permits these longer delayed imaging times.

Imaging Device

All images are obtained on a gamma camera.

Discussion

Both the gamma camera and the rectilinear scanner have been used for imaging; each has advantages and disadvantages, which are discussed elsewhere.[11] Gamma cameras have replaced rectilinear scanners in most institutions. Images obtained from a rectilinear scanner are shown in Figures 6-2, 6-13A, 6-15A, and 6-16.

Collimator

Our choice in collimators is the pinhole collimator for anterior neck imaging and an I-131 parallel-hole collimator for the chest, abdomen, and extremities.

Discussion

The construction of an acceptable I-131 collimator is compromised by competing factors, such as septal penetration and sensitivity. A high-energy parallel-hole collimator minimizes septal penetration at the expense of sensitivity, and a medium-energy collimator increases sensitivity at the expense of greater septal penetration. We have found the latter to be an acceptable compromise, except when significant residual radioactivity is present, such as in the thyroid region. In these situations a "star pattern" (Fig. 6-3) is seen secondary to septal penetration, and this radioactivity may obscure other areas of I-131 uptake. Pinhole collimator images eliminate this problem and offer magnified, higher resolution images of the neck-bed region or any other specific region of interest.

Energy Peak and Window

An energy peak of 364 Kev is used with a 20% window.

Images

The standard imaging series to be employed depends on whether the patient has known thyroid metastasis or has a high likelihood of thyroid metastasis. For patients with no known metastasis, images may be obtained of the anterior neck and chest, posterior neck and chest, and posterior abdomen and pelvis. If a lesion is detected on any of these images, then the series is extended. For those patients with known metastasis or a high index of suspicion, the whole body is imaged, excluding hands and feet.

Markers

Preceding each of the images, the patient is positioned and radioactive markers are placed on various anatomical landmarks within the field of view (*e.g.,* suprasternal notch, chin, thyroid cartilage, xyphoid, shoulder, and iliac crest) and recorded on a separate hard copy film. Without moving the patient, the radioactive markers are removed, and the image is obtained for the desired time (counts) on a second hard copy film. After developing the films, the film with the markers is superimposed on the film with the patient's images.

Duration of Imaging

Each image obtained for 20,000 total counts or 45 minutes, whichever occurs first.

Discussion

As the duration of imaging increases, the ability to detect areas of radioiodine uptake improves because the signal/noise ratio is improved. At all times, however, balance between lesion detection, patient comfort, and cost must be maintained. An example of how the combination of duration of imaging and collimator can affect lesion detection is shown in Figure 6-4.

QUANTITATION

Quantitative measurements may be performed concomitant with the I-131 neck and chest scintigraphy to aid in the selection of ablative or therapeutic doses of I-131. Several approaches have been reported;[12–14] an overview of the approach of Benua and Leeper[13,15] is presented below.

As is generally the case in radiation therapy, the treatment is limited not by the radiation dose to the target (tumor), but to some other "critical organ." However, the selection of a therapeutic dose of I-131 for the treatment of metastatic thyroid disease has largely been an arbitrary judgment applied to the population as a whole and not to the individual. An alternative approach, that used by Benua and Leeper at Memorial Sloan Kettering Cancer Center, determines a treatment dose of I-131 based on estimates of the radiation exposure to an assumed critical organ, which in this case is the blood (bone marrow). A dose is selected to keep the total radiation exposure to the blood within a "safe" limit. In order to calculate the radiation dose to the assumed critical organ, a model is required which incorporates the data of biological uptake and clearance, radioactive decay, and other physical characteristics of the I-131 in the various compartments comprising the model. The particular model used by this group was developed in the late 1940s.[16]

According to the Benua and Leeper method, the radiation dose to the blood (marrow) from the beta and gamma emissions of I-131 is calculated assuming a two-compartment model (*i.e.,* blood and a generalized whole body). Because of the limited range of beta particles, it can be assumed that their energy is totally absorbed within the same compartment from which they are emitted. Hence, the beta component of the radiation dose to the critical region is derived only from the radioactivity which is contained within the whole blood compartment. To a lesser extent, the critical organ (bone marrow) is also irradiated by the gamma emissions from the I-131 that is distributed throughout the whole body. This part of the model assumes that *both* the bone-marrow radioactivity and the whole-body radioactivity are homogeneously distributed. An *average* gamma dose can then be calculated using a formula that incorporates a geometrical factor based on the patient's height and weight.[17]

The biokinetics of orally administered I-131 varies from patient to patient. In order to perform the calculations using this model, it is necessary to know what percentage of the administered I-131 resides in the blood and whole-body compartments as a function of time. To obtain this data, measurements are performed over the course of at least 5 days following the administration of a known dose (1–5 mCi [37–185 MBq]) of I-131. Blood samples are drawn into 5-ml heparinized vials at periodic intervals (*e.g.,* 4 hr, 24 hr, 48 hr, and 72 hr) and stored for counting at the completion of the study. Two independent techniques are employed in the measurement of the whole-body component, one based on a direct measurement of the gamma radiation from the patient, and one inferred from a collection of the patient's urine. In the former, a large, well-shielded, uncollimated sodium iodide (Tl) scintillation detector is used to measure the gamma radiation emanating from the patient who is located at a reproducible distance from the probe (10–15 feet). Whole-body counting is performed immediately after dosing, and then at 2-, 4-, and 24-hour intervals for at least 4 days or until the residual activity is less than 2% of the administered dosage or until the clearance rate approximates physical decay.

The whole-body retention is also measured indirectly by determining the total radioactivity excreted in the urine. Complete urine collection is performed over 24-hour intervals beginning from the time of dosing. Aliquots of urine are counted against a standard at the completion of the study to determine the total I-131 excreted each day. The residual whole-body radioactivity at the end of each day is inferred from the difference between the value for the previous day and the percentage of the dosage excreted during the last 24-hour interval. Computer programs are available to integrate these retention curves (Fig. 6-5A and B) and to compute the dose (in rads/mCi) to the blood using the Marinelli formula.[16] The treatment dose is then that quantity of I-131 that will deliver a dose to the blood of 200 rads (2Gy) — demonstrated to be the "largest safe dose" that does not produce clinically significant bone marrow suppression.[13,18] Based upon their clinical experience, Benua and Leeper have added an additional constraint on the treatment dose, namely that the projected whole-body retention at 48 hours must not exceed 120 mCi (4.44 GBq) or 80 mCi (2.96 GBq) in the presence of diffuse lung metastases.

Radiation Absorbed Doses

As noted in Figure 6-5, the clearance of I-131 can have and usually does have significant variability. This together with other variable factors (*e.g.,* percent uptake of I-131 in thyroid bed, percent uptake in metastasis, and residual times) make it very difficult to give singular estimates of radiation absorbed doses for neck and chest scintigraphic dosages that are meaningful to any individual patient. However, calculations of estimates can be made based on the schema of the Medical Internal Radiation Dose (MIRD) Committee for calculating the absorbed dose from biologically distributed radionuclides.

OBSERVATION AND INTERPRETATIONS

When asked to read a neck and chest exam, the frequent response of untrained physicians is either "It's beyond me" or "It's easy — just look for the hot spots." Neither of these responses is appropriate.

Like most scintigraphs or radiographs, an I-131 neck and chest exam is read in three steps:

1. Ensuring quality of technique
2. Observation
3. Interpretation*

The first step, ensuring quality technique, is often difficult and can be disastrous if overlooked. A well-written procedure should be available, and the physician must attempt to ensure quality by questioning the patient before beginning the exam as to his or her compliance with thyroid hormone withdrawal; reviewing TSH blood levels; questioning the technologist regarding window settings, exposure intensity, duration of imaging, counts, and any other factors affecting the image acquisition/recording process; and reviewing the films for the placement of marker sources. Without thus ensuring the quality of technique, steps two and three may have reduced or no value.

The second step, observation, is simple and involves describing areas of increased radioactivity. (The physician who states that the I-131 neck and chest scintigraphy is "easy" typically only performs this step.) The third step, interpretation, involves the individual evaluation of each area of increased radioactivity for etiology. This task is facilitated by using five tools:

An understanding of normal distribution of radioiodine with time
Knowledge of normal variants
Knowledge of artifacts

*The authors created this division as a tool to aid students while recognizing that the steps are not always distinct.

Knowledge of the rare false-positives
Techniques to distinguish all of the above from metastasis

Each of these is illustrated within the atlas section of this chapter: normal distribution of radioactivity in Figures 6-6 through 6-12; normal variants in Figures 6-13 through 6-17; artifacts in Figure 6-18; false-positives in Figures 6-29 to 6-32. Techniques to distinguish the etiologies of I-131 radioactivity are noted in the *Comment* section of the respective figures.

Without these tools, the I-131 neck and chest scintigraphy will be seen as "beyond me," and the procedure, rather than the interpreter, will often be unfairly criticized. If one learns and uses these tools, we believe the physician will discover that the I-131 neck and chest scintigraphy is neither "easy" nor "beyond" him or her, but rather very valuable and rewarding.

REFERENCES

1. Hurley JR, Becker DV: The use of radionuclide in the management of thyroid cancer. In Freeman LM, Weissman HS (eds): Nuclear Medicine Annual, pp 329–384. New York, Raven Press, 1983
2. Baum S, Vincent NR, Wu SY: Atlas of Nuclear Medicine Imaging pp 145–146. New York Appleton-Century-Crofts, 1981
3. Sisson JC: Applying the radioactive eraser: I-131 to ablate normal thyroid tissue in patients from whom thyroid cancer has been resected. J Nucl Med 24:743, 1983
4. Hilts SV, Hellman D, Anderson J, et al: Serial TSH determination after T_3 withdrawal or thyroidectomy in the therapy of thyroid carcinoma. J Nucl Med 20:928, 1979
5. Vagenakis AG, Braverman LE, Azizi F et al: Recovery of pituitary function after withdrawal of prolonged thyroid suppression therapy. N Engl J Med 293:681, 1975
6. Gosling BM: Effect of a low iodine diet on I-131 therapy in follicular thyroid carcinoma. J Endo 64:30, 1975
7. Hamburger JI: Diuretic augmentation of I-131 uptake in inoperable thyroid cancer. N Engl J Med 280:1091, 1969
8. Marvca J, Santner S, Miller K et al: Prolonged iodine clearance with a depletion regimen for thyroid carcinoma: Concise communication. J Nucl Med 25:1089, 1984
9. Waxman A, Ramanna L, Chapman N et al: The significance of I-131 scan dose in patients with thyroid cancer: Determination of ablation. Concise Communication. J Nucl Med 22:861, 1981
10. Bekerman C, Gottschalk A, Hoffer P et al: Optimal time for I-131 total body imaging to detect metastatic thyroid carcinoma. J Nucl Med 15:477, 1974
11. Kirchner P, Brown T, Bidani N et al: I-131 whole body surveys: Tomographic vs gamma camera imaging. J Nucl Med 21:p 30, 1980
12. Thomas SR, Maxon HR, Kereiakes JG et al: Quantitative external counting techniques enabling improved diagnostic and therapeutic decisions in patients with well-differentiated thyroid cancer. Radiology 122:731, 1977
13. Benua SR, Cicale NR, Sonenberg M et al: The relation of radioiodine dosimetry to results and complication in the treatment of metastatic thyroid cancer. Am J Roent Rad Ther Nucl Med 87:171, 1962
14. Maxon HR, Thomas SR, Hertzberg VS et al:Relation between effective radiation dose and outcome of radioiodine therapy for thyroid cancer. N Engl J Med 309:937, 1983
15. Leeper RD, Shimaoka K: Treatment of metastatic thyroid cancer. Clin Endo Metab 9:383, 1980
16. Marinelli ID, Quimby EH, Hine GJ: Dosage determination with radioactive isotopes. Am J Roent Rad Therapy, 50:260, 1948
17. Loevinger R, Holt JG, Hine GJ: Internally Administered Isotopes. Radiation Dosimetry, pp 801–873. New York, Academic Press, 1956
18. Cember H: Introduction to Health Physics, pp 183. Oxford, Pergamon Press, 1969
19. Snyder J, Gorman C, Scanlon P: Thyroid remnant ablation: Questionable pursuit of an ill-defined goal. J Nucl Med 24:659, 1983
20. Balachandran S, Sayle BA: Value of thyroid carcinoma imaging after therapeutic doses of radioiodine. Clin Nucl Med 6:162, 1981
21. Tyson JW, Wilkinson RH, Witherspoon LR et al: False positive I-131 total body scan. J Nucl Med 15:1052, 1974
22. Grossman M: Gastroesophageal reflux: A potential source of confusion in technetium thyroid scanning: Case report. J Nucl Med 18:548, 1977
23. Berquist TH, Nolan NG, Stephens DH et al: Radioisotope scintigraphy in diagnosis of Barrett's esophagus. Am J Roent 123:401, 1975
24. Dhawan VM, Kaess KP, Spencer RP: False positive thyroid scan due to Zenker's diverticulum. J Nucl Med 19:1231, 1978
25. Salvatore M, Gallo A: Accessory thyroid in the anterior mediastinum: Case report. J Nucl Med 16:1135, 1975
26. Lin DS: Thyroid imaging-mediastinal uptake in thyroid imaging. Sem Nucl Med 13:395, 1983
27. Jackson G, Graham W, Flickinger F et al: Thymus accumulation of radioactive iodine. Penn Med 82:37, 1979
28. Wiseman, J: Bony metastases from thyroid carcinoma or contamination? Clin Nucl Med 9:363, 1984
29. Abdel-Dayem H, Halker K, Sayed ME: The radioactive wig in iodine-131 whole body imaging. Clin Nucl Med 9:459, 1984
30. Kirk G, Schulz E: Post-laryngectomy localization of I-131 at tracheostomy site on a total body scan. Clin Nucl Med 9:409, 1984
31. DeGroot LJ, Larsen PR, Refetoff S et al: The Thyroid and its Diseases, p 634. New York, John Wiley & Sons, 1984
32. Massin J, Savoie J, Garnier H et al: Pulmonary metastasis in differentiated thyroid carcinoma. Cancer 53:982, 1984
33. Samaan N, Schultz PN, Haynie TP et al: Pulmonary metastasis of differentiated thyroid carcinoma: Treatment results in 101 patients. J Clin Endocrinol Metab 65:376, 1985
34. Froment J: Les metastases respiratoires du cancer thyroiden. Nancy, France, These, 1969

35. Beierwaltes WH, Nishiyama RH, Thompson NW et al: Survival time and "cure" in papillary and follicular thyroid carcinoma with distant metastases: statistics following university of Michigan therapy. J Nucl Med 23:561, 1982

36. Maheshwari YK, Hill CS, Haynie TP et al: I-131 therapy in differentiated thyroid carcinoma. Cancer 47:664, 1981

37. Castillo LA, Yeh SDJ, Leeper RD, Benua RS: Bone scans in bone metastasis from functioning thyroid carcinoma. Clin Nucl Med 5:200, 1980

38. Parker LN, Wu SY, Kim DD et al: Recurrence of papillary thyroid carcinoma presenting as a focal neurological deficit. Arch Int Med (in press) 146:1985, 1986

39. Fernandez- Ulloa M, Maxon HR, Mehta S et al: Iodine-131 uptake by primary lung adenocarcinoma, misinterpretation of I-131 scan. JAMA 236:857, 1976

40. Acosta J, Chitkara R, Kahn F et al: Radioactive iodine uptake by a large cell undifferentiated bronchogenic carcinoma. Clin Nucl Med 8:368, 1982

41. Wu SY, Kollin J, Coodley E et al: I-131 total-body scan: Localization of disseminated gastric adenocarcinoma. Case report and survey of the literature. J Nucl Med 25:1204, 1984

42. Michael BE, Forouhar FA, Spencer RP: Medullary thyroid carcinoma with radioiodide transport—effects of iodine-131 therapy and lithium administration. Clin Nucl Med 10:274, 1985

43. Yeh EL, Meade RC, Ruetz PP: Radionuclide study of *struma ovarii.* J Nucl Med 14:118, 1973

44. Wu SY, Brown T, Milne N et al: Iodine 131 total body scan-extrathyroid uptake of radioiodine. Semin Nucl Med 16:82, 1986

Atlas Section

Figure 6-1

Value of Delayed Imaging

(A) The 24 hour I-131 image of the chest shows one, and possibly two, subtle focal areas of increased radioactivity in the pericardial area *(arrowhead)* and a focal lesion just superior and contiguous to the fundus of the stomach *(arrow);* however, both are partially obscurred by cardiac blood pool activity. ***(B)*** The 48-hour image shows two well-defined focal areas of radioactivity in the same area as the two subtle lesions *(arrowheads)* seen in the 24-hour image, and better definition of the lesion superior to the fundus of the stomach *(arrow).* The latter was not esophageal (for the distinction, see Figs. 6-13and 6-14. Markedly reduced cardiac blood pool radioactivity is readily apparent. (*N*, neck bed; *S*, stomach, *B*, bowel radioactivity) ***Comment:*** Delayed images (48 hours or longer postdosing) improve sensitivity for lesion detection, because blood pool radioactivity decreases with time while target (lesion) radioactivity may increase, remain unchanged, or decrease at a slower rate than background. This results in a better target-to-background ratio.

A

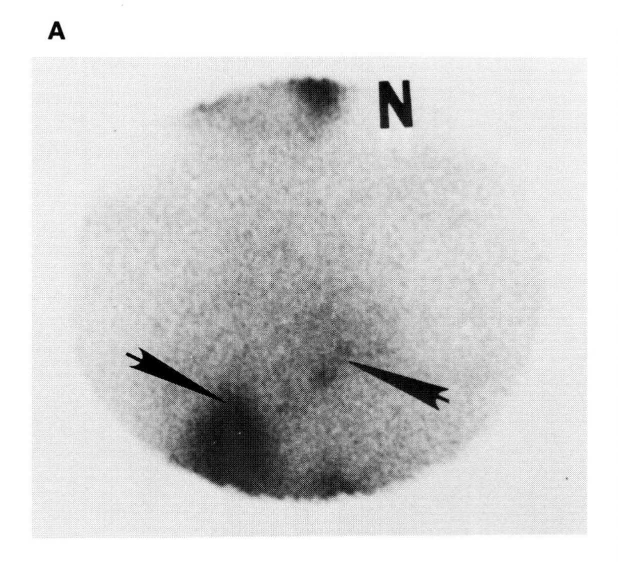

B

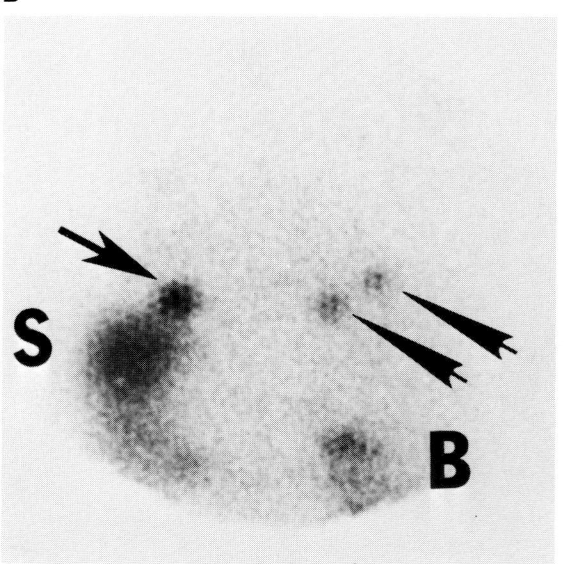

Figure 6-2 ***Image Obtained from a Rectilinear Scanner***

(See the section, "Imaging Technique, Imaging Device.")

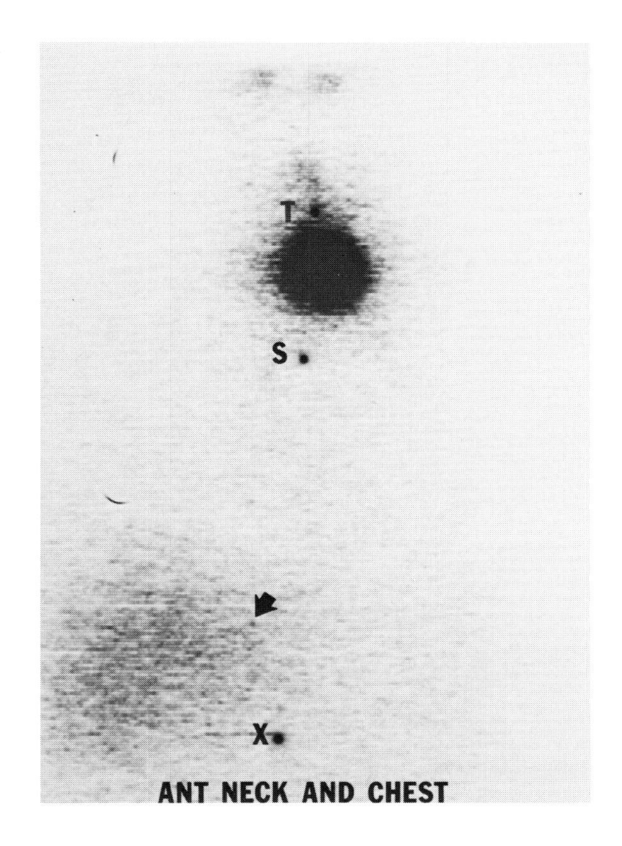

Figure 6-3 ***Septal Penetration***

The "star pattern" above represents septal penetration and may obscure other areas of I-131 uptake. Pinhole or high-energy collimators may eliminate or reduce this problem, respectively. (See the section, "Imaging Technique, Collimator.")

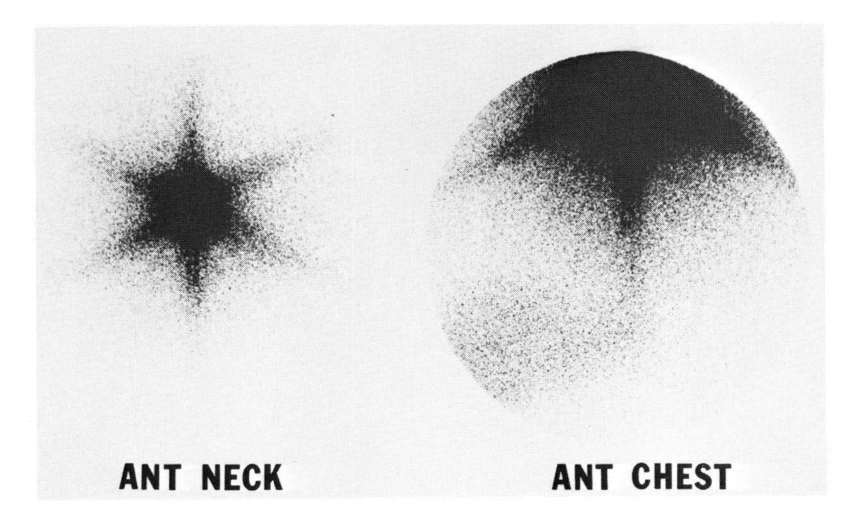

Figure 6-4 ***Value of Longer Imaging Time and Collimator***

(A) Three focal areas of residual neck bed radioactivity are visible on this I-131 study performed with a parallel-hole collimator for a total imaging time of 0.75 hours. Chin *(C)*, suprasternal notch *(S)*, and xyphoid *(X)* are noted with cobalt-57 markers. ***(B)*** Imaging was immediately repeated with the same camera and other technical factors, except the imaging time was twice as long (1.5 hours) and a pinhole collimator was used. Three additional areas of uptake are demonstrated *(arrows)*. ***Comment:*** Technical factors such as longer imaging times or pinhole collimation may improve lesion detection. Failure to standardize may cause a misdiagnosis, such as "newly" visualized lesions suggesting metastatic foci.

A

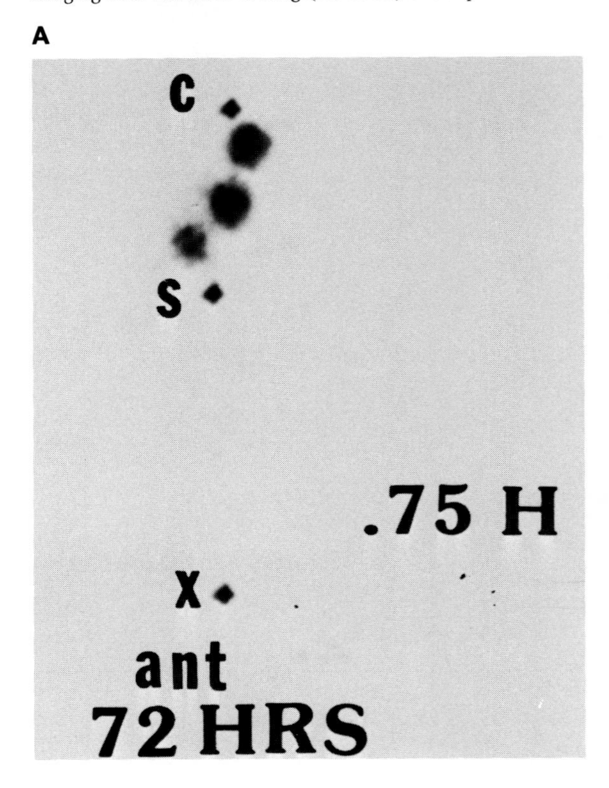

B

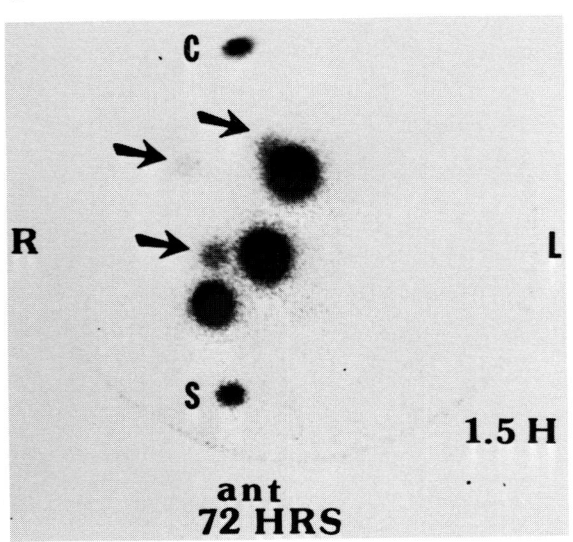

Figure 6-5 *Quantitative Clearance Curves*

(*A* and *B*). The clearance curves of radioiodine are measured by blood, whole body probe, and whole body urine for two patients. ***Comment:*** For a given dosage, the patient with the slower clearance ***(B)*** would have an expected higher radiation dose to the bone marrow than the patient with the faster clearance ***(A)***. Thus, the patient in ***B*** would have a lower calculated I-131 therapy dosage.

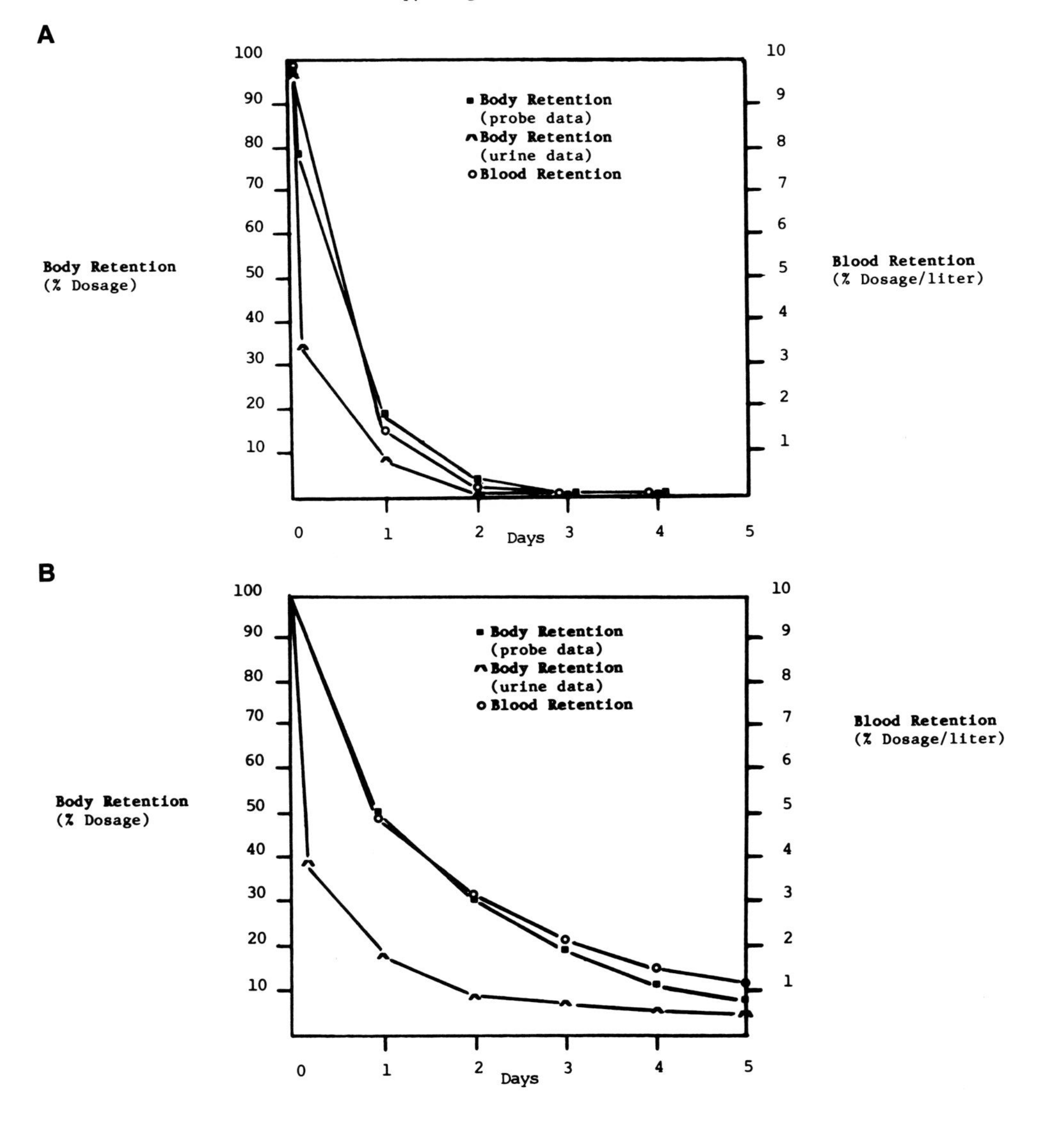

Figure 6-6

Normal Salivary and Nasopharyngeal Uptake

This anterior image of the head demonstrates normal radioiodine uptake in the parotid glands *(short arrows)* and submandibular salivary glands *(long arrows).* Radioactivity in the midline *(thick arrow)* represents oropharyngeal activity. The faint radioactivity that outlines the head is secondary to residual blood pool radioactivity, and the head is hyperextended. (The radioactivity in the midline below the two long arrows is discussed in Fig. 6-17.) ***Comment:*** Radioiodine is normally taken up in the salivary glands and may be secreted into the oropharynx.

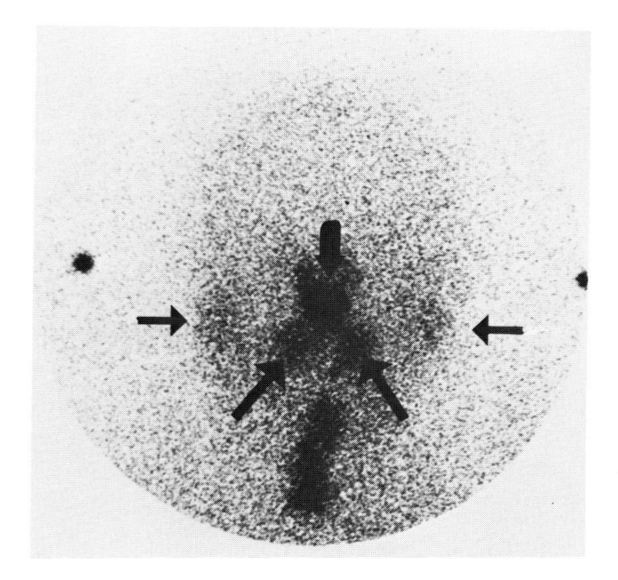

Figure 6-7

Normal-Nasal Uptake

A right lateral view of the head and neck was performed 72 hours after a diagnostic dose of I-131. Nasal radioactivity is demonstrated, which is consistent with normal uptake in minor salivary glands in the tip of the nose *(curved arrow).* Normal radioiodine uptake is noted in the oropharynx *(O),* parotid gland *(P),* and thyroid bed *(N).* ***Comment:*** Nasal uptake of I-131 should be recognized as a normal finding. The configuration is typically circular or oval, and the degree of radioiodine uptake may vary from background to greater than the intensity of radioactivity in the parotid gland and/or mouth.

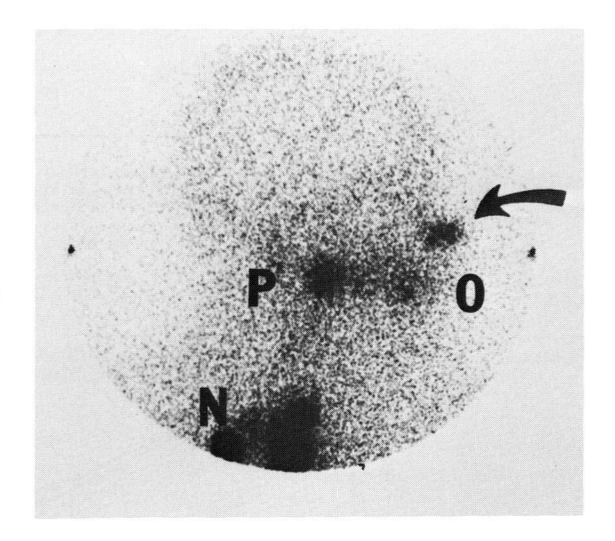

Figure 6-8 ***Normal Blood Pool Uptake***

This anterior I-131 neck and chest image was obtained 24 hours after dosing with 1.0 mCi (37 MBq) of I-131 orally. Increased radioiodine is noted in the cardiac region *(curved arrow)* and represents normal cardiac blood pool activity at 24 hours; this area cleared on the 72-hour image (not shown). The chin *(C)*, suprasternal notch *(S)*, and xyphoid *(X)* are noted with cobalt markers. *r*, right; *l*, left. ***Comment:*** The physician should be alert for and recognize blood pool activity. Although it is normal and should not be mistaken for diffuse metastasis, blood pool radioactivity may mask metastasis not only in the cardiac region but also in other areas. Delayed views must be repeated on subsequent days until the blood pool has cleared sufficiently (see Fig. 6-1).

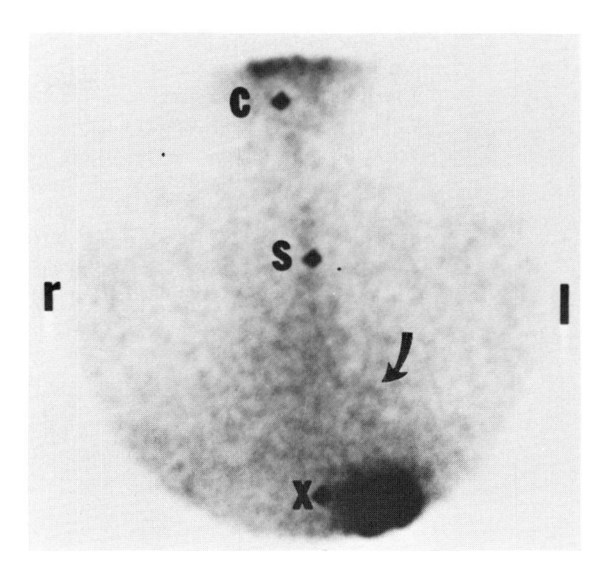

Figure 6-9 ***Normal Uptake—Residual Functioning Thyroid Tissue in Thyroid Bed Region***

On this anterior pinhole image of the thyroid region, the suprasternal notch is at the bottom of the field of view, and the chin lies just outside the top of the field of view. Minimal but definite radioactivity is present in the "thyroid bed" region. (The term "thyroid bed" is a colloquialism that refers to the normal anatomic position of the thyroid.) Although a typical representation of residual functioning thyroid tissue postoperatively, this radioactivity may also represent metastatic disease. (*r*, right; *l*, left) ***Comment:*** Residual functioning thyroid tissue is typically present after thyroid surgery, and ablation (destruction) with 29 mCi (1.07 GBq) to 150 mCi (5.6 GBq) is often performed to potentiate the future visualization and treatment of metastasis and/or to treat any remaining cancerous foci within the residual visualized area. No consensus exists among nuclear therapists regarding what represents adequate ablation (acceptable residual functioning thyroid tissue).[3,19] Factors that affect this decision include criteria such as the visual estimate or quantitative (*i.e.*, uptake) amount of radioactivity, the size of the area of radioactivity, and the histology of the original primary tumor. (Further reading is available.[3,19])

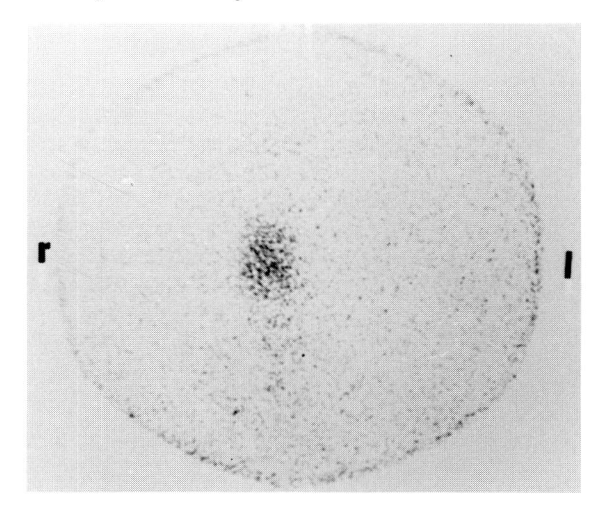

Figure 6-10

Normal Background Radioactivity in an Ablated Thyroid Bed

This anterior neck and chest image was obtained 72 hours after the oral administration of 5 mCi (185 MBq) of I-131 and showed no detectable residual thyroid bed radioactivity following a subtotal thyroidectomy and I-131 neck bed ablation for well-differentiated adenocarcinoma of the thyroid. No cardiac blood pool radioactivity is seen. Normal stomach activity is noted *(arrow),* and cobalt-57 markers are present over the lower aspect of the shoulders. (*C,* chin; *T,* thyroid cartilage, *S,* suprasternal notch) ***Comment:*** Before the above findings are interpreted as an ablated thyroid bed, the physician should confirm that the TSH is elevated.

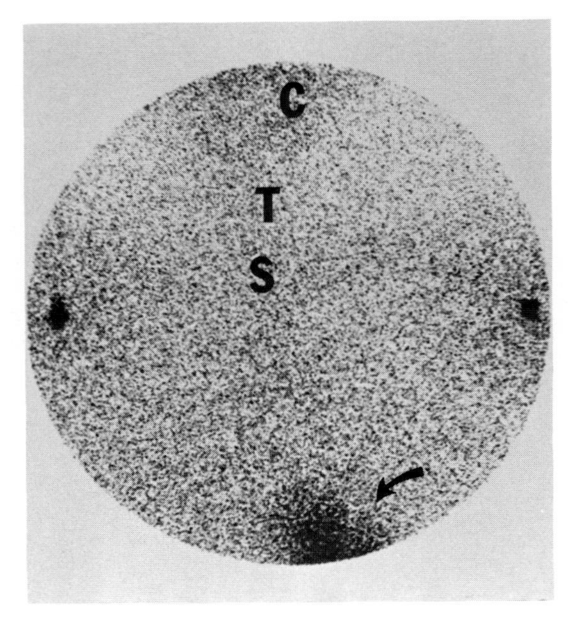

Figure 6-11

Normal Uptake—Abdomen

I-131 image of the abdomen shows normal stomach *(ST),* bowel *(B)* and urinary bladder *(UB)* radioactivity. The xyphoid *(X)* and right iliac crest *(i)* are noted with cobalt markers. ***Comment:*** In normal patients, radioiodine is secreted by gastric mucosa and passes into the intestinal lumen where it may be reabsorbed. Radioiodine is also removed from the blood by the kidneys and excreted into the urinary tract. Serial imaging or imaging following bowel cleanings or voidings may be necessary in order to distinguish these normal entities from metastases, such as in overlying bone or liver.

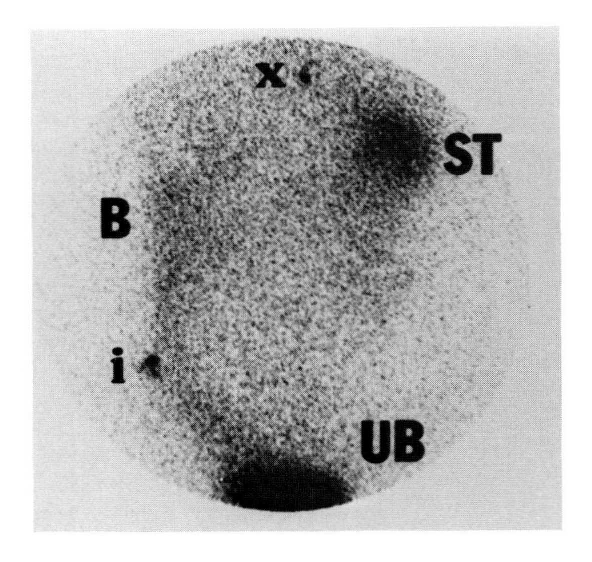

Figure 6-12 *Normal— or Abnormal?— Liver Uptake*

Anterior abdominal image was performed 10 days after a therapeutic dosage of I-131 and shows diffuse, homogeneous tracer uptake in the liver *(arrow)*. Markers denote xyphoid *(X)*, umbilicus *(U)*, and skin *(S)*. This radioactivity most likely represents radioiodinated compounds from the metabolism of radiolabeled triiodothyronine, thyroxine, or related chemicals, which were synthesized with radioiodine by either residual functioning normal thyroid tissue or functioning thyroid carcinoma.

Comment: (1) Occasionally minimal liver radioactivity may be present on images obtained within 24 hours of dosing (diagnostic or therapeutic), and this radioactivity represents blood pool. On images obtained from 24 to 96 hours after dosing, the liver is not tyically visualized; however, if residual functioning normal or abnormal thyroid tissue is present, then the liver typically is visualized on images performed 96 hours or more after dosing. (2) Functioning liver metastases usually appear as focal areas of increased radioactivity. To our knowledge, diffuse functioning liver metastasis from thyroid carcinoma has not been reported. (3) Images may be performed from 7 to 14 days after ablative or therapeutic dosages of I-131 to detect metastasis not demonstrated on images obtained after diagnostic dosages at 24 to 96 hours.[20]

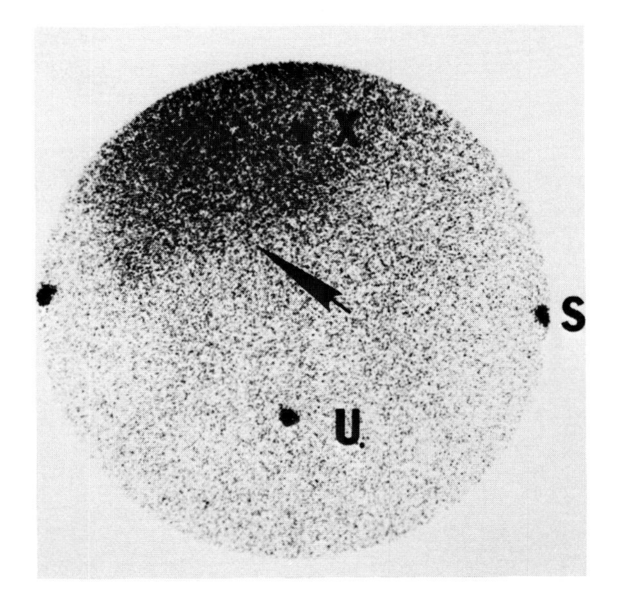

Figure 6-13 ***Normal Variant: Oral Pharyngeal/Esophageal Uptake***

(***A*** and ***B***). I-131 rectilinear scans are of the neck region using identical scan parameters. The multiple areas of radioiodine concentration seen in ***A*** are not visualized in ***B***, which was obtained after the patient drank water. ***(C)*** In a barium swallow study of the same patient, barium can be seen in a dilated cervical esophagus, which corresponds to the pyriform sinuses and the areas of abnormal I-131 accumulation seen in *A*. Markers represent the levels of thyroid cartilage *(TC)* and suprasternal notch *(SSN)*. ***Comment:*** Although location, appearance, and/or symmetry will often help to distinguish normal radioactivity within the oropharynx, neck, or mediastinum from abnormal radioactivity, additional manipulations may be required. Oral fluids may rinse the area of the radioactivity; oblique images may localize the radioactivity as deep (*i.e.*, possible esophagus) or superficial (*i.e.*, possible bone metastasis); serial images may demonstrate a change in configuration which is atypical for metastasis; and images after lemon juice gargle may "washout" radioactivity in salivary glands. (Tyson JW et al: False positive I-131 total body scan. J Nucl Med 15:1052, 1974)

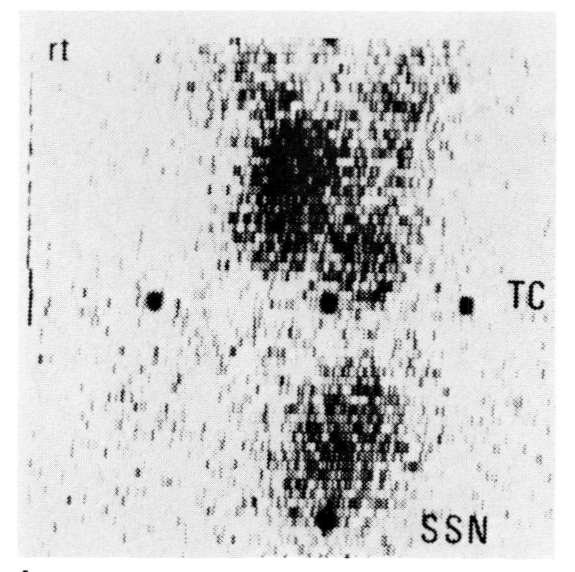

A

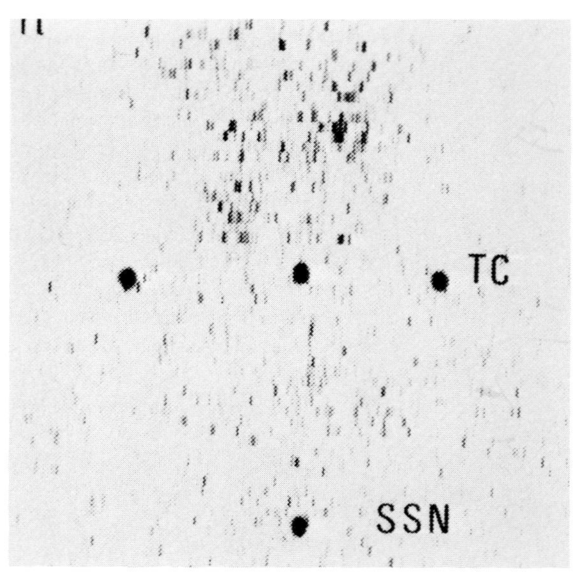

B

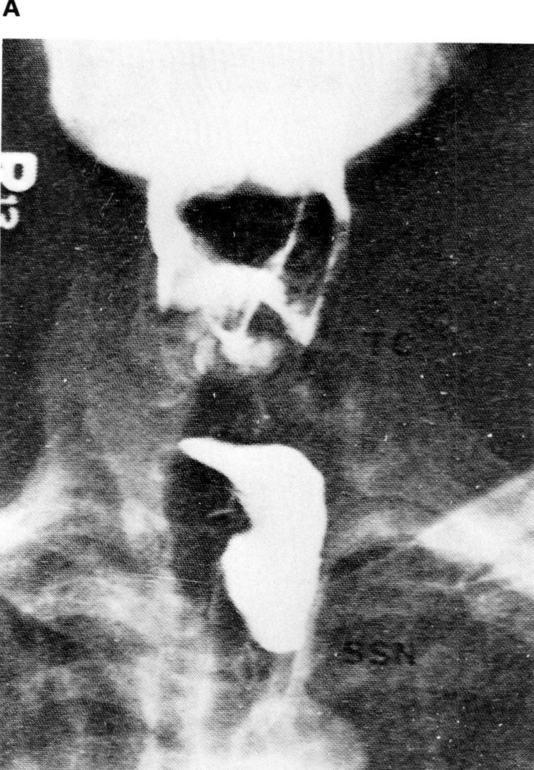

C

Figure 6-14 ***Normal Variant—Esophageal Radioactivity***

Although these thyroid images were obtained with Tc-99m O_4^-, they further emphasize the potential pitfall of esophageal radioactivity. ***(A)*** Anterior image of the thyroid performed with a pinhole collimator demonstrates a large linear area of radioactivity inferior to the right lobe of the thyroid *(arrow)*. ***(B)*** Obtained with a parallel-hole collimator, this image demonstrates that the radioactivity extends to the xyphoid region *(large arrow)*. After the patient drank several glasses of water, this radioactivity cleared. The patient had achalasia, and the area of radioactivity was believed to be Tc-99m O_4^-, which came from the salivary glands and was swallowed into the dilated esophagus. (*X*, xyphoid; *S*, suprasternal notch; *ST*, stomach) ***Comment:*** I-131, like Tc-99m O_4^-, may appear in the esophagus via several mechanisms. First, the I-131 that appears in the salivary glands may be secreted into the oropharynx and subsequently swallowed into the esophagus. Second, I-131 that is secreted by the gastric mucosa into the stomach may reflux into the esophagus.[22] Third, ectopic gastric mucosa in a Barrett's esophagus may "secrete" I-131.[23] Although the esophageal radioactivity from any of the above mechanisms should clear rapidly, stasis may occur with Zenker's diverticulum[24] or achalasia. As emphasized in the previous case, further manipulations may be necessary in order to distinguish normal from abnormal accumulations of radiotracer.

A

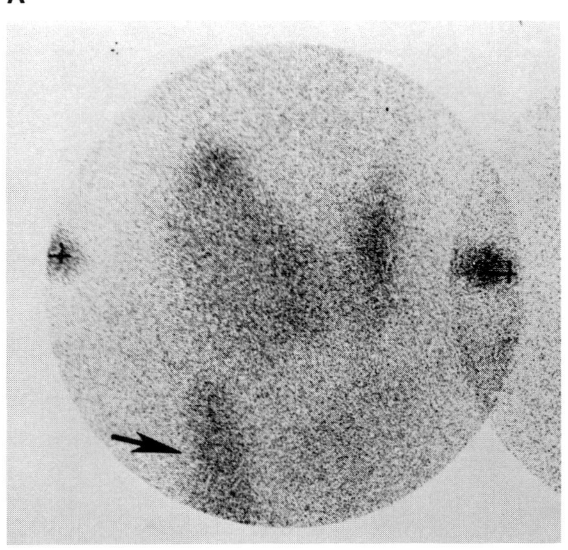

B

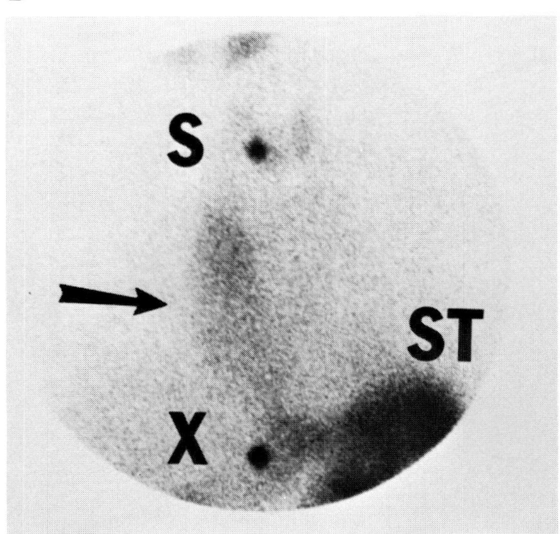

Figure 6-15 *Normal Variant—Substernal/Mediastinal Thyroid*

(A) On an I-131 rectilinear scan, the "*v*" represents the suprasternal notch, and the parallel lines represent the contours of the clavicle. Radioiodine uptake is noted in the normal thyroid bed region above the suprasternal notch; however, a large area of radioiodine is seen below the suprasternal notch. This area corresponds to mediastinal widening on chest radiograph ***(B)*** and was confirmed surgically as normal anterior mediastinal thyroid tissue.
Comment: This case emphasizes an important normal variant which could easily be confused with lung or mediastinal metastasis. We are unaware of any additional manipulations of radioiodine images that would have resolved this issue.[26] (Salvatore M, Gallo A: Accessory thyroid in the anterior mediastinum: Case report. J Nucl Med 16:1135-1136, 1975)

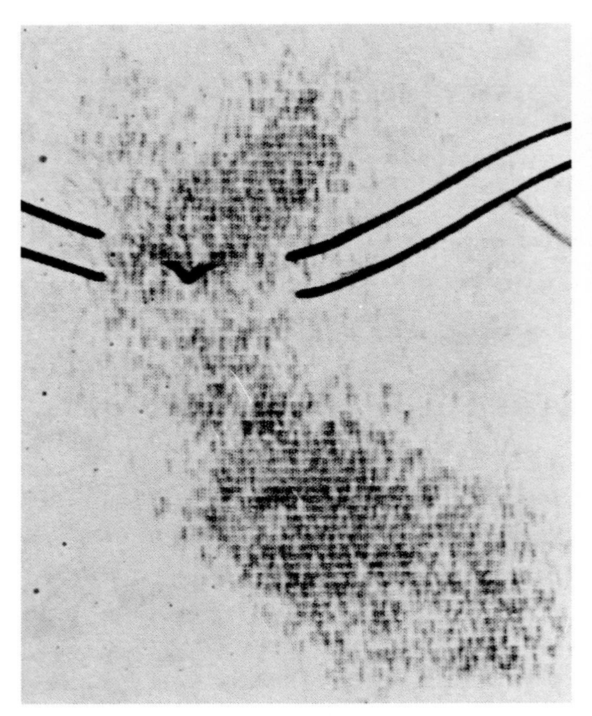

A

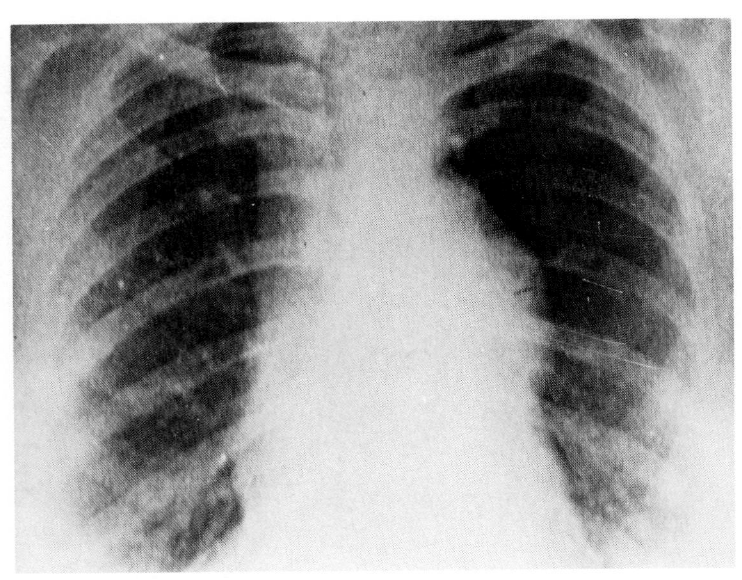

B

Figure 6-16 *Normal Variant—Possible Thymus*

This I-131 scan demonstrates a focal area of radioactivity in the region of the right lobe of the thyroid. Because subsequent surgery demonstrated thymic tissue in this region with no normal or metastatic thyroid tissue, the focal area of radioiodine was interpreted as radioiodine uptake in thymus. ***Comment:*** Although this case does not conclusively prove that thymic tissue takes up radioiodine, a case reported by Jackson and co-workers[27] indicated the presence of radioiodine in the thymus by autoradiography. These two findings together should alert the physician to the possibility of thymic radioiodine uptake. (Jackson G et al: Thymus accumulation of radioactive iodine. Penn Med 82:37, 1979)

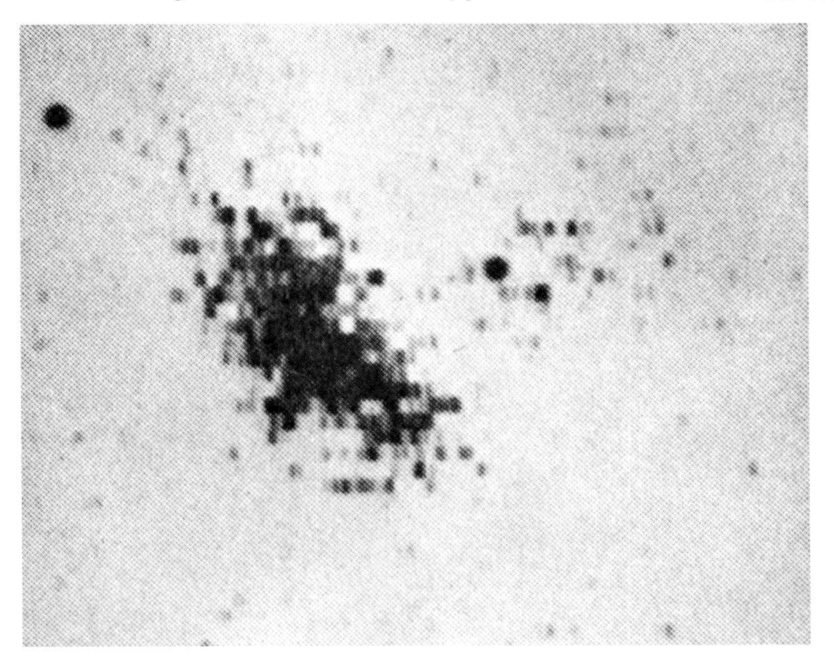

Figure 6-17 *Normal Variant—Thyroglossal Duct Remnant*

An I-131 image of the neck bed was performed with a pinhole collimator. An elongated area of radioactivity is present in the midline at the level of the thyroid cartilage *(T)* and is consistent with functioning thyroid tissue within the thyroglossal duct remnant *(arrowhead).* Two small areas of residual radioactivity are noted in the neck bed *(arrows);* the chin *(C)* and suprasternal notch *(S)* are noted with cobalt markers. ***Comment:*** The diagnosis of thyroglossal duct remnant is suggested by the midline location and the linear/oval distribution of radioactivity. Although one can exclude esophageal activity with techniques described in Figure 6-13, one may not be able to differentiate thyroglossal duct remnant from local lymph node metastasis.

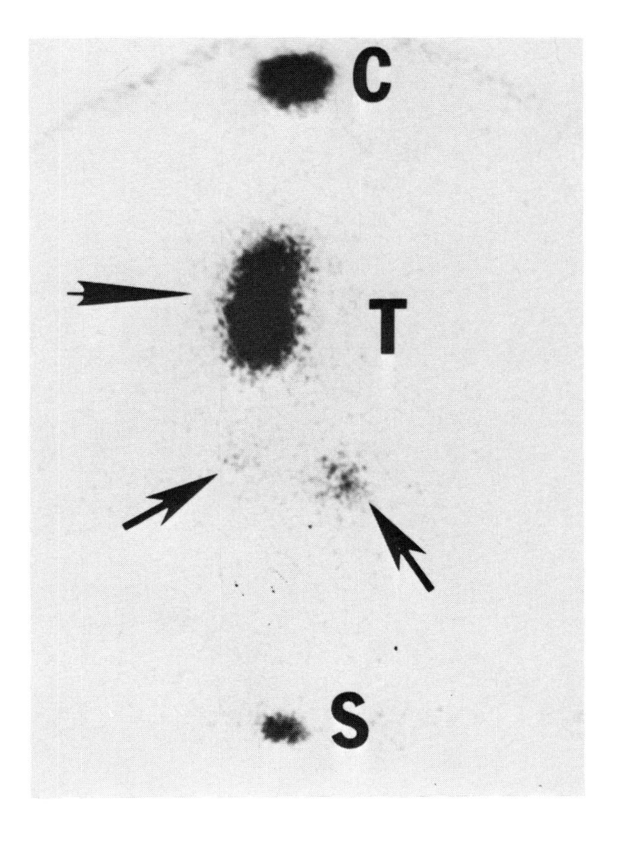

Figure 6-18 *Artifacts*

These cases demonstrate various types of artifacts. (*A* and *B*). Nasal contamination. These images are whole-body anterior I-131 neck and chest scintigraphs of the same patient; however, *B* was obtained after a shower. The multiple areas of radioactivity that washed off were due to a heavily contaminated handkerchief stored in the left hip pocket. Marked increased paranasal radioactivity is present, and cross-contamination by the hands has occurred. Urine contamination may give an identical presentation. (The etiology of the radioactivity which remained after the shower was not available.) ***(C)*** Radioactive wig. This anterior I-131 image of the head shows normal nasopharyngeal radioactivity and abnormal radioactivity in the region of the calvarium, attributable to contaminated hair by radioactivity from perspiration. (*D* and *E*). Radioactive tracheostomy tube. Anterior image ***(D)*** of the thyroid neck region was performed with a pinhole collimator. The radioactivity in the superior aspect represents salivary and oropharyngeal radioactivity. A focal area of radioactivity *(arrow)* is present in the thyroid bed region and corresponds to the area of the patient's tracheostomy tube. After removal of the tube, the radioactivity in the neck bed disappeared, and an image of the removed tracheostomy tube only demonstrates radioactivity in the tube ***(E)***. The radioactivity of the tube may be secondary to contamination from normal radioactivity in salivary secretions. ***Comment:*** Artifacts can commonly be attributed to contamination or sequestration of radioiodine that has been eliminated from the body through common (*e.g.*, urine, salivary secretions) or less common (*e.g.*, nasal and pulmonary secretions, perspiration) mechanisms. (***A*** and ***B:*** Wiseman J: Bony metastases from thyroid carcinoma or contamination? Clin Nucl Med 9:363, 1984. ***C:*** Abdel-Dayem H et al: The radioactive wig in iodine-131 whole body imaging. Clin Nucl Med 9:459, 1984. ***D*** and ***E:*** Kirk G, Schulz E: Post-laryngectomy localization of I-131 at tracheostomy site on a total body scan. Clin Nucl Med 9:409, 1984.)

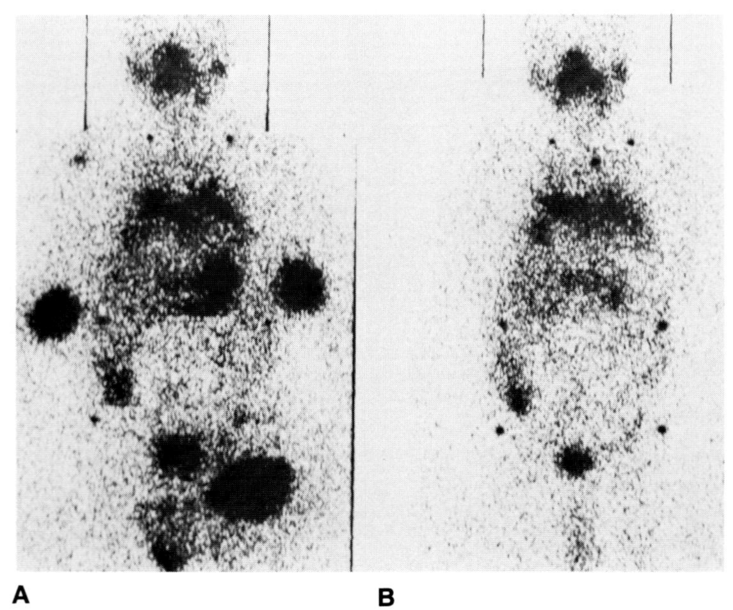

A **B**

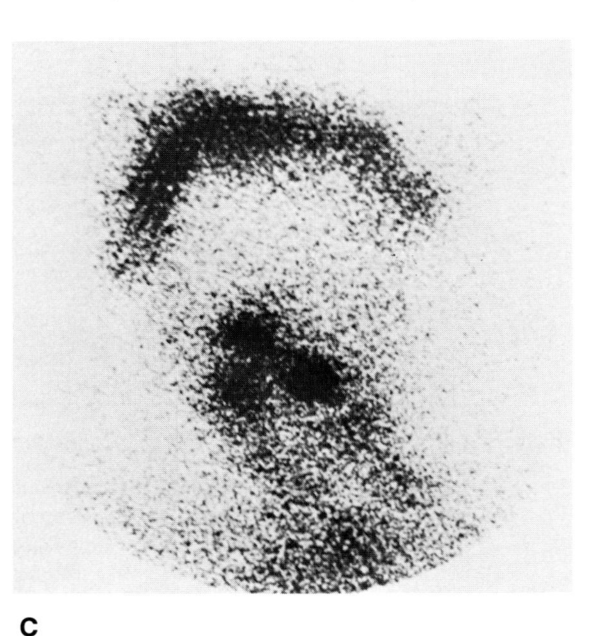

C

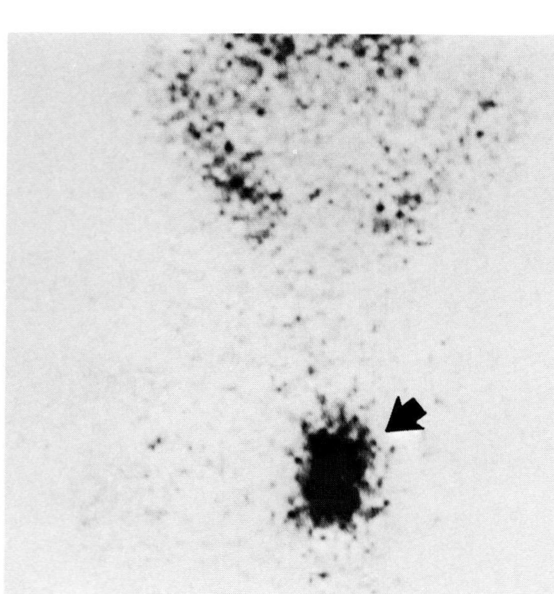

D

E

Figure 6-19 ***Metastasis— Cervical Lymph Node***

Anterior I-131 image performed with a pinhole collimator shows multiple focal areas of radioiodine within and outside the thyroid bed. Because of the number and distribution, these areas were interpreted as highly suggestive of cervical lymph node metastasis. (*C*, chin; *T*, thyroid cartilage; *S*, suprasternal notch) ***Comment:*** Focal areas of I-131 uptake in the neck region (but outside of the thyroid bed) are strong evidence for cervical lymph node metastasis. Although normal thyroid tissue may be present outside of the thyroid region (lateral aberrant tissue), this occurs infrequently and does not typically involve a larger number of foci of radioactivity.[31] Even if all foci were within the thyroid bed, this pattern would be atypical for normal, residual, functioning postoperative thyroid tissue.

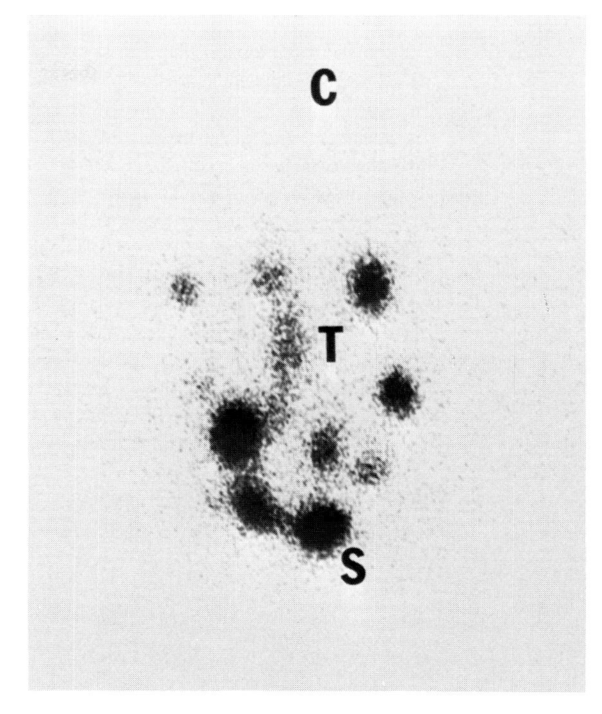

Figure 6-20 ***Metastasis— Mediastinal Lymph Node***

An I-131 image of the neck and chest region shows normal stomach radioactivity *(ST),* cardiac blood pool radioactivity *(C),* liver blood pool *(L)* radioactivity, and three focal areas of increased tracer radioactivity in the mediastinum *(arrows).* These areas were interpreted as thyroid metastasis to mediastinal lymph nodes. Cobalt markers denote the levels of the suprasternal notch. (*S*, suprasternal notch; *X*, xyphoid) ***Comment:*** This is an example of a scintigraphic appearance of metastasis to the mediastinal lymph nodes, which is a frequent location for metastasis from papillary carcinoma.

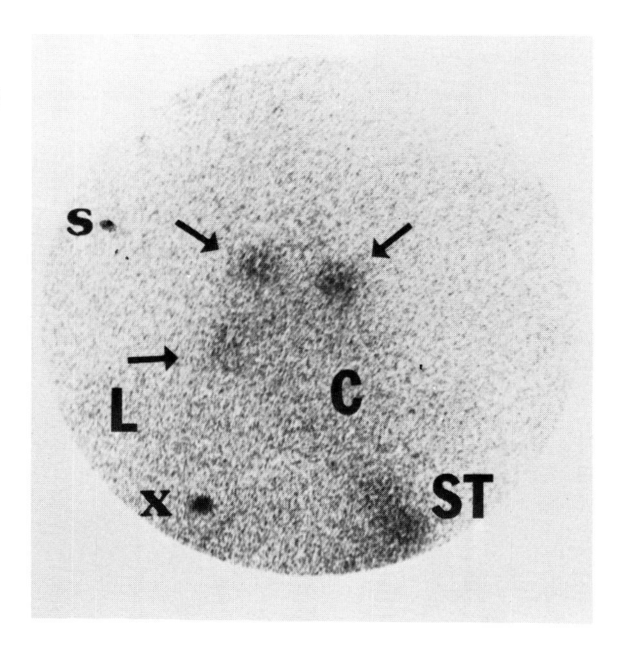

Figure 6-21 ***Value of Comparing Scans***

(A) Anterior I-131 image of the neck and chest in a patient with mixed papillary follicular thyroid carcinoma. Multiple focal areas of radioiodine are noted in the thyroid bed, mediastinum, and hilum. The radioiodine in the thyroid region may represent normal residual thyroid tissue and/or metastasis; the radioiodine in the mediastinum and hilum is highly suggestive of metastatic thyroid carcinoma.

(B) A follow-up scan performed one year after radioiodine therapy also demonstrates multiple focal areas in the thyroid bed, mediastinum, and hilum. However, a comparison with ***A*** demonstrates a change in the pattern, with new areas of radioiodine uptake in the thyroid bed and cervical area *(curved arrows),* and decreased or resolved areas of radioiodine uptake in the thyroid bed and mediastinum *(arrowheads).* These changes could not be attributed to technique. The new areas suggest progression, and the decreased/resolved areas suggest response to the radioiodine therapy. (*C,* chin; *S,* suprasternal notch; *X,* xyphoid, are noted without cobalt markers)

Comment: This case emphasizes a well-accepted code of practice: images must be compared with previous exams when available. The implication of comparative changes, however, can be ambiguous. For example, persistent radioiodine is not necessarily indicative of therapy failure. Neither is decreased or resolved uptake necessarily evidence of response to therapy, since it might also represent only loss of the function of I-131 uptake. Other modalities (*i.e.,* chest x-ray, computed tomography, magnetic resonance imaging) are recommended for following measurable lesions.

A

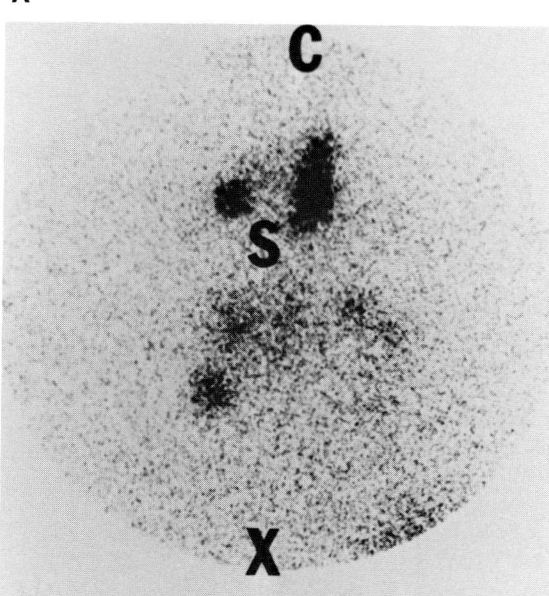

B

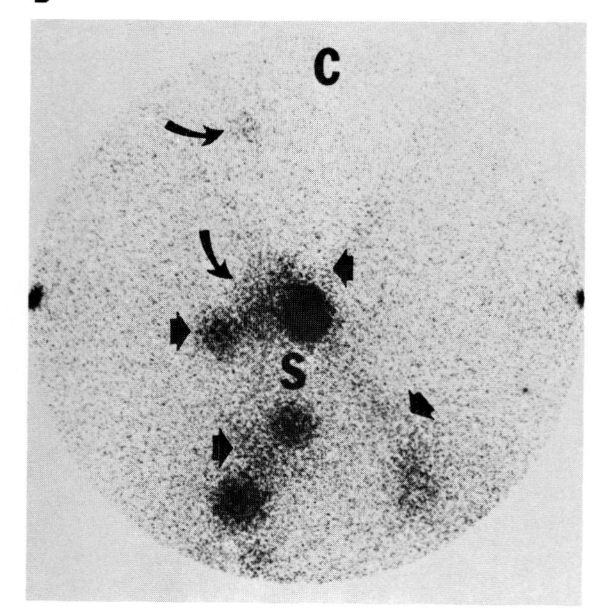

Figure 6-22 ***Pulmonary Metastasis—Scintigraphic Patterns***

Pulmonary metastasis from well-differentiated adenocarcinoma of the thyroid is reported to occur in 2% to 20% of patients[32] and may occur from either a papillary or follicular primary lesion. The I-131 scintigraphic patterns of functioning pulmonary metastasis range from solitary to multiple lesions, a macronodular (focal) to micronodular (diffuse) pattern, and regional to total involvement of the lungs. To aid in the description of the scintigraphic findings, this figure illustrates a proposed scintigraphic classification. Several of the scintigraphic patterns are shown in Figures 6-23, 6-24, and 6-25. ***Comment:*** Although the various classifications are semiarbitrary and may overlap clinically, these classifications and other aspects of pulmonary metastasis may have prognostic and therapeutic implications. A more favorable prognosis has been suggested when the metastasis takes up I-131,[33] when the chest radiograph is normal,[33] and when the scintigraphic pattern of I-131 uptake is micronodular rather than macronodular.[32] The concomitant presence of mediastinal lymph node involvement may portend a worse prognosis,[32,34] and younger patients do better than older patients.[1]

The scintigraphic classification may also have therapeutic implications. Speculatively, patients with micronodular metastasis treated with radioiodine may be at higher risk for developing radiation pneumonitis, pulmonary fibrosis, and subsequent respiratory compromise in comparison to patients with macronodular disease, because the micronodular pattern, all other factors being equal, may deliver more radiation dose to the normal adjacent lung parenchyma. Also speculatively, patients with regional disease (whether macronodular or micronodular) may tolerate pulmonary complications better than patients with total lung involvement, because the uninvolved lung would be spared. Further study is needed to clarify or confirm these examples of prognostic and therapeutic implications.

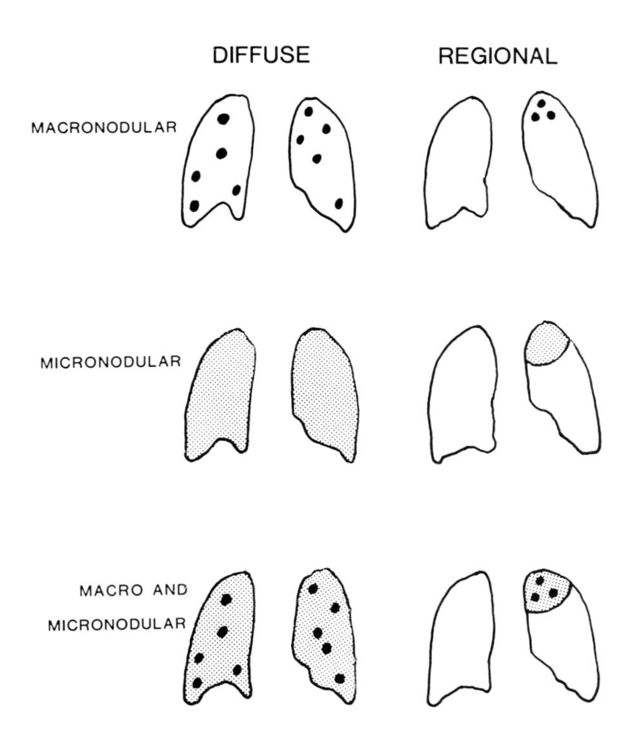

Figure 6-23 ***Pulmonary Metastasis — Entire Lung Involvement of Diffuse (micronodular) Disease***

Anterior I-131 chest image demonstrates diffuse, markedly increased uptake throughout the entire lungs, which was interpreted as compatible with micronodular pulmonary metastasis from the patient's known papillary thyroid carcinoma. The chest radiograph also demonstrated diffuse miliary lung nodules consistent with metastatic thyroid carcinoma. The level of the thyroid cartilage *(T)* and suprasternal notch *(S)* are noted. The cobalt-57 markers, which are at the edge of the field of view, represent the level of midchest. The two small focal areas of radioactivity in the region of the suprasternal notch represent functioning tissue in the thyroid bed. ***Comment:*** This pattern of diffuse radioiodine uptake has been reported to occur in 38% of patients with functioning pulmonary metastasis;[32] when this is present, the physician should be cautious in selecting a dosage of radioiodine for therapy. The dosage should maximize destruction of the pulmonary metastasis while minimizing the risk of radiation pneumonitis and pulmonary fibrosis. The quantitative clearance of radioiodine described earlier may be helpful in selecting therapeutic radioiodine dosages. This case also demonstrates that I-131 neck and chest scintigraphy may be of value in the work-up of radiographic miliary lung nodules, because I-131 uptake would be highly suggestive of metastatic thyroid carcinoma.

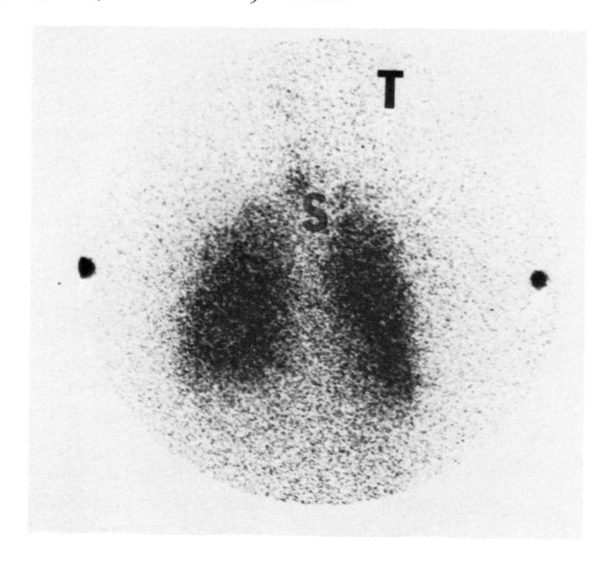

Figure 6-24 ***Pulmonary Metastasis — Focal (macronodular)***

This posterior I-131 chest image demonstrates a single focal area of radioactivity in the right lower lung area. Lateral and oblique images (not shown) demonstrated that the radioactivity was centrally located (and not over spine) and thus not in bone; delayed imaging confirmed the persistence of radioactivity, and this made normal variant or an artifact less likely (see Figs. 6-13, 6-14, and 6-18). Transmission images* demonstrated further that the area was located above the diaphragm and thus not in liver; a solution of Tc-99m sulfur colloid given orally outlined the esophagus, which excluded this already unlikely organ as the site, and computed tomography and magnetic resonance imaging were noncontributory. Based on these findings, the focal area was interpreted as highly suggestive of thyroid metastasis. Although one cannot differentiate whether this is cardiac, pericardial, or pulmonary in location, a pulmonary site is most likely based upon prevalence alone. The level of the seventh cervical spine is noted, and a cobalt marker is present on both edges of the image noted. ***Comment:*** This case demonstrates not only the typical appearance of focal (macronodular) regional pulmonary metastasis, but also some of the manipulations used to differentiate a pulmonary metastasis from normal variants, artifacts, or bone metastasis. When possible, pulmonary metastasis should be differentiated from bony metastasis because the latter has a worse prognosis, which in turn may encourage more aggressive therapy (*i.e.,* more frequent or higher therapeutic dosages).[35]

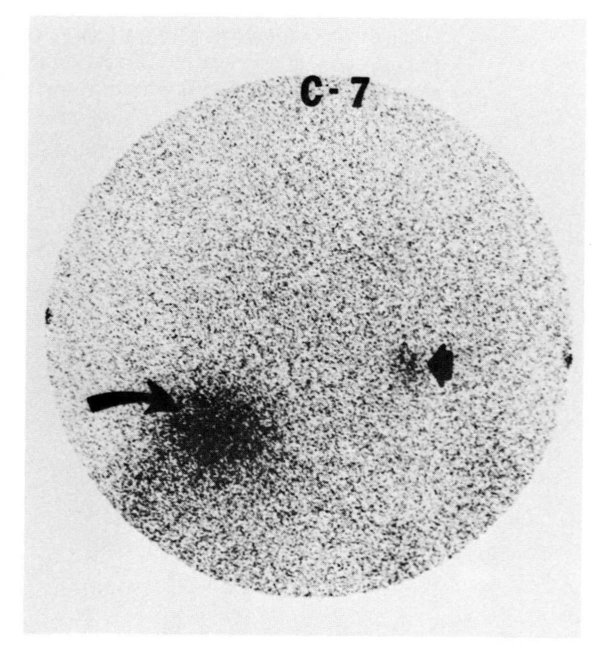

* Transmission images may be obtained by positioning the patient between a large uniform flood source of Tc-99m O_4^- and the gamma camera. The differential attenuation of the gamma rays by the lung and liver should outline the upper limits of the diaphragm. Lesions above the diaphragm are not in liver.

Figure 6-25 ***Pulmonary Metastasis—Entire Lung Involvement of Diffuse (micronodular) and Focal (macronodular)***

The patient has been shifted slightly to his left for this anterior chest and abdominal image, and the accompanying diagram should aid in identification of findings. This image demonstrates increased radioiodine in the thyroid bed, diffuse moderately increased radioiodine throughout most of both lungs (which is indicative of micronodular metastasis), and faint but definite focal areas of increased radioiodine throughout both lung fields. The focal areas of radioactivity were attributed to pulmonary macronodular metastasis, rather than bony metastasis, because no focal areas suggesting bony metastasis were present outside of the lung fields. Trace normal radioactivity is noted in the liver. ***Comment:*** This case suggests that both pulmonary patterns may occur together. Whether these patterns represent two distinct types of pulmonary metastasis or two different stages of pulmonary metastasis is not known.

A

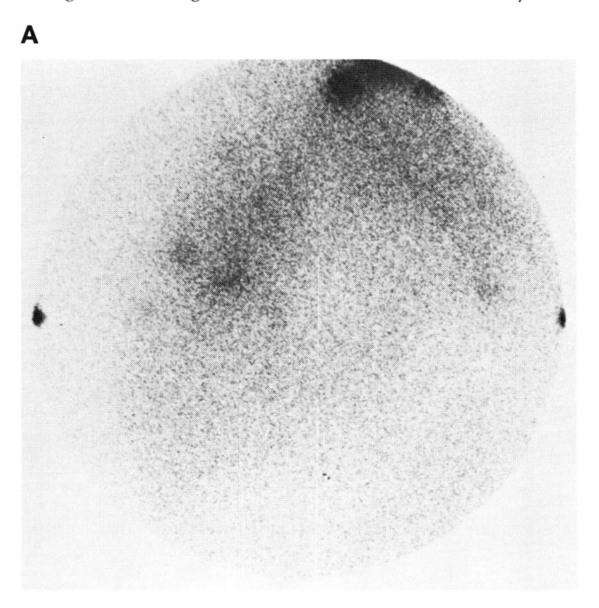

B

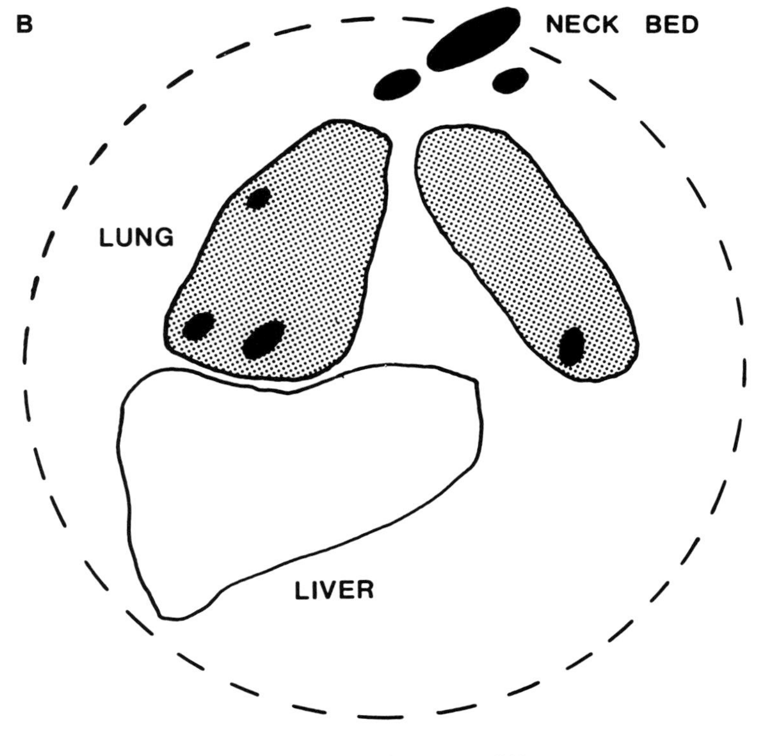

Figure 6-26 *Metastasis—Bone*

Anterior *(A)* and lateral *(B)* I-131 images of the left lower extremity revealed focal increased tracer uptake in both the left distal femur and proximal tibia *(arrows)*. These abnormalities persisted following repeated washings of the area and were localized on multiple projections to within bone. These areas were still visualized on follow-up study one year later. Radiographs and bone scintigraphy of these areas were normal. In the anterior view, cobalt markers are noted at the level of the knees; in the lateral view, cobalt markers are at the levels of the hip and ankle. ***Comment:*** Thyroid metastasis to bone has been reported to occur in as many as 9% of patients,[36] but it is rare in children. More frequently it is observed in follicular than papillary primaries, and it has a more severe prognosis. Appropriate manipulations should be performed to eliminate artifacts.

A

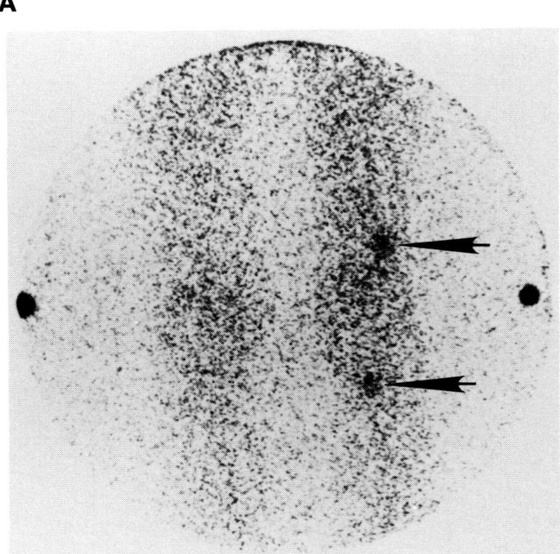

B

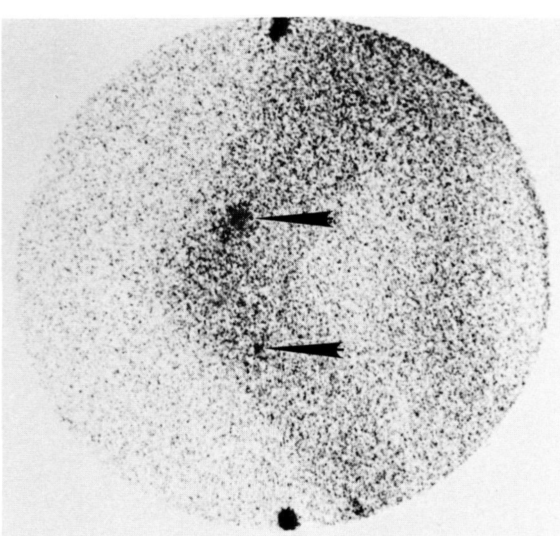

Figure 6-27 ***Neck and Chest Survey vs. Bone Scintigraphy in the Detection of Bone Metastasis***

(A) I-131 image of the pelvis shows multiple focal areas of increased tracer activity in the region of the pelvic bone (*small arrows* and *arrowheads*). The areas of increased I-131 uptake noted by the small arrows have no definite corresponding abnormality on bone scintigraphy (***B*** and ***C***), and the areas of increased I-131 uptake noted by the *arrowheads* correspond approximately to both "hot" and "cold" lesions on bone scintigraphy (***B***, *arrows*). The right *(R)* and left sides at the level of the iliac crests are noted with cobalt markers. ***Comment:*** I-131 scintigraphy has been reported to be significantly more sensitive than bone scintigraphy in detecting bone metastasis from thyroid carcinoma, and in fact bone scintigraphy may have a sensitivity as low as 50%.[37] Thus, although bone scintigraphy may be of value in the evaluation of thyroid metastasis, the failure to demonstrate an abnormality on bone scintigraphy does not prove that an abnormality on I-131 scintigraphy is extraskeletal. Likewise an abnormality on bone scintigraphy may be secondary to bony thyroid metastasis, even though the I-131 scintigraphy is normal in the corresponding area, because not all thyroid metastasis are functioning and will take up I-131.

A

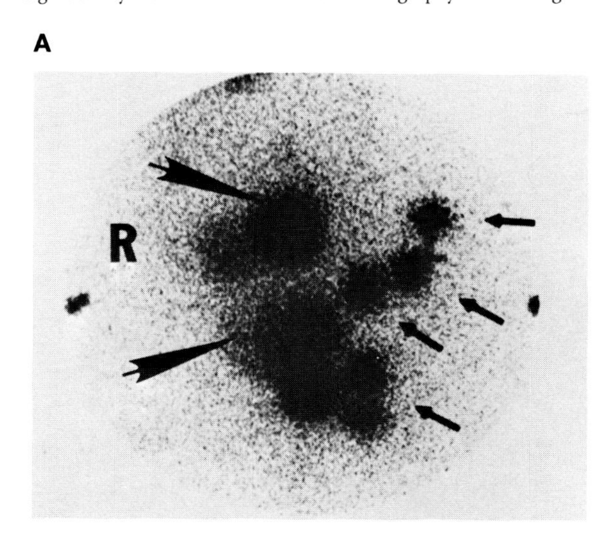

B C

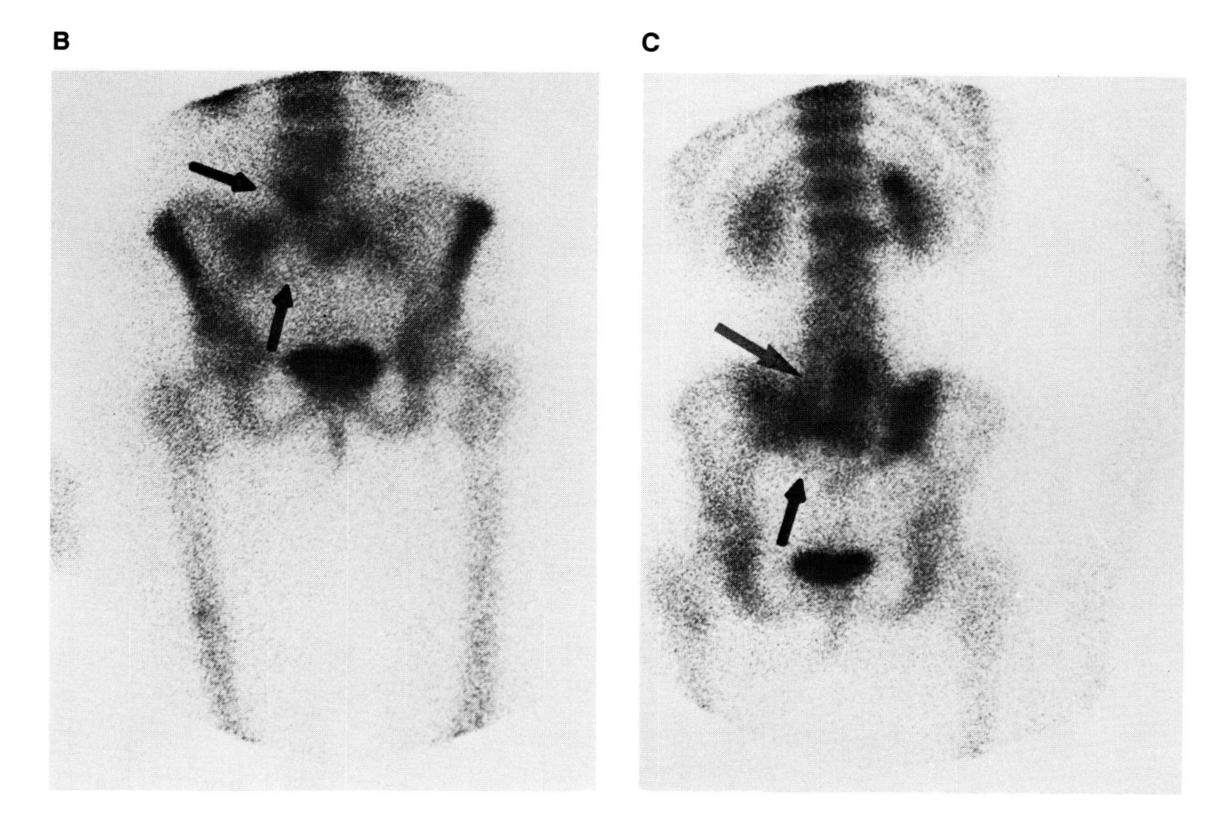

Figure 6-28 ***Metastasis—Brain***

(***A*** and ***B***) These coronal tomographic whole body I-131 images were obtained using a Phocon camera. An abnormal focal area of radioiodine ***(B)*** is present in the head region, and brain involvement is demonstrated in the multiple tomographic images. Metastatic intracerebral papillary thyroid carcinoma was confirmed surgically, and after partial excision of the mass, follow-up images demonstrated markedly reduced radioiodine uptake. Radioactivity is also present in the thyroid *(T)* and gastrointestinal *(G)* region. ***Comment:*** Thyroid metastasis to the brain is infrequent, and since I-131 irradiation of brain metastasis has been associated with intracerebral bleeding, the therapist may consider either reduction of the I-131 therapeutic dosage or surgical removal. In differentiating brain from bone metastasis, tomographic scintigraphy or planar images from multiple projections may help to localize the metastasis. However, other modalities, such as radiographs, bone scintigraphy, and computed tomography, may be necessary. (Parker LN, Wu SY, Kim DD, Kollin J: Recurrence of papillary thyroid carcinoma presenting as a focal neurological deficit. Arch Intern Med 146:1985, 1986; copyright 1986, American Medical Association)

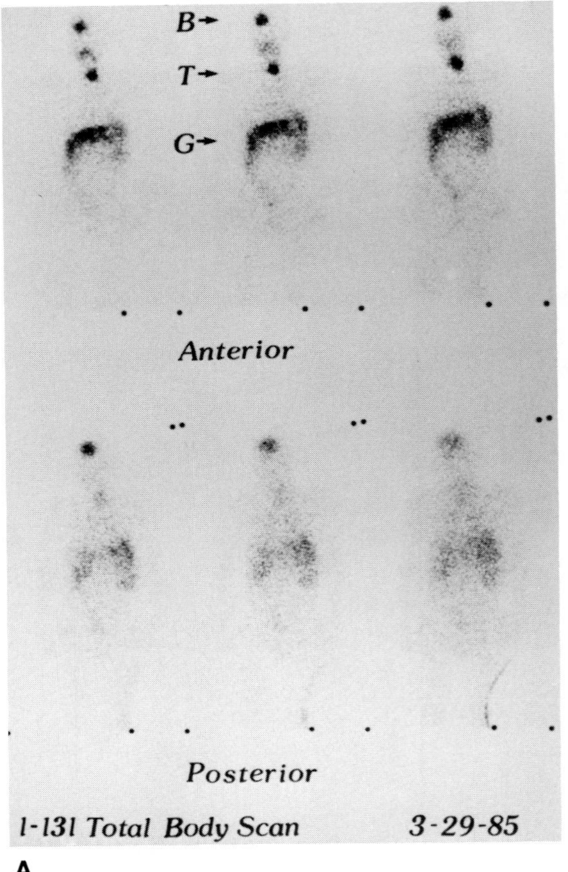

A

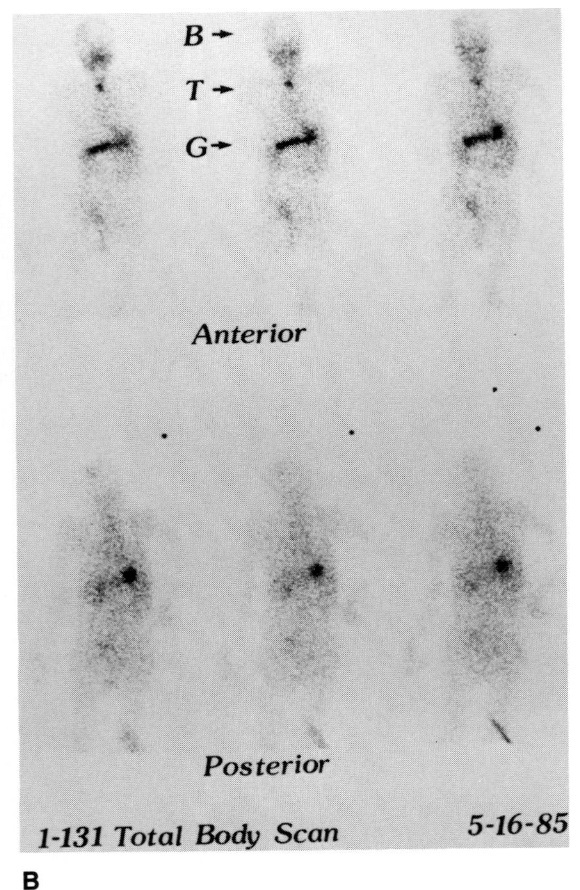

B

A

C

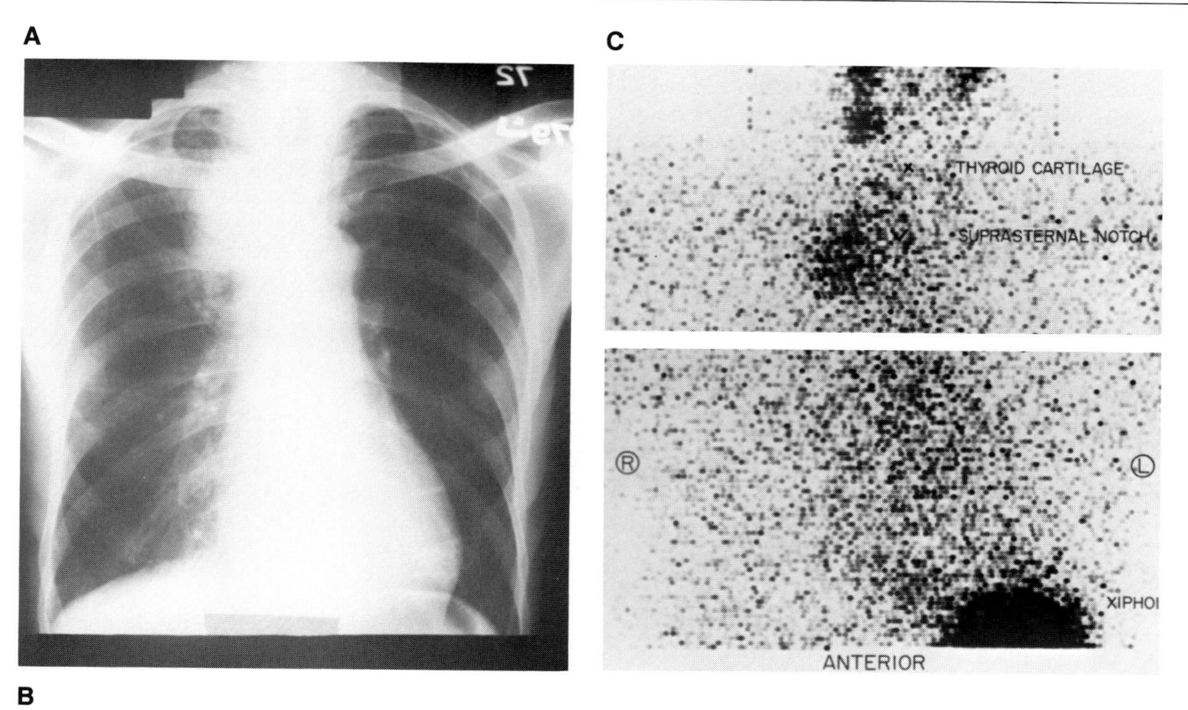

B

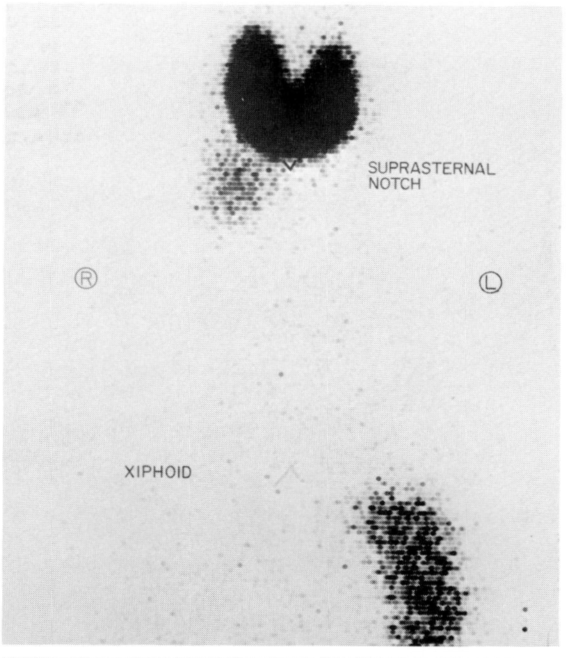

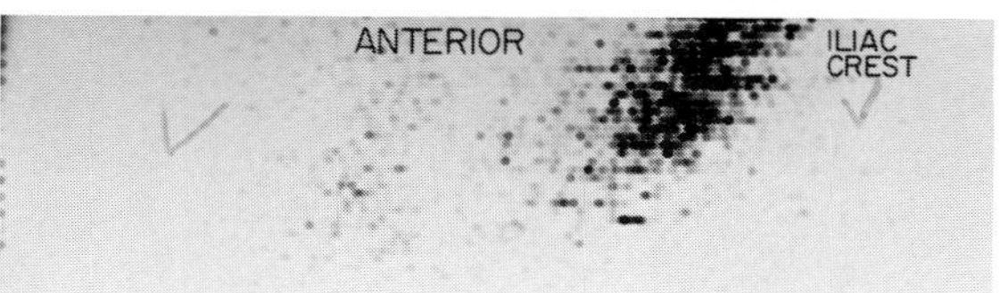

Figure 6-29 ***Primary Lung Adenocarcinoma***

(A) A 54-year-old male had a right upper lobe lung mass as demonstrated on the chest radiograph. Because the biopsy of this mass indicated papillary adenocarcinoma, metastatic thyroid carcinoma was suspected, and an I-131 scan following TSH injection was performed. *(B)* This study shows homogenous I-131 uptake in the thyroid gland as well as in the mass. After total thyroidectomy, extensive pathological exam of the thyroid gland revealed no primary thyroid carcinoma. A repeat I-131 scan *(C)* showed persistent uptake in the lung mass. Postmortem exam of the lung mass revealed identical histopathology to the antemortem biopsy specimen; mucicarmine stain was positive on the lung tumor but negative on control sections of another known papillary carcinoma of the thyroid, and the neoplasm was believed to be a primary lung adenocarcinoma. ***Comment:*** When a neoplasm takes up I-131, well-differentiated thyroid carcinoma is almost always the primary lesion; however, occasionally other neoplasms may take up radioiodine (Figs. 6-30, 6-31, and 6-32). Although the mechanism for radioiodine uptake in lung carcinoma is unclear, in this case and in others it was proposed that the mitotic cells may have been pluripotential and may have developed receptors capable of radioiodine affinity or uptake.[39,40] (Fernandez-Ulloa M et al: Iodine-131 uptake by primary lung adenocarcinoma. JAMA 236:857, 1976)

Figure 6-30

Metastatic Gastric Adenocarcinoma

Whole-body coronal tomographic I-131 images in a 65-year-old black male with metastatic gastric adenocarcinoma show numerous areas of abnormal tracer uptake throughout the chest, shoulders, spine, liver, and pelvis. At autopsy these metastases were moderately differentiated mucin-secreting adenocarcinomas of the stomach. (*S*, right shoulder; *L*, liver; *P*, pelvis) ***Comment:*** This case demonstrates another nonthyroidal neoplasm that may take up radioiodine and that could potentially be treated with radioiodine. The mechanism of I-131 uptake in this case is not known. Wu and co-workers have speculated that although the primary gastric tumor and normal gastric mucosa may have initially showed near-background radioactivity, cell dedifferentiation of the metastatic cells may have resulted in longer retention of radioiodine and subsequent visualization at 72 hours. (Su SY et al: I-131 total-body scan: Localization of disseminated gastric adenocarcinoma. J Nucl Med 25:1204, 1984)

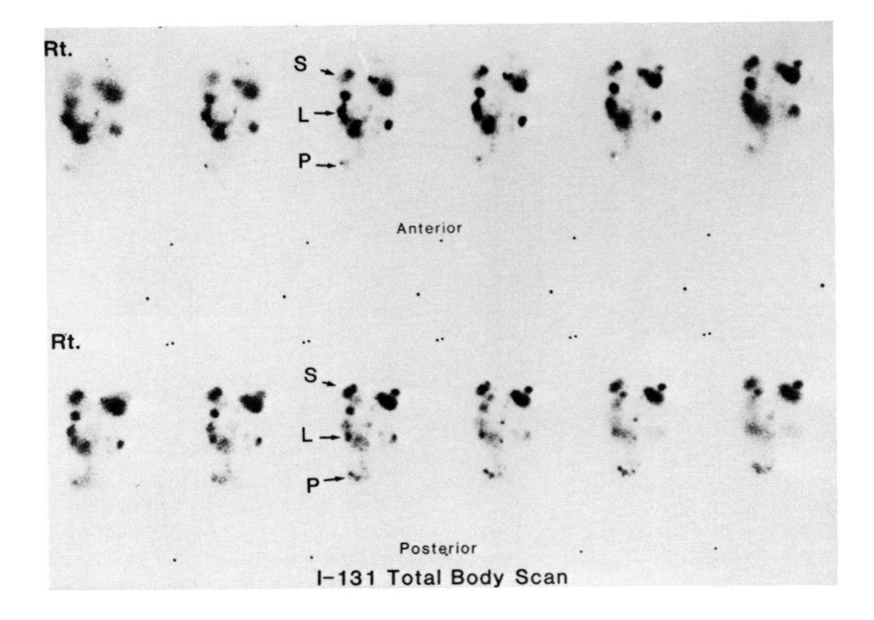

Figure 6-31 ***Medullary Thyroid Carcinoma***

I-131 images of posterior head ***(A)*** and posterior chest ***(B)*** show multiple focal areas of radioiodine uptake in the skull, spine, and right humerus. At postmortem, extensive histopathological sections showed no evidence of neoplasm other than medullary thyroid carcinoma. ***Comment:*** I-131 neck and chest scintigraphy has been generally believed to be of little value in the evaluation or treatment of medullary carcinoma, because I-131 uptake was not suspected in light of the neoplasm's cell origin. This case documents that medullary thyroid carcinoma can take up I-131 with a potential for radioiodine therapy. The frequency of mechanism of I-131 uptake in this neoplasm remains undetermined. (Michael BE, Forouhar FA, Spencer RP: Medullary thyroid carcinoma with radioiodide transport—effects of iodine-131 therapy and lithium administration. Clin Nucl Med 10:274, 1985)

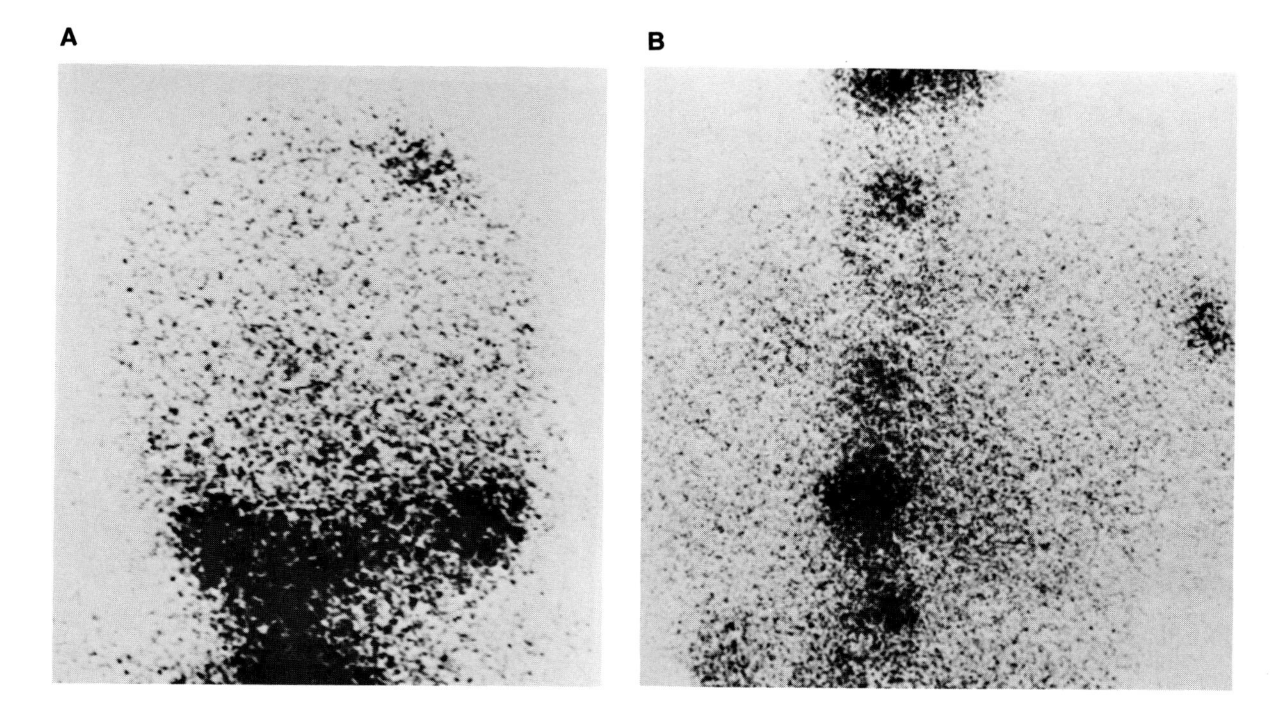

Figure 6-32 ***Struma Ovarii***

(A) An overlay of a patient's I-131 rectilinear scan onto her pelvic radiograph demonstrates I-131 uptake in the pelvic region. This uptake corresponded to a 10-cm pelvic mass, which was an infarcted teratoma composed mostly of active thyroid tissue (struma ovarii). A measurement of the 24-hour I-131 uptake performed over the mass (with the bladder catheterized) was 17%. ***(B)*** The pathological specimen and the corresponding *in vitro* I-131 image of this specimen. The I-131 had been administered 9 days prior to surgery.

Comment: Although struma ovarii itself is very rare and the likelihood of its simultaneous occurrence with thyroid carcinoma is even more remote, pelvic or abdominal I-131 uptake in struma ovarii could be mistaken for metastatic thyroid carcinoma. Other rare etiologies of I-131 uptake include Warthin's tumor, metastatic salivary gland tumors, papillary meningioma, and fungal infection.[44] (Yeh EL, Meade RC, Ruetz PP: Radionuclide study of struma ovarii. J Nucl Med 14:118, 1973)

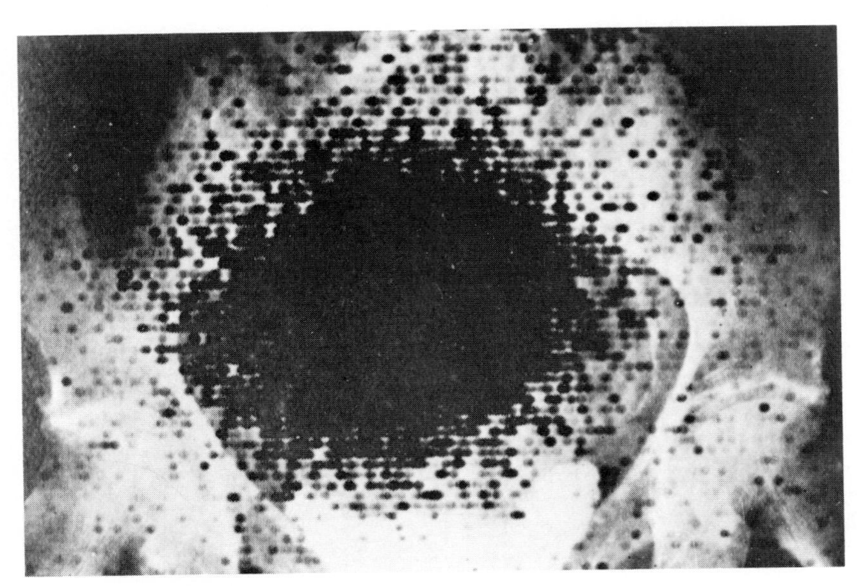

A

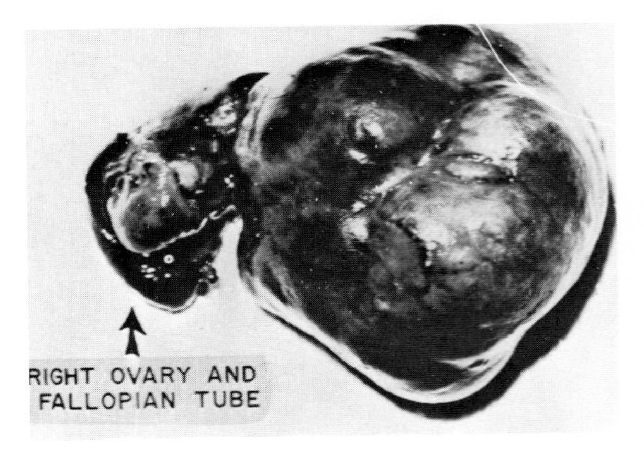

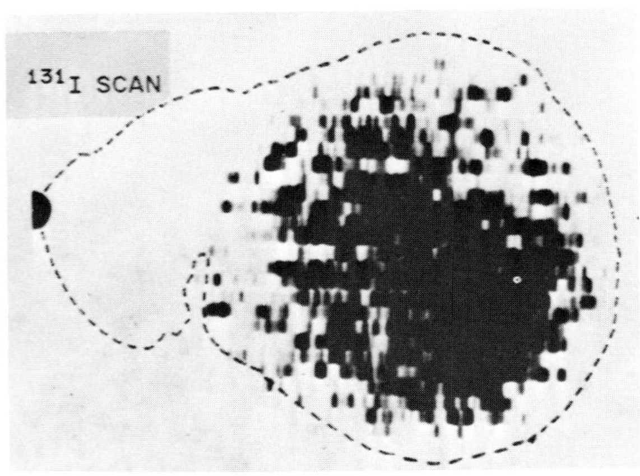

B

Table 6-1 *Proposed Use of TSH Blood Levels*

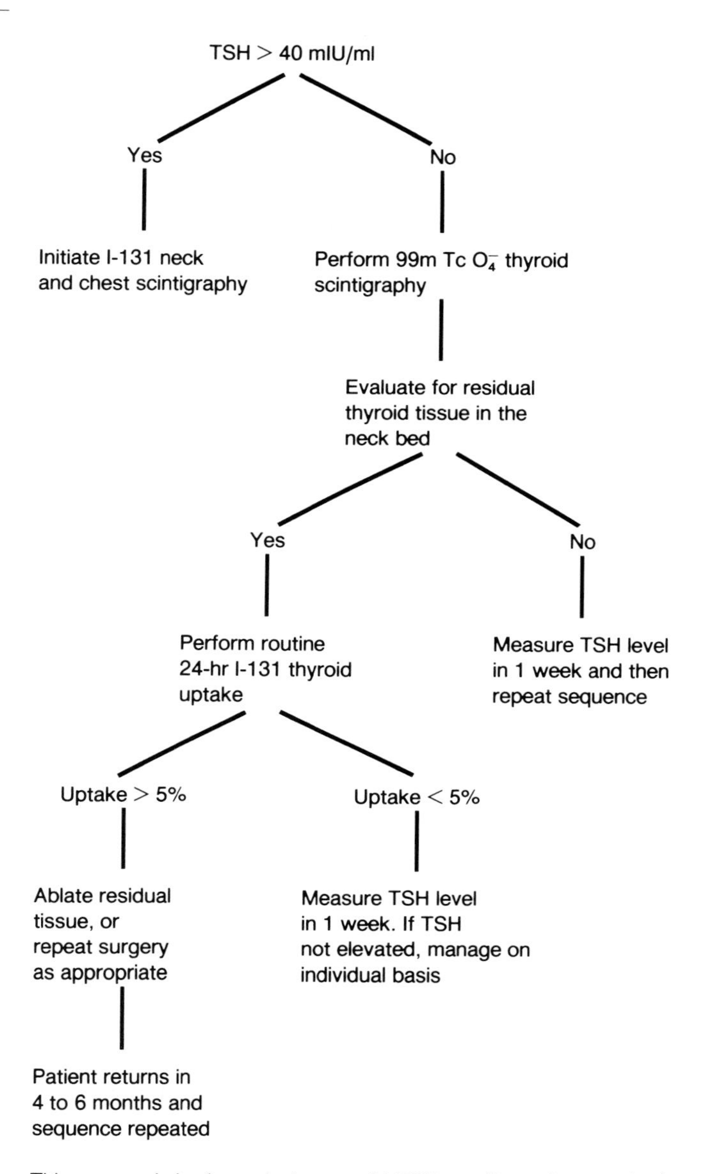

This approach is dependent on rapid TSH results and no marked functioning metastasis (*e.g.*, diffuse lung metastasis that could suppress TSH).

Michel Meignan
François Begon
Pierre Galle

Chapter 7 FLUORESCENT THYROID SCANNING

Fluorescent thyroid scanning produces images that reflect the distribution of stable iodine (I-127) without injection of radioactive material. It also allows *in vivo* estimation of the amount of stable iodine in the thyroid. The method was first suggested in 1968 by Hoffer[1] but has not been widely used. It was only recently that certain technological advances improved image quality and the accuracy of measurements sufficiently to permit x-ray fluorescent imaging to be applied to human thyroid studies under satisfactory conditions.[2-10]

TECHNIQUES

Principles of X-Ray Fluorescent Imaging

The basic principle of x-ray fluorescent imaging is simple (Fig. 7-1): a beam of x-rays or gamma rays is aimed at a particular point of an organ. If the energy of the incident rays is higher than the energy of the inner shell target atom electron, then the latter is ejected, creating a vacancy that is quickly filled by an outer shell electron. The energy lost in transit by this electron generates either a fluorescent photon or an Auger electron. The energy of fluorescent photons is characteristic of target atoms and therefore permits the identification of these atoms. They can also be detected quantitatively, since the number of characteristic photons is directly dependent on the concentration of the element being sought. Unfortunately, tissue generally contains atoms with low atomic numbers, whose characteristic photons have a low energy yield and are therefore mostly absorbed by the tissue, even in an organ as near the surface as the thyroid. In addition, the fluorescent yield in elements with low atomic numbers is fairly low. For all these reasons, many elements of biological interest cannot be studied by the fluorescent technique.

However, fluorescent imaging is very suitable for the *in vivo* study of iodine in the human thyroid. Several factors provide satisfactory conditions for this purpose, including the concentration of iodine in the thyroid, the fact that the gland is near the surface, the good fluorescent yield of the characteristic iodine K x-ray (86%),[11] and the energy these rays generate (K_α=28.5 keV and K_β=32.4 keV), which is not excessively absorbed in the soft tissues.

Imaging Procedure

To obtain an image, the thyroid gland is irradiated from an outside source. The fluorescent iodine x-ray is detected by a detector and a spectrometer. The unit combining the detection device and irradiation source is part of the mobile arm of a scanner. Scanning the front surface of the neck permits fluorescent imaging of the iodine atoms in the thyroid.

Obtaining good quality pictures of the thyroid is a tricky operation. First, the intensity of the fluorescent emission is low, and second, there is important scattering by Compton diffusion of the americium photons on the soft tissue and cartilage of the neck. The aim is therefore to create conditions ensuring the highest possible fluorescent photon flux and permitting the clearest possible differentiation between K_α rays and background noise. The number of the emitted fluorescent photons depends on the energy of the photons and the intensity of the incident photon flux striking the iodine atoms. The energy of the incident photons must be higher than, but also as close as possible to, 33.2 keV, which is the energy of the iodine K-shell absorption peak. Under these conditions, the probability of the interaction between the incident rays and the K-electrons in this shell increases.

Several types of sources for the generation of exciting photons have been suggested. Of the gamma ray emitters, the most commonly used is americium-241 (half-life of 476 years, $E\gamma = 59.8$ keV).[11] Aubert and colleagues recently irradiated the thyroid with an x-ray generator, which released a continuous spectrum of photons whose energy was adjustable between 1 and 79 keV. The beam was filtered through 1 mm of copper.[12] This enabled delivery of a larger photon flux than achievable by gamma ray generators, and thus increased the method's sensitivity. Nevertheless, this polychromatic beam can produce scattered photons whose energy may seriously hamper detection of the fluorescent ray. Monochromatic americium gamma rays therefore seem preferable to polychromatic beams. Individualization of K_α rays from the background necessitates the choice of a high-resolution semiconductor detector with or without a collimator. Acute spectrometry is required with a window analyzer set at 1 to 2 keV, and there should be a second spectrometric channel for subtraction of background noise.

The above considerations have led to the commercialization of two types of devices (Fig. 7-2). Both use Am-241 as a source of activity ranging from 1 to 2 Ci. They differ in their exciting beam source, which may be single or multiple. If multiple, the sources are concentrically arranged around the detector. The source and detector unit are mounted on the arm of a conventional scanner. We use one such system equipped with 20 sources of 1 Ci americium each.[13] They are individually beamed by a collimator to a focal point and the sources are separated by a lead septum. The detector used is a silicon–lithium 25-mm diameter crystal. It is cooled to the temperature of liquid nitrogen (77° K) so that the material behaves as an insulator and produces a large number of ion pairs when a photon interaction deposits a very small amount of energy.

Energy resolution, measured on a linear source filled (line spread function [LSF]) with iodine and placed on diffusing material, is 0.867 keV (Full Width Half Maximum [FWHM]) at 28.5 keV. Spatial resolution is satisfactory. At the focal point, the FWHM of the LSF is 0.47 inch. To obtain an image, the detection unit is placed 1 inch away from the skin detection area. Thyroid scanning lasts 13 minutes for a good quality image. A hyperpure germanium (HPGe) detector has recently been proposed.[13] The improved efficiency of HPGe allows the use of collimators with higher resolution, which results in improved image quality. In addition, this detector allows the simultaneous study of the distribution of technetium-99m.

PHYSIOLOGY OF INTRATHYROIDAL IODINE POOL AND DETERMINATION OF STABLE IODINE CONTENT IN THYROID

The fluorescent technique enables *in vitro* thyroid iodine content to be measured during imaging. In the normal thyroid gland the major form in which iodine is found is iodothyronines (triiodothyronine [T_3], reverse triiodothyronine [rT_3], and thyroxine [T_4]) and their precursors, the iodothyrosines (monoiodothyrosine [MIT] and diiodothyrosine [DIT]), which are mainly contained in the thyroglobulin molecule; about 20% of iodine is represented by other organic iodocompounds and inorganic iodine. There is a correlation between the amount of thyroglobulin stored and the degree of iodination of protein. Therefore the stable iodine content represents an estimation of the amount of thyroid tissue present, which is necessary to maintain euthyroidism.[8]

Up until now, iodine content had been determined by indirect methods using radioactive isotopes[14] which measure the exchangable iodine pool rather than the true iodine content pool (two values that often differ in pathology.)[14] Determination of the true iodine pool by fluorescent scanning requires the construction of a standard curve by recording the number of impulses obtained while scanning a thyroid phantom that contains increasing amounts of iodine; the phantom may be placed, for instance, in a Plexiglas block in order to simulate the soft tissue. In patients, the number of impulses considered to be of thyroid origin corresponds to the number recorded by thyroid scanning, minus the background noise defined on the second spectrometric beam. The iodine pool is determined by reference to the standard curve constructed from the phantom.

Although simple in principle, quantitative analysis is difficult in practice. This is because the thickness of the tissue covering the gland can cause errors by attenuating the fluorescent ray which has a low K_α energy and a 2-cm half-attenuation layer in soft tissue. It has been suggested that skin thickness should be measured and the measurement corrected. This is feasible if the attenuation ratio of the K_α and K_β iodine rays is studied as a function of skin thickness and calculated for each subject.[15] More recently, Gollnick and coworkers suggested injecting patients with 1 μCi of iodine-123 and then calculating the depth of the thyroid from the attenuation ratio of the x-rays and gamma I-123 rays.[16] Some investigators, like Patton,[17] established two standard curves, for thin and fat subjects respectively. Others, including ourselves,[18] considered the detection area as homogeneous up to 1.5 cm from the focal area on each side and made no geometric corrections. In practice, the errors in iodine measurement caused by changes in skin thickness are much smaller than the variations observed in various disease states.

Finally, two other causes of error must also be mentioned. First, morphologic abnormalities of the thyroid can make the background noise connected with the scatter of americium vary considerably from one point of the gland to another. Second, iodine atom distribution may also be very heterogeneous. However, both the heterogeneity and morphologic variations are difficult to simulate.

Despite all the limitations, excellent correlations have been obtained between *in vitro* doses and *in vivo* measurements by authors like Patton[13] who found that $r = 0.99$. It would therefore be possible to estimate total iodine content by this method with 15% accuracy. Digitalization of fluorescent scanning has been used by Aubert and colleagues and Begon and co-workers and is valuable for background subtraction.[12,19]

ESTIMATED RADIATION ABSORBED DOSE

For the centering and scanning techniques in current clinical use, the patient doses recorded by thermoluminescent dosimeters placed on a Rando[20] radiotherapy phantom are 8 millirads for the thyroid, 10 millirads for the skin (center of the field and lower angles of the scanning area), and 2 millirads for the bone marrow. Since the dose conveyed by the neutrons and fluorescent rays is negligible,[18,21] the radiation absorbed by the thyroid during fluorescent scanning is essentially delivered by the Am-241 gamma photons, but such irradiation is very slight and extremely localized without any gonadal irradiation (of interest in pregnant patients). Tc-99m O_4^- delivers more than 100 millirads and I-131 more than 50 rads to the thyroid for imaging purposes. In addition, unlike I-131 and Tc-99m O_4, the estimated radiation absorbed dose to the gonads and whole body from a fluorescent study is negligible.

DISCUSSION

Hoffer was the first to show the usefulness of fluorescent scanning in 1971, when he published a report covering 159 patients.[22,23] Since then, several series of clinical results have been published.[3,10,24–26] The interest of this type of examination became evident during iodine flooding to diagnose iodine contamination in subjects undergoing thyroid-suppressive treatments and in the evaluation of hyperthyroidism induced by iodine overloading. The present technique also appears to be an aid in the etiological diagnosis of cold nodules on radionuclide scans as well as of toxic nodules, and also in the prognosis and observation of the evolution of subacute and autoimmune thyroiditis. It could also be used to decide an iodine therapy in some goiters in endemic regions with low-iodine diets. The slight, exclusively local irradiation to which it exposes the patient allows thyroid examination in pregnant and lactating patients and small children. However, most of the indications are for pathophysiological studies, and fluorescent scanning is not extensively used for routine clinical practice.

Clinical Indications For Fluorescent Scintigraphy
Evaluation of thyroid in pregnant, lactating women
Evaluation of nodules
Evaluation of autonomous nodules
Evaluation of patients who have had iodine flooding or thyroid suppression
Evaluation for hemiagenesis
Follow-up of patients with iodine-induced thyrotoxicosis (*i.e.*, drugs such as amiodarone)
Evaluation and follow-up of patients with thyroiditis
Evaluation of thyroid in pediatric patients (measurement of iodine pool in children's goiter)
Follow-up of radioiodine therapy in patients with hyperthyroidism

This method nevertheless has the advantage of allowing the observation of changes in thyroid iodine content during the various thyroid complaints. It is the only noninvasive method that gives this information in a short time. Comparison of the results obtained by fluorescent and radioactive tracer scanning, including any discrepancies that may occur, can subsequently prove of great pathophysiological importance. However, it must be kept in mind that x-ray fluorescence cannot distinguish the functionally important thyroglobulin from the other fractions of intrathyroidal iodine.

Nevertheless, some investigators believe that the great variability of thyroid iodine content in both normal and pathologic subjects will make it difficult to assess the diagnostic value of absolute iodine measurement, especially as the technical difficulties involved in such measurement are added obstacles. This would be particularly true in patients with large goiters where K_α x-rays from the deepest regions of the gland are not detected. However the excellent reproducibility of the fluorescent method easily justifies longitudinal studies, and the variations observed in diseases are far greater than the variations due to technical problems. Therefore this method has interesting pathophysiological implications. The results of fluorescent thyroid scanning and of the estimation of thyroid iodine content will be now discussed in normal and abnormal conditions.

Normal Thyroid Iodine Content

Image quality is almost as good as in I-131 or I-123 scintigraphy (Fig. 7-3). In normal subjects, stable iodine distribution is homogeneous. Measurable iodine is found only in thyroid tissue which contains follicles with colloid. The thyroid scan with x-ray fluorescence can be considered a mapping of colloid-bearing follicles regardless of their functional state.[24] The iodine intrathyroid pool measured by fluorescence scanning is extremely variable in a normal population. According to Palmer, the mean value is 10.3 ± 2.2 mg for men and 8.2 ± 2.5 mg for women.[2] Wahner found a mean of 15.5 ± 6 mg, as did Rougier and colleagues.[4,5] Despite the large scatter of the results, the method's reproducibility is good (variations of less than 10% in longitudinal studies during more than 100 days).[24] Even though some lower values have been reported, 10 mg to 15 mg seems to be the appropriate range in normal human subjects, as measured by kinetic analysis with radioiodine. The iodine concentration in normal subjects has been evaluated by this technique to be between 0.37 mg/g and 1.8 mg/g of thyroid tissue, using ultrasound to measure the thyroid volume.[9] Ermans and associates have shown that above 0.25 mg/g there is no change in the relative proportion of iodo compounds in the thyroid.[27] This is the critical limit under which thyroid decompensation is met.

Iodine Flooding

Detection of states of iodine flooding is one of the main purposes of fluorescent scanning. Even after the administration of iodine contrast drugs, very satisfactory scans of these states are still possible under conditions which would make scintigraphy impracticable, even if considerable Tc-99m activity were to be injected (Fig. 7-4). The same applies to patients treated with Lugol's solution or with iodine-containing drugs such as amiodarone. The response of the thyroid iodine pool after an oral iodine loading has been studied by Pavoni and co-workers[28] and Wahner and colleagues.[4] An oral dose of 0.08 mg I/kg induces an early increase of the iodine pool (at 4 hours) followed by a discharge explained by the Wolft–Chaikoff effect at 10 hours, and by a new increase within 48 hours. The iodine content then returns to its previous level. The increase in thyroid iodide is similar whether the single dose given is 1 mg or 100 mg, and this is presumably due to the uptake-blocking effect of the iodine. The return to previous levels, however, is slower with the higher dose. The detection of iodine contamination is usually made by measuring urinary excretion of iodine. However, the urinary excretion may normalize long before the thyroid iodine content has normalized.

Cold Nodules

Numerous ways of diagnosing malignant lesions have been suggested in cases in which nodules are cold on the radioactive tracer scan. These methods also include thallium-201 imaging, echography, and thermography. Fluorescent scanning can also prove helpful in this respect. This type of *in vivo* examination of thyroid nodules has made it possible to show that cancerous thyroid tissue has a relatively low iodine content.[29] In a study of 42 patients before surgery, Patton studied the ratio of iodine content in the cold nodules to that in a corresponding area of the contralateral lobe.[17] *In vivo* analysis enabled him to show that malignant nodules contained less iodine than normal tissue with an iodine content ratio below 0.6. In a recent retrospective analysis of 150 cold nodules with histological diagnosis the same investigator additionally demonstrated that benign nodules could have iodine content ratios lower or higher than this cut-off value.[30] Since 47% of the nodules with iodine content ratio below 0.6 were malignant, but 99% with iodine content ratio above 0.6 were benign, Patton states that an iodine content ratio of greater than 0.6 can be used to identify nodules that have a high potential for benignancy.[30]

Similar results have been reported by Tang Fui and co-workers.[31] They studied 39 patients with cold nodules; 12 patients had thyroid cancer. Iodine was undetectable in all but one of 12 patients with thyroid cancer. However, this also occurred in 8 of 27 patients with benign nodules. We have reported a case of colloid cyst (Fig. 7-5).[3] The nodule was cold on the I-131 scan but filled with stable iodine.

The frequency of cold nodules with high iodine

content differs from one study to another. Hoffer studied 61 patients with nodules that were cold on the fluorescent scan, and all except 2 were devoid of stable iodine. The two exceptions were, respectively, a case of colloid cyst and one of encapsulated carcinoma.[22] Wahner explored seven nodules that remained cold in response to Tc-99m, and they were all devoid of stable iodine; three were benign, and the other four cancerous.[4] Perhaps these results reflect the great scatter of intrathyroid iodine values, with low values sometimes overlapping into the normal zone, as Wahner pointed out. From all these data one can conclude that fluorescent scanning statistically excludes malignant lesions if iodine is found in a cold nodule, but the converse is not indicative of malignancy. So far, no observations based on fluorescent scanning of thyroid metastases have been published. This method can probably not detect deep metastases because of the low energy of the rays projected, but in theory detection of cervical adenopathy should be possible, although this would depend on the degree of tissue differentiation.

Hot Nodules

Hot nodules can have different stable iodine content. Some nontoxic hot nodules contain nondetectable amounts of stable iodine. Some toxic nodules present with a high amount of iodine and some do not. These differences can be explained by the fact that newly formed follicles growing into nodules may present different functional defects. One of the larger studies is by Thrall and associates.[32] In 12 subjects with solitary, autonomously functioning adenomas or toxic adenoma, they showed that the stable iodine concentration in the thyroid was relatively uniform; for although the nodule bound far more Tc-99m than did healthy tissue, its iodine content was not very different. In addition, in two of these patients with no extranodular radioactive isotope binding, fluorescent imaging was able to visualize extranodular tissue, a result already reported by Hoffer and Gottschalk in three patients.[22] The clarity of contralateral lobe visualization also varied.[32] Wahner never observed it, but such visualization was always very clear in the four patients we studied (Fig. 7-6).[3] It was thus possible to see the contralateral lobe in toxic, autonomously functioning adenomas without any injection of thyroid-stimulating hormone (TSH). Comparison of two images obtained on the same day, one with tracers and the other by fluorescence, enabled direct confirmation of the diagnosis without further examination of the patient (Fig. 7-7). In addition, the fluorescent method eliminates the risk arising from the injection of bovine protein TSH.

Differences observed in the stable iodine content of the contralateral thyroid parenchyma may be explained in the two ways mentioned by Thrall and colleagues[32]: The patient may have been imaged before the gland had had time to empty itself of iodine; or TSH may not have been completely suppressed. As the authors suggested, the impossibility of revealing extranodular tissue by isotope scintigraphy may reflect the latter's relative insensitivity in distinguishing nonexistent from greatly attenuated functions.

Hypothyroidism

In secondary hypothyroidism, thyroid iodine content remains normal. In contrast, primary hypothyroidism is accompanied by the disappearance of intrathyroidal iodine. In Hoffer's 13 patients, the thyroid was almost entirely devoid of iodine, and the same applied to the six patients we studied (Fig. 7-8).[3] Fluorescent scanning can detect iodine-induced hypothyroidism (iodide goiter) and can be used as an aid in the diagnosis and the treatment of this disease.

Thyroiditis

A low thyroidal iodine content also is generally encountered in thyroiditis; all of Hoffer's hypothyroid patients with Hashimoto's thyroiditis had a low thyroidal iodine content. But although values can be low in both euthyroidal and hyperthyroidal subjects, they also may be normal or high.[23] Comparable data were reported by Wahner. Due to the insidious occurrence of hypothyroidism in some thyroiditis (such as atrophic autoimmune thyroiditis), fluorescent scanning could be helpful in such patients. A low intrathyroidal iodine pool must lead the clinician to be careful about the risk of hypothyroidism and to press the other investigations in order to begin substitutive therapy. Fragu and colleagues[7] studied 13 patients with de Quervain's subacute thyroiditis. The iodine content was about 2.5 times lower than normal values when thyroiditis had developed in a normal thyroid and still decreased after clinical remission. Thyroid iodine content started to increase after the elevation of serum TSH but remained low for 1 year after complete remission. Iodine treatment increased the iodine content, which returned to initial values faster than in normal subjects after withdrawal of this therapy. In the six patients we studied, the thyroid was devoid of iodine during the initial phase.[3] Rappoport and associates[33] described the evolution of one case of subacute thyroiditis, in which a return to the normal iodine level was observed 10 months after normalization of not only an I-125 scintigram but also the hormonal levels. In two cases of focal thyroiditis, Hoffer showed a reduction of the iodine pool in the nodular lesions, whereas iodine content in the rest of the gland was normal. Paradoxically, one of our subjects who was examined in the hyperthyroid state of de Quervain's subacute nodular thyroiditis had accumulated iodine in the nodule, but there was none in the rest of the thyroid (Fig. 7-9). One month later, Tc-99m scintigraphy showed the reappearance of a residual cold nodule in the thyroid.

Thus, fluorescent imaging during thyroiditis enables the evolution of the disease to be traced and the physiopathology of hormonogenetic disorders to be studied. It can also help to suggest surgical intervention when cancer of the thyroid is suspected because of a nodule that appears during thyroiditis.

Goiter

Even if the total amount of iodine is increased in nontoxic goiter (24 ± 11 mg),[5] this amount is not proportional to goiter size, for some of the large goiters contained relatively little stable iodine. The average iodine concentration is significantly below the normal (288 ± 109 μg/g).[9] During iodide treatment in children, the mean concentration increased with a slight reduction of TSH. In adults, a decreased responsiveness to thyrotropin-releasing hormone (TRH) is observed. Fluorescent scanning is useful in diagnosing the cases of sporadic goiter with low thyroid iodine content, particularly in children. These cases can be treated with small doses of iodine which induce shrinkage of the goiter.[9] The treatment with thyroid hormones would further deplete the iodine stores.

Goiter is the disease in which the most discrepancies arise between radioactive tracer and fluorescent scanning (Figs. 7-10 and 7-11), and the iodine replacement rate probably varies from one region of the thyroid to another during the disorder.[34] This expresses the chemical and functional heterogeneity of endemic or simple goiter. During the growth of the goiter, epithelial cells replicate and generate new follicles with a wide range of different structures and function (high or low iodine turnover).

Hyperthyroidism

Results in the use of the fluorescent technique in nontreated cases of hyperthyroidism are conflicting. Neither Hoffer[23] or Wahner,[4] who studied 15 and 5 patients, respectively, found any difference between the distribution of radioactive iodine and Tc-99m or of stable iodine. The level of the intrathyroid iodine pool was either normal or, more often, high. In a series of 40 patients, Okerlund found an intrathyroidal iodine mean value of 14.1 ± 2.9 mg.[35] This is far above the normal value in women (8.2 ± 2.2 mg). Conversely, our results for 10 patients showed that the iodine pool may vary from an elevated to a frequently undetectable level.[3] Accordingly in untreated thyrotoxic patients, Rougier and colleagues also found a wide range in the thyroid iodine content (16.5 ± 19.5 mg).[5] Twenty percent of the patients exhibited a value above the upper limit of the normal range whereas 22% of the patients were below the lower limit. The variations in the intrathyroidal pool are demonstrated in Figure 7-12.

In vitro studies have shown a reduction in intrathyroidal iodine during hyperthyroidism,[36] and *in vivo* a normal level for such iodine was reported by Heedman using spectrophotometry.[34] Robertson, who applied neutron activation analysis, found low values which returned to normal after the disease was cured.[37] This reduction appears to reflect considerable hormonogenesis. Okerlund showed that the intrathyroidal iodine level was inversely correlated to the triiodothyroxine/thyroxine ratio.

These results might lead one to think that in human hyperthyroidism there is a connection between the iodine content and the T_3/T_4 production ratio. This relationship was not, however, verified *in vitro* by Larsen,[38] and the total iodine pool is probably only an indirect indication of the pool involved in hormonogenesis.

There have been few reports about fluorescent examination of treated hyperthyroid cases. In this connection Okerlund, who periodically examined patients taking synthetic antithyroid drugs, found that all those who still had a high intrathyroidal iodine pool value after 2 years of treatment frequently had a relapse after radioactive iodine therapy. He showed that some of them had practically no iodine left in their thyroid 2 months after cessation of treatment.[35] He raised the question of whether fluorescent scanning would make it possible to predict which patients would develop hypothyroidism after treatment and which would not. Rougier and co-workers[5] found two groups of patients with different evolution of the thyroid iodine content under carbimazole therapy. Patients with high iodine content experienced a decrease under therapy. Patients with low iodine content before treatment experienced an increase at recovery. When hyperthyroidism was due to iodine overloading, fluorescent scanning showed a rise in the thyroid iodine pool. Due to its high iodine content (38% of iodine by weight) amiodarone, a frequently used antiarrythmic and antianginal drug, is responsible for a large portion of iodine-induced thyrotoxicosis.[10] Under this drug, the remaining euthyroid patients accumulate about four times as much iodine as normals (mean of 35 mg, ranging from 20–100 mg) whereas some dysthyroid patients do not. It is felt that the inability of these glands to accumulate the drug or the presence of low thyroidal iodine content before treatment could be of prognostic significance. Therefore, as recommended by Jonckeer,[26] if a patient taking amiodarone for 3 months presents with a thyroid content lower than 20 mg, then it is strongly suspected that he may develop either hyper- or hypothyroidism, and the patient should be very closely examined and observed. Additionally, fluorescent scanning is a useful tool for the follow-up of these cases, because thyroid content evolves in close parallel with the thyroid state in treated patients or in patients recovering spontaneously.

The serial measurement of thyroid iodine content was recently proven useful by Lee and associates[39] as a diagnostic predictor of the outcome of radioiodine therapy in patients with hyperthyroidism. Following radioiodine therapy there is a rapid fall of total iodine content, the nadir of which occurs around the third month post-therapy. The chance of early hypothyroidism within 12 months of radiotherapy is about 82% if thyroid iodine content is less than 2 mg, but only 14% if thyroid iodine content is 2 mg or more.

Pitfalls

Several pitfalls have been described in connection with fluorescent scanning. They may occur in adenopathy, when lymphography fills a gland with iodine,[40] or by the visualization of other fluorescent x-ray isotopes close to the iodine K_α ray. In one of our cases, we observed the presence of a glass jacket button containing barium, which has a K_α ray of 32 keV (Fig. 7-13).

REFERENCES

1. Hoffer PB, Jones WB, Crawford RD: Fluorescent thyroid scanning: A new method of imaging the thyroid. Radiology 90:343, 1968

2. Palmer DW, Deconinck F, Swann SJ et al: Low cost intrathyroidal iodine quantification with a fluorescent scanner. Radiology 119:733, 1976

3. Meignan M, Galle P: Exploration thyroidienne par fluorescence x. Nouv Presse Med 7:13, 1978

4. Wahner HW, Sweet RA, McConahey WM et al: Fluorescent thyroid scanning. A method based on stable iodine measurements. Mayo Clin Proc 53:151, 1978

5. Rougier P, Fragu P, Aubert B et al: Mesure du contenu en iode intrathyroidien par fluorescence x. Intérêt et applications. Path Biol 29:31, 1981

6. Tadros TG, Maisly MN, Ng Tang Fui SC et al: The iodine concentration in benign and malignant thyroid nodules measured by x-ray fluorescence. Brit J Radiol 54:626, 1981

7. Fragu P, Rougier P, Schlumberger M et al: Evolution of thyroid ^{127}I stores measured by fluorescence in subacute thyroiditis. J Clin Endocrinol Metab 54:169, 1982

8. Jonckheer MH: Stable iodine and thyroid function. In Jonckheer MH, Deconinck F (eds):X Ray Fluorescent Scanning of the Thyroid, p 100. Boston, Martinus Nijboff Publishers, 1983

9. Leisner B: Endemic non toxic goiter. In Jonckheer MH, Deconinck F (eds): X Ray Fluorescent Scanning of the Thyroid, p 100. Boston, Martinus Nijboff Publishers, 1983

10. Leger AF, Fragu P, Rougier P et al: Thyroid iodine content measured by x ray fluorescence in Amiodarone induced thyrotoxicosis: Concise communication. J Nucl Med 24:582, 1983

11. Hoffer PB: Fluorescent thyroid scanning. Am J Roentgenol 105:721, 1969

12. Aubert B, Fragu P, DiPaola M et al: Application of x-ray fluorescence to the study of iodine distribution and content in the thyroid. Eur J Nucl Med 6:407, 1981

13. Patton JA, Brill AB: Simultaneous emission and fluorescent scanning of the thyroid. J Nucl Med 19:464, 1978

14. Degroot LJ: Kinetic analysis of iodine metabolism. J Clin Endocrinol Metab 26:149, 1966

15. Kaufman L, Shames D, Powell M: An absorption correction technique for in vivo iodine quantitation by fluorescent excitation. Invest Radiol 8:167, 1973

16. Gollnick DA, Greenfield MA: The in vivo measurement of the total iodine content of the thyroid gland by x-ray fluorescence. Radiology 126:197, 1978

17. Patton JA, Hollifiels BB, Leeg S et al: Differentiation between malignant and benign solitary thyroid nodules by fluorescent scanning. J Nucl Med 17:17, 1976

18. Meignan M, Marinello G, Galle P: Exploration thyroidienne par fluorescence x. Application clinique et étude dosimétrique. J Fr Biophys Med Nucl 1:13, 1979

19. Begon F, Dubrulle C, Boui C: A computerized fluorescent system for thyroid imaging. In Raynaud C (ed): Nuclear Medicine and Biology. Proceedings of the Third World Congress of Nuclear Medicine and Biology, p 2589. Paris, Pergamon Press, 1982

20. Lanzi LH: The Rando Phantom and its Medical Applications. Chicago, University of Chicago, Department of Radiology, 1973

21. Venkataraman G, Jayaraman S: Radiation hazards from ^{241}Am sources used in thyroid studies. J Nucl Med 17:408, 1976

22. Hoffer PB, Gottschalk A: Fluorescent thyroid scanning. Scanning without isotopes. Initial clinical results. Radiology 99:117, 1971

23. Hoffer PB, Bernstein J, Gottschalk A: Fluorescent techniques in thyroid imaging. Semin Nucl Med 1:379, 1971

24. Jonckheer MH, Wahner HW: Clinical usefulness of x ray fluorescence thyroid iodine quantitation and scanning. In Jonckheer MH, Deconinck F (eds): X Ray Fluorescent Scanning of the Thyroid, p 163. Boston, Martinus Nijboff Publishers, 1983

25. Fragu P, Schlumberger M, Aubert B et al: Thyroid iodine content measurement helps for the diagnosis of hyperthyroidism with undetectable radioiodine uptake. In Jonckheer MH, Deconinck F (eds): X Ray Fluorescent Scanning of the Thyroid, p 145. Boston, Martinus Nijboff Publishers, 1983

26. Jonckheer MH: Thyroid iodine content measured by x ray fluorescence in amiodarone induced thyrotoxicosis. J Nucl Med 25:536, 1984

27. Ermans AM, Kinthaert J, Camus M: Defective intrathyroidal iodine metabolism in non toxic goiter inadequate iodation of thyroglobulin. J Clin Endocrinol and Metab 28:1307, 1968

28. Pavoni P, Raganella L, Di Luzio S et al: Feasibility of in vivo XRF dynamic study of the thyroid following stable iodine administration. In Jonckheer MH, Deconinck F (eds):

X Ray Fluorescent Scanning of the Thyroid, p 100. Boston, Martinus Nijboff Publishers, 1983

29. Le Blanc AD, Bell RL, Johnson PC: Measurement of I-127 concentration in thyroid tissue by x-ray fluorescence. J Nucl Med 14:816, 1973

30. Patton JA, Sandler MP, Partain CL: Prediction of benignancy of the solitary "cold" thyroid nodule by fluorescent scanning. J Nucl Med 26:461, 1985

31. Tang Fui Ng SC, Maisey MN: A stationary x ray fluorescent system for measuring thyroid iodine concentrations. In Jonckheer MH, Deconinck F (eds): X Ray Fluorescent Scanning of the Thyroid, p 46. Boston, Martinus Nijboff Publishers, 1983

32. Thrall JH, Bruman KD, Gillin MT et al: Solitary autonomous thyroid nodules. Comparison of fluorescent and pertechnetate imaging. J Nucl Med 18:1064, 1977

33. Rappoport B, Block MB, Hoffer PB et al: Depletion of thyroid iodine during subacute thyroiditis. J Clin Endocrinol Metab 36:610, 1973

34. Heedman PA, Jacobson B: Thyroid iodine determined by x-ray spectrophotometry. J Clin Endocrinol Metab 24:246, 1964

35. Okerlund MD: The clinical utilization of fluorescent scanning of the thyroid. In Kaufman L, Price DC (eds): The Medical Applications of Fluorescent Excitation Analysis, p 149. Orlando, Florida, CRC Press, 1979

36. Gutman AB, Benedict EM, Baxter B et al: The effect of administration of iodine on the total iodine, inorganic and thyroxine content of the pathological thyroid gland. J Biol Chem 97:303, 1932

37. Robertson I, Bodou K, Hooper MJ: Total thyroidal content of iodine in thyrotoxic patients measured by in vivo neutron activation analysis. Clin Endocrinol 5:151, 1976

38. Larsen PR: Thyroidal triiodothyronine and thyroxine in "Graves" disease: Correlation with presurgical treatment, thyroid status and iodine content. J Clin Endocrinol Metab 41:1098, 1975

39. Lee GS, Sandler MP, Patton JA et al: Serial thyroid iodine content in hyperthyroid patients treated with radioiodine. Clin Nucl Med 11:115, 1986

40. Nijensohn E, McCartney W, Hoffer PB: Identification of iodine in a supra clavicular lymph node following lymphography. A fluorescent artifact. J Nucl Med 14:179, 1973

Atlas Section

Figure 7-1 ***Fluorescence Principle***

The incident photon ejects a K-shell target electron. Electronic rearrangement produces fluorescent x-rays (Ter-Pogossian M: Physical Aspects of Diagnostic Radiology. New York, Hoeber, 1969) (Meignan MA, Galle P: Principles and clinical application of thyroid scanning by x-ray fluorescence. Clin Nucl Med 5:71 – 80, 1980)

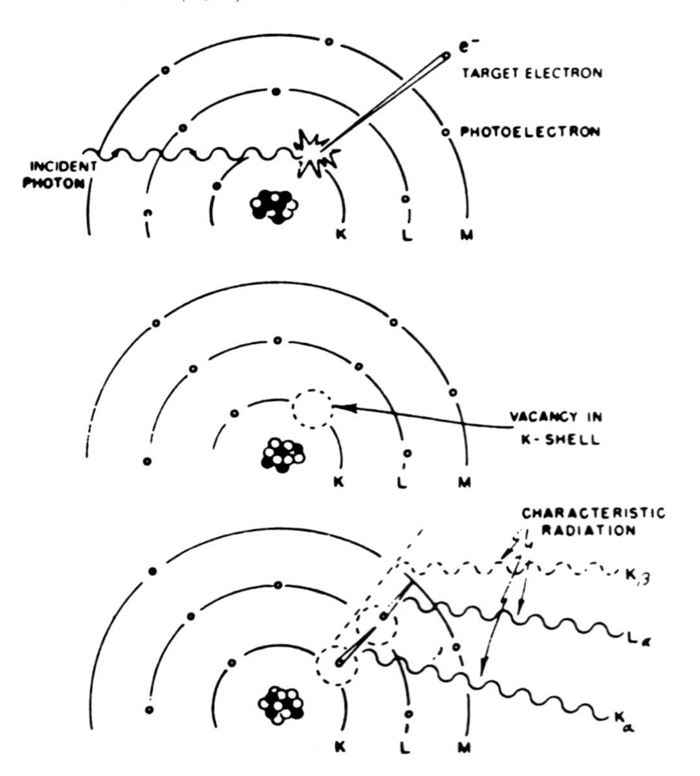

Figure 7-2 ***Two Devices Commonly Used for Fluorescent Thyroid Scanning***

Commercially available fluorescent thyroid scanning devices may use either a single exciting beam source *(lower image)* or multiple concentrically arranged exciting beam sources *(upper image)*. Liquid nitrogen is necessary for cooling the semiconductor to its critical temperature. (Meignan MA, Galle P: Principles and clinical application of thyroid scanning by x-ray fluorescence. Clin Nucl Med 5:71–81, 1980)

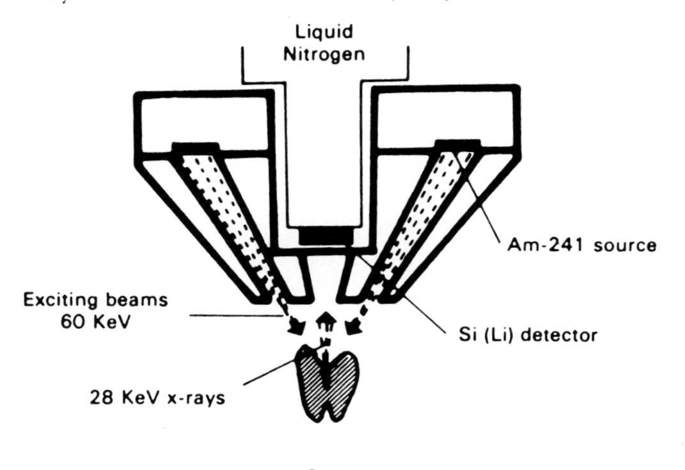

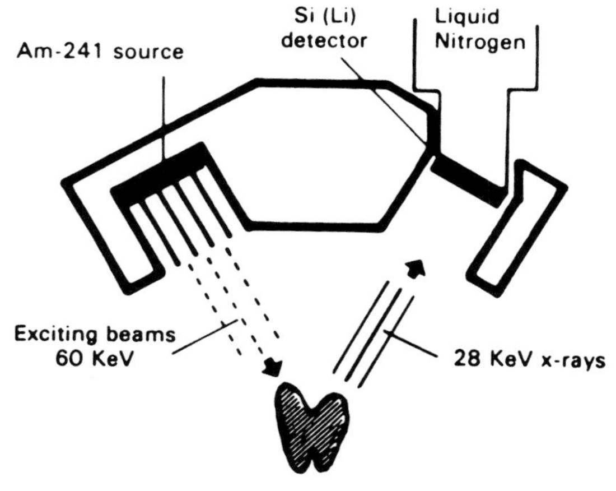

Figure 7-3

Normal Patient

I-131 scintigraphy *(left)* and fluorescent scan *(right)*. (Meignan MA, Galle P: Principles and clinical application of thyroid scanning by x-ray fluorescence. Clin Nucl Med 5:71 - 81, 1980)

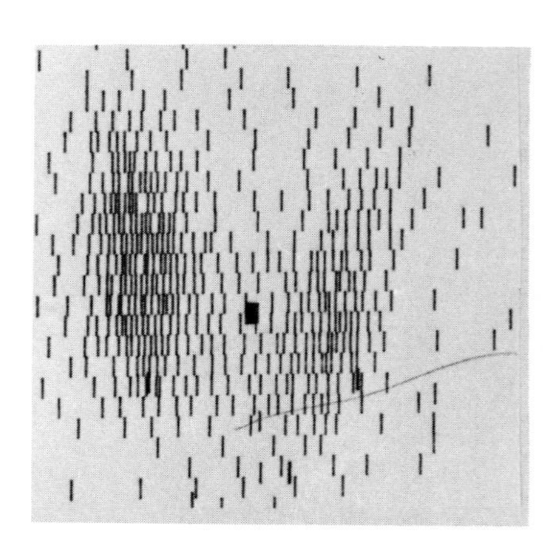

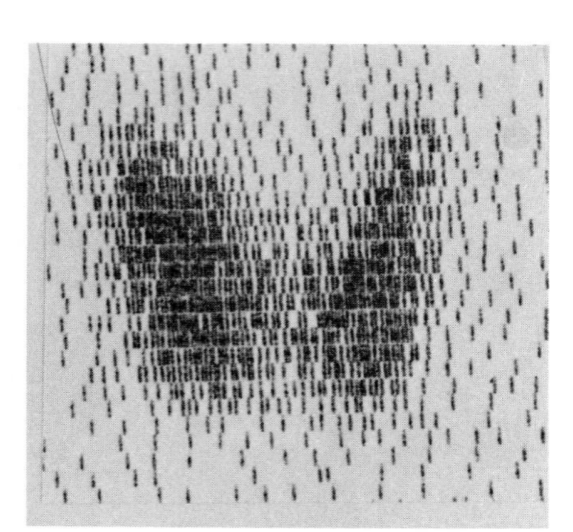

Figure 7-4

Iodine Flooding After a Lymphography

The Tc-99m scan *(left)* demonstrates no trapping, and the fluorescent scan *(right)* demonstrates excellent iodine content. This is what one may see after iodine flooding, such as after lymphography. Detection of iodine contamination is one of the main purposes of fluorescent thyroid scanning. (Meignan MA, Galle P: Principles and clinical application of thyroid scanning by x-ray fluorescence. Clin Nucl Med 5:71 - 80, 1980)

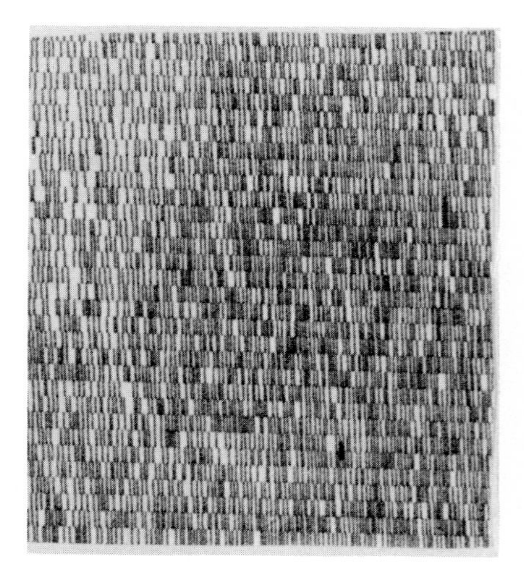

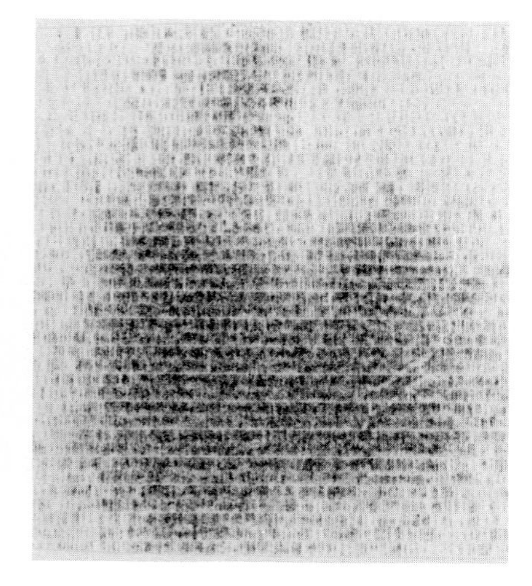

Figure 7-5 ***Colloid Cyst***

The nodule is cold on the I-131 scan *(left, arrow)* but filled with stable iodine *(right, arrow)*. Fluorescent scanning statistically excludes malignant lesions if iodine is found in a cold nodule. (Meignan MA, Galle P: Principles and clinical application of thyroid scanning by x-ray fluorescence. Clin Nucl Med 5:71-80, 1980)

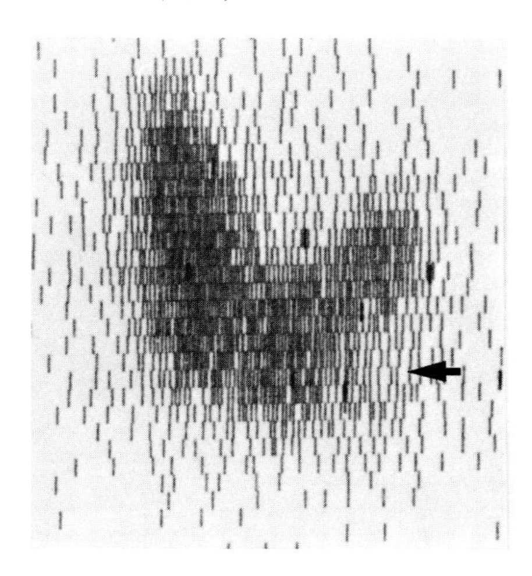

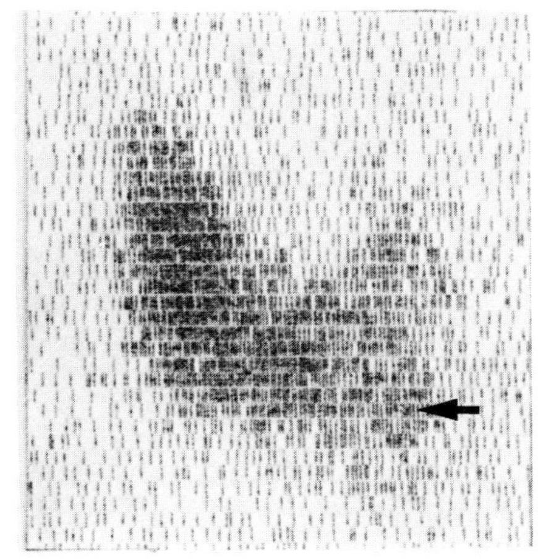

Figure 7-6 ***Autonomous Adenoma***

The I-131 thyroid scan *(left)* demonstrates radioactivity in the autonomous adenoma *(arrows)*, but no radioactivity in the region of the contralateral lobe *(arrowhead)*. The contralateral lobe is clearly demonstrated on the fluorescent scan *(right arrows)*. (Meignan MA, Galle P: Principles and clinical application of thyroid scanning by x-ray fluorescence. Clin Nucl Med 5:71-80, 1980)

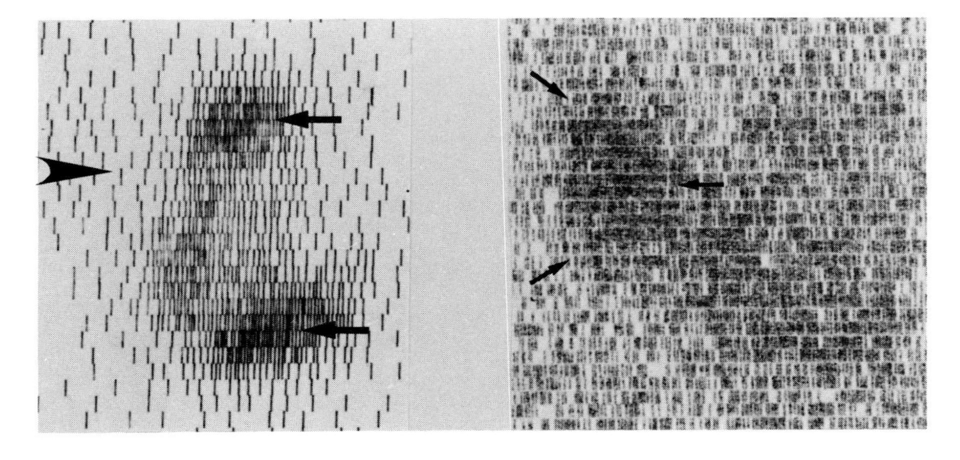

Figure 7-7 ***Toxic Adenoma***

The I-131 thyroid scan ***(A)*** demonstrates a hyperfunctioning adenoma *(outlined area).* A fluorescent scan performed on the same day ***(B)*** demonstrates the presence of the left lobe and thyroid tissue in the right lobe superior to the adenoma *(arrows),* which suggests that these areas are suppressed and that the adenoma is autonomous and most likely toxic. A repeat I-131 scan after TSH administration ***(C)*** confirmed the findings on fluorescent scan. ***Comment:*** Fluorescent scan may help avoid repeated examinations of the patient. (Meignan MA, Galle P: Principles and clinical application of thyroid scanning by x-ray fluorescence. Clin Nucl Med 5:71–80, 1980)

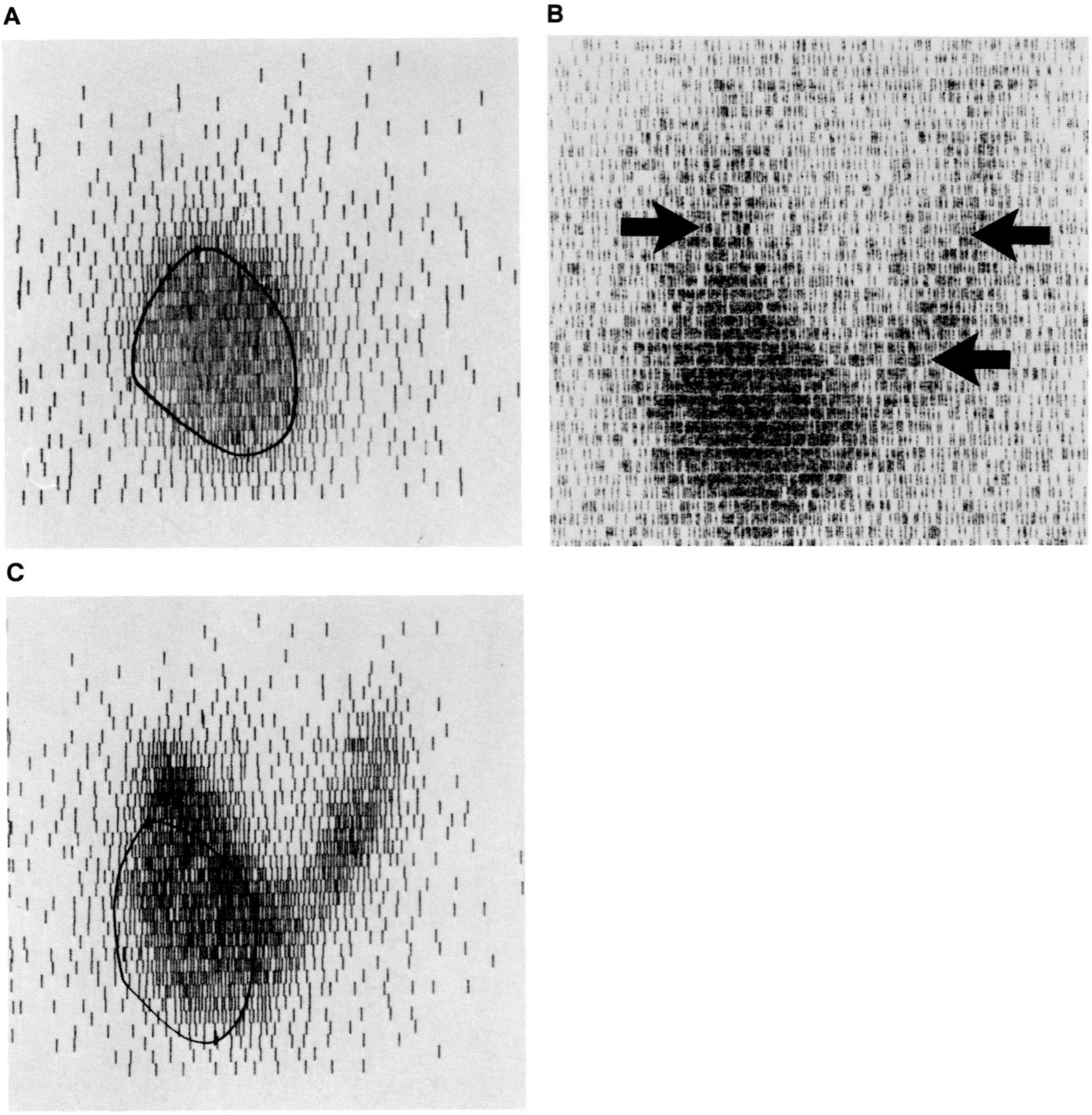

A

B

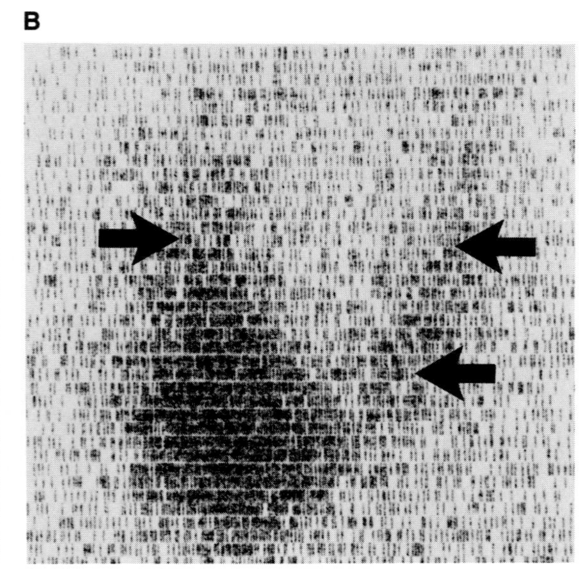

C

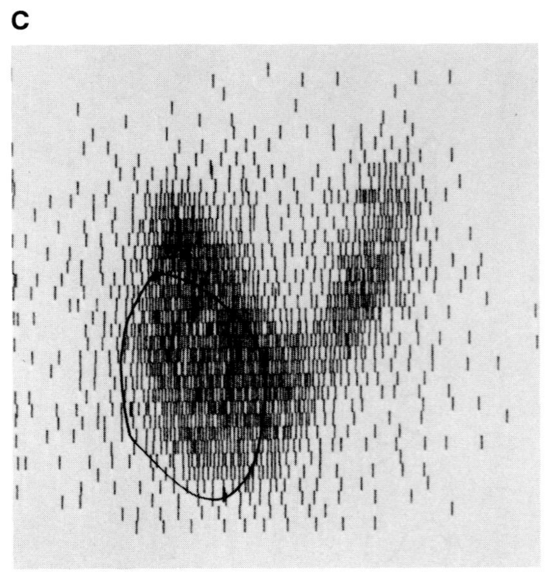

Figure 7-8 ***Primary Hypothyroidism***

I-131 thyroid scan *(left)*; fluorescent scan *(right)*. Although the thyroidal iodine pool is depleted, a slight uptake of I-131 is detectable after a long imaging time. This difference could be explained by the different sensitivities of the two imaging techniques; the low residual iodine pool probably is under the limit of detectability of the fluorescent scan. An organification defect could also be present. In secondary hypothyroidism fluorescent scanning is normal and in iodine-induced hypothyroidism the iodine pool is increased. (Meignan MA, Galle P: Principles and clinical application of thyroid scanning by x-ray fluorescence. Clin Nucl Med 5:71-80, 1980)

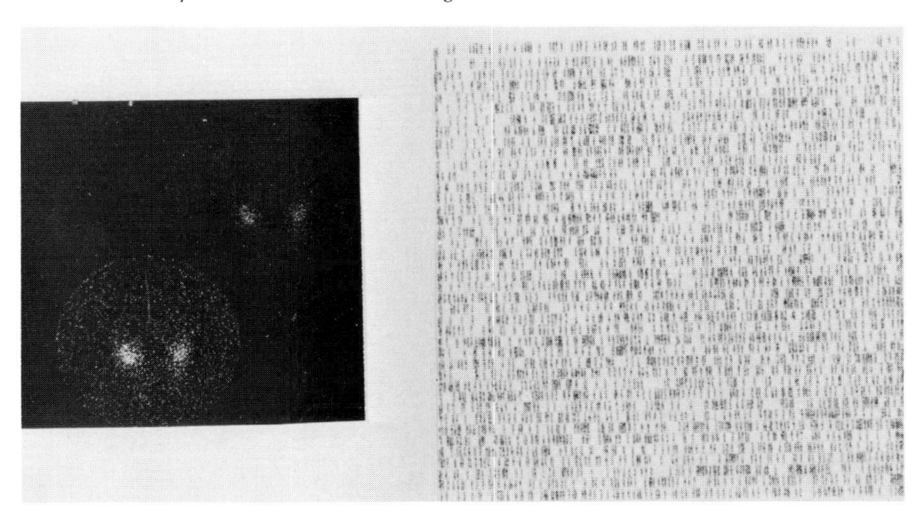

Figure 7-9 ***Nodular Subactue Thyroiditis (Hyperthyroidism)***

The Tc-99m scan *(left)* demonstrates no trapping compatible with subacute thyroiditis. The fluorescent scan *(right)* shows stable iodine in one area, which corresponded to a palpable nodule. This suggests the nodule is benign; and one month later the nodule disappeared. ***Comment:*** The above was attributed to a cystic nodule in a thyroid with subacute thyroiditis. It is proposed that the iodine in the cyst was the only pool of stable iodine remaining. (Meignan MA, Galle P: Principles and clinical application of thyroid scanning by x-ray fluorescence. Clin Nucl Med 5:71-80, 1980)

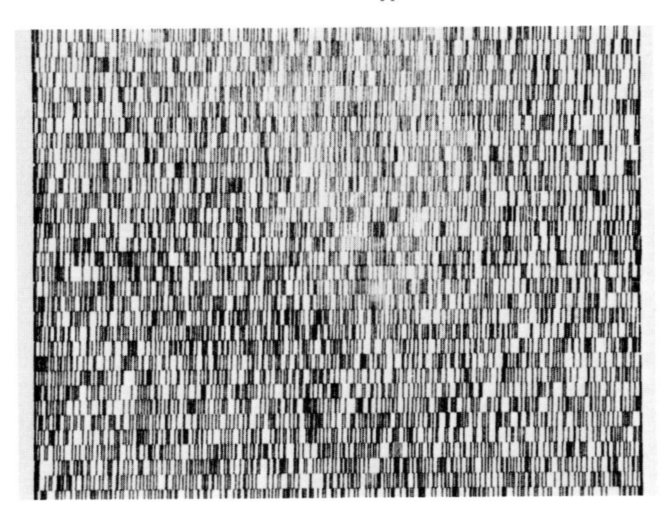

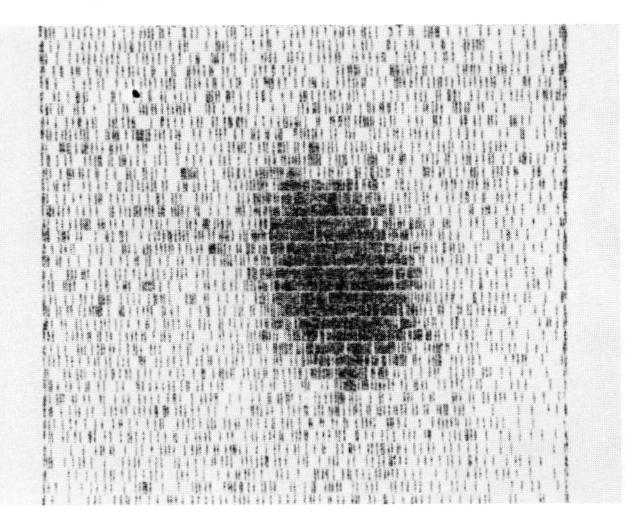

Figure 7-10 ***Goiter***

The I-131 scan *(left)* and the fluorescent scan *(right)* demonstrate one pattern for thyroid goiter with variable distribution of I-131 uptake and stable iodine, respectively. Note the relatively similar pattern of distribution. (Meigan MA, Galle P: Principles and clinical application of thyroid scanning by x-ray fluorescence. Clin Nucl Med 5:71 - 80, 1980)

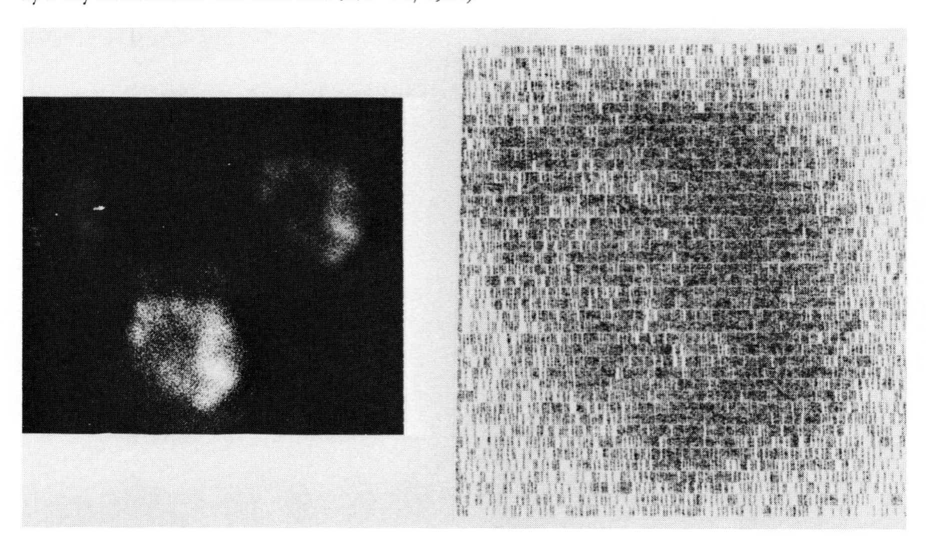

Figure 7-11 ***Goiter***

Tc-99m scan *(left)* demonstrates variable trapping in the right lobe *(arrows)*, and the fluorescent scan *(right)* demonstrates variable levels of stable iodine *(arrowheads)*. Note the relatively different pattern of distribution, expressing the typical chemical heterogeneity of goiters. (Meignan MA, Galle P: Principles and clinical application of thyroid scanning by x-ray fluorescence. Clin Nucl Med 5:71 - 80, 1980)

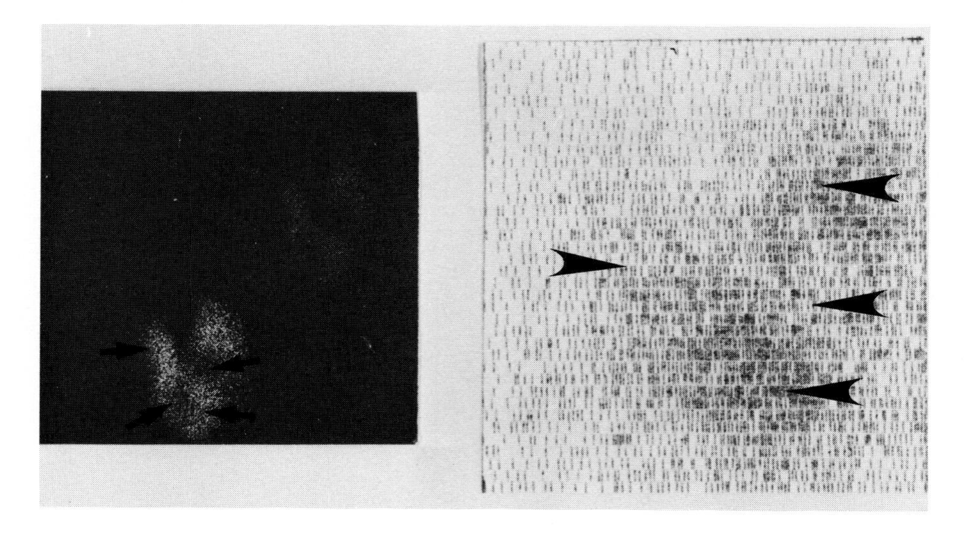

Figure 7-12 ***Hyperthyroidism***

I-131 scans *(upper and lower left)* and fluorescent scans *(upper and lower right)* in two different patients demonstrate the wide range in intrathyroid iodine content. (Meignan MA, Galle P: Principles and clinical application of thyroid scanning by x-ray fluorescence. Clin Nucl Med 5:71–80, 1980)

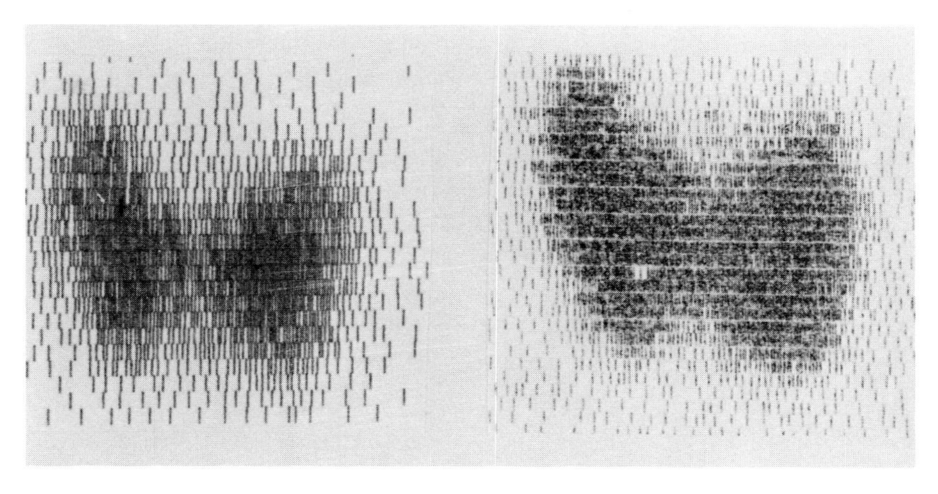

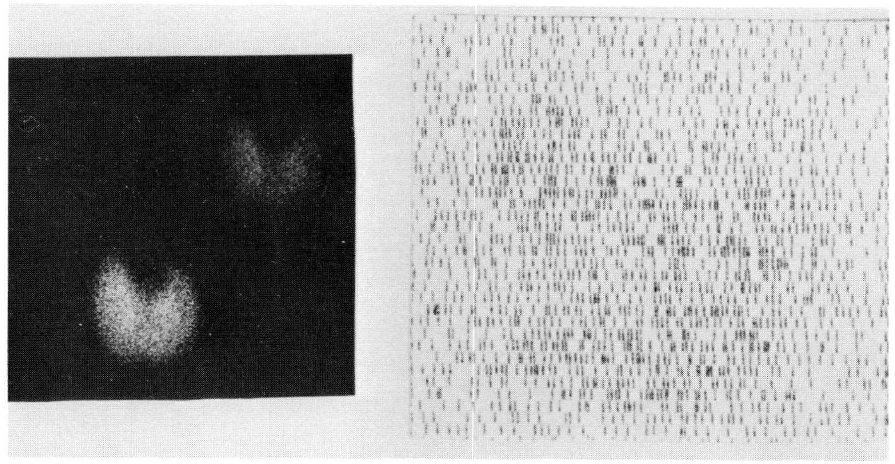

Figure 7-13 ***Artifact***

(A) Fluorescent scan of the thyroid and a glass jacket button *(arrow)*. ***(B)*** Fluorescent scan of the button containing barium. The pitfall of including other elements with detectable K_α rays must be kept in mind. (Meignan MA, Galle P: Principles and clinical application of thyroid scanning by x-ray fluorescence. Clin Nucl Med 5:71–80, 1980)

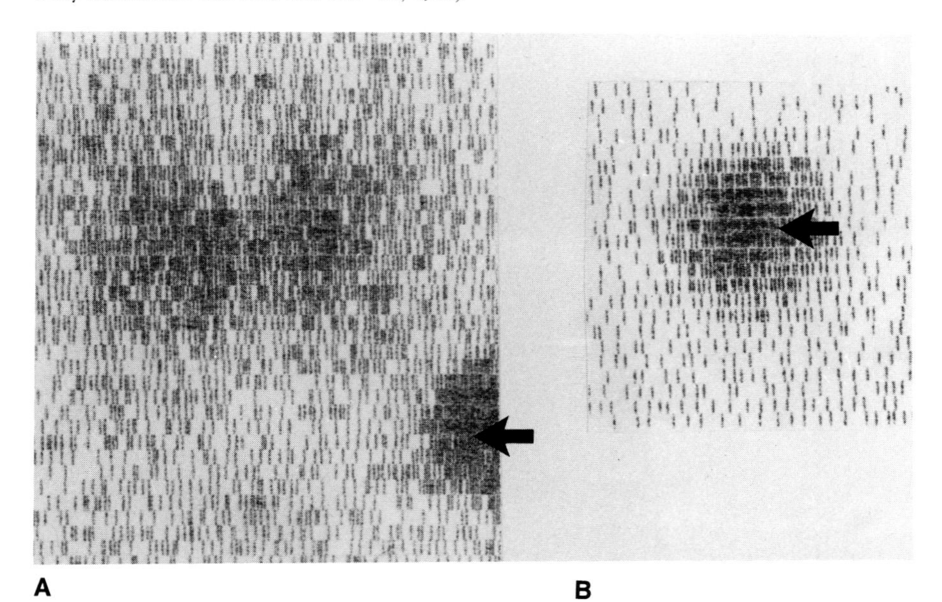

Leonard Wartofsky
Yolanda C. Oertel

Chapter 8 FINE NEEDLE ASPIRATION OF THYROID NODULES

The critical question for the clinician managing the problem of a patient with a solitary thyroid nodule is whether there is sufficient risk of malignancy to warrant consideration for thyroidectomy. Tumors elsewhere in the body often have been approached initially by some type of biopsy procedure to obtain a pathologic diagnosis. Earlier reports of thyroid biopsy for this purpose employing Tru-Cut or Vim – Silverman needles failed to generate enthusiasm for the technique because of the associated morbidity, fear of spreading neoplastic cells along the needle track, and, more often than not, the acquisition of insufficient tissue for definitive diagnosis.

The subject of this chapter, fine needle aspiration and its use to obtain cells for cytologic examination, must be clearly differentiated from the biopsy methods just mentioned, which were intended to obtain tissue for histologic evaluation.

Diagnostic evaluation of the solitary thyroid nodule usually will include radionuclide imaging to determine whether or not a nodule is functioning, and sonography to indicate whether the nodule is solid or cystic in nature. Although nonfunctioning ("cold") and solid lesions have a greater likelihood of being malignant than do functioning or cystic lesions, these criteria alone will not serve as an indication for thyroidectomy because of the unfavorable risk/benefit ratio. Over the past several decades, numerous other factors had been evaluated for their discriminatory value in predicting which patients with a thyroid mass should go to surgery, but the improved yield was minimal until the recent introduction, or rather, popularization of the fine needle aspiration technique.

The technique is, in fact, not a new one and has been in use in selected medical centers in the United States for over 50 years, but it has enjoyed more widespread application in Scandinavia, particularly during the past 20 years. The value of the technique lies in its ability to establish a diagnosis of either benign or malignant disease and by so doing eliminate the necessity in a large number of patients for an exploratory thyroidectomy with its attendant morbidity.

To use the technique reliably, there are certain mandatory requirements in regard to the performance of the procedure and the expertise of the pathologist interpreting the specimen. The Atlas is intended to provide guidelines for the technique of fine needle aspiration and to illustrate characteristic cytologic findings for the most common disorders of the thyroid gland encountered in clinical practice.

We are grateful to the expert technical staff of our two respective insitutions who helped support our excellent results in needle aspiration, to Mrs. Estelle Coleman for editorial assistance and preparation of the manuscript, and to Ms. Barbara Neuburger for assistance with the photomicrographs.

MATERIALS AND METHODS

The Operator

Virtually any physician could learn how to perform needle aspiration of the thyroid gland. However, the technique is deceptively simple. There has to be commitment, willingness to pay attention to detail, and dexterity comparable to that needed for a venipuncture combined with a knowledge of the anatomy of the neck, a careful physical examination of the thyroid, and the localization of the thyroid lesion in question. It is invaluable to observe an experienced individual perform the technique. Then a role reversal should occur as the experienced physician watches the beginner perform the procedure.

We cannot overemphasize the importance of obtaining a good sample and smearing it correctly. The pathologist's interpretation of the smears is only as good as the sample provided. Although the technique of acquisition of the specimen is critical to success, the subsequent interpretation of the smears is obviously of equal importance. One definite advantage to having a pathologist do the procedure is that the slides can be immediately previewed for their adequacy. Thus, if material insufficient for interpretation is obtained, repeated aspirations can be performed until satisfactory, thereby obviating the need for return visits by the patient in order to repeat the procedure.

In the thyroid clinic at Walter Reed Hospital, the aspiration is done by one of our staff or fellows with a cytotechnologist on hand to assist in the immediate preparation of smears, with duplicate slides smeared, stained, and then reviewed by the physician doing the aspiration. If adequate cellular material is obtained, the procedure is terminated and all the materials are taken to the cytopathology laboratory for further preparation and interpretation. With minimal training, an internist, surgeon, or nuclear medicine specialist could become adept enough at the technique to ensure acquisition of material suitable for interpretation, while still relying on the expertise of the laboratory for optimal preparation of slides and diagnosis. In that way the patient is not caused any undue inconvenience. Although needle guidance by ultrasound may be useful (especially for small or posteriorly placed nodules), in practice it is rarely employed.

Materials Required

Needle aspiration of the thyroid is performed with the following materials:

- 22-23 – gauge, 1- to 1½-inch disposable needles with clear hubs
- 10 ml disposable (Luer-Lok) syringes
- Syringe holder
- Iodine and alcohol wipes for skin sterilization
- Sterile surgical towel or drape (optional)

Technique

Patient Preparation

The patient is counseled regarding the indications, technique, and risks of the procedure. A signed, witnessed consent form may be advisable.

The patient is placed in a supine position on an examining table with the neck hyperextended over a pillow in much the same position as employed for a thyroid scintiscan, resulting in moderate relaxation of the cervical musculature and reasonable comfort for the patient. The neck is examined in this position, making careful note of the location of the nodule(s) in question. Nodules in either the left or right lobe of the thyroid gland can be approached equally well from either side of the patient, so the operator should stand on whichever side is conducive to his or her greatest comfort, based upon the operating space, facilities, and his or her technical assistance.

The skin may be prepared with alcohol wipes alone or with three wipes of 10% providone – iodine, followed by a final wipe with an alcohol-saturated sponge. Attempts to maintain a truly aseptic technique are really not necessary and will vary from center to center. Although some workers wear sterile surgical gloves, these tend to interfere with the sensitivity of nodule palpation. It is actually only the left hand (of a right-handed operator) that is used to fix the nodule in place and needs to remain sterile, whereas the right hand operates the (nonsterile) syringe holder. The infiltration of a local anesthetic is not recommended because the discomfort of the 22-23 – gauge aspiration needle is not significantly greater than that from the 27-gauge needle used for the infiltration of Xylocaine. Moreover, the subsequent aspiration of Xylocaine into the syringe with the specimen will hemolyze cells and alter the findings.

Sampling Procedure

Before inserting the needle into the thyroid, we have found it useful to introduce into the syringe approximately 2 cc of air, which is used at the termination of the procedure to expel the aspirated material out of the needle hub. The nodule is immobilized with one hand (usually the left) between the index and middle fingers. The needle tip is quickly but gently passed into the nodule without exerting any negative pressure (retraction) on the plunger. Once it is clear that the needle is within the nodule, the pistol grip of the syringe holder then is retracted to create a vacuum within the needle and the syringe barrel(Fig. 8-1). The nee-

Figure 8-1 ***Technique of Needle Aspiration***

(A) A 10-ml (preferably) or 20-ml syringe with 1- to 1½-inch clear hub 22–23 gauge needle is locked into aspiration pistol. ***(B)*** Needle is inserted into nodule without exerting suction. ***(C)*** Needle is quickly but gently moved forward and back within the nodule while retracting handle to create suction. ***(D)*** The handle is released to stop suction. ***(E)*** The needle is withdrawn from the nodule. ***(F)*** The aspirate is expressed onto a glass slide.

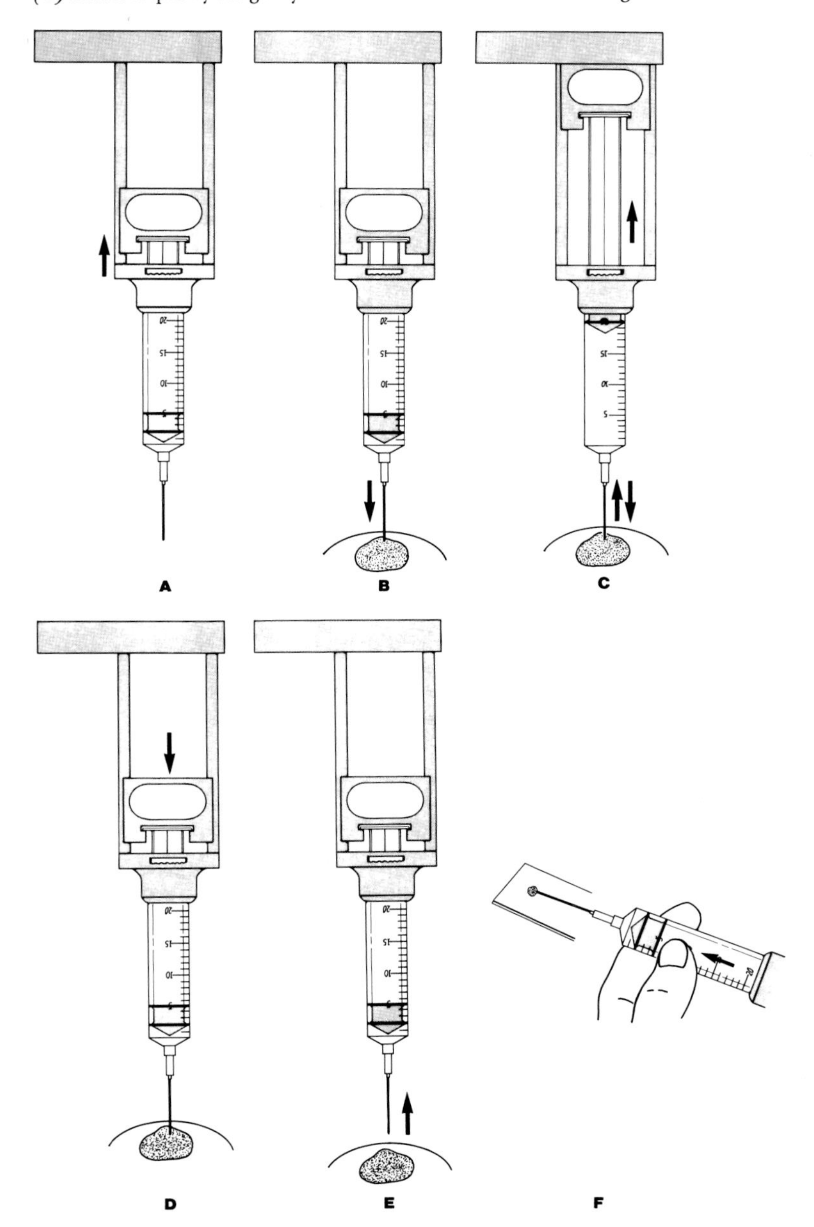

dle tip is quickly but gently moved forward and backward, five to ten times over an approximate 1 mm to 2 mm excursion, which serves to free up cellular material. Some operators prefer to alternately squeeze and release the trigger of the syringe holder during needle insertion. Sampling of multiple sites greatly increases the likelihood of obtaining adequate cellular material. We aspirate a minimum of two times (even if the lesion is 5 mm in diameter or less). The only instance in which we do only one aspiration is when the lesion is cystic and it collapses after the fluid is evacuated, leaving no palpable residual lesion to reaspirate.

Once a drop of bloody material is seen to enter the needle hub, the suction is terminated and the pressure on the pistol grip of the holder is released in preparation for withdrawal of the needle. This is perhaps the most critical technical aspect of the procedure, because if the needle is withdrawn while still maintaining suction, the cellular material will be drawn into the barrel of the syringe and made very difficult or impossible to readily retrieve, smear, and stain, and very likely the sampling will have to be repeated.

Slide Preparation

For convenience in expressing the aspirate onto the slides, the syringe and needle may be disengaged from the syringe holder (some prefer to omit this maneuver). If a few ml of air were introduced into the syringe prior to the sampling, this air can now be used to express the material in the needle onto a slide by putting some pressure on the plunger of the syringe. If there is insufficient air in the barrel to do this, then the needle is first detached, the syringe is filled with air, and the material is then expressed out onto one slide or several slides, depending on how much aspirate was obtained. The technique will vary depending upon the type of material aspirated. Slides with one frosted end are preferable in order to facilitate labeling. When only a drop of aspirate is obtained, it is immediately smeared with a hemacytometer cover glass in much the same manner as employed for preparation of a peripheral blood smear.

The drop is allowed to run on the slide along one edge of the hemacytometer cover glass and then the cover glass is drawn with a smooth even movement along the slide, resulting in a smear of usually a half to one inch in length. Visible clumps or particles should be further smeared out by pressing gently with the flat surface of the coverslip. This technique will differ when a more liquid aspirate, either blood or cyst fluid, is obtained. Cyst fluid will vary in appearance from chocolate brown (hemorrhagic) to serum-like and often viscous, to clear and colorless (parathyroid cyst). Visible particles in the aspirate should be smeared and fixed as usual, but the remainder of the fluid should be immediately centrifuged, after which the sediment is smeared, fixed, and stained. When a very bloody aspirate is obtained, it may be difficult to ascertain whether it represents fresh hemorrhage within a nodule or entry of the needle into a thyroidal blood vessel. The former might be more likely if an ultrasound examination was available beforehand and indicated a cystic or mixed solid–cystic structure. In either case, it may be helpful to separate out any cellular material in the aspirate from the blood before it clots. This can be done by expelling the blood onto glass slides and then employing 2×2 gauze pads to soak up the liquid while endeavoring to leave cellular particles on the slide. The particles are then pressed with the flat surface of a cover glass and smeared as just described.

Methods for fixation of the smear vary from one institution to another. Most workers prefer to let the smears air-dry (an air blower can be used to hasten this), and then stain them with a modified Wright's stain, the May-Grünwald-Giemsa (MGG) stain, or the commercially available Diff-Quik. Others prefer to fix the smears with alcohol or a cytofixative spray and then stain with hematoxylin-eosin or by the Papanicolaou technique. All of the hematological stains produce good staining of cytoplasmic detail and colloid, whereas the Papanicolaou and hematoxylin-eosin stains yield greater nuclear detail. Both fixation/staining techniques can be employed if there is an adequate number of smears.

Visual Interpretation and Diagnosis

Application of the fine needle aspiration technique may be very helpful in the diagnosis of a number of benign and malignant disorders of the thyroid gland, but it is important to recognize the limitations of the procedure. The major deficiency which precludes greater practical utility of the aspiration method is the very large propensity for false-negative diagnosis due to "missed" lesions or inadequate aspirations; these cases are usually related to improper technique. It should be apparent that the real value of the procedure lies in obtaining a specific diagnosis. Since the absence of findings, such as malignant cells, does not rule out malignancy, a cytology report simply stating "no malignant cells seen" is ambiguous and virtually worthless; such a report provides little help in the decision-making process for management of a given thyroid mass. Equally obvious is the fact that the reliability of the cytologic interpretation is directly related to the training, experience, and interpretative skill of the pathologist. We would be reluctant to place faith in someone's interpretations unless that individual had undergone special training or apprenticeship in thyroid cytology.

The clinician may expect to receive reports from an experienced pathologist, which will classify the specimen into one of four specific categories: benign, malignant, indeterminate, or inadequate. *Benign* lesions may be characterized as suggestive of adenomatoid nodule, colloid nodule, colloid goiter, Hashimoto's disease, or granulomatours thyroiditis, but will rarely go beyond these specific diagnoses, except to describe the nature and type of cells seen. *Malignant* readings will be indicated as papillary carcinoma, medullary carcinoma, anaplastic carcinoma, malignant lymphoma, or metastatic carcinoma. *Indeterminate* readings may indicate some degree of suspicion for malignancy, but not enough to warrant more specific categories or cell types (*e.g.,* follicular neoplasia). These specimens should not be confused with inadequate ones, and may in fact be very cellular but lacking enough features to call one way or another. For example, a highly cellular smear could be read as "suggestive of follicular neoplasm; indeterminate for malignancy." This is because the diagnosis of follicular carcinoma rests on the demonstration of capsular or vascular invasion seen only on histologic sections and not visible on cytologic smears. (Recently, some investigators have been measuring the DNA content and ploidy level of the nuclei to distinguish between benign and malignant follicular cells; aneuploidy indicates malignancy.) Finally, the *inadequate for diagnosis* category includes smears with abundant blood or dysmorphic debris, but insufficient cellular material to provide any interpretation.

The result of obtaining an interpretation in these four categories may be briefly summarized as follows: Benign readings indefinitely defer surgery and give the patient and physician peace of mind for continued conservative management. A reading of malignant or suspicion of malignancy is grounds for immediate surgery and appropriate subsequent follow-up care. A reading in the indeterminate category may lead either to surgery or continued conservative management, depending on the risk factors for malignancy. Those patients conservatively managed may be subsequently re-aspirated for confirmation and reassurance. The patient could escape thyroidectomy unless the nodule enlarges while the patient is taking thyroid hormone. Results in the inadequate category demand another attempt at aspiration as long as the indication for the initial attempt was a valid one. Subsequent failures to obtain adequate material for interpretation may suggest a very acellular lesion, such as colloid-filled nodule, but a surgical procedure is required for definitive diagnosis.

CHARACTERISTIC CYTOPATHOLOGY

Cystic Lesions

The aspirated fluid of cystic lesions (Col. Fig. 8-1) may be chocolate brown (suggesting old hemorrhage), straw-colored and viscous, or very rarely purulent. Hemorrhagic cyst fluid may show some cellular atypia and amorphous debris; the degenerative nature of the lesion is further indicated by the presence of numerous multinucleated giant cells and hemosiderin-laden macrophages. If particularly viscous fluid is encountered, the needle may be withdrawn and replaced with a larger bore (*e.g.,* 18-gauge) needle to facilitate drainage of the cyst. Clear, colorless fluid has been described in parathyroid cysts, and such fluid might be sent for parathyroid hormone assay to confirm this possibility. Cyst fluid and follicular cells in an aspirate of a cervical lymph node are evidence of metastatic papillary carcinoma.

Inflammatory Disorders

Acute (suppurative) thyroiditis is an extremely rare disorder usually associated with infection elsewhere and sepsis. The aspirate will demonstrate numerous polymorphonuclear leukocytes and necrotic debris, but few follicular cells. Rarely, the infection is a complication of an underlying undifferentiated carcinoma, and care must be taken to adequately sample indurated or nodular regions of the gland to exclude this possibility.

Subacute (de Quervain's or granulomatous) thyroiditis is marked clinically by pain and tenderness of the thyroid, extremely low radioiodine uptake, and hyperthyroidism in approximately one third of cases. The aspiration is frequently very uncomfortable for these patients. The aspirate reveals clusters of epithelial cells (which may be in various stages of degeneration), histiocytes, and multinucleated giant cells. The cytoplasm of the intact follicular cells may be relatively scant, granular, and contain vacuoles (Col. Fig. 8-2).

Painless thyroiditis is a clinical designation for a disorder which usually presents with hyperthyroidism, a small goiter, absence of stigmata of Graves' disease, and a low radioiodine uptake as seen in de Quervain's thyroiditis. This disorder has been recognized with increasing frequency following pregnancy; it is associated with high titers of thyroid autoantibodies and is called "postpartum thyroiditis." The aspirate is similar to that seen in chronic lymphocytic thyroiditis.

Hashimoto's (chronic lymphocytic) thyroiditis is an autoimmune disorder occurring predominantly in women, which is marked by circulating antithyroid antibodies and a spectrum of thyroidal changes—from lymphocytic infiltration and goiter to progressive fibrosis and atrophy—often culminating in so-called "idiopathic" myxedema. Early on, the follicular epithelium may be hyperplastic and highly cellular with clusters of oxyphilic Hürthle (Askanazy) cells, plasma cells, and lymphocytes. The Hürthle cells may be in confluent sheets, leading to difficulties differentiating such a process from Hürthle cell neoplasm. The lymphocytes are often very abundant and may range from reactive immunoblastic forms to small, mature lymphocytes. Histiocytes containing phagocytized debris may be seen, as well as follicular cells with cytoplasmic vacuoles. Colloid is generally scant to absent, and with progressive fibrosis and atrophy the gland may become nodular and less likely to yield cellular material (Col. Figs. 8-3 and 8-4).

Although the changes of Hashimoto's thyroiditis may on occasion be localized, caution must be exercised in the case of aspirates of cold nodules, because patchy lymphocytic infiltrates may occur in association with thyroid carcinoma, and malignant lymphoma (see later discussion) often arises amidst a background of lymphocytic thyroiditis.

Goiter

Nontoxic goiter includes several types of benign thyroid enlargement, such as colloid goiter, endemic (iodine deficiency) goiter, and multinodular or adenomatous goiter. The cytologic characteristics will vary depending upon the etiology of the goiter, but some generalizations are possible. Large amounts of colloid will be seen in colloid goiter, with clusters to sheets of small follicular cells with round, regular nuclei and few nucleoli (Col. Figs. 8-5–8-9). Colloid will stain blue (cyanophilia) with MGG stain and an orange green with Papanicolaou. The larger multinodular goiters will have degenerating nodules as well, which may show clear yellow to chocolate brown fluid on aspiration with numerous hemosiderin-laden macrophages. Absorption of the fluid on the slide with gauze (as described previously) often will leave granular material usually representing degenerating follicular cells with colloid debris, or occasionally calcified particles.

Toxic goiter rarely presents a sufficient clinical diagnostic dilemma to warrant needle aspiration. Cold nodules may occur in a Graves' gland, however, and probably represent a greater risk for malignancy in that they presumably arise in the face of hyperthyroxinemia. The cytologic features of the uncomplicated Graves' thyroid include numerous clusters of hyperplastic follicular cells with vacuoles that are endocytosed "colloid droplets." The colloid itself will be scant due to the hyperactive secretory process, and moderate to numerous lymphocytes will be seen. These glands are extremely vascular and the aspirate will contain many erythrocytes.

Neoplasms

Follicular adenoma cannot be differentiated from *follicular carcinoma* by cytologic examination alone; the distinction requires multiple histologic sections through the capsule to exclude the capsular invasion as well as vascular invasion seen in carcinoma. The typical findings are of clusters to sheets of monotonous follicular cells with enlarged nuclei forming "rosettes" or "neoplastic follicles." The follicular pattern is most apparent in those adenomas classified histologically as microfollicular (fetal) or embryonal (trabecular). The nuclei are enlarged, relatively uniform, and the cytoplasm will be delicate and scant. When sheets of cells are obtained, there will be moderately dense crowding and overlapping of the nuclei, and nucleoli may or may not be prominent. While some of these nodules may represent adenomatoid nodules and will demonstrate moderate amounts of colloid on the smear, most follicular adenomas are primarily cellular and hence colloid will be scant to absent on the aspirate (Col. Figs. 8-10–8-12).

Hürthle cell tumors are a variant of follicular neoplasms which may be either benign (Hürthle cell adenoma) or malignant (Hürthle cell carcinoma). The Hürthle (oxyphilic; Askanazy) cell is a metaplastic follicular cell with increased mitochondrial and oxidative enzyme activity; the cells are large with abundant, finely granular eosinophilic cytoplasm, and enlarged hyperchromatic round to oval nuclei, slightly eccentric, often with a macronucleolus. Colloid is present in variable amounts but is frequently scant (Col. Figs. 8-13–8-16).

Hürthle cells may be seen commonly in Hashimoto's thyroiditis and also in nodular goiters. An aspirate showing sheets or large clusters of Hürthle cells with prominent nucleoli indicates that the diagnosis is Hürthle cell neoplasm rather than one of the latter disorders more likely to have smaller groups of Hürthle cells (usually with inconspicuous nucleoli), along with the other findings typical of the primary process. This is complicated by the fact that Hürthle cell tumors may occur in a Hashimoto's gland. As with other follicular neoplasms, malignancy cannot be ruled out on the basis of the cytologic examination, and recognition of a Hürthle cell tumor generally represents an indication for surgery.

Papillary carcinoma is the most common thyroid cancer. The typical histologic features of papillary fibrovascular stalks with malignant epithelium and psammoma bodies (calcification) correlate with the usual cytologic features. Papillary fragments may be seen with some nuclear palisading or monolayer sheets of cells with occasional remnants of a follicular pattern. Intranuclear inclusions may be seen; these so-called "pseudonucleoli" are pale rounded holes in the nucleus representing cytoplasmic invaginations. These sometimes are seen in medullary carcinoma and metastatic melanoma. The nuclei vary in size, and nucleoli sometimes may be observed. The cytoplasm is often abundant and dense, but sometimes may be clear to foamy with sharper borders than are usually seen in a follicular neoplasm. Multinucleated macrophages are common; the aforementioned psammoma bodies with their concentric lamellae are not frequently seen (Col. Figs. 8-17–8-22).

Medullary carcinoma is a neoplasm derived from the parafollicular cells (C cells) of the thyroid which secrete calcitonin (CT). The tumor may occur on either a familial or sporadic basis, and patients will have elevated levels of CT in their blood, either in the basal state or after provocative stimulation tests. Histologically, the tumors are typically pleomorphic with a stroma containing amyloid. Similar features are seen on cytologic examination (Col. Figs. 8-23–8-27). The smears are very cellular. The shape of the cells will range from round to oval cells of varying size with eccentric nuclei, to larger polygonal cells and spindle-shaped cells. The nuclei may vary in size and shape. Binucleation is frequently seen, but nucleoli are very seldom observed. Intranuclear cytoplasmic inclusions may be present. The cytoplasm of some cells will show red pink granules with MGG stain and will be cyanophilic with Papanicolaou stain. Colloid is generally absent. When present, the amyloid is primarily extracellular and might be confused with colloid, but occasionally can be seen intracellularly; the amorphous clumps are pale-staining and will stain metachromatically with alkaline congo red and appear fluorescent with thioflavine T.

Anaplastic carcinoma is the most undifferentiated form of thyroid malignancy and has an almost uniformly dismal prognosis. Aspiration of adequate cellular material may be difficult due to inflammatory changes, necrosis, and sclerosis. The cells are markedly atypical with wide variation in size and shape of both the cells and the nuclei. Some binucleate cells may be seen, and there are prominent nucleoli (Col. Figs. 8-28–8-30). The *giant cell* type is the most bizarre, with much necrotic material mixed with giant and spindle cells. The *small cell* type may be difficult to differentiate from thyroid lymphoma.

Lymphoma may be a primary tumor of the thyroid and can present as a single nonfunctioning nodule or as more diffuse involvement. This disorder is to be distinguished from Hodgkin's disease and other non-Hodgkin's lymphomas which may invade the thyroid gland from the adjacent cervical nodes; typically Reed–Sternberg cells can be seen in an aspirate from Hodgkin's disease. It may be difficult to distinguish lymphoma from Hashimoto's thyroiditis; the absence of epithelial cells is more compatible with lymphoma. The problem is compounded by the fact that primary lymphoma often arises in postmenopausal females with underlying Hashimoto's disease and positive titers of antithyroid antibodies in their serum. In such cases, the cytologic features will be marked by masses of lymphomatous cells with the pleomorphic background typical of Hashimoto's disease. A malignancy superimposed on thyroiditis should be suggested by the presence of clumps of primitive types of cells (such as transformed lymphocytes and poorly differentiated lymphoblasts); nonetheless, accurate cytologic diagnosis may be difficult. The features of primary lymphoma in the absence of background Hashimoto's disease are more clear-cut and include marked cellularity with monotonous aggregates of lymphocytes without germinal center cells, and the absence of Hürthle cells, normal follicular cells, or colloid. The nuclei are round and fairly regular and the cytoplasm is cyanophilic (Col. Figs. 8-31 and 8-32).

Metastatic carcinoma to the thyroid gland occasionally may be confirmed by needle aspiration cytology. The tumors which more commonly appear in the thyroid include breast cancer, renal cell carcinoma, malignant melanoma, and bronchogenic carcinoma. There may be a history of one of these lesions or the diagnosis may be first considered on the basis of the presentation with a metastasis to the thyroid (Col. Figs. 8-33 and 8-34).

DISCUSSION

Analysis of Results

Analysis of the results of fine needle aspiration cytology largely rests on the frequency of false negatives or false positives for thyroid malignancy. As mentioned early in this chapter, successful results depend upon the technical skill of the operator and the interpretative expertise of the cytopathologist. At most centers, rates of both false positives and false negatives are highest after initial utilization of the technique, and then these fall with progressive experience. In competent hands, there should be virtually a 0% incidence of false positives, and approximately 2% to 8% of false negatives. It is clear that the best results are obtained when a trained pathologist does the entire study, that is, performs the aspiration, makes the smears, and reads the slides. This significantly reduces the frequency of inadequate samplings and poorly smeared or fixed slides.

The overall outcome of fine needle aspiration will vary greatly depending upon the method of reporting, for example, whether or not indeterminate results are included and whether surgical confirmation of biopsies was obtained. Given this potentially wide variation, a very approximate breakdown of average yields after fine needle aspiration can be given: 5% to 7% inadequate, 50% benign (including false negatives), 30% suspicious or indeterminate, and 15% definitely malignant. Published results will vary depending upon the disposition of the indeterminate or suspicious lesions (*i.e.,* whether or not they are operated upon).

Outcome and Consequences of Aspiration Cytology

Early operation and appropriate follow-up therapy are recommended for all patients with malignant aspirates. While operation for all patients with indeterminate readings is probably not justifiable on the basis of frequency of cancer, most should be advised to have surgery unless there are compelling clinical data or contraindications to surgery that would warrant a conservative approach. Those aspirates which are inadequate must be repeated at a later date, or else the patients can go directly to operation. Nonfunctioning nodules which are clearly benign on aspiration are generally treated with thyroxine in replacement to suppressive dosage and observed. Those nodules which fail to show any significant shrinkage are candidates for repeat aspiration at a future date to confirm the initial impression.

Even when the clinical circumstances weigh heavily toward the likelihood of malignancy, we and our surgeons prefer to obtain a preoperative needle aspiration. A more definitive preoperative diagnosis permits specific discussions with the patient of the potential extent of the surgery and of the postoperative management. We have had cases in which a definite diagnosis of thyroid cancer was obtained by needle aspiration while the intraoperative frozen section was equivocal; foreknowledge of the tissue diagnosis prepared the surgeon to pursue lymph node exploration, biopsy, and appropriate dissection, rather than terminating the surgical procedure prematurely.

There has been some debate in the literature regarding the role of fine needle aspiration in the patient with a history of radiation to the head or neck and a thyroid abnormality by either palpation or scintiscan. When such patients are operated on, malignant lesions are found in 30% to 40%, but often in a location remote from the cold nodule of original concern. Hence, scan or palpation abnormalities in an irradiated gland should be viewed as a marker for pathology and a negative needle aspiration of one area does not preclude cancer in another. As long as this is borne in mind, needle aspiration may be performed with positive results, serving to prepare the patient and surgeon, whereas negative results will be of little value.

Fine needle aspiration brings some definite advantages to the hitherto prolonged and costly evaluation of the patient with a thyroid nodule. The method is simple, relatively inexpensive, and will significantly accelerate the work-up. Its use has resulted in a doubling of the yield of malignancies found at surgery, while at the same time reducing the number of (unnecessary) thyroidectomies for benign disease. That taken together with the more direct and rapid evaluation that aspiration permits equals a saving in health care dollars of between $500 to $1000 per patient.

Thus, results of numerous recent reports are indicating that fine needle aspiration cytology is assuming an increasingly dominant (and early) role in the diagnosis and management of thyroid nodules. While the technique appears to be affording our patients earlier specific selection for either medical or surgical therapy, time will tell whether such management has had a significant impact on thyroid cancer morbidity and mortality statistics. It is hard to believe that the effect will not be a salutary one; hence, fine needle aspiration with cytologic examination should maintain an important role in the evaluation of thyroid masses for a long time to come.

BIBLIOGRAPHY

Block MA, Dailey GE, Robb JA: Thyroid nodules indeterminate by needle biopsy. Am J Surg 146:72–78, 1983

Compagno J, Oertel J: Malignant lymphoma and other lympho-proliferative disorders of the thyroid gland. Amer J Clin Patho 64:1–11, 1980

Frable MA, Frable WJ: Fine-needle aspiration biopsy revisited. Laryngoscope 92:1414–1418, 1982

Galvan G: Thin needle aspiration biopsy and cytological examination of hypofunctional, "cold" thyroid nodules in routine clinical work. Clin Nucl Med 2:413–421, 1977

Gershengorn MC, McClung MR, Chu EW et al: Fine needle aspiration cytology in the preoperative diagnosis of thyroid nodules. Ann Intern Med 87:265–269, 1977

Gharib H, Goellner JR, Zinsmeister AR et al: Fine-needle aspiration biopsy of the thyroid: The problem of suspicious cytologic findings. Ann Intern Med 101:25–28, 1984

Hamberger B, Gharib H, Melton LH III et al: Fine-needle aspiration biopsy of thyroid nodules: Impact on thyroid practice and cost of care. Am J Med 73:381–384, 1982

Kini SR, Miller JM, Hamburger JI: Cytopathology of Hürthle cell lesions of the thyroid gland by fine needle aspiration. Acta Cytol 25:647–652, 1981

Kini SR, Miller JM, Hamburger JI: Problems in the cytologic diagnosis of the "cold" thyroid nodule in patients with lymphocytic thyroiditis. Acta Cytol 25:506–512, 1981

Lowhagen T, Guanberg PD, Lundell G et al: Aspiration biopsy cytology (ABC) in nodules of the thyroid gland suspected to be malignant. Surg Clin North Am 59:3–18, 1979

Miller JM, Hamburger JI, Kini S: Diagnosis of thyroid nodules: Use of fine-needle aspiration and needle biopsy. JAMA 241:481–484, 1979

Rosen IB, Wallace C, Strawbridge HG, Walfish PG: Reevaluation of needle aspiration cytology in detection of thyroid cancer. Surgery 90:747–756, 1981

Soderstrom N, Telenius-Berg M, Akerman M: Diagnosis of medullary carcinoma of the thyroid by fine needle aspiration biopsy. Acta Med Scand 197:71–76, 1975

Sprenger E, Lowhagen T, Vogt-Schaden M: Differential diagnosis between follicular adenoma and follicular carcinoma of the thyroid by nuclear DNA determination. Acta Cytol 21:528–530, 1977

Stavric GD, Karanfilski BT, Kalamaras AK et al: Early diagnosis and detection of clinically non-suspected thyroid neoplasia by the cytologic method. A critical review of 1536 aspiration biopsies. Cancer 45:340–344, 1980

Walfish PG, Hazani E, Strawbridge TG et al: Combined ultrasound and needle aspiration cytology in the assessment and management of hypofunctioning thyroid nodule. Ann Int Med 87:270–274, 1977

Willems JS, Lowhagen T: Fine-needle aspiration cytology in thyroid disease. Clin Endocrinol Metab 10:247–266, 1981

Stephen M. Manier
Douglas Van Nostrand
Ralph W. Kyle
Sue H. Abreu

Chapter 9 RENAL TRANSPLANT SCINTIGRAPHY

This chapter is designed to aid the physician who is learning to interpret renal transplant scintigraphy. The text consists of

- A basic discussion of camera and computer techniques for the acquisition of Iodine-131 (I-131) Hippuran (OIH) sequential images, flow images, and technetium-99m diethylenetriaminepentaacetic acid (Tc-99m DTPA) sequential images
- A brief discussion of quantitative methods for evaluating renal flow and function (kidney to aortic ratio, effective renal plasma flow, and excretion index)
- A table for the visual description of the flow images, Tc-99m DTPA images, and I-131 Hippuran images
- A table relating differential diagnosis to the visual description of the first scan performed within 48 hours of transplantation
- A table relating differential diagnosis to visual description of scans performed serially.

The atlas section presents figures that are representative of the more common disease entities presented in the tables.

The authors wish to thank Ms. Mary Sue Mood for her assistance in the preparation of this manuscript.

This chapter *does not* replace a nuclear medicine textbook; discuss all reported computer quantitative methods for transplant scintigraphy; discuss renal transplant scintigraphy with iodine-123 fibrinogen, indium-111–labeled white cells, Tc-99m sulfur colloid, indium-111–labeled platelets, or gallium-67; or discuss quality control considerations such as the radiochemical impurities of I-131 Hippuran.

We recognize that inevitable and serious limitations such as over-simplifications and omissions, occur in any such chapter and that an understanding of renal transplant scintigraphy must be based on a comprehensive understanding of physiology and time–activity curves. Nevertheless, we believe this chapter will be a valuable aid to the student and teacher of renal transplant scintigraphy and may stimulate a more in-depth understanding of renal transplant physiology and scintigraphy, as well as further independent study.[1-5]

TECHNIQUE

Patient Preparation

Patients are instructed to drink three 8-ounce glasses of water 30 minutes prior to the study unless they are on fluid restriction. After emptying the urinary bladder, the patient is placed in the supine position under the scintillation camera so that the renal transplant, ureter, and urinary bladder are in the field of view.

Camera Image Acquisition

For I-131 Hippuran images, 150 μCi (5.5 MBq) of I-131 orthoiodohippurate (I-131 OIH) is given intravenously. Sequential 2-minute analog film images are acquired for 24 minutes followed by prevoid, postvoid, and injection-site images.

For a flow study, 15 mCi (555 MBq) of Tc-99m DTPA is given intravenously using a bolus injector. Analog film images are acquired at 3 seconds per frame for 48 seconds.

Two-minute sequential Tc-99m DTPA images are acquired for 24 minutes followed by pre- and postvoid images of the bladder. Additional images may be requested by the nuclear medicine physician.

Computer Image Acquisition

The I-131 OIH study is acquired at 15-second frames for 24 minues.

The Tc-99m DTPA flow study is acquired at 1-second frames for 60 seconds.

Sequential Tc-99m DTPA images are acquired at 15-second frames for 24 minutes.

QUANTITATIVE ANALYSIS

Many computer applications have been used to quantitate renal transplant flow and function. The three that will be discussed here are effective renal plasma flow, excretory index, and kidney-aortic ratio.

Hippuran

Selected frames are used to choose regions of interest over the renal cortex (*straight arrow* in Figure 9-1) and adjacent background (*curved arrow* in Fig. 9-1). A background-subtracted time–activity curve of the renal cortex is then made (Fig. 9-4E).

ERPF Study

Effective renal plasma flow (ERPF) can be calculated from the value of a single plasma sample of I-131 OIH drawn from peripheral venous blood 44 minutes after injection and the use of a regression equation as proposed by Tauxe.[6]

Excretory Index

The excretion index (EI) measures the efficiency of tubular cells in "moving" extracted Hippuran from the cells into the collecting system. The index is a ratio of I-131 OIH excreted in the urine during the first 35 minutes, divided by the expected amount of I-131 OIH excreted in the urine for a specific ERPF. The expected amount of I-131 OIH excreted in the urine for a specific value of ERPF is determined empirically by Kontzen.[7]

$$\text{EI} = \frac{\text{Actual \% dose of I-131 OIH in urine}}{\text{Expected \% dose of I-131 OIH in urine for ERPF determined by patient blood sample}}$$

The numerator of the above equation is equal to the sum of the percent dose in voided urine and percent dose in bladder after voiding, as the following equation demonstrates:

$$\text{EI} = \frac{\text{\% Dose in voided urine} + \text{\% Dose in bladder after voiding}}{\text{Expected \% dose of I-131 OIH in urine for ERPF determined by patient blood sample.}}$$

The percent dose in voided urine is calculated as follows:

$$\text{\% dose in voided urine} = \frac{\text{Urine counts/ml} \times \text{Urine volume}}{\text{Standard counts for dose administered}}$$

Percent dose of I-131 OIH in the bladder after voiding is calculated by using these equations:

$$\text{Volume of residual urine (ml)} = \frac{\text{Voided urine volume} \times \text{Postvoid bladder counts}}{\text{Prevoid bladder counts} - \text{Postvoid bladder counts}}$$

$$\text{\% Dose left in bladder after voiding} = \frac{\text{\%Dose in voided urine} \times \text{Volume of residual urine}}{\text{Voided urine volume}}$$

The proposed relationship of ERPF and EI in normals and in various disease states of renal transplants is shown in Figure 9-2.[8]

Kidney–Aortic Ratio

Selected frames showing the aorta and kidney are used to choose regions of interest. A constant-sized rectangular region of interest is placed over the distal abdominal aorta just above the iliac artery bifurcation (Fig. 9-3, *straight arrow*). An irregular region of interest is manually drawn to include the entire kidney (Fig. 9-3, *curved arrow*). Time–activity curves are made of both regions of interest. The slope of the initial rise of each curve is calculated by means of a least-squares-fit program.[9] The ratio of these slopes gives the kidney–aortic (K/A) ratio (Fig. 9-4*B*), which is one of many estimates of renal blood flow. The K/A ratios in nor-

Table 9-1
Kidney–Aortic Ratio

	Average	*Range*	*Number of Studies*
Normal transplants	0.87	0.64–1.16	23
Acute tubular necrosis	0.27	0.11–0.40	10
Stable but abnormal function	0.54	0.27–0.92	14
Terminal failure	0.16	≤ 0.23	9

(Kirchner P et al: Clinical application of the kidney to aortic blood flow index (K/A ratio). Contrib Nephrol 11:120–126, 1978)

mals and in various disease states are shown in Table 9-1 as reported by Kirchner.[9]

VISUAL DESCRIPTION AND INTERPRETATION

The visual description of renal transplant scans is shown in Table 9-2. The visual interpretation of the first renal transplant scan performed within 48 hours of transplantation is shown in Table 9-3. The visual interpretation of serial scans is shown in Table 9-4.

(Text continues on page 207.)

Table 9-2
*Primer for Visual Description of Renal Scans in Renal Transplants**

First Pass: Blood Flow (ml/min)	*DTPA: Glomerular Filtration Function*	*Hippuran: Tubular Function*	*Collecting System*	*Other*
How Good is Bolus? 1. Time from initial appearance of aortic activity to peak aortic activity should be < 3 sec. 2. Aortic activity should disappear and reappear during recirculation/blood pool images. *How Good is Flow?* 1. Peak kidney activity occurs 3–6 sec after peak aortic activity. 2. Peak kidney activity should be ≥ peak aortic activity.	*How Good is Filtration?* 1. Maximum ratio of parenchymal to peak background activity should be > 3. 2. Time to peak activity should be 3–5 min. 3. Time of first activity in collecting system is < 5 min. 4. Rate of washout of activity from parenchyma over 24 min is very subjective and partly dependent on state of hydration. 5. Background washes out with time.	*How Good is Tubular Function?* 1. Maximum ratio of parenchymal to peak background activity should be > 4. 2. Time of peak activity should be 1.5–4.5 min. 3. Time of first activity in collecting system should be 3–6 min. 4. Washout† from parenchyma should be almost complete by 24 min. However, rate is very subjective and partly dependent on state of hydration. Hippuran washes out faster than DTPA. 5. Total activity in bladder and collecting system relative to parenchyma at 24 min suggests excretion‡ efficiency; very subjective. 6. Total activity in bladder, collecting system, and parenchyma suggests extraction§ efficiency and/or ERPF; very subjective.	*Is Anatomy Normal?* 1. Evaluate renal pelvis, ureter, and bladder for size, configuration, displacement, course, and so on. *Is Urinary Obstruction Present?* 1. Measure rate of washout of activity from renal pelvis or ureters. Very variable and dependent on state of hydration. 2. Check dilatation of renal pelvis or ureter.	1. Evaluate kidney for size, configuration, defects, etc. 2. Evaluate flow in region of parenchymal defect. 3. Evaluate for any urine activity in region of renal defect (e.g., collecting system). 4. Evaluate extra-renal cold or hot areas.

* *Each portion of study to be evaluated separately.*

† *Washout is a function of input (ERPF and extraction efficiency) and output (excretion efficiency).*

‡ *Excretion efficiency: See text under "excretion index."*

§ *Extraction efficiency: Ability of tubular cells to remove I-131 Hippuran from blood.*

(Dubovsky EV: Primer and atlas for renal transplant scintigraphy. Clin Nucl Med:10:Part I [52–62], Part II [118–133], 1985)

Table 9-3
Primer for Visual Interpretation of First Renal Transplant Scan*

Description					*Implications*
Flow	*DTPA*	*Hippuran*	*Collecting System*	*Other*	
Normal	Normal	Normal	Normal		Normal (Fig. 9-4)
Absent (or trace)	No activity (or trace)	No activity (or trace)	Not visualized	Photopenic defect in area of entire kidney	Renal artery occlusion/ thrombosis (Fig. 9-5) Renal vein occlusion/ thrombosis Hyperacute rejection (rare) (Fig. 9-5)
Variable. (Immediately after surgery, it may be supernormal returning to normal or mildly reduced by 48 hrs.)	Markedly reduced peak, although time to peak may be normal. Parenchyma may have little to no washout while background increases with time.	Good but markedly delayed peak. Typically no collecting system activity. Increasing parenchymal activity with time. ERPF and extraction good, but excretion very poor.	Little or no activity		Acute tubular necrosis (Fig. 9-7)
Minimally to moderately reduced	All parameters minimally to moderately reduced or delayed	All parameters minimally to moderately reduced or delayed			Acute rejection (cellular); very rare in first 2 days (Fig. 9-8)
Early visualization of inferior vena cava					Arteriovenous fistula (Fig. 9-6)
				Abnormal tracer accumulation outside transplant collecting system	Urine extravasation/ urinoma (Figs. 9-12, 9-13, 9-14) Wound dehiscence (Fig. 9-15) Urine reflux in native ureter (Fig. 9-17) Uterus: menses (Fig. 9-6), pregnancy, fibroid (Fig. 9-16)
				Intrarenal photopenic defect	Infarct (Fig. 9-22) or emboli Hematoma Segmental ischemia or acute tubular necrosis (see acute tubular necrosis)

(continued)

Table 9-3
Primer for Visual Interpretation of First Renal Transplant Scan* (continued)

Description					*Implications*
Flow	*DTPA*	*Hippuran*	*Collecting System*	*Other*	
				Extrarenal photopenic defect	Hematoma (Fig. 9-21) Urinoma; may fill in with radioactivity over time (Figs. 9-12, 9-13, 9-14) Edema Bowel Bladder
Normal	Normal	Normal	Prolonged renal pelvic transit time and/or "columning" without dilatation		State of hydration; dehydration may cause delayed clearance Gravity: ureter may have gravity-dependent portion. Upright postvoid image may resolve this. Obstruction (Fig. 9-10): Postoperative edema (resolution within 1-2 weeks)
Variable	Variable	Variable	Prolonged renal pelvic transit time with or without dilatation of ureter		Obstruction: Ureteral blood clot (Fig. 9-11) Extrinsic mass (*i.e.*, hematoma) Inadequate ureteral implantation Acute rejection of ureter

* *Performed within 48 hours after transplantation*
(Dubovsky EV: Primer and atlas for renal transplant scintigraphy. Clin Nucl Med:10:Part I [52-62], Part II [118-133], 1985)

Table 9-4
Primer for Visual Interpretation of Serial Renal Transplant Scans

Change in Flow, DTPA, Hippuran	*Collecting System*	*Other*	*Implication*
No change (normal function)			Normal
No change (trace or absent function)			Renal artery occlusion/thrombosis (Fig. 9-5) Renal vein occlusion/thrombosis Hyperacute rejection (Fig. 9-5)
No change (present but abnormal function)			Acute tubular necrosis (Fig. 9-7) Acute rejection that is stable on treatment
Improvement			Acute tubular necrosis (Fig. 9-7) Acute rejection with good response to treatment
Deterioration			Acute rejection (Fig. 9-8) Pyelonephritis (*e.g.,* bacterial, cytomegalic virus) Chronic rejection (humoral) (Fig. 9-9); not typical in first 10 days Glomerulonephritis; may be membranous or proliferative and does not typically occur in first 10 days Drug toxicity
		Abnormal tracer accumulation outside collecting system	Urine extravasation/urinoma (Figs. 9-12, 9-13, 9-14) Wound dehiscence (Fig. 9-15) Urine reflux in native ureter (Fig. 9-17) Uterus: menses (Fig. 9-6), pregnancy, fibroid (Fig. 9-16)
		New or progressive intrarenal photopenic defect	Infarct (Fig. 9-22)/emboli Abscess Hematoma Segmental ischemia or acute tubular necrosis (see acute tubular necrosis)
		Extrarenal photopenic defect	Urinoma (Figs. 9-12, 9-13, 9-14) Lymphocele (Figs. 9-18, 9-19); may appear photopenic on initial images after injection, but may become "warm" and "disappear" on delayed imaging. They typically appear at 1 - 2 weeks and may be confused with urinoma (Figs. 9-18, 9-19). A rim of increased activity may also be seen (Fig. 9-20). Abscess
		Resolving extrarenal photopenic defect	Hematoma (Fig. 9-21) Edema Bowel Bladder
	Prolonged renal-pelvic transit time (see Table 9-3)		(See Table 9-3.)

(Dubovsky EV: Primer and atlas for renal transplant scintigraphy. Clin Nucl Med:10:Part I [52 - 62], Part II [118 - 133], 1985)

Table 9-5
Radiation Absorbed Dose Estimates

Organ	*TC-99m DTPA* (Rad/mCi)*	*I-131 Hipporan† (Rad/mCi)*
Bladder (critical organ)	0.55	7.8
Kidney	0.04	0.07
Ovaries	0.019	0.16
Testes	0.01	0.16
Whole body	0.016	

* *(Modified from Hauser W et al; Technetium-99m-DTPA: A new radiopharmaceutical for brain and kidney scanning. Radiology 94:679, 1970; and Harbert JC et al: Absorbed dose estimates from radionuclides. Clin Nucl Med 9:210, 1984)*

† *(Modified from Harbert JC et al: Absorbed dose estimates from radionuclides. Clin Nucl Med 9:210, 1984; and Henk JM, Cottrall MF, Taylor DM: Radiation dosimetry of the ^{131}I-Hippuran renogram. Br J Radiol 40:327, 1967)*

RADIATION ABSORBED DOSES

Estimates of radiation absorbed doses for Tc-99m DTPA and I-131 Hippuran are given in Table 9-5. The critical organ is the one that typically receives the largest radiation absorbed dose.

REFERENCES

1. Kirchner PT, Eckelman WC et al: Renal transplant evaluation. In Kirchner PT, Eckelman WC et al (eds): Nuclear Medicine Review Syllabus, pp 359–366. New York, Society of Nuclear Medicine, 1980
2. Britton KE, Maisey MN: Renal radionuclide studies. In Maisey MN, Britton KE, Gilday DL (eds): Clinical Nuclear Medicine pp 93–133. Philadelphia, WB Saunders, 1983
3. Kirchner PT, Rosenthall L: Renal transplantation evaluation. Sem Nucl Med 12:370–378, 1982
4. Tauxe WN, Duborsky EV: The kidney. In Rocha AFG, Harbert JC (eds): Text of Nuclear Medicine Clinical Applications, pp 344–375. Philadelphia, Lea and Febiger, 1979
5. Taylor AT: Quantitative renal function scanning: A historical and current status report on renal radiopharmaceuticals. In Freeman LM, Weissmann HS (eds): Nuclear Medicine Annual 1980, pp 303–340. New York, Raven Press, 1980
6. Tauxe WN, Maher FT, Taylor WF: Effective renal plasma flow: Estimation from theoretical volumes of distribution of intravenously injected 131-I-orthoiodohippurate. Mayo Clin Proc 46:524, 1971
7. Kontzen FN, Tobin M, Dubovsky EF, Tauxe WN: Comprehensive renal function studies: Technical aspects. J Nucl Med Tech 5:81, 1977
8. Diethelm AG et al: Dubovsky EV, Whelchel JD et al: Diagnosis of impaired renal function after kidney transplantation using renal scintigraphy, renal plasma flow, and urinary excretion of Hippurate. Ann Surg 191:604–615, 1980
9. Kirchner P et al: Clinical application of the kidney to aortic blood flow index (K/A ratio). Contrib Nephrol 11:120–126, 1978
10. Hauser W, Atkins HL, Nelson KG et al: Technetium-99m-DTPA: A new radiopharmaceutical for brain and kidney scanning. Radiology 94:679, 1970
11. Harbert JC, Pollina R et al: Absorbed dose estimates from radionuclides. Clin Nucl Med 9:210, 1984
12. Henk JM, Cottrall MF, Taylor DM: Radiation dosimetry of the ^{131}I-Hippuran renogram. Br J Radiol 40:327, 1967

Atlas Section

Figure 9-1

Regions of interest of renal parenchyma *(straight arrow)* and background *(curved arrow)* on Hippuran images. (Dubovsky EV: Primer and atlas for renal transplant scintigraphy. Clin Nucl Med:10:Part I [52-62], Part II [118-133], 1985)

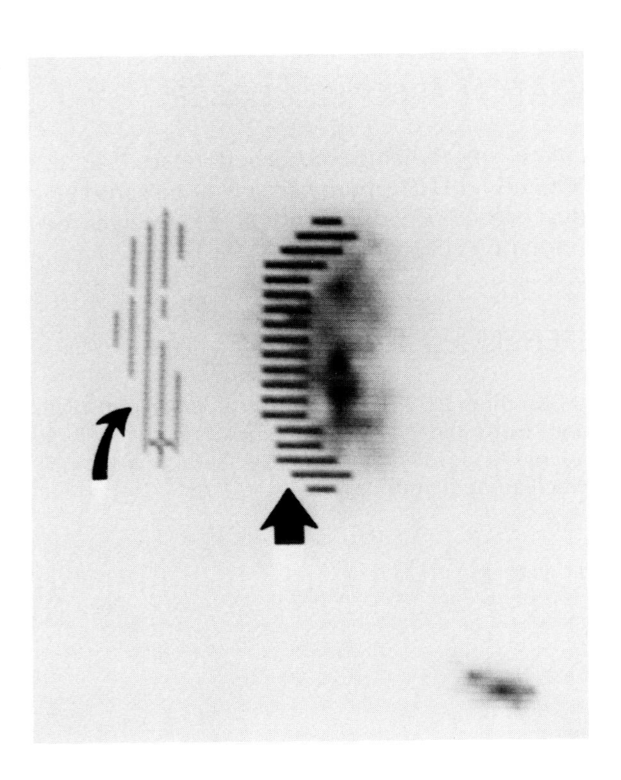

(Atlas continues on page 209.)

THE COLOR PLATES HAVE BEEN MADE POSSIBLE IN PART BY A GENEROUS GRANT FROM THE FLINT LABORATORIES, INC., DEERFIELD, ILLINOIS.

Color Figure 8-1

Adenomatoid Nodule with Cystic Degeneration

Sheets of follicular epithelial cells mixed with numerous hemosiderin-laden macrophages. The hemosiderin pigment stains blue to black. Many red blood cells are seen in the background. (Diff-Quik ×80)

Color Figure 8-2

Granulomatous Thyroiditis

There is a striking difference between the follicular epithelial cells, which show round nuclei and very delicate and scant cytoplasm, and the epithelioid cells, forming a granuloma, which show elongated ovoid nuclei in a palisade arrangement. Numerous red blood cells lie in the background. (Diff-Quik ×80)

Color Figure 8-3

Hashimoto's Thyroiditis

"Lymphoid tangle" with crushed and stretched-out nuclear chromatin is seen in the center of this photomicrograph. Adjacent small groups of follicular cells show a slight variation in nuclear size and in the amount of cytoplasm. (Diff-Quik ×80)

Color Figure 8-4

Hashimoto's Thyroiditis

Photomicrograph is of a group of oncocytic, mitochondrion-rich (Hürthle) cells surrounded by lymphoid cells. (Diff-Quik × 160)

Color Figure 8-5

Colloid and Benign Follicular Epithelial Cells

The thicker clumps of colloid are stained deep blue, thinner colloid appears purplish pink. The follicular cells are arranged in groups of variable sizes. (Diff-Quik ×40)

Color Figure 8-6

Colloid Nodule, Colloid Goiter

The colloid appears as overlapping irregular ribbons. (Diff-Quik × 16)

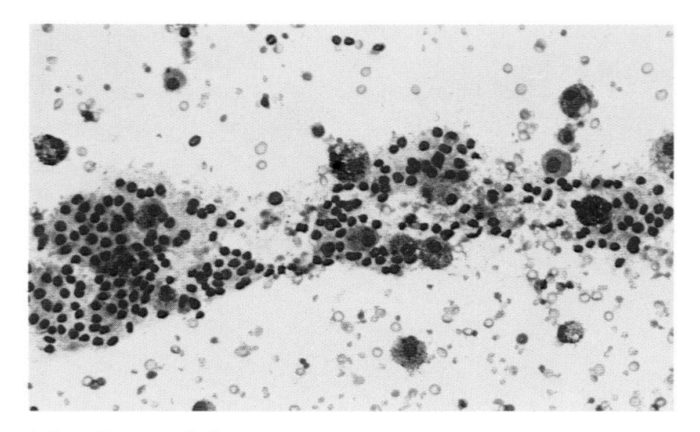

Color Figure 8-1.

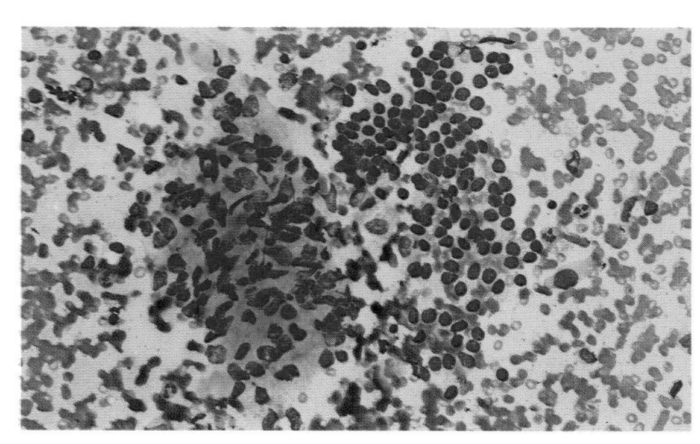

Color Figure 8-2.

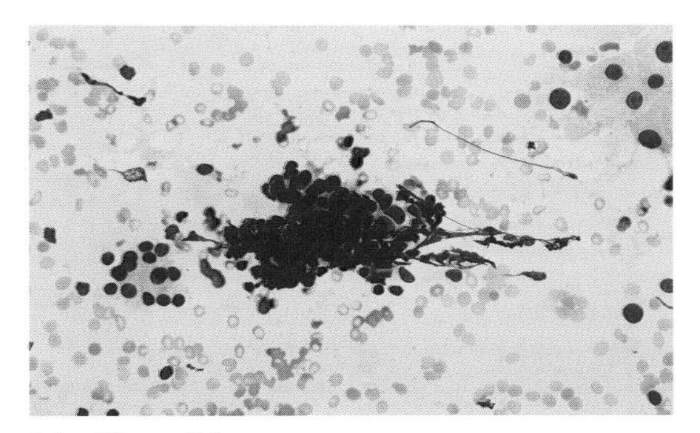

Color Figure 8-3.

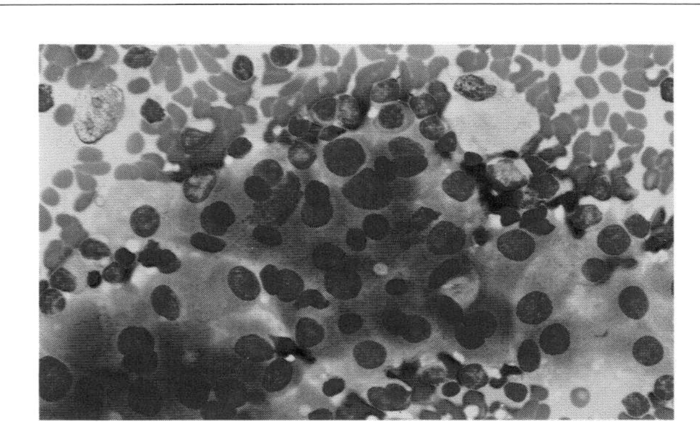

Color Figure 8-4.

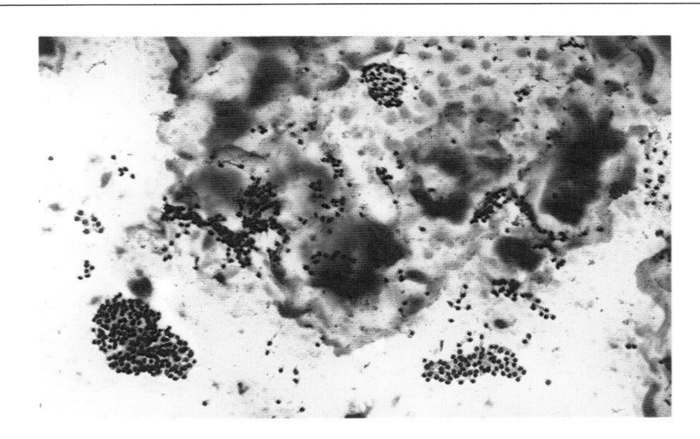

Color Figure 8-5.

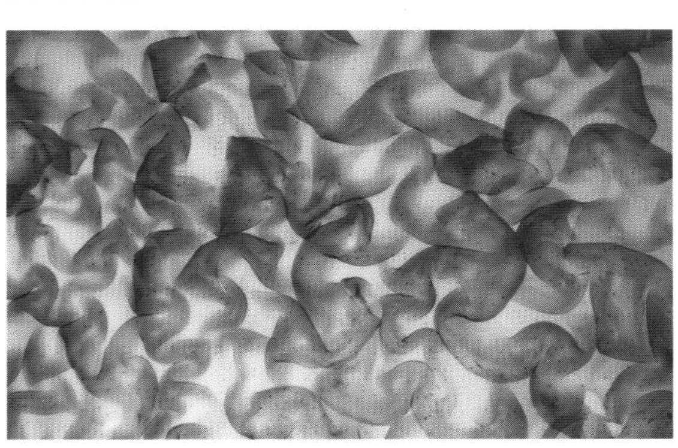

Color Figure 8-6.

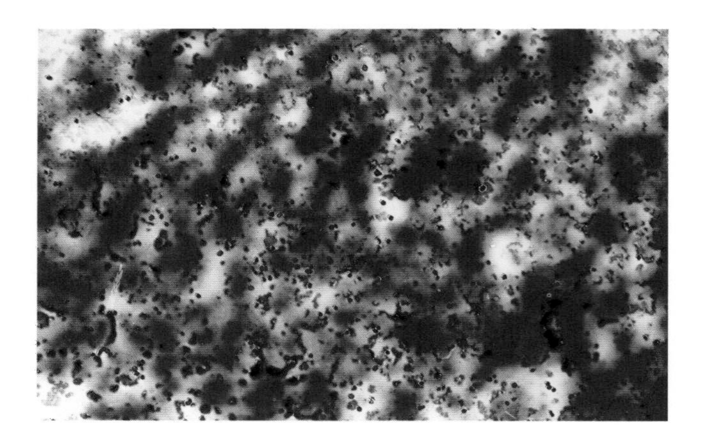

Color Figure 8-7

Colloid Nodule, Colloid Goiter

Colloid resembles an irregular lattice. (Diff-Quik ×40)

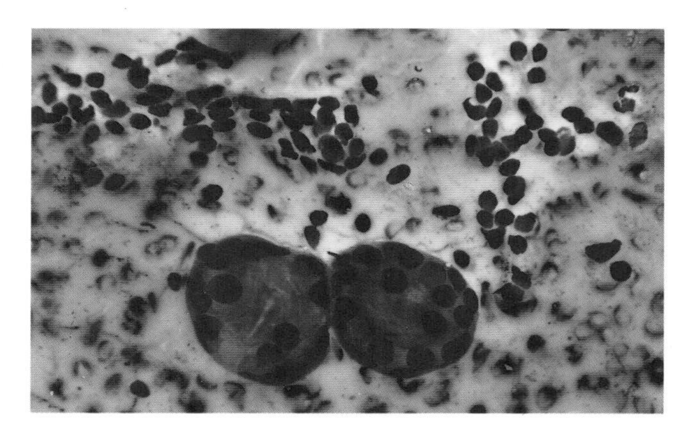

Color Figure 8-8

Follicular Epithelial Cells

Two different types of arrangement: sheets and "spherules" (Non-neoplastic follicles). Note thin, pale pink colloid in the background as well as some red blood cells. (Diff-Quik ×160)

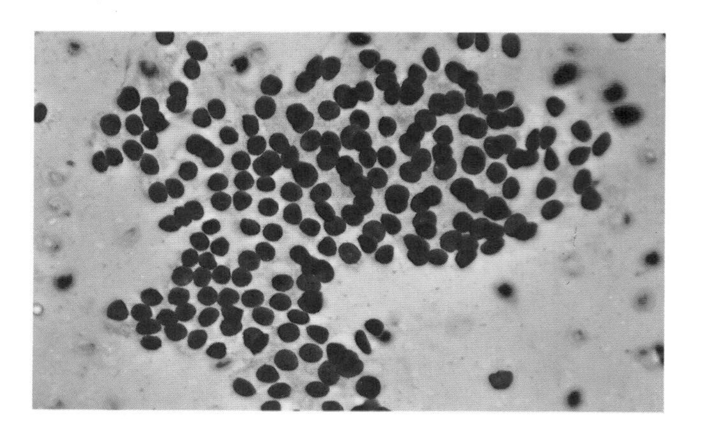

Color Figure 8-9

Follicular Epithelial Cells

Note that the nuclei are fairly regular. No nucleoli are evident. Thin pink colloid is observed in the background. (Diff-Quik ×160)

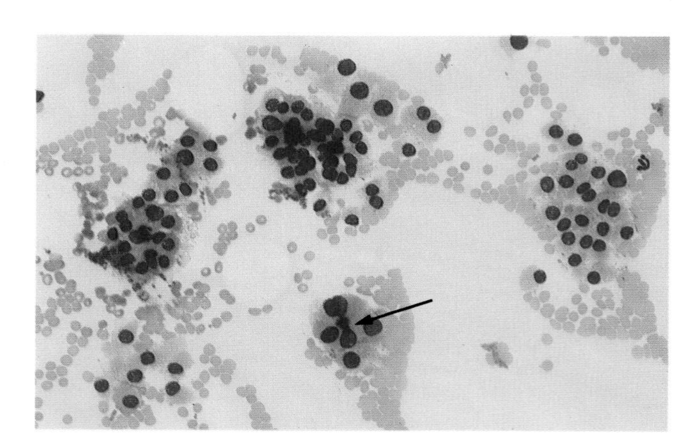

Color Figure 8-10

Follicular Neoplasia

The group of cells in the center of the field are forming "neoplastic follicles." The cells are arranged around dark blue, dense colloid (*arrow*). The cells on the left side of the field are forming "rosettes." (Diff-Quik ×80)

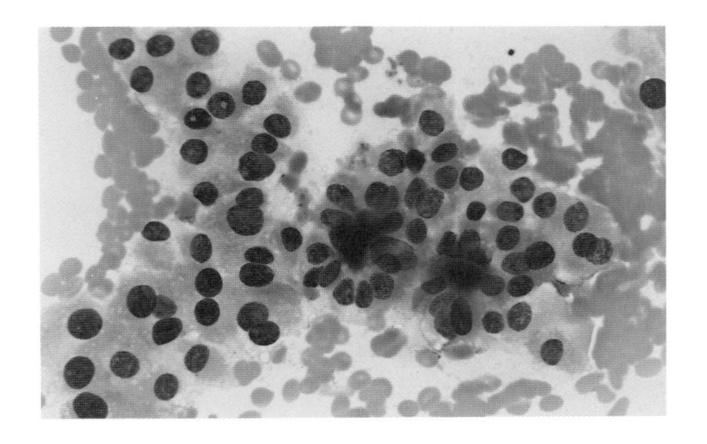

Color Figure 8-11

Follicular Neoplasia

On the left the cells tend to form cords or tubules. Those on the right are forming "neoplastic follicles." (Diff-Quik ×160)

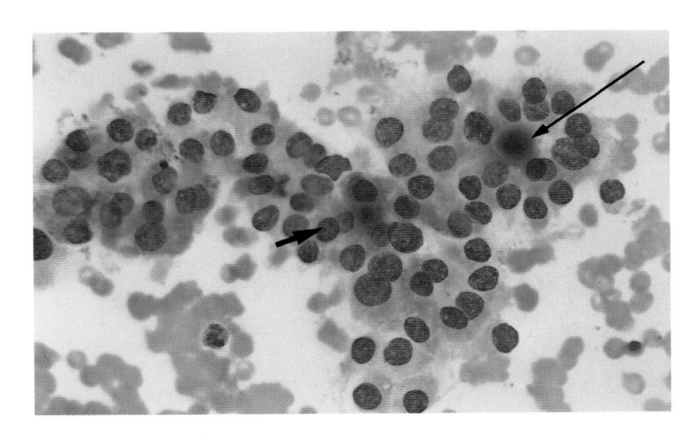

Color Figure 8-12

Follicular Neoplasia

Cells with enlarged nuclei, readily visible nucleoli (*short arrow*), and pale blue cytoplasm are arranged around dense, dark blue colloid (*long arrow*) forming "neoplastic follicles," (Diff-Quik ×160)

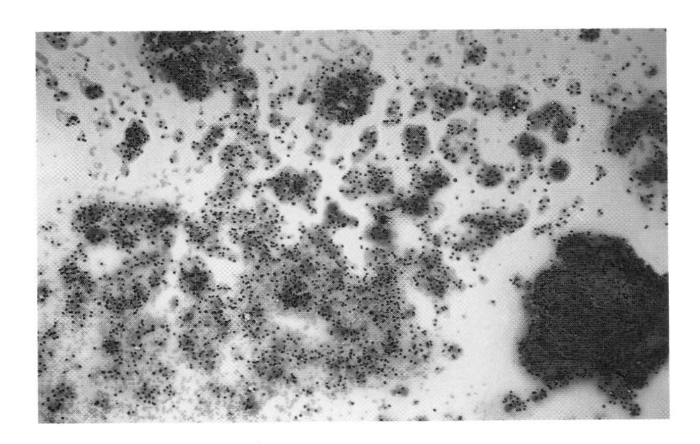

Color Figure 8-13

Hürthle Cell Neoplasia

The smears are extremely cellular ("tumor cellularity"). (Diff-Quik ×16)

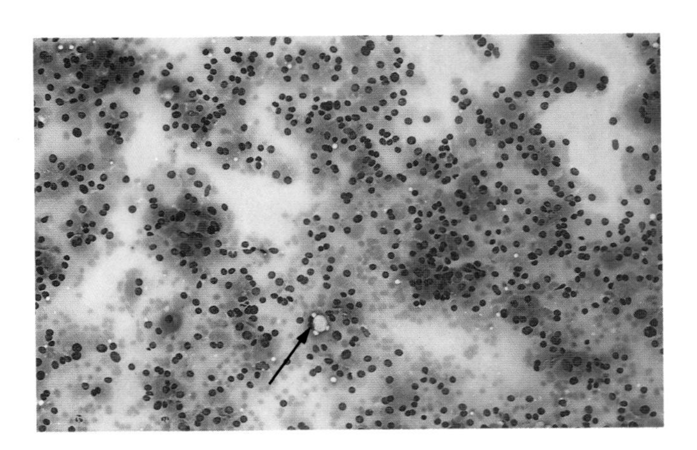

Color Figure 8-14

Hürthle Cell Neoplasia

The neoplastic cells are arranged predominantly in irregular sheets and groups. Some of the cells tend to form follicles with empty lumina (*arrow*). The cytoplasm of the cells shows variable shades of bluish pink. (Diff-Quik ×40)

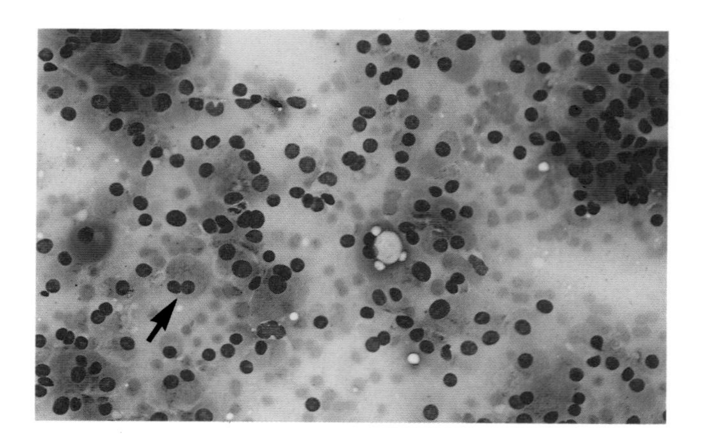

Color Figure 8-15

Hürthle Cell Neoplasia

Notice "empty follicle" in the center of the field. Binucleation (*arrow*) is also evident. (Diff-Quik ×80)

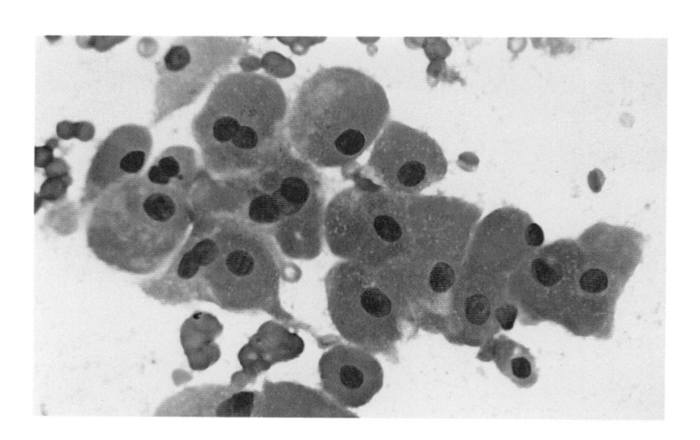

Color Figure 8-16

Hürthle Cell Neoplasia

The cells show very abundant, dense bluish pink granular cytoplasm with well-demarcated cell borders. The nuclei are round and eccentric, and nucleoli are prominent. Binucleation is very common. (Diff-Quik ×160)

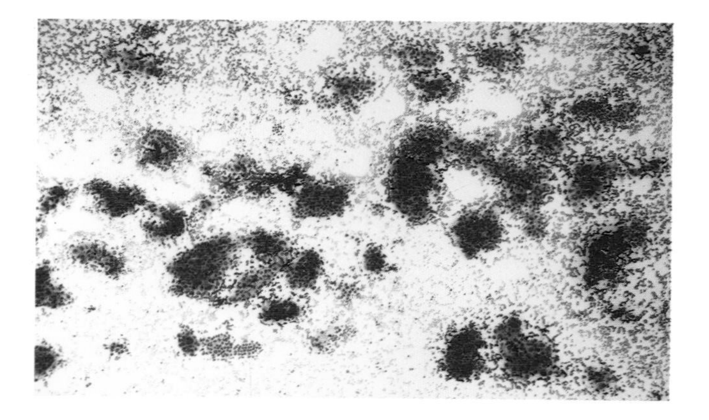

Color Figure 8-17

Papillary Carcinoma

This very low magnification depicts the so-called tumor cellularity. Many irregular tissue fragments of variable sizes are observed. (Diff-Quik ×16)

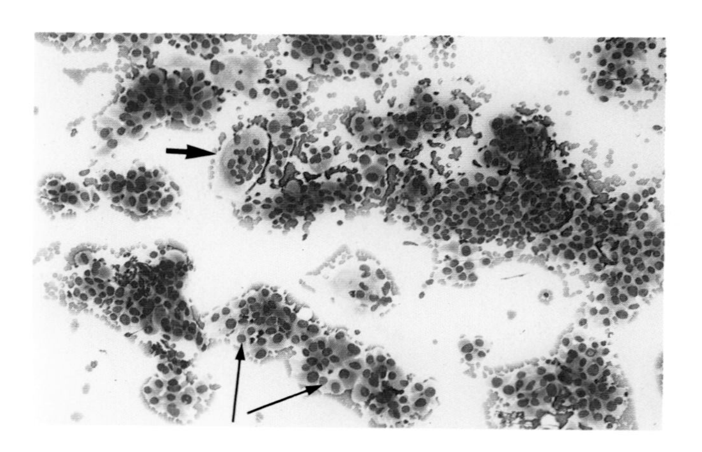

Color Figure 8-18

Papillary Carcinoma

Many enlarged follicular epithelial cells are arranged in irregular tissue fragments (long arrows). Some multinucleated histiocytes are easily detected (short arrow). (Diff-Quik ×40)

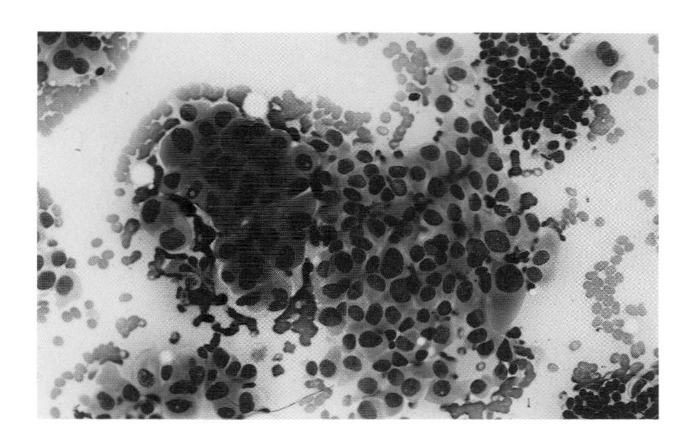

Color Figure 8-19

Papillary Carcinoma

The nuclei of the neoplastic follicular cells are enlarged and vary in size. This is quite evident if one compares their size to that of the red blood cells seen in the background and that of the nuclei of normal follicular cells observed in the upper right corner. The cytoplasm of the neoplastic cells is abundant, dense, and bluish pink. (Diff-Quik ×80)

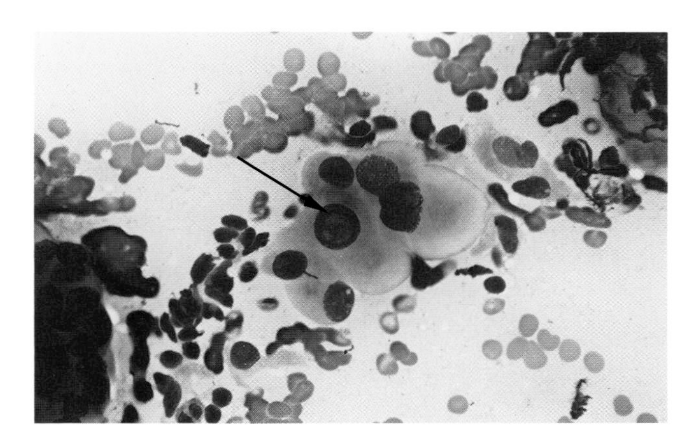

Color Figure 8-20

Papillary Carcinoma

Small group of neoplastic cells in the center shows escalloped borders and an "intranuclear inclusion" (*arrow*) (Diff-Quik ×160)

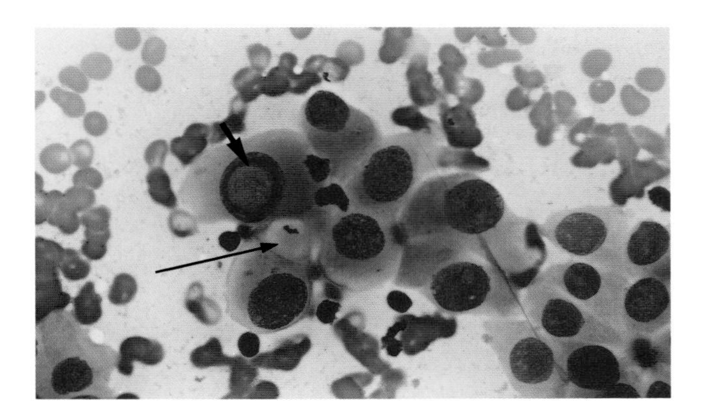

Color Figure 8-21

Papillary Carcinoma

Sheet of neoplastic cells with enlarged nuclei (compare in size to red blood cells in the background) and one intranuclear inclusion (*short arrow*) are observed. The adjacent cell shows a cytoplasmic vacuole (*long arrow*). Notice the dense cytoplasm and the well-demarcated cell borders. (Diff-Quik ×200)

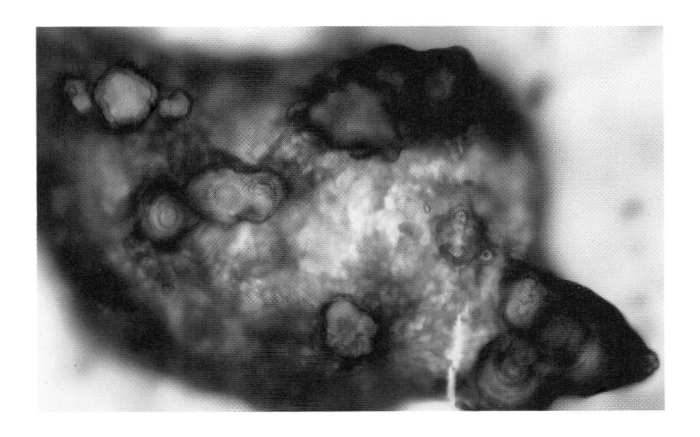

Color Figure 8-22

Papillary Carcinoma

Multiple concentric lamellae (psammoma bodies) are seen, with some colloid in the background. (Diff-Quik ×160)

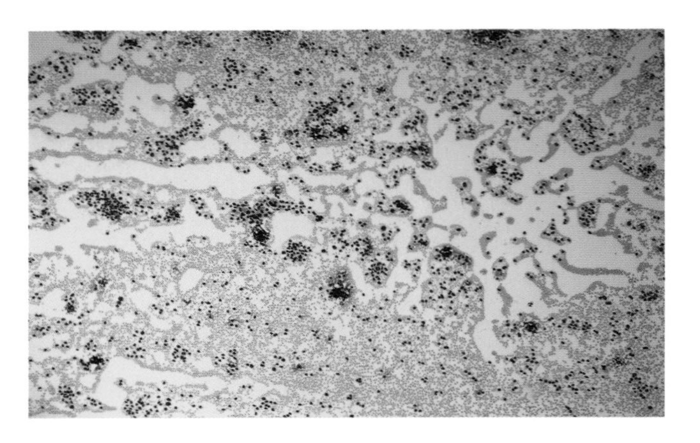

Color Figure 8-23

Medullary Carcinoma

The smears are very cellular (tumor cellularity). However, no large tissue fragments are observed (compare with those of Hürthle cell neoplasia, Fig. 8-13, and papillary carcinoma, Fig. 8-17, at the same magnification). (Diff-Quik ×16)

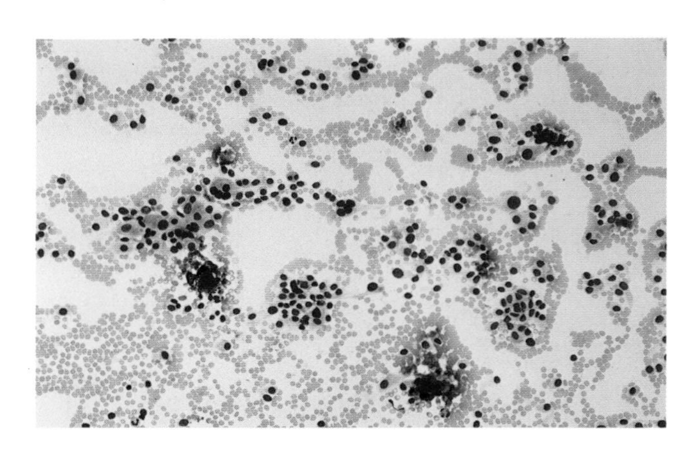

Color Figure 8-24

Medullary Carcinoma

Many neoplastic cells lying singly or in loosely cohesive groups. Some variation in nuclear size is apparent even at this low magnification. (Diff-Quik ×40)

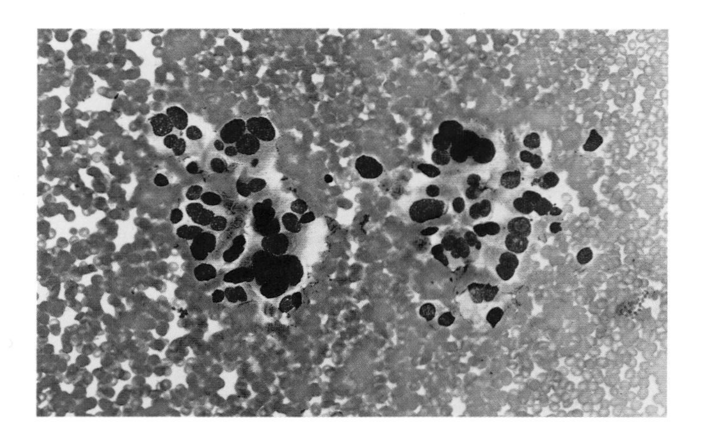

Color Figure 8-25

Medullary Carcinoma

Loosely cohesive cell clusters show marked nuclear atypia. (Diff-Quik ×80)

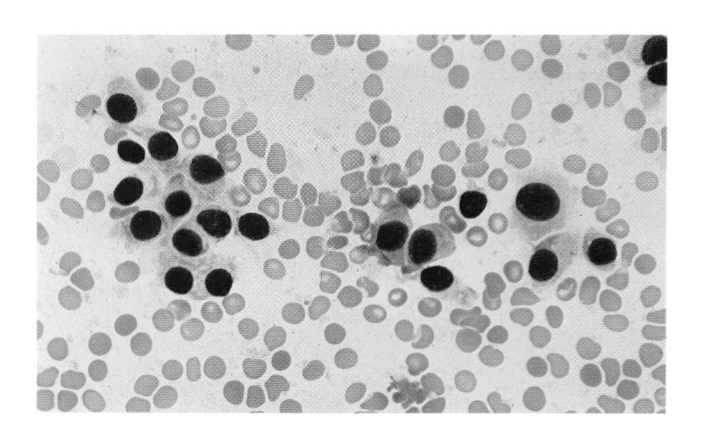

Color Figure 8-26

Medullary Carcinoma

The cells are loosely cohesive and have a plasmacytoid appearance. The two cells in the center show some pink cytoplasmic granules. (Diff-Quik ×160)

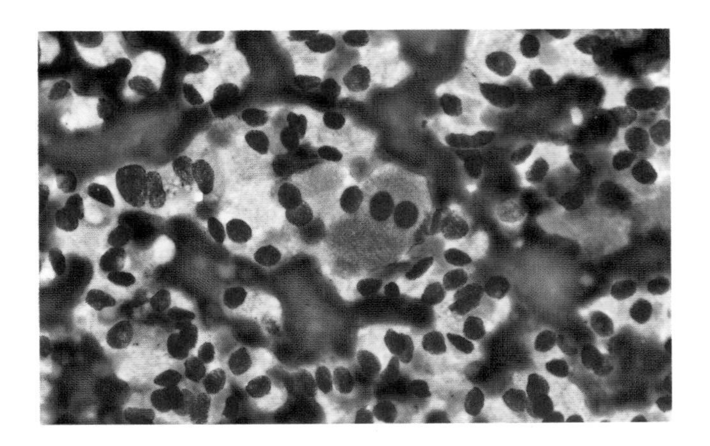

Color Figure 8-27

Medullary Carcinoma

The binucleated cell in the center shows the pink calcitonin granules. (Diff-Quik ×160)

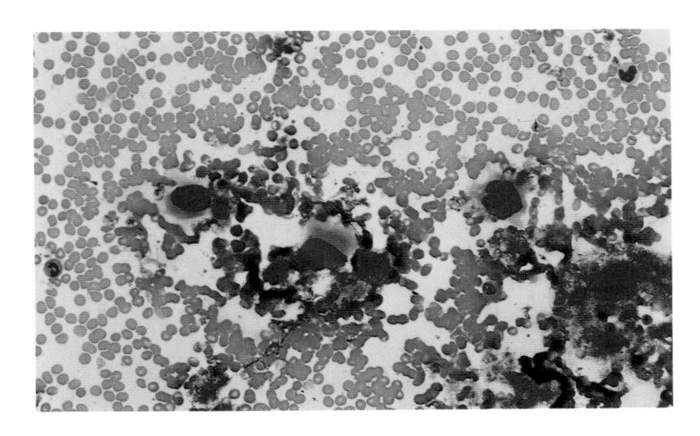

Color Figure 8-28

Anaplastic Carcinoma

A few bizarre cells, amorphous debris, and red blood cells are seen. (Diff-Quik ×80)

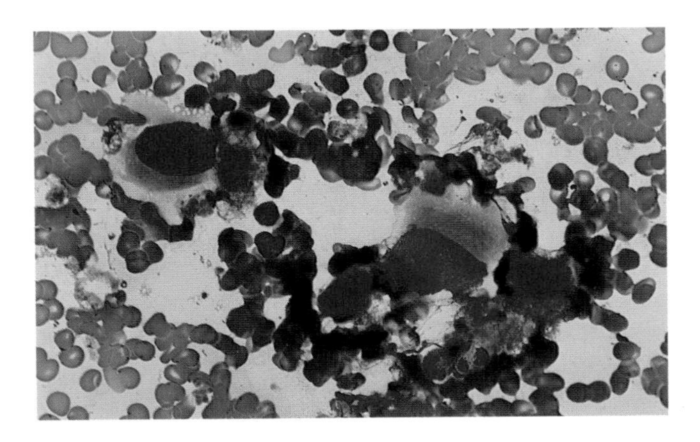

Color Figure 8-29

Anaplastic Carcinoma

This view is part of the same field as Figure 8-29, but at higher magnification. (Diff-Quik ×160)

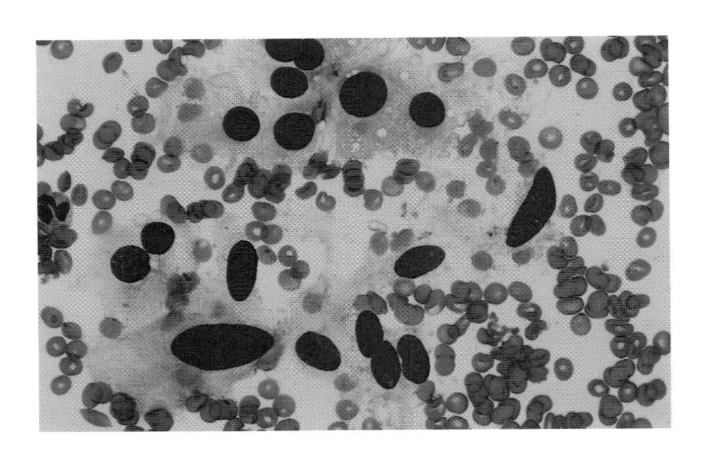

Color Figure 8-30

Anaplastic Carcinoma

Notice marked variation in nuclear size and shape. (Diff-Quik ×160)

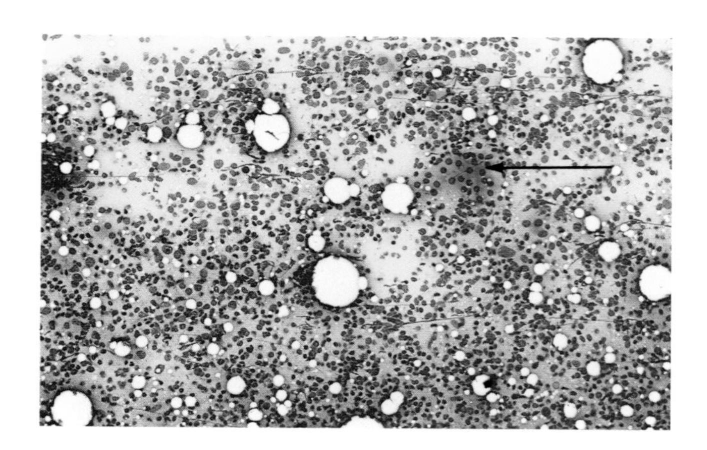

Color Figure 8-31

Malignant Lymphoma

One group of follicular epithelial cells (*arrow*) is seen slightly to the right of center. The rest of the field shows innumerable lymphoid cells. (Diff-Quik ×40)

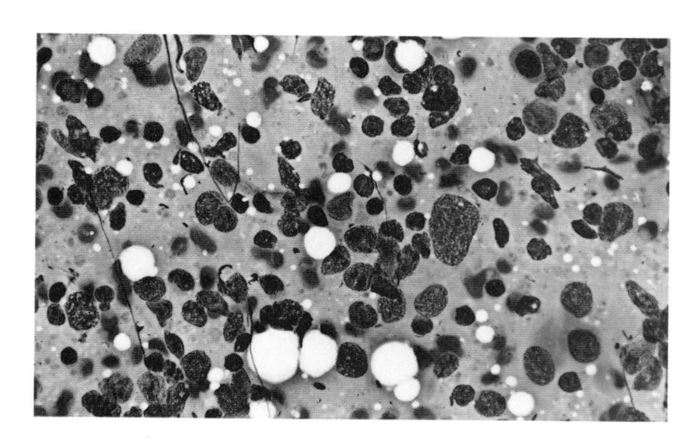

Color Figure 8-32

Malignant Lymphoma

The two cells showing bluish cytoplasm, lower right, are most likely follicular cells. Numerous large lymphoid cells with variation in nuclear size (some bizarre) and prominent nucleoli are observed. (Diff-Quik ×160)

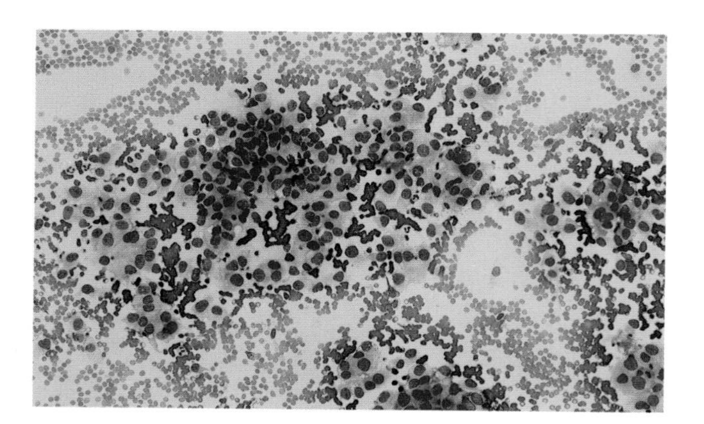

Color Figure 8-33

Renal Cell Carcinoma

Large irregular groups of epithelial cells (tumor cellularity) are seen in a hemorrhagic background. (Diff-Quik ×40)

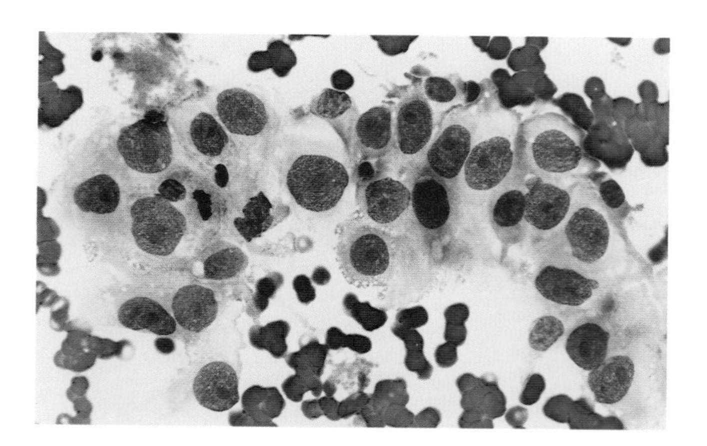

Color Figure 8-34

Renal Cell Carcinoma

These cells are quite different from those of primary thyroid neoplasia (compare with previous figures at same magnification). The nuclei show marked variation in size and shape and the nucleoli are extremely large. The cell borders are very visible, but the cytoplasm appears very delicate to clear. (Diff-Quik ×160)

(Continued from page 208.)

Figure 9-2 ***Relationship of Excretion Index (EI) to Effective Renal Plasma Flow (ERPF) in Various Disease Entities***

{Reproduced with permission from Tauxe and *Annals of Surgery* (8)} (Diethelm AG, et al: Diagnosis of Impaired Renal Function after Renal Transplantation using Renal Scintigraphy, Renal Plasma Flow, and Urinary Excretion of Hippurate. Ann Surg 191:604-605, 1980)

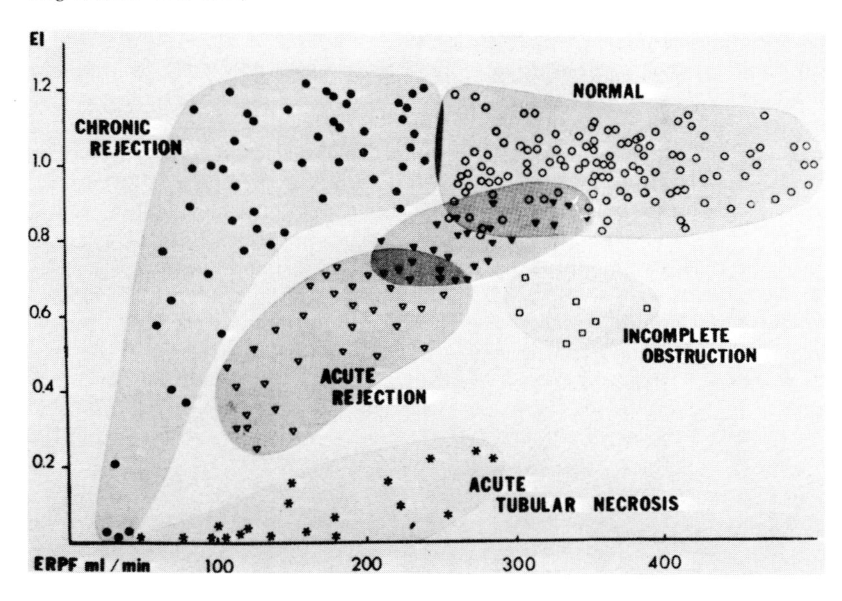

Figure 9-3

Regions of interest of renal parenchyma *(curved arrow)* and aorta *(straight arrow)* on flow images. (Dubovsky EV: Primer and atlas for renal transplant scintigraphy. Clin Nucl Med:10:Part I [52-62], Part II [118-133], 1985)

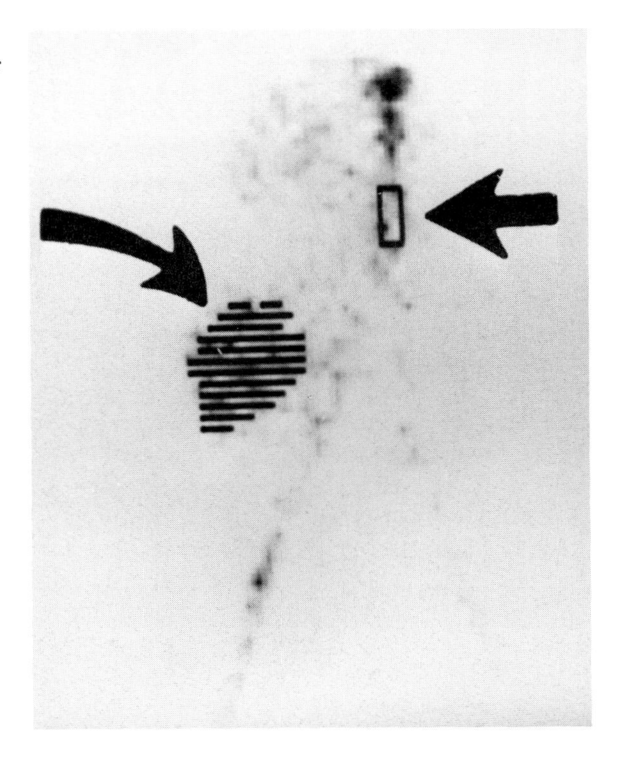

Figure 9-4 ***Normal Tc-99m DTPA Renal Transplant Study****

(A) Time from initial appearance of aortic activity (0-to-3-second frame) to peak aortic activity (3-to-6-second frame) is 3 seconds with rapid clearance of aortic activity. Peak renal activity occurs within 3 to 6 seconds of peak aortic activity, and the peak renal activity equals or exceeds peak aortic activity. Thus, the bolus and renal flow are excellent. (The authors recognize that all times are very approximate.) ***(B)*** Computer-generated time-activity curves show a rapid rise and fall of both aortic *(squares)* and renal *(pluses)* activity. The rate of rise of the kidney curve divided by the aortic cruve will give a quantitative index related to blood flow (ml/min). This is a normal kidney-aortic ratio (K/A). ***(C)*** DTPA images show maximum ratio of parenchymal to background activity (peak) to be greater than 3 (in this patient it is approximately 8) and to occur at 2 to 4 minutes. The renal pelvis is first seen at 0 to 2 minutes. Excellent parenchymal washout is noted over 24 minutes. Background decreases with time. The renal collecting system is normal with excellent washout of the renal pelvis. ***(D)*** I-131 OIH images show maximum ratio of parenchymal to background activity (peak) to be greater than 4 and to occur at 2 to 4 minutes. Renal pelvis activity is first seen at 2 to 4 minutes. Excellent parenchymal washout is noted over 24 minutes. Total activity in bladder and collecting system relative to parenchyma at 24 minutes suggests excellent excretion efficiency. Total activity in bladder, collecting system, and parenchyma suggests excellent extraction and ERPF. The arm injection site shows no subcutaneous infiltration of radiotracer. Partial infiltration of the radiotracer may cause the above parameters to be unreliable. ***(E)*** The time-activity curve shows a rapid upslope with peak activity at 3 minutes and a rapid downslope. The ERPF was 451, 481, and 428 ml/min on three separate determinations; the corresponding excretion indices (EI) were 0.85, 0.80, and 0.94. (Dubovsky EV: Primer and atlas for renal transplant scintigraphy. Clin Nucl Med:10:Part I [52-62], Part II [118-133], 1985)

* Variable photographic intensities may be apparent on many of the 16-frame format images because of photographic factors.

A

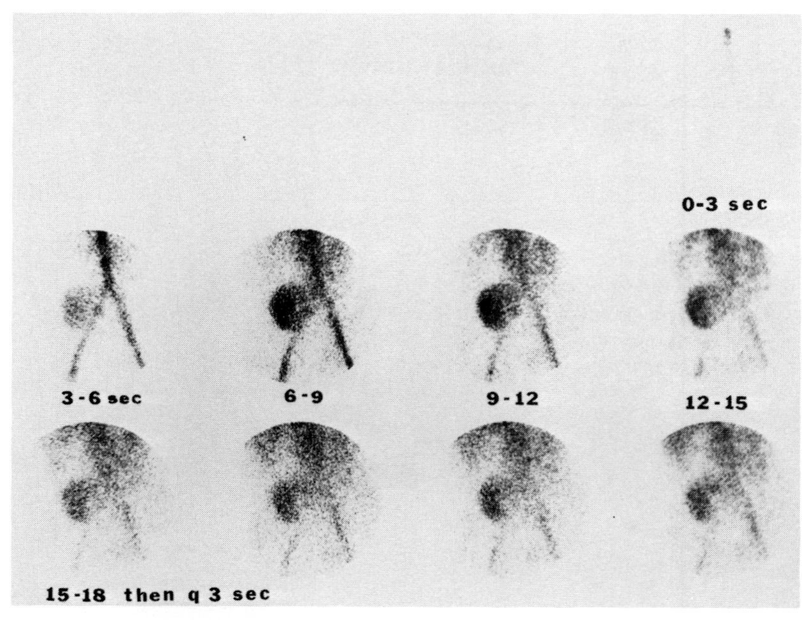

B

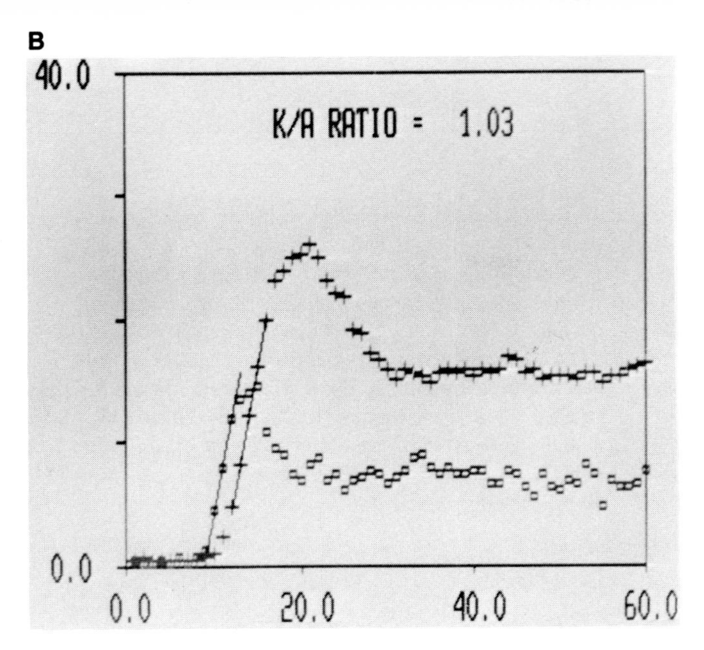

40.0
K/A RATIO = 1.03
0.0
0.0
20.0
40.0
60.0

C

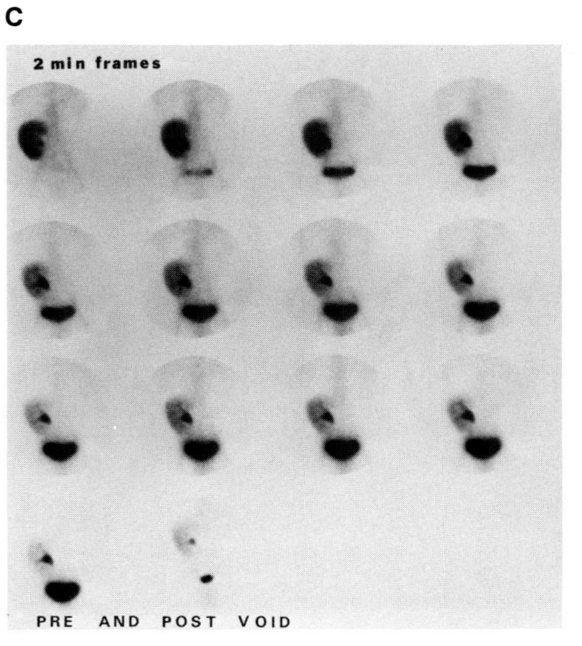

2 min frames
PRE AND POST VOID

D

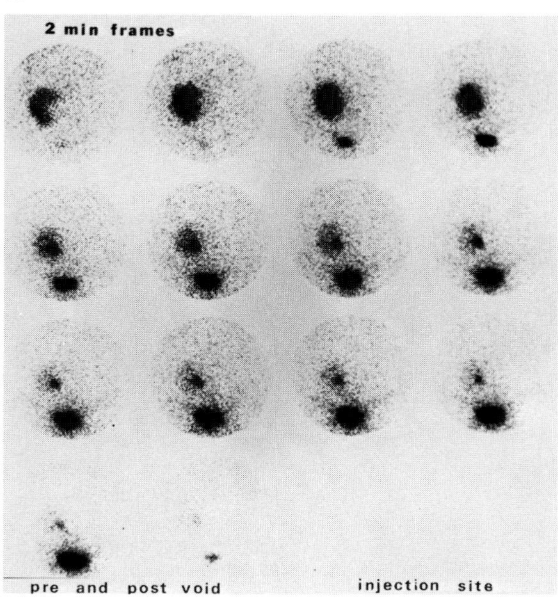

2 min frames
pre and post void
injection site

E

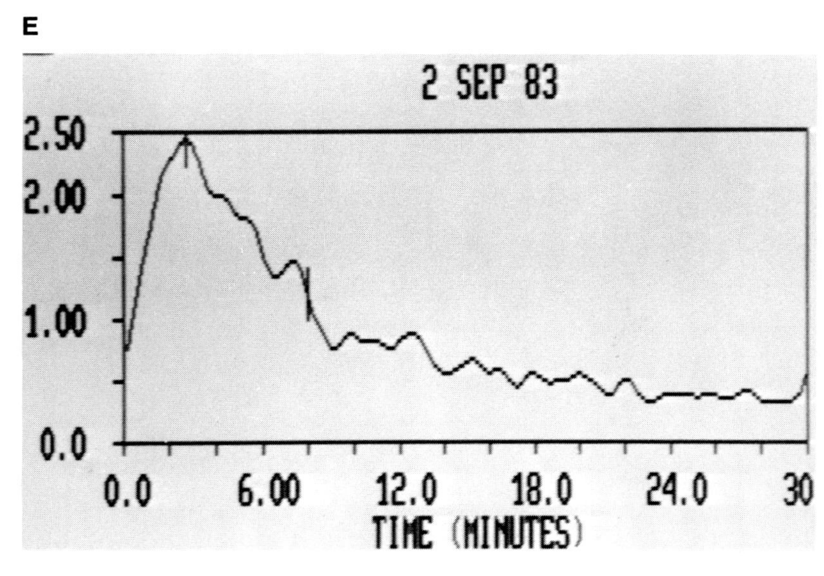

2 SEP 83
2.50
2.00
1.00
0.0
0.0
6.00
12.0
18.0
24.0
30
TIME (MINUTES)

Figure 9-5 ***Renal Artery Thrombosis***

(A) This Tc-99m DTPA renal transplant flow study was performed one day after surgical repair of renal artery stenosis. The time from initial aortic activity to peak aortic activity is ≤ 3 seconds, which denotes a good bolus. No kidney activity is seen, reflecting markedly reduced or absent kidney blood flow. (A small area in the pelvis shows increased flow of unknown etiology. See Figs. 9-6 and 9-16.) ***(B)*** A representative image from the sequential Tc-99m DTPA images shows an oval photon deficiency in the location of the entire renal transplant with a rim of increased activity around the kidney *(arrows)*. No glomerular filtration function or collecting system activity is seen. The etiology of the rim of increased radioactivity may be perirenal hyperemia. A small area of increasing activity *(arrowhead)* is noted in the pelvis. The etiology is unknown but may represent active hemorrhage. The transplanted kidney was removed and revealed renal artery thrombosis with diffuse hemorrhagic necrosis of the kidney. This pattern has also been reported in hyperacute rejection. (Dubovsky EV: Primer and atlas for renal transplant scintigraphy. Clin Nucl Med:10:Part I [52-62], Part II [118-133], 1985)

A

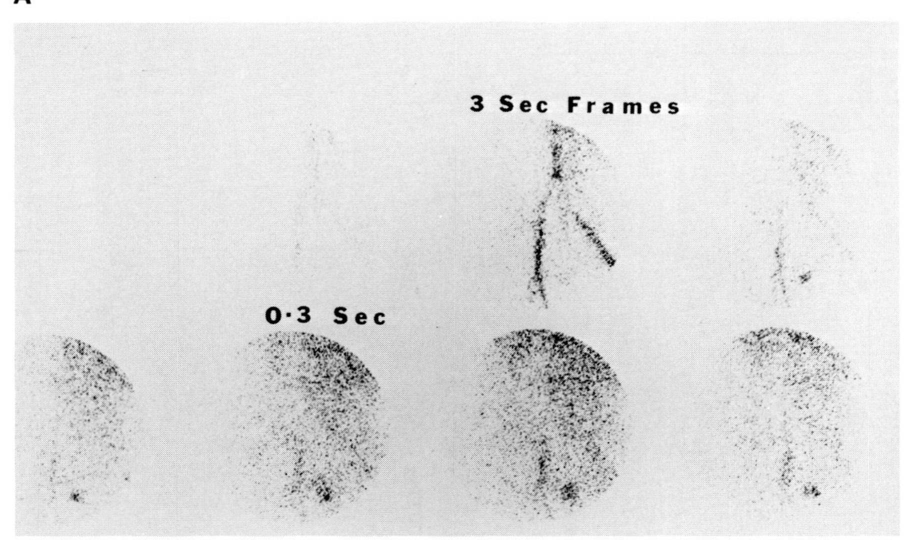

B

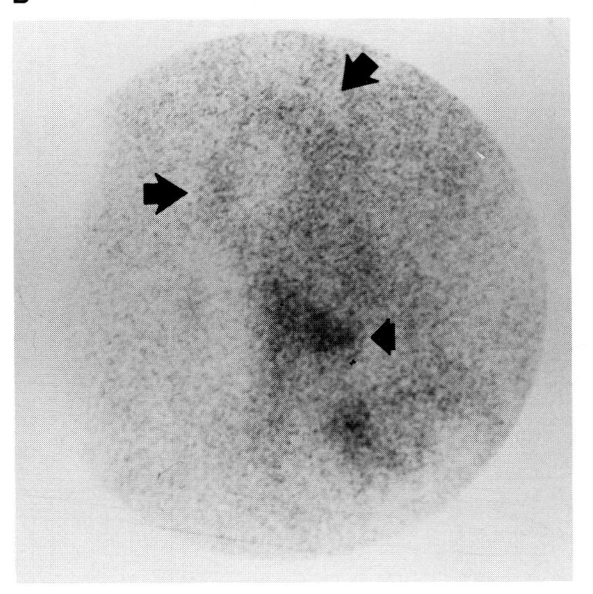

Figure 9-6 ***Arteriovenous Fistula***

(A) This is a Tc-99m DTPA study performed one day after a renal transplantation in a patient with a bruit and thrill over the renal graft area. After injection in the antecubital vein, the Tc-99m DTPA flow study showed simultaneous visualization of the aorta *(small arrow)* and inferior vena cava *(arrowhead)*. Computer images (not shown) at a 1-second frame rate showed rapid sequential visualization of aorta, renal hilum, and inferior vena cava before peak renal parenchymal activity was seen. Subsequent surgery confirmed that the patient's hypogastric artery had been anastomosed to a renal transplant vein, which had apparent communication with other renal transplant veins. *(B)* A Tc-99m DTPA flow study after surgical correction shows resolution of the AV fistula with normal sequential visualization of the aorta, kidney, and inferior vena cava. Incidentally noted during the flow phase of ***A*** is a highly vascular area in the pelvis *(long arrow)*. This activity was attributed to the patient's menstruating uterus because sequential serial images demonstrated the activity to be superior to the bladder, the patient was menstruating, and the area of activity had resolved by the time ***B*** was performed, after cessation of menses. (See Fig. 9-16.) (Dubovsky EV: Primer and atlas for renal transplant scintigraphy. Clin Nucl Med:10:Part I [52–62], Part II [118–133], 1985)

A

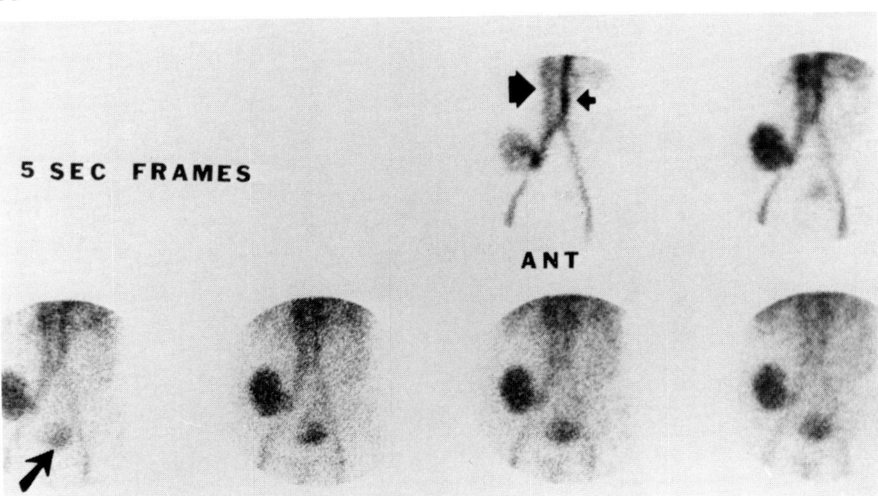

B

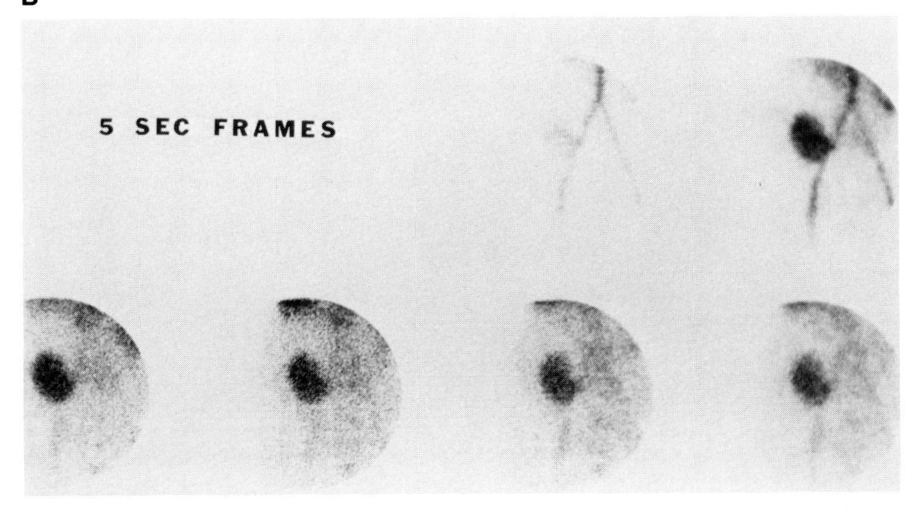

Figure 9-7 *Acute Tubular Necrosis*

(A) This Tc-99m DTPA flow study was performed one day after transplant and demonstrates the time from initial aortic activity (0-to-3-second frame, *arrow*) to peak aortic activity (3-to-6-second frame, *arrowhead*) to be ≤ 3 seconds, which denotes a good bolus. Peak kidney activity (6-to-9-second frame, *curved arrow*) occurs 3 seconds after peak aortic activity and exceeds peak aortic activity. These parameters denote good kidney blood flow. *(B)* The Tc-99m DTPA sequential study shows a markedly reduced kidney-to-background activity ratio throughout the study with no definite peak, no collecting system activity, and no parenchymal washout with time. Although poorly demonstrated because camera intensities varied in each column, soft-tissue activity does increase with time. *(C)* The I-131 OIH study shows a very good but markedly delayed peak on the 22-to-24-minute frame *(arrow)*. There is no collecting system activity and parenchymal activity increases with time. The excretion efficiency is very poor, but the ERPF and extraction efficiency are very good. *(D)* K/A ratio is below normal, suggesting slightly reduced flow. *(E)* The I-131 OIH curve shows progressive accumulation of radioactivity. *(F–P)* Serial scans were performed 14 days *(F–J)* and 56 days *(K–O)* post-transplantation. The flow studies (*A*, *F*, and *K*) show progressive improvement of flow. The K/A ratios (*D*, *I*, and *N*) also show improvement. The DTPA images (*B*, *G*, and *L*) show improvement; however, parenchymal washout on *L* has not returned to normal. The latter was retrospectively attributed to mild acute rejection confirmed on biopsy several days after the scan. The I-131 OIH studies (*C*, *H*, and *M*) show marked improvement in excretion efficiency, which is confirmed in *P*. Although it is visually difficult to determine if the ERPF has improved, *P* shows improved ERPF as determined by blood sample. *E*, *J*, and *O* show the time–activity curve returning toward normal. (Dubovsky EV: Primer and atlas for renal transplant scintigraphy. Clin Nucl Med:10:Part I [52–62], Part II [118–133], 1985)

A

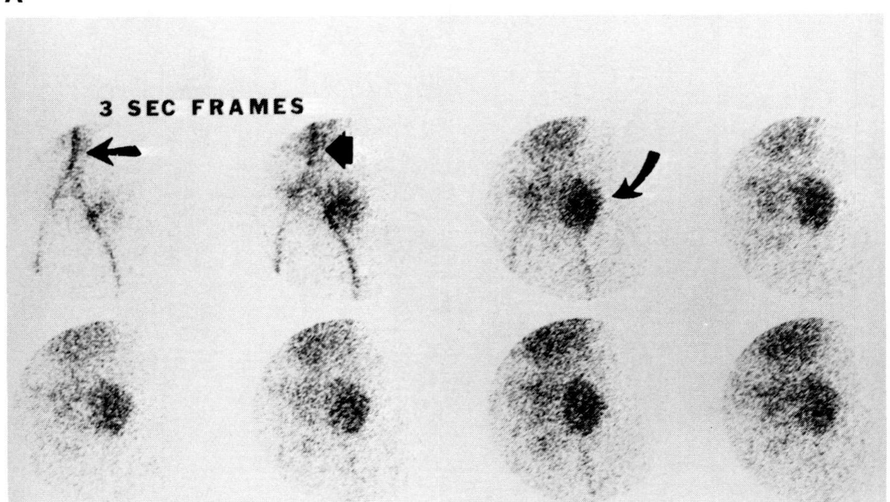

B

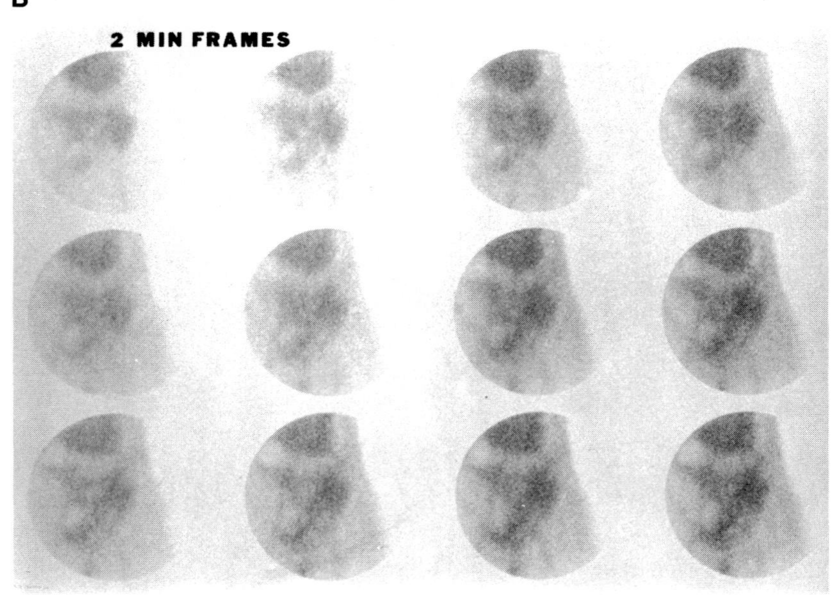

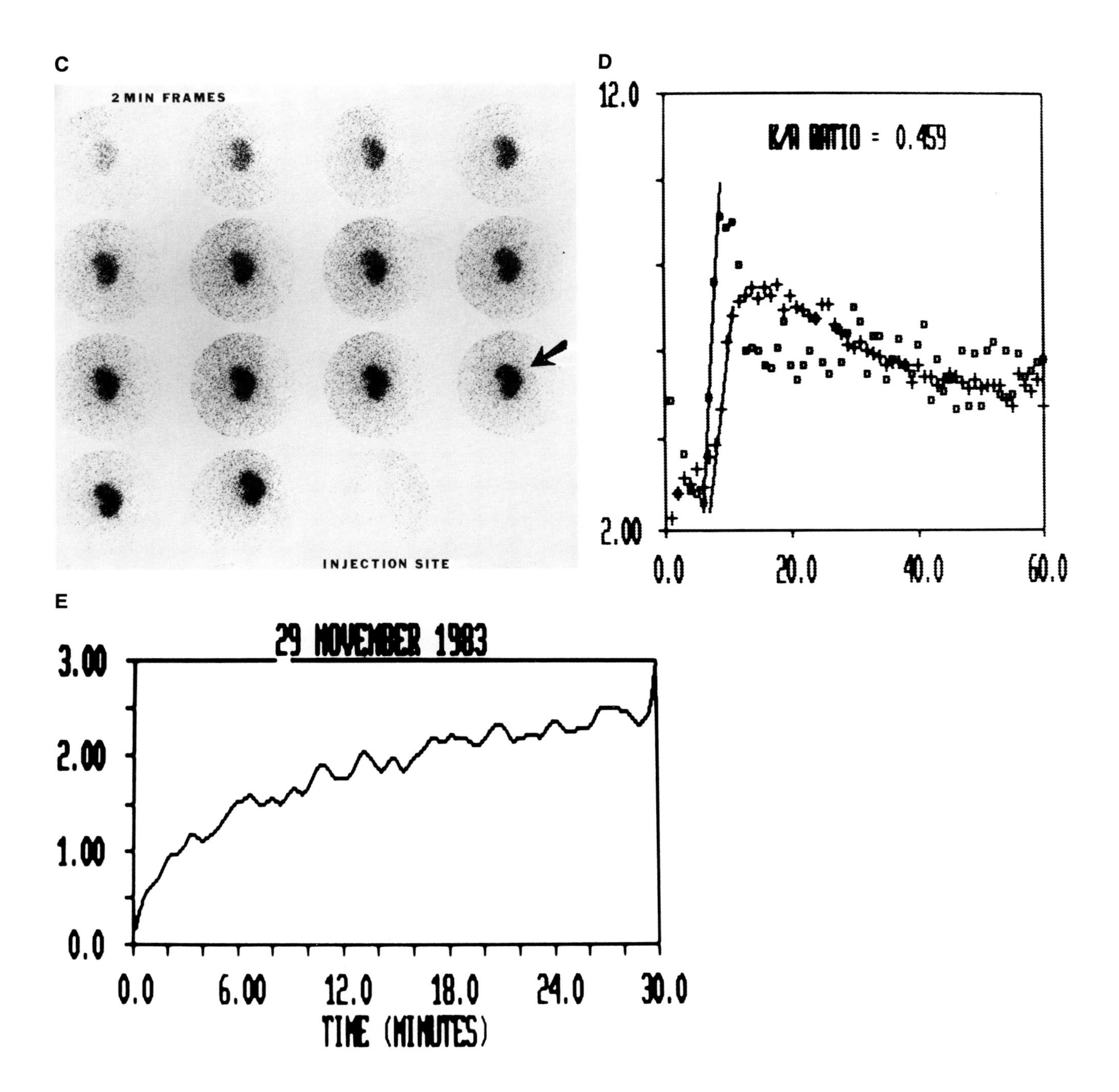
C
2 MIN FRAMES
INJECTION SITE
D
K/A RATIO = 0.459
12.0
2.00
0.0
20.0
40.0
60.0
E
29 NOVEMBER 1983
3.00
2.00
1.00
0.0
0.0
6.00
12.0
18.0
24.0
30.0
TIME (MINUTES)

F

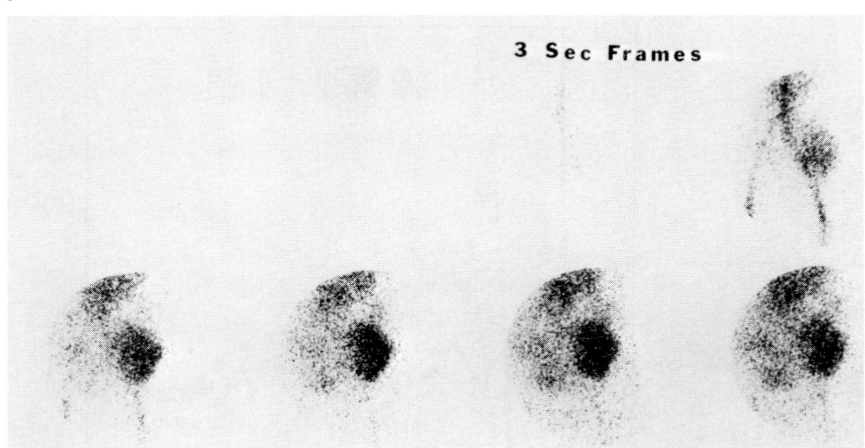

G

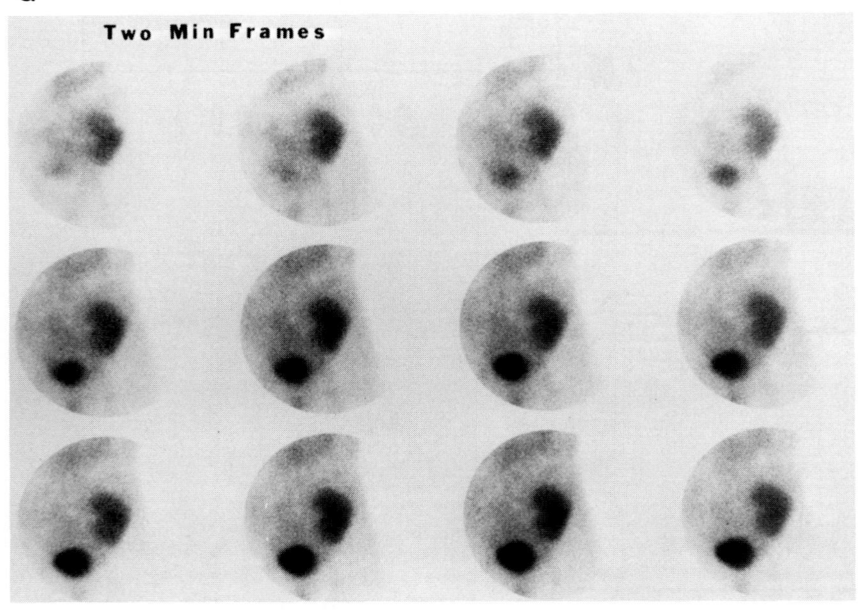

H

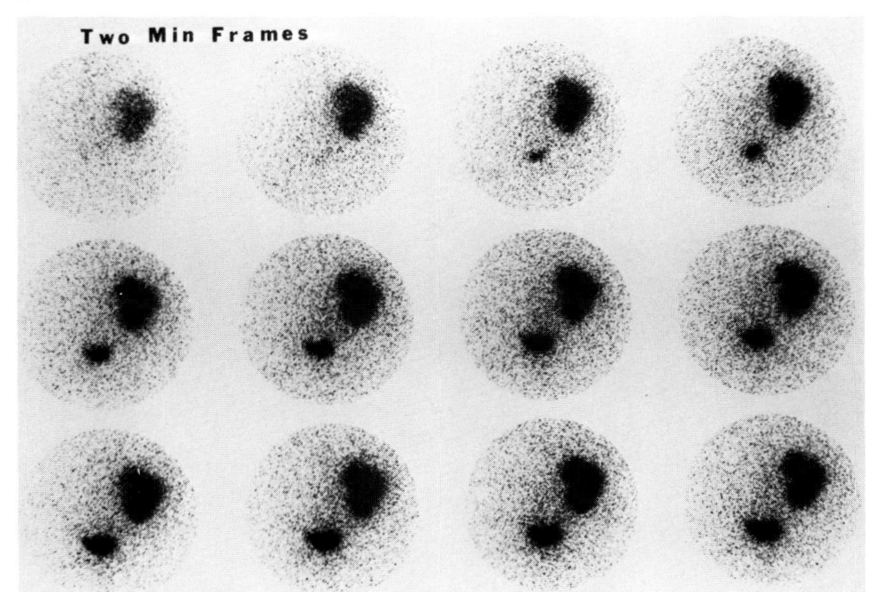
Two Min Frames

I

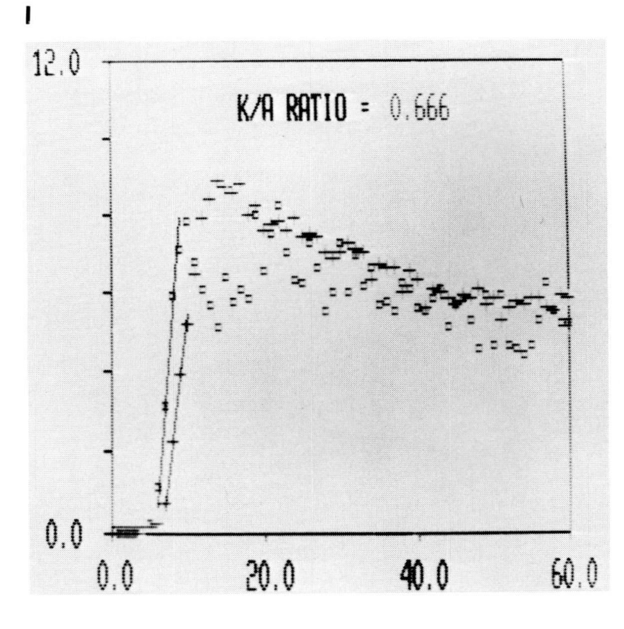
K/A RATIO = 0.666
12.0
0.0
0.0
20.0
40.0
60.0

J

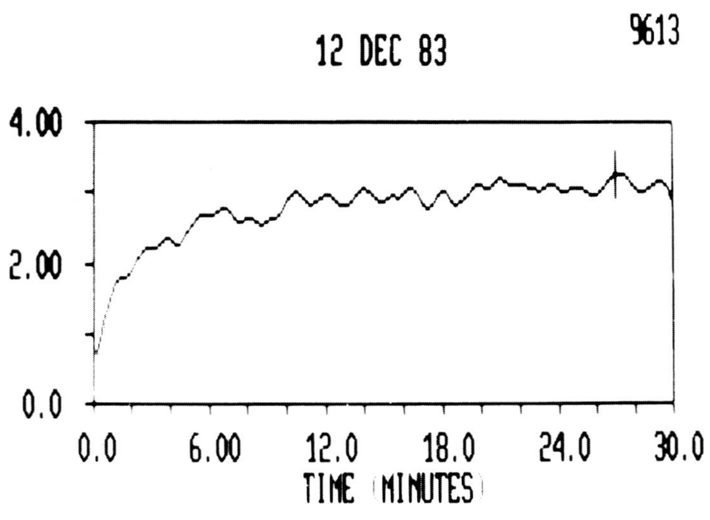
12 DEC 83
9613
4.00
2.00
0.0
0.0
6.00
12.0
18.0
24.0
30.0
TIME (MINUTES)

K

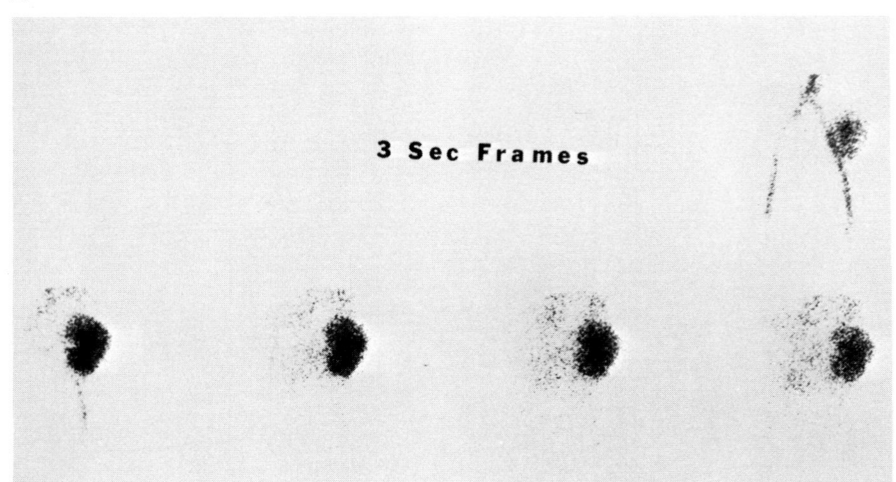

L

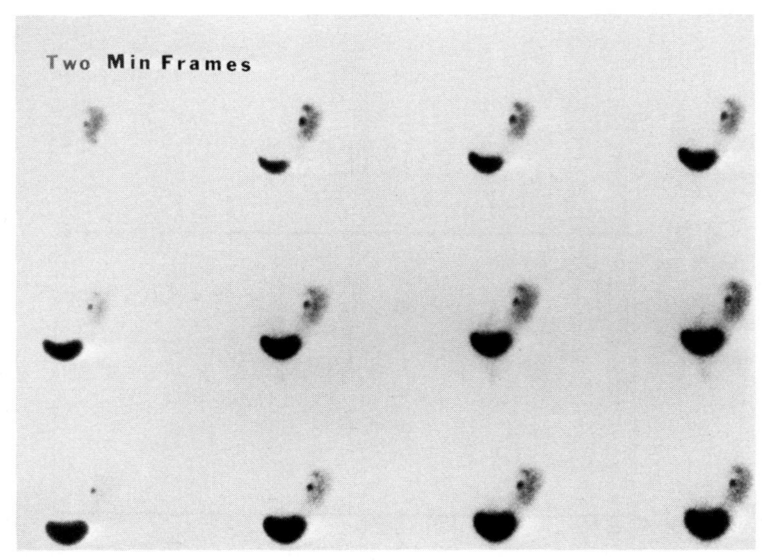

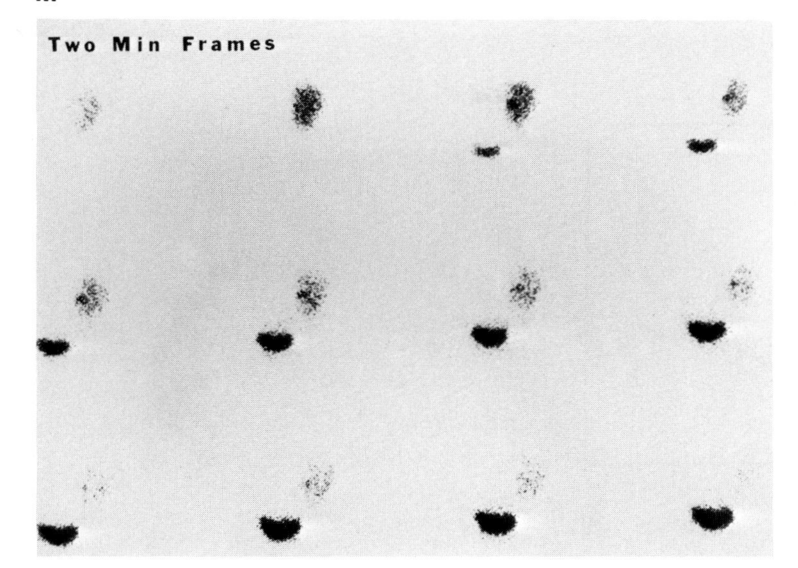

N

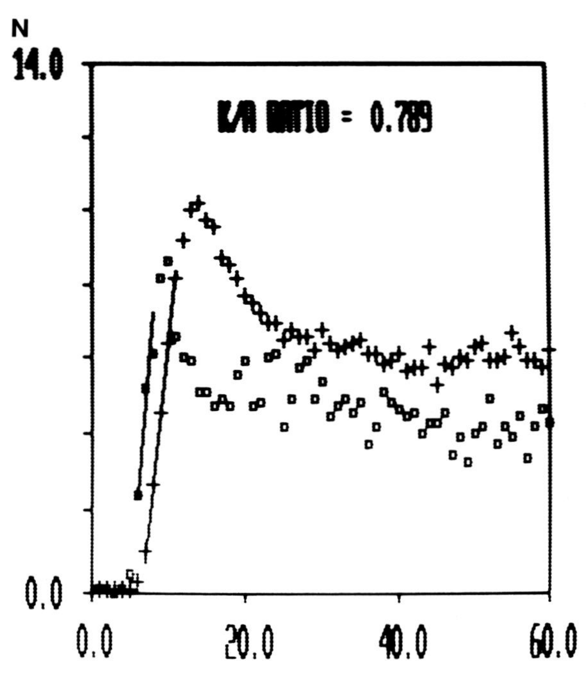

O

23 JAN 84

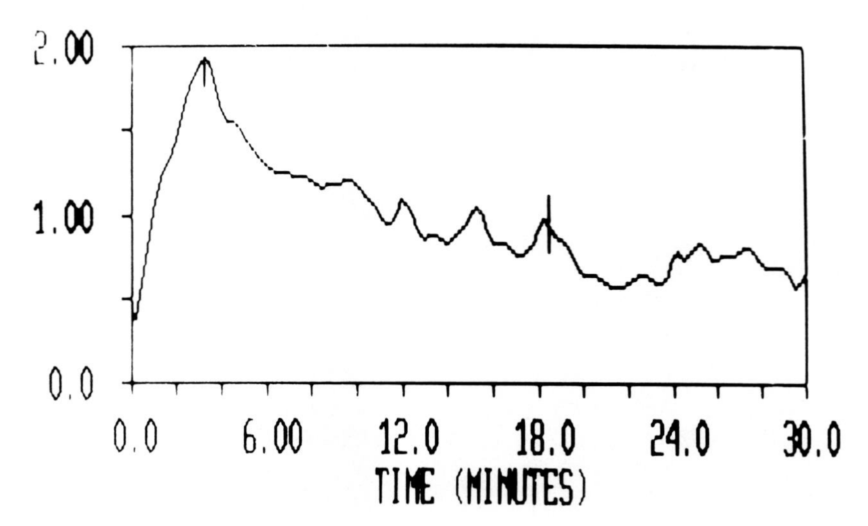
2.00
1.00
0.0
0.0
6.00
12.0
18.0
24.0
30.0
TIME (MINUTES)

P

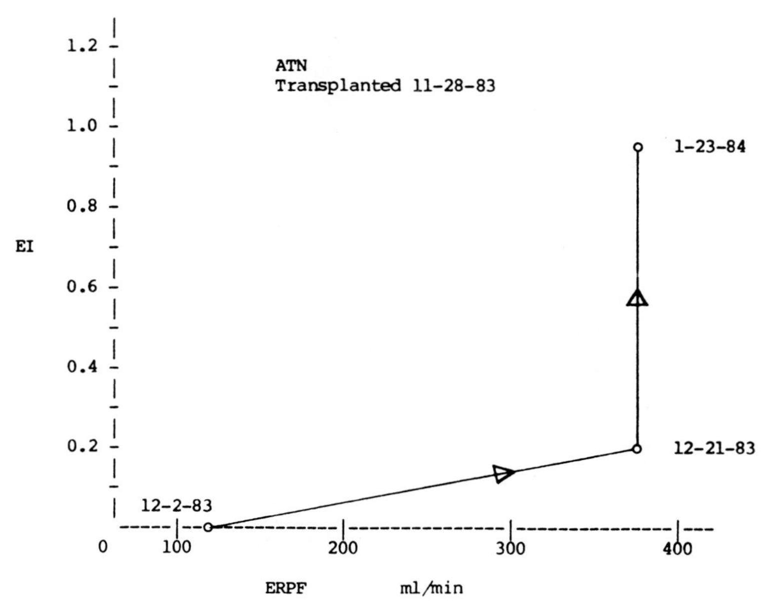
1.2
1.0
0.8
0.6
0.4
0.2
EI
ATN
Transplanted 11-28-83
1-23-84
12-21-83
12-2-83
0
100
200
300
400
ERPF ml/min

Figure 9-8 *Acute Rejection*

(A) This Tc-99m DTPA flow study was obtained 4 weeks after transplantation. The time from initial aortic activity (0-to-3-second frame, *arrow*) to peak aortic activity (3-to 6-second frame, *arrowhead*) is 3 seconds, which denotes a good bolus. Peak kidney activity (12-to-15-second frame, *curved arrow*) occurs 9 seconds after peak aortic activity (3-to-6-second frame) and is less than peak aortic activity. These parameters denote reduced renal blood flow. *(B)* The sequential Tc-99m DTPA study shows reduced peak activity occurring at the normal time (4 minutes). Time of first definite collecting system activity is seen on the 4-to-6-minute frame *(arrow)*. There is reduced parenchymal washout over 24 minutes and background activity changes little with time. *(C)* The I-131 OIH study shows normal, but delayed peak parenchymal activity at 6 through 24 minutes. First definite collecting system activity occurs at 6 to 8 minutes *(arrow)*. There is initial parenchymal tracer accumulation in the first 6 minutes, but the parenchymal activity remains constant through the remainder of the study. Extraction and ERPF are good; however, excretion efficiency is only fair. *(D)* The K/A ratio is reduced at 0.39. *(E)* The Hippuran curve shows initial parenchymal tracer accumulation in the first 6 minutes, which remains constant through the remainder of the study. *(F)* This is the same patient approximately 6 weeks later. Tc-99m DTPA flow study shows time from initial aortic activity (0-to-3-second frame, *arrow*) to peak aortic activity (3-to-6-second frame, *arrowhead*) to be 3 seconds, which denotes a good bolus. No kidney activity is identified; this denotes markedly reduced renal blood flow. *(G)* The Tc-99m DTPA study shows a markedly reduced peak activity, no definable peak time, no collecting system activity, and no parenchymal washout. Background activity changes little with time. *(H)* The I-131 OIH study shows minimal progressive accumulation of activity with a reduced peak occurring at 24 minutes or later and no collecting system activity. The ERPF, extraction efficiency, and excretion efficiency are very poor. (Photographic intensities are variable.) *(I)* The K/A ratio has decreased to 0.16. *(J)* The Hippuran curve shows slight progressive parenchymal tracer accumulation. *(K)* This graph displays all available K/A ratios by dates and shows a continuous decline in ratios. *(L)* This graph displays the relationship of estimated ERPF and EI, which were available on three studies. (Dubovsky EV: Primer and atlas for renal transplant scintigraphy. Clin Nucl Med:10:Part I [52-62], Part II [118-133], 1985)

A

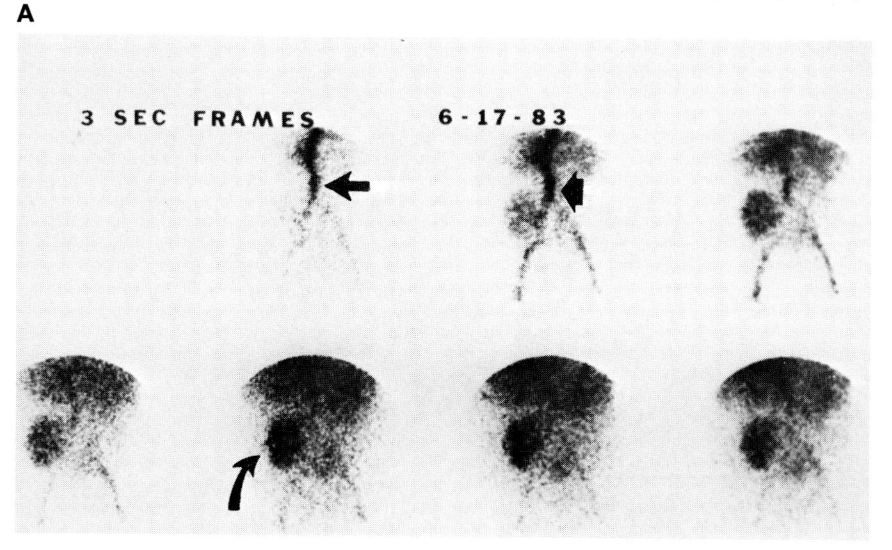

B

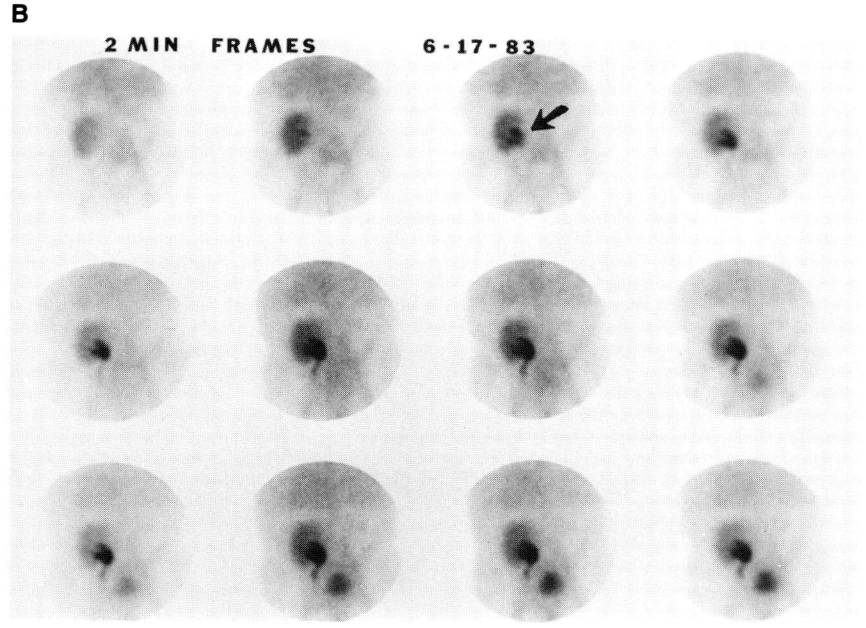

C

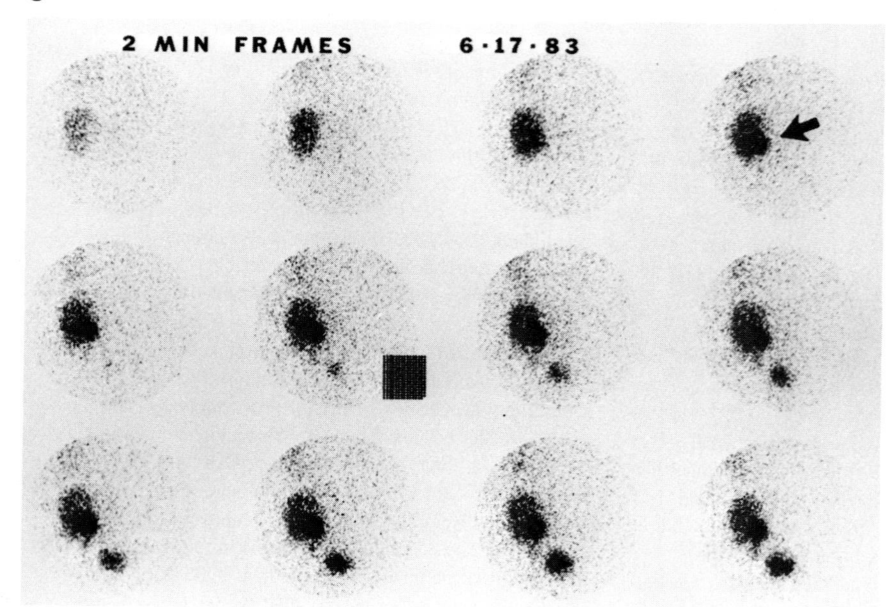

2 MIN FRAMES
6·17·83

D

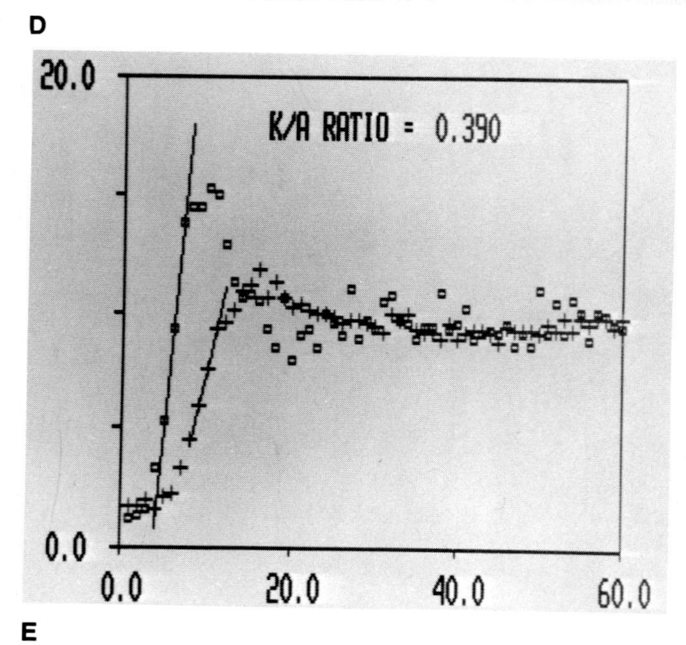

K/A RATIO = 0.390
20.0
0.0
0.0
20.0
40.0
60.0

E

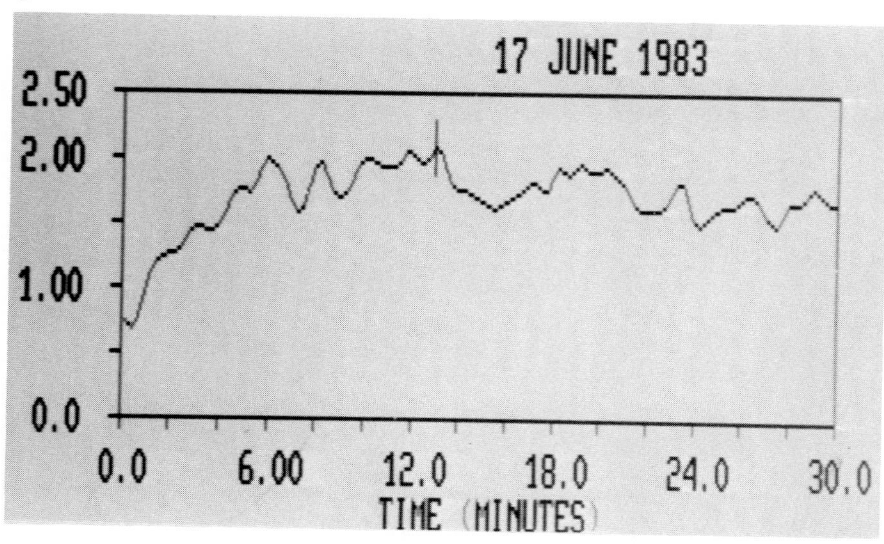

17 JUNE 1983
2.50
2.00
1.00
0.0
0.0
6.00
12.0
18.0
24.0
30.0
TIME (MINUTES)

F

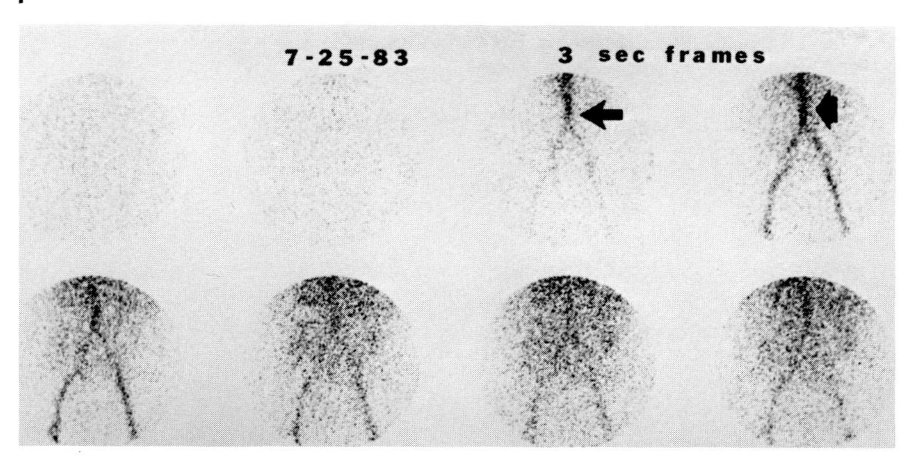

G

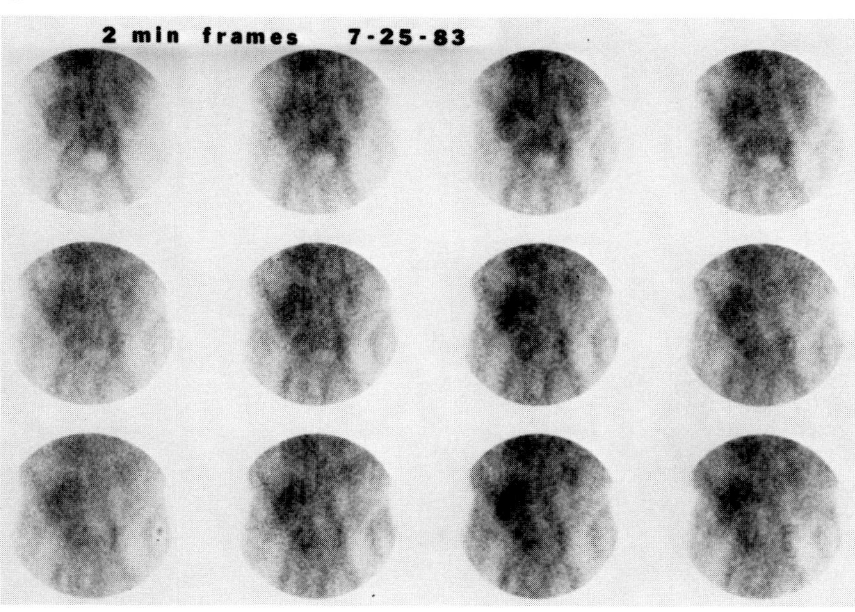

H

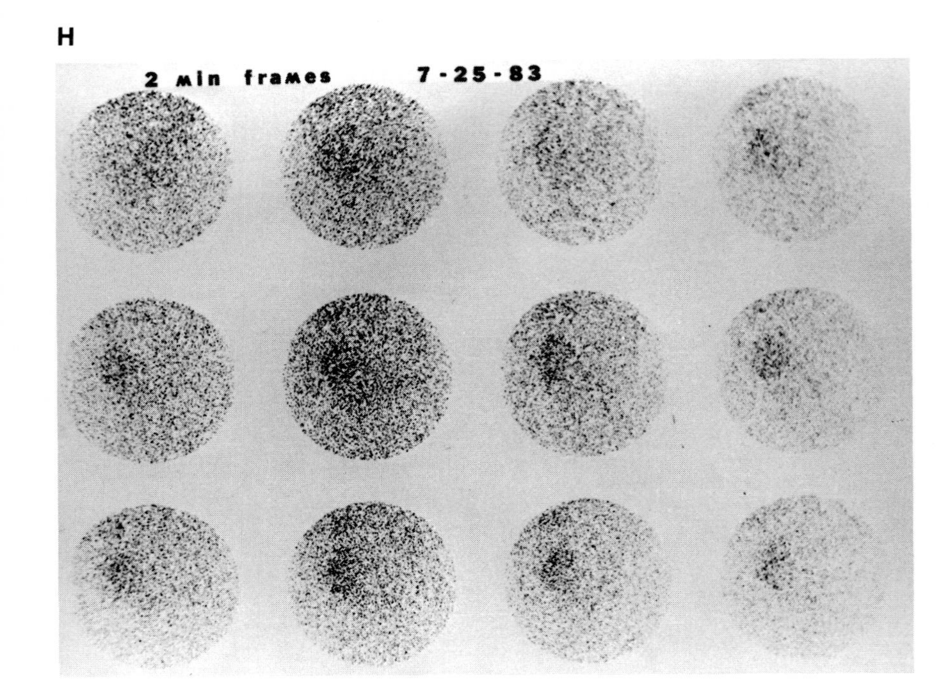
2 min frames 7-25-83

I

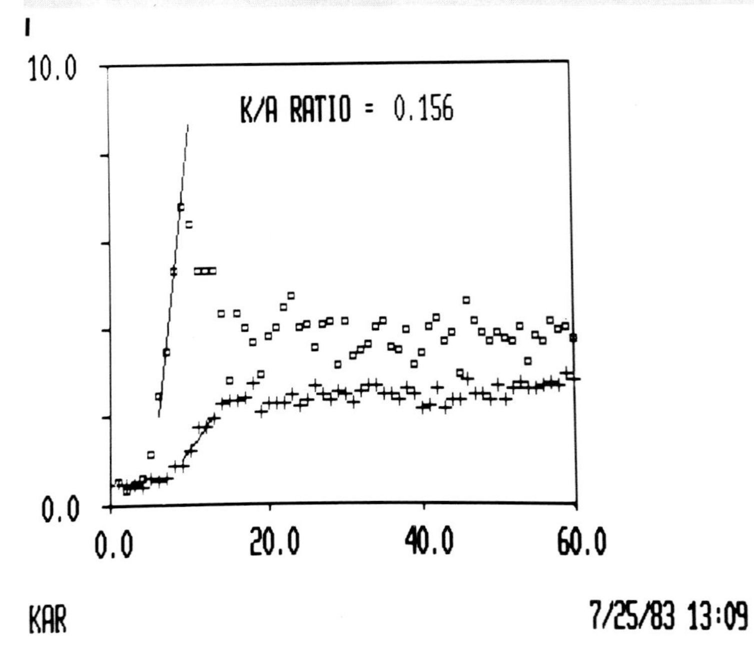
K/A RATIO = 0.156
10.0
0.0
0.0
20.0
40.0
60.0
KAR
7/25/83 13:09

J

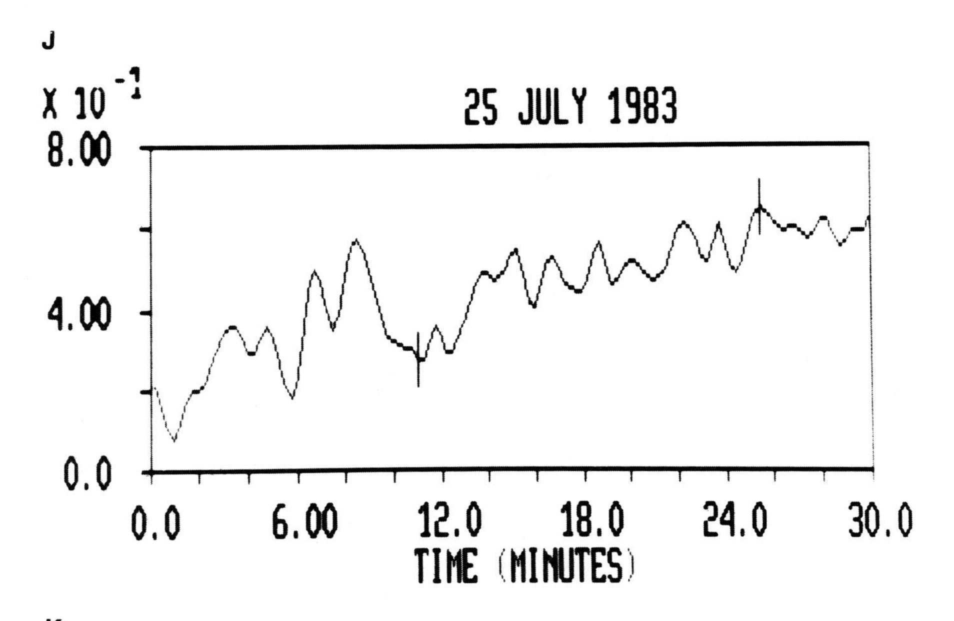
X 10^-1
25 JULY 1983
8.00
4.00
0.0
0.0
6.00
12.0
18.0
24.0
30.0
TIME (MINUTES)

K

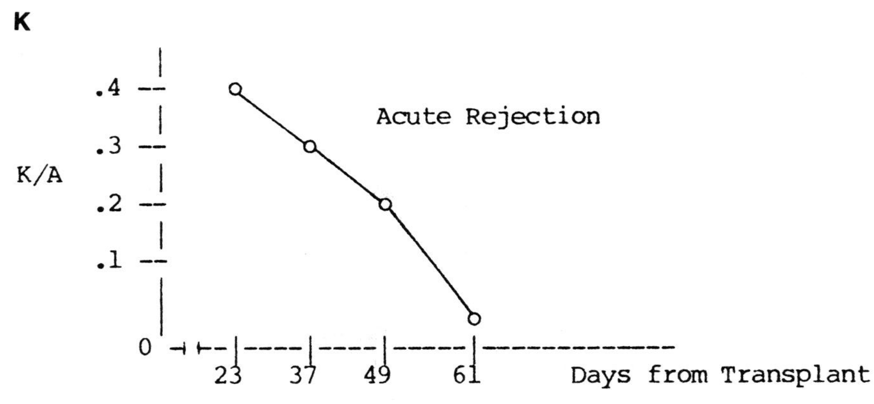
.4
.3
.2
.1
0
K/A
Acute Rejection
23
37
49
61
Days from Transplant

L

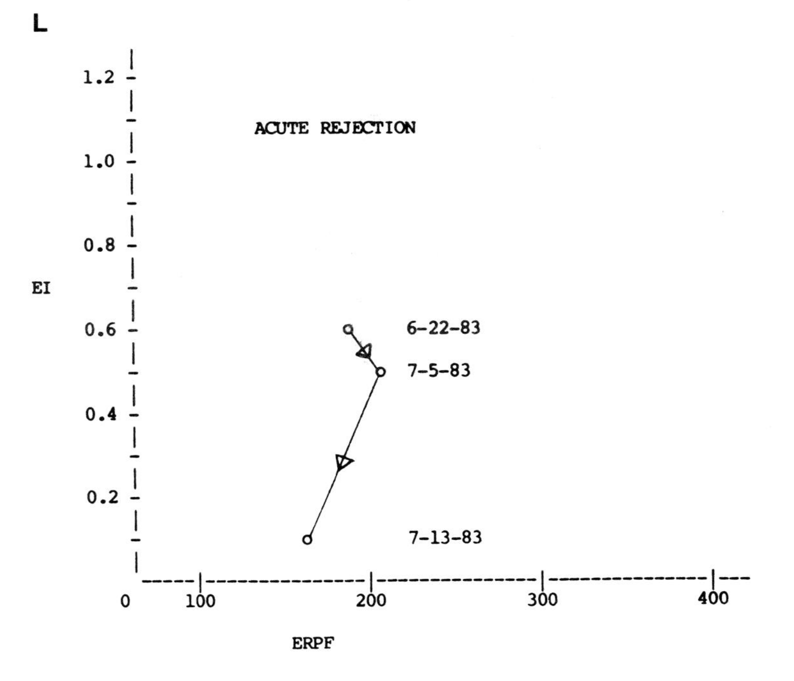
1.2
1.0
0.8
0.6
0.4
0.2
EI
ACUTE REJECTION
6-22-83
7-5-83
7-13-83
0
100
200
300
400
ERPF

Figure 9-9 ***Chronic Rejection***

(A) Scans *1* (9-17-72), *2* (10-30-74), *3* (4-8-75), and *4* (9-9-76) correspond to four Hippuran studies shown in the graph in image **B.** Scan 1 shows excellent peak parenchymal activity on the 0-to-3-minute image, with progressive deterioration of peak activity in scans 2–4. There is a normal time to peak parenchymal activity seen on the 0-to-3-minute image in scan 1. Scan 2 shows mild deterioration in time to peak, which occurs on the 3-to-6-minute image, and scan 3 shows no further visual delay in time to peak. In scan 1, the time of first activity in the collecting system is normal and occurs at 3 to 6 minutes. Scans 2 and 3 show progressive delay in the time of first collecting system activity. Scan 1 shows normal parenchymal washout of tracer. Scans 2–4 show progressive deterioration in washout. Scan 1 showa a normal ERPF and excretion efficiency. Scans 2–4 show a progressive decline in ERPF and excretion efficiency. Scan 1 shows excellent excretion efficiency. Scans 3 and 4 suggest progressive decline in excretion efficiency, which is confirmed in the EI versus ERPF graph. The three graphs to the right of the images are sequential time–activity curves which show progressive decrease in peak activity and delay in time to peak of the kidney (*K* in image *A*). The bladder (*B* in image *A*) *(B)* activity curves demonstrate decreasing rate and amount of bladder activity on sequential studies. (Dubovsky EV: Primer and atlas for renal transplant scintigraphy. Clin Nucl Med 10:Part I [52–62], Part II [118–133], 1985)

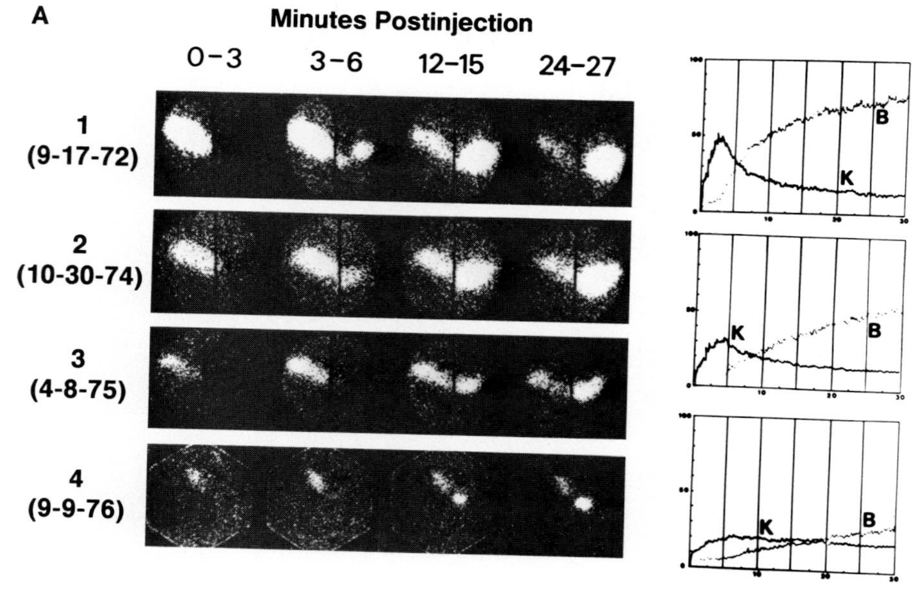

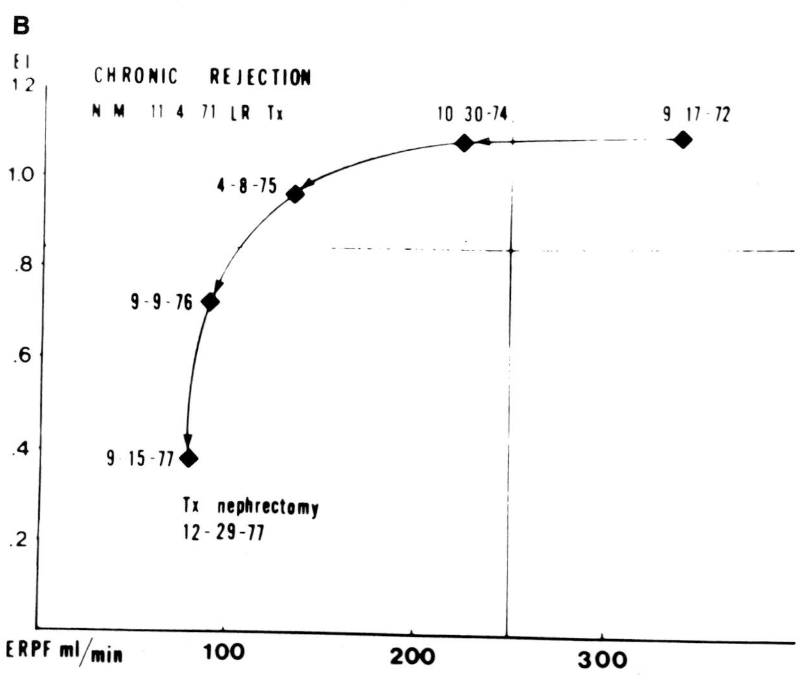

Figure 9-10 ***Partial Ureteral Obstruction — Postoperative Edema***

(A) This patient was imaged one day after renal transplant. Tc-99m DTPA study shows columning of the ureter to the level of the ureterovesical junction *(arrow)* with only partial ureteral clearance on the erect postvoid film *(curved arrow)*. The patient was well hydrated. ***(B)*** A follow-up study 11 days later shows more rapid clearance of tracer from the ureter *(arrows)* with significant reduction in ureteral columning on the erect postvoid image *(curved arrow)*. There was no operative intervention between the two studies. The partial obstruction of the distal ureter is believed to be due to transient postoperative edema at the ureterovesical junction. ***Comment:*** When significant ureteral columning is seen on the erect postvoid image, a partial obstruction of the distal ureter is suggested. Although visually one cannot differentiate the various etiologies of obstruction (Tables 9-3 and 9-4), the time of appearance of ureteral columning postoperatively and its subsequent course may be of value. Postoperative edema occurs early and resolves with time. It is not an infrequent finding. [Dubovsky EV: Primer and atlas for renal transplant scintigraphy. Clin Nucl Med:10:Part I [52-62], Part II [118-133], 1985)

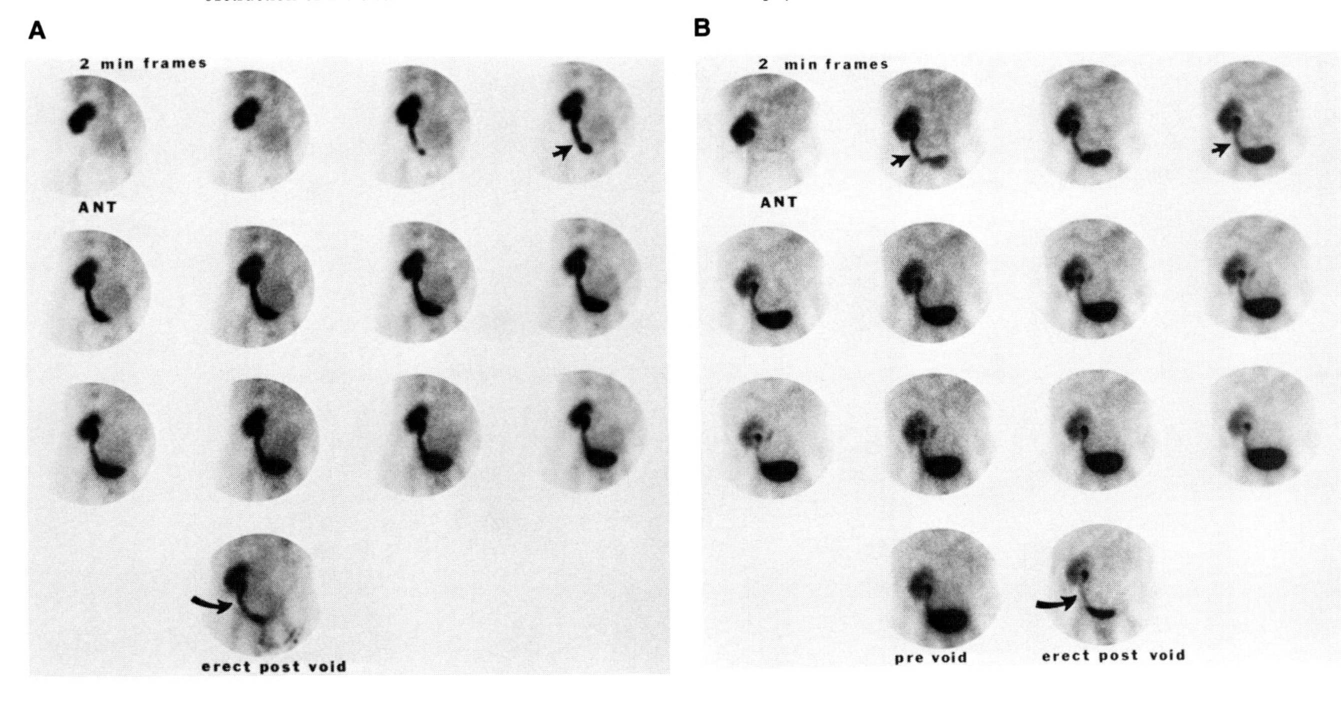

Figure 9-11 ***Collecting System Obstruction***

(A) This DTPA study was performed shortly after renal biopsy which was complicated by hematuria. The sequential images show progressive tracer accumulation in a dilated renal pelvis *(arrowhead)* with an irregular and poorly filled bladder *(arrow)*. The renal parenchymal washout is delayed. ***(B)*** The erect postvoid image shows poor drainage of the dilated renal pelvis *(arrowhead)* with nonvisualization of the ureter. The above findings are highly suggestive of a partial ureteropelvic junction obstruction. Again, the bladder configuration is irregular *(arrow)*. Because multiple blood clots were present in the urine at the time of the above study and follow-up renograms showed resolution of obstructive findings when urine blood clots were no longer present, the suspected obstruction and irregular bladder configuration were attributed to the blood clots. (Dubovsky EV: Primer and atlas for renal transplant scintigraphy. Clin Nucl Med:10:Part I [52–62], Part II [118–133], 1985)

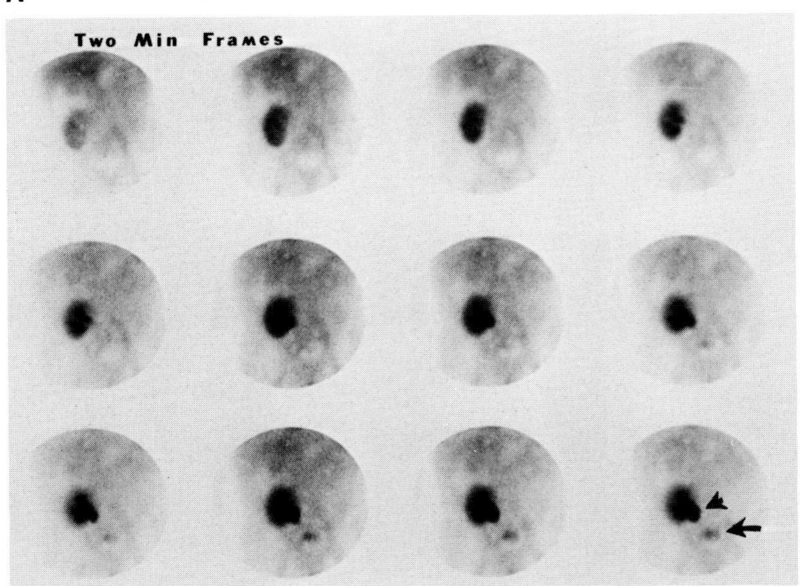

B

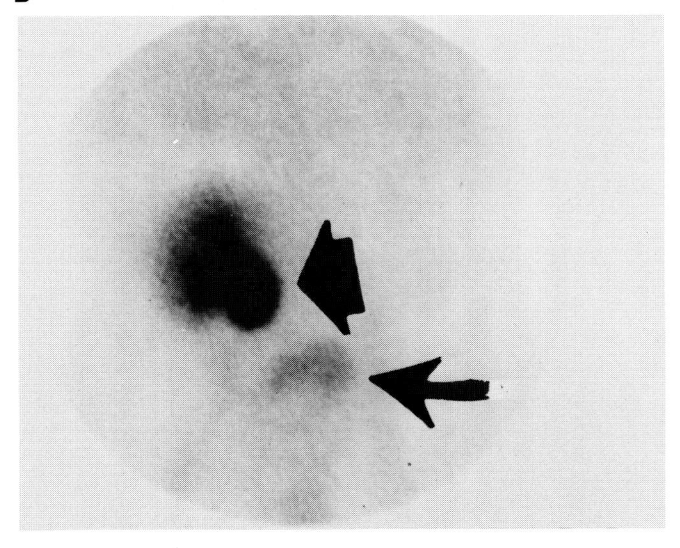

Figure 9-12 ***"Hot" Urinoma***

This is a Tc-99m DTPA sequential study in a patient 10 days after renal transplantation. Progressive tracer accumulation appears in an irregular outline near the region of the bladder *(arrow)*. A urine leak at the ureterovesical anastomosis was found at surgery. ***Comment:*** This case demonstrates the typical scintigraphic appearance of a urinoma: a hot, irregular accumulation of radioactivity in the pelvis extrinsic to the normal renal collecting system. The most common location of urinary leak is the ureterovesical anastomosis. Leakage may be a complication of ischemia of the ureter, rejection of the distal ureter, or incompetence of a suture line. (Dubovsky EV: Primer and atlas for renal transplant scintigraphy. Clin Nucl Med:10:Part I [52-62], Part II [118-133], 1985)

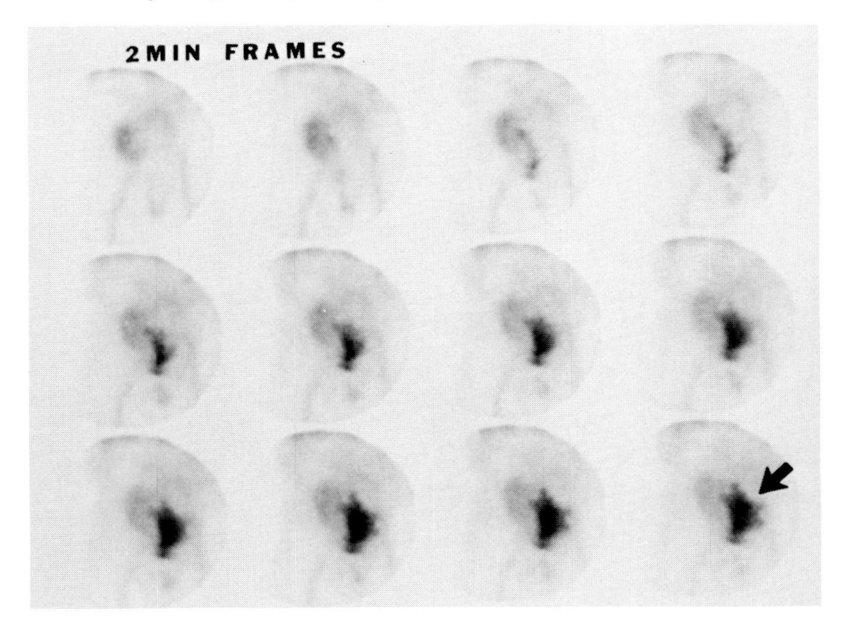

Figure 9-13 ***"Hot" Urinoma: Value of Delayed Views***

(A) This Tc-99m DTPA sequential study (from the same patient as in Fig. 9-22) shows a photon deficiency in the pelvis on the initial image *(arrowheads)*. This may be urine-filled bladder, urinoma, lymphocele, or hematoma. Routine sequential images show a distorted ureter, bladder, or urinary extravasation *(arrow)*. ***(B)*** The patient was unable to void, and a 60-minute delayed image showed abnormal tracer distribution throughout the pelvis and abdomen *(arrows)*, which is diagnostic of urinary extravasation. The curved arrow is the renal transplant. A urinary leak at the ureterovesical anastomosis was found at surgery. A Tc-99m DTPA study, performed 10 days after surgical correction, showed normal bladder filling and emptying without urine extravasation. ***Comment:*** Delayed views should be considered whenever the urinary bladder fails to fill and empty normally, an abnormal ureteral configuration exists, or a focal deficiency persists after voiding. Occasionally, the extravasation of radiotracer may only be visualized on a 24-hour delayed image. In these cases, I-131 Hippuran, which has a longer half-life, may be of value. (Dubovsky EV: Primer and atlas for renal transplant scintigraphy. Clin Nucl Med:10:Part I [52-62], Part II [118-133], 1985)

A

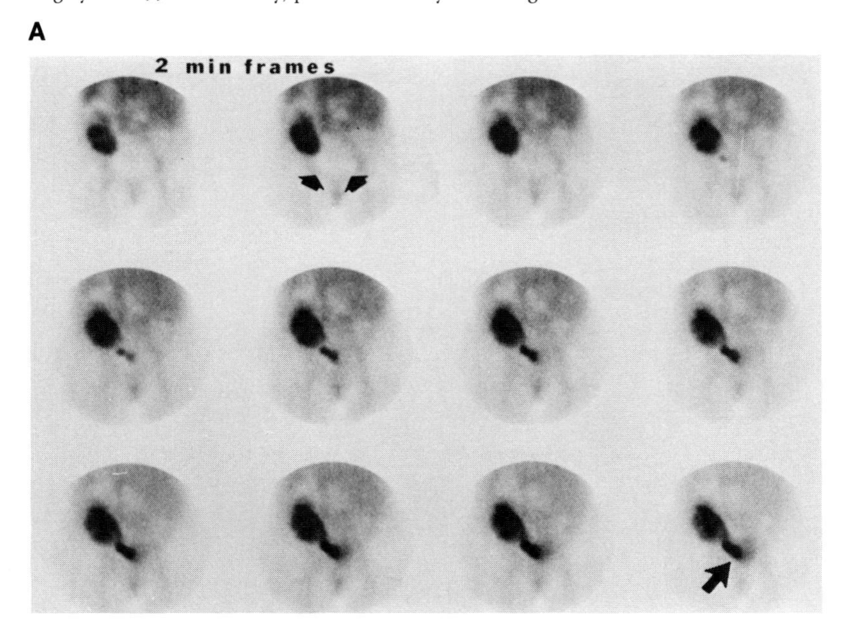

B

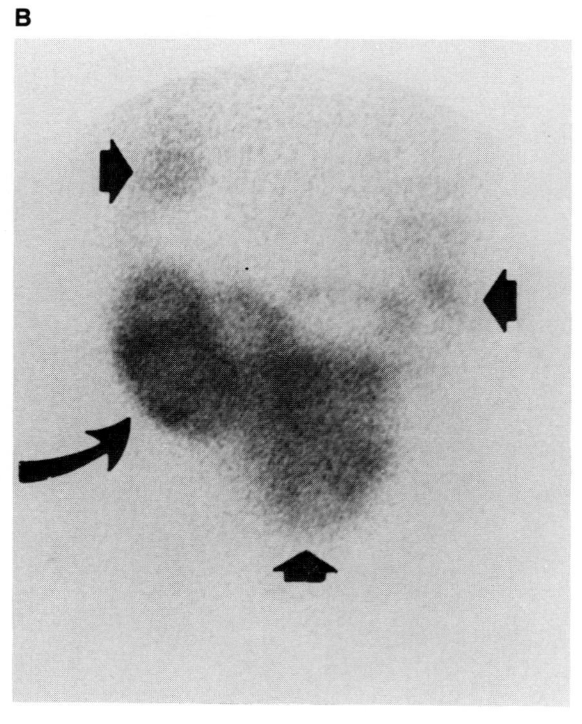

Figure 9-14 *"Cold" to "Hot" Urinoma*

(A) This Tc-99m DTPA study made 3 weeks after renal transplant shows a large photon-deficient area superior and lateral to the kidney, on the 4-minute image *(arrow)*. This area and the region inferior to the ureter show subtle, progressive increase in activity on the 30-minute image (***B****, arrows*). ***(C)*** The postvoid image shows a dramatic increase in activity superior and lateral to the kidney in the initial "cold" areas *(arrows)*. Surgery revealed a urinoma adjacent to a 2 mm bladder perforation that was 2 cm from the ureterovesical anastomosis and was attributed to necrosis from electrocautery. ***Comment:*** Urinomas may occur not only in the pelvic region, but also superior to the kidney. In addition, a urinoma may initially appear as a "cold" defect and become a "hot" area on sequential routine images. (Dubovsky EV: Primer and atlas for renal transplant scintigraphy. Clin Nucl Med:10:Part I [52-62], Part II [118-133], 1985)

A

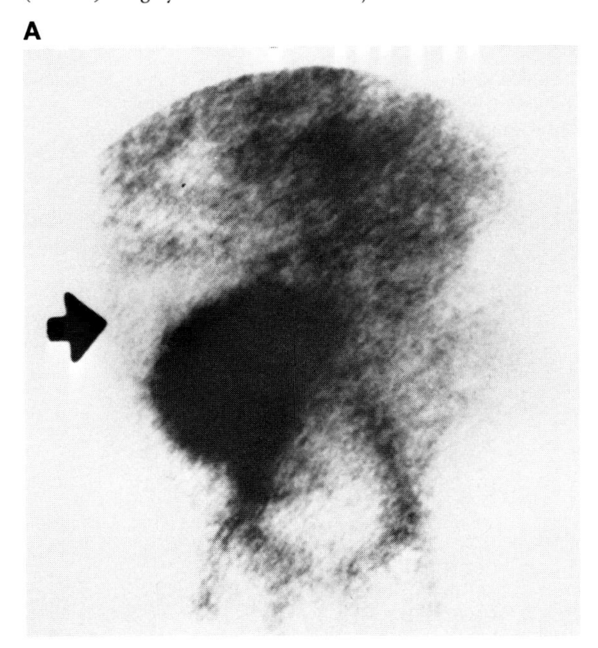

B

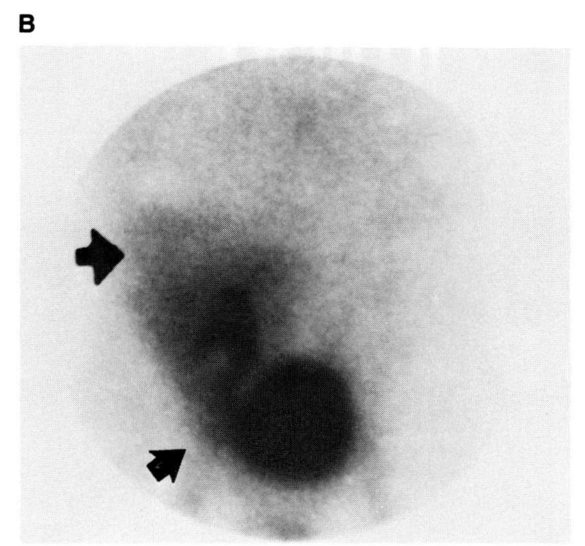

C

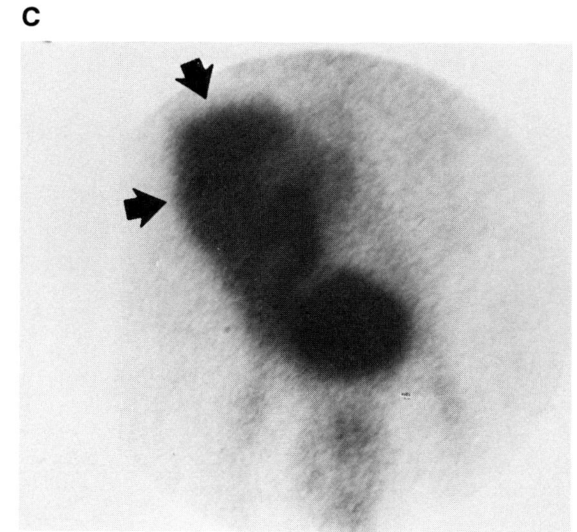

Figure 9-15 ***Wound Dehiscence***

Sequential images from a Tc-99m study show a broad, ill-defined area of increased tracer activity projecting over the pelvic region *(arrows)*. The area extends from the region of the kidney to the bladder and increases in intensity with time. Physical examination of the patient revealed wound dehiscence corresponding to the area of abnormal tracer activity. Oblique and lateral images confirmed the anterior location of the abnormality. As healing occurred, serial studies demonstrated less intense tracer activity. ***Comment:*** Lateral or oblique views may help localize an area of tracer activity, thereby distinguishing a urinoma (typically retroperitoneal) from other entities. This case also demonstrates that an area of abnormal tracer activity which increases with time is not always a urinoma.

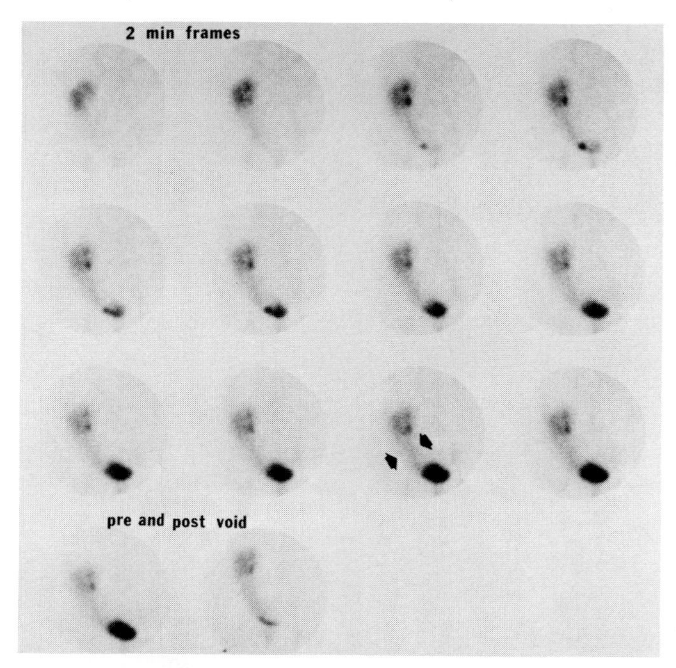

Figure 9-16 ***Uterine Leiomyomata***

The Tc-99m DTPA sequential images show a focal area of increased tracer activity *(solid arrows)* with a relatively photo-deficient center. This area of radioactivity is present in the first 2-minute image and does not appear to increase in activity with time. The central area "fills in" with time. The distorted bladder configuration *(open arrow)* suggests an extrinsic mass. At the time of the renal transplantation, the uterus was noted to be enlarged with leiomyomata. ***Comment:*** The uterus may be a cause of an area of increased radiotracer, and this must be distinguished from early bladder activity (uterine activity may appear prior to renal excretion of radioactive urine) or urine extravasation (uterine activity does not appear to increase with time). The uterine activity may represent hyperemia secondary to pregnancy, cyclic hormonal changes (secretory or menstrual phases), or uterine neoplasia. The etiology of the relative photon deficiency is uncertain.

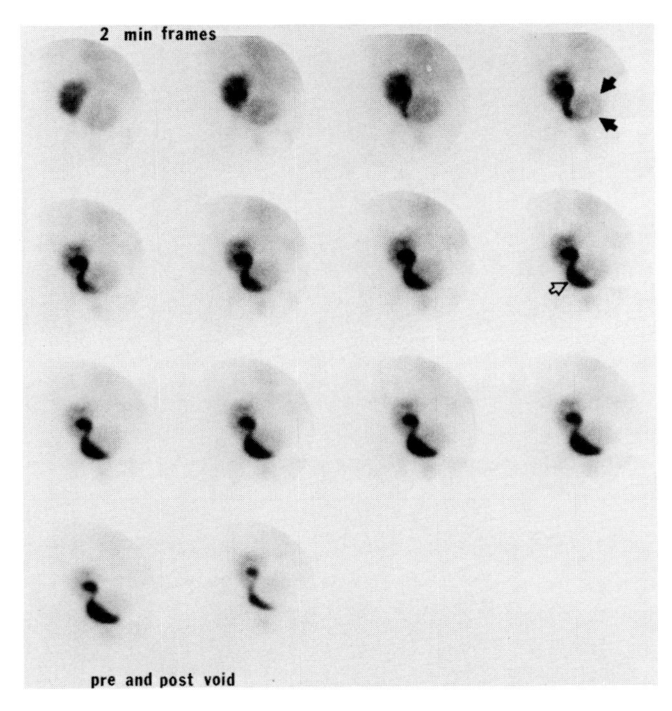

Figure 9-17

Reflux of Urine into Native Ureter

This Tc-99m DTPA study shows a linear collection of tracer above and to the left of the bladder *(closed arrow)*. This area increases in intensity through 24 minutes and then decreases with voiding *(open arrow)*. We believe this activity represents reflux of radioactive urine into the left native ureter because of the following: the patient was nine months post-transplantation; this finding was not present on studies performed one week before or one week after the exam; and the radioactivity had the above location, configuration, and changes with voiding.

Comment: Reflux into a native ureter should not be confused with a urinoma. Location, configuration, changes with voiding, and time from transplantation may help distinguish this entity from a urinoma.

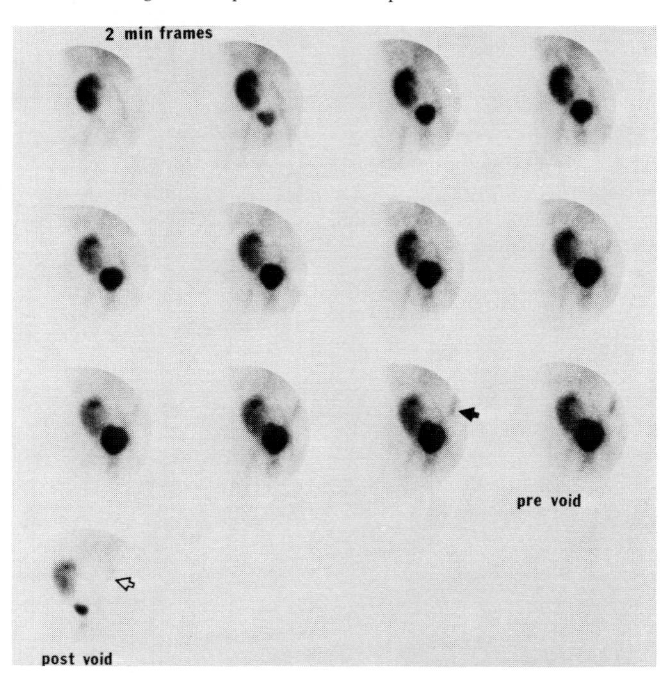

Figure 9-18

Lymphocele

This Tc-99m DTPA study was performed 3 weeks after renal transplantation. The 10-minute image shows the urinary bladder *(large arrow)* displaced to the left, suggesting a non-delineated pelvic mass. An ill-defined, curvilinear photon deficiency is present along the lateral margin of the renal cortex *(small arrow)*. These findings 3 weeks post-transplantation suggest lymphocele. A pelvic lymphocele which extended from the area of the bladder along the superior and lateral margin of the kidney was surgically drained. ***Comment:*** Pelvic lymphoceles may not present as clearly defined photon deficiencies; the only abnormality may be bladder distortion. (Dubovsky EV: Primer and atlas for renal transplant scintigraphy. Clin Nucl Med:10:Part I [52-62], Part II [118-133], 1985)

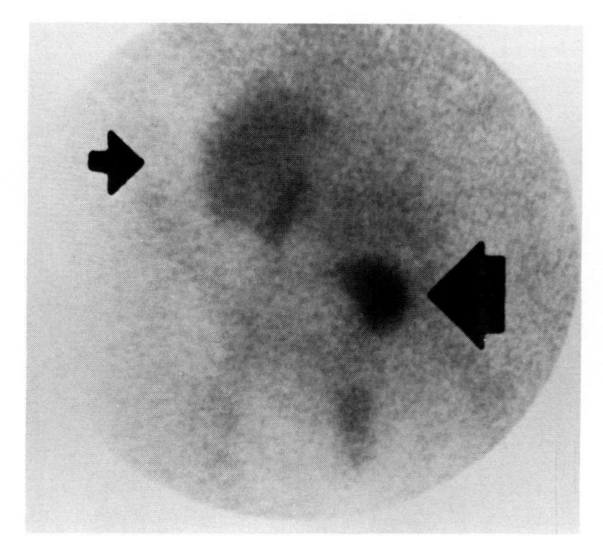

Figure 9-19 *"Warm" Lymphocele*

These are the 2-minute ***(A)***, 26-minute ***(B)***, and postvoid ***(C)*** images of a Tc-99m DTPA study performed 3 weeks after renal transplantation. In ***A***, a large, sharply marginated, round photon deficiency lies inferior to the kidney *(arrow)*. In ***B***, the "cold" area *(arrow)* becomes "hotter," which implies tracer accumulation; this observation is more certain when it is compared to the unchanging photon-deficient stomach in the left upper quadrant *(arrowhead)*. ***C*** also confirms the mild "filling in" of the "cold" pelvic defect *(arrow)*. In both ***B*** and ***C***, the bladder configuration suggests the presence of an extrinsic mass. A suggestion of a "rim" of increased radiotracer is noted around the defect. (This finding is better demonstrated in the case shown in Fig. 9-20). Sonography showed a well-defined hypoechoic area suggesting a urinoma or lymphocele. Needle aspiration and surgical drainage revealed a lymphocele with 550 ml of clear xanthochromic fluid.
Comment: Lymphoceles typically present as relatively photon-deficient areas on radionuclide studies. They occasionally accumulate radiotracer and usually occur several weeks after surgery. (Dubovsky EV: Primer and atlas for renal transplant scintigraphy. Clin Nucl Med:10:Part I [52–62], Part II [118–133], 1985)

A

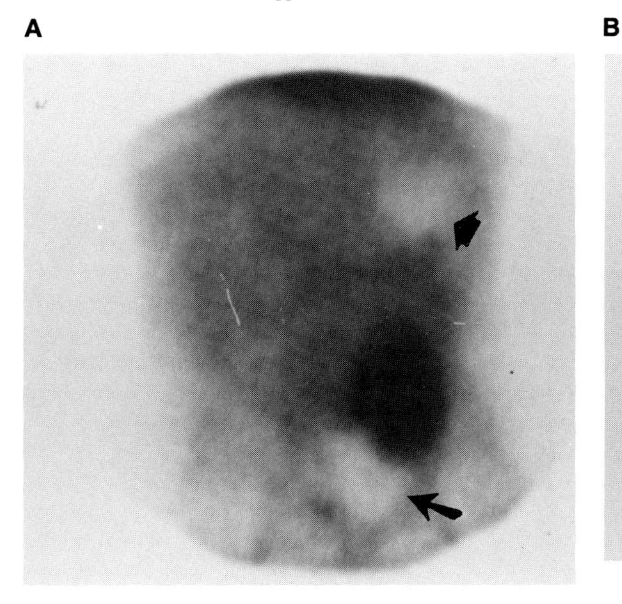

B

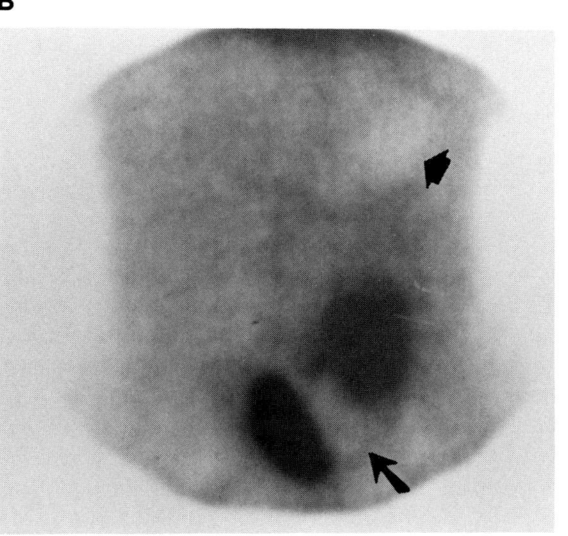

C

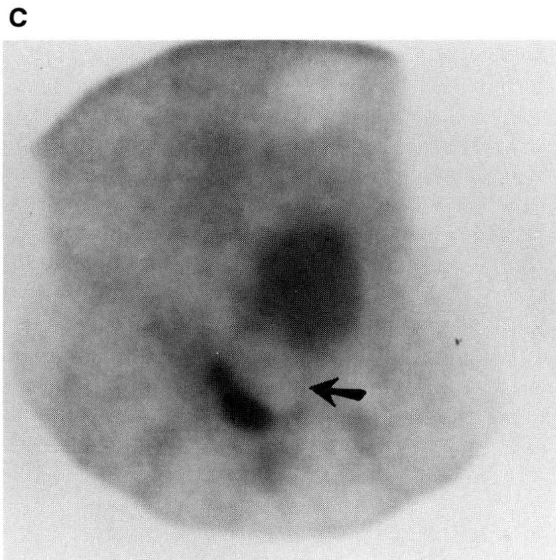

Figure 9-20 ***Rimmed Lymphocele***

This image was obtained approximately 2 weeks after renal transplantation and 16 minutes after injection of Tc-99m DTPA. In the pelvis, a rim of increased tracer activity is noted slightly superior to and to the left of the bladder *(arrow)*. The center of the area has radioactivity approximately equal to or slightly less than background. The distorted bladder configuration suggests an extrinsic mass. Aspiration of this area confirmed the presence of a lymphocele. ***Comment:*** Lymphoceles can demonstrate a rim of increased radioactivity, which should not be confused with urine extravasation. The etiology is not known; however, the authors have observed that this rim of activity resolved after aspiration of the lymphocele. Possible etiologies could include hyperemia, compressed normal tissue, or Compton scatter. (Dubovsky EV: Primer and atlas for renal transplant scintigraphy. Clin Nucl Med:10:Part I [52 - 62], Part II [118 - 133], 1985)

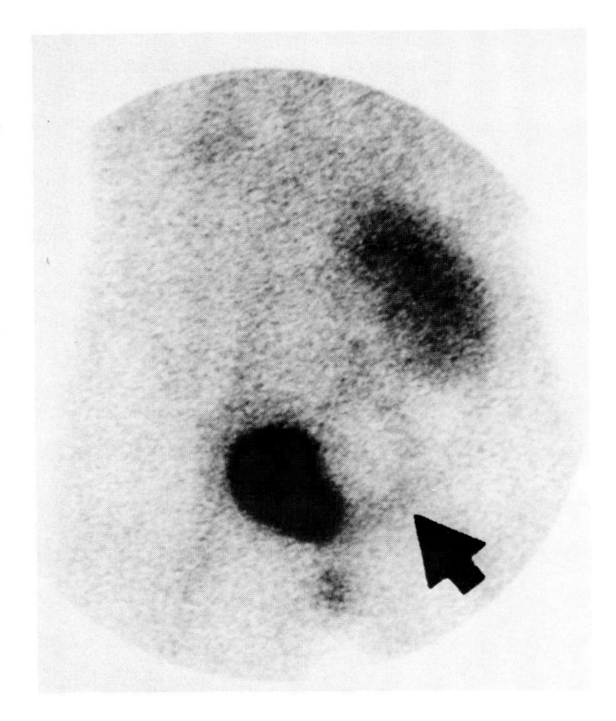

Figure 9-21 ***Perirenal Hematoma***

A Tc-99m DTPA study was performed one day after renal transplantation, and this image was obtained 14 minutes after radionuclide injection. A large photon deficiency *(arrows)* surrounds the kidney, extends into the pelvis, and deforms the bladder *(curved arrow)*. An 800-ml perirenal and pelvic hematoma with active hemorrhage was found at surgery. (Dubovsky EV: Primer and atlas for renal transplant scintigraphy. Clin Nucl Med:10:Part I [52-62], Part II [118-133], 1985)

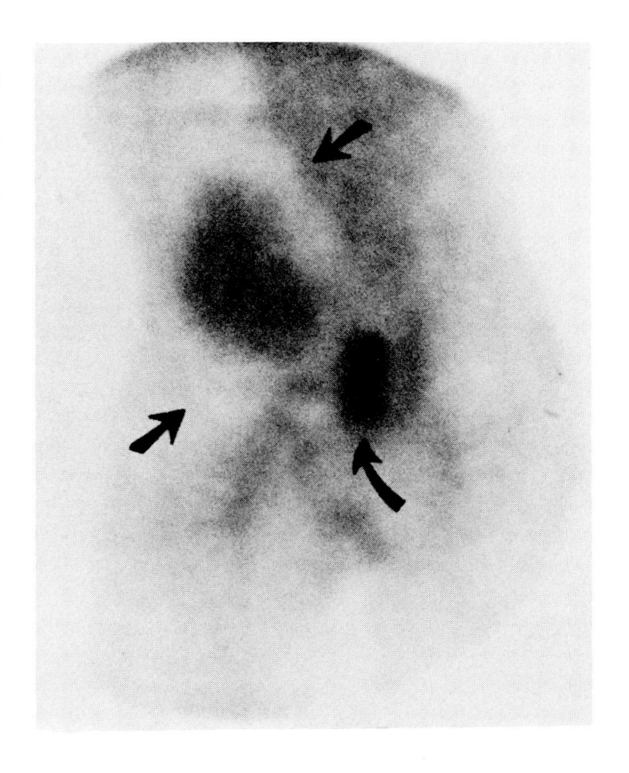

Figure 9-22 ***Renal Infarction***

This is a Tc-99m DTPA flow study 3 days after renal transplantation in a patient with oliguria. Renal flow images show a large area of decreased activity in the upper pole *(large arrow)* with a more subtle wedge-shaped area of decreased activity in the lower pole laterally *(small arrow)*. Angiography showed thrombosis of one upper pole renal artery and suggested infarction in the lower pole area. (Dubovsky EV: Primer and atlas for renal transplant scintigraphy. Clin Nucl Med:10:Part I [52-62], Part II [118-133], 1985)

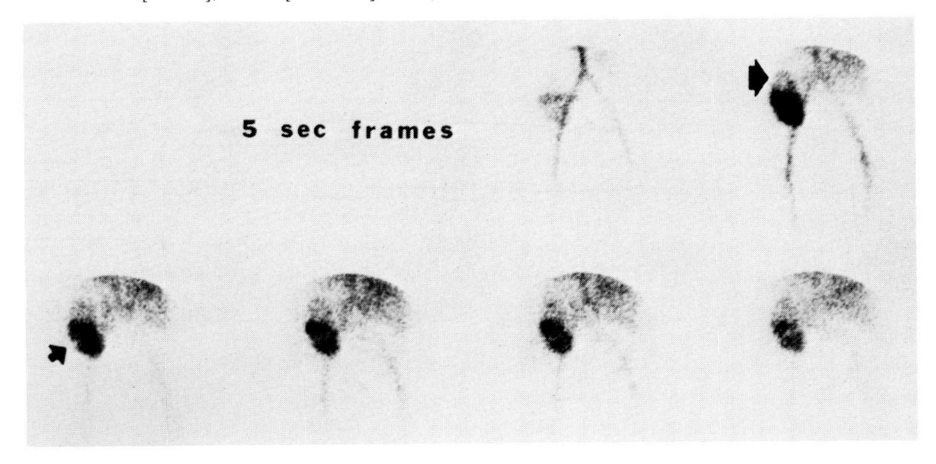

Figure 9-23 *Vascular "Blush" of Undetermined Etiology*

(A) This flow study was performed 11 days after renal transplantation. Irregular areas of increased radioactivity are seen in the abdomen *(arrow)* immediately after the appearance of radioactivity within the aorta. The activity in these areas and the aorta decreased rapidly. ***(B)*** In the DTPA sequential images, the radioactivity *(arrowhead)* is slightly greater than soft-tissue activity and washes out at a rate similar to other soft tissues. Renal scans performed prior to this exam showed a progressive increase in intensity of the vascular "blush" up to the present exam, and follow-up exams showed a continual decrease in intensity. The vascular blush was no longer present 6 months post-transplantation. At the time of the exam shown in this figure, the patient was being treated for suspected acute rejection. ***Comment:*** The etiology of this vascular "blush" is undetermined but is not infrequently seen. We hypothesize it may be increased mesenteric or omental blood flow. Its relationship to digestion, drugs, and rejection is not known; however, it does not appear to be related to surgery, since we have seen it years after surgery, and it can appear and resolve within 2 weeks. (Dubovsky EV: Primer and atlas for renal transplant scintigraphy. Clin Nucl Med:10:Part I [52–62], Part II [118–133], 1985)

A

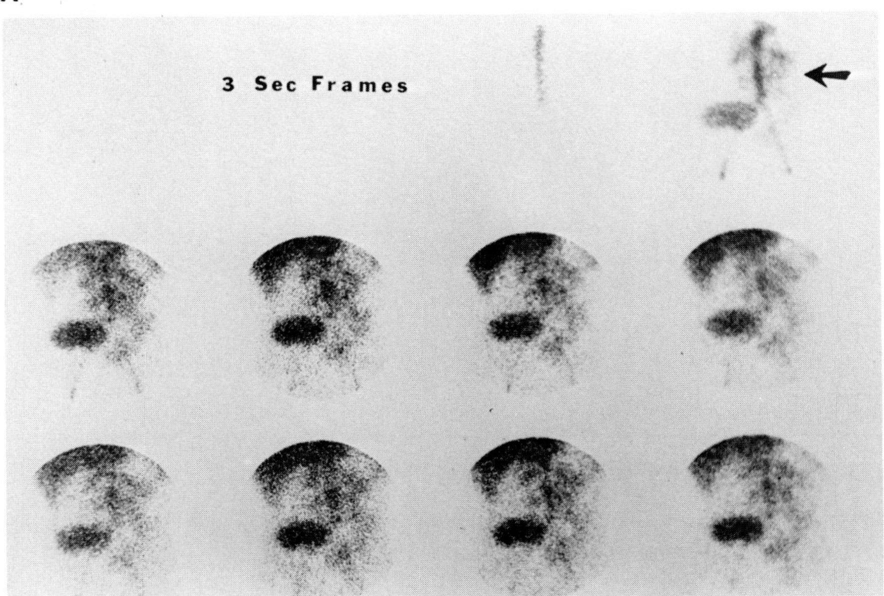

B

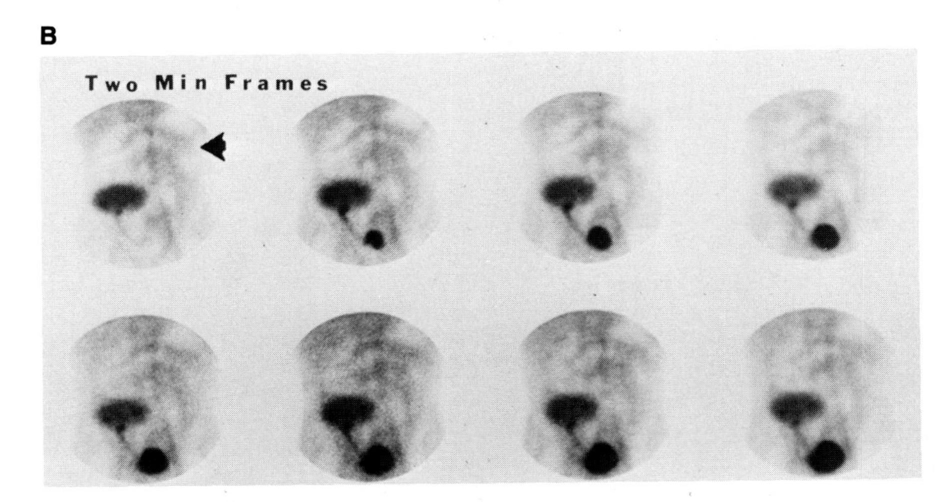

Robert Lisbona
Vilma Derbekyan
Javier A. Novales-Diaz

Chapter 10 Tc-99m RED BLOOD CELL VENOGRAPHY IN DEEP VENOUS THROMBOSIS OF THE LOWER LIMB

Deep vein thrombosis (DVT) of the leg is a common disease that can be clinically elusive. When the condition is suspected, objective testing is frequently essential to complement the impressions of the physical examination. Among the widely diverse testing methods available, technetium-99m red blood cell (RBC) venography ranks high as being both simple and sensitive. The advantages and limitations of Tc-99m RBC venography in the clinical setting are reviewed in detail in this chapter.

TECHNIQUE

Scrupulous technique is essential in the performance of Tc-99m RBC venography, since poor labeling, faulty positioning, and inferior imaging methods can result in nondiagnostic examinations or even misleading tests.

Optimal labeling of the blood pool is a prerequisite for a diagnostic venographic study. Labeling of the red blood cells with efficiencies in excess of 90% may be achieved through a combination of *in vivo* and *in vitro* tagging methods. In our department, the patient is pretreated with 3.4 mg of stannous chloride (cold stannous pyrophosphate) administered intravenously in an antecubital vein. Twenty minutes after injection, 10 ml to 20 ml of whole blood is withdrawn from the patient in a heparinized syringe that already contains 20 mCi of Tc-99m pertechnetate. The mixture of whole blood and technetium is then incubated within the syringe at room temperature for 10 minutes, and is periodically swirled. The radioactive blood is then reinjected as a bolus through an antecubital vein. In individuals with difficult veins, less ideal but sometimes adequate tagging can also be obtained by using a full *in vivo* labeling technique.

Scanning begins only 5 minutes following the injection of radionuclide to allow equilibrium of the tagged red blood cells. Meticulous imaging is necessary to optimize the interpretation of the study. Two million counts are accumulated for each view through the low-energy, general-purpose collimator of a large-field-of-view Anger camera. Anterior images of the lower abdomen and thighs are obtained with the patient in the supine position. Posterior views of the thighs, knees, and calf regions, with the patient in the prone position, are also registered to maximize the visualization of the lower superficial femoral, popliteal, and deep calf veins. An examination is complete only if all views are obtained, because each helps to assess a different portion or aspect of venous morphology. Careful and symmetrical positioning of the lower extremities is mandatory, because even slight degrees of rotation of one leg can interfere with the side-by-side comparison of both limbs.

When possible, prior to imaging, the patient should be stripped of leg bandages or stockings to avoid compression of the low pressure calf veins. We also discourage imaging of the calf veins with the patient supine and the camera below the stretcher, because the weight of the limb against the hard surface of the scanning table may distort the major veins. If the patient can only lie supine, then the calves should be raised at the ankles for the posterior views, just enough to clear the bed while the camera is positioned beneath the patient.

Sustained contractions of the calf muscles can alter the appearance of the veins and cause false-positive studies. Unilateral tensing of the muscles can be particularly troublesome because it results in an asymmetrical appearance of the calf veins. Hyperextension of the knee can also cause pinching of the popliteal vein. The patient should therefore be instructed by the technologist to keep a relaxed muscle tone.

PHYSIOLOGIC BASIS OF Tc-99m RBC VENOGRAPHY

Although the labeled cells are distributed throughout the vascular system, nuclear imaging of the lower limb with Tc-99m RBCs will provide selective visualization of the deep veins of the leg because of the preferential distribution of blood in these large capacity channels. The arterial vessels, with a cross-sectional area only one-third to one-half that of accompanying veins and a commensurate blood mass, are eclipsed by the individual deep venous channels. The superficial venous system, which normally has little reservoir function, is also generally overshadowed by these same deep veins. Normally there is intense and crisp definition of the deep veins of the legs on the nuclear venogram. In thrombophlebitis, the nascent and growing clot, primarily composed of fibrin and red cells interspersed with platelets and white elements, propagates along major venous pathways and tributaries. As it grows, it occludes the lumen and reduces the blood content of the affected veins. Thrombosis may then become manifest by the absence of a vein or the poor definition of the venous shadow in comparison to the contralateral companion vein. Also, congestion from the vascular obstruction and local inflammation and edema likely contributes to distortion of the image outline of thrombosed veins. The deranged venous circulation of the limb can also be appreciated on occasion by visualizing collateral channels, which serve as alternative pathways of drainage in extensive DVT. The saphenous veins may also be accentuated because they carry increased venous return.

Small nonobstructive clots or floating thrombi will not significantly depress the blood content of veins and, as such, are unlikely to be detected on blood pool venography because of their lack of physiological disturbance.

ESTIMATED RADIATION ABSORBED DOSE

The nature of the Tc-99m emissions and the short 6-hour half-life of the radionuclide result in a tolerable radiation burden. The total body dose from the procedure ranges from 0.018 to 0.019 rads/mCi of tracer injected, whereas the red bone marrow load averages 0.222 rads/mCi. The gonadal dose to the ovaries and testes is 0.022 and 0.012 rads/mCi, respectively.

VISUAL DESCRIPTION AND INTERPRETATION

Normal Anatomy

The venous system of the leg is composed of four groups of veins. One major grouping is the deep system, consisting of the plantar veins of the foot and three paired veins below the knee: the anterior and posterior tibials and the peroneal veins. These three paired veins merge to form the popliteal vein on the dorsal side of the knee. This latter channel continues in the thigh as the superficial femoral vein (or, simply, the femoral) into which the profunda femoris and greater saphenous veins drain to ultimately form the external iliac vein. In the pelvis, external and internal iliac veins fuse to form the common iliac that joins the contralateral companion vein to give rise to the inferior vena cava (IVC).

A second group of veins consists of the superficial channels in the subcutaneous tissue of the leg. The lesser saphenous vein is located posteriorly in the calf and empties into the popliteal vein. The greater saphenous system extends the entire length of the medial aspect of the lower extremity and drains into the femoral channels.

Communicating veins, or *perforators,* that connect the superficial and the deep venous system, form another group of leg veins. They are short vessels with valves that normally permit flow only from the superficial to the deep veins. Perforators are most numerous in the foot and the leg and less evident in the thigh.

The last major group of veins of the leg are the muscular veins. They are represented in the calf by the gastrocnemius and soleus veins. The gastrocnemius veins usually empty into the popliteal vein, while the soleal veins drain into the posterior tibial and peroneal veins. In the thigh, the profunda femoris is the major route of drainage of the thigh musculature and is a tributary of the superficial femoral vein.

Normal Tc-99m RBC Venogram

In normal and well-positioned patients Tc-99m venography results in a symmetrical and equally intense delineation of the proximal deep venous system of the lower limbs with excellent visualization of the popliteal, superficial femoral, external iliac, and common iliac veins (Fig. 10-1). The internal iliac veins and other deep pelvic veins do not normally visualize. The profunda femoris is not always apparent.

In the calf, the peroneal and posterior tibial veins of the deep system are consistently demonstrated, although individual members of each pair cannot be resolved. The anterior tibials are generally not identified.

The sural veins draining the medial aspect of the gastrocnemius muscle are fequently seen in normal studies. They are generally sharply defined, but in some patients, particularly the elderly, debilitated, or bedridden, these veins may be slender, irregular, and only faintly seen. The veins may also be emptied by spasms or contractions of the gastrocnemius muscle. Visualization of the lateral surals is a variable and less common phenomenon. The soleal muscular venous plexus is not necessarily seen in all normal subjects. Prominent perforators can be resolved and are recognized by their horizontal course, which might cause the longitudinal veins to appear whiskered. The greater saphenous vein may image faintly in a normal individual, while the lesser saphenous vein is generally not evident.

Normal Artifacts and Variants

Imaging artifacts are usually those of compression of the venous system (Fig. 10-2) or those induced by muscular contractions (Figs. 10-3 and 10-4). The interpreter must also be familar with normal variants. For instance, the normal common iliac veins may be symmetrically attenuated in the obese patient or in patients with ascites (Fig. 10-5); their course may also be meandering at times. If sufficiently tortuous, the iliac veins, although patent, may be dissimilar (Fig. 10-6). This asymmetry, in the absence of other abnormalities on the Tc-99m RBC venogram, is generally an innocent finding because iliac thrombosis is almost always associated with DVT in the femoral region. A pedal injection will verify the nature of this normal variant if venous disease is still suspected on clinical grounds. A notch may also be noted at the most proximal aspect of the left common iliac vein and is probably related to the bifurcation of the aorta (Fig. 10-7). More rarely, it can be seen on the right side.

In the thigh, splitting of the superficial femoral veins can be recognized (Fig. 10-8). It is often bilateral and is sometimes most evident on the posterior view. An occult duplication of the femoral vein may degrade its appearance on the anterior scan. Close scrutiny of the posterior views of the thigh usually discloses the duplication. Otherwise, the split can be documented by pedal intravenous injection of the radiotracer.

If the femoral veins are at different depths in the thigh, their appearance may be asymmetrical on the anterior image because of the greater attenuation of the deeper vein (Fig. 10-9). On the posterior views of the thigh, however, the disparity reverses and the attenuated vein of the anterior image should now be more intense than its contralateral companion.

In the calf, the posterior tibial veins may be congenitally absent (Fig. 10-10). At other times, the distal one-third of either the peroneal or posterior tibial veins can be quite thin relative to the more proximal part of the channel. This finding represents a slender beginning to the normal vein.

Abnormal Tc-99m RBC Venogram

Thrombophlebitis will alter the appearance of the normal Tc-99m RBC venogram and will be identified by some characteristic features. The most definitive portrayal for DVT on Tc-99m RBC venography is the nonvisualization of a vein that usually can be detected on the nuclear images (Figs. 10-11 and 10-12). This finding reflects the presence of total or near-total occlusion of the vessel by extensive clotting. At times, this nonvisualization is of a segmental nature only (Fig. 10-13). A significant increase in background activity may be noted also when major veins are thrombosed and not delineated (Fig. 10-14). This activity is generally a manifestation of an extensive network of smaller, superficial collateral vessels.

A second reliable sign of thrombosis is asymmetry, as manifested by attenuated activity and/or poor definition of a vein or portion thereof, as compared to its equivalent in the opposite limb (Figs. 10-15–10-17). Suspicions for bilateral thrombosis should be raised when companion vessels in both limbs are ill-defined and at variance from the normal venogram (Figs. 10-18 and 10-19). Pedal injections of radiotracer can be used for further assessment of these vessels. Soft tissue swelling, collateral vessels, and intense visualization of the greater saphenous vein are frequent ancillary signs in patients with DVT. They are insufficient in themselves, however, to establish the diagnosis, as they may be features of other pathologies, such as the postphlebitic syndrome or valvular incompetence.

DISCUSSION

Deep venous thrombosis of the leg is a common condition with suspected yearly incidence of at least 2.5 million cases in the United States.[1] The clinical diagnosis of the disease is notoriously unreliable, since many of its signs and symptoms are nonspecific or muted. Objective testing will exclude DVT in 45% to 65% of patients.[2-4] Up to 50% of patients with DVT have no clinical manifestations of the condition.[3] In the face of these diagnostic difficulties, it is almost always necessary to supplement the clinical impressions with more definitive approaches in order to reach a suitable therapeutic decision. A correct diagnosis is of particular significance, since there are risks to treatment and the complications of thrombolytic therapy must always be weighed against the hazards of pulmonary embolization from untreated lower limb DVT. In fact, 80% to 95% of pulmonary emboli, many of which are potentially fatal, originate from the leg veins.[1-5] Only a minority of emboli arise from the upper limb, the deep pelvic veins, the renal veins, or the right heart.

There are many advantages to blood pool venography in the clinical setting.[6-9] It is a simple, short, and immediate test, which is accessible to all nuclear medicine departments for the diagnosis of lower limb DVT. Tc-99m is readily available and most advantageous for imaging, and the labeling of the red blood cell is straightforward. The test is not hazardous, is easily tolerated, and may be frequently repeated.Tc-99m RBC venography permits the detection of DVT above and below the knee region with high levels of sensitivity and specificity. The sensitivity of the Tc-99m RBC venographic technique for documenting DVT in the leg has been reported from various centers to vary from 77% to 100% (an average of 89%), while specificity ranges from 71% to 94% (a mean of 86%).[6-13] Even when inconclusive, blood pool venography is extremely valuable as a road map for any subsequent contrast venographic procedure because it will highlight regions of concern.

Tc-99m RBC venography results in the global assessment of the deep venous system of the lower limbs. DVT in the leg is frequently multicentric and the static high-resolution radionuclide images are ideally suited to localizing the sites and extent of thrombosis because the entire venous system of both limbs is imaged simultaneously. The technique can also be used without undue patient discomfort to monitor the course of the disease and its response to thrombolytic therapy. A serial comparison of the status of individual veins is possible at all times. When treatment fails, propagation of the thrombosis can readily be documented (Figs. 10-20 and 10-21). With successful therapy, Tc-99m RBC venography will frequently show improvement of the DVT, often to the point of total resolution of the initial process (Figs. 10-22 – 10-25). On other occasions, the geographical distribution of the abnormalities will be unchanged as the disease becomes chronic despite therapy.

Because some patients will fail to resolve their DVT, we routinely obtain baseline reference studies after the treatment of acute DVT. These reference scans ease future difficulties in differentiating an acute from a chronic event, because recurrent phlebitis will show a deterioration of the image from the baseline post-therapy examination (Fig. 10-26). As a general rule, however, blood pool venography, on a single examination, will not distinguish acute from chronic thrombosis. Obliterated venous columns, ill-defined pathways, and a collateral network may be features of both acute or chronic DVT on the venographic study (Fig. 10-27). Nonetheless, an abnormal Tc-99m RBC venogram in an acutely symptomatic patient with no previous complaints referable to the leg is strongly suggestive of acute DVT. If, however, the patient manifests no signs or symptoms of the disease, then remote DVT or phlebothrombosis is a possibility.

It is important to bear in mind that clots that only slightly narrow a venous lumen will generally be overlooked on blood pool venography (Fig. 10-28). Also, thrombosis of the muscular veins is usually problematic on the Tc-99m RBC venogram; the study is generally blind to events in the deep femoral vein, for example, and therefore the clinical presentation should guide the diagnostic and therapeutic management of the patient (Fig. 10-29). Thrombosis of the sural veins, which is hard to document by most techniques, proves difficult to diagnose with Tc-99m RBC venography owing to the variable appearance of the vein on the nuclear venogram. A flagrant absence of the vein should suggest the diagnosis in the appropriately symptomatic patient (Fig. 10-30). It is wise, however, to try and confirm this diagnosis with contrast venography. Thrombosis of the soleal plexus will also be missed on blood pool venography. This condition, however, is of lesser clinical importance, as it is rarely associated with serious complications unless it spills over into the longitudinal veins of the calf.

A raging cellulitis of the calf region can be confounding (Fig. 10-31). Soft tissue swelling from the inflammatory process may, of itself, compress the veins and disturb their normal appearance. A complicating DVT is to be suspected only if the veins are also irregular and/or interrupted, in addition to being attenuated. Nonvisualization of a vein also indicates DVT associated with cellulitis. Similar caution in interpretation is recommended in the presence of marked edema of the limbs.

Perivascular lesions may also compress veins, obliterating their lumens and causing an abnormal blood pool venogram (Figs. 10-32 – 10-35). Such extrinsic compression is most frequently seen in the pelvis where aneurysm, abscess, or tumor may obstruct and mask the venous channels. It can also be noted in the popliteal fossa as a result of a Baker cyst or local hematomas. Scintigraphically, the appearance on Tc-99m RBC venography may be indistinguishable from DVT. It has been our repeated experience, however, that isolated abnormalities of the pelvic or popliteal veins rarely represent DVT and are most often due to extrinsic phenomena. Extrinsic defects can also be noted at other sites of the deep venous system and again manifest as a local abnormality with an otherwise normal blood pool venogram (see Fig. 10-35).

As discussed earlier, covert duplication of the deep venous system can also be disturbing. A subtle duplication of the femoral vein, particularly when unilateral, may lead to asymmetries and a difficult interpretation on Tc-99m RBC venography. The posterior views of the thigh, however, generally give enough detail to clarify the bifid nature of the vein. In addition, the split may be readily demonstrated by the pedal injection of radiotracer (Figs. 10-36 and 10-37).

Thrombophlebitis of the leg and pulmonary embolization are intimately related. Since many pulmonary emboli can silently complicate DVT, a lung scan would be indicated in the presence of a positive blood pool venogram even when chest symptoms are absent. The documentation of pulmonary embolism in patients with DVT of the leg is relevant because this complication of thrombophlebitis may necessitate longer term anticoagulant therapy.

Tc-99m RBC venography is also a useful and noninvasive test for the comprehensive assessment of patients with pulmonary embolization. In 74% of patients with symptomatic pulmonary embolization, the nuclear venogram will reveal the presence of DVT and a source of embolization.[14] In many, the leg thrombosis will not be suspected clinically (Fig. 10-38). The potential for lung embolism in leg DVT is somewhat dictated by the level of DVT: the more proximal the process the greater the incidence of embolization. Yet, interestingly, up to 35% of patients with pulmonary embolism will show evidence of DVT confined to the calf.

REFERENCES

1. Sherry S: The problem of thromboembolic disease. Semin Nucl Med 7:205, 1977

2. Nicolaides AN, Kakkar VV, Field ES et al: The orgin of deep vein thrombosis: A venographic study. Brit J Radiol 44:653, 1971

3. De Nardo GL, De Nardo SL: Diagnosis of thrombophlebitis. Medical monograph. Arlington Heights, IL, Amersham Corporation, 1978

4. Hull R, Hirsch J, Sackett DL et al: Cost effectiveness of clinical diagnosis, venography and noninvasive testing in patients with symptomatic deep vein thrombosis. N Engl J Med 304:1561, 1981

5. Moser KM, Fedullo PF: Imaging of venous thromboembolic with labeled platelets. Semin Nucl Med 14:188, 1984

6. Lisbona R: Radionuclide blood pool imaging in the diagnosis of deep vein thrombosis of the leg. In Freeman LM, Weissman H (eds): Nuclear Medicine Annual. New York, Raven Press, 1986.

7. Lisbona R, Derbekyan V, Novales-Diaz J, Rush CL: 99mTc-Red blood cell venography in deep venous thrombosis of the lower limb: An overview. Clin Nucl Med 10:208, 1985

8. Lisbona R, Stern J, Derbekyan V: 99mTc-Red blood cell venography in deep vein thrombosis of the leg: A correlation with contrast venography. Radiology 143:771, 1982

9. Lisbona R, Leger J, Stern J et al: Observations on 99mTc-erythrocyte venography in normal subjects and in patients with deep vein thrombosis. Clin Nucl Med 6:385, 1981

10. Beswick W, Chmiel R, Booth R et al: Detection of deep venous thrombosis by scanning of 99mTechnetium labeled red cell venous pool. Br Med J 1:82, 1979

11. Kempi V, Van der Linden W: Diagnosis of deep vein thrombosis with in-vitro Tc-99m-labeled red blood cells. Eur J Med 6:5, 1981

12. Kempi V: Radionuclide diagnosis of deep vein thrombosis, pp 41–61. Monograph. Göteborg, 1985

13. Fogh J, Levin Nielsen SV, Vitting K, et al: The diagnostic value of angioscintigraphy with 99mTc-labeled red blood cells for detection of deep vein thrombosis. Nucl Med Commun 3:172, 1982

14. Lisbona R, Rush C, Lepanto L: Technetium-99m red blood cell venography of the lower limb in symptomatic pulmonary embolization. Clin Nucl Med 12:93, 1987

Atlas Section

Figure 10-1

Normal Tc-99m RBC Venogram

There is a crisp and symmetrical visualization of the deep venous system of both lower limbs at the expense of the arterial and superficial venous compartments. Anterior views *(upper left, center)* are obtained over the pelvis and thighs. A posterior view *(upper right)* of the thigh region is registered to assess the distal femoral veins. Posterior views *(lower)* over the knee and calf areas result in optimal definition of popliteal and calf vessels. (*IVC,* inferior vena cava; *A,* aorta; *I,* iliac vein; *F,* superficial femoral vein; *P,* popliteal vein; *Pe,* paired peroneal veins; *T,* paired posterior tibial veins; *S,* sural vein, *So,* soleal vein) ***Comment:*** The deep veins of the leg generally eclipse their companion arteries, which in cross section are only one third to one half the size of the veins and contain proportionally less radioactive blood. Similarly, the blood volume content of the superficial venous compartment is also modest, so that it also is usually overshadowed by the deep venous system. (Lisbona R, Derbekyan V, Novales-Diaz JA, Rush CL: Tc-99m red blood cell venography in deep venous thrombosis of the lower limb—An overview. Clin Nucl Med 10(3):208, 1985)

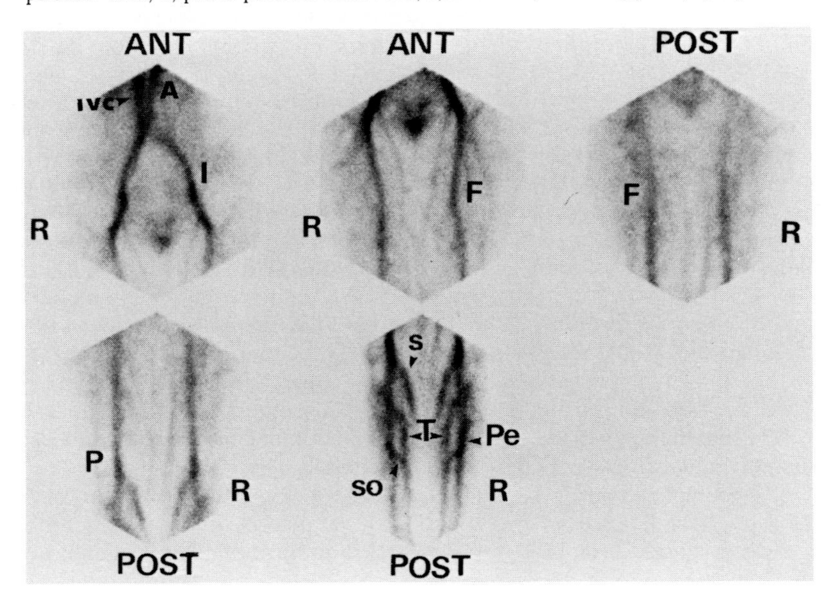

Figure 10-2 ***Compression Artifact***

Although the calf veins are normal when the patient is imaged prone *(left),* an artifact is induced when the posterior views are obtained with the patient supine and the legs resting over a pillow *(right).* ***Comment:*** The deep veins are lacking in smooth muscle and form a high-capacity, but low-pressure, system which is extremely susceptible to external forces. (Lisbona R, Derbekyan V, Novales-Diaz JA, Rush CL: Tc-99m red blood cell venography in deep venous thrombosis of the lower limb—An overview. Clin Nucl Med 10(3):208, 1985)

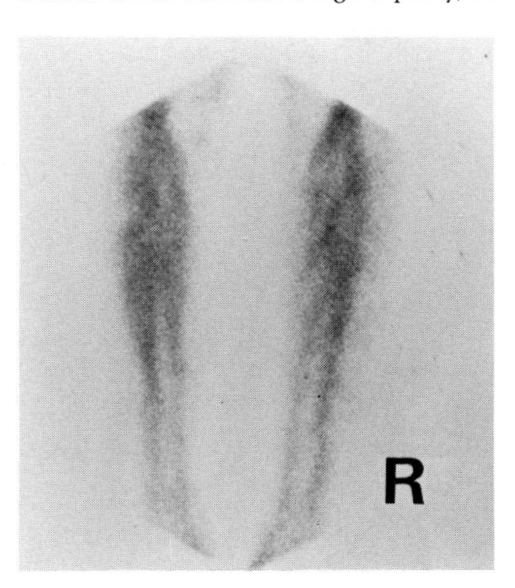

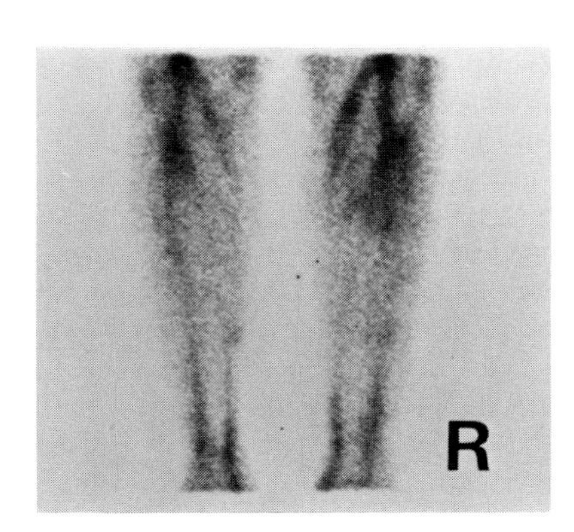

Figure 10-3 ***Contraction Artifact***

The effect of muscular contractions on the appearance of the Tc-99m RBC venogram can be appreciated in these views of a normal individual. The normal veins *(left)* partially empty and blur during moderate contractions *(center)* and are fully drained of blood with sustained contractions *(right).* ***Comment:*** The calf muscles act as a pump to promote venous return from the deep veins of the lower limbs. Circulatory dynamics in the leg are such that contraction of the calf musculature, which can generate pressures of 100 to 200 mm Hg, compresses the deep veins to propel blood centrally toward the heart. Competent valves in the perforators prevent any reflux into the superficial channels. When the muscles relax, the pressures within the deep compartment drop below that of the superficial one, so that blood is directed across the perforating veins and into the deep channels. Gravitational forces, intrathoracic pressures, and the suction effect of the right ventricle at end diastole also influence the dynamics of flow in the leg veins. (Lisbona R: Radionuclide blood pool imaging in the diagnosis of deep vein thrombosis of the leg. In Freeman LM, Weissman H (eds): Nuclear Medicine Annual. New York, Raven Press, 1986)

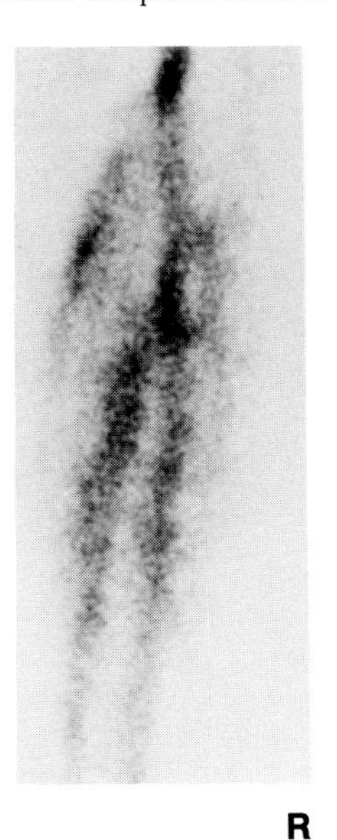

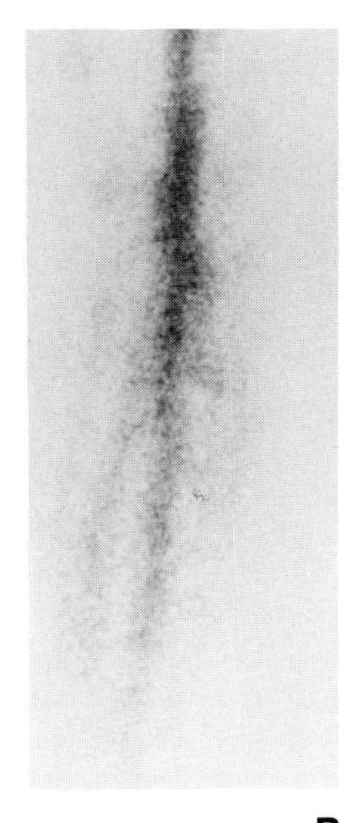

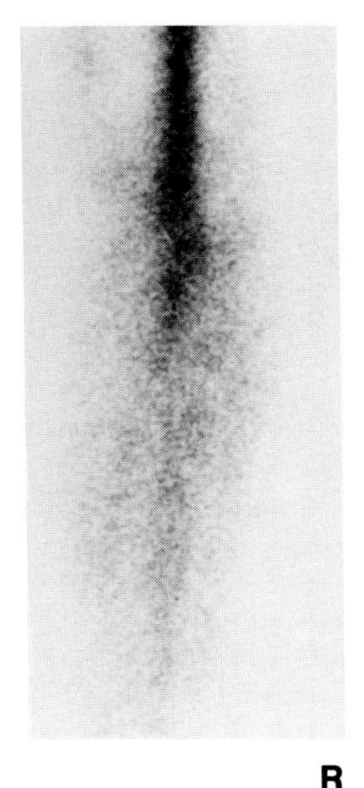

Figure 10-4 *Contraction Artifact*

In this normal individual, the calf veins have different appearances. Due to unilateral tensing of the left calf musculature, there is emptying of the sural veins *(arrowhead)* and blurring of the tibioperoneal trunk at the site of insertion of the soleus muscle *(open arrow).* Distally, the tense musculature attenuates the more peripheral aspects of the veins *(closed arrows),* in contrast to the relaxed right calf where all venous segments image normally. ***Comment:*** The appearance in the left calf is quite characteristic for tensed gastrocnemius and soleus muscles. It should not be misconstrued for DVT, because the abnormalities are confined to sites of muscular bundles. Tensing of the muscles can interfere with the side-by-side comparison of the limbs on the blood pool venogram. The technologist must ensure that the patient is relaxed at the time of imaging. (Lisbona R: Radionuclide blood pool imaging in the diagnosis of deep vein thrombosis of the leg. In Freeman LM, Weissman H (eds): Nuclear Medicine Annual. New York, Raven Press, 1986)

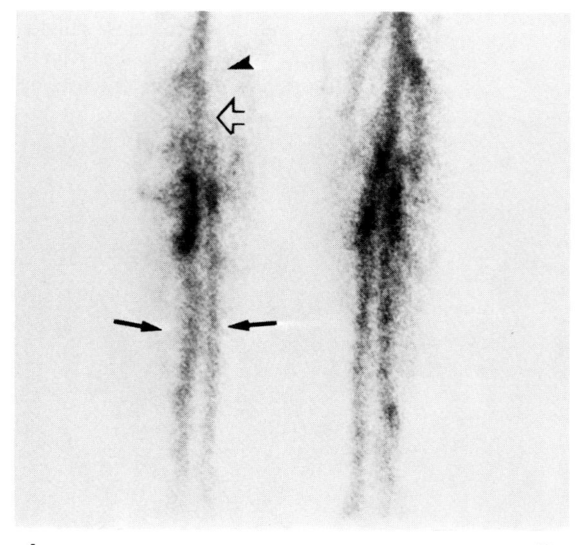

Figure 10-5 *Normal Variant*

The iliac veins are equally attenuated in this image of a normal but obese patient. ***Comment:*** This is a common variant in the pelvic area. With bilateral iliac DVT, one may expect collateral vessels and irregularity or interruption of the iliac channels alongside their faint appearance. (Lisbona R, Derbekyan V, Novales-Diaz JA, Rush CL: Tc-99m red blood cell venography in deep venous thrombosis of the lower limb—An overview. Clin Nucl Med 10(3):208, 1985)

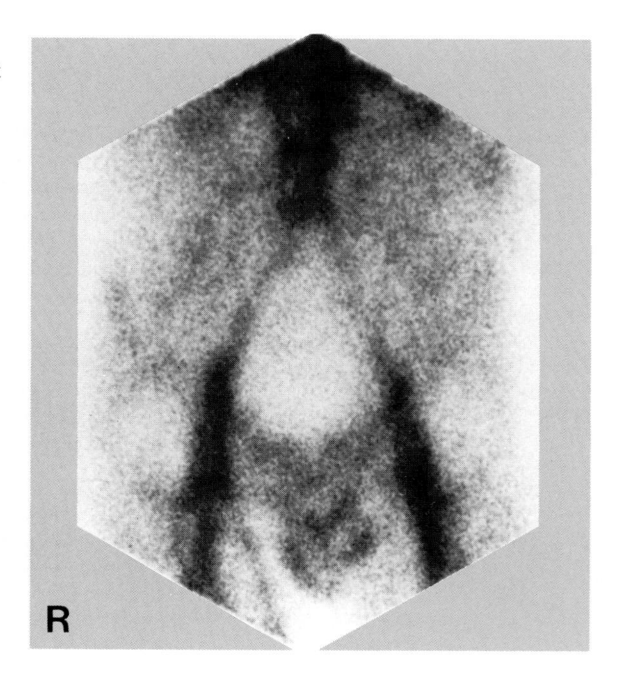

Figure 10-6

Normal Variant

The iliac veins are dissimilar, with dim visualization of the left channel on the Tc-99m RBC venogram *(left)*. A pedal injection of radiotracer confirms the normalcy of the left iliac vein *(right)*. ***Comment:*** The normal iliac veins may vary in intensity on the blood pool venogram because one may have a different course in the pelvis causing it to appear fainter. This is a common and benign variant if the asymmetry is confined to the level of the iliac veins. In the event of iliac vein thrombosis, involvement of the proximal femoral vein is almost always present, resulting in an asymmetrical appearance of the femoral vessels as well. (Lisbona R, Derbekyan V, Novales-Diaz JA, Rush CL: Tc-99m red blood cell venography in deep venous thrombosis of the lower limb—An overview. Clin Nucl Med 10(3):208, 1985)

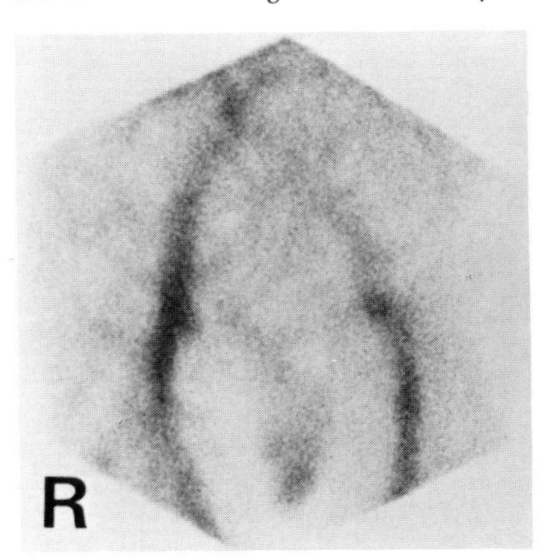

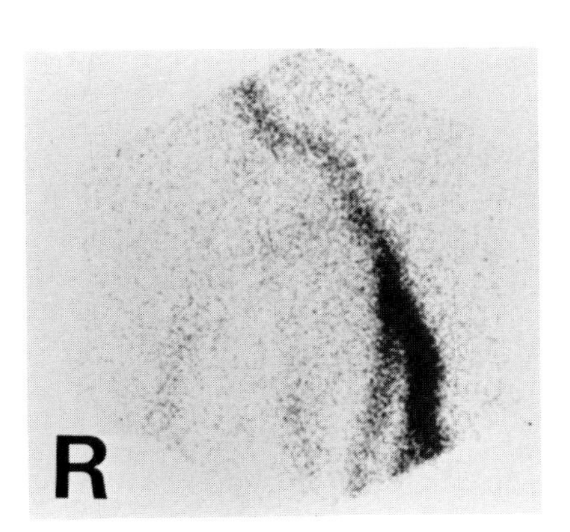

Figure 10-7

Normal Variant

A notch may be noted in the proximal aspect of the left iliac vein *(arrowhead)*. ***Comment:*** This minor asymmetry of the iliac veins is likely due to some extrinsic compression of the vein at the aortic bifurcation. Occasionally, the notch is right-sided.

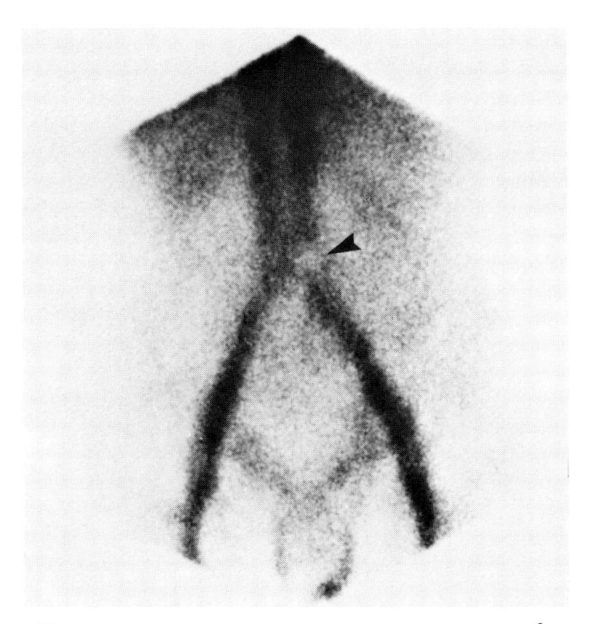

Figure 10-8 ***Normal Variant***

The femoral veins are split or duplicated *(arrowheads)* on both anterior *(left)* and posterior *(right)* images of the thigh. ***Comment:*** An obvious split of the femoral veins does not interfere with the interpretation of the study. A covert duplication, however, may be troublesome because the femoral vein may then have an unusual appearance on the anterior view. The posterior views of the thigh should always be closely scrutinized for the possibility of a subtle duplication of the vein. (See also Figs. 10-36 and 10-37.)

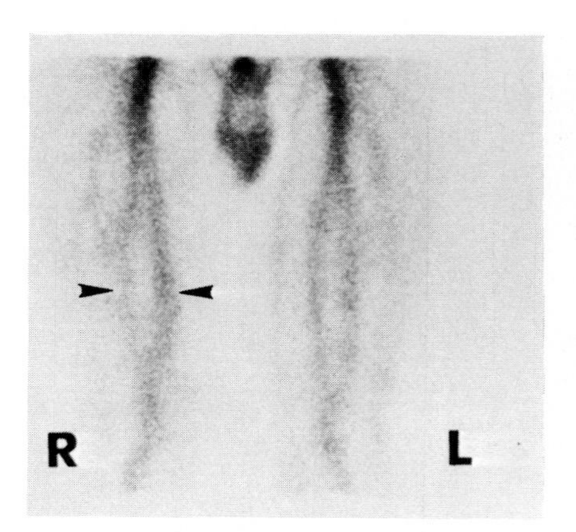

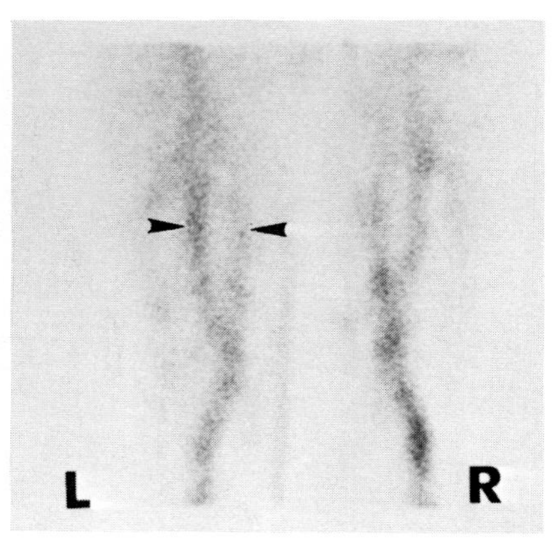

Figure 10-9 ***Normal Variant***

These images of an obese patient show unequal portrayal of the femoral veins on the anterior view *(top)* of the thighs with poorer definition of the left femoral channel *(arrowhead)*. On the posterior view *(bottom)* of the thigh, however, the left femoral vein *(arrowheads)*, which is normal, visualizes to better advantage than the right femoral vein. ***Comment:*** The normal femoral veins may appear different on the anterior images if right and left veins course at unequal depth in the thigh. A posterior view of the thigh is essential to take that eventuality into account, and the examination is not ready for interpretation unless the posterior image is registered. These asymmetries may also be more focal. (Lisbona R, Derbekyan V, Novales-Diaz JA, Rush CL: Tc-99m red blood cell venography in deep venous thrombosis of the lower limb—An overview. Clin Nucl Med 10(3):208, 1985)

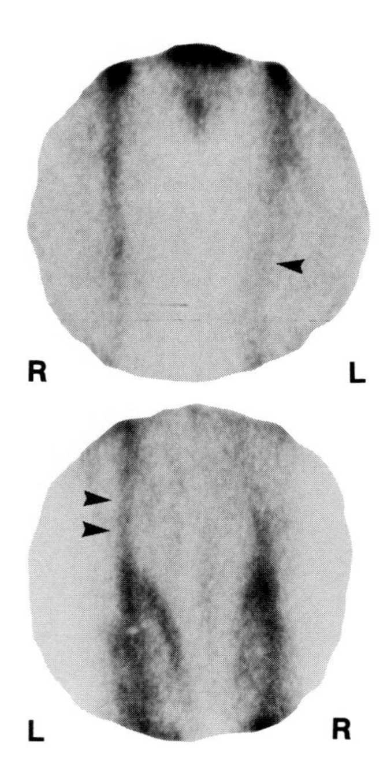

Figure 10-10 *Normal Variant*

Drainage of the calf occurs through a predominant, peroneal venous pathway. The posterior tibial veins are congenitally absent in this patient and consequently are not seen on the blood pool *(A)* or contrast venograms *(B)*. ***Comment:*** This variant can be recognized when the plantar arch veins drain into the peroneal channels. The oblique course of the veins *(arrowheads)* directly connecting the plantar arch and the peroneal veins confirms the diagnosis.

A

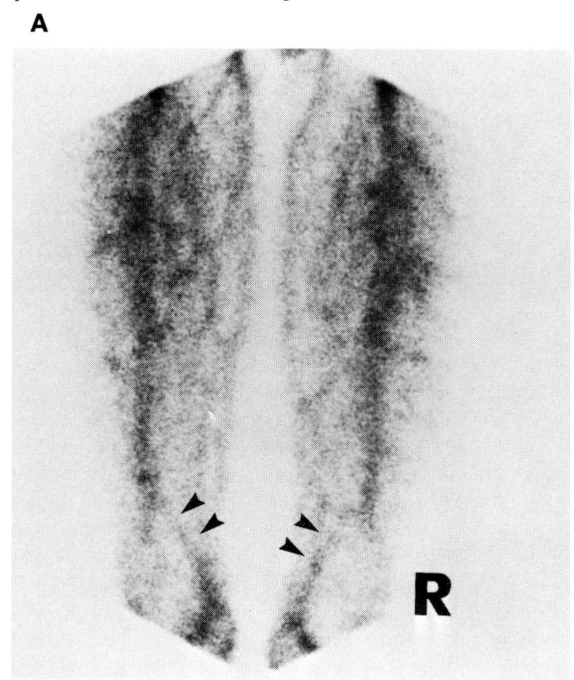

B

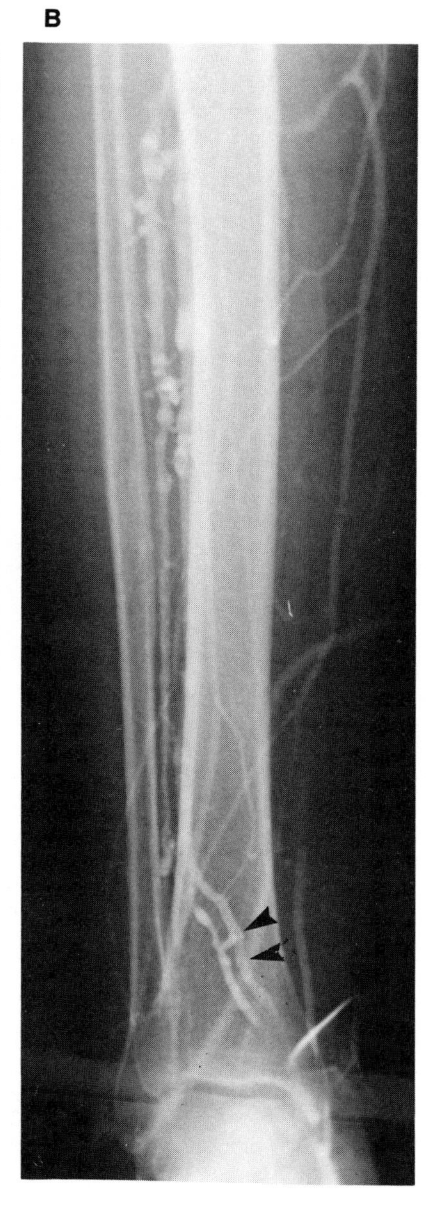

Figure 10-11 ***Occlusive Thrombosis in a 45-Year-Old Male with Extensive DVT***

The elongated clots noted on the anterior phlebogram *(A)* almost totally occlude the iliofemoral column, thus obliterating the outline of the left iliac vein and large portions of the left femoral vein on the Tc-99m RBC venogram *(B)*. Some femoral activity is noted on the nuclear study *(arrowhead)*, where the femoral lumen on the contrast study shows greater patency *(arrow)*. ***Comment:*** The final portrayal of a vein on Tc-99m RBC venography depends on its blood content and hence radioactivity. With occlusive thrombosis, the vein vanishes on the nuclear venogram as the clot displaces the blood. Lesser degrees of obstruction by a clot will manifest on the venogram as a detectable venous outline, albeit abnormal, due to the presence of residual blood and radioactivity around the thrombus. (See Figs. 10-12 – 10-14.) (Lisbona R: Radionuclide blood pool imaging in the diagnosis of deep vein thrombosis of the leg. In Freeman LM, Weissman H (eds): Nuclear Medicine Annual. New York, Raven Press, 1986)

A

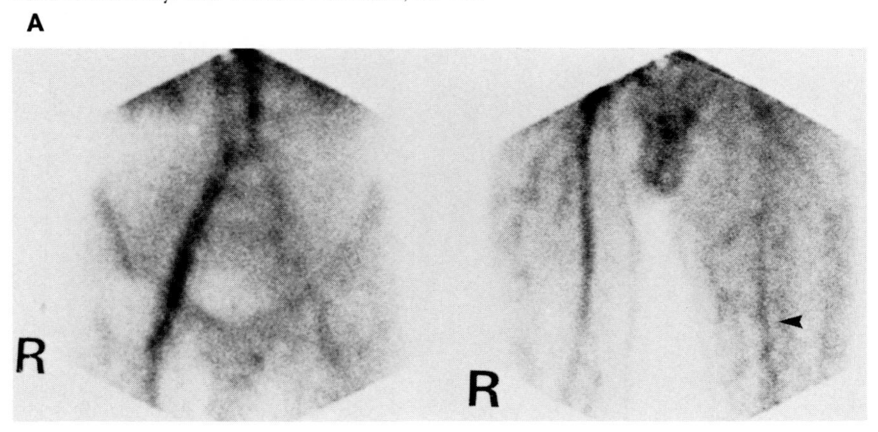

B

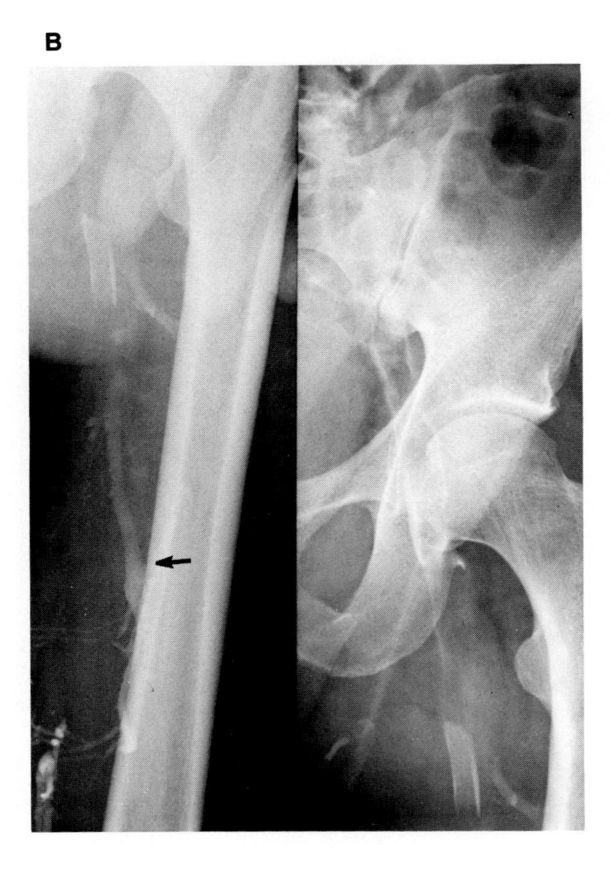

Figure 10-12 ***Occlusive Calf Thrombosis***

The left posterior tibial veins fail to visualize in their entirety, thus indicating left-sided calf DVT with occlusive thrombosis of those veins. (Lisbona R, Derbekyan V, Novales-Diaz JA, Rush CL: Tc-99m red blood cell venography in deep venous thrombosis of the lower limb—An overview. Clin Nucl Med 10(3):208, 1985)

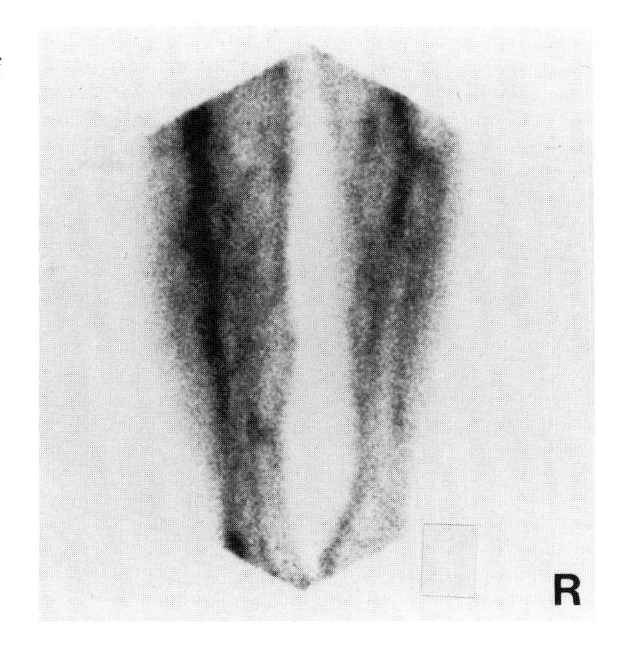

Figure 10-13 ***Occlusive Segmental DVT***

The left calf is swollen and collateral vessels *(closed arrows)* are present. The longitudinal veins are absent in their mid-portion, indicating regional occlusive DVT. Distally, the deep venous system is patent *(arrowheads),* as is the tibioperoneal trunk *(open arrow).*

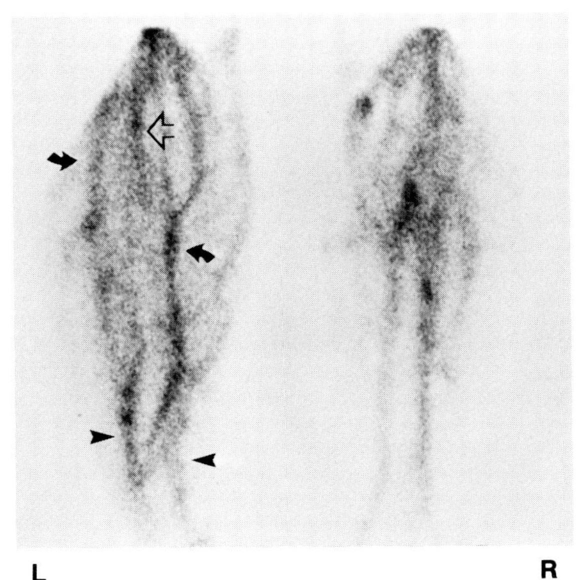

Figure 10-14 ***The "Dirty" Calf***

The longitudinal veins of the calf *(left)* are not visualized because of DVT. The intense level of background activity is due to a myriad of smaller, superficial collateral vessels, which can be appreciated on the contrast venogram *(right)*. ***Comment:*** A high degree of soft tissue activity usually implies severe thrombotic occlusion of the deep venous system. Frequently, only the superficial pathways are visualized on the contrast venogram. (Lisbona R, Derbekyan V, Novales-Diaz JA, Rush CL: Tc-99m red blood cell venography in deep venous thrombosis of the lower limb—An overview. Clin Nucl Med 10(3):208, 1985)

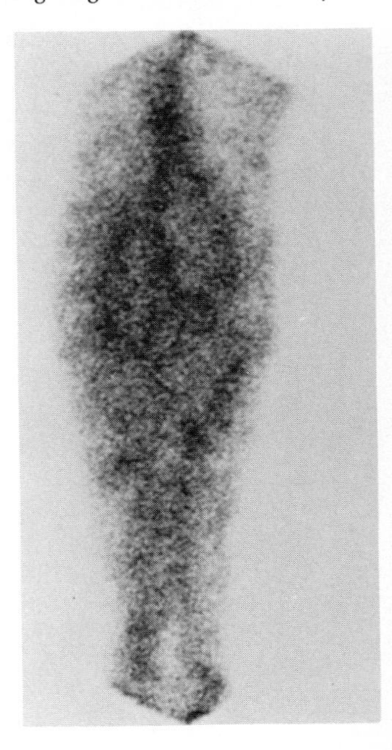

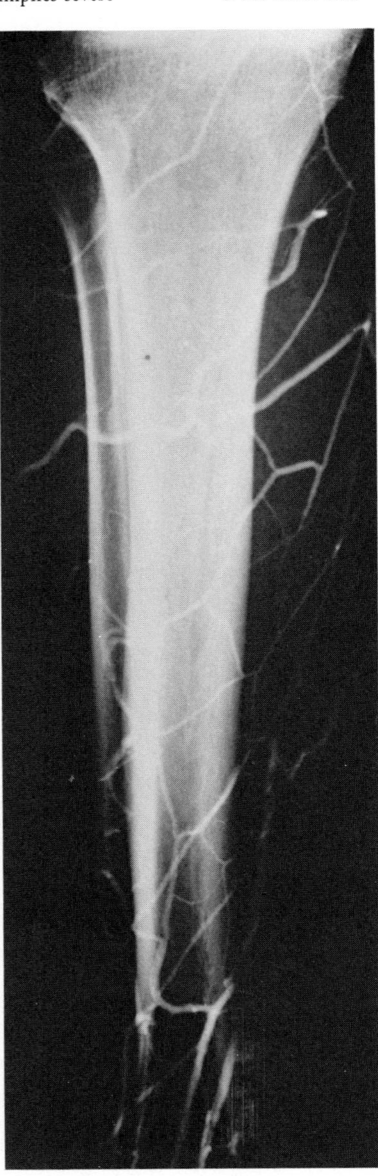

Figure 10-15 ***Femoral DVT, Attenuated Pattern***

The entire right femoral vessel shows diffuse attenuation and thinning in contrast to the left one, indicating thrombophlebitis. ***Comment:*** Diminished activity along a venous pathway is another sign of DVT. Persistent visibility of the venous outline, however, generally implies greater patency of the lumen than when the vein is totally absent on the nuclear venogram. (See also Figs. 10-16 and 10-17.)

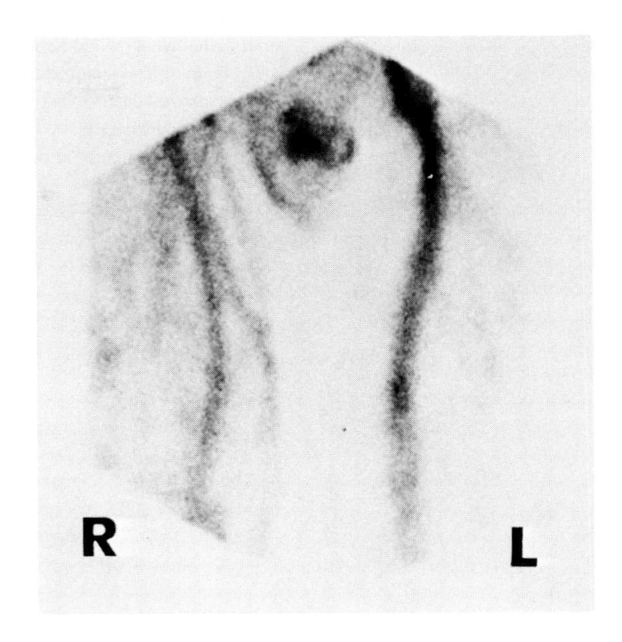

Figure 10-16 ***Calf DVT, Attenuated Pattern***

The longitudinal veins of the right calf lack sharpness and are irregular and of small caliber when compared to the normal left side. This is consistent with thrombosis along the right posterior tibial and peroneal pathways. (Lisbona R: Radionuclide blood pool imaging in the diagnosis of deep vein thrombosis of the leg. In Freeman LM, Weissman H (eds): Nuclear Medicine Annual. New York, Raven Press, 1986)

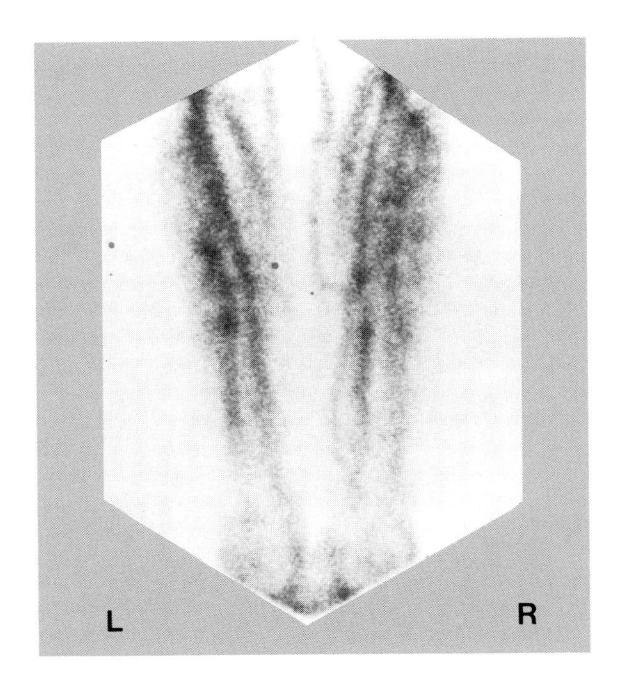

Figure 10-17 ***Segmental Femoral DVT***

The left distal femoral vein *(left, arrowheads)* is poorly defined on the nuclear venogram in comparison to the right vessel. The contrast study *(right)* documents thrombotic abnormalities to the level of the valve. The portion of the vein above the valve *(arrows)* is normal both on the nuclear and contrast venograms. ***Comment:*** Localized DVT will induce limited but discernible changes on the Tc-99m RBC venogram. (Lisbona R, Derbekyan V, Novales-Diaz J, Rush CL: Tc-99m red blood cell venography in deep venous thrombosis of the lower limb—An overview. Clin Nucl Med 10(3):208, 1985)

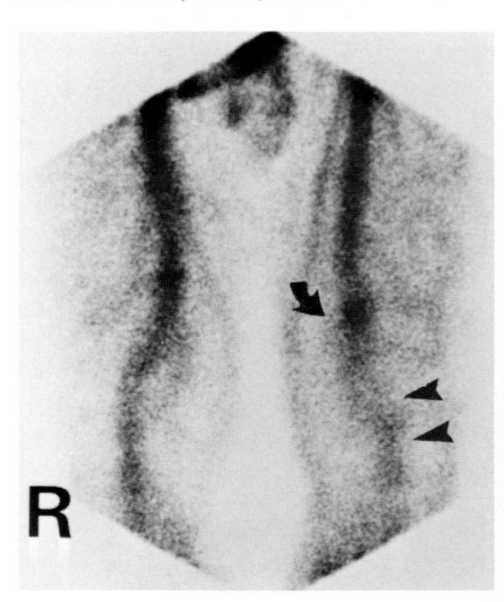

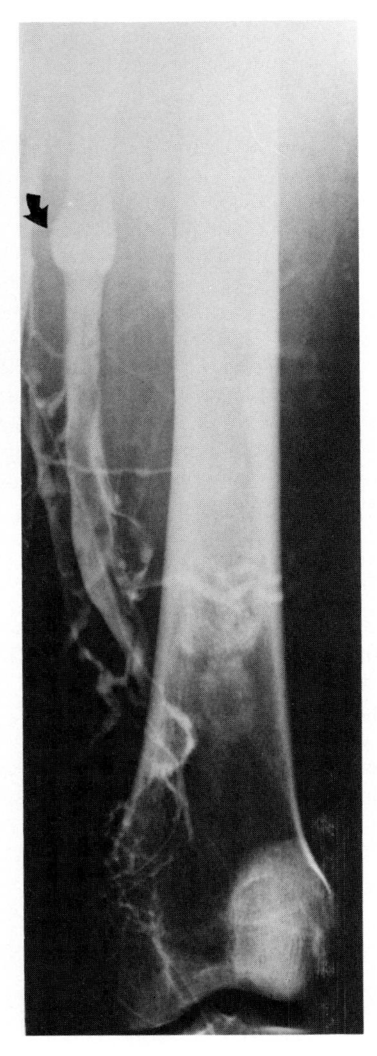

Figure 10-18 ***Bilateral Femoral DVT***

Bilateral femoral thrombosis on the Tc-99m RBC venogram *(left)* is evidenced by the irregularity and poor definition of both femoral channels. A left-sided pedal injection of radiotracer *(right)* confirms the abnormality of the left femoral vein. ***Comment:*** A side-by-side comparison of the legs is fruitful in documenting unilateral thrombosis on the nuclear venogram. If both sides vary from the norm, however, bilateral DVT is to be suspected. Also on occasion, pedal intravenous injections of radiotracer may disclose an extensive network of collaterals that may be underestimated on Tc-99m RBC venography. These discrepancies are related to the more dynamic nature of the pedal study which reflects patterns of flow, while the Tc-99m RBC venogram depicts the blood content of the veins at equilibrium. See also Fig. 10-19. (Lisbona R, Derbekyan V, Novales-Diaz JA, Rush CL: Tc-99m red blood cell venography in deep venous thrombosis of the lower limb—An overview. Clin Nucl Med 10(3):208, 1985)

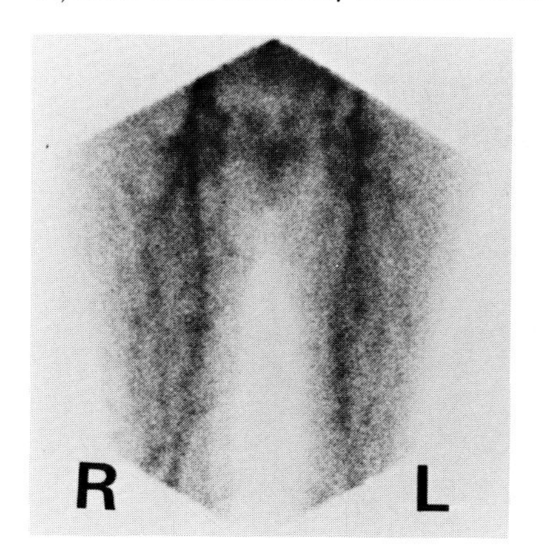

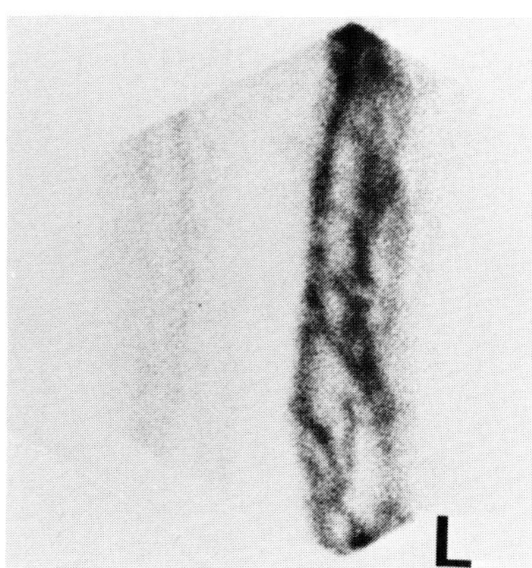

Figure 10-19 ***Bilateral Calf DVT***

The architecture of the calf veins is grossly distorted in both legs, indicating the presence of bilateral DVT.

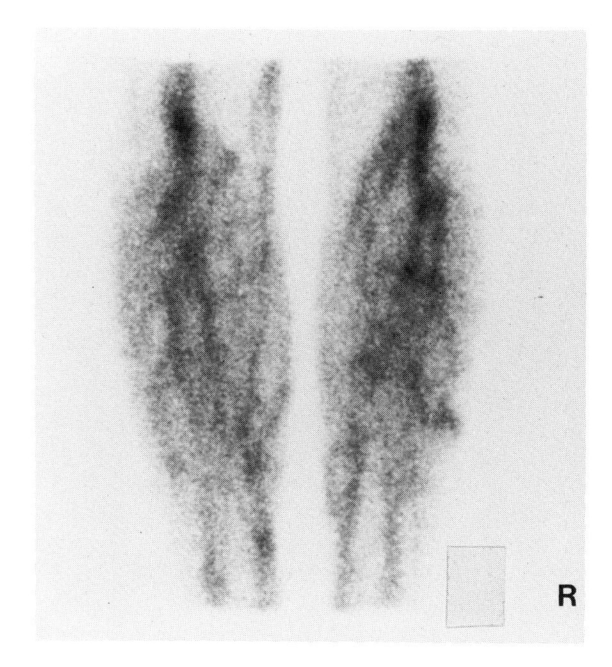

Figure 10-20 ***Progressive DVT***

(A) The Tc-99m RBC venogram shows evidence of right popliteal and calf vein DVT; these venous channels are ill-defined relative to the normal left side. ***(B)*** Three weeks later, despite anticoagulation therapy, the DVT has progressed to involve the right femoral and iliac veins. ***Comment:*** Propagation of thrombosis can be accurately mapped by serial blood pool venograms. This is a difficult clinical endeavor. See also Fig. 10-21. (Lisbona R, Derbekyan V, Novales-Diaz JA, Rush CL: Tc-99m red blood cell venography in deep venous thrombosis of the lower limb—An overview. Clin Nucl Med 10(3):208, 1985)

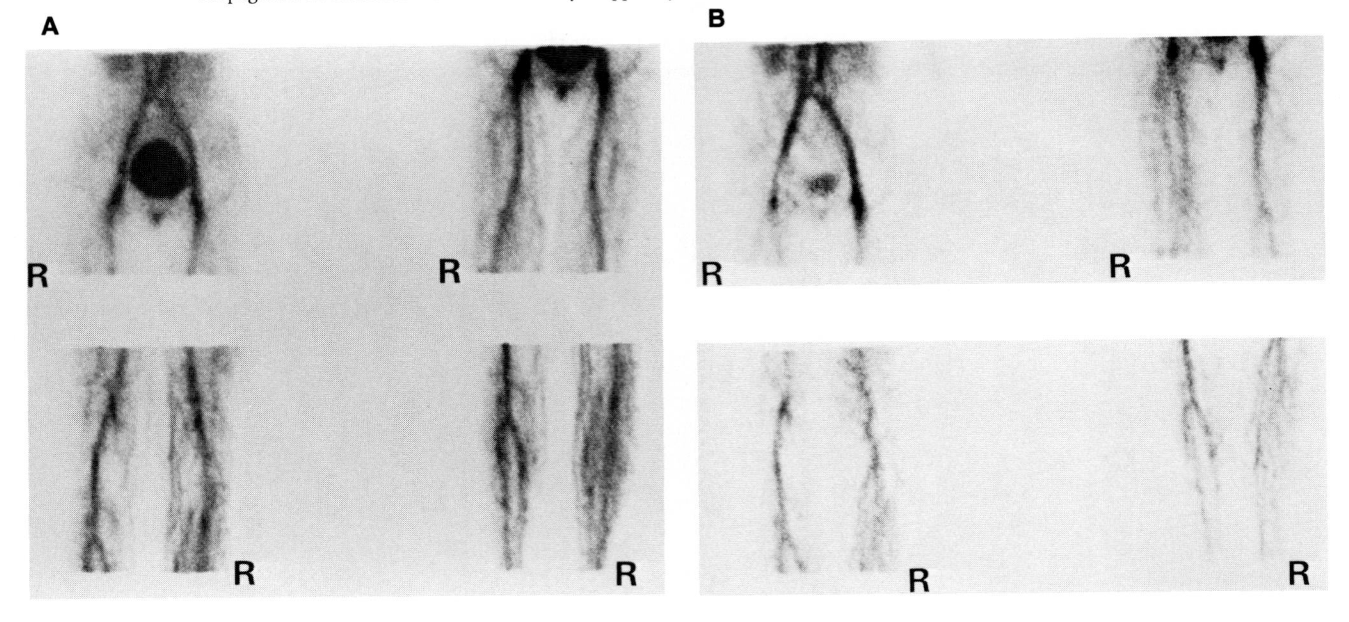

Figure 10-21 ***Progressive DVT***

(A) This elderly patient was admitted to the hospital with left-sided DVT involving the calf veins and femoral vessel. *(B)* Despite anticoagulation therapy, extensive DVT developed in the opposite limb and there was no resolution of the initial DVT. (Lisbona R: Radionuclide blood pool imaging in the diagnosis of deep vein thrombosis of the leg. In Freeman LM, Weissman H (eds): Nuclear Medicine Annual. New York, Raven Press, 1986)

A

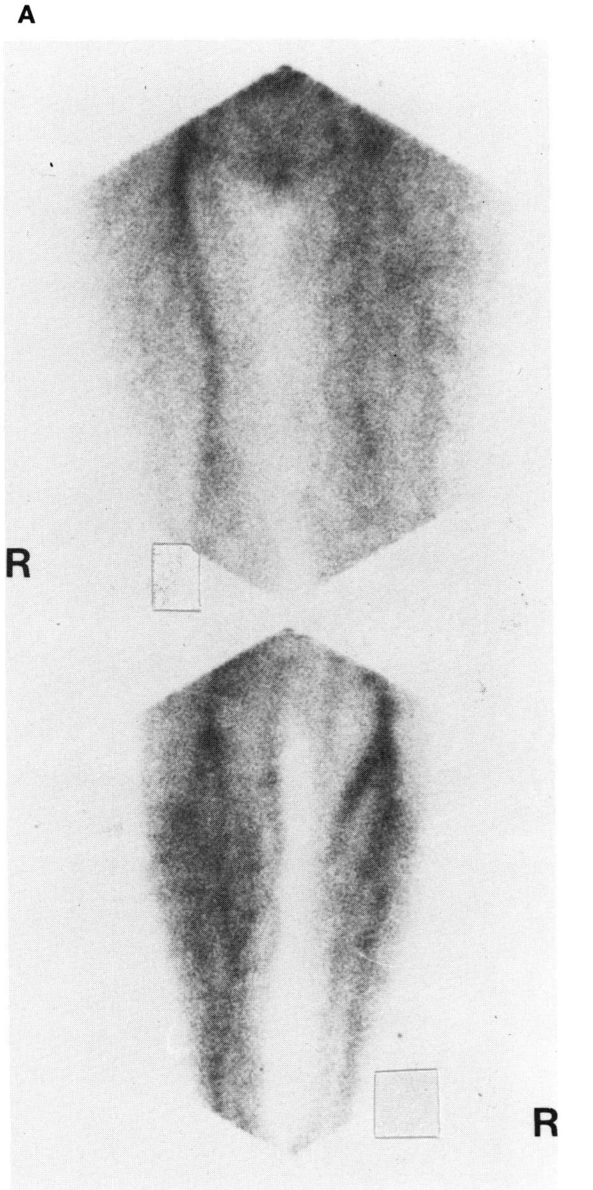

B

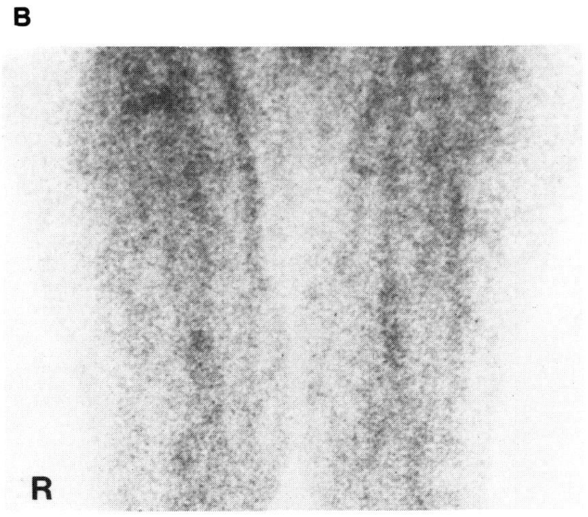

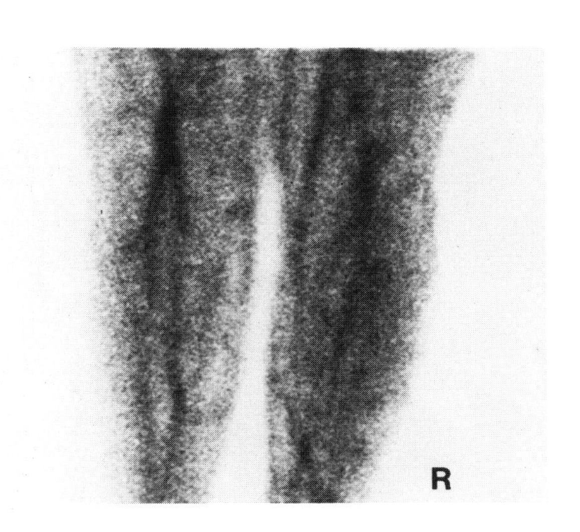

Figure 10-22 ***Resolving DVT***

(A) Prior to therapy, there is evidence of extensive left-sided DVT involving the iliofemoral and calf veins. *(B)* After ten days of heparin therapy, the deep venous system of the left leg reverted to normal. ***Comment:*** Tc-99m RBC venography is a unique, noninvasive screening test in the evaluation of DVT in that it permits the serial assessment of the status of individual veins, either proximal or distal, when the patient undergoes treatment. See also Figs. 10-23 - 10-25. (Lisbona R, Derbekyan V, Novales-Diaz JA, Rush CL: Tc-99m red blood cell venography in deep venous thrombosis of the lower limb—An overview. Clin Nucl Med 10(3):208, 1985)

A

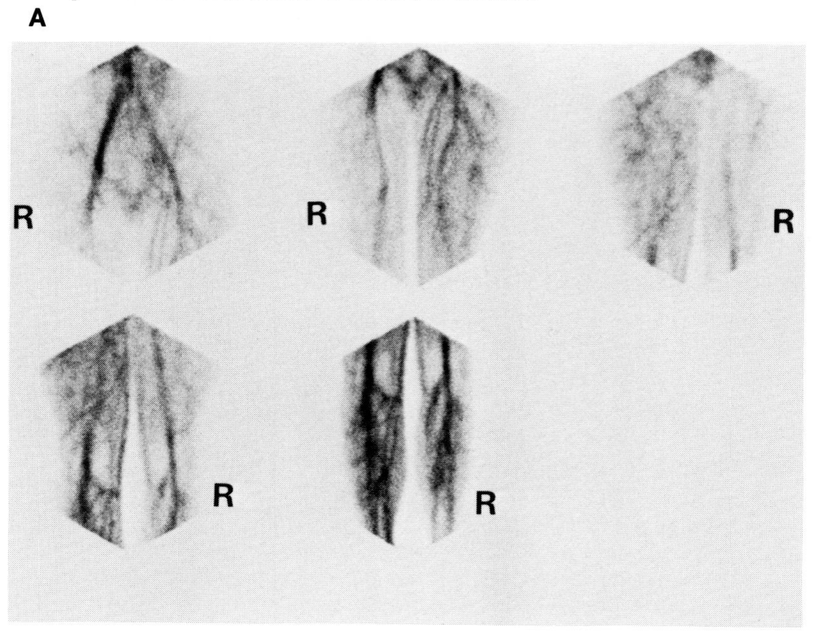

B

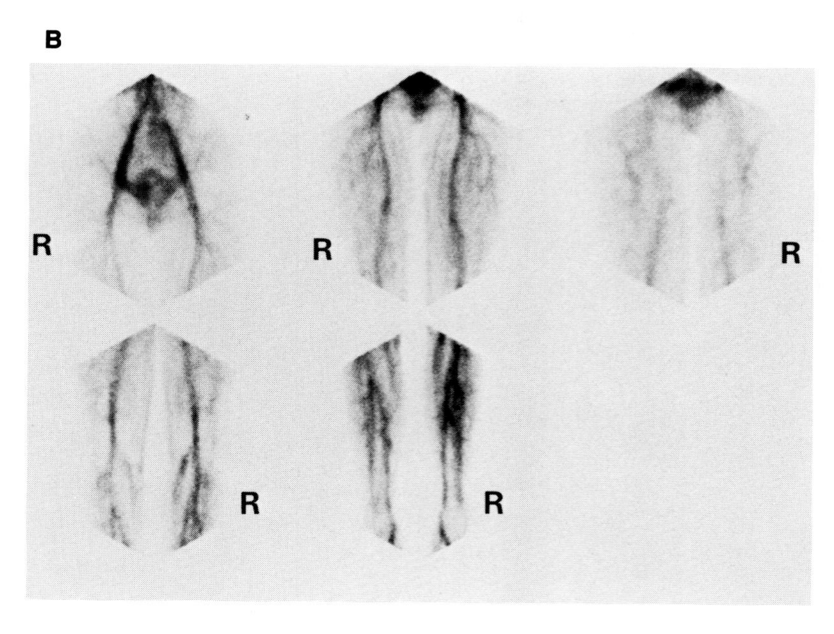

Figure 10-23 ***Resolving Iliac DVT***

With therapy, the left iliofemoral thrombosis *(left)* undergoes resolution and the iliac channel *(right)* recanalizes. (Lisbona R: Radionuclide blood pool imaging in the diagnosis of deep vein thrombosis of the leg. In Freeman LM, Weissman H (eds): Nuclear Medicine Annual. New York, Raven Press, 1986)

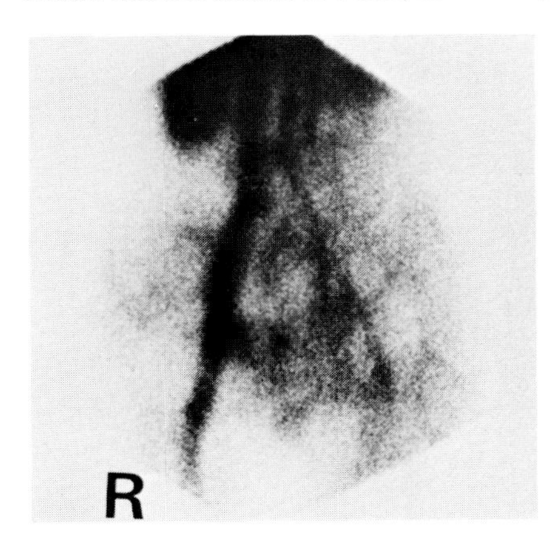

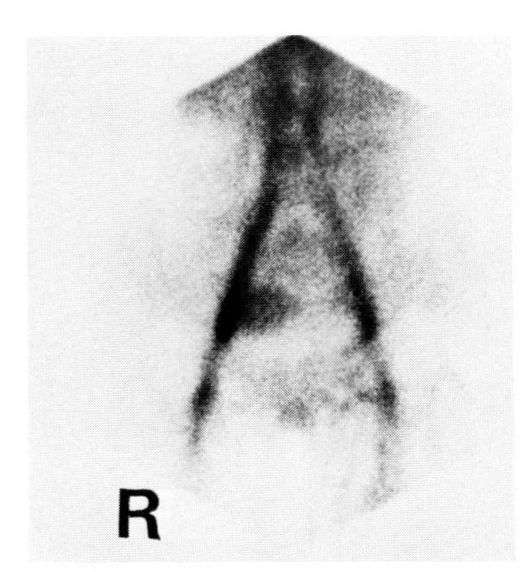

Figure 10-24 ***Resolving DVT***

On admission *(left)*, there is poor definition and thinning of the left femoral and iliac veins consistent with iliofemoral DVT. Following treatment *(right)*, the left femoral vein displays a normal appearance and even its bifid nature can now be appreciated. (Lisbona R: Radionuclide blood pool imaging in the diagnosis of deep vein thrombosis of the leg. In Freeman LM, Weissman H (eds): Nuclear Medicine Annual. New York, Raven Press, 1986)

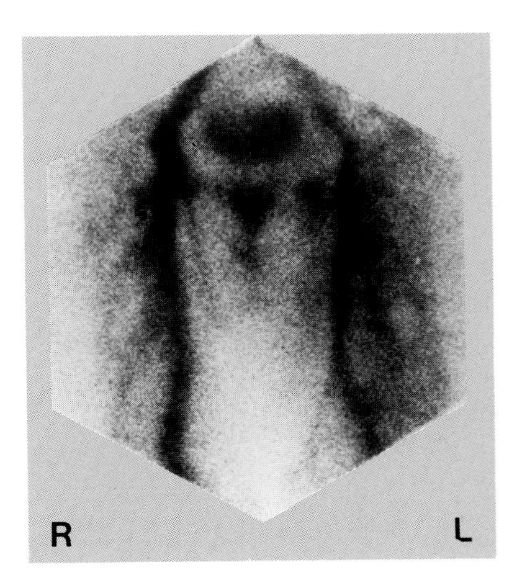

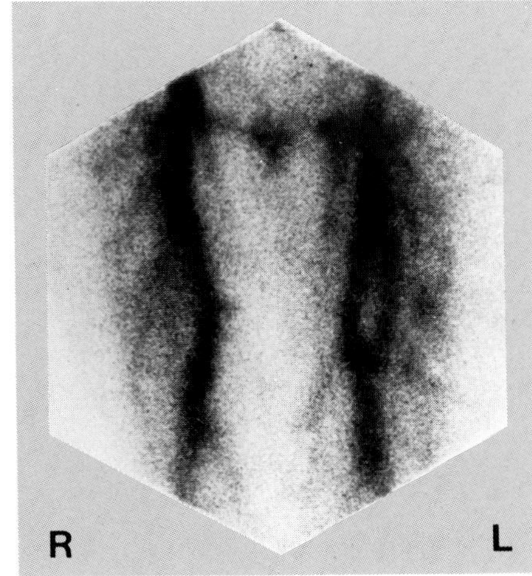

Figure 10-25 ***Resolving Calf DVT***

The absence of the left peroneal veins and abnormal portrayal of the distal left posterior tibial veins confirmed clinical suspicions of left calf DVT in this patient *(left)*. After intravenous anticoagulation therapy, the calf veins returned to normal *(right)*. (Lisbona R, Derbekyan V, Novales-Diaz JA, Rush CL: Tc-99m red blood cell venography in deep venous thrombosis of the lower limb—An overview. Clin Nucl Med 10(3):208, 1985)

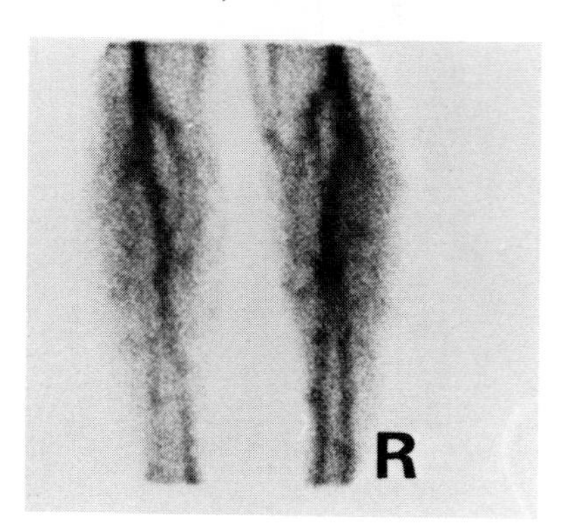

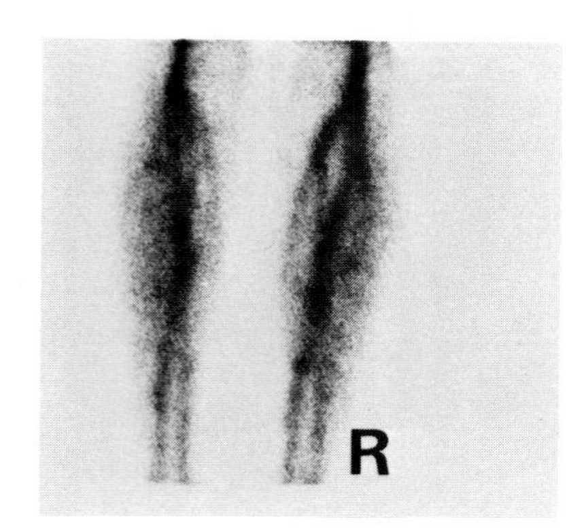

Figure 10-26 ***Acute DVT Superimposed upon Chronic Changes***

(A) In this patient with recurrent bouts of DVT, extensive residual abnormalities of the deep venous system are disclosed in the left calf and left femoral region, as well as in the right calf. ***(B)*** When the patient is reexamined due to new symptomatology, greater distortion of the left femoral vein *(arrow)* is noted with nonvisualization of a communicating vein *(single arrowhead)* and a segment of the greater saphenous vein *(arrowheads)*. These findings point to a diagnosis of acute DVT superimposed on chronic changes. ***Comment:*** Follow-up scans should be obtained after therapy in all patients. They can serve as baseline studies for any future evaluation of the patient and thus ease the interpretation of new scintigrams. (Lisbona R, Derbekyan V, Novales-Diaz JA, Rush CL: Tc-99m red blood cell venography in deep venous thrombosis of the lower limb—An overview. Clin Nucl Med 10(3):208, 1985)

A

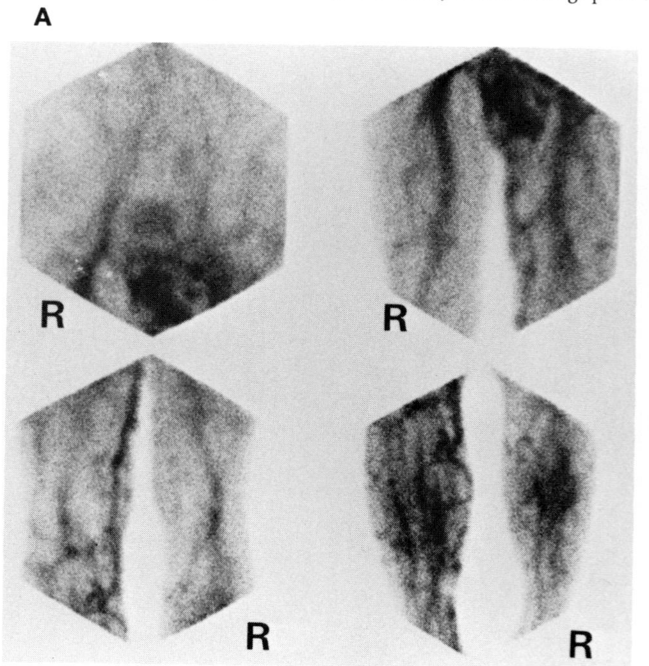

B

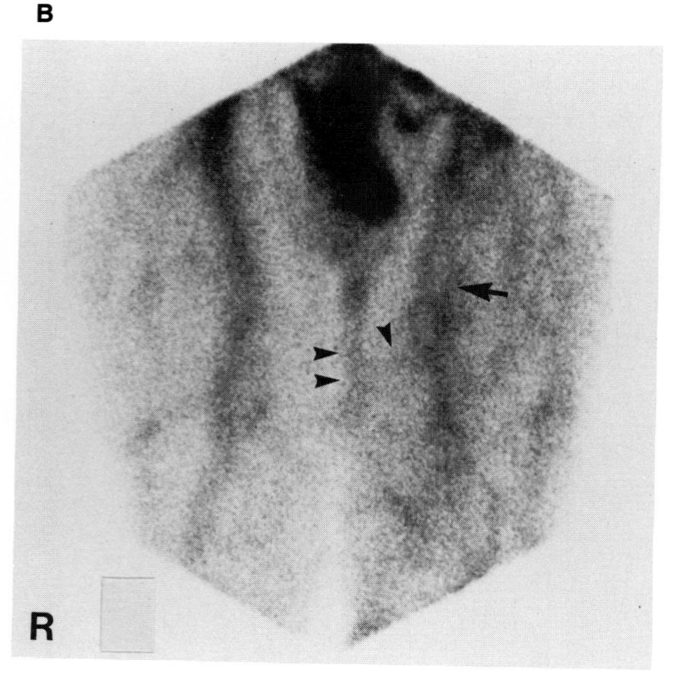

Figure 10-27 ***Chronic DVT***

The blood pool venogram is grossly aberrant, with chronic residual abnormalities persisting in both limbs of this patient with multiple remote episodes of phlebitis. ***Comment:*** Absent and ill-defined venous channels and a collateral network can be features of acute or chronic DVT. As with most screening tests for DVT, Tc-99m RBC venography cannot date the age of thrombosis. The most extensive collateralization is nevertheless usually seen in cases of chronic, proximal, unresolved occlusive thrombosis. (Lisbona R, Derbekyan V, Novales-Diaz JA, Rush CL: Tc-99m red blood cell venography in deep venous thrombosis of the lower limb—An overview. Clin Nucl Med 10(3):208, 1985)

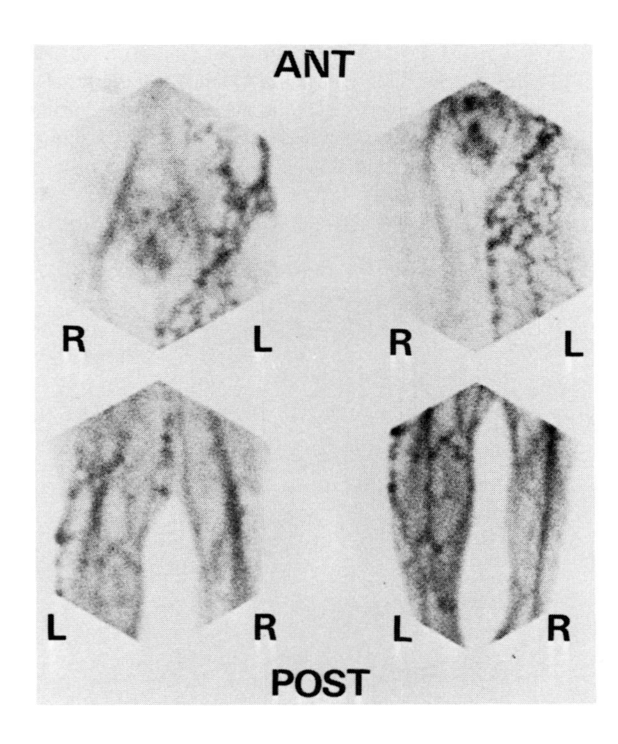

Figure 10-28 ***False-Negative Tc-99m RBC Venogram***

Although the nuclear venogram *(A)* is unremarkable, the contrast venogram *(B)* discloses floating clots *(arrowhead)* in the right popliteal channel. ***Comment:*** Smaller thrombi or floating clots may displace too little blood to affect the portrayal of veins on Tc-99m RBC venography and will, therefore, result in false-negative examinations. (Lisbona R, Derbekyan V, Novales-Diaz JA, Rush CL: Tc-99m red blood cell venography in deep venous thrombosis of the lower limb—An overview. Clin Nucl Med 10(3):208, 1985)

A

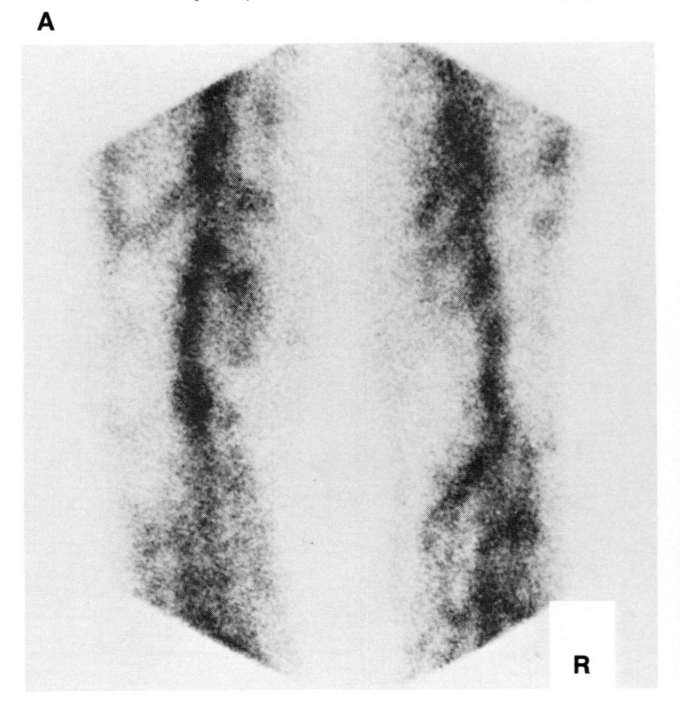

B

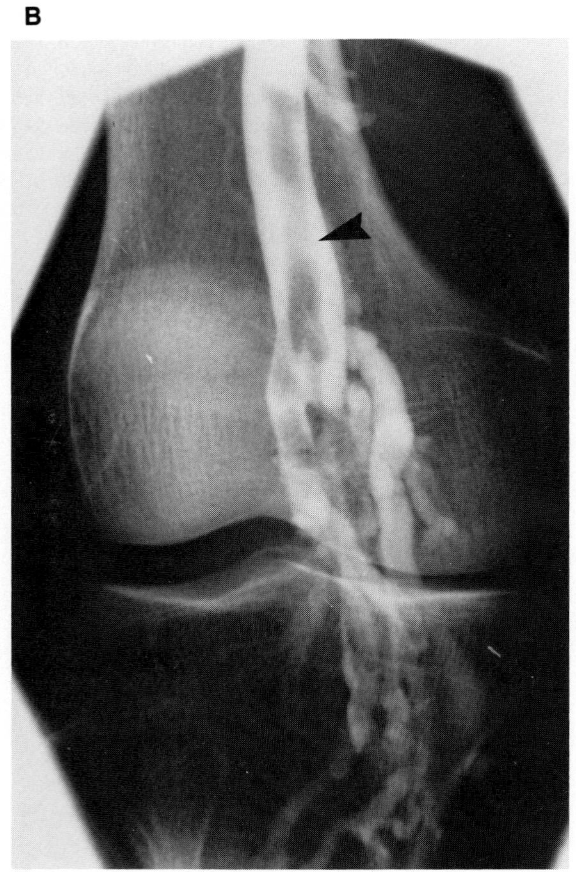

Figure 10-29 ***Thrombosis of the Deep Femoral Vein***

(A) The Tc-99m RBC venogram is unremarkable in this 47-year-old patient with recent right hip arthroplasty. *(B)* The contrast venogram confirms the normalcy of the right superficial femoral vein. However, a filling defect *(arrowheads),* which represents a clot, is noted in the right deep femoral vein. ***Comment:*** The nuclear venogram is usually blind to events in the deep femoral vein, since this channel does not necessarily visualize on the study, even in normals. Deep femoral vein thrombosis can also be a difficult diagnosis on the contrast venogram, since the vein will be opacified in only 50% of cases.

A

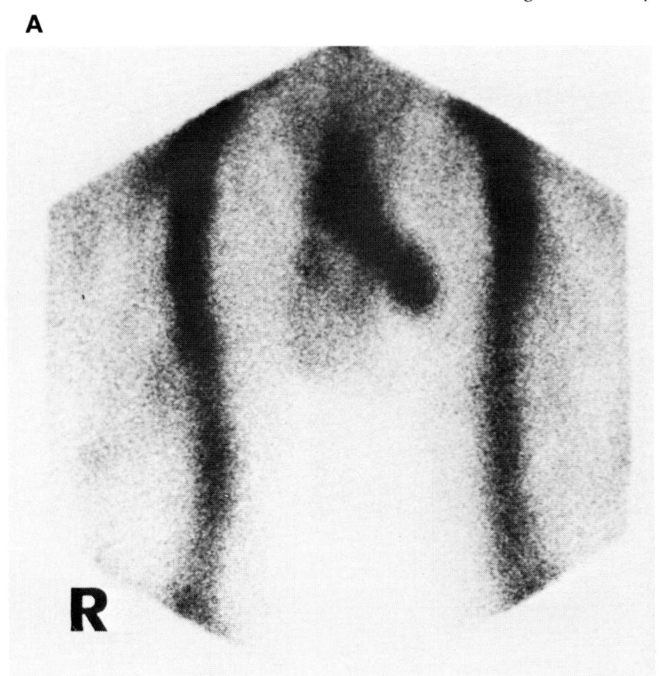

B

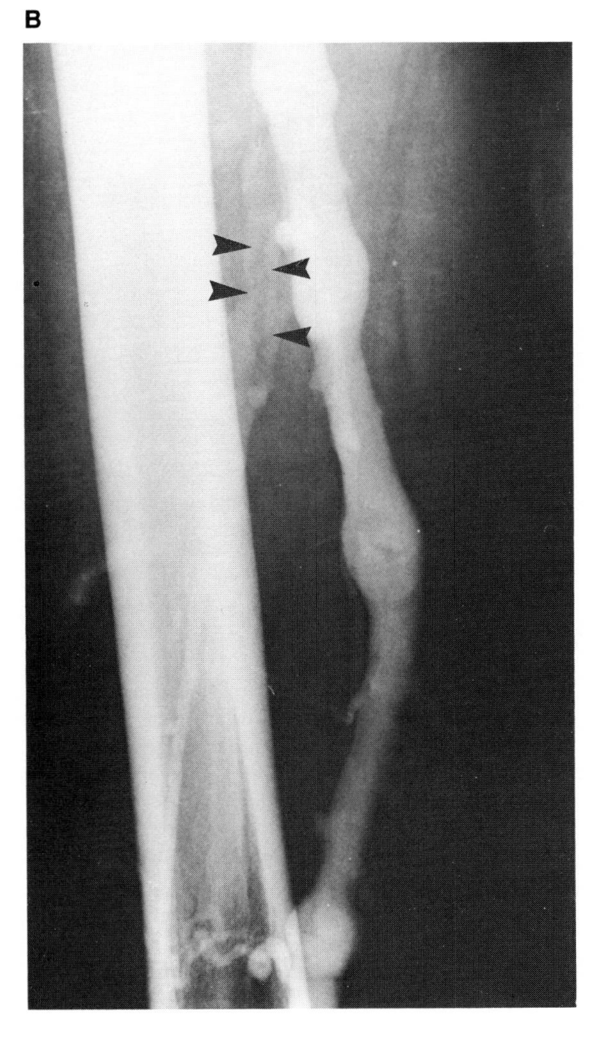

Figure 10-30 ***Sural Phlebitis***

There is flagrant asymmetry on the posterior view of the knees of this patient, with total nonvisualization of the right sural vein *(left).* On contrast venography *(right),* the head of an occlusive thrombus is seen to protrude in the right popliteal vein *(arrowhead),* thus establishing a diagnosis of sural phlebitis. ***Comment:*** The detection of sural phlebitis is a difficult task with most techniques. On Tc-99m RBC venography, the variable appearance of the normal vein makes the interpretation unreliable. Nevertheless, flagrant asymmetry of the sural veins on the nuclear venogram should suggest the diagnosis in the patient with appropriate symptoms. However, it is wise to try and confirm the abnormality by contrast venography. (Lisbona R, Derbekyan V, Novales-Diaz JA, Rush CL: Tc-99m red blood cell venography in deep venous thrombosis of the lower limb—An overview. Clin Nucl Med 10(3):208, 1985)

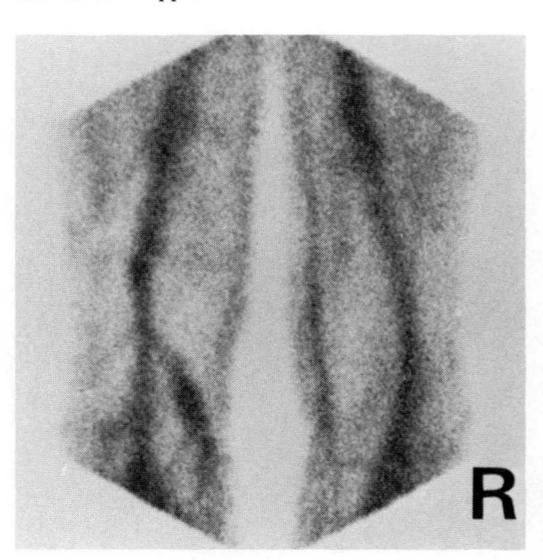

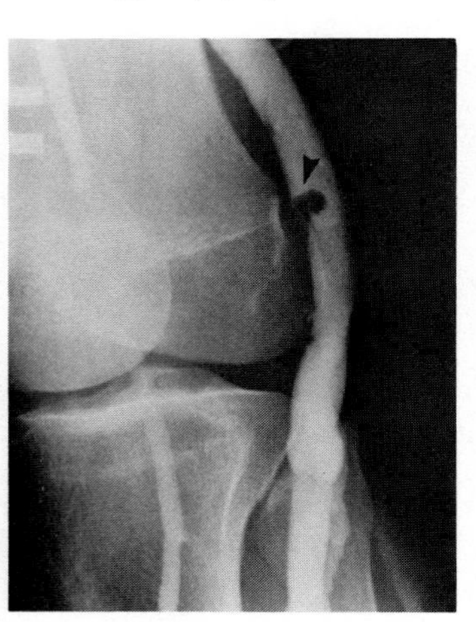

Figure 10-31 ***Cellulitis***

This patient with right-calf cellulitis and no DVT shows evidence of soft tissue swelling on the nuclear venogram. Also, the longitudinal veins are markedly attenuated but still maintain an integral outline. ***Comment:*** A flaring cellulitis uncomplicated by DVT can diffusely attenuate the intensity of venous channels; these should, however, still be seen and should still maintain a sharp outline. When DVT is associated with cellulitis, the vessels become irregular and interrupted. However, the distinction between a simple rampant cellulitis and one with superimposed DVT may not always be simple to make.

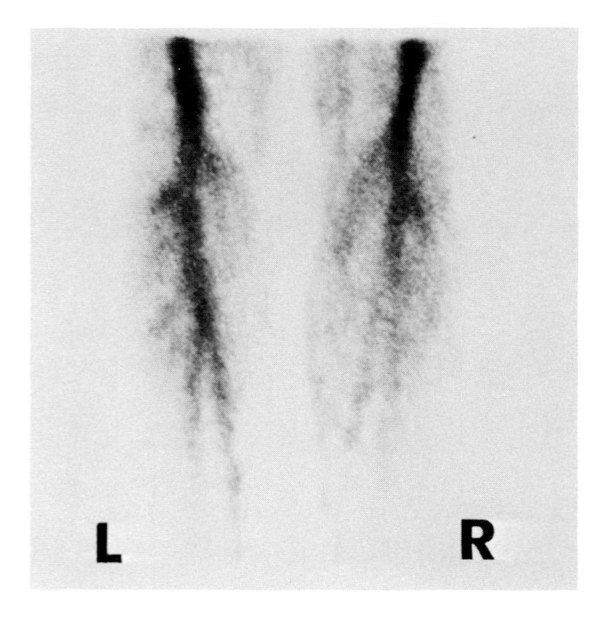

Figure 10-32 ***Extrinsic Compression***

(A) On Tc-99m RBC venography, the right iliac vein fails to visualize as a result of pelvic lymphoma, which compresses and masks the vein. The nuclear venogram is otherwise normal. ***(B)*** Ultrasound confirms the presence of a lymphomatous mass *(curved arrow)* about the iliac vein *(arrows)* and absence of DVT, since the vein is free of clots. ***Comment:*** Although selective abnormalities of iliac veins on the nuclear venogram may mimic thrombosis, they generally result from extrinsic compression of the low-pressure veins by tumor, abscess, hematoma, or aneurysm. Iliac DVT is almost always accompanied by femoral thrombosis. The distinction between extrinsic obstruction and venous thrombosis may sometimes be difficult, however, both clinically and scintigraphically. (See also Figs. 10-33–10-35.) (Lisbona R: Radionuclide blood pool imaging in the diagnosis of deep vein thrombosis of the leg. In Freeman LM, Weissman H (eds): Nuclear Medicine Annual. New York, Raven Press, 1986)

A

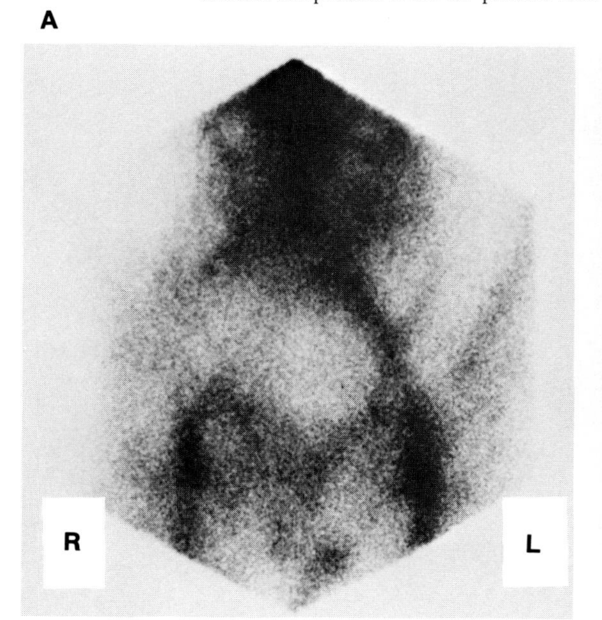

B

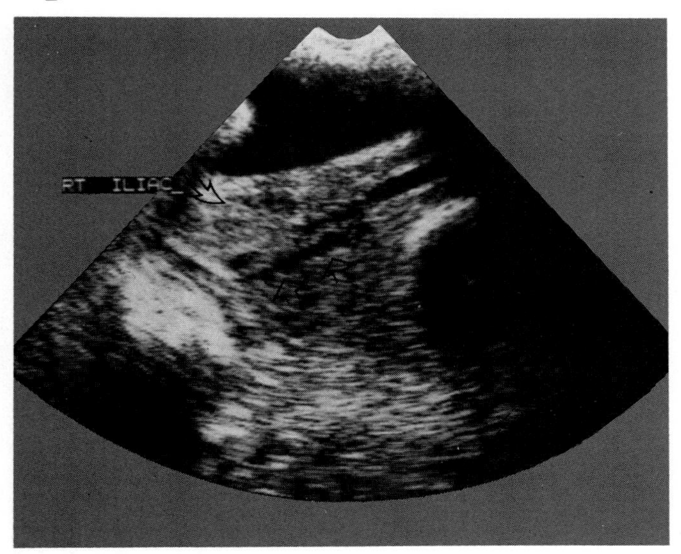

Figure 10-33 ***Extrinsic Compression of the Iliac Vein***

This elderly male patient sustained a large left-sided pelvic hematoma which extended into the thigh. ***(A)*** The left iliac and very proximal femoral veins fail to visualize on the nuclear study. ***(B)*** A contrast venogram obtained on the same day shows no intrinsic abnormality of the left-sided vessels to suggest thrombosis.

A

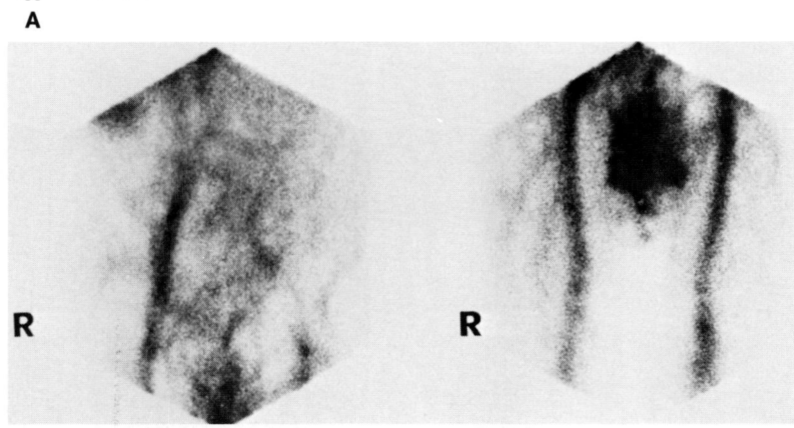

B

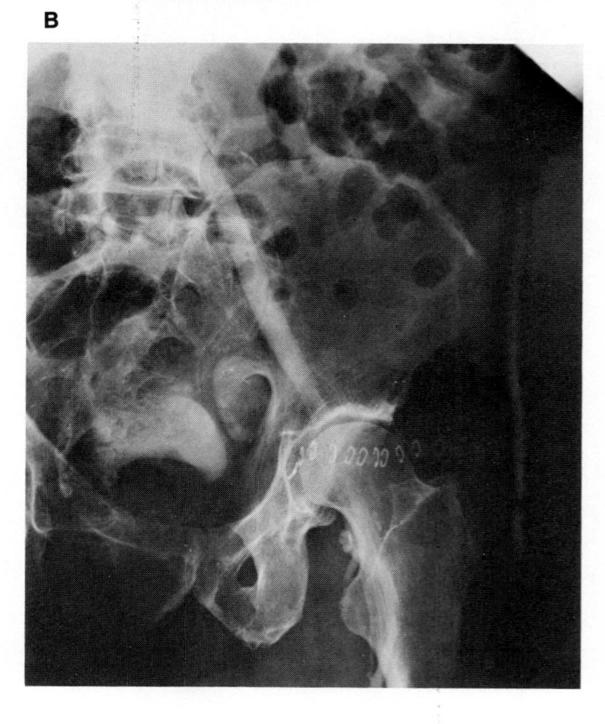

Figure 10-34 *Extrinsic Compression of Veins*

A focal abnormality of the right popliteal vein *(upper left, open arrow)* is the result of compression of the vein by a Baker cyst, as seen on the contrast arthrogram *(right, closed arrow)*. The appearance of the right popliteal vein returns to normal *(lower left)* when the cyst is displaced manually. ***Comment:*** Solitary lesions of the venous system that are confined to the popliteal fossa rarely represent thrombosis. They usually reflect extrinsic compression of the vein from popliteal cysts or local hematomas. (Lisbona R, Derbekyan V, Novales-Diaz JA, Rush CL: Tc-99m red blood cell venography in deep venous thrombosis of the lower limb—An overview. Clin Nucl Med 10(3):208, 1985)

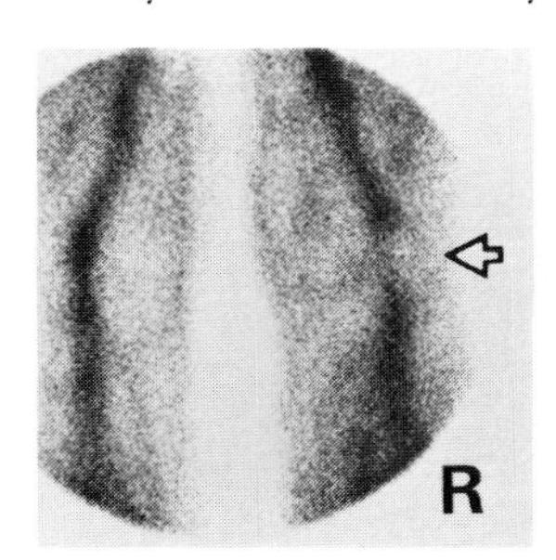

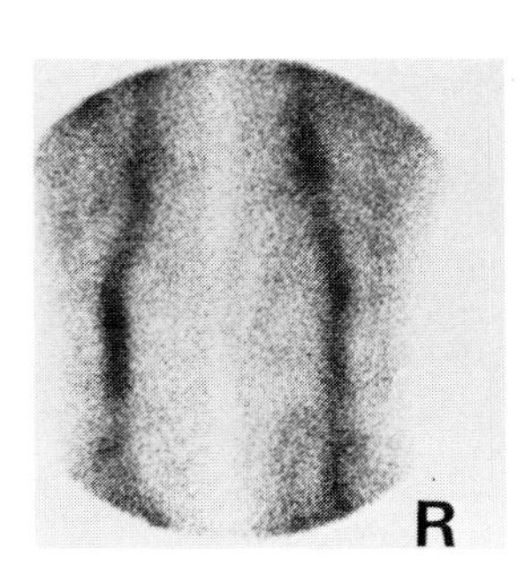

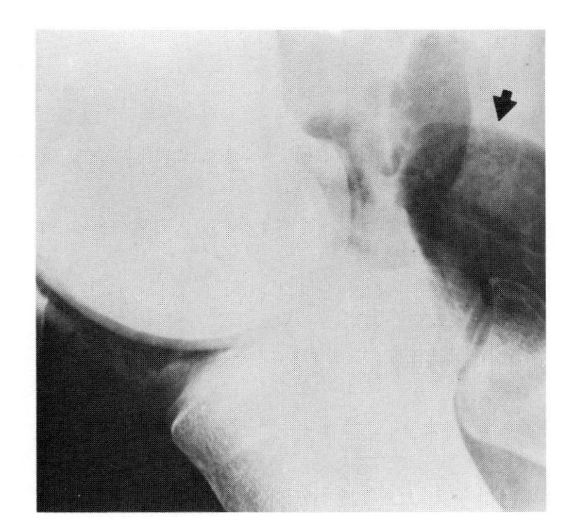

Figure 10-35 ***Extrinsic Compression***

The large calcific mass evident on the radiographs *(A)* pinches the femoral vein *(arrowhead)* and narrows its lumen, as can be appreciated on the venogram *(B)*.

A

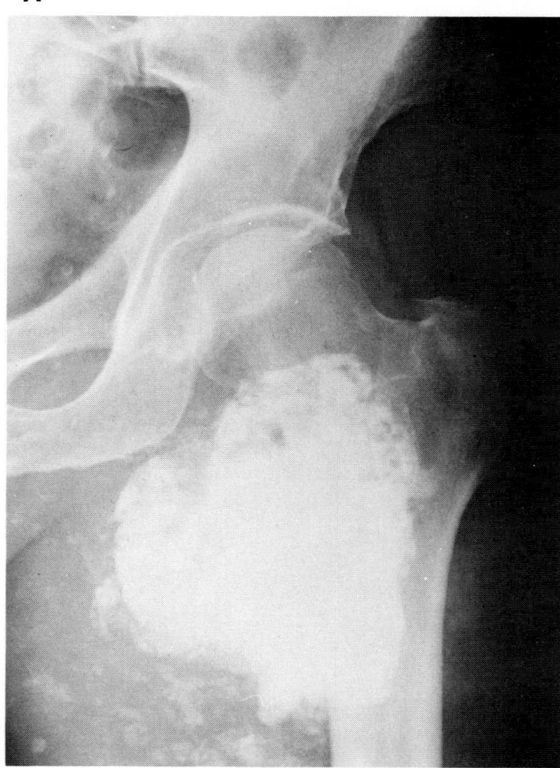

B

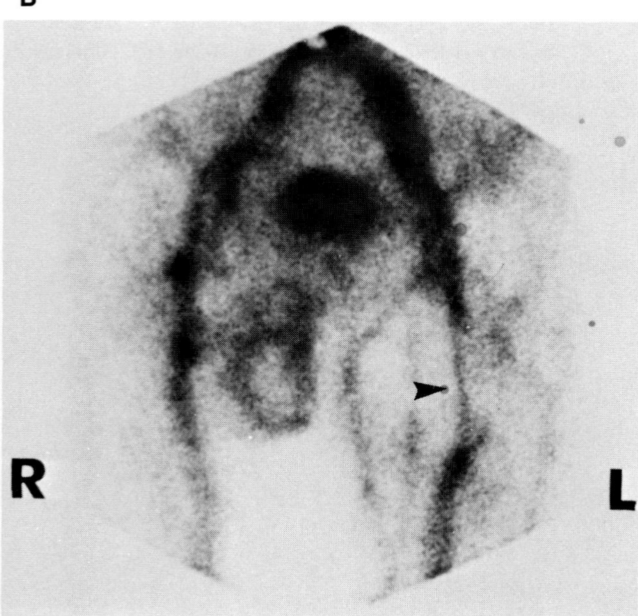

Figure 10-36 *Occult Duplications*

Both femoral veins are poorly delineated on the anterior images *(upper row), (arrows).* Splitting of the right femoral vein *(arrowheads)* is noted on the posterior views of the lower thighs *(lower row),* which explains the vein's unusual appearance on the anterior views. A covert split of the left femoral vein should then also be suspected to explain its lack of sharpness on the Tc-99m RBC venogram. A left-sided pedal injection of radiotracer with the patient supine *(lower right)* confirms the duplication of the left femoral vein. ***Comment:*** Hidden duplications of the femoral veins, particularly when unilateral, may be troublesome on interpretation. Duplications should be suspected if the mid-segment of the vein lacks definition on the anterior images. Close scrutiny of the posterior views may confirm the split. If not, a pedal injection of radiotracer will document the duplication. See also Fig. 10-37. (Lisbona R, Derbekyan V, Novales-Diaz JA, Rush CL: Tc-99m red blood cell venography in deep venous thrombosis of the lower limb—An overview. Clin Nucl Med 10(3):208, 1985)

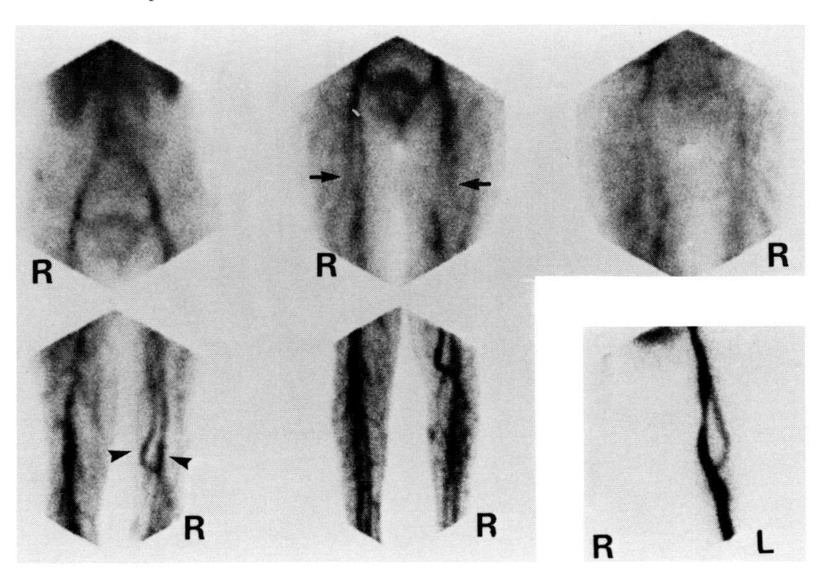

Figure 10-37 *Unilateral Occult Duplication of Femoral Vein*

The abnormal short segment in the proximal left femoral vein on Tc-99m RBC venography *(left)* is not due to DVT, but is the result of a subtle split of the vessel, which is readily demonstrated on pedal injection of radiotracer *(right).* The posterior views of the thighs in this patient did not reveal the duplication. (Lisbona R: Radionuclide blood pool imaging in the diagnosis of deep vein thrombosis of the leg. In Freeman LM, Weissman H (eds): Nuclear Medicine Annual. New York, Raven Press, 1986)

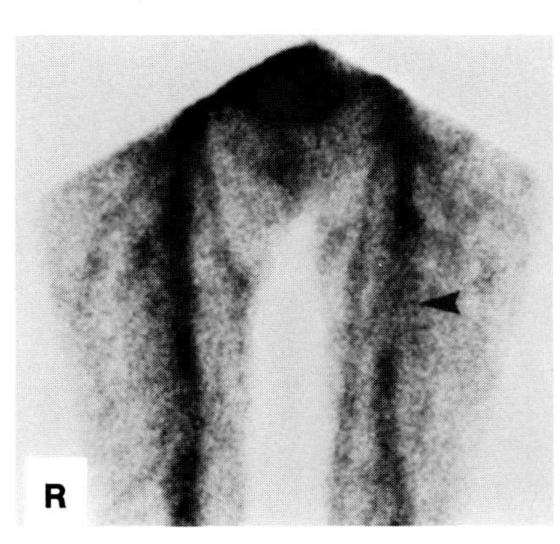

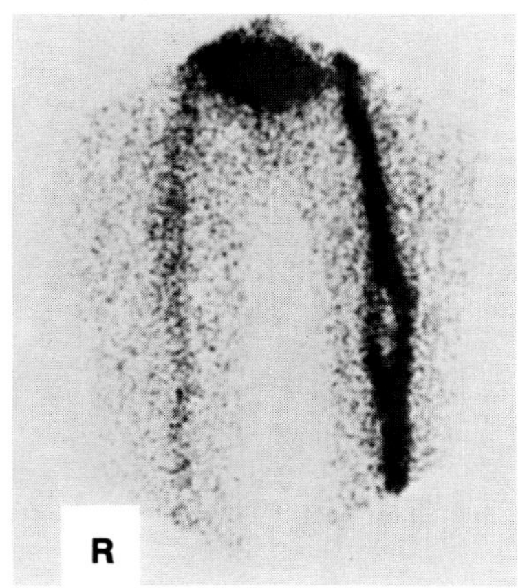

Figure 10-38 ***Pulmonary Embolization and DVT***

(A) Multiple segmental defects of perfusion are noted on the lung scan. The ventilation study was normal. *(B)* A nuclear venogram of this patient with pulmonary embolization reveals evidence of right calf and popliteal DVT. ***Comment:*** In our experience, 74% of patients with symptomatic pulmonary embolization will show evidence of DVT on the nuclear venogram. In many, the DVT will not be suspected clinically.

A

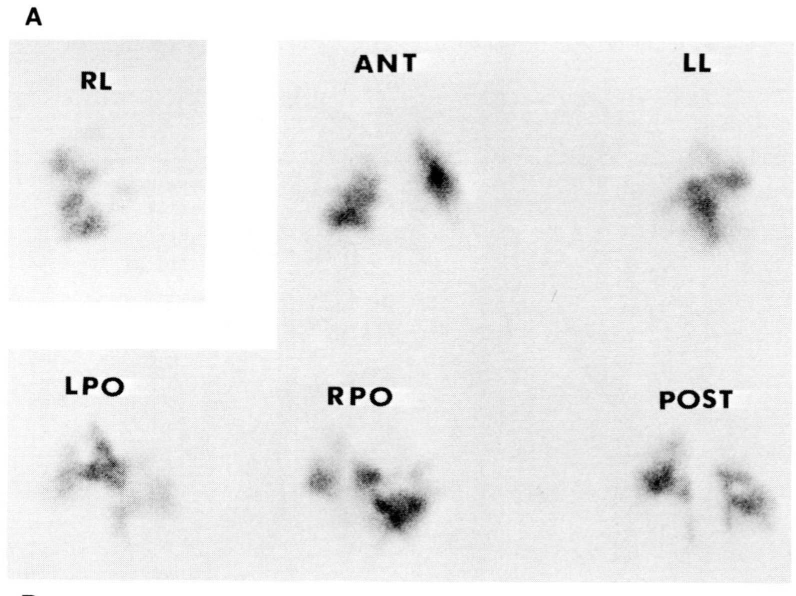

B

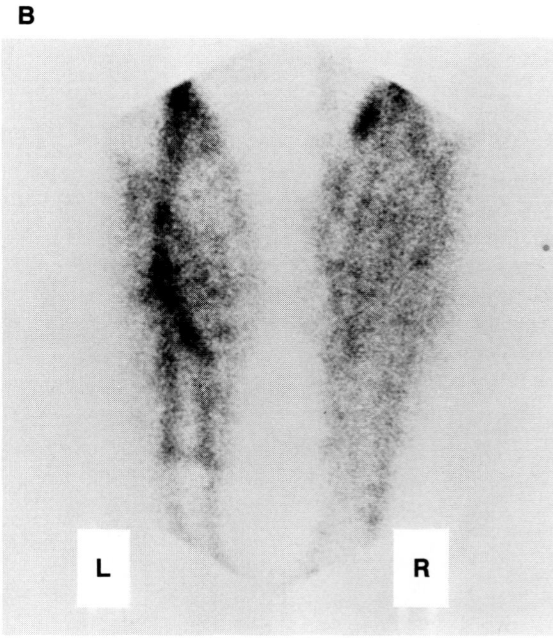

Douglas F. Eggli
Julio E. Garcia

Chapter 11 RADIONUCLIDE IMAGING OF THE ACUTELY PAINFUL SCROTUM

Testicular imaging using technetium-99m pertechnetate was first performed in 1973 to evaluate the acutely painful scrotum preoperatively. Initially only perfusion images were employed,[1] but subsequently the current technique of flow and blood pool imaging was used.[2,3] Since that time scintigraphy has become the primary imaging modality when clinical examination is inconclusive. Scintigraphic imaging of the scrotum has achieved a sensitivity of 96% and specificity of 98% with an overall accuracy of 96% as compared to clinical evaluation which has a sensitivity of 66%, a specificity of 96%, and an accuracy of 88%.[4] In pooled data, clinical examination alone has a 20% false-negative rate.

The clinical application of testicular imaging is the evaluation of the acutely or subacutely painful scrotum. The main objective of testicular imaging is to determine if the etiology of the pain is surgical (torsion of the spermatic cord) or inflammatory. Scintigraphic imaging is not the primary modality for the evaluation of chronic or painless disorders of the scrotum. Ultrasound has greater specificity in this arena and is the screening test of choice.[5] The scintigraphic imager, however, should recognize the patterns associated with chronic disease. Imaging of the acutely painful scrotum is a legitimate emergency, since testicular viability after torsion of the spermatic cord is dependent on the number of hours from the onset of pain to surgical reduction of the torsion.

PHYSIOLOGY OF THE TESTIS

A basic familiarity with the embryology and blood supply of the testis is essential to the understanding of the developmental abnormality that predisposes to torsion. In the sixth to eighth week of fetal life, the primitive germ cells begin to form the gonadal cords in the retroperitoneum. A dense fibrous connective tissue layer (the tunica albuginea) forms, covering the evolving testis. The appendix testis is formed by the regression of the paramesonephric (müllerian) duct cranially, while the appendix epididymis develops from the remnant of the mesonephric (wolffian) duct not incorporated into the ductus deferens.

By the eighth week of gestation a urogenital mesentery develops connecting the testis and mesonephros (developing duct system) to the posterior coelomic wall. This mesentery persists caudally to become the caudal genital ligament (gubernaculum), which extends from the site where the inguinal canal will develop to the lower pole of the testis. As the fetus develops, the gubernaculum remains fixed and anchors the testis in place as the fetus grows cranially, effectively "dragging" the testis and its aortic blood supply from midabdomen to the inguinal canal by preventing cranial migration of the testis with its site of orgin.

The tunica vaginalis forms as an outpouching of

the peritoneal lining covering the developing testis. This peritoneal lining, along with the muscular and fascial layers of the abdominal peritoneum, "descend" with the testis through the inguinal ring into the developing scrotal pouch. The testis and the gubernaculum remain outside the vaginal process, preserving their retroperitoneal location. The testis is anchored inferiorly by the gubernaculum and posteriorly through the attachments of the epididymis to the posterior scrotal wall. Normally, the tunica vaginalis only covers the testis anteriorly.

Abnormalities of descent and attachment of the testis predispose it to torsion. The primary abnormality leading to torsion of the spermatic cord is the "bell-clapper" testis. In the bell-clapper anomaly, the tunica vaginalis completely surrounds the testis, epididymis, and spermatic cord, preventing normal posterior and inferior anchoring of the testis. The testis and its vascular bundle are suspended freely like the clapper of a bell between the layers of the tunica. This abnormality is almost always bilateral. Other related but more subtle abnormalities predisposing to torsion are an unusually capacious tunica vaginalis, absence of the ligaments attaching the epididymis to the posterior scrotal wall, abnormal length and loose attachment of the intravaginal portion of the spermatic cord, abnormal descent of the testis, or, as in many cases, a combination of the above (Fig. 11-1). The incidence of torsion in maldescent is tenfold higher (6.3%) than if descent was normal (0.63%),[6] but maldescent has a very low incidence. Torsion of an incompletely descended testis above the inguinal ring is very difficult to detect scintigraphically.

The male genitalia has a dual blood supply that is important to understand and recognize scintigraphically. The testes and scrotum have separate blood supplies. The vessels of the spermatic cord include the testicular artery, which arises from the abdominal aorta just below the orgin of the renal arteries; the cremasteric artery, which arises from the inferior epigastric artery; and the deferential artery, which arises from either the internal iliac or vesical artery. The scrotal blood supply includes the superficial external pudendal artery arising from the femoral artery; the anterior scrotal artery from the deep external pudendal, which, in turn, arises from the femoral artery below the superficial external pudendal artery; and the posterior scrotal artery from branches of the internal pudendal artery, which arises from the internal iliac artery.

Many of the controversies associated with the flow portion of the scintigraphic examination arise from the failure to make the distinction between the blood supply to the testis through the spermatic cord and the blood supply to the scrotum, which is predominantly through the pudendal vessels. Scintigraphically, the vessels of the spermatic cord enter the scrotum at a steeper and more vertical axis than the scrotal (or pudendal) vessels, which enter the scrotum more horizontally (Fig. 11-2).

TECHNIQUE

Imaging Procedure

The scrotum is positioned over a groin shield especially designed for testicular imaging (Fig. 11-3B) so that the testes lie over the shield and the median raphe of the scrotum is midline. The shield is designed to permit visualization of both the iliac and femoral vessels while shielding the scrotum from thigh activity. The penis is taped up in the midline over the low abdominal wall so that its blood pool does not interfere with imaging. The physician interpreting the study should examine the patient's scrotum to be able to describe the physical findings as part of the examination and to guarantee that the testes overlie the shield. If testis redux is present, the examiner should gently retract the testes over the shield. *Testis redux* is the involuntary contraction of the cremasteric muscle in the presence of torsion or inflammation, which retracts the testis superiorly. Ten to fifteen mCi of Tc-99m O_4^- (200–250 μCi/kg with a minimum dosage of 1.5–2 mCi) are injected intravenously 15 to 20 minutes after blocking the thyroid with potassium perchlorate (6–8 mg/kg). In adults and teenagers, a standard or large-field-of-view camera with a general-purpose, low-energy collimator is adequate for imaging. In smaller children, a converging collimator is recommended. True rather than electronic magnification is most desirable. In infants and the smallest children, a pinhole collimator may be necessary for adequate magnification and resolution. Radionuclide angiogram images are obtained at 3-to-5-second intervals for one minute, followed by an immediate tissue phase image containing 500 to 750 K counts (300–500 kilocounts in small children using a converging collimator, 100–150 kilocounts with a pinhole collimator). Magnification should be employed to limit the field of view from just below the umbilicus to approximately the junction of the upper and middle thirds of the femur. This should permit visualization of the iliac and femoral vessels. A subsequent marker image with a cobalt marker should identify the symptomatic testis. A delayed tissue phase image should be obtained following the marker image. The entire examination should be completed within 15 to 20 minutes of injection.

Computer Acquisition

The flow study can be obtained on computer using a 64 × 64 pixel matrix in word mode with one-second frames over the first minute after injection. Tissue phase, marker, and delayed tissue phase images should be obtained on a 128 × 128 matrix or larger in word mode for the same total counts as the tissue phase film images. With larger matrices, scrotal skin may be resolved separate from the testes. Word mode is used because the vessels during the angiogram phase and bladder in the tissue phase may contribute a significant number of counts to any pixel. If there is pixel overflow from one of these structures, low level activity in scrotal skin and ischemic testes may be lost in the digital image. If true magnification is employed for the film images, no additional magnification should be necessary on the digital images.

Computer Analysis

Perfusion curves and quantitative data are not typically obtained from scrotal scintigrams. The main role of computerization for the study is to prevent loss of the study. In many clinics, scrotal scintigraphy is not commonly performed and radiopharmaceutical dosages vary depending on the size or body surface area of the patient. This oftens leads to uncertain film exposure techniques.

PHYSIOLOGIC MECHANISM OF Tc-99m PERTECHNETATE

Pertechnetate is used for testicular imaging because of its ready availability, requiring no significant time for preparation or quality control. It serves primarily as a first-pass and immediate blood pool agent. From this point of view, any number of soluble nonparticulate radiopharmaceuticals could be used, but because all of them lengthen preparation time before the examination can be performed, they reduce the clinical utility of the examination. It is only in the late or delayed tissue phase images that the unique physiology of free pertechnetate plays any role in the examination. Pertechnetate has the same charge and similar ionic radius as iodide. As such it rapidly leaves the vascular space and distributes throughout extracellular fluid. It is this property of pertechnetate that may lead to the detection of minimal late perfusion in incomplete torsion. Like iodide, technetium pertechnetate is actively concentrated in the thyroid, salivary glands, stomach, and choroid plexus. This activity can be blocked by competitive inhibition pretreating with perchlorate. The primary route of excretion is urinary, accounting for the appearance of the bladder in the late tissue phase images.

ESTIMATED RADIATION ABSORBED DOSE

Tc-99m pertechnetate is administered intravenously as a sodium salt. Radiation doses in rads/mCi without blocking the thyroid and stomach are thyroid 0.13, stomach 0.25, colon 0.06, marrow 0.019, ovaries 0.022 and testes 0.009.[7] Radiation exposures to the thyroid, stomach, and colon are further reduced by prior administration of perchlorate.

VISUAL DESCRIPTION AND INTERPRETATION

The first step in reviewing the scintigraphic examination should be to ensure the technical adequacy of the images. In the flow (angiogram) portion of the study, aortic, iliac, and femoral perfusion are evaluated along with perfusion to the scrotum itself. The properly positioned patient may demonstrate a small segment of aorta just above the bifurcation at the top of the field of view. Femoral vessels just below the scrotum should be at the bottom of the field of view. The properly designed and positioned groin shield should not obscure either the iliac or femoral vessels. The scrotum and testes should overlie the shield, with the median raphe positioned in the midline.

In the arterial phase of the examination, symmetry of scrotal perfusion should be noted with reference to the side of symptoms. The opposite testis is the reference point. Spermatic and pudendal trunks should be separately identified if present. Careful examination of the iliac vessels near the origin of the spermatic cord vessels for a vascular cut-off or nubbin sign may be extremely useful in the diagnosis of spermatic cord torsion or missed (late) torsion. Perfusion is usually categorized as increased, normal, or decreased for each of the vascular trunks indentified, the epididymis, the testis, and the overlying scrotal skin. If perfusion is increased, an attempt should be made to determine if the route is predominantly by way of the spermatic cord or pudendal/scrotal trunk. When the distinction can be made with confidence, the differential diagnosis can be narrowed significantly.

In the tissue phase, testis, epididymis, testicular mediastinum, and scrotal skin are evaluated along with any residual hyperemia along with course of the spermatic cord or pudendal trunk. The location of both testes within the scrotum must be known with absolute certainty if their perfusion status is to be determined. Physical examination of the patient with marker views provide this information. An attempt should be made to localize any abnormal tracer accumulations (either increased or decreased) to the testis itself, the epididymis, or scrotal soft tissues.

Several questions must be answered systematically when reviewing the tissue phase images. Initially, does the symptomatic testis have normal, increased, or decreased blood pool compared to the opposite testis and thigh activity? Does this change on the delayed (or late) tissue phase images obtained after the marker view(s)? The opposite testis may not always be normal; if it is not, the thigh activity just lateral to the scrotum is a useful alternate reference point. Testicular activity should be approximately equal to or slightly less than thigh activity. Is the epididymis normal (not seen above scrotal background), or does it have increased activity? Is the epididymis in the normal location, or is it displaced? Are there any focal areas of abnormal tracer accumulation within the hemiscrotum? Where are they in relation to the testis? Specifically, is there any nontestis photon deficiency in the hemiscrotum? How large is it and where is the testis in relation to it (*i.e.*, is the testis displaced by or in the middle of the photon deficiency)? Is the scrotal skin normal or hyperemic? If a rim or ring is present around the testis, it should be carefully described. Is it thin and fairly uniform or is it thick, shaggy, and irregular? Is the testis inside the rim round or oval-shaped, or is it complex and irregular? Is the testis itself inside the ring less, equal, or greater in activity than the opposite side? Applications of such observations to specific diagnoses are found throughout the atlas section of this chapter.

DISCUSSION

Acute torsion of the spermatic cord is a surgical emergency. The surgical treatment is uncomplicated, but long-term results remain poor. Delay of definitive treatment is the dominant factor in the poor outcome. "Castration by neglect"[8] is too often the outcome, with two-thirds of surgical reductions developing some late testicular atrophy. Atrophy will occur in some patients after as little as 4 hours of ischemia, and will be inevitable by 8 to 10 hours after complete torsion of the spermatic cord. Fully half of the patients with viable testes after surgical reduction of a torsion develop abnormal seminal analyses; immunological injury to the normal testis as a result of contralateral torsion is the probable mechanism. This mechanism has been demonstrated in rats[9] and postulated in humans.[10]

The cases presented in the atlas represent a spectrum of the abnormalities which might be detected by testicular imaging, with special attention given to testicular torsion and the acute diseases which may mimic it clinically. The clinical presentation of torsion, missed torsion, torsion of the testicular appendages, and inflammatory disease may overlap, and diagnosis may be confusing clinically. Scintigraphic testicular imaging is a useful and accurate method of aiding clinical diagnosis.

Overall, as previously noted, the test is very accurate at determining the etiology of the acutely painful scrotum. The accuracy should approach 100% for the diagnosis or exclusion of acute torsion, and the time involved should not significantly delay surgical reduction of the acute torsion. A significant number of suspected torsions may be relegated to a nonsurgical category, sparing the expense and potential morbidity of general anesthesia and a surgical procedure. Similarly, the number of missed diagnoses and delayed treatments may be reduced. The acutely painful scrotum should share the status accorded pulmonary embolus, in which the indication for the scintigraphic test is the suspicion of the diagnosis.

REFERENCES

1. Nadel NS, Gitter MH, Hahn LC, Vernon AR: Preoperative diagnosis of testicular torsion. Urology 1:478, 1973
2. Mishkin F, Lawrence, D: Radionuclide imaging in scrotal abnormalities (abstr). J Nucl Med 15:518, 1974
3. Heck LL, Coles JL, Van Hove ED, Riley TW: Value of 99-Tc pertechnetate imaging in evaluation of testicular torsion (abstr). J Nucl Med 15:501, 1974
4. Tanaka T, Mishkin FS, Datta NS: Radionuclide imaging of scrotal contents. In Freeman LM, Weissman HS (eds): Nuclear Medicine Annual, pp 195–221. New York, Raven Press, 1981
5. Nachtsheim DA, Scheible FW, Gosink B: Ultrasound of testis tumors. J Urol 129:978, 1983
6. Williamson RCN:Torsion of the testis and allied conditions. Br J Surg 63:465, 1976
7. Harbert JC, Pollina R: Absorbed dose estimates from radionuclides. Clin Nucl Med 9:210, 1984
8. Thomas WEG, Williamson RCN: Diagnosis and outcome of testicular torsion. BR J Surg 70:213, 1983
9. Harrison RG, Lewis-Jones DI, Morendo de Marval MJ et al: Mechanism of damage to the contralateral testis in rats with an ischemic testis. Lancet 2:723, 1981
10. Nagler H, White RD: The effect of testicular torsion on contralateral testes. Presented at the annual meeting of the American Urological Association meeting. Kansas City, Missouri, May 16–20, 1982
11. Mendel JB, Taylor GA, Treves T et al: Testicular torsion in children: Scintigraphic assessment. Pediatr Radiol 15:110, 1985
12. Majd M: Personal communication.
13. Chen DCP, Holder LE, Melloul M: Radionuclide scrotal imaging: Further experience with 210 patients. Part II: Results and discussion. J Nucl Med 24:841, 1983
14. Angell JC: Torsion of the testicle: A plea for diagnosis. Lancet 1:19, 1963
15. Corriere JN: Horizontal lie of the testicle: A diagnostic sign in torsion of the testis. J Urol 107:616, 1972
16. Holder LE, Melloul M, Chen D: Current status of radionuclide scrotal imaging. Semin Nucl Med 11:232, 1981
17. Dunn EK, Macchia RJ, Soloman NA: Scintigraphic pattern in missed testicular torsion. Radiology 139:175, 1981
18. Barker K, Raper FP: Torsion of the testis. BR J Urol 36:35, 1964
19. Skogland RW, McRoberts JW, Ragde H: Torsion of the spermatic cord: A review of the literature and an analysis of 70 new cases. J Urol 104:604, 1970
20. Smith, GI: Cellular changes from graded testicular ischemia. J Urol 73:355, 1955

21. Krarup T: The testis after torsion. Br J Urol 50:43, 1978

22. Bartsch G, Frank S, Marberger H et al: Testicular torsion: Late results with special regard to fertility and endocrine function. J Urol 124:375, 1980

23. Rolnick D, Kawanoue S, Szanto P et al: Anatomic incidence of testicular appendages. J Urol 100:755, 1968

24. Skogland RW, McRoberts JW, Ragde H: Torsion of testicular appendages: Presentation of 43 new cases and a collective review. J Urol 104:598, 1970

25. Gilday DL, Hitch D, Shandling B et al: Testicular imaging for testicular torsion in pediatric surgery. J Nucl Med 17:553, 1976

26. Holland JM, Graham JB, Ignatoff JM: Conservative management of twisted testicular appendages. J Urol 125:213, 1981

27. Chen DCP, Holder LE, Melloul M: Radionuclide scrotal imaging: Further experience with 210 patients. Part I: Anatomy, pathophysiology, and methods. J Nucl Med 24:735, 1983

28. Kohler FP: On the etiology of varicocele. J Urol 97:741, 1967

29. van der Vis-Melsen MJE, Baert RJM, van der Beek FJ et al: Sensitivity of scrotal scintigraphy in the diagnosis of varicocele. Clin Nucl Med 7:287, 1982

Atlas Section

Figure 11-1 *Anatomy of the Testis*

(A) A normal scrotum is illustrated in the left lateral projection. The separation between the visceral and parietal layers of the tunica vaginalis is normally a potential space; the gap has been exaggerated in this case to emphasize anatomic relationships. Note that the testis is normally a retroperitoneal organ with both layers of the tunica vaginalis anterior to it. The epididymis is attached to the posterolateral margin of the testis. Both are normally anchored to the posterior scrotal wall by the testicular mesorchium. The testicular hilum is half-way along the posterior margin of the testis. *(B)* Bell-clapper deformity. In this bilateral congenital abnormality, the tunica vaginalis completely invests the testis. The normal posterior mesorchial anchor is absent, allowing the testis to twist on its

A

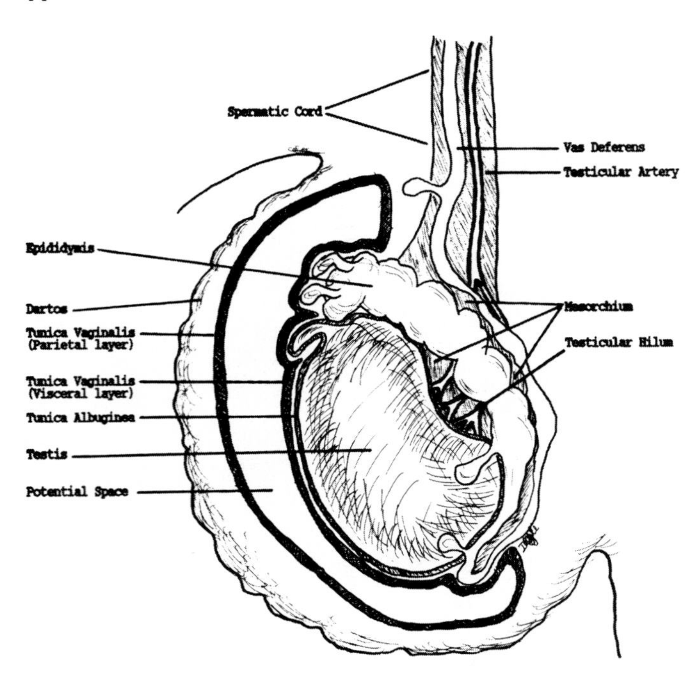

vascular pedicle. The posterior midtesticular insertion of the testicular artery (testicular hilum) results in the abnormal horizontal lie of the testis that is diagnostic of bell-clapper deformity. *(C)* Abnormal posterior fixation. Abnormal fixation is a rare cause of torsion. In this case, as a result of a redundant mesorchium, the testis is poorly attached to the epididymis which in turn is loosely attached to the posterior scrotal wall, allowing rotation. The scrotal mesorchium is the connective tissue that attaches the testis to the epididymis, the epididymis to the intrascrotal portion of the vas deferens, and the entire complex to the posterior scrotal wall. Normally this is a firm attachment.

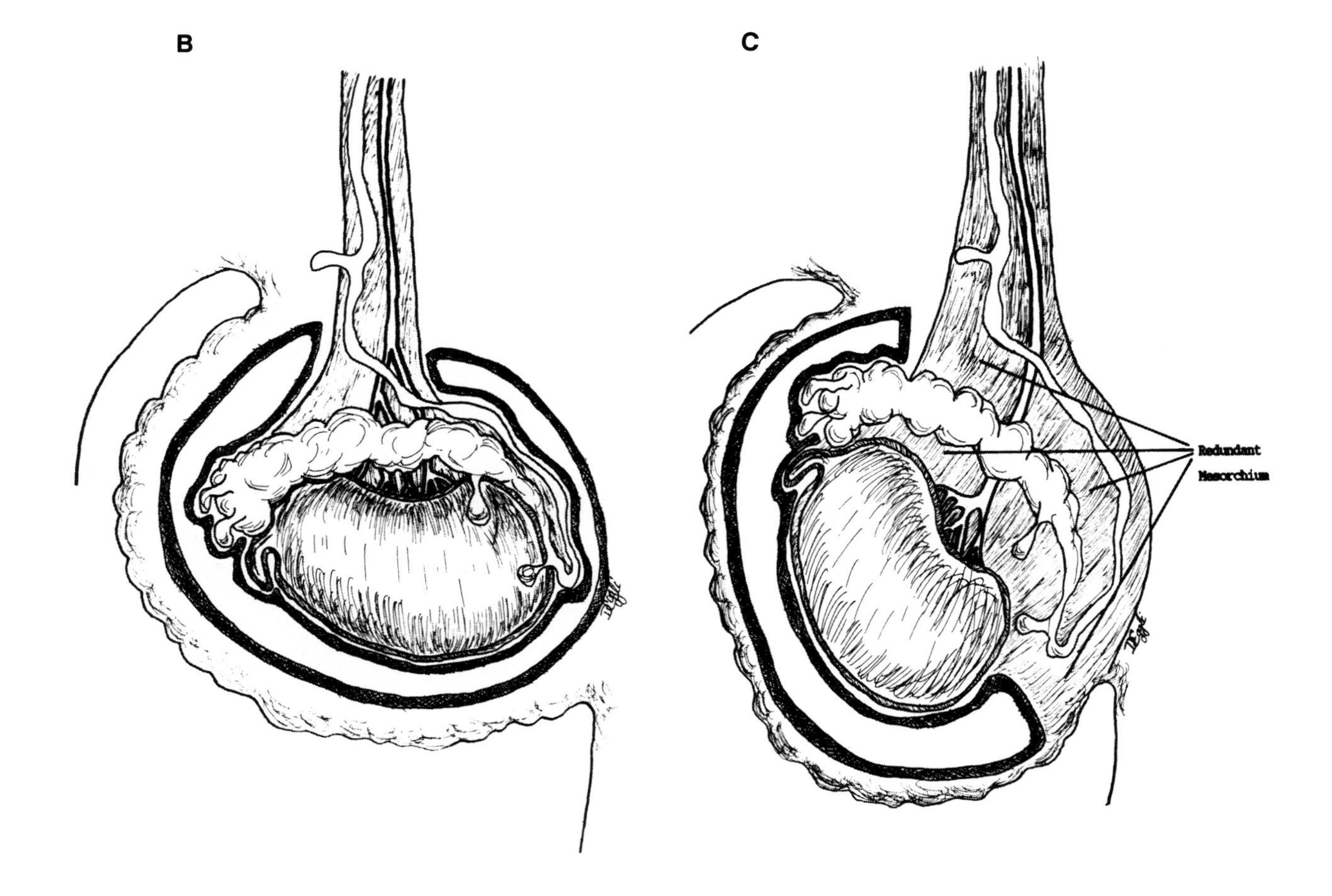

Figure 11-2 ***Vascular Anatomy***

The dual arterial supply of the male genitalia is demonstrated in relation to the scrotum, bony pelvis, and vessels of origin. Arterial anatomy is demonstrated on the left and venous anatomy of the spermatic cord is demonstrated on the right side of the pelvis. Scrotal venous drainage is not demonstrated, as it does not play a significant role in the scintigraphic evaluation of testicular torsion. Note that the vascular trunk of the spermatic cord enters the scrotum more superiorly and more vertically than the pudendal vessels supplying the scrotal walls, which enter the scrotum more horizontally and laterally. Individual spermatic cord vessels are not resolved scintigraphically, but are seen as a vascular trunk.

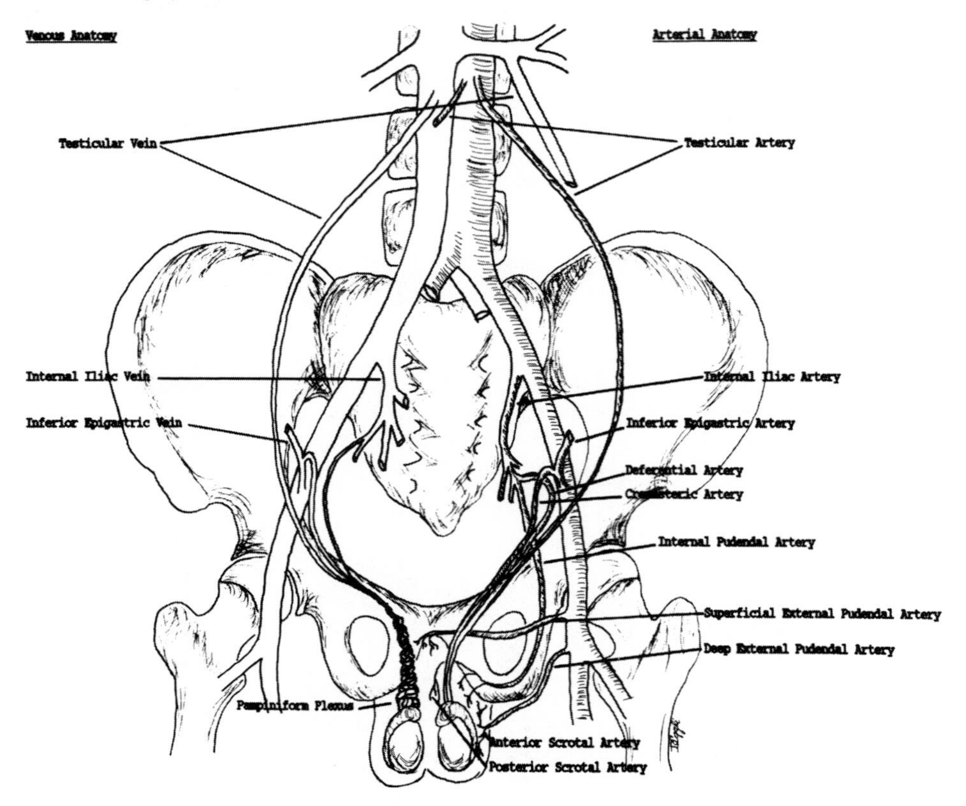

Figure 11-3 ***Normal Testes of a 34-Year-Old Man with Persistent Pain One Week after Vasectomy***

Physical examination was normal. ***(A)*** Flow study: Flow images obtained at 5 sec/frame demonstrate a normal flow pattern to the scrotum. Normally distinct spermatic cord and scrotal vascular groups are not identified. A uniform capillary and venous phase blush of the scrotum is recognizable. ***(B)*** Tissue phase: The image demonstrates normal right and left testes *(open arrows)* overlying a custom-made groin shield. The shield is designed not to overlie vascular structures of the groin or pelvis. A curved or semicircular cut-out along the top margin of the shield allows placement of the shield at the base of the scrotum with the testes overlying the shield. ***Comment:*** If there is cremasteric spasm and upward retraction of the symptomatic testis (testis redux, a common physical examination finding in torsion of the spermatic cord), the examiner can gently position the symptomatic testis over the shield. In our experience, there has not been a patient whose testes could not be positioned over the shield by the examiner. Similar experiences with the use of a groin shield even in the smallest children have been reported at Children's Hospital in Boston[11] and Children's Hospital National Medical Center in Washington, D.C.[12] Chen and colleagues[13] report a 26% improvement in diagnostic accuracy and/or confidence using a groin shield. Activity in normal testes is approximately equal to that of adjacent thigh and should be symmetric left to right. In the early part of tissue phase, the bladder *(Bl)* is commonly seen as an oval or rounded photon deficiency above the base of the penis. Likewise, a prominent blood pool blush of the penis is common (not present in this image), which is why the penis is taped up in the midline over the lower abdomen. If more delayed tissue phase images are obtained, the bladder may begin to fill in with activity excreted by the kidney (as in *C*). (*I*, iliac vessels; *P*, pudendal vessels; *F*, femoral vessels)

(C) Marker view. Position of normal and symptomatic testes is always confirmed by physical examination in the scanning position. A cobalt marker is always used to confirm the location of the symptomatic testis. All examinations shown in this chapter have had the location of the testes delineated by physical examination and confirmed with marker views. This should be routine. Note that the bladder *(Bl)* is beginning to fill in with Tc-99m O_4^- excreted by the kidneys.

A

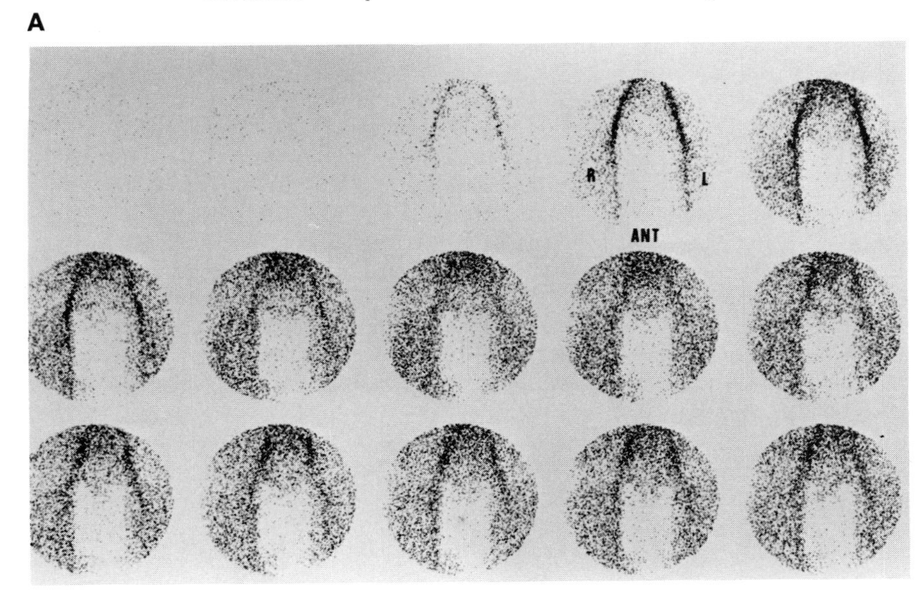

B

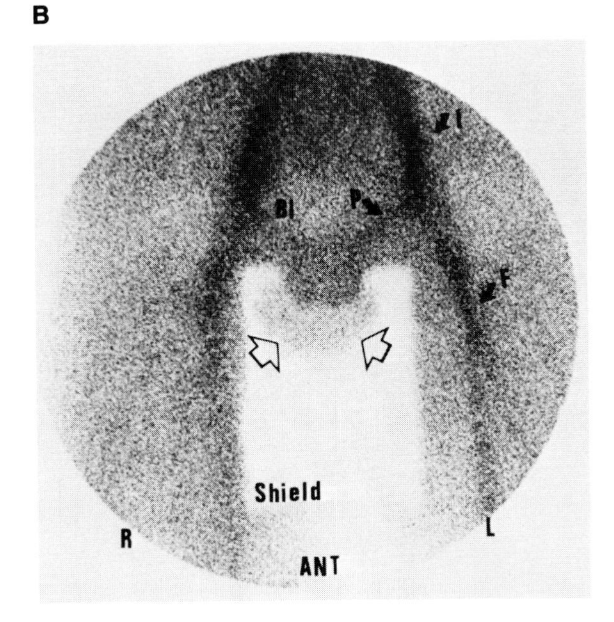

C

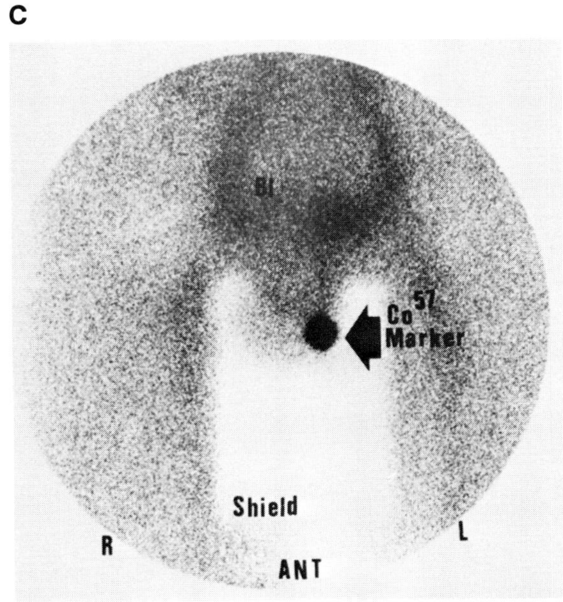

Figure 11-4 *Torsion of the Spermatic Cord*

A 28-year-old man presented with acute onset of severe left scrotal pain of approximately 6 to 8 hours' duration. The patient was nauseated on the scan table. Physical examination revealed a mildly enlarged, firm, indurated, exquisitely tender left testis. There was no prior history of testicular pain. Abnormal lie was not detected on the unaffected side. ***(A)*** Flow study: A nubbin sign is demonstrated on the left in the third flow image *(closed arrow).* A small amount of normal flow through the spermatic cord is seen on the right *(curved arrows).* Subtle reduction in flow can be seen over the left hemiscrotum *(open arrow);* however, areas of absent or reduced perfusion are better seen on the tissue phase image. ***(B)*** Tissue phase image: Open arrows mark the location of the right and left testes on physical examination. There is essentially no perfusion in the left testis in the initial tissue phase image. An oval area of increased tracer is present in the midline above the testes at the base of the penis *(arrowheads);* this is a normal blush at the base of the penis, due to the venous blood pool of the paired corpora cavernosa of the penis. The large curved arrow ***(A)*** marks an artifact from the examiner's hand. It was necessary to gently retract the testes over the shield, due to testis redux. ***(C)*** Late tissue phase. A small amount of Tc-99m O_4^- is now seen in the left testis *(open arrows mark right and left testes).*

Comment: The bell-clapper deformity will produce an abnormal lie on the unaffected side, as the deformity is virtually always bilateral. The abnormal lie can be detected on physical examination in approximately a third of all cases and is diagnostic of the deformity.[6,14,15] Pain is present in the vast majority of patients with torsion of the spermatic cord, except in the neonate in whom pain is usually absent. Nausea and vomiting may be present in as many as 40% of affected individuals.[6] Urinary tract symptoms are uncommon and urinary sediments are unremarkable. Low-grade fever may be present in as many as 20%. The peak incidence of testicular torsion occurs during adolescence with approximately 65% of cases occurring in children from ages 12 through 18.[6] The cumulative risk of torsion for a male up to age 25 is 1 in 160.[6] Torsions have been reported to occur from birth into the eighth decade of life, but torsion after the age of 40 is rare. In neonates, the torsion usually occurs *in utero.* The mechanism of neonatal torsion is different from that in the bell-clapper deformity. In neonatal torsion, the entire scrotal contents are involved in the torsion, including both layers of the tunica vaginalis.[6]

The significance of the nubbin sign in ***A*** is the same as with any vessel cut-off on radionuclide angiography and represents vascular occlusion, in this case as a result of the torsion of the spermatic cord.[16]

Torsions producing profound testicular ischemia have been reported from as little as 120° torsion to as much as 1440° (four complete turns).[6] Initially after torsion, venous return is occluded by the twisted vascular pedicle. Arterial perfusion continues until venous congestion chokes off the arterial supply within the tough fibrous sheath of the spermatic cord. The fibrous sheath of the spermatic cord permits only limited expansion of volume, effectively producing a compartment syndrome. It is the initial venous occlusion that leads to the firm swollen testis, the "woody induration" described by Williamson.[6] The pattern of late and very reduced activity in the testis seen in this patient suggests that the torsion may be less than 360° and that there is the likelihood of testicular salvage. At surgery in this patient, the torsion was only 270°.

A

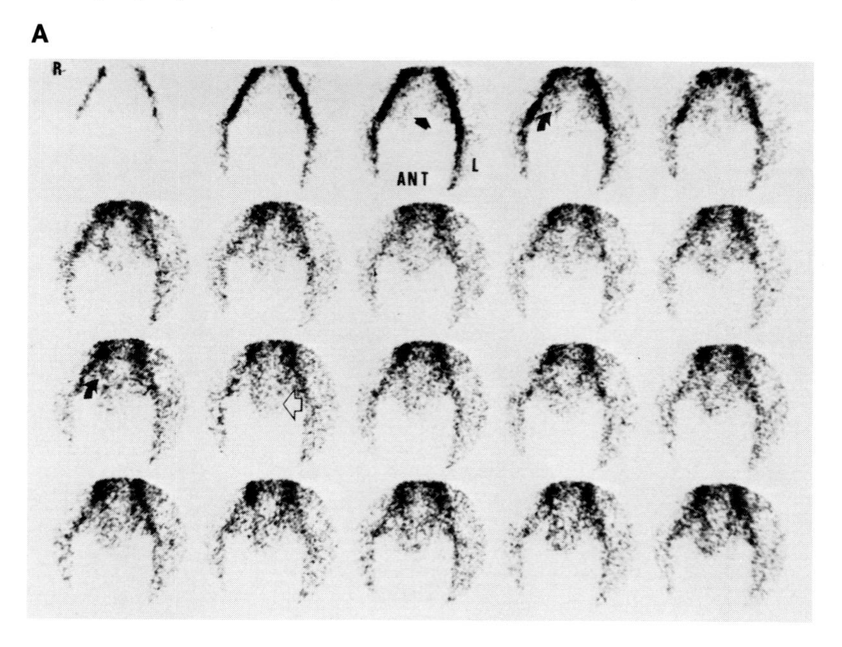

B

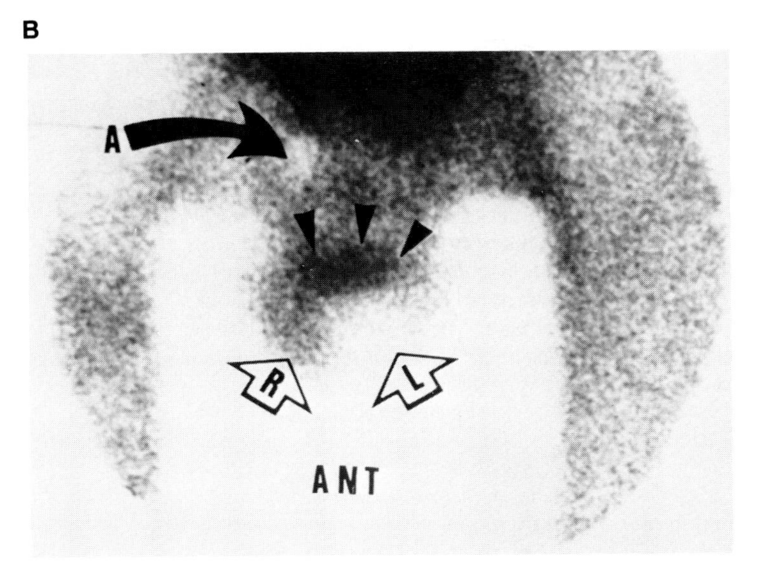

C

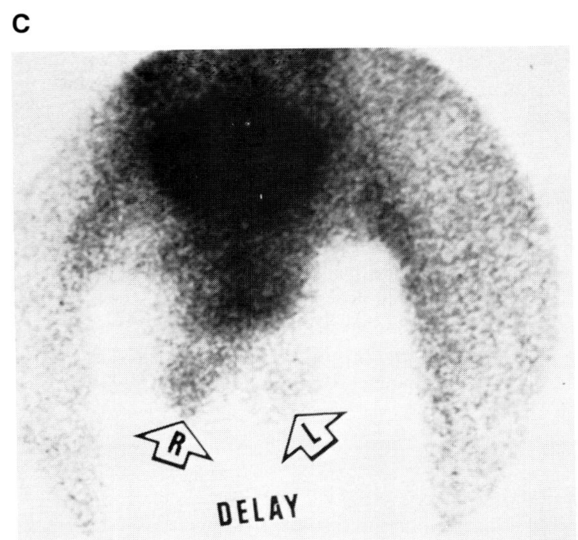

Figure 11-5 ***Testicular Torsion (imaged without a groin shield)***

A 19-year-old man presented after sudden onset of severe but intermittent left testicular pain accompanied by nausea and vomiting over the four hours prior to presentation. The patient was afebrile with a normal urinalysis and white blood cell count of 11.1 K. Physical examination demonstrated an enlarged, firm, tender left testis. ***(A)*** Flow study: A faint amount of normal perfusion is seen on the right side *(straight arrows)*. Although more difficult to see on the unshielded examination, perfusion is reduced on the left. *(Curved arrow marks left testis.)* ***(B)*** Tissue phase: Images are more difficult to interpret without a groin shield, but demonstrate decreased perfusion in the left hemiscrotum *(curved arrow)*. Activity on the right side is normal for an unshielded examination. *(Straight arrow marks the right testis.)* ***(C)*** Marker view confirms that the area of reduced photon accumulation is the symptomatic left testis *(cobalt dot is centered on the left testis)*. The reduced activity on the left compared to right is probably more easily seen on this image than on the initial tissue phase image. Bladder *(Bl)* activity is again noted on the delayed tissue phase image, which is normal, and normal slightly increased activity is seen at the base of the penis *(P)*.

A

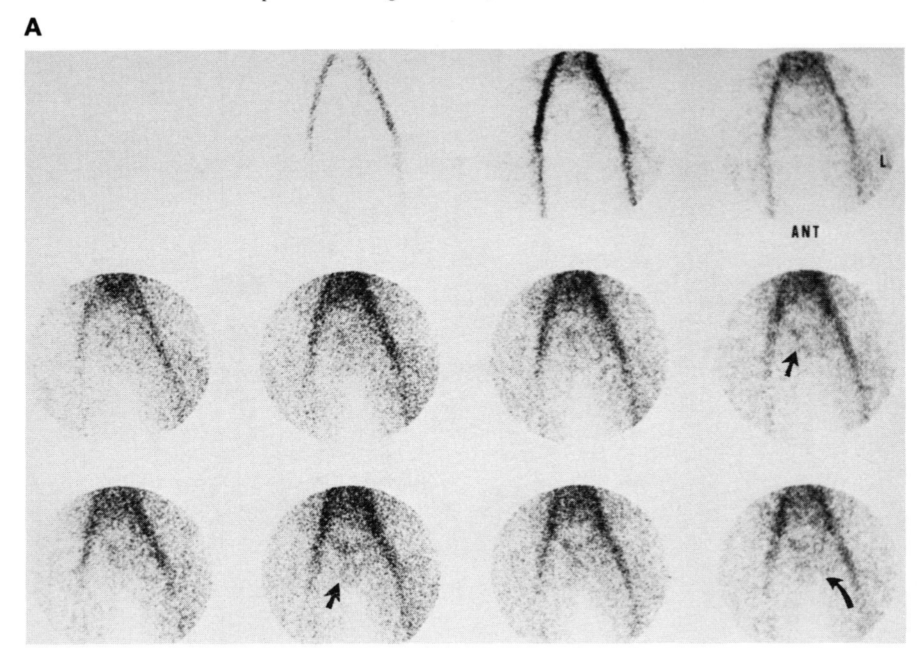

B

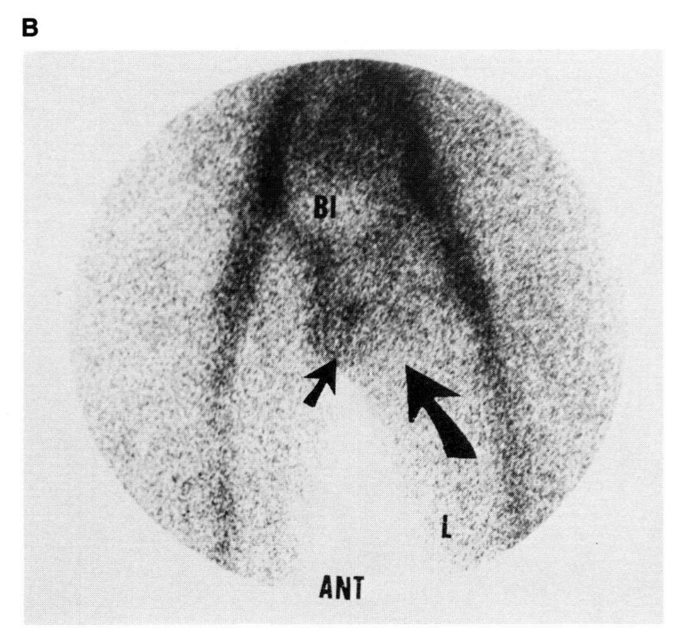

C

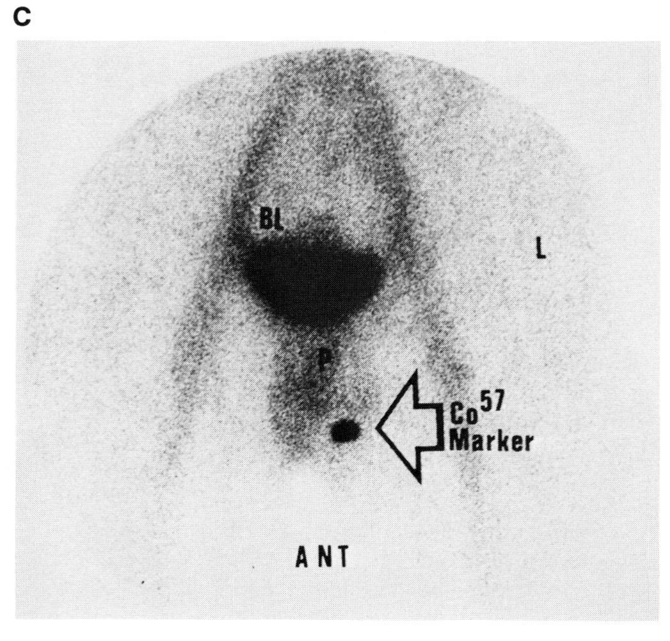

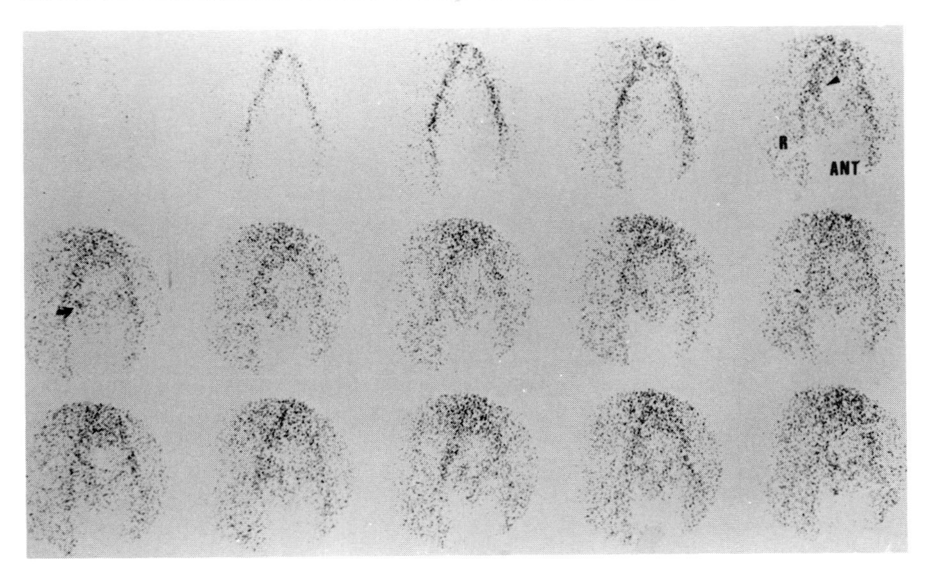

A

Figure 11-6

Missed Torsion

A 29-year-old man presented six days prior to this examination with right scrotal pain, which was felt clinically to represent epididymitis. He was initially treated with antibiotics, based on the clinical diagnosis. He returned on Day two and again on Day six with persistent pain and was ultimately referred for scintigraphy to evaluate slowly resolving epididymitis (or epididymitis unresponsive to therapy) *versus* missed torsion. ***(A)*** Flow study: A subtle nubbin sign is present *(upper right, arrowhead)* along with a minimal amount of scrotal perfusion through the pudendal trunk. *(Curved arrow marks right pudendal trunk.)* ***(B)*** Tissue phase: The image demonstrates the "rim sign" [17] of missed torsion and reduced tracer accumulation in the symptomatic testis. ***Comment:*** Missed torsion typically is demonstrated by normal to mildly increased flow. Spermatic cord vessels and scrotal vessels are not usually well-delineated. Typically, when vessels are seen, they are pudendal/scrotal *(curved arrow in A)* rather than spermatic cord vessels. A nubbin sign may be seen in missed torsion as well as in acute torsion.[13,16] It may be very difficult for physicians who do not routinely see large numbers of scrotal scans to distinguish between spermatic cord and pudendal vessels. Spermatic cord vessels originate higher and have a steeper course to the scrotum than the pudendal supply, which originates lower and has a somewhat more horizontal course. However, when only one of the two supplies is visualized, determining its origin can be extremely difficult.

The hyperemic rim around the torsed testis is due to dartos hyperemia in response to the torsion. The testis itself usually demonstrates reduced tracer accumulation in comparison to the uninvolved testis. This case demonstrates both the rim sign and the reduced tracer in the torsed testis. It should be noted that the ability to detect a reduction in testicular activity (compared to the uninvolved testis) inside the rim will depend on the intensity of the activity in the rim in relation to the size of the testis (*i.e.,* if the testis is relatively small and the rim relatively intense, the central photon deficiency may not be detectable).

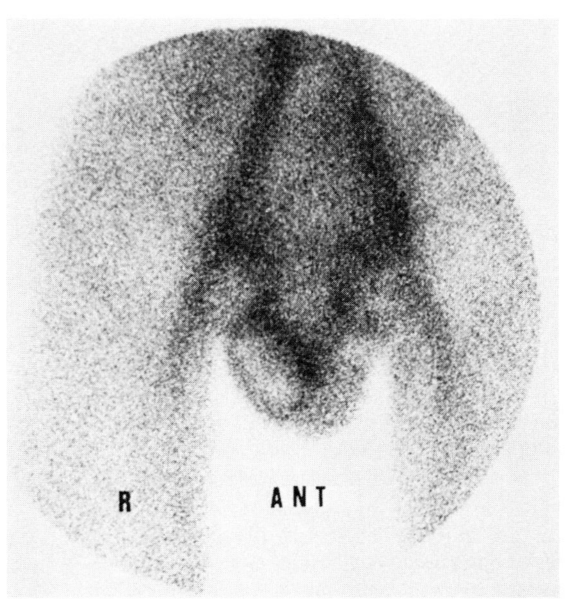

B

The rim of the missed torsion is usually fairly thin, uniform, and complete. Medially in the upper pole, some hyperemia of the scrotal mediastinum may appear to thicken the rim. A thick, shaggy, irregular, or discontinuous (broken) rim should suggest other possible diagnoses such as epididymitis, epididymo-orchitis, abscess, or tumor. The flow study may add specificity to the diagnosis. It should also be noted that not all missed or late torsions will have a well-developed rim sign, and a few testes with the pattern of a missed torsion may be salvageable, at least with respect to endocrine function. This does not, however, diminish the importance of recognizing the scintigraphic pattern. The vast majority of missed torsions will demonstrate a typical rim sign. Late phase viability is low, less than 20%,[18,19] and the condition is an operative one, regardless of ultimate viability due to the previously discussed, immunologically mediated injury to the contralateral testis (see discussion of Figure 11-7 for additional detail).

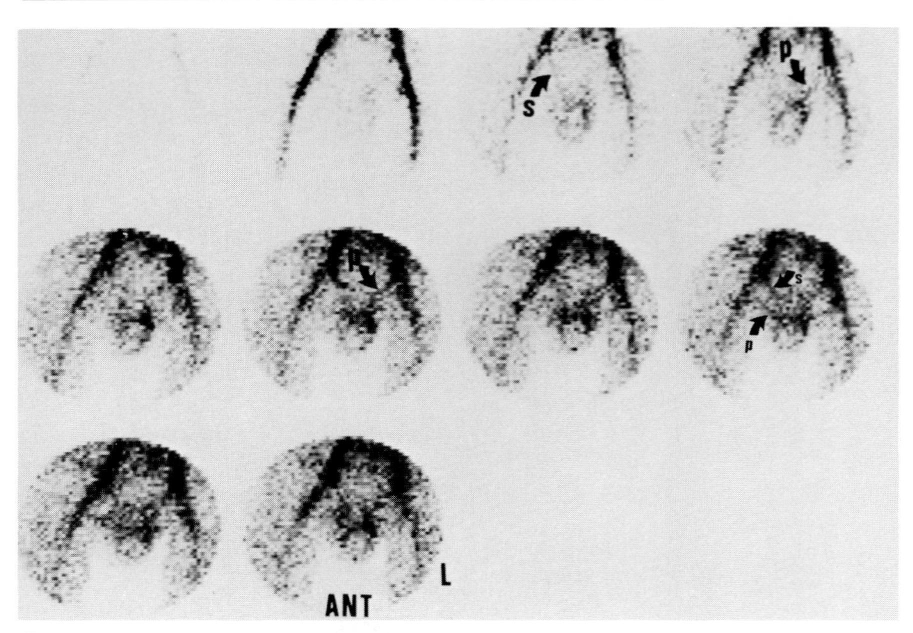

A

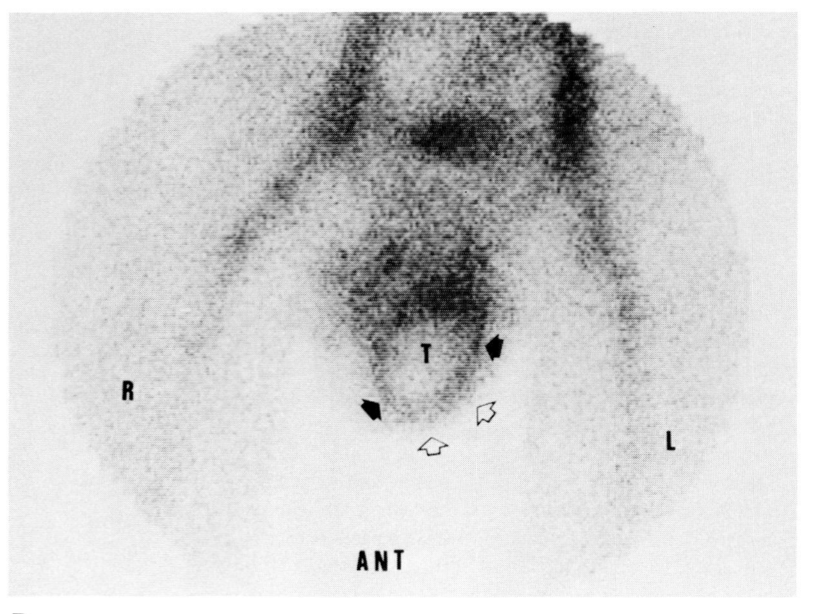

B

Figure 11-7

Missed Torsion

A 15-year-old youth presented with left scrotal pain. He had been kicked in the groin three days prior to the onset of pain. He had the additional history of similar pain on several prior occasions, resolving spontaneously. ***(A)*** Flow study: Prominent hyperemia is noted in the left hemiscrotum in the capillary and venous phases of the flow study. There is, however, very little increased flow seen in the main vascular trunks. Only a faint amount of perfusion is seen in the left pudendal group *(p)*. A faint normal spermatic cord vascular trunk is seen on the right *(s)*. Note the slightly different site of origin and course of the spermatic cord vessels and the pudendal group. ***(B)*** Tissue phase: The tissue phase image again demonstrates the characteristic thin, uniform, continuous rim of the missed torsion *(closed arrows)* with some apparent thickening superiorly and medially from mediastinal hyperemia. The left testis *(T)* does not clearly demonstrate reduced activity compared to the right side. The scrotal skin line *(open arrows)* reveals normal overlying scrotal skin, commonly seen in missed torsions if camera resolution is adequate. ***Comment:*** Fully a third of patients presenting with torsion of the spermatic cord have had prior similar incidents of pain.[6,19] This is felt to represent recurrent, spontaneously resolving torsion or intermittent torsion.

Hyperemia in the scrotum may be fairly prominent in the flow study of a patient with missed torsion, but the vessels themselves may remain difficult to detect. When increased flow is present in missed torsion, it is by way of the pudendal vessels. Increased flow should not be seen in the spermatic cord vessels.

Experimental data show that severe damage to spermatogenic cells occurs within 4 hours of severe ischemia. Leydig (hormonal function) cells are more tolerant of ischemia but are irreparably damaged after 10 to 12 hours of profound ischemia.[20] Clinically, viability of the testis after surgical reduction is rare more than 10 to 12 hours after onset of symptoms, in the absence of spontaneous reduction.[6] Viability, however, does not imply recovery of spermatogenic function. Removal of the nonviable testis is indicated to preserve spermatic function on the uninvolved side. Retention of the infarcted testis causes immunologically mediated reduction in fertility.[8,10,11,21,22]

Figure 11-8 ***Anatomy of the Testicular Appendages***

The location of the five testicular appendages is diagramed in relation to the testis, epididymis, and vas deferens. There may be two or more appendices of the epididymis.

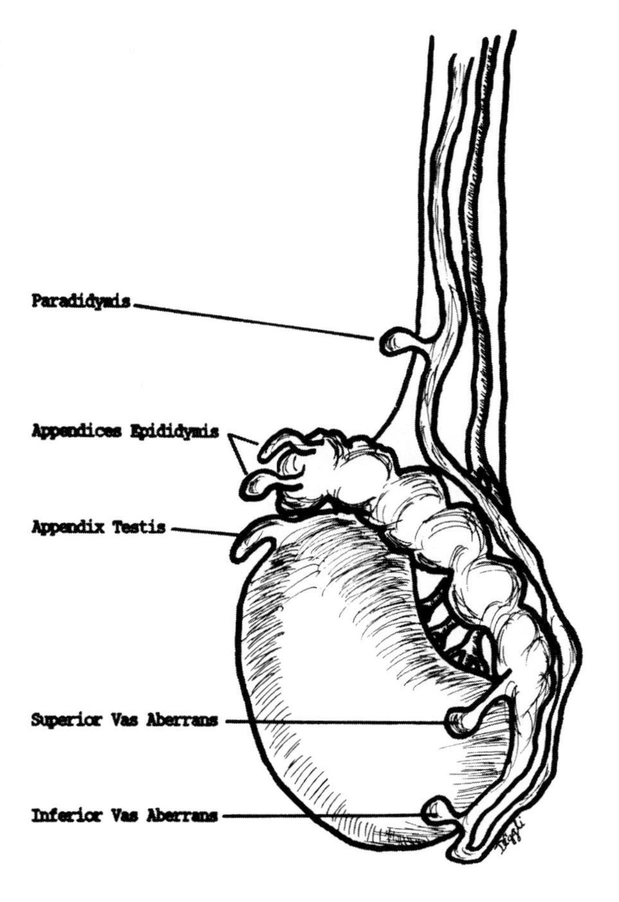

Figure 11-9 *Torsion of a Testicular Appendage*

A 10-year-old boy with pain and swelling of the left testis for five days was referred with a presumptive clinical diagnosis of missed torsion. (***A*** and ***B***) Flow study: The radionuclide angiogram demonstrates mildly increased flow to the base of the left hemiscrotum *(arrowheads)* through pudendal vessels *(curved arrow)* predominantly in the venous phase, probably due to inflammation of overlying dartos. ***(C)*** Tissue phase: There is a semilunar area of increased activity at the upper pole of the left testis *(curved arrow)*. Normal testes were present bilaterally and have normal tracer accumulation in tissue phase *(open arrows)*. ***Comment:*** The testis has five appendages: the appendix testis, the appendix epididymis, paradidymis, superior vas aberrans, and inferior vas aberrans. The appendix testis is present at least unilaterally in 92% of males, the appendix epididymis in 34%.[23] Only those appendages which are pedunculated (as opposed to sessile) are subject to torsion, that is, 82% of appendices testis and 100% of appendices epididymis. Torsion of the appendix testis accounts for 92% to 95% of all torsed appendages; nearly all of the remainder are torsions of the appendix epididymis. The other three appendages account for a few case reports.[24] Clinically, torsions of the appendix testis and appendix epididymis are identical and, due to anatomic proximity, are indistinguishable by scan. On physical examination, there may be a palpable nodule at the upper pole of the involved testis and a "blue dot" may be present deep in the soft tissues when the scrotum is transilluminated.[6,22] In comparison, although pain is present in equal frequency to patients with torsion of the cord, pain is less severe and nausea and vomiting are rare.[6] There is more likely to be inflamed skin overlying torsed appendages (60%) than with torsion of the cord (36%).[6] Abnormal testicular lie is rarely seen. In children ages 13 and under (except for neonates), torsion of the appendages occurs with equal frequency as torsion of the spermatic cord. The peak incidence falls between ages 7 to 14 years, with 85% of cases occurring in the peak age range.[6,22] Torsions of the appendages are rare over the age of 18.

This is the typical pattern of focal accumulation seen in torsion of the appendages. This pattern is well recognized by pediatric nuclear imagers and has been reported by Mendel at Children's Hospital in Boston[11] and by Gilday at the Hospital for Sick Children in Toronto.[25] However, the pattern is not present in all cases of torsed appendages. In a series of 14 cases from Children's Hospital in Boston, 7 demonstrated abnormality in both phases, 2 were abnormal on flow study only, and 5 were normal.[11]

It is important to recognize the pattern of torsion of the testicular appendages, as management is conservative and nonoperative. Holland reported the results of conservative management in 23 patients with torsion of the testicular appendages based on physical findings.[26] Twenty of the 23 were pain-free within a week, most in 2 to 5 days. Only three required surgery for persistent pain. Five additional patients who were treated surgically for presumed torsion of the spermatic cord were also found to have torsion of the appendix testis.

A

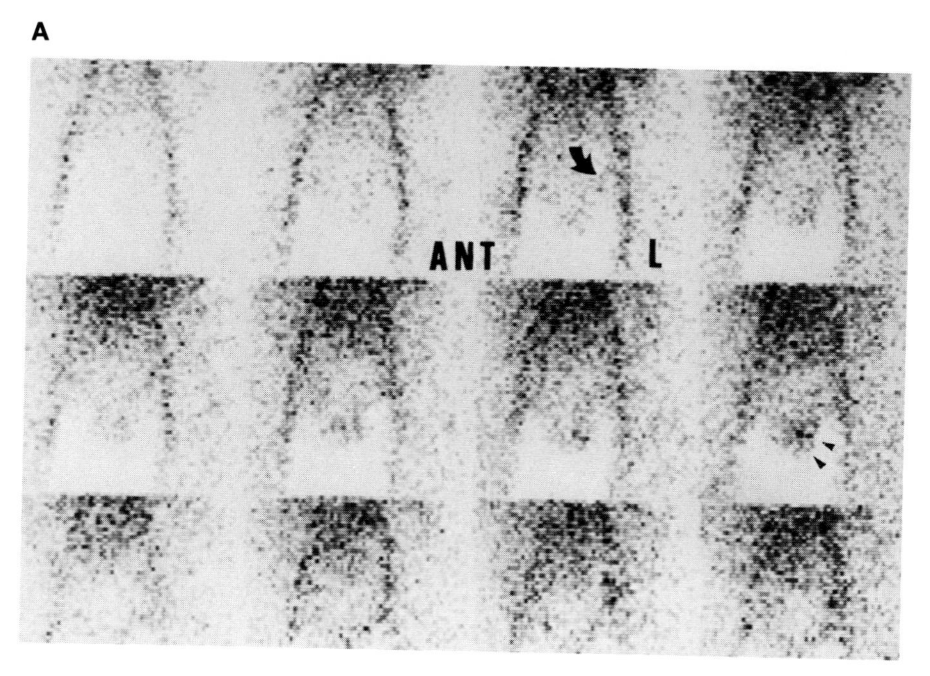

B

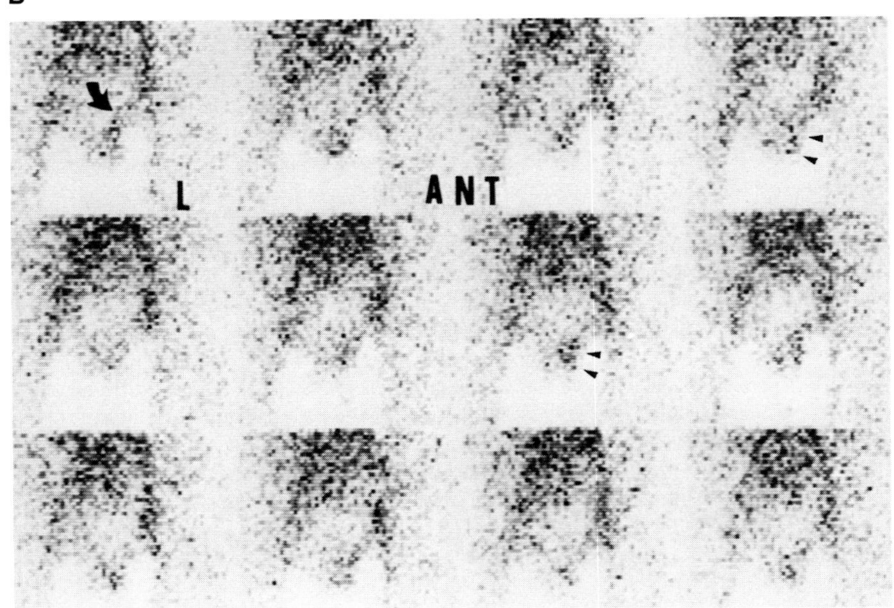

C

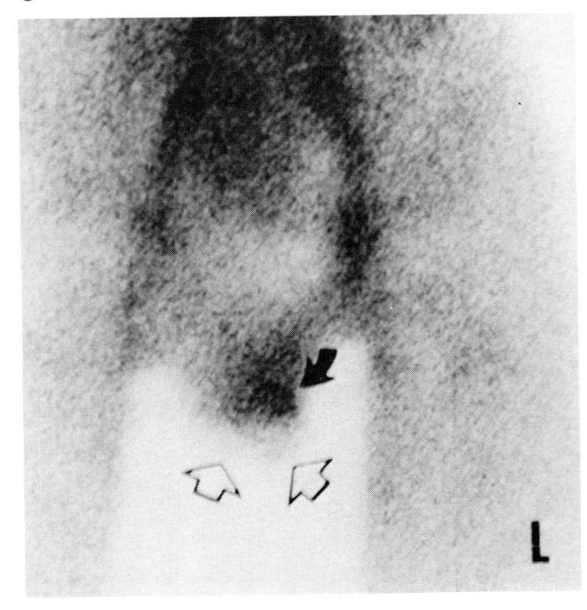

Figure 11-10 *Traumatic Torsion of a Testicular Appendage*

The patient is a young adult with persistent scrotal pain after groin trauma. ***(A)*** Flow study: There is a focal area of increased perfusion just at or above the upper pole of the right testis *(arrow)*. ***(B)*** Tissue phase: A focal comma-shaped area of increased tracer is present superior and lateral to the right testis *(closed arrow)*. Testes *(open arrows)* are normal bilaterally, confirmed by physical examination and markers. ***Comment:*** It may be difficult to distinguish by scintigraphy between torsion of the appendages and focal epididymitis in the head of the epididymis. Clinical context should permit differentiation of these two distinctly different populations. Urinary symptoms are uncommon and urinary sediments are normal in torsion of the appendages.[6] Age of presentation is also useful. Epididymitis is relatively uncommon in prepubertal males 7 to 14 years old, in whom torsion of the appendages predominantly occurs. Prepubertal children with inflammatory disease more commonly have epididymo-orchitis of viral origin, a different scan pattern from that seen in torsion of the appendix testis. The scan pattern of viral epididymo-orchitis usually shows mildly increased flow with tissue phase images demonstrating a mild to moderate diffuse increase in activity on the symptomatic side, involving testis and epididymis uniformly. Bacterial epididymitis in a prepubertal boy is distinctly uncommon. When it does occur, it is usually in association with a congenital genitourinary tract abnormality, most commonly ectopic ureteral insertion into the vas deferens or epididymis. Typical bacterial epididymitis and epididymo-orchitis first appear in the adolescent with sexual activity and peak incidence is in the late teenage and early adult years. Conversely, torsion of the testicular appendages is rare (except with a history of trauma, as in the presented case) in the age range in which epididymitis commonly occurs.

A

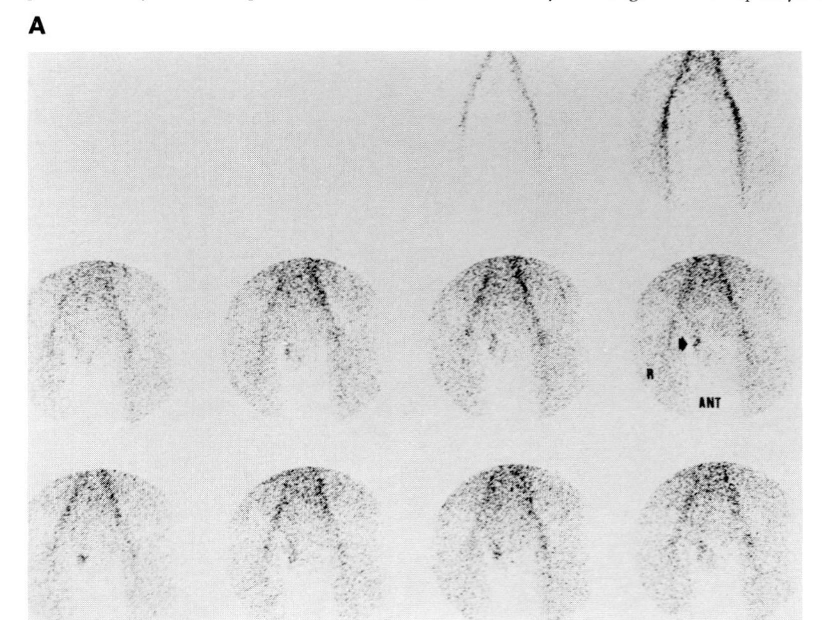

B

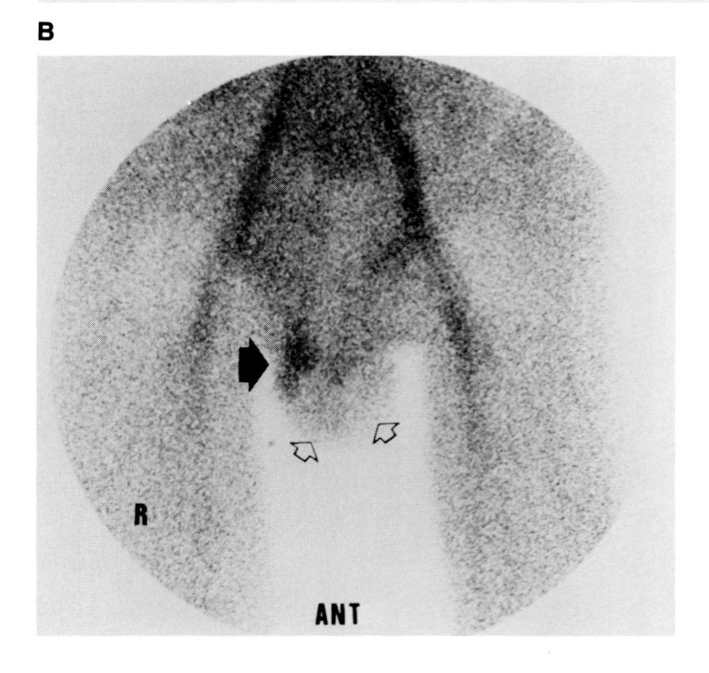

Figure 11-11 ***Traumatic Epididymitis***

A 43-year-old man presented with a history of trauma to the right scrotum one week earlier. ***(A)*** Flow study: Increased perfusion through the right spermatic cord *(curved arrow)* is demonstrated, beginning in the early arterial phase of perfusion. The increased flow is distributed curvilinearly along the lateral margin of the hemiscrotum, along the course of the epididymis *(arrowheads)*. The right testis *(open arrow)* is normal and seen medial to the inflamed epididymis. ***(B)*** Tissue phase: Hyperemia of the spermatic cord *(curved arrow)* persists in tissue phase, as does the hyperemia of the epididymis again seen as a crescent of increased tracer *(arrowheads)* along the lateral margin of an otherwise normal right testis *(open arrow)*. ***Comment:*** Traumatic, reactive, and infectious forms of epididymitis have an identical scintigraphic appearance in the absence of other complications. The study illustrates the classic scintigraphic appearance of epididymitis with hyperemia involving the head, body, and tail of the epididymis as well as the spermatic cord vessels, sparing the adjacent testis. In distinction to missed torsion, the pattern of epididymitis is a lateral hyperemic crescent or half circle, rather than a complete rim.

A

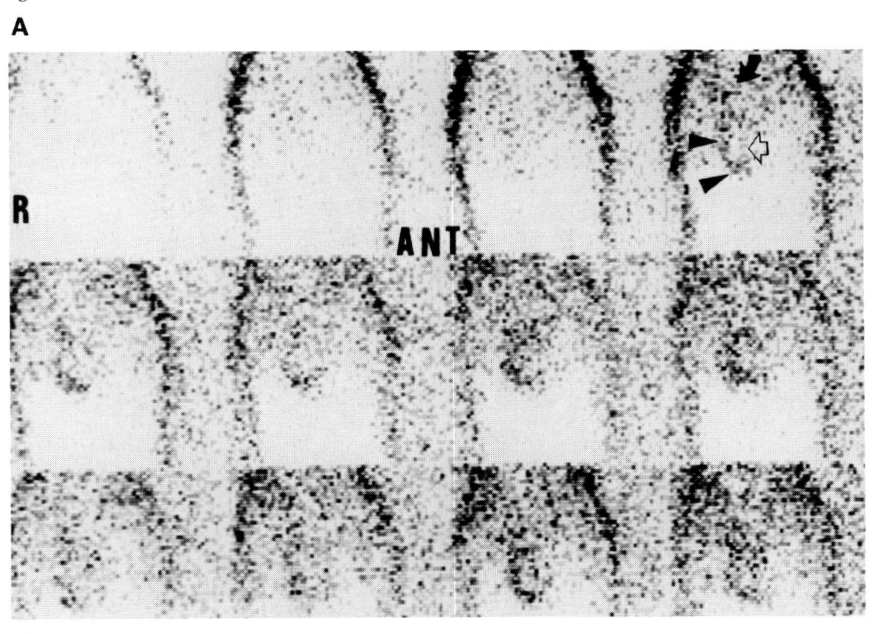

B

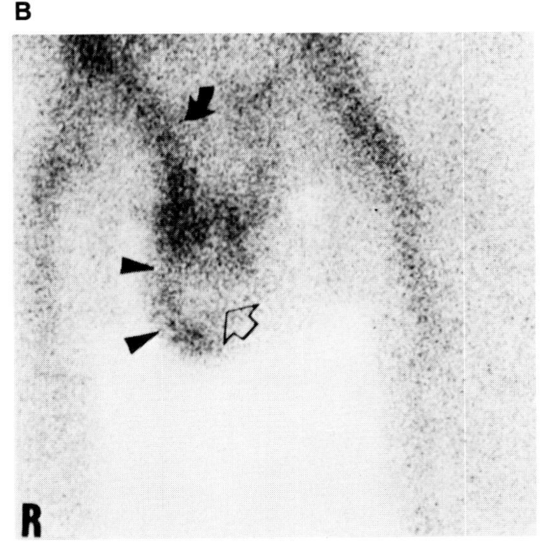

Figure 11-12 *Epididymo-orchitis*

A 24-year-old man presented to the emergency room with acute onset of right testicular pain. He had no history of trauma. ***(A)*** Flow study: The radionuclide angiogram demonstrates diffusely increased perfusion to the entire right hemiscrotum *(arrowheads)* through spermatic cord vessels *(curved arrow)*. ***(B)*** Tissue phase: Hyperemia of the spermatic cord and head of the epididymis persist in tissue phase *(curved arrows)*. There is diffusely increased tissue phase tracer accumulation in the right testis *(open arrow)* as compared to the normal left side. A reactive hydrocele is present lateral and superior to the right testis *(arrowheads)* and is indicated by a curvilinear photon deficiency. ***Comment:*** Acute onset of pain is not the most common presenting complaint in inflammatory disease. The onset of pain is usually more gradual in infectious epididymitis and epididymo-orchitis. A more significant systemic illness is usually present. Fever is more likely to be present and less likely to be low-grade, often greater than 38°C. The white blood cell count is likely to be significantly elevated. Urinary symptoms (dysuria and frequency) are common. Pyuria and significant bacteriuria are likely to be present on examination of the urine.[6] The yield on urinalysis and culture can be significantly increased by prostate massage.

This case points out that clinical history may be misleading and that there may be a significant overlap in the clinical presentation of testicular torsion and infection.

Reactive hydroceles are commonly present in inflammatory disease of the scrotum and often limit the physical examination of the scrotum. Although not seen on all scans and not detected on all physical examinations, reactive hydroceles (some small) are present at surgery in virtually all cases of inflammatory disease[6,22,23] and are often present on ultrasound examinations. Torsions of the testicular appendages also commonly develop significant reactive hydroceles and are often missed clinically because the reactive hydroceles obscure the diagnostic clinical findings.

A

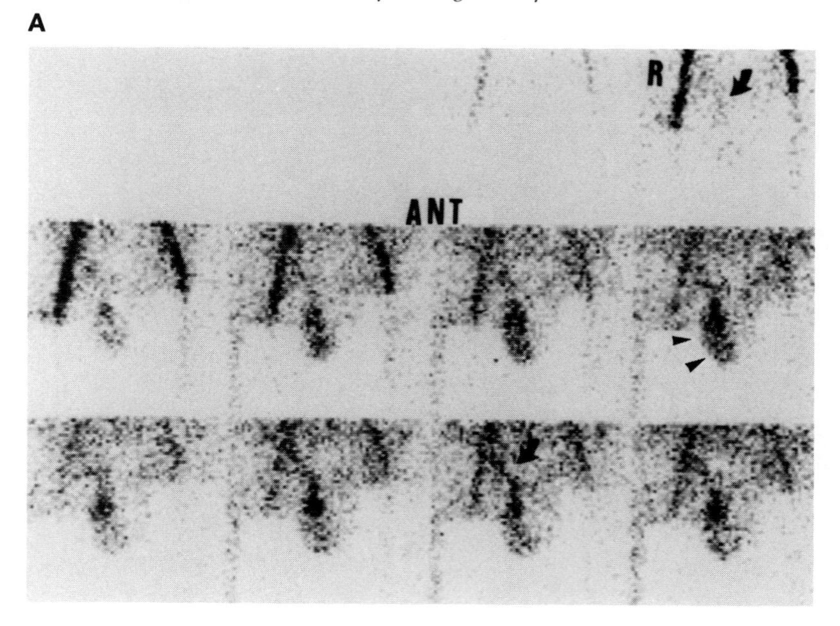

B

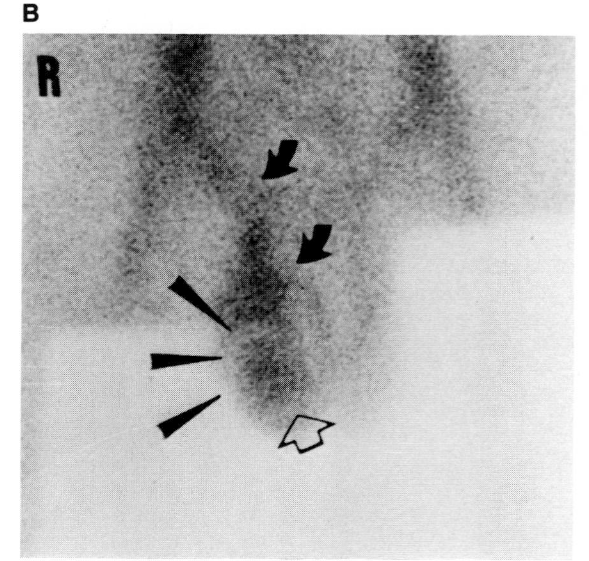

Figure 11-13 *Orchitis with Focal Epididymitis*

The patient is a 68-year-old receiving antibiotic treatment for clinically diagnosed epididymitis who returned with persistent pain and progressive swelling of his left hemiscrotum. ***(A)*** Flow study: Prominent increased flow through the spermatic cord vessels to the head of the epididymis is present in the early portion of the radionuclide angiogram *(arrows)*. In the late phase of the flow study, the hyperemia extends to involve the majority of the left hemiscrotum. Significant tortuosity of the iliac arteries is incidentally noted. ***(B)*** Tissue phase: Hyperemia of the left spermatic cord and head of the epididymis *(curved arrows)* persists in the tissue phase of the study. The body and tail of the epididymis *(straight arrows)* are, however, normal. This may be due to the prior antibiotic therapy. The left testis is diffusely hyperemic compared to the uninvolved right side *(open arrows mark right and left testes)*. Note again the presence of a small reactive hydrocele *(arrowheads)* medial to the left testis. ***Comment:*** Epididymitis usually responds to antibiotic therapy and rest within 5 to 7 days after institution of therapy.[6] If the patient is unresponsive to treatment, then abscess formation, infarction secondary to infection, missed torsion of the spermatic cord, and slowly resolving epididymitis are in the differential diagnosis. Scrotal scintigraphy is a pertinent examination to obtain early in the course of a patient's reevaluation.

This patient has orchitis as well as epididymitis. Clinically, orchitis was not initially present. The continued pain and progressive swelling of the left scrotum on therapy are compatible with the development of orchitis on treatment. Abscess, infarction, and missed torsion are excluded.

A

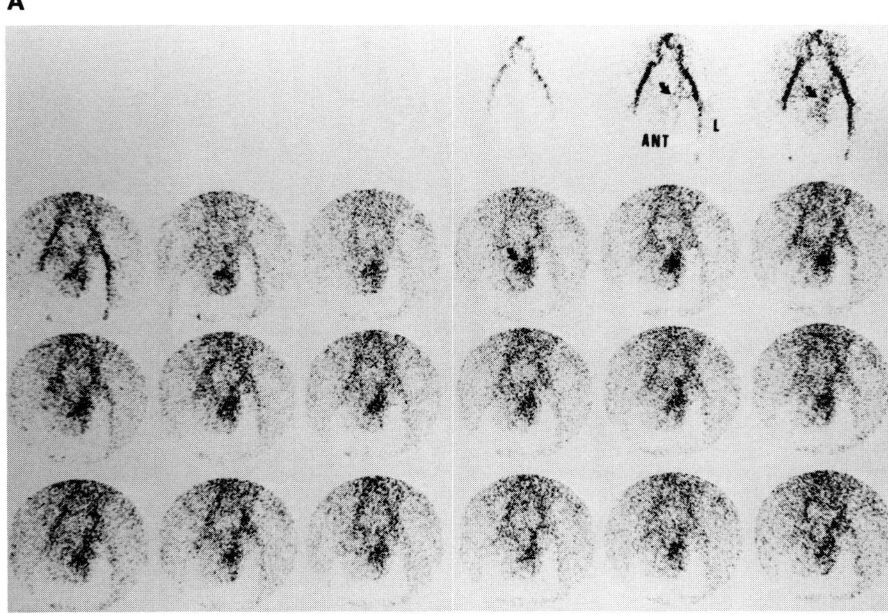

B

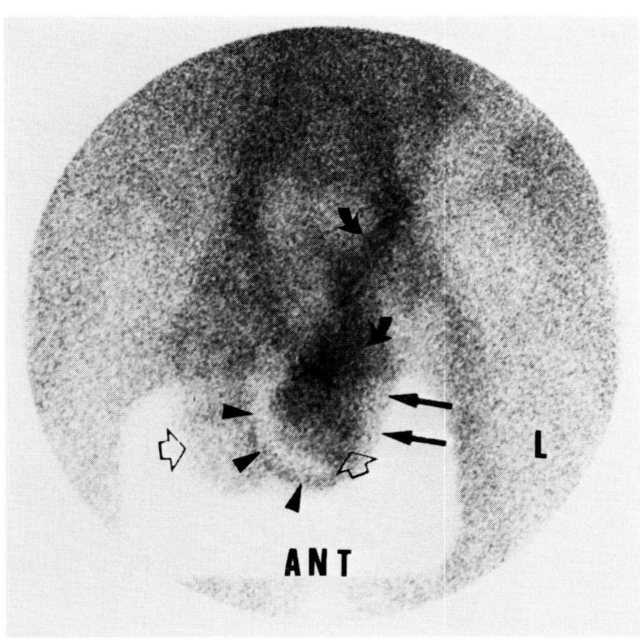

Figure 11-14 *Testicular Abscess*

A 36-year-old man experienced increasing pain, swelling, and constitutional symptoms of fever and malaise for a week after orchiopexy for torsion of the right testis. On examination in the clinic, the patient was toxic. ***(A)*** Flow study: Markedly increased perfusion to the right hemiscrotum through the right spermatic cord *(curved arrow)* is identified. A subtle photon deficiency is suggested centrally on the right *(straight arrow)*. ***(B)*** Tissue phase: Markedly increased activity is present in the right testis with two irregular photon deficiencies present centrally *(arrowheads)*. Mildly increased tracer is also present in the uninvolved left testis *(open arrow)*; this was felt to be reactive rather than infectious. ***(C)*** Ultrasound confirmed the presence of an abscess. *(Margins of the right testis are marked with arrowheads.)* The abscess was demonstrated to be multilocular. The ultrasound demonstrated the typical pattern of an abscess *(open arrow)*: a complex mass with generally decreased echo texture, irregular anechoic areas with smaller areas which are echogenic or have a mixed echo texture, compatible with cavitation and debris. The section shown is transverse through a loculus of the abscess in the midtestis.

Comment: This pattern should not be mistaken for the rim sign of missed torsion. The rim, although complete, is thick and irregular with two photon deficiencies within a very enlarged right testis. Although hyperemia of the torsed testis may be seen in the flow study of a missed torsion, this magnitude of hyperemia in the spermatic cord (as opposed to pudendal trunk) should not be seen. Also, the study should never be interpreted out of clinical context. This patient was far too sick for an uncomplicated missed torsion. The combination of clinical presentation, markedly increased spermatic cord perfusion on the flow study, marked persistent hyperemia in the tissue phase with an irregular multilocular photon deficiency, and the shaggy irregular rim suggest abscess formation. Surgery confirmed a multilocular abscess. The testis was not infarcted; however, there was too much tissue destruction by the abscess to permit salvage and orchiectomy was performed. The left testis was normal at surgery.

A

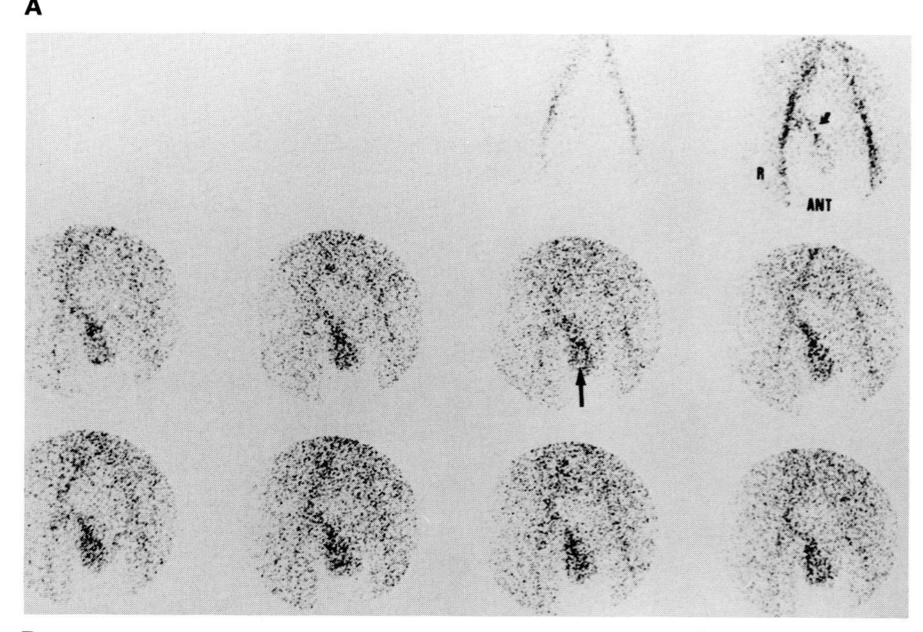

B

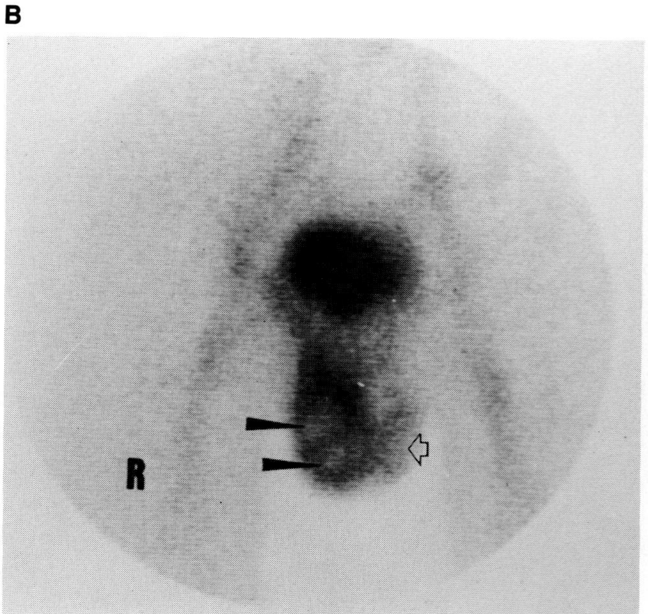

C

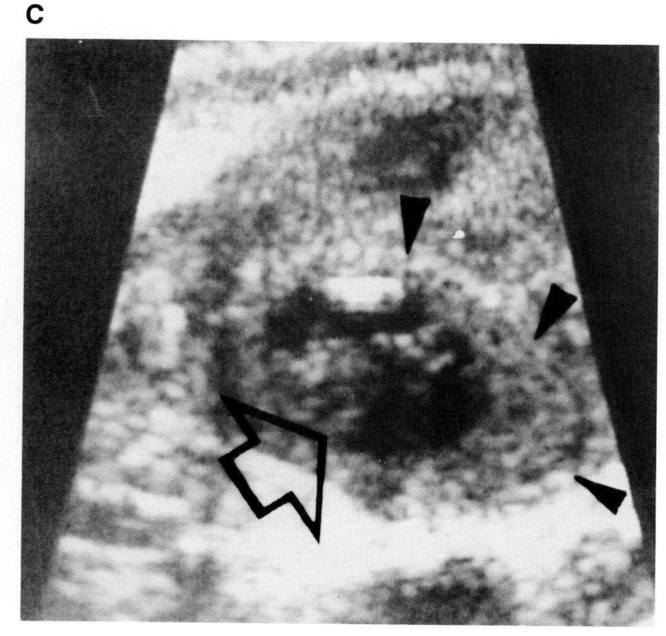

Figure 11-15 *Traumatic Hematoma/Hematocele of the Scrotum*

A 21-year-old man experienced blunt trauma to the scrotum with subsequent swelling and marked tenderness, predominantly on the right side. ***(A)*** Flow study: There is minimally increased flow to the left hemiscrotum *(arrowheads)* through the spermatic cord *(curved arrow),* which is related to a post-traumatic inflammatory process. ***(B)*** Tissue phase: Symmetric, mildly increased activity is present in both testes *(open arrows),* due to reactive hyperemia resulting from the trauma. The right testis is displaced to the left and the left testis is displaced superiorly by the large hematoma *(curved arrow)* in the right hemiscrotum, which is seen as a large photon deficiency. Location of the testes was confirmed by physical examination. Thinning of the right scrotal wall *(thin arrows)* correlates with obliteration of the scrotal rugae present on examination. Active soft tissue extravasation/bleeding is suggested in the right groin *(arrowheads),* which correlates with a large hematoma readily apparent on physical examination.

Comment: The right testis is literally floating in a lake of blood contained between the layers of the tunica vaginalis. The blood which accumulates between the layers of the tunica vaginalis is in the same location as a hydrocele. Scintigraphically, hematocele/hematoma may be indistinguishable from a hydrocele. Clinically, however, differentiation is usually not difficult. Unlike hydrocele, transillumination of a hematoma is usually impossible. Obliteration of the scrotal rugae suggests fluid under tension. Although this finding is typically seen in hematoma/hematocele, it may also be seen in large-volume hydroceles. The blood in the hematocele usually comes from an intratesticular hematoma that subsequently ruptures through the tunica albuginea. The presence of a hematocele implies the probability of significant testicular injury. Hematocele is a surgical condition, requiring scrotal exploration to determine if a repair of the laceration of the tunica albuginea is necessary.

A

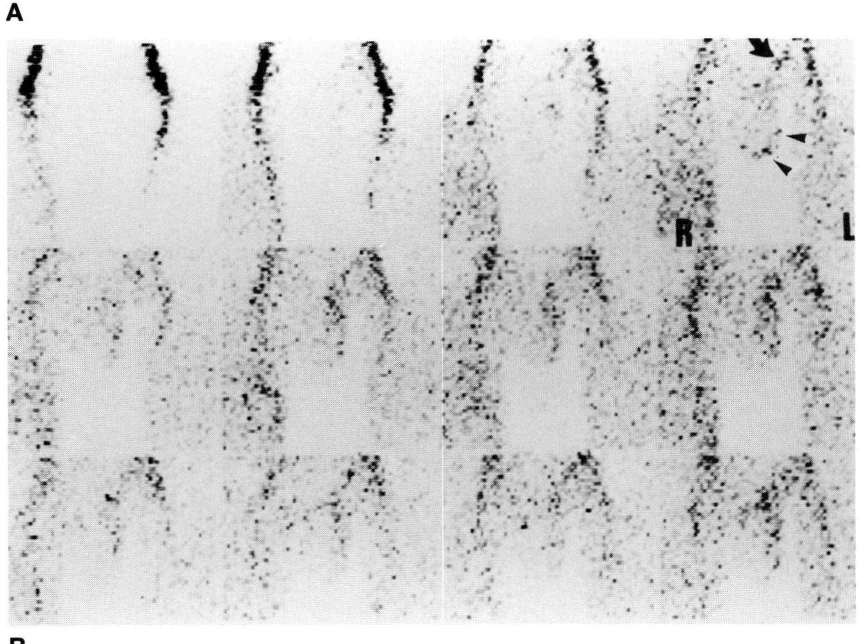

B

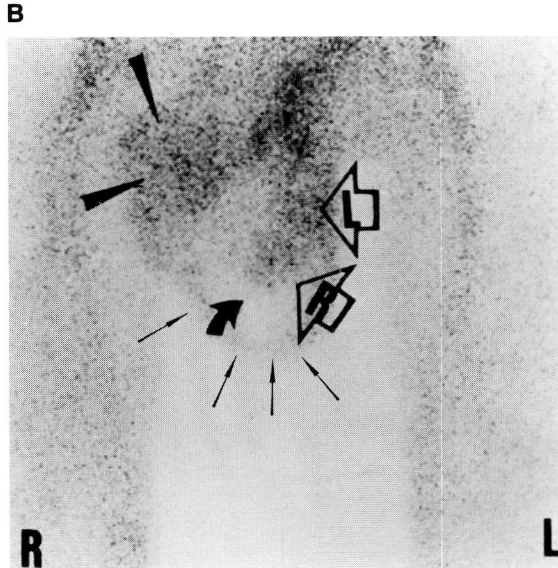

Figure 11-16 *Inguinal Hernia*

A 21-year-old man with acute onset of left scrotal pain and swelling 3 weeks earlier presented with an enlarged left hemiscrotum. He had no prior history of trauma. The flow study was normal and is not included here. This tissue phase image demonstrates a photon-deficient area in the left hemiscrotum, extending superiorly to the left inguinal ring *(arrowheads)*. The testes were normal on physical examination *(arrows)*, but were displaced to the right by the herniated bowel loop. ***Comment:*** Differential diagnosis would include hydrocele and hematocele. The demonstration of a normal testis excludes torsion of the spermatic cord. Clinical history excludes hematocele. A "tail" on the photon-deficient mass, extending to the level of the inguinal ring, should suggest the possibility of an inguinal hernia. Inguinal hernias occur through a patent processus vaginalis. The processus vaginalis is patent in nearly all newborns, 60% of two-year-olds, and 20% of adult males.[27]

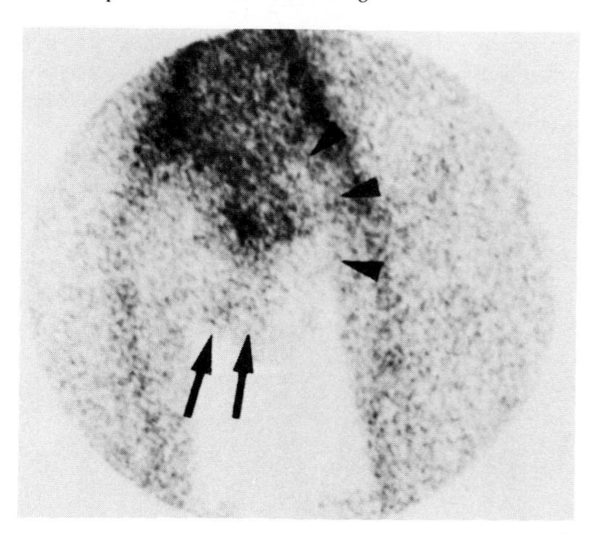

Figure 11-17 *Hydrocele*

A 19-year-old youth presented with a complaint of swelling and moderate pain in the right hemiscrotum. He had no history of trauma. Symptoms had been increasing gradually. ***(A)*** Flow study: Scrotal perfusion is normal. ***(B)*** Tissue phase: The tissue phase image demonstrates a large half moon-shaped band of photopenia lateral to the right testis *(arrowheads mark overlying scrotal skin)*. Testes *(open arrows)* show normal symmetric activity. ***Comment:*** Lack of abnormal findings in the flow study or in the testis and epididymis suggests uncomplicated primary hydrocele. The hydrocele will transilluminate on physical examination. In secondary hydrocele, the cause (*e.g.*, epididymitis, torsion, trauma) may be apparent from the scan.

A

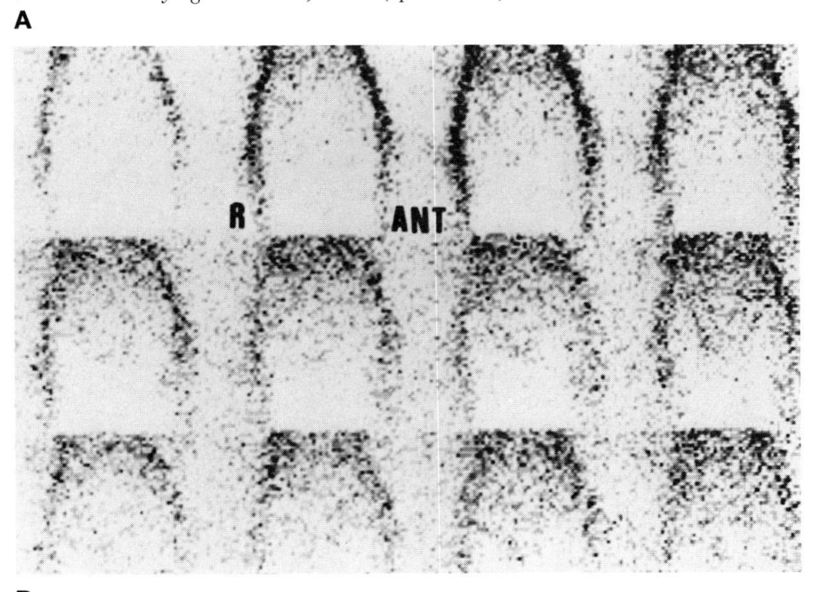

B

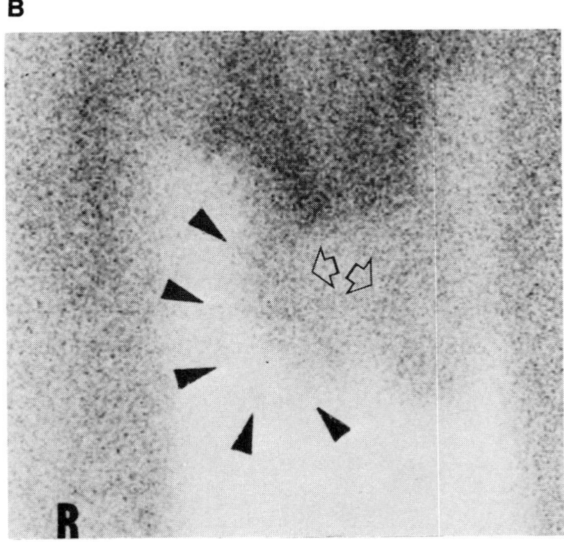

Figure 11-18 ***Testicular Tumor***

A 23-year-old man noticed swelling of the left testis after groin trauma. On presentation, the left testis was markedly enlarged, hard, and painless. Overlying scrotal skin was normal. The right testis was normal to palpation. The scintigraphic examination was performed without a groin shield. ***(A)*** Flow study: Scrotal perfusion is normal. ***(B)*** Tissue phase: The left testis *(arrowheads)* is enlarged with mildly increased and somewhat mottled activity. No fluid collections are noted around the testicle. The right testis *(open arrow)* is normal for an unshielded examination. ***Comment:*** Testicular radionuclide imaging is not the primary imaging modality for imaging of testicular tumors; ultrasound is the study of choice. However, clinical history can be misleading in the case of testicular tumors, and occasionally a tumor may present as an acutely or subacutely painful scrotum. An acute event such as minor trauma may cause the patient to notice an enlarged testis and assume an acute onset. Testicular tumors may also present with acute pain and swelling, as a result of hemorrhage into a necrotic area of tumor. As such, it is important to recognize the pattern in tumors. In the case of hemorrhage into a tumor, the pattern on scan may be similar to the abscess in Figure 11-14. However, clinical findings for a toxic patient with high fever, elevated white blood cell count, urinary symptoms, and pyuria are likely to be absent. The degree of pain experienced by the patient on physical examination of the symptomatic testis is typically far less than would be expected in an abscess, or may even be absent. The testis should be hard on palpation. At surgery, this patient was found to have an embryonal carcinoma of the left testis.

A

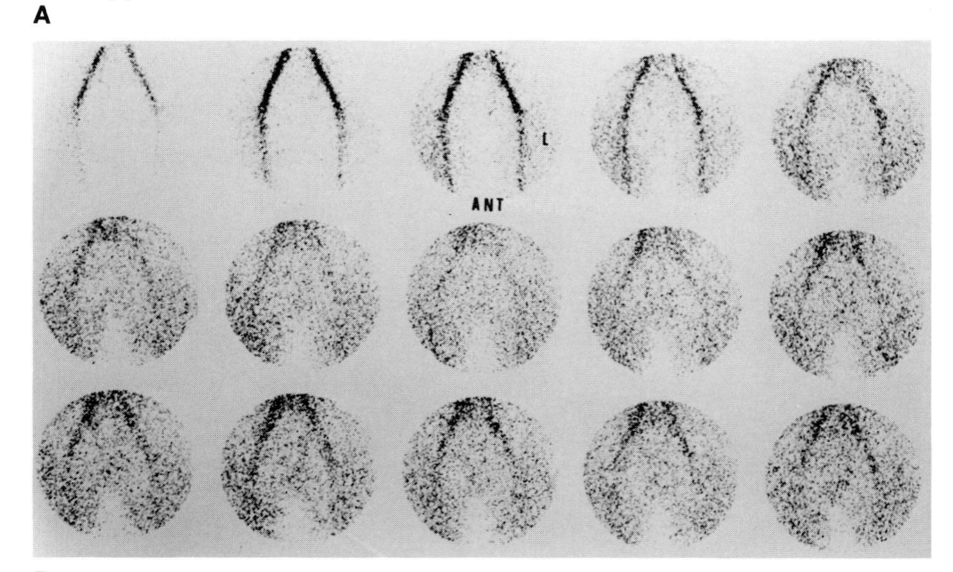

B

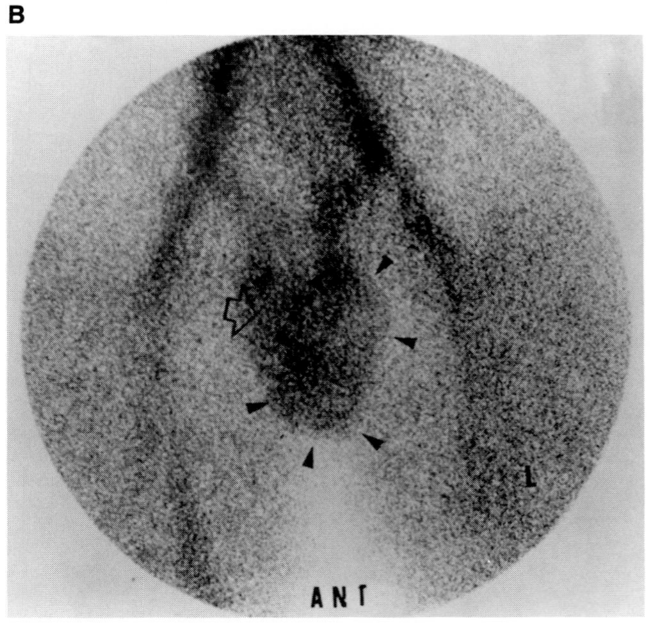

Figure 11-19 ***Varicocele, Flow Study***

The study was obtained as part of a male infertility evaluation. Five-second flow images demonstrate focally increased flow in the left hemiscrotum during the venous phase. Thin arrows in the 0-to-5-second frame mark the iliac arteries in early arterial phase. The larger arrow in the 15-to-20-second image marks the appearance of the varicocele in the early venous phase of the flow study. (Mali WPTM, Oei HY, Arndt JW et al: Hemodynamics of the varicocele, Part I. Correlation among the clinical, phlebographic and scintigraphic findings. J Urol 135:483, 1986; copyright by Williams & Wilkins)

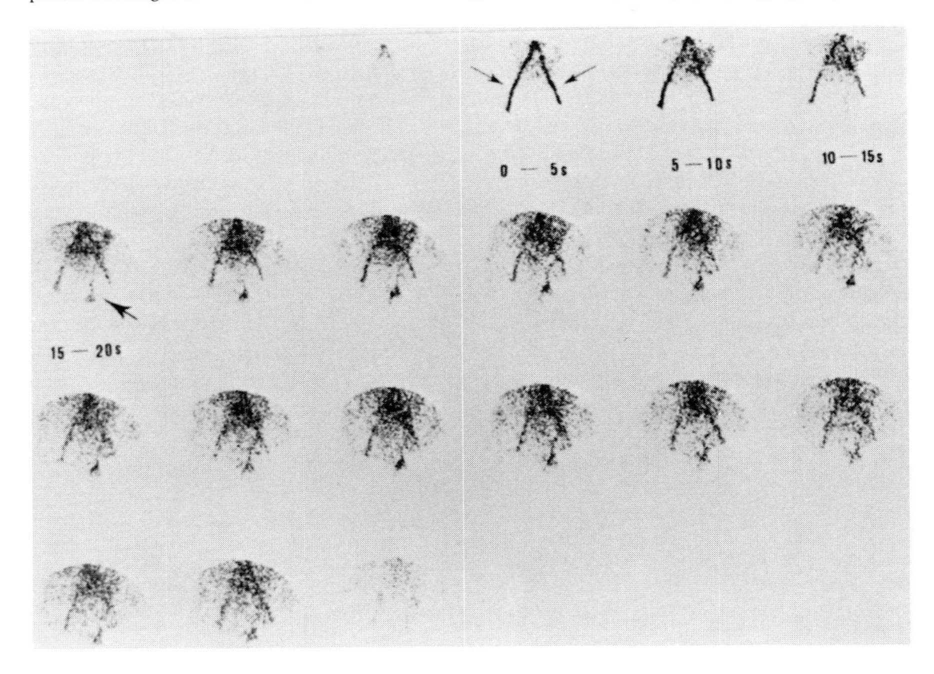

Figure 11-20 *Varicocele, Tissue Phase*

(A) Normal image is contrasted with ***(B),*** a static image demonstrating persistent blood pool in the left hemiscrotum *(arrows)* compatible with varicocele. ***(C)*** is a schematic representation of ***(B)***. ***Comment:*** A varicocele is a distended, tortuous, engorged pampiniform plexus within the scrotum. The common physical examination description of a varicocele is a "bag of worms." The pampiniform plexus is formed from the testicular, deferential, and cremasteric veins. These veins contain valves that normally prevent the retrograde flow and stasis of venous blood. When these valves become incompetent, a varicocele can form. Virtually all varicoceles (98%) form on the left side. Relative physiologic obstruction of the left testicular vein due to its right-angle insertion into the left renal vein and the compression of the left renal vein by the mesenteric vessels as it crosses the aorta have been postulated as contributing factors in the development of varicoceles.[28] Male infertility patients have a twofold higher incidence of varicocele than the general population (39% compared to 20%) and surgical correction improves fertility in two thirds of the patients reported.[29] Sensitivity of the scintigraphic examination for the detection of varicocele is 93% compared to 57% for physical examination alone.

Varicocele should not present a problem in the diagnosis of acute spermatic cord torsion. Varicoceles are spermatic cord origin structures and as such would have occluded perfusion.

There is potential for a large varicocele to confuse the interpretation of epididymitis. The late appearance of the varicocele in the venous phase of the flow study should be a useful differential diagnostic finding. The increased perfusion in epididymitis appears early in the arterial phase of the flow study.

Typically varicoceles can be distinguished clinically, as they do not present with an acute pain syndrome. If a patient with varicocele develops epididymitis, the arterial phase of the study should be positive, leading to a correct diagnosis. (Mali WPTM, Oei HY, Arndt JW et al: Hemodynamics of the varicocele. Part I. Correlation among the clinical, phlebographic and scintigraphic findings. J Urol 135:483, 1986; copyright by Williams & Wilins)

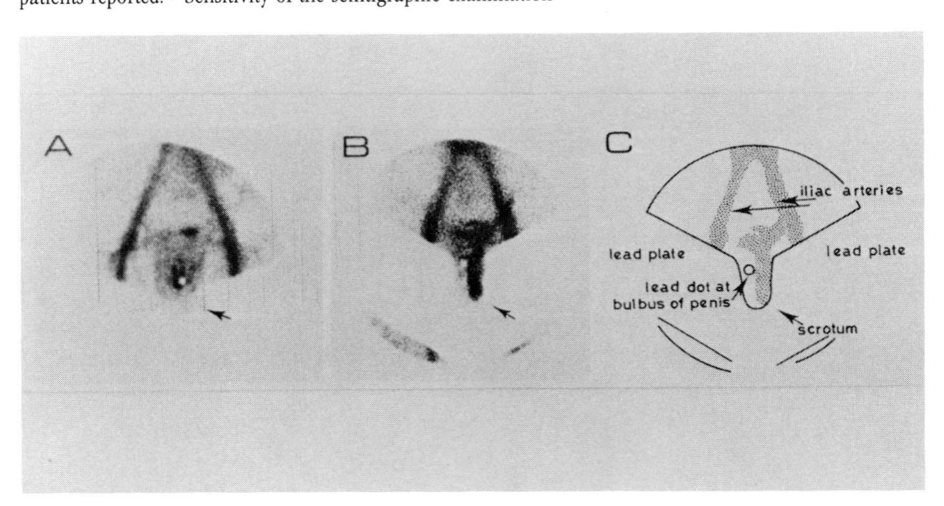

Michael S. Kipper

Chapter 12 INDIUM-111 WHITE BLOOD CELL IMAGING

A full decade has now passed since Thakur and associates[1] first reported the use of Indium-111 labeled white blood cells (In-111 WBC) for abscess detection in dogs. Ten years of experience have served to more clearly define the applicability and accuracy of the technique as well as to better delineate the limitations and pitfalls of leukocyte imaging.

This chapter examines the technique of cell labeling, the physiologic mechanism(s) of the radiopharmaceutical (including radiation dosimetry), and the spectrum of clinical utility and presents an in-depth guide to interpretation of the normal (negative), abnormal, and variant studies seen in clinical practice. The atlas examples are from the author's personal experience. Final diagnoses are based upon surgical findings, biopsy results, autopsy, or definitive correlative tests. These cases should give the reader an appreciation for the diversity of pathology and variants encountered in the acute care setting.

TECHNIQUE

Leukocyte Labeling

The following protocol for leukocyte labeling is based on the method of Thakur.[2] Fifty ml to 80 ml of the patient's venous blood is drawn (using a 19-gauge butterfly IV and three-way stopcock) into two syringes, each containing 250 IU preservative-free heparin and 10 ml hetastarch (Fig. 12-1). After gentle mixing the syringes are turned on their hubs and allowed to stand for 45 minutes. This allows the erythrocytes to settle out and results in a supernatant rich in leukocytes and platelets. The supernatants are combined and centrifuged at 150 g for 8 minutes. This results in a leukocyte button (predominantly polymorphonuclear) and a platelet-rich plasma (PRP) supernatant. The PRP is removed and centrifuged at 450 g for 8 minutes, yielding a platelet-poor plasma (PPP) supernatant which is saved for subsequent use. The leukocyte button is resuspended in 6 ml of sterile saline. To this suspension In-111 oxine (approximately l mCi) is added dropwise with gentle swirling. The mixture is incubated at room temperature for 30 minutes (gently swirled three to five times) and then recentrifuged at 150 g for 8 minutes. The radioactivity in the button and the supernatant are counted to determine labeling efficiency and the supernatant is discarded. The labeled white cell button is then resuspended in 10 ml of PPP and injected intravenously into the patient.

This procedure takes approximately 2 hours and routinely yields a labeling efficiency of 80% or better.

Practical considerations:

No patient preparation is required.

Hypotonic erythrocyte lysis is not indicated, because it may damage WBCs.

- Density-gradient leukocyte separation is not necessary for routine studies, although it can be used to produce a pure population of polymorphonuclear leukocytes (PMN).
- Children can be studied using reduced volumes of blood (as little as 20 ml) and a reduced amount of radiopharmaceutical (dosage based on body weight or body surface area).
- Donor blood can be used in leukopenic patients (*e.g.*, WBC count less than 3000), but must be carefully screened.
- Standard adult dosage is 500 μCi (18.5 MBq) In-111 WBC.
- In-111 oxine is now commercially available and does not require an investigational new drug (IND) protocol.
- Labeling can be satisfactorily performed in-house or by a central radiopharmacy.

Imaging Procedure

Scanning is performed on a large-field-of-view gamma camera with a medium-energy collimator. Camera characteristics determine the most practical method; these alternatives are acceptable:

- 700 counts/cm^2 information density over the liver with all other images for the same time. Some authors have chosen 500 counts/cm^2.
- 400 to 600 K over liver or spleen with remainder of images same
- Ten minutes per image or minimum of 100 K per image

A 20% window centered over the 173 keV and 247 keV photopeaks of In-111 may be used. Routine views include those of anterior and posterior chest, abdomen, and pelvis; additional views are obtained as required. Optimal imaging time is 18 to 24 hours after injection of labeled WBC.

Practical considerations:

- Early views (*i.e.*, 4–6 hours) may be very helpful in suspected inflammatory bowel disease.
- Sensitivity may be as low as 33% at 1 to 4 hours after injection.[3]
- Combined imaging with technetium-labeled radiopharmaceuticals can be satisfactorily accomplished.
- A small percentage (less than 5%) of patients require imaging beyond 24 hours (personal experience).
- The value of single photon emission computed tomography (SPECT) imaging has yet to be established.

PHYSIOLOGIC MECHANISM(S) OF RADIOPHARMACEUTICAL

In-111 oxine (8-hydroxyquinoline) is routinely used for neutrophil labeling. There are reports on the use of other chelating agents (acetylacetone and tropolone); however, the author has had experience only with In-111-oxine WBC. Oxine is a lipophilic ligand that chelates bivalent and trivalent metal ions. The In-111 oxine complex is lipid-soluble and easily diffuses across the cell membrane. Once intracellular, the In-111 dissociates and binds firmly to components in the cytoplasm. Oxine then diffuses out of the cell.[4] The labeled cells maintain their viability and function and progressively accumulate at sites of inflammation/abscess over the subsequent 24 to 48 hours.

Practical considerations:

- Oxine may be toxic in excessive amounts; however, current methods employ safe concentrations.
- In-111 oxine labels lymphocytes, platelets, and erythrocytes. Their numbers, relative to polymorphonuclear leukocytes, should not routinely interfere with scan interpretation.

ESTIMATED RADIATION ABSORBED DOSE

Estimated radiation doses received from 500 μCi of In-111 labeled leukocytes are given in Table 12-1.[5,6]

Practical considerations:

- Use a maximum of 500 μCi In-111 per study in adults.
- For children, the recommended dosage is 10 to 12 μCi/kg body weight, up to a maximum of 500 μCi.[7]
- Lymphocytes incidentally labeled with In-111 in mixed leukocyte preparations for abscess localization are killed by the radiation and pose no long-term risk.[8]

Table 12-1
Radiation Dosimetry (based on 500 μCi In-111 WBC)

Organ	Rads
Whole body	0.25
Liver	1.5–2.5
Spleen	9
Bone marrow	0.1–0.15
Testes	0.03
Ovary	0.07

VISUAL DESCRIPTION AND INTERPRETATION

Labeled leukocytes normally localize to the liver, spleen, and bone marrow as early as 2 to 4 hours after reinjection and prominently by 24 hours. Transient pulmonary activity may be seen in the first few hours and is in part due to WBCs reversibly damaged in the labeling process (including embolization of cell aggregates). In general, the spleen will demonstrate the most intense uptake. Liver activity is slightly less than that of the spleen, while marrow uptake, although easily visualized, is the least intense.

Scan interpretation is based on comparison with the intensity of uptake at physiological sites of accumulation (liver, spleen, marrow). Additional factors include focal versus diffuse uptake, movement with time, and change in intensity over time.

Usually, abscesses are of equal or greater intensity as compared to the spleen, focal rather than diffuse, and demonstrate neither movement nor diminution of activity with time.

In my experience, abdominal foci equal to the liver generally indicate a significant inflammatory process, but infrequently require surgical intervention. Abdominal activity equal to or less than marrow uptake rarely indicates significant inflammation, a notable exception being peritoneal dialysis catheters.[9] Of note, positive uptake less than that in the spleen remains a controversial area and must be interpreted in each patient based on all available clinical information.

Practical considerations:

- Healing, noninfected wounds show minimal uptake 48 to 72 hours after surgery.
- Diffuse pulmonary activity is uncommonly associated with an infectious process.[10]
- Recent stomas may demonstrate modest uptake localized to the abdominal wall.
- Intestinal activity due to swallowed leukocytes tends to move with time.
- Intense uptake may occur in recent hematomas.[11]
- Genitourinary tract activity most often indicates an infectious process, since In-111 is not excreted in the urine.
- Localized foci of accumulation are frequently seen at tracheostomy sites, nasogastric tube sites, gastrostomy tube sites, and at the distal tip of indwelling vascular catheters.
- Intramuscular injection sites often demonstrate increased uptake.

DISCUSSION

In-111 WBC imaging is now a practical, reliable method for investigating the patient with a suspected abscess or inflammatory process. Multiple factors must be considered in assessing the utility or any imaging procedure; these include sensitivity, specificity, accuracy, ease of performance, patient preparation, completion time, technical limitations, radiation exposure, comparison with other modalities, potential pitfalls, and cost.

Sensitivity. Most large series (including our own) report sensitivities of 90% or better. Those reporting lower values often contain many cases of chronic bone infection. In fact, many authors favor gallium-67 citrate for all types of chronic infection (*i.e.*, more than 2 weeks' duration). Arguments in favor of gallium in this subset of patients seem to be anecdotal or based on small numbers of patients. Theoretical reasons for the preferential use of gallium in chronic infection include "walling off" of the abscess with subsequent reduction of white cell access and subpar leukocyte function in the setting of chronic infection. Datz, Thorne, and Christian have recently reported no significant reduction in sensitivity in chronic infections.[12] Their data were based on retrospective review of more than 300 studies. Other factors that might reduce the sensitivity of the In-111 WBC scan include antibiotic therapy, steroids, hyperglycemia, uremia, and hyperalimentation. Again, Datz, Thorne, and Taylor could find no significant reduction in the sensitivity of the study due to these factors.[13] The sensitivity of the In-111 WBC study is expected to be low in those processes which generate a modest white cell response (*e.g.*, parasitic infections, fungal infections, and bacterial endocarditis).

Specificity. Early reports of specificity were all in the 90% to 100% range. Although experience has brought this figure closer to 90%, the advantage compared to gallium remains clear. The normal routes of excretion for gallium are the kidneys (first 24 hours) and gastrointestinal (GI) tract, often resulting in significant gallium activity in these organs, which can be misleading. Labeled leukocytes are not normally found in either the kidneys or GI tract, obviating this problem. Also, migration of labeled leukocytes to recent surgical wounds is significantly greater with gallium. More importantly, with rare exception, In-111 WBC do not accumulate at tumor sites. Table 12-2 is a listing of false-positive and false-negative studies reported to date. In many instances, careful patient evaluation, assessment of the intensity and/or focality of the uptake, or delayed imaging can prevent an erroneous interpretation. In the chest, focal uptake is much more likely to represent a true inflammatory process than is diffuse uptake.[13,14] In the abdomen, swallowed WBCs can usually be distinguished from a true inflammatory focus due to their tendency to move with time.

Table 12-2
False-Positive and False-Negative Results With In-111 WBC Imaging

False Positive	*False Negative*
Chest	Chronic infections (especially in bone)
Congestive heart failure	Aortofemoral graft
Adult respiratory distress syndrome	Left upper quadrant abscess
Pulmonary embolism	Infected pelvic hematoma
Embolized cells	Retroperitoneal abscess
Cystic fibrosis	Hepatic abscess
	Empyema
Abdomen/Pelvis	Pancreatic abscess
Accessory spleen(s)	Renal abscess
Colonic accumulation (*e.g.*, multiple enemas, swallowed WBCs)	Splenic abscess
	Postulated
Renal transplant rejection	Patients on antibiotics
Gastrointestinal hemorrhage	Hyperalimentation
Vasculitis	Hyperglycemia
Ischemic bowel disease	Hemodialysis
Post-CPR	Steroids
Uremia	
Post-radiation therapy	
Wegener's granulomatosis	
Acute lymphocytic leukemia	
Miscellaneous	
Intramuscular injections	
Histiocytic lymphoma	
Cerebral infarction	
Arthritis	
Skeletal metastasis	
Thrombophlebitis	
Hematoma (including subdural)	
Hip prosthesis	
Cecal carcinoma	
Postsurgical pseudoaneurysm	

Gastrointestinal hemorrhage may be obvious historically or in conjunction with a falling hematocrit. Accessory spleen(s) can be confirmed with liver/spleen scanning. Most of the miscellaneous causes are isolated case reports, with the exception of hematomas and intramuscular injection sites.

Accuracy. Most authors report an accuracy of greater than 85% for the In-111 WBC study. The study is most appropriately used in a patient suspected of harboring an abscess/inflammatory focus without localizing signs. A positive In-111 WBC study suggests the best confirmatory test (*e.g.*, CT, ultrasound) which, if positive, directs appropriate intervention. A negative In-111 WBC study is strong evidence against an abscess/inflammatory process. However, if clinical index of suspicion remains high, gallium scanning should be considered. Under these conditions the clinical value of the test is very high. Figure 12-2 is an algorithm for testing which the author finds generally helpful. However, patient individualization remains the key in selecting the highest yield study.

Ease of Performance. Indium-111 labeling of leukocytes is now approved without an IND permit by the United States Food and Drug Administration. Labeling can be done in central nuclear pharmacies or in the nuclear medicine department at most hospitals. Requirements include a sterile work area, equipment outlined in Figure 12-1A, and a qualified technologist. The labeling process takes approximately 2 hours, routinely results in a labeling efficiency of greater than 80%, and requires almost no patient preparation. Many modifications of Thakur's original method exist and each appears to work satisfactorily. All share certain common characteristics: minimization of cell handling and time out of body, gentle centrifugation(s), and sterile technique.

Patient Preparation. As dicussed above, minimal patient preparation is required. The author has successfully studied patients with WBC counts as low as 2500; below this level, I recommend that either gallium or donor leukocytes be used. Donor cells currently must be ABO- and Rh-matched, VDRL-tested, and screened for hepatitis as well as acquired immune deficiency syndrome (AIDS). Children may be studied; however, the volume of blood drawn and the radiopharmaceutical dose should be appropriately reduced (see discussion of Estimated Radiation Absorbed Dose). One advantage of labeled leukocytes over gallium is the lack of excretion through the GI tract of the In-111 WBC. This obviates the need for cathartics. Patients on dialysis can be studied; if possible, postdialysis seems preferable due to minimization of uremia.

Completion Time. The vast majority of In-111 WBC studies are completed within 24 hours. In the author's experience less than 5% of studies require repeat imaging beyond this time. Earlier imaging (*e.g.*, 4–6 hours after reinjection) may be positive, but negative early views do not predict a negative study at 24 hours. Early views may be extremely valuable in the critically ill patient (if positive), and in suspected bowel disease (due to subsequent movement of the labeled leukocytes). In many instances, equivocal accumulations are clarified by imaging at 48 hours. This is especially true with pulmonary activity, which is often of no clinical significance.

Technical Limitations. Indium-111 decays by electron capture and emits gamma rays with energies of 173 keV and 247 keV. This necessitates the use of a medium-energy collimator, making portable studies suboptimal at best. A physical half-life of 2.8 days, while beneficial for delayed imaging, theoretically could present some problems in performing combined studies with Tc-99m-labeled pharmaceuticals. In practice, this presents no major problem. Liver or bone scans can be performed immediately after labeled leukocyte studies and In-111 WBC images can be obtained subsequent to technetium injection by using only the 247 keV photopeak. To date, there is little reported experience with SPECT and In-111 WBC. Use of the current recommended dose (500 μCi or less) suggests that SPECT may be of limited value. Significant improvement in sensitivity seems unlikely and more precise localization of abnormal foci can be achieved with additional views (*e.g.*, lateral or oblique).

Estimated Radiation Absorbed Dose. It is important to note that the usual administered dosage of gallium (3–10 mCi) results in a significantly higher whole body dose compared to In-111 WBC (1.5–7.5 rads/5 mCi versus 0.25 rads/500 μCi).[6] The spleen is the critical organ in an In-111 WBC study and may receive up to 10 rads exposure.

Concern has been raised that labeling with In-111 may have detrimental effects on human lymphocytes.[15] However, Thakur and McAfee, after careful analysis, concluded that the mean survival time of irradiated peripheral blood lymphocytes is extremely short; moreover, the increased burden of chromosomal aberrations is extremely small, the risk of lymphoid malignancy appears to be small and acceptable, and lymphocytes incidentally labeled with In-111 in mixed leukocyte preparations are killed by the radiation and pose no long-term risk.[8]

Comparison with Other Modalities. In-111 labeled leukocytes compare quite favorably to other imaging modalities. Although the design of any study comparing radionuclide imaging, CT, and ultrasound presents logistical problems, reasonable data are available. A report by McDougall and colleagues comparing In-111 WBC and ultrasound documented a sensitivity of 89% for WBC versus 64% for ultrasound.[16] Knochel and coworkers studied 170 patients with CT, ultrasound, and In-111 WBC and found accuracies of 96%, 90%, and 92%, respectively.[17]

Potential Pitfalls. Potential pitfalls exist at each step along the way in In-111 WBC studies, from patient selection to image interpretation. These have been covered in part in previous sections but warrant repeating and elaboration.

In patients with WBC counts of less than 2500 it is recommended that donor WBC or gallium be used. Suspected chronic bone infection, parasitic or fungal infection, or those infectious processes which do not generate a significant leukocyte response are best evaluated with other modalities.

Careless handling of the cells, prolonged labeling or time out of body, and improper separation or incubation times may result in a reduced labeling efficiency.

Sensitivity may be as low as 33% in patients imaged 1 to 4 hours after reinjection.[3] Images obtained at 24 hours only may misplace the site of abnormality in some cases of inflammatory bowel disease because of movement of labeled leukocytes; in suspected cases, earlier views can be helpful. Activity from swallowed WBC may also move with time and this tendency can be used to separate a true abscess/inflammatory process from migrating leukocytes within the bowel.

The technique presented labels a mixed-cell population including PMNs, mononuclear cells (lymphocytes, monocytes), erythrocytes, and platelets. This will occasionally result in lymph node visualization, blood pool activity (*e.g.*, heart), and may partially explain visualization of sites of hemorrhage or hematoma. Virtually any process which results in white cell accumulation can show uptake; these sites are interpreted based on intensity of uptake, patient history, and physical examination. False-negative studies are uncommon; however, one must be aware of potential problems in the area of the liver and spleen, chronic bone infections, and dialysis catheters (these may show only modest uptake, yet require removal). False-postive studies can be minimized if one has a good understanding of the normal distribution of labeled leukocytes, and if one correlates the scan findings with history and physical examination, and uses confirmatory tests such as CT, ultrasound, or other nuclear studies.

Cost. In-111 WBC studies cost approximately the same as gallium surveys, in general, somewhere between ultrasound and CT. In today's cost-conscious environment, results in 24 hours (In-111 WBC) versus 48 to 72 hours (gallium) may be important.

ACKNOWLEDGMENTS

The author wishes to thank Dr. Ted Gay, Department of Infectious Disease, University of California, San Diego for his review of this manuscript and his suggestions. The author also wishes to thank Sandra Barkwell for her preparation of this manuscript.

REFERENCES

1. Thakur ML, Coleman RE, Mayhall CG: Preparation and evaluation of 111-Indium-labeled leukocytes as an abscess imaging agent in dogs. Radiology 119:731, 1976

2. Thakur, ML: Indium-111: A new radioactive tracer for leukocytes. Exp Hematol (Suppl) 5:145,1977

3. Datz FL, Jacobs J, Baker W et al: Decreased sensitivity of early imaging with In-111 oxine-labeled leukocytes in detection of occult infection: concise communication. J Nucl Med 25:303, 1984

4. Thakur ML, Segal AW, Louis L et al: Indium-111 labeled cellular blood components: Mechanism of labeling and intracellular location in human neutrophils. J Nucl Med 18:1022, 1977

5. Thakur ML, Lavender JP, Arnot RN et al: Indium-111-labeled autologous leukocytes in man. J Nucl Med 18:1012, 1977

6. McDougall IR: The use of white blood cell scanning techniques in infectious disease. In Remington JS, Swartz MN (eds): Current Clinical Topics in Infectious Diseases, pp 130–152. New York, McGraw-Hill, 1983

7. Gainey MA, McDougall IR: Diagnosis of acute inflammatory conditions in children and adolescents using Indium-111 oxine white blood cells. Clin Nucl Med 9:71,1984

8. Thakur ML, McAfee JG: the significance of chromosomal aberrations in Indium-111-labeled lymphocytes. J Nucl Med 25:922, 1984

9. Kipper SL, Steiner RW, Witztum KF et al: In-111-leukocyte scintigraphy for detection of infection associated with peritoneal dialysis catheters. Radiology 151:491,1984

10. Cook PS, Datz FL, Disbro MA et al: Pulmonary uptake in Indium-111 leukocyte imaging: Clinical significance in patients with suspected occult infections. Radiology 150:557, 1984

11. Wing VW, Van Sonnenberg E, Kipper S et al: Indium-111-labeled leukocyte localization in hematomas: A pitfall in abscess detection. Radiology 152:173,1984

12. Datz FL, Thorne D, Christian PE: Effect of chronicity of infection on the sensitivity of the Indium-111-labeled leukocyte scan. J Nucl Med 27:914,1986

13. Datz FL, Thorne D, Taylor A: Evaluation of factors that could potentially decrease the sensitivity of the Indium-111-labeled leukocyte scan. J Nucl Med 26:914,1986

14. McAfee JG, Samin A: Indium-111 labeled leukocytes: A review of problems in image interpretation. Radiology 155:221, 1985

15. tenBerge RJM, Natarajan AT, Hardeman MR et al: Labeling with indium-111 has detrimental effects on human lymphocytes: Concise communication. J Nucl Med 24:615, 1983

16. McDougall IR, Goodwin DA, Silverman P et al: Comparison of indium-111 white cell scans and ultrasonography for detection of intra-abdominal abscess. Clin Nucl Med 4:558, 1980

17. Knochel JQ, Koehler PR, Lee TG et al: Diagnosis of abdominal abscesses with computed tomography, ultrasound, and Indium-111 leukocyte scans. Radiology 137:425, 1980

Atlas Section

Figure 12-1 *Procedure for Leukocyte Labeling*

(A) The total cost per patient for materials for labeling leukocytes is under ten dollars. *(B)* Drawing set-up: Two 60-ml syringes, each containing heparin plus hetastarch are attached to a three-way stopcock and 19-gauge butterfly intravenous needle. *(C)* Blood drawing: Fifty ml to 80 ml (half in each syringe) are drawn from antecubital vein. *(D)* Settling of erythrocytes: Syringes are left to stand on their hubs for 45 to 60 minutes. *(E)* After erythrocytes settle, plasma is rich in leukocytes as well as platelets. Pinkish hue of plasma will be seen and is due to presence of erythrocytes. *(F)* Supernatants are expressed from each syringe into a single, sterile centrifuge tube. *(G)* Centrifugation: Initial centrifugation is at 150 g for 8 minutes. This results in a leukocyte button and platelet-rich plasma (PRP). *(H)* Product of first centrifugation: A red color in the leukocyte button will be seen due to contaminating erythrocytes. *(I)* The PRP is removed and centrifuged at 450 g for 8 minutes, yielding platelet-poor plasma (PPP), which is saved for later use. *(J)* Resuspension/tagging: The leukocyte button is resuspended in 6 ml of saline, and In-111 oxine is added dropwise. The suspension is incubated for 30 minutes at room temperature with periodic gentle swirling. *(K)* Final centrifugation: After incubation, the mixture is centrifuged at 150 g for 8 minutes. This results in radioactive supernatant and labeled leukocytes *(red pellet)*. Each is counted to determine labeling efficiency. *(L)* The labeled WBC button is then resuspended in 10 ml of PPP and reinjected. (Thakur ML: Indium-111: A new radioactive tracer for leukocytes. Exp Hematol (Suppl) 5:145, 1977)

A

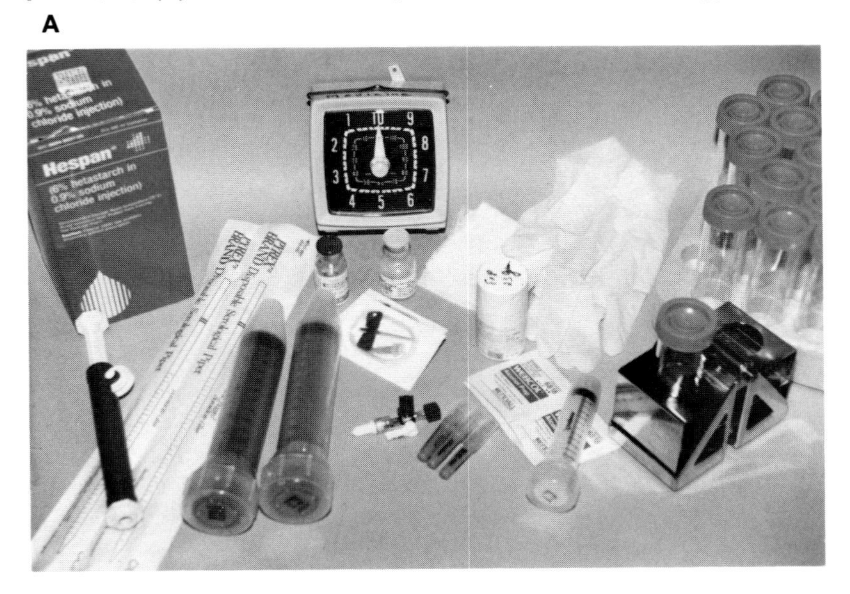

B

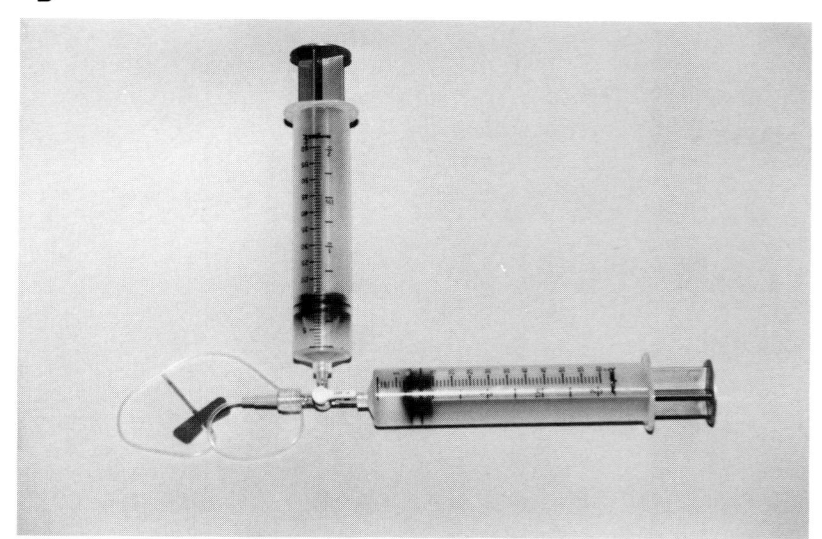

C

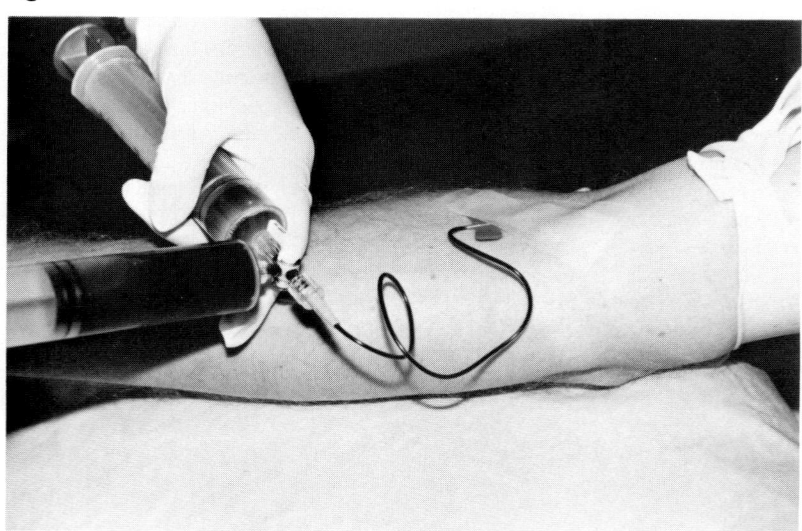

D

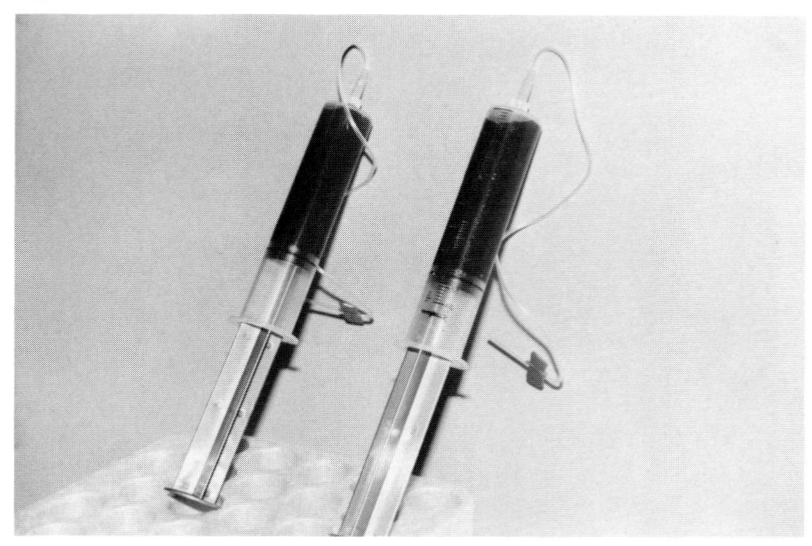

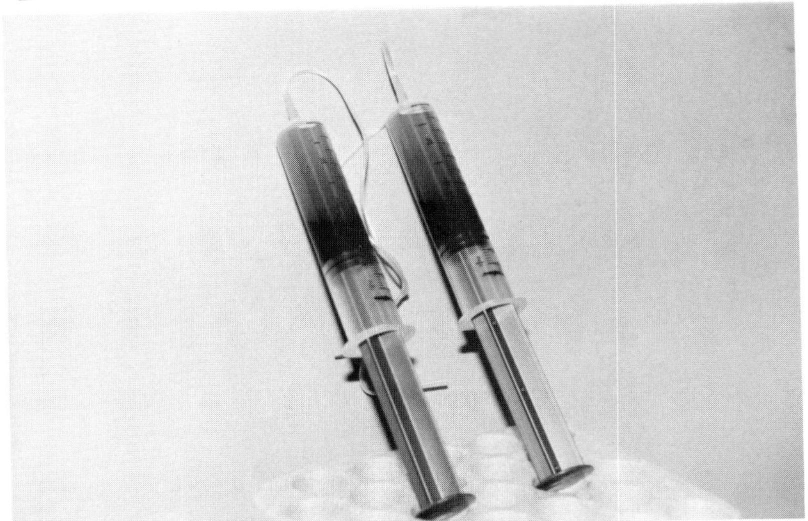

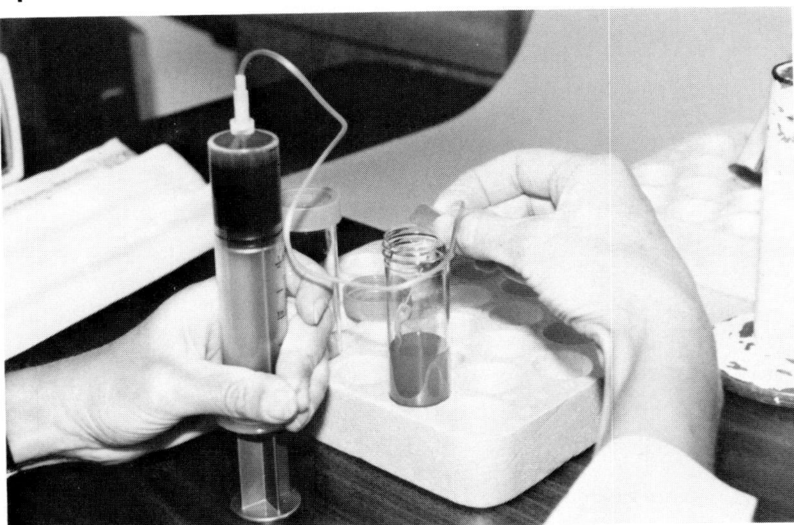

G

H

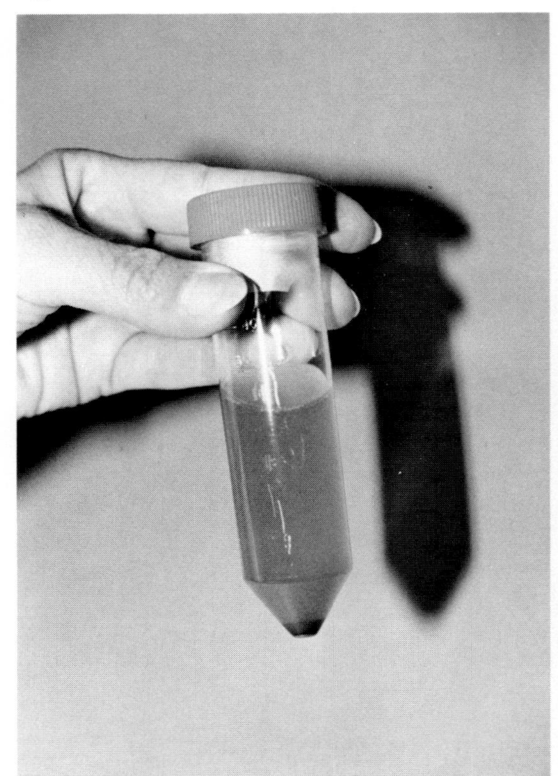

I

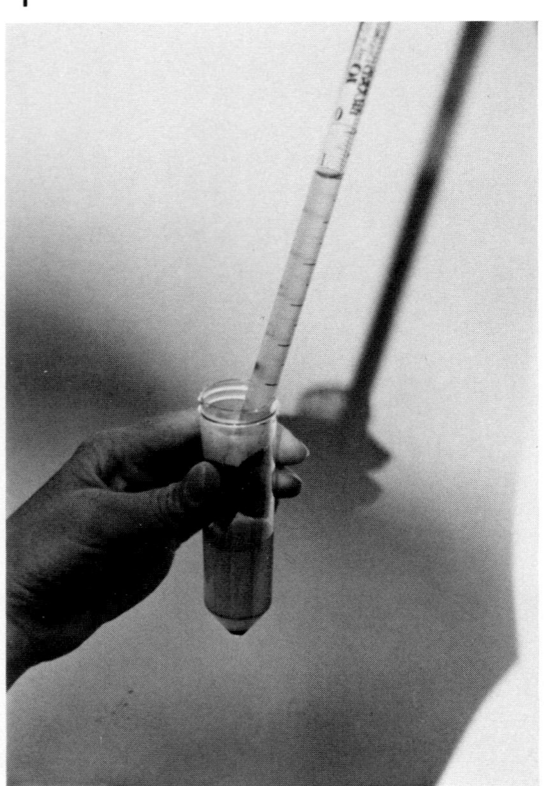

J

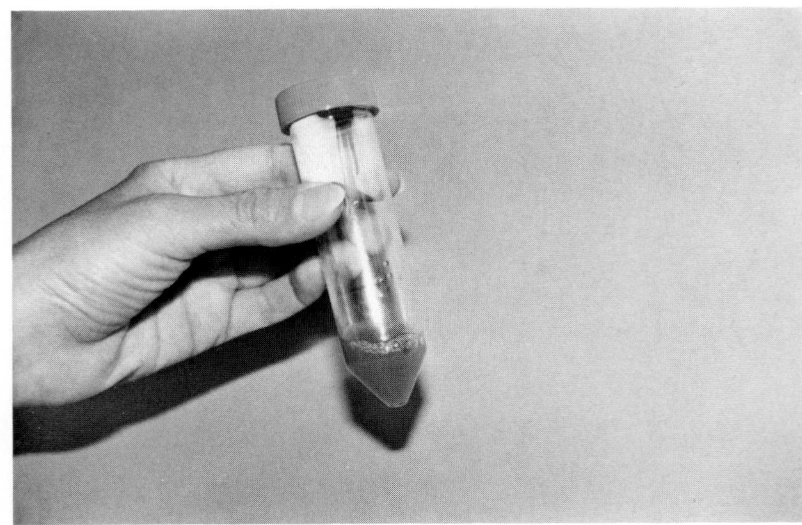

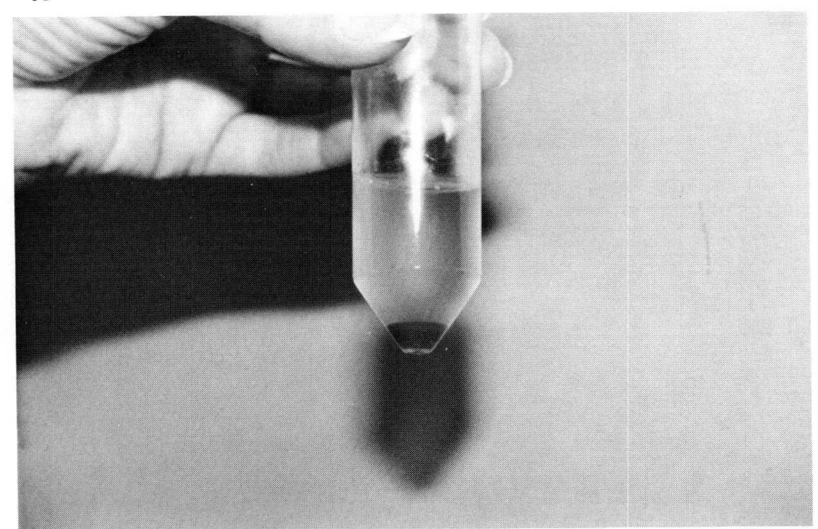

L

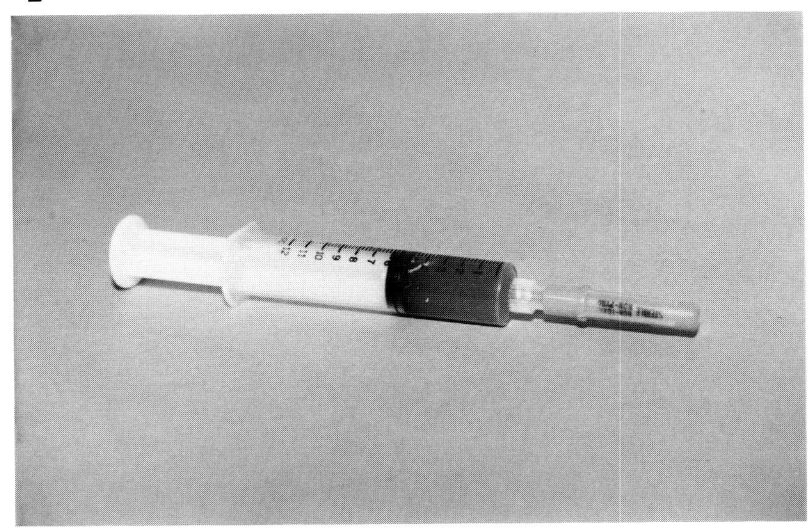

Figure 12-2 *Algorithm for Use of In-111 WBC scan.*

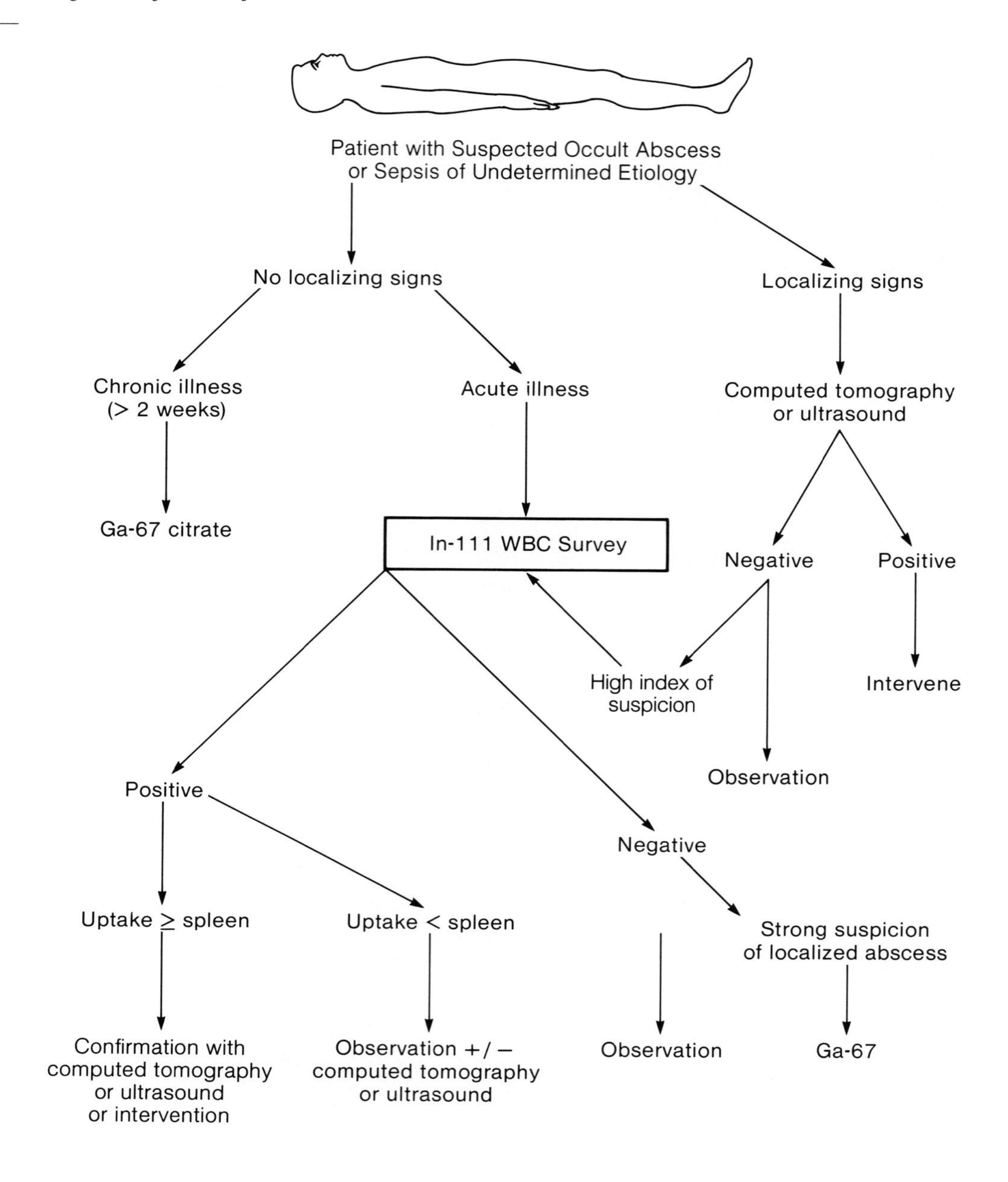

Figure 12-3 *Normal Scans*

(A) Anterior views. Note liver, spleen, and marrow activity.
(B) Posterior views. Note intensity of splenic uptake. (*H*, head; *C*, chest; *A*, abdomen; *P*, pelvis)

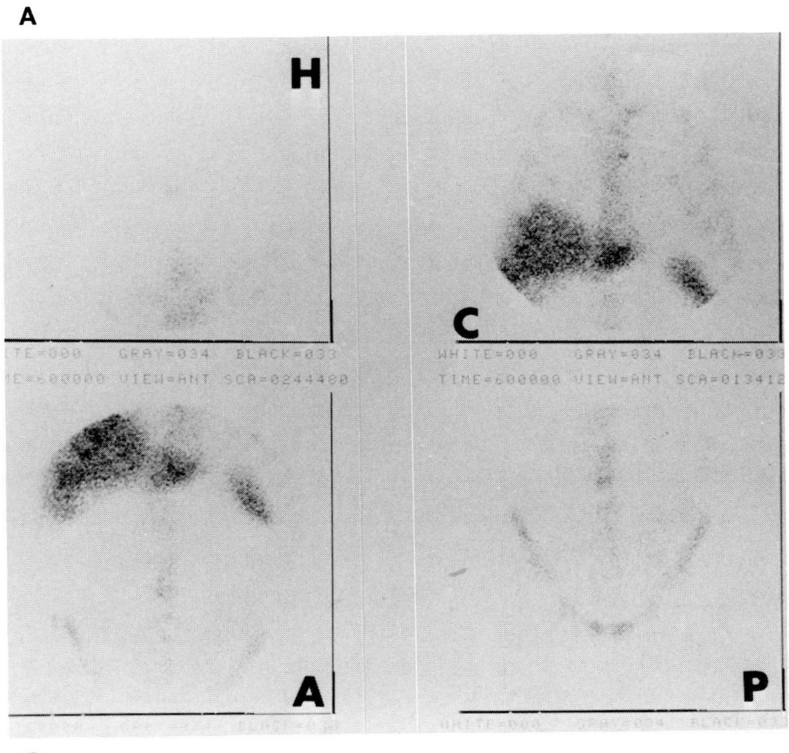

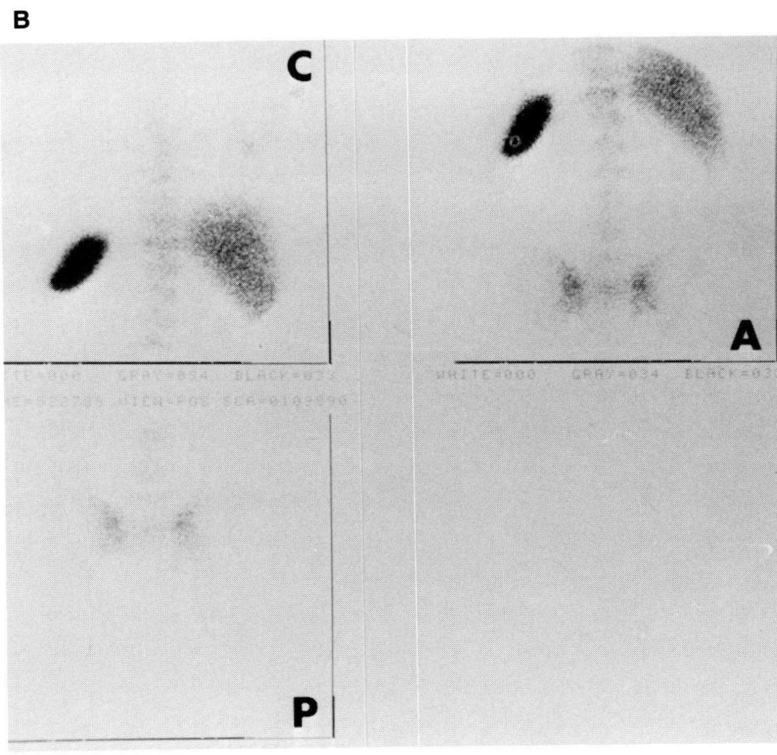

Figure 12-4 ***Abdominal Abscess***

Anterior abdominal view demonstrates typical findings for abscess *(arrow)*. Focus is well-delineated and of greater intensity than the spleen (superior to abscess).

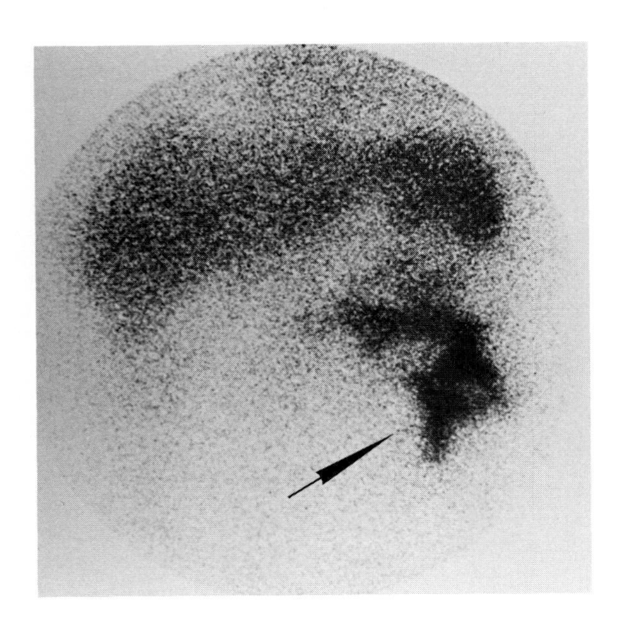

Figure 12-5 ***Hepatic Abscess***

(A) Anterior view from a liver scan demonstrates large focal defect in right lobe *(arrow)*. The patient presented with fever and abnormal liver function tests. ***(B)*** Anterior abdominal view from an In-111 WBC study. Large arrow points to the intrahepatic abscess which fills in the defect seen on the liver scan. Note second focus *(small arrow)* inferior to liver. This proved to be a walled-off bowel perforation and the source of infection to the liver. (Kipper MS, Williams RJ: Indium-111 white blood cell imaging. Clin Nucl Med 8(9):449–455, 1983)

A

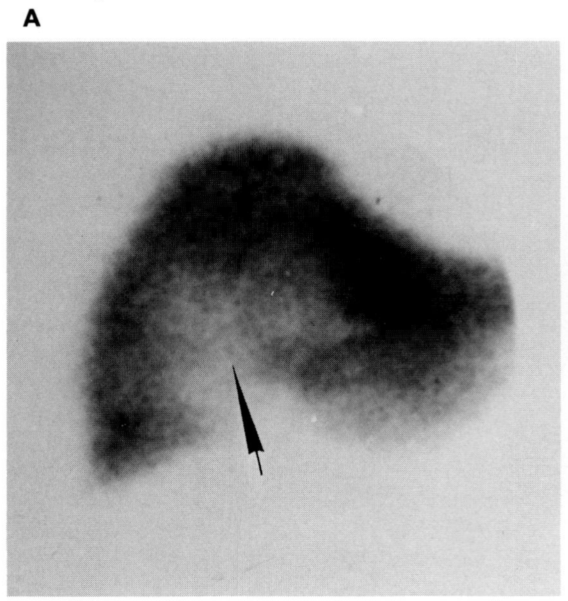

B

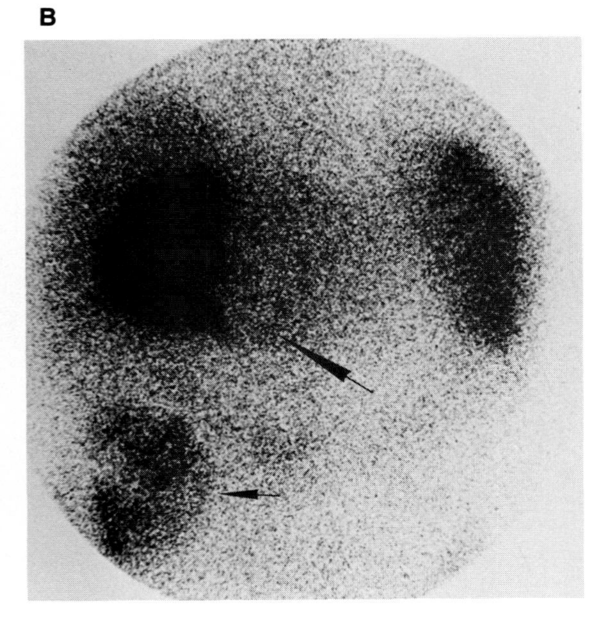

Figure 12-6 ***Infected Pancreatic Pseudocyst***

Anterior abdominal view demonstrates increased uptake in area of infection *(large arrow).* Small arrow points to the liver, which is much less intense than the abscess. Scattered foci of uptake below abscess are secondary to internal drainage of labeled leukocytes into bowel prior to surgery. (Kipper MS, Williams RJ: Indium-111 white blood cell imaging. Clin Nucl Med 8(9):449-455, 1983)

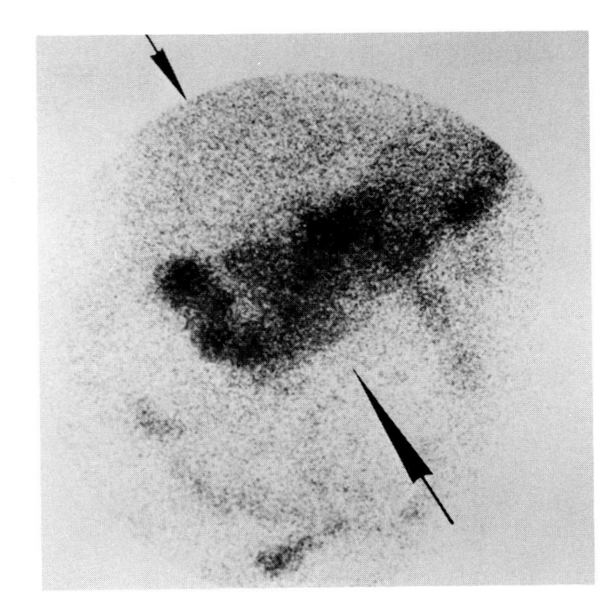

Figure 12-7 ***Myocardial Abscess***

Arrow points to abscess on anterior chest view. Patient died of this complication of endocarditis. ***Comment:*** In-111 WBC studies are negative in uncomplicated subacute bacterial endocarditis. (*R*, right; *L*, left)

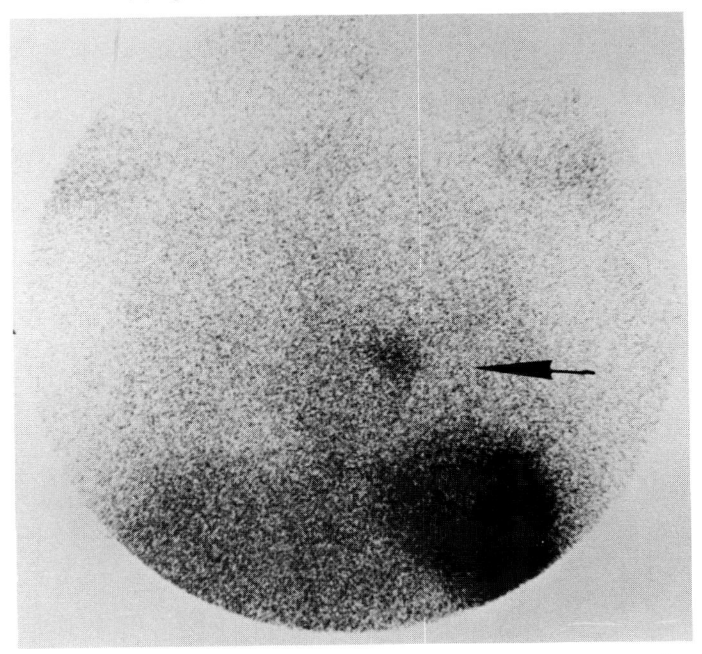

Figure 12-8 ***Loculated Pus Within Peritoneum (postappendectomy)***

(A) Anterior abdomen/pelvis view demonstrates diffuse right-sided activity. Large arrow points to collection of WBC. Small arrow denotes liver. *(B)* Pelvic view was taken to enhance abnormality. Arrow points to large area of purulence.

A

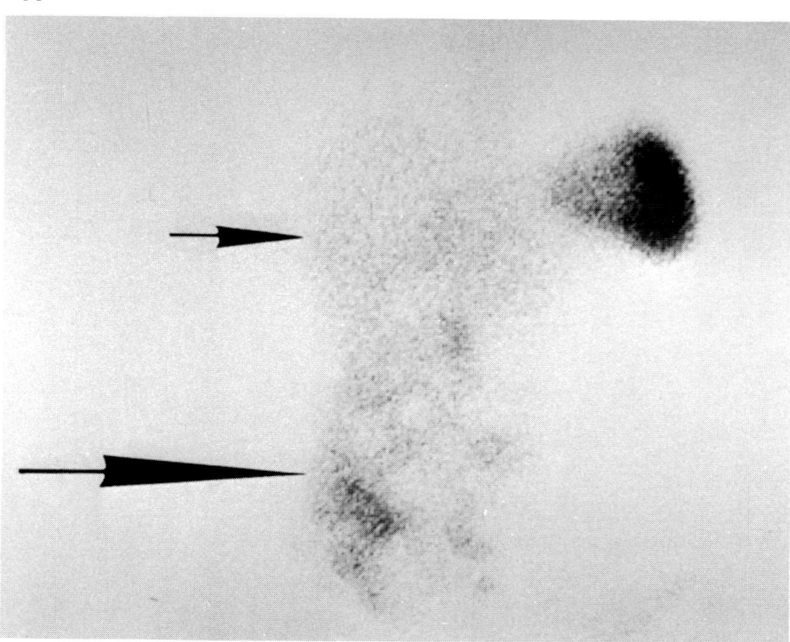

B

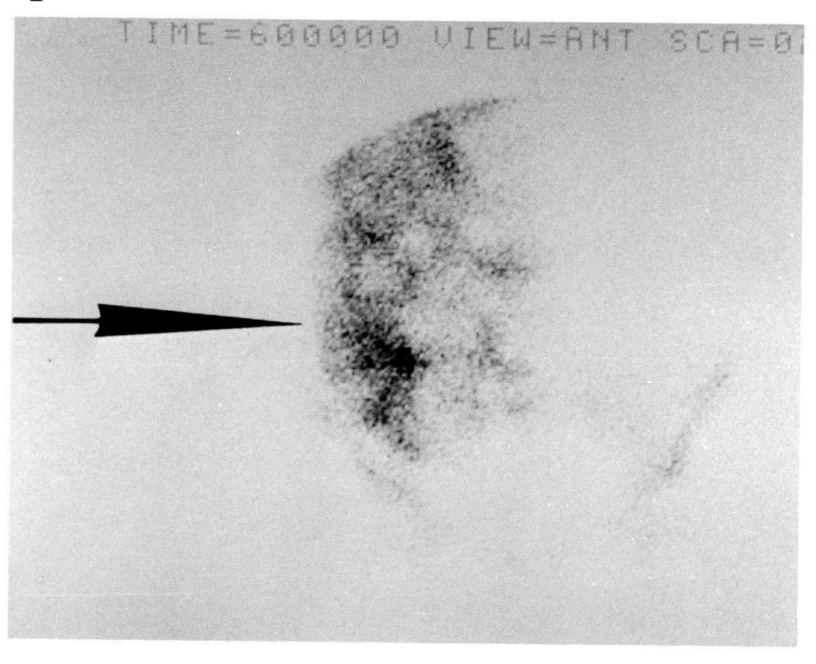

Figure 12-9 ***Wound Abscess***

On anterior pelvic view, arrow points to linear area of wound abscess with pocket of pus inferiorly. ***Comment:*** Uncomplicated, healing wounds show minimal activity 48 to 72 hours after surgery.

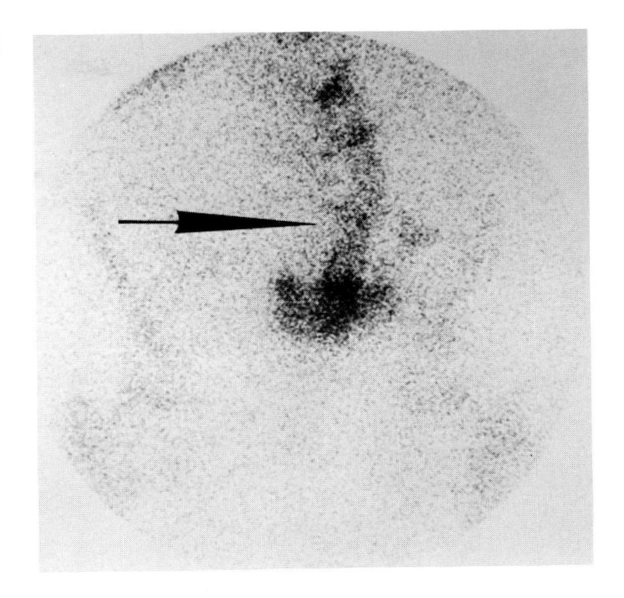

Figure 12-10 ***Pelvic Abscess***

Anterior abdomen *(A)* and pelvis *(P)* views. Arrow points to large focus in lower pelvis. Note intensity is greater than spleen. Also note absence of confusing bowel activity, which commonly occurs with gallium.

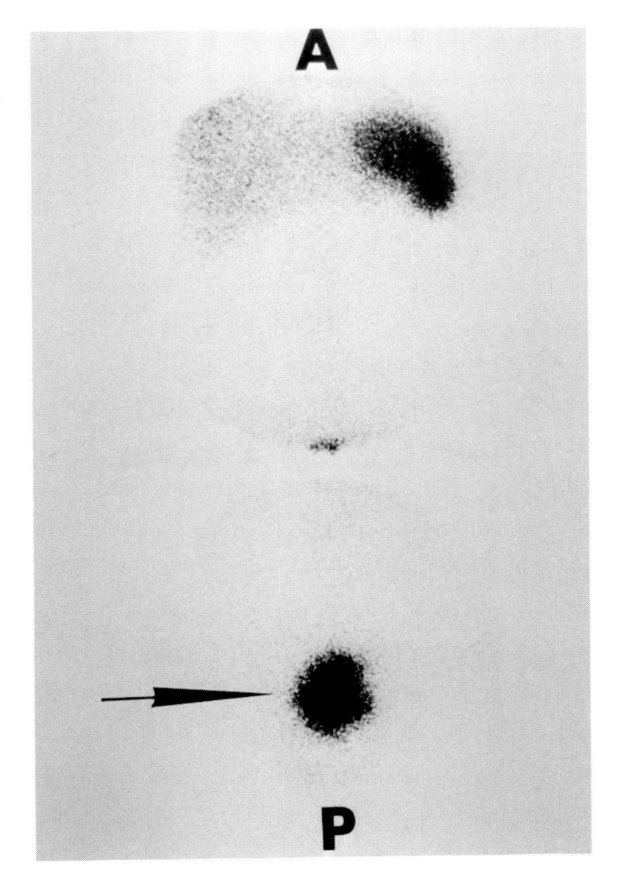

Figure 12-11 ***Pelvic and Paracolic Gutter Abscess***

Anterior chest *(C)* and abdomen/pelvis *(A/P)* views. Arrow points to area of abscess. Note intensity relative to spleen.

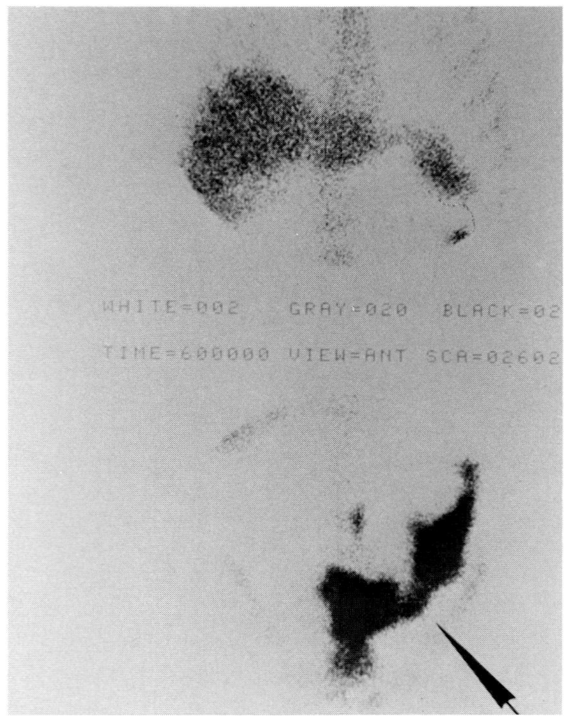

Figure 12-12 ***Suppurative Cholecystitis***

Anterior image of the abdomen demonstrates a gangrenous gallbladder *(arrow)*.

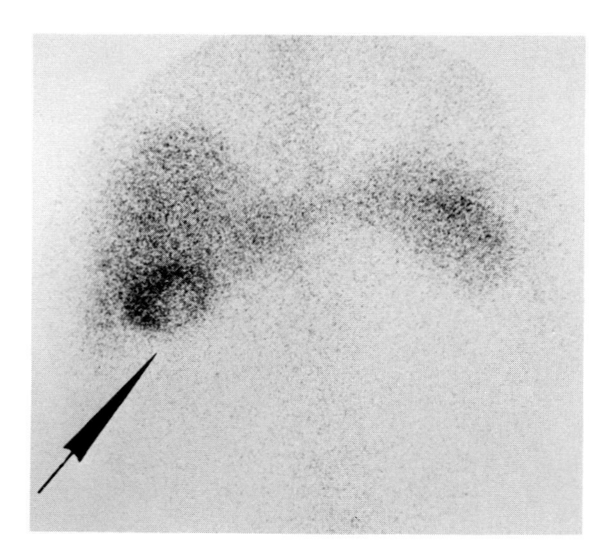

Figure 12-13 ***Osteomyelitis and Soft Tissue Abscess***

(A) Bone-scan image of left proximal femur. Arrow points to "cold defect." Note irregular, increased uptake both proximal and distal. *(B)* Corresponding In-111 image. Large arrow points to proximal left femur, which was surgically proven to be osteomyelitis. Small arrow points to surgically proven soft tissue abscess, medial to femur.

A

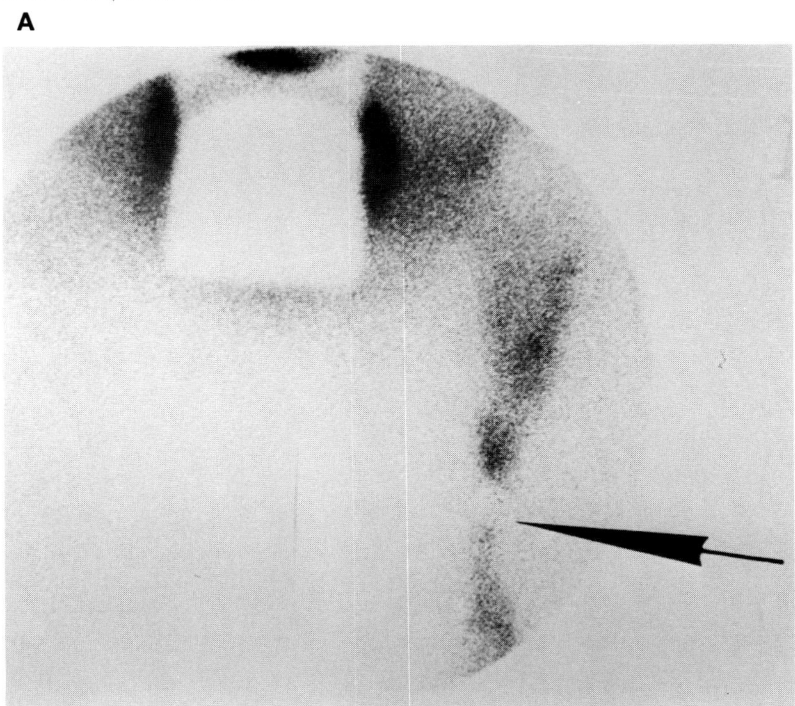

B

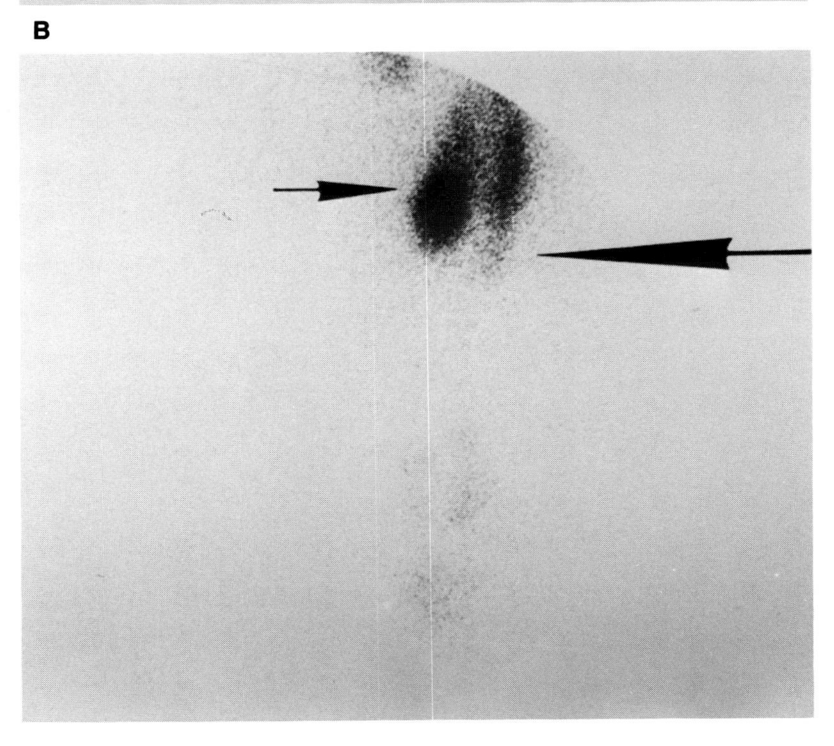

Figure 12-14 ***Renal Abscess***

(A) Posterior pelvic view from In-111 scan. Arrow points to area of moderately increased uptake (equal to liver) in right lower quadrant. ***(B)*** Image after administration of DTPA. Arrow points to "cold defect" in the inferior pole of the right kidney. This was shown to be a renal abscess. ***Comment:*** Renal infections often show only mild to moderate uptake of labeled leukocytes.

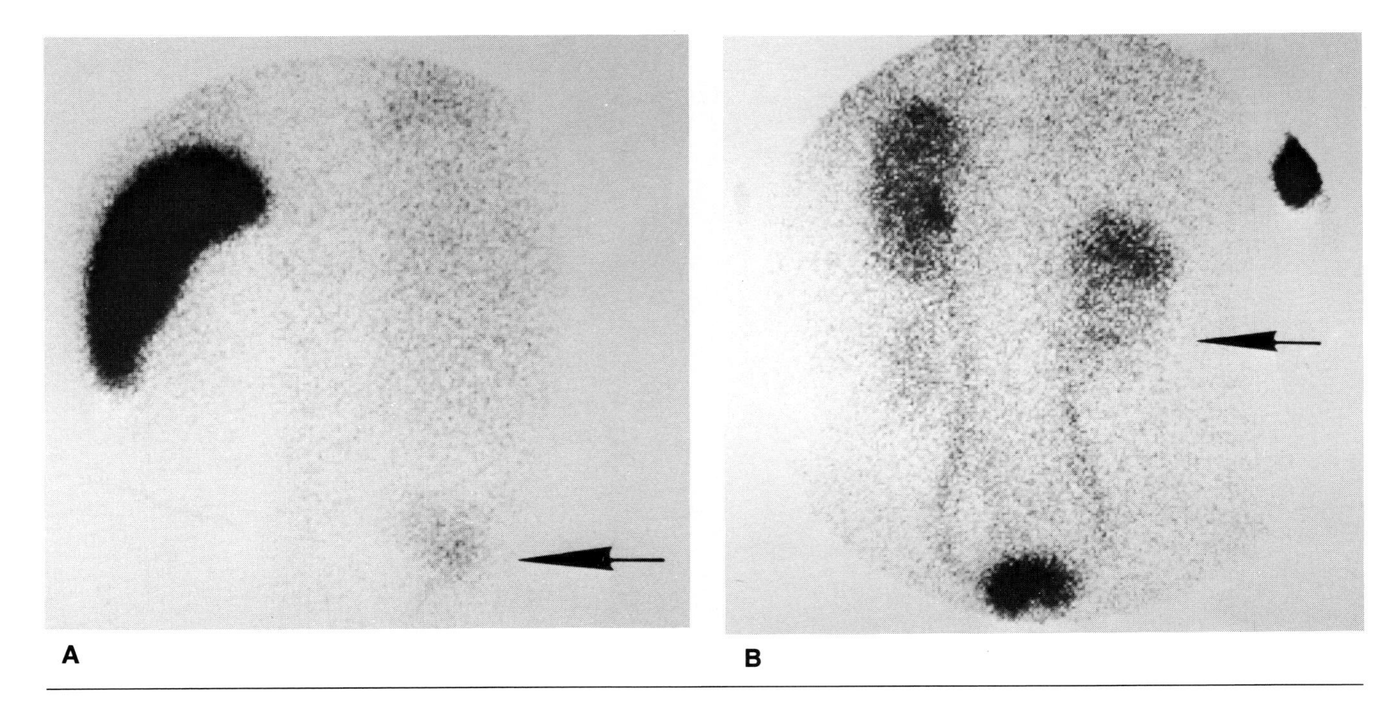

Figure 12-15 ***Pancreatitis***

In the anterior abdominal view, the large arrow points to a midline focus of increased uptake, the intensity of which is between that of liver and spleen. This generally means pancreatitis without abscess. Small arrows point to liver *(L)* and spleen *(S)*.

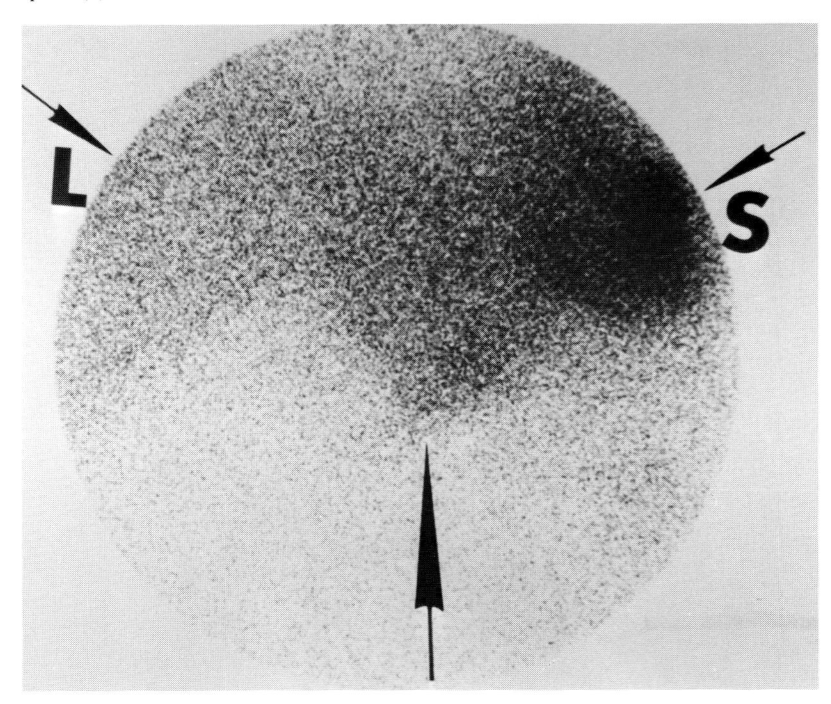

Figure 12-16 ***Pancreatic Abscess***

In the anterior abdominal view, the arrow points to an abscess, which is equal in intensity to spleen. Compare this scan to the one in Figure 12-15. (*R*, right; *L*, left)

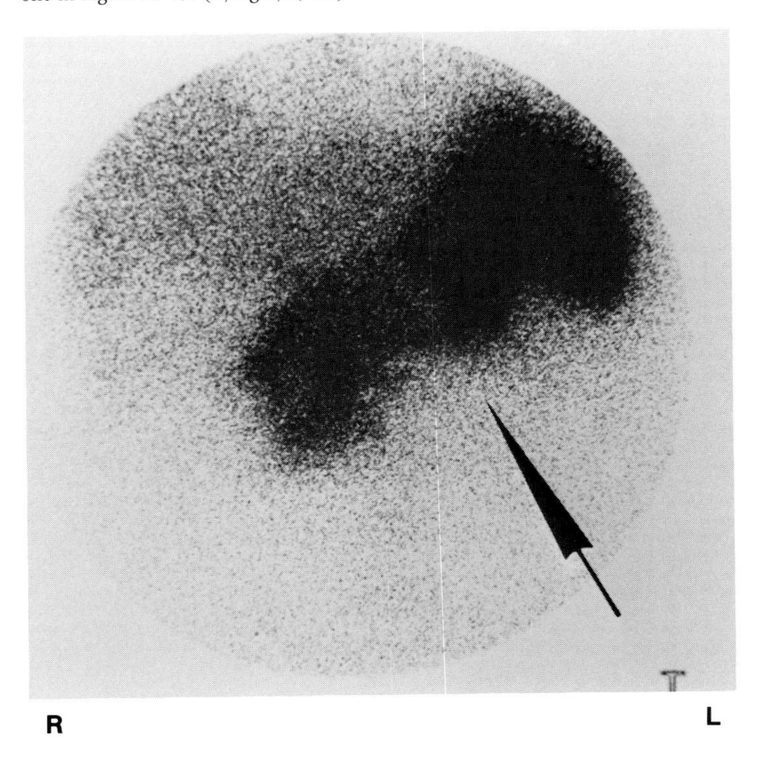

Figure 12-17 ***Diverticulitis***

In the anterior pelvic view, the arrow points to focus in left lower quadrant. ***Comment:*** Diverticular abscesses are generally more intense than uncomplicated diverticulitis.

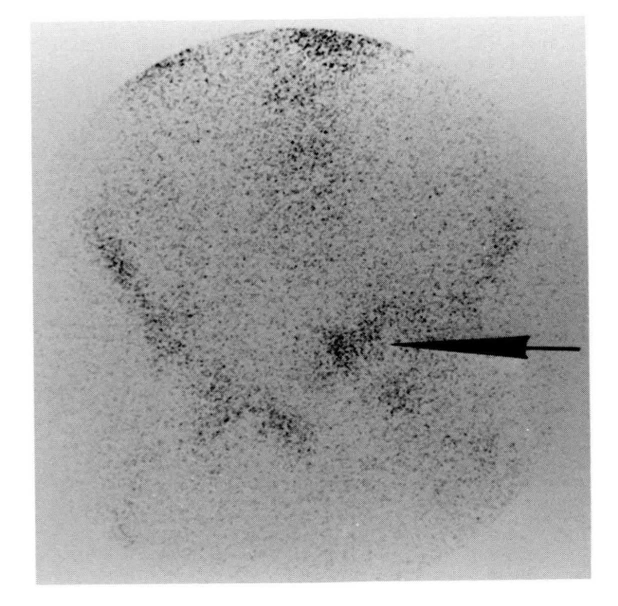

Figure 12-18 ***Colitis***

In the anterior abdominal view, the arrows point to uptake in ascending and transverse colon. ***Comment:*** This study is often helpful to assess extent of bowel involvement.

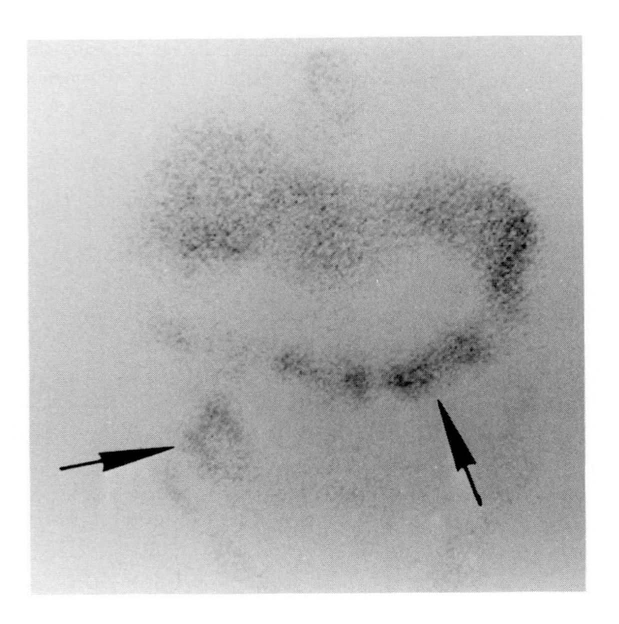

Figure 12-19 ***Appendicitis***

In this anterior abdominal view the arrow points to "inflamed appendix"; intensity of uptake is approximately equal to the liver.

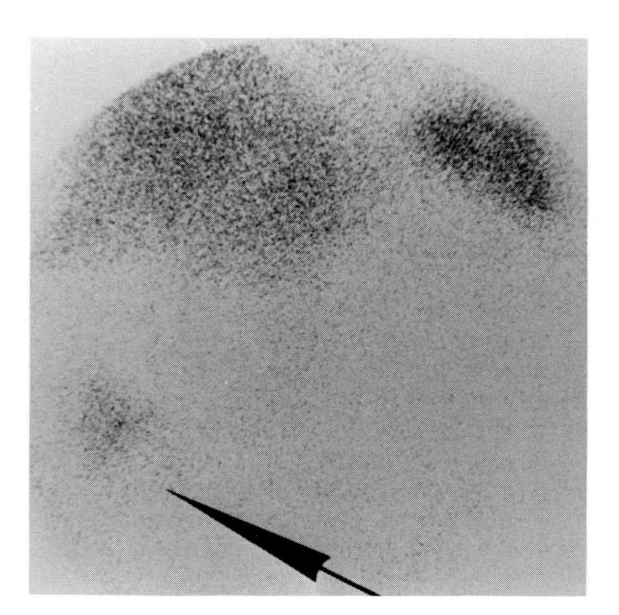

Figure 12-20 ***Multiple Abscesses and Infected Right Pleural Effusion***

(A) Posterior chest view. Large arrow points to left upper-quadrant abscess. Small arrow points to right pleural effusion. The patient had just undergone a Whipple procedure and therefore the spleen is absent. ***(B)*** Anterior chest view. Small arrow denotes a second, midline, anterior abscess. Large arrow points to modest actvity at tracheostomy site. Again note right pleural activity.

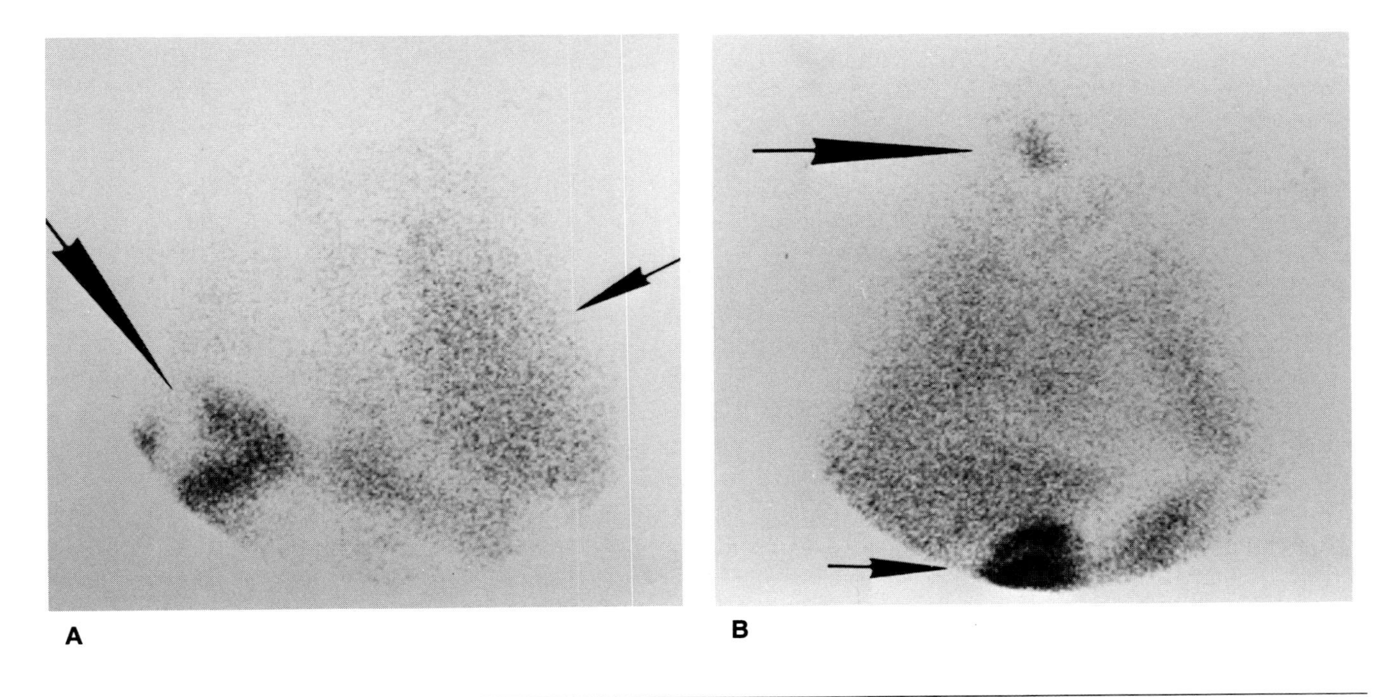

Figure 12-21 ***Empyema and Pneumonia***

In this posterior view of the chest, the large arrow points to left-sided empyema. Small arrow points to right-sided pneumonia.

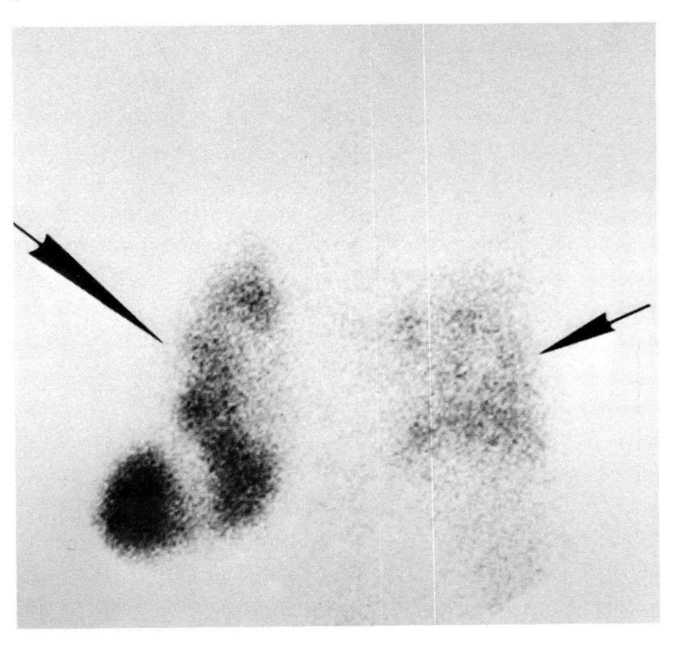

Figure 12-22 ***Felty's Syndrome***

(A) Posterior chest view. Arrows point to lower lobes and increased uptake in "rheumatoid lung." ***(B)*** Posterior abdomen view. Note impressive splenomegaly *(arrow)*. The patient had a leukocyte count of 2800, yet scan demonstrated a septic right knee focus. (Kipper MS, Williams RJ: Indium-111 white blood cell imaging. Clin Nucl Med 8(9):449–455, 1983)

A

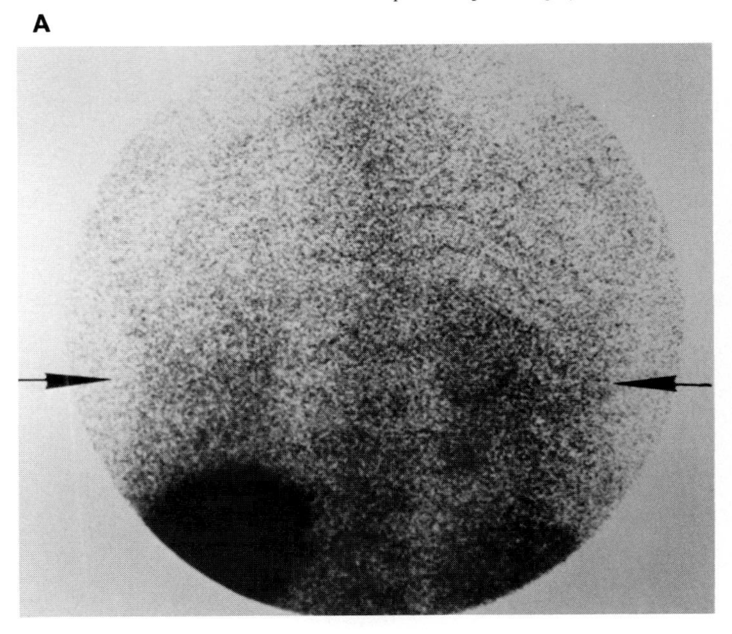

B

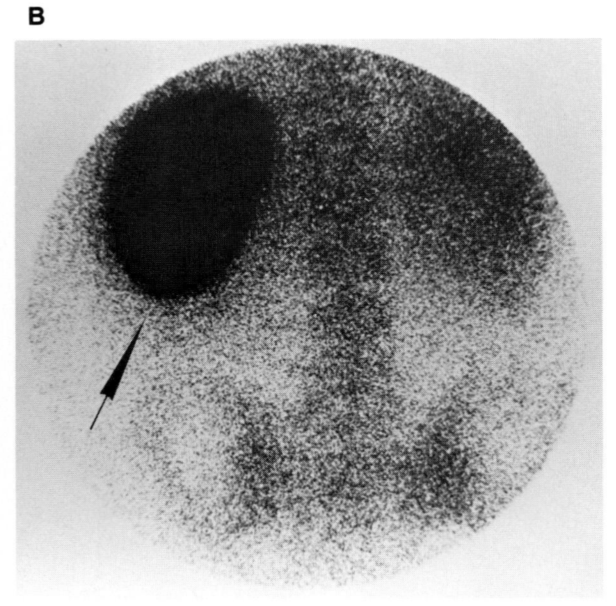

Figure 12-23 ***Vascular Graft Infection/Abscess***

In the anterior pelvic view, the small arrows point to uptake (often modest in this condition) in infected aortoiliac grafts. Large arrow points to femoral abscess.

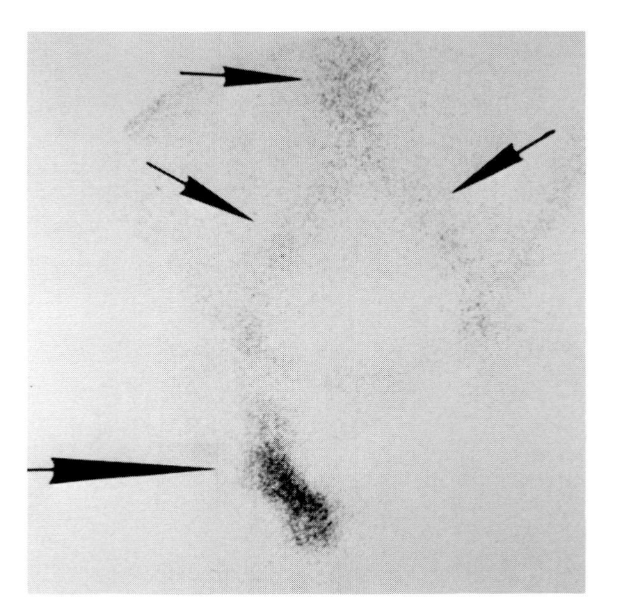

Figure 12-24 ***Colostomy Site***

Modest, circular uptake at a stoma is demonstrated on this anterior abdominal view *(arrow)*. ***Comment:*** This degree of uptake is normal in the first few days after surgery.

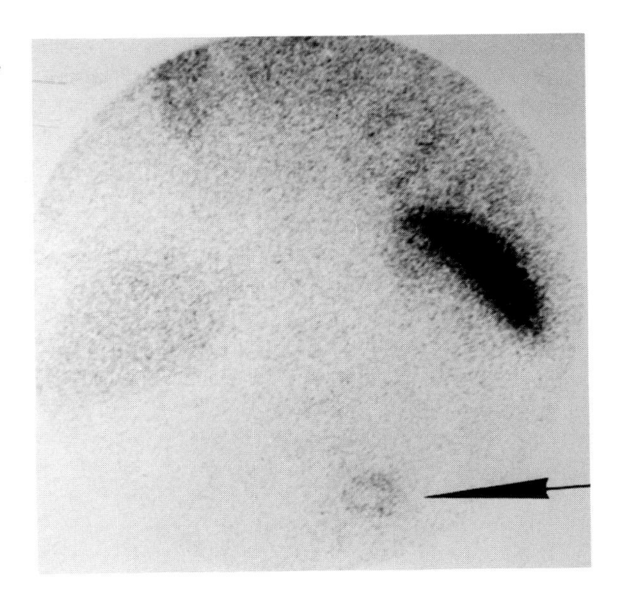

Figure 12-25 ***Lymph Node Uptake***

In this anterior pelvic view, the arrows denote inguinal node activity. ***Comment:*** Nodes are occasionally visualized and are usually of modest intensity.

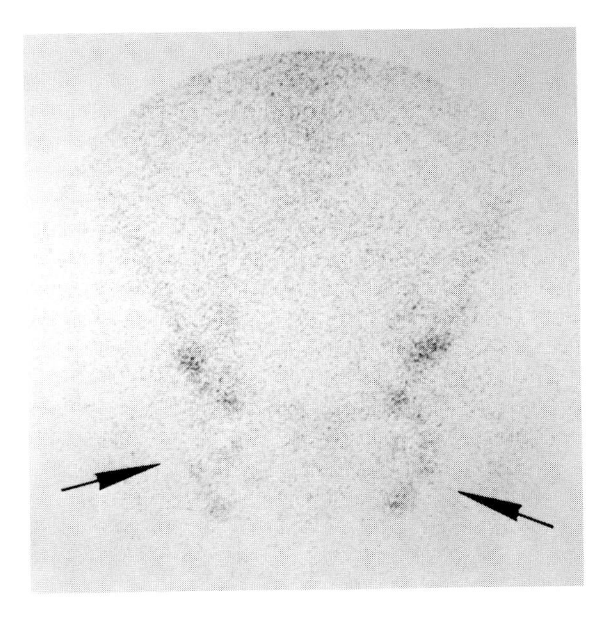

Figure 12-26 ***Drain Infection***

In this anterior abdominal view, the arrow points to curvilinear focus in left midquadrant. This proved to be an infected drain. ***Comment:*** Surgical drains, beyond a few days, may show only modest uptake and indicate infection. A similar degree of uptake is often seen in infected peritoneal dialysis catheters. Note absence of spleen.

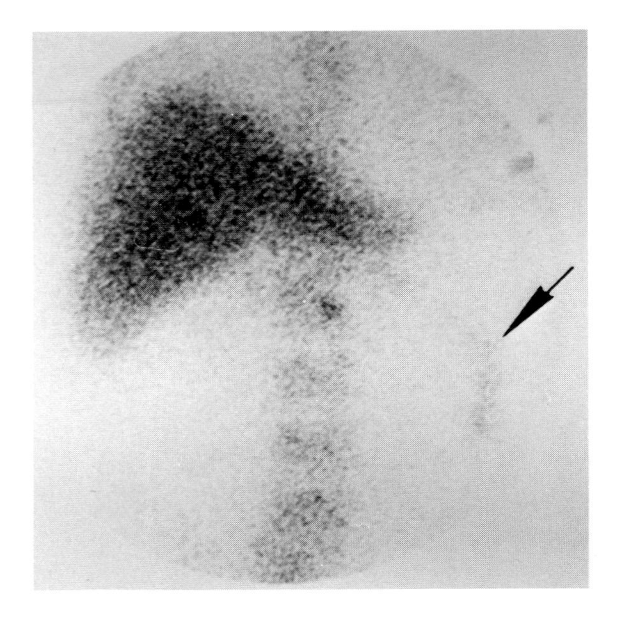

Figure 12-27 ***Cold Defect***

In this posterior pelvic view, the arrow points to absent activity in left sacroiliac region. This proved to be due to a lytic process secondary to tumor involvement. ***Comment:*** It is important to remember to look for symmetry in areas of normal leukocyte accumulation.

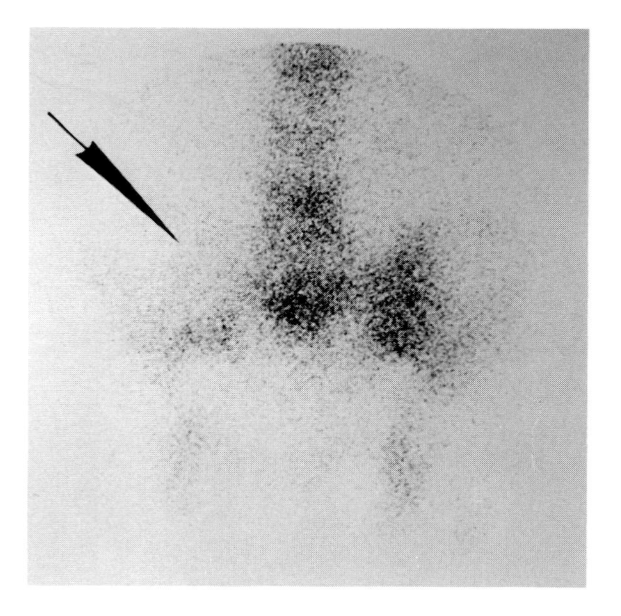

Figure 12-28 ***Subclavian Vein Infection***

In this anteior view of the chest, the arrow points to left subclavian vein abscess (secondary to catheter). ***Comment:*** Uncomplicated catheters are not visualized. Note intensity of uptake in comparison to that in Figure 12-33.

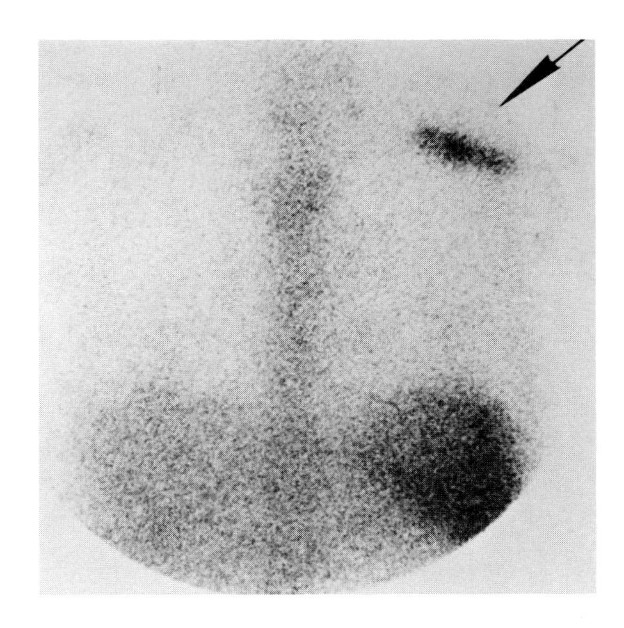

Figure 12-29 ***Mycotic Aneurysm***

In this anterior view of the legs, the arrow points to very modest activity in left popliteal area. This was treated by surgical resection. ***Comment:*** Fungal and parasitic infections usually generate minimal WBC response.

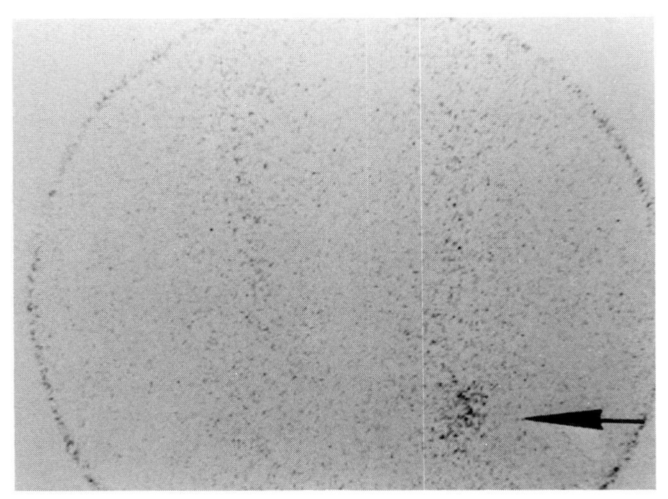

Figure 12-30 ***Interstitial Nephritis***

In this posterior abdominal view, the arrows point to bilateral, modest renal uptake, which was the clinical diagnosis at discharge. Compare to Figure 12-14.

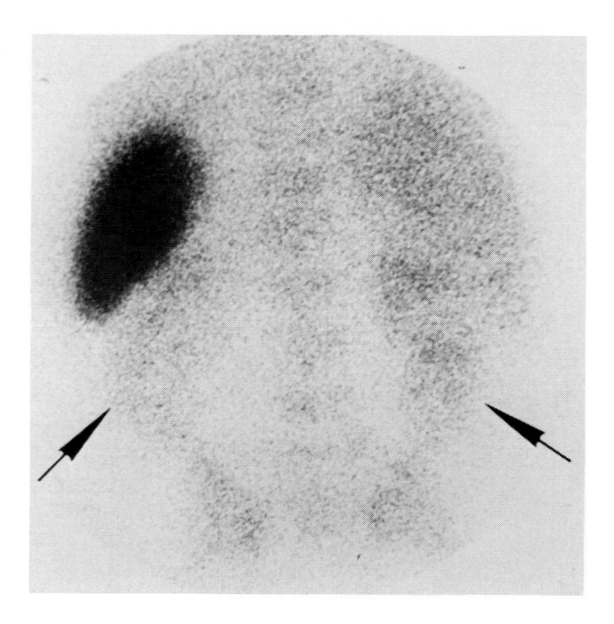

Figure 12-31 ***Ischemic Bowel Disease***

In this anterior abdominal view, intense uptake is noted throughout the large bowel, and this activity may be difficult to distinguish from inflammatory bowel disease owing to marked leukocyte response. This patient died shortly after scan and the postmortem examination revealed necrotic bowel.

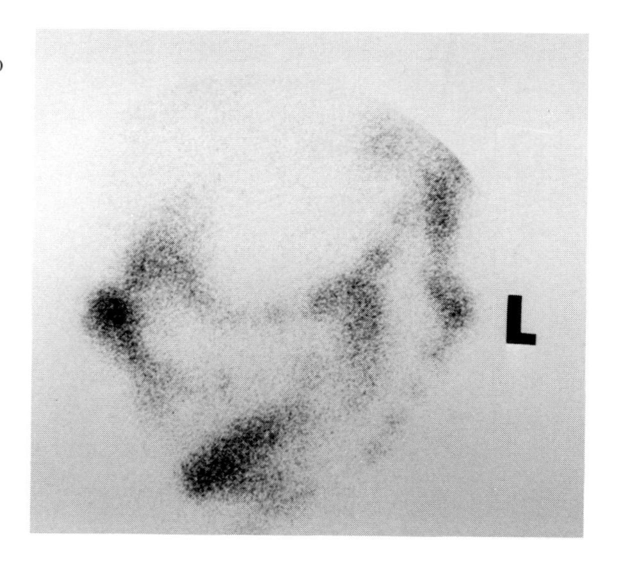

Figure 12-32 ***Intramuscular Injection Sites***

In this posterior pelvic view, the arrows point to bilateral injection sites. Note that intensity of uptake may be significant.

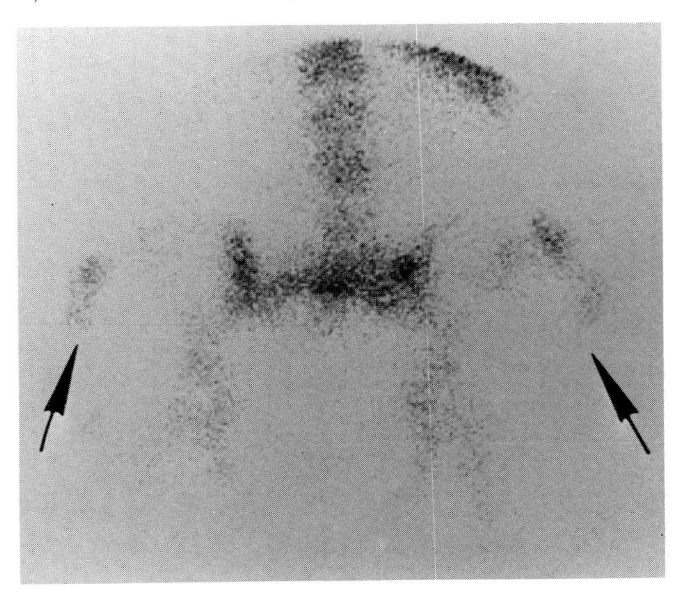

Figure 12-33 ***Tube Sites***

In this anterior view of the head and upper chest, the large arrow points to nasogastric tube site. Small arrow at left points to subclavian catheter. Arrow with *t* points to tracheostomy site. Subclavian activity is likely due to injection through the catheter (see Fig. 12-36).

Figure 12-34 ***Hematoma***

In this anterior chest view, the arrows point to right subclavian vein hematoma, proven by venography. ***Comment:*** Hematomas may be as intense as abscesses.

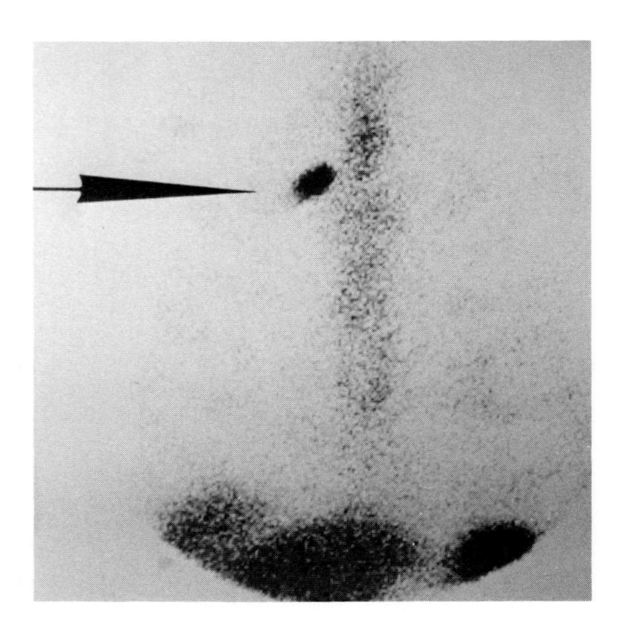

Figure 12-35 ***Splenic Hematoma***

(A) Posterior abdominal view. Arrow points to large "cold" defect in spleen. ***(B)*** Posterior view from subsequent liver/spleen scan. Arrow points to identical defect.

A

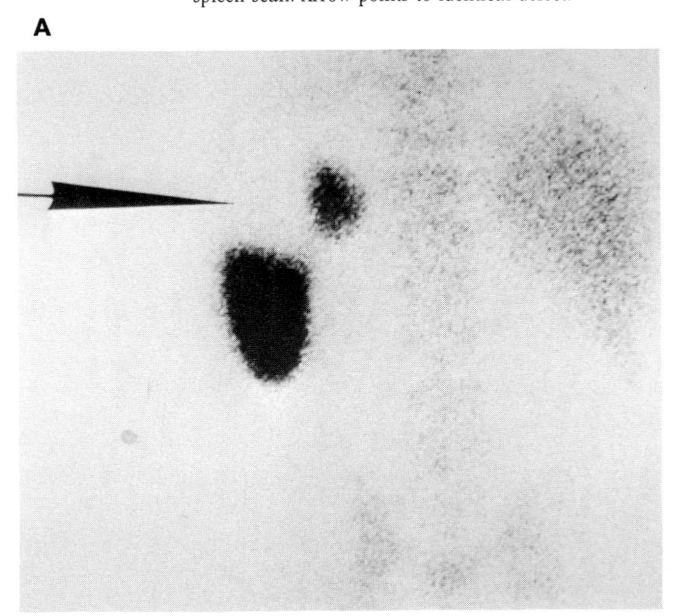

B

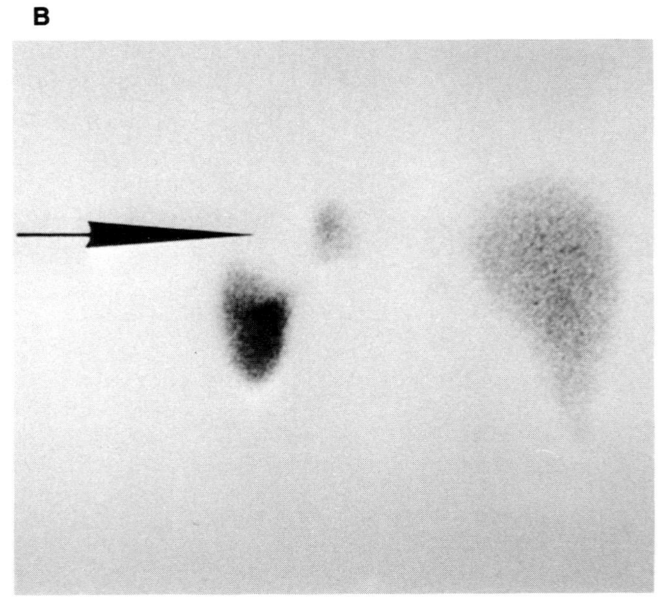

Figure 12-36 ***Catheter Injection***

In this anterior head/chest view, the arrow points to a curvilinear focus of uptake in the right neck. This radioactivity is felt to be secondary to platelet and white cell adherence at time of injection. This injection was made through an indwelling jugular catheter. Focal increase at distal tip is likely due to "trapping" of WBC and platelets.

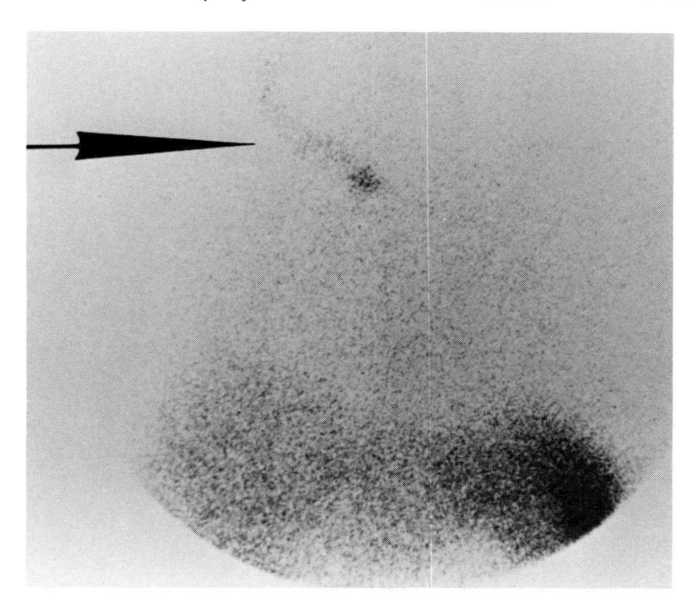

Figure 12-37 ***Sinusitis/ Swallowed WBC***

(A) Anterior head view demonstrates increased uptake in sinuses *(arrow)*. Focal area below sinuses is tracheostomy site. *(B)* Posterior pelvic view demonstrates uptake in rectosigmoid region *(arrow)*. This is due to swallowed WBC.

A

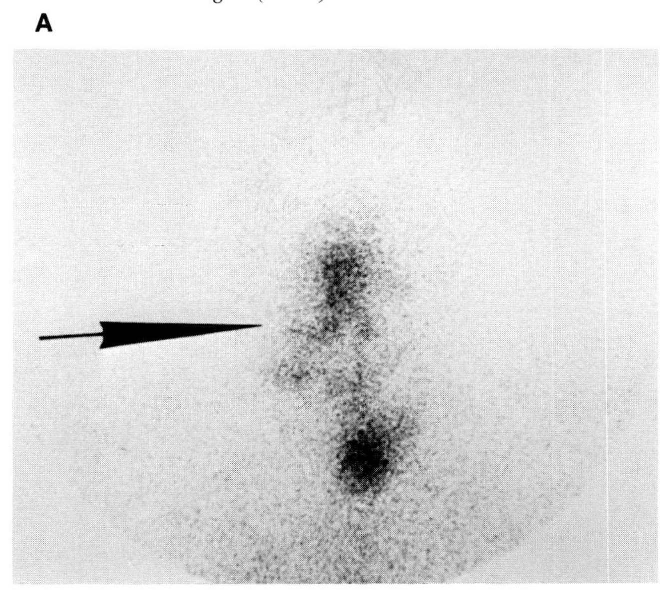

B

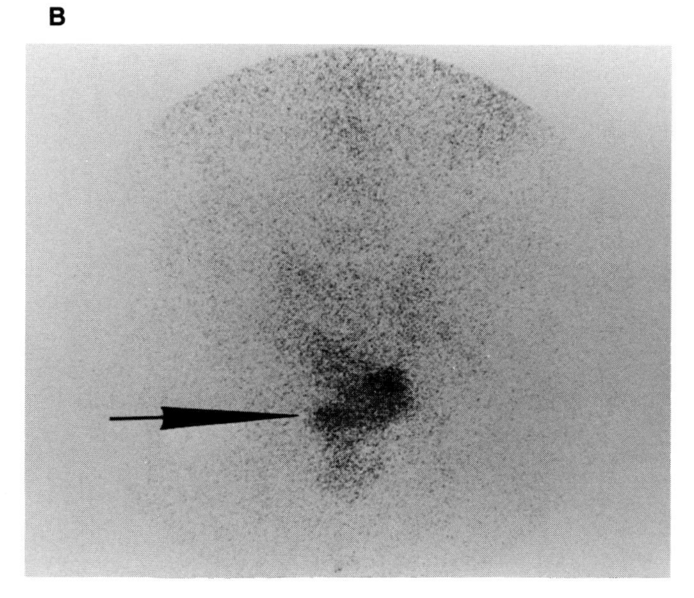

Figure 12-38 ***Thrombophlebitis***

In this anterior pelvic view, the arrow points to left femoral thrombophlebitis. Forty-eight hours later, the patient developed pulmonary emboli.

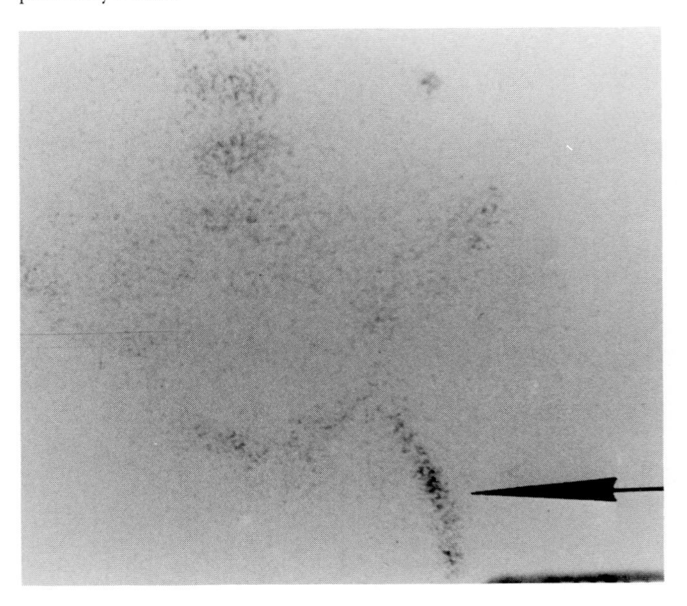

Figure 12-39 ***Hemorrhagic Cyst***

Anterior pelvic view demonstrates a focus of modest uptake *(arrow),* which was determined to be a hemorrhagic, uninfected cyst at surgery.

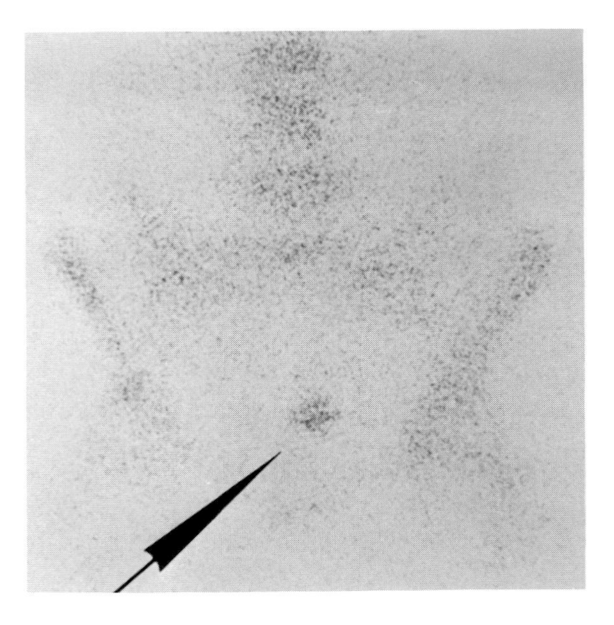

John R. Sty
Robert G. Wells
S. Bert Litwin

Chapter 13 PEDIATRIC NUCLEAR CARDIOLOGY

The original use of radiopharmaceuticals for evaluating the human cardiovascular system began with the work of Blumgart and Weiss in 1927.[1] Although their work was limited clinically because they had to use radium products, these investigators determined the circulation time in humans by injecting a radium salt into an arm vein and timing the appearance of the radioactivity within the artery in the opposite arm with a Wilson cloud chamber. In 1948 the technique was revived by Prinzmetal and colleagues[2] who used a Geiger-Müller counter and sodium-24.

Since the early seventies there has been a renaissance in nuclear cardiology. The acceptance of nuclear medicine techniques has grown considerably because of the advances in instrumentation and a growing interest among cardiologists and cardiovascular surgeons. With the development of high-resolution scintillation cameras, dedicated computer systems, and sophisticated software, radionuclide techniques have taken their place with electrophysiologic, biochemical, and echocardiographic methodologies.

The major categories of nuclear cardiac examination in children at the clinical level include radionuclide angiocardiography, gated blood pool imaging, myocardial imaging and particle imaging. Although these categories are similar to those used in evaluation of adult patients, they differ in the performance of the examination, the type of abnormalities investigated, and, consequently, the information obtained. All of these examinations are sensitive, noninvasive, accurate, and available at a clinical level.

RADIONUCLIDE ANGIOCARDIOGRAPHY

Qualitative Radionuclide Angiocardiography

Qualitative radionuclide angiocardiography emerged in the 1960s. Since the earlier studies performed by Bender and Blau in 1963, there has been a steady improvement in technique.[3] The development of the Anger scintillation camera, combined with the introduction of technetium-99m pertechnetate, initiated the diagnostic field of pediatric nuclear cardiology. The major reason for developing radionuclide angiocardiography was to offer a means other than cardiac catheterization to visualize the heart, lungs, and great vessels. High-quality images have contributed to the wide acceptance of this technique.

Imaging Technique

Radionuclide angiocardiography is simple to perform, but to obtain diagnostic images, meticulous attention must be given to the technical aspects of the procedure. The site of injection is generally the right external jugular vein in the infant or median antecubital vein in an older child. Tc-99m pertechnetate is administered at a dosage of 8.3 mCi/m^2, with a minimum dosage of 2 mCi. A small volume with high specific activity is used. Potassium perchlorate (6 mg/kg) is given orally prior to the injection of the radiopharmaceutical to block uptake of pertechnetate by the thyroid and to hasten its excretion. Data must be recorded at a rate that permits observation of the temporal sequence of chamber filling, but for long enough that each frame contains enough data to achieve the required spatial resolution. The optimum frame rate depends on heart rate. In general, a frame rate of 4 to 6 per second is used. Analog images are satisfactory, but computer processing may be necessary to summate, enhance, or smooth the images.

Estimated Radiation Absorbed Doses

One estimate of radiation absorbed doses for Tc-99m 0_4 is given in Table 13-1.

Normal Images

In the anterior view, the radiopharmaceutical enters the right atrium from the superior vena cava within 2 seconds of injection. The bolus is diluted with unlabeled blood at the junction of the innominate vein with the right subclavian vein, which at times appears as an indentation in the superior vena cava. Approximately 0.5 to 1 second later the right ventricle is demonstrated; its shape varies with the axis of the heart. The cardiac axis will depend on the particular cardiac malformation being investigated.

The radiopharmaceutical then enters the main pulmonary artery. The infundibulum of the right ventricle and the pulmonary valve are narrower than the main pulmonary artery; this is usually not noticeable in children under age 4. The right and left pulmonary arteries then fill; the left pulmonary artery is usually more distinguishable from the surrounding structures. The tracer then enters the peripheral pulmonary circulation; the right lung normally appears more dense than the left. The lung is visualized within 2 to 5 seconds after the injection. The left atrium is not well-visualized on the anterior view. During the dextrophase of the cardiac cycle, the left ventricle is demonstrated as a photon-deficient void. During the levophase, the left ventricle usually appears as a crescent or an oval structure (Fig. 13-1).

Clinical Applications

The clinical applications of qualitative radionuclide angiocardiography are broad-based. Diagnostic information can be obtained concerning obstruction of the superior vena cava and other great veins, cardiac tumors, intracardiac shunts, pulmonary hypertension, etiology of heart failure, postoperative cardiovascular reconstruction, and cardiac valvular and chamber abnormalities[4–15] (Figs. 13-2–13-8).

For example, in obstruction of the superior vena cava, there is interruption of the normal flow pattern into the right heart, which is frequently associated with demonstrable collateral circulation. In right-to-left shunts there is early appearance of radiopharmaceutical in the left heart and aorta concomitant with visualization of the lung. At times, the level of the right-to-left shunt may be detected. For example, in tricuspid atresia, which has an associated opening in the atrial septum (Fig. 13-3), the radiopharmaceutical may flow from the right atrium into the left atrium and left ven-

Table 13-1
Estimates of Radiation Absorbed Doses (RAD/mCi)

Radiopharmaceutical	*Target Organ*	*Other*	*Red Marrow*	*Ovaries*	*Testicular*	*Whole Body*
Tc-99m pertechnetate*	Stomach 0.25	Thyroid unblocked 0.13	0.019	0.022	0.009	0.014
Tc-99m red blood cells†	Heart 0.075–0.081	Lungs 0.048–0.064	0.033–0.041	0.022	0.012	0.018–0.019
Thallium-201‡	Renal medullae 1.2	Heart 0.50		0.47	0.52	0.21
Tc-99m pyrophosphate§	Bladder 0.2	Bone 0.07	0.038	0.03	0.015	0.02

* *(Modified from MIRD Dose Estimate Report No. 8: Summary of current radiation dose estimates to normal humans from 99m-Tc as sodium pertechnetate. J Nucl Med 17:74, 1976)*

† *(Modified from Malmud L: Dosimetry of 99m-Tc-labeled blood pool scanning agents. Clin Nucl Med 3:421, 1978)*

‡ *(Modified from Atkins HL, Budinger TF, Lebowitz E et al: Thallium-201 for medical use. J Nucl Med 18:133, 1977)*

§ *(Modified from Graham LS, Krishnamurthy GT, Blahd BT: Dosimetry of skeletal-seeking radiopharmaceuticals. J Nucl Med 15:496, 1974)*

tricle without visualization of the right ventricle. In right-to-left shunts at the ventricular level, as in the tetralogy of Fallot, the radiopharmaceutical flows from the right ventricle to the left ventricle and exits the aorta and pulmonary artery with minimal lung visualization. In left-to-right shunting, the radionuclide angiocardiogram typically demonstrates persistent high levels of radioactivity in the lung and/or right ventricle because of early recirculation. With large-volume left-to-right shunts, the left-sided cardiac chambers may not be visible due to recirculation and poor clearance of tracer from the lung.

Radionuclide angiocardiography can be used in the postoperative evaluation of transposition of the great vessels (Fig. 13-7). One form of corrective surgery for this anomaly includes the placement of an intracardiac baffle (Mustard procedure), which helps drain off caval and coronary sinus blood into the left ventricle and pulmonary artery, as well as pulmonary venous blood into the right ventricle and aorta. Residual shunting through the atrial baffle or obstruction of the systemic or pulmonary veins by the baffle are complications that can be documented from assessment of flow patterns.

Patency or occlusion of shunts constructed to palliate children with cyanotic heart disease can be determined by careful assessment of pulmonary dilution curves or by pulmonary perfusion imaging. For example, in a functioning Blalock–Taussig shunt (subclavian artery to pulmonary artery anastomosis) or Waterston shunt (ascending aorta to right pulmonary artery anastomosis), the "radionuclide dilution curve" obtained over the lung ipsilateral to the anastomosis reflects a greater and earlier recirculation than in the contralateral lung. The status of a Glenn shunt (superior vena cava to right pulmonary artery anastomosis) can be easily evaluated with pulmonary perfusion imaging or radionuclide angiography (Fig. 13-8).

Radionuclide angiocardiography is rapid, safe, and minimally invasive, yet yields sufficient data to answer specific clinical questions. It can be used to exclude or to guide cardiac catheterization angiography. However, the technique also has limitations: it has limited anatomic resolution and does not measure intracardiac pressures or oxygen saturations. These are important parameters in evaluating pediatric cardiac disorders. Rapid technical advances and increased experience in real-time ultrasound have resulted in the use of this technique rather than radionuclide evaluation in many cases.

Quantitative Radionuclide Angiocardiography

Quantitative radionuclide angiocardiography provides a variety of useful, noninvasive measurements in pediatric cardiac disease. The more frequent procedures include left-to-right shunt quantification and ejection fraction determination. Both the gamma camera and the cardiac probe have proven useful. Quantitative analysis allows time–activity curves to be derived from regions of interest created over selected vascular structures. The radionuclide time–activity curves are analogous to dye dilution curves and are evaluated in a similar manner. With these curves, quantification of cardiac shunts, cardiac output and ejection fraction, and mean pulmonary blood flow can be determined.

Left-to-Right Shunt Quantification

Simple, complex, empirically derived, and totally explicit model-based methods have all been developed to calculate the size of an intracardiac shunt by external counting and generating and analyzing a regional time–activity curve from the lung. The principle of locating, quantifying, and calculating the recirculation of radiopharmaceutical due to a left-to-right shunt is highly satisfactory with all of these methods.[16–18]

Imaging Technique. Patient preparation is not necessary for quantitative radionuclide angiocardiography and the dosage of radiopharmaceutical is identical to that for qualitative radionuclide angiocardiography. The patient is placed in a supine position. A butterfly-type needle is inserted into a vein that is close to the central circulation (external jugular vein or median antecubital vein). A single compact bolus of radiopharmaceutical is necessary for an accurate temporal sequence and the quantification of small shunts. Technical success greater than 95% should be achieved with experience. The saline flush is limited to 2 ml in a premature infant, but up to 15 ml may be needed in the adolescent. Because crying and breath-holding increase pulmonary vascular resistance, it is important not to inject the tracer if the child is vigorously crying. Such Valsalva-type maneuvers can invalidate the examination by prolonging or interrupting the intravascular flow of the bolus of radiopharmaceutical.

The normal radionuclide angiocardiogram shows the sequential passage of tracer from the injection site to the superior vena cava, right atrium, right ventricle, lung, left atrium, left ventricle, and aorta. The time sequence of tracer flow varies with the age of the child. With a left-to-right intracardiac shunt, the radionuclide angiocardiogram will show persistence of high level activity in the lung and in the right heart chambers due to early recirculation. The left heart and aorta become more difficult to visualize as the magnitude of the left-to-right shunt increases.

To detect and quantify a left-to-right shunt, regions of interest are created over the superior vena cava and the right lung. These regions must be accurately identified, so that there is a definite separation of the lung from the cardiac chambers. The regions of interest in the lung must not extend to the lung periphery because respiratory motion will falsely produce fluctuations in the time–activity curve. Once the regions of interest are chosen, time–activity curves are generated from each region. The curve from the superior vena cava is used to assess the integrity and compactness of the bolus. If the transit time of the bolus in the superior vena cava is greater than 3 seconds, invalid data are obtained unless deconvolution analysis is applied.

The normal pulmonary time–activity curve (Fig. 13-9) is composed of an initial segment of no activity as the radiopharmaceutical resides in the right heart. Activity then sharply increases as the radiopharmaceutical reaches the vessels of the lung. This is followed by a rapid decline as it returns to the left heart. Following this decline, there is a second peak of lesser amplitude and greater duration than the first peak that corresponds to the normal recirculation of the radiopharmaceutical bolus as it returns to the lung from systemic circulation.

The child with a left-to-right shunt shows significant alterations of the pulmonary time–activity curve because of the early recirculation of systemic blood by way of the intracardiac shunt. The curve shows an abnormal early reappearance of radioactivity that interrupts the normal exponential decrease in counts after the first peak. The degree of recirculation is proportional to the shunt flow (Fig. 13-10).

Computer Analysis. A gamma variate represents an idealized representation of a single passage of radiopharmaceutical through the cardiopulmonary circulation without recirculation. A gamma variate is fit to the actual pulmonary time–activity histogram and is then subtracted from the original pulmonary time–activity histogram, yielding another curve that represents shunt and normal systemic recirculation. A second gamma variate is then fit to the second histogram. In this manner two areas are defined: area A1, proportional to pulmonary blood flow (Q_p), and area A2, proportional to shunt blood flow. Area A1 minus area A2 is proportional to the systemic blood flow (Q_s). The ratio $A1/(A1 - A2)$ is derived, representing the $Q_p : Q_s$ ratio (Figs. 13-9 and 13-10).

With this technique, $Q_p : Q_s$ ratios can be accurately calculated when the shunt is greater than 1.2 : 1 and less than 3.0 : 1. A significant advantage of the gamma variate technique is that if only a proportion of the pulmonary time–activity curve is available, it is possible to fit a gamma variate to that part of the curve and extrapolate the remainder.[16–19]

Detection of Level of Shunt. In older children, careful selection of regions of interest over the superior vena cava, right atrium, or right ventricle and generation of time–activity histograms can, at times, define the level at which the left-to-right shunt occurs. This is technically difficult and only possible if there is no overlap of the cardiac chambers. For example, in total anomalous pulmonary venous connection to the superior vena cava, it may be possible to detect a second peak of radioactivity in the superior vena cava due to recirculation at this level. Similarly, there is evidence of recirculation on the curve obtained from the region of interest over the right atrium in an atrial septal defect (ASD). Recirculation peaks will also occur over the right ventricle and lung with an ASD. With a ventricular septal defect (VSD), the right atrial curve generally will not show recirculation, but recirculation is detected in the lung and the right ventricle. In children with a patent ductus arteriosus (PDA), early recirculation will be observed over the lung but not over the right atrium or ventricle (Fig. 13-11). Inspection of the radionuclide angiocardiogram can also show coronary artery fistulae which are an uncommon cause of left-to-right shunts[20] (Fig. 13-12, A and B).

Ventricular Ejection Fraction

Ventricular function in children can be accurately analyzed with a scintillation camera and computer system. The first-pass radionuclide angiocardiogram allows for noninvasive measurement of the right and left heart performance. Theoretically, this examination is ideally suited for children because the time for data acquisition is brief, usually less than 30 seconds. Examination is limited to older children, however, because statistical accuracy cannot be achieved with the minimal quantities of radiopharmaceutical used in newborns. Also, because of small heart size, it is difficult to separate the cardiac chambers in infants.

Imaging Technique. The first-pass method of measuring ventricular performance entails a study of the initial transit of the radiopharmaceutical bolus through the central circulation. Patient cooperation is required for a relatively short period of time. Careful positioning of the child beneath the scintillation camera is essential so that separation of the right and left ventricles, the aortic outflow, and the left atrium can be achieved. The optimal position is with the child supine and a 10° to 15° right anterior oblique angulation of the detector and a 5° to 10° caudal tilt. The choice of collimator is critical, but a compromise is ordinarily met between maximal count efficiency (sensitivity) for improved statistics and sufficient resolution.

The injection technique and radiopharmaceutical dosage are identical to that used for left-to-right shunt detection and quantification. As with left-to-right shunt evaluation, analysis of the data from the first-pass study is largely dependent on the application of the indicator dye dilution principal. The theoretical assumption of this technique is that there is uniform mixing of tracer with the cardiac blood volume such that changes in count rate are directly proportional to the change in ventricular volume. By virtue of the temporal and anatomic separation of the radiopharmaceutical within each cardiac chamber, quantification of the right and left ventricular performance can be obtained in patients without an intracardiac shunt. The technique using first-pass transit for determining ejection fraction requires a framing rate of at least 20 frames per second.

The time–activity curve that is generated from the first-pass technique must contain counts originating exclusively from the ventricles and be free of counts arising from the noncardiac background structures (Figs. 13-13 and 13-14). This requires that the data analysis be confined only to that portion of the recorded data which corresponds to the time when the radiopharmaceutical resides in a specific ventricular chamber.

Computer Analysis. The left ventricular ejection fraction (LVEF) can be determined either as an average of several individual beats or from the summation of the cardiac cycles formed by adding several beats frame by frame. Each peak of maximal activity corresponds to end-diastolic volume and each valley of minimal activity represents the end-systolic volume. In order to correct background contribution to the left ventricular time–activity curve, two regions of interest are generated with a light pen or comparable flagging system. One region indicates the left ventricle and the other corresponds to the noncardiac background. Usually the background region of interest assumes the shape of a horseshoe surrounding the left ventricle. From these regions of interest simultaneous time–activity curves are generated. Each curve represents the counts accumulated within a defined region of interest at a specific time. It is necessary to digitally smooth the background curve to minimize statistical variation. It should then be normalized to the left ventricular curve by multiplying each point of the background curve by the ratio of the area of the left ventricular region of interest to the area of the background region of interest. The normalized background time–activity curve is then subtracted point for point from the left ventricular time–activity curve using the formula:

$$\text{LVEF} = \frac{\text{end-diastolic counts} - \text{end-systolic counts}}{\text{end-diastolic counts} - \text{background counts}}$$

The same principle is used to determine the right ventricular ejection fraction (RVEF). Background correction is less critical for RVEF determination because there is very little noncardiac activity during the dextrophase of the cardiac cycle.

One advantage of determining ejection fraction by the radionuclide technique over contrast angiography or echocardiography is that no assumption is made regarding the shape of the ventricle. The ejection fraction is based on the change in the count rate, which is directly proportional to change in chamber volume. In addition, pulmonary transit time, cardiac output, and blood volume determination can be incorporated into the study to supply additional information.

The radionuclide technique does have disadvantages. The determination of ejection fraction is interactive; that is, the physician has to make the decision when determining the regions of interest. In very small children, the data will not accurately reflect ejection fraction, since low count rates are produced by the small chamber size and dose limitations.[21–29]

Clinical Applications. The assessment of performance of both the right and left ventricles at rest and during exercise has become increasingly important in adults. The same is true in children, particularly as more of them undergo surgery for congenital heart anomalies and many now reach adulthood. Left ventricular dysfunction in children may be due to anomalous origin of the left coronary artery, left ventricular outflow abnormality, long-term transfusion cardiomyopathy, or myocarditis. Evaluation of right heart dysfunction is particularly applicable in congenital heart disease involving that chamber, such as tetralogy of Fallot.[30–36]

In children with myocarditis or acute rheumatic carditis, assessment of biventricular performance at rest may allow appropriate categorization of patients in terms of therapy and prognosis. In some circumstances, it may be useful to combine first-pass studies with thallium-201 perfusion imaging.

The symptoms of congestive heart failure may complicate sickle cell crisis. Differentiation of right ventricular and left ventricular dysfunction can be measured using these techniques. Since therapy for sickle cell crisis includes administration of large volumes of fluid, determination of left ventricular performance can have major therapeutic applications.

In children with cystic fibrosis, the major hemodynamic burden falls on the right ventricle due to pulmonary hypertension. Right heart failure and right ventricular hypertrophy frequently are difficult to identify prior to overt cardiac decompensation, and these usually occur late in the course of the disease. Abnormalities in right ventricular performance are predictive of subsequent cardiopulmonary decompensation. Radionuclide angiocardiography, with the assessment of ventricular function, can provide a means of selecting appropriate patients with cystic fibrosis for therapy and for following the hemodynamic responses to therapy.

In patients with obstructive valvular heart disease, the determination of ventricular performance during exercise may allow identification of an appropriate time for corrective surgery and provide an objective means for evaluating the results of surgical procedures. For example, resting right ventricular dysfunction has been identified in patients with mitral stenosis. This hemodynamic abnormality suggests pulmonary arterial hypertension. Following mitral valve replacement, there is usually a significant improvement in RVEF.

In patients with aortic stenosis, the major hemodynamic burden is on the left ventricle. Exercise radionuclide studies have been used in conjunction with electrocardiography to assess the relationship between the ventricular geometry and the functional myocardial reserve. Many circumstances and lessons learned from adults with acquired heart disease can be applied to children.

GATED BLOOD POOL IMAGING

In adults, gated blood imaging is the most frequently performed radionuclide study of the cardiovascular system and the results of these investigations have been extensively published. In children, this procedure is less widely employed because the overall incidence of heart disease is relatively less.

Imaging Technique

Several criteria must be met before gated blood pool imaging can be used effectively as a diagnostic technique. The child's heart rate must be relatively constant during the period of data acquisition, because an arrhythmia significantly interferes with the accuracy of the examination. There must be no patient motion and diaphragmatic excursions should be minimized. The triggering signal, usually the R-wave of the QRS complex, must have a fixed relation to the mechanical events of the cardiac cycle. The radiopharmaceutical must be contained within the intravascular compartment during the time interval of the study. The count density in the images must be sufficient to provide adequate spatial resolution. A sufficient number of views must be recorded to ensure that each cardiac chamber is maximally visualized and is separated from the others. This latter criterion is more difficult to achieve in children because of the variation in cardiac size and position and the abnormal relationships of the ventricles that are found in many forms of congenital heart diseases.

The gated blood pool technique, unlike the first-pass method, requires complete mixing of tracer in the vascular compartment before the study commences. Consequently, all of the cardiac chambers are simultaneously visualized in various positions, depending on the angle of view (Fig. 13-15). The ratio of radioactivity in the cardiac blood pool to that of the pulmonary blood pool determines the level of certainty with which the cardiac chambers can be identified.

More routinely used blood pool tracers include human serum albumin (HSA) or *in vivo* labeled red blood cells (RBC), both labeled with Tc-99m. HSA has a larger distribution in space when compared to that of labeled RBC, particularly in the liver and to a lesser extent in the lung. Therefore, images recorded with HSA will have a lower heart-to-lung ratio than those with RBC. Even with Tc-99m-labeled RBC, however, a ratio of only 3 : 1 can be achieved.[37–45]

Both the single- and multicrystal scintillation cameras are capable of recording equilibrium blood pool studies in children. The single-crystal gamma camera, with either a parallel-hole or conversion collimator, is the most commonly employed instrument. The triggering signal used to record the information is the R-wave of the electrocardiogram. The signal has several features that make it desirable:

- It is easy to obtain.
- The signal of the size and shape of the R-wave occurs only once during a complete cardiac cycle.
- It bears a fixed relationship to the mechanical events of the cardiac cycle.

Using the R-wave as a reference, this method records a series of images during the cardiac cycle. Because of the limited number of counts in a single cardiac cycle, it is necessary to produce a series of blood pool images (500 to 2000 heart cycles), which are then superimposed in a single recording. Thus, the gated blood pool study represents a summation of a large number of cardiac cycles, more or less equally divided using the R-wave as a point of reference. This summated image is usually temporally divided into 16 to 32 frames. The duration of each frame is dependent on the R-to-R interval of the electrocardiogram. Each view requires approximately 8 to 20 minutes to acquire, depending on the age of the patient. Because of this lengthy imaging time, young children frequently require mild sedation.

Computer Analysis

Analysis of data from the gated blood pool study is performed in two steps: first, qualitative inspection by display of the data in a continuous cinematic format, and second, quantification of function by computer analysis. Because the externally monitored count rate is proportional to the changes in ventricular volume, ventricular volume can be accurately quantified by analyzing time–activity curves obtained from the ventricle. Whatever methodology is used to acquire the data, the final product is an image that has been digitized and stored in a computer. Data recording programs from most manufacturers of computer systems allow the operator to select different recording parameters in order to tailor the examination. The most common frame duration required for adequate studies is approximately 40 milliseconds (msec); however, this may be lengthened or shortened depending on the R-to-R interval. In addition, correction for arrhythmias is usually possible by specifying the maximal acceptable variation in the R-to-R interval. Usually, gate tolerance (maximal acceptable range of R-to-R interval) is approximately 100 msec. This rejection system helps to prevent distortion of data from cardiac beats of aberrant duration. If premature beats are frequent, greater than 10%, or if other arrhythmias are present, the examination cannot be accurately performed. The end of the study is usually determined by a preset number of cardiac cycles or time. Most studies require one million counts.

It is difficult at times to accurately identify the entire left ventricular contour because radiopharmaceutical is present in all cardiac chambers simultaneously. The free wall of the left ventricle, the apex, and the septum can be identified without difficulty under normal circumstances. It is challenging to identify the atrioventricular valve plane on an unprocessed image. This difficulty can be overcome by creating an image of the stroke volume that is obtained by subtracting the end-systolic from the end-diastolic image. Because only the pixels viewing the ventricles increase at diastole, stroke-volume measurements will delineate the atrioventricular plane. In order to generate the stroke-volume image, end-diastolic and end-systolic frames need to be identified. This can be accomplished from information obtained from the count rate versus the time curve generated from the left ventricle. Once the stroke-volume image has been produced, a left ventricular region of interest limited to the entire left ventricle can be drawn using the end-diastolic frame to indicate the septal, apical, and free wall of the left ventricle, and the stroke-volume image used to identify the atrioventricular valve plane. The amount of activity that is contributed from background is calculated from a horseshoe region of interest surrounding the left ventricle. To determine if the background is correct, a time–activity curve is generated. If only the lung is included in the area, a straight line is obtained.

Gated blood pool imaging is superior to the first-pass technique in children for several reasons:

- The injection technique is not critical.
- Low count rates are compensated for by increased imaging time.
- Specialized investigations, such as assessment of ventricular response to stress or drug administration, are possible without additional injections.

Estimated Radiation Absorbed Dose

One estimate of the radiation absorbed dose for Tc-99m-label RBC is noted in Table 13-1.

Clinical Applications

Gated blood pool studies can be used in children to diagnose and manage many congenital and acquired abnormalities, including single ventricle, idiopathic hypertrophic subaortic stenosis (Fig. 13-16), valvular aortic stenosis (Fig. 13-17), congenital ventricular diverticulae, ventricular aneurysms (Fig. 13-18), drug-induced cardiomyopathies, myocarditis (Fig. 13-19), and cardiac function abnormalities due to chronic lung disease.

Gated blood pool examinations are useful in the management of adolescents with known acquired valvular heart disease or congenital malformations. The examination permits noninvasive evaluation of ejection fractions at rest and with exercise, assessment of functional myocardial reserve, and detection of regional wall abnormalities such as might occur with unsuspected coronary artery disease in patients with rheumatic valvular disease.

Reliable measurements of ventricular function prior to surgery for congenital or acquired valvular disease may be useful indicators of potential response to therapy. In a study of patients with ASD, Liberthson and colleagues[46] used radionuclide imaging to assess right ventricular function. Right ventricular dilatation was present in all patients preoperatively. Preoperatively, nine patients had normal right ventricular wall motion and 11 had diffuse right ventricular hypokinesis. Follow-up studies showed a reduction in right ventricular size. The patients with hypokinesia, all of whom were symptomatic, had a mean preoperative ejection fraction of 34%, compared to the 64% in patients with normal wall motion. Seven patients had clinical evidence of cardiac decompensation. All of these patients improved symptomatically after surgery, and one became totally asymptomatic. Atrial fibrillation persisted in the five patients so affected. Significant dilatation remained, however, in all of the patients who had persistent right ventricular hypokinesia. The mean postoperative RVEF was 47%, a significant increase over the preoperative management.

Right ventricular function is closely related to the clinical course before and after surgical correction of an ASD. Radionuclide imaging is often helpful in assessing older symptomatic patients, adults and children, with ASD. Right ventricular function may improve postoperatively when hypocontractility is present but cannot be expected to return to normal. Similar studies indicate that gated blood pool imaging can be useful in the long-term follow-up of patients with aortic regurgitation and/or stenosis. For example, patients with a normal ejection fraction and left ventricular hypertrophy due to aortic stenosis have a better long-term prognosis than those with a reduced ejection fraction.

In children with viral myocarditis, the clinical value of exercise gated blood pool examination can be anticipated from the study of Das and co-workers.[47] They performed gated blood pool imaging at rest and with exercise to determine left ventricular functional reserve in six asymptomatic men who previously had myocarditis and in six age-matched controls. The patients were all studied 2 years after the bout of myocarditis. All patients had been healthy prior to their illness. The resting mean LVEF was 58%, as compared with 65% in the control subjects. Three patients had values below 50%. Mean exercise values were 53% in the patients and 75% in the controls. Three of the patients actually had a decline in the ejection fraction with exercise, and the other three had only a slight increase. No regional wall abnormalities were observed at rest or during exercise in any case. All patients had normal exercise electrocardiograms, chest radiographs, and echocardiograms. This experience, combined with that obtained in adults with coronary artery disease, indicates that exercise gated blood pool imaging is valuable in screening asymptomatic children with a history of viral myocarditis for abnormal ventricular function, especially those who wish to participate in vigorous athletics.

MYOCARDIAL IMAGING

Historically, a variety of radiopharmaceuticals have been used to evaluate the human myocardium. These include potassium-43, cesium-129, ytterbium-81, thallium-201, and technetium-99m complexes. The most common tracers employed in clinical practice are thallium-201 and technetium-99m pyrophosphate. Their clinical value has been supported by numerous investigations in adult patients.

Because the spectrum of heart disease in children differs markedly from that in adults, myocardial imaging is an infrequent procedure in children. Although the clinical applications of Tl-201 myocardial scintigraphy are less defined in children, preliminary experience suggests several potential applications:

- To detect fixed changes in left ventricular myocardial perfusion and regional perfusion reserve
- To assess early right ventricular volume or pressure overload, or both, or right ventricular hypertrophy due to noncardiac disease
- To detect cardiomyopathy due to systemic disease
- To evaluate the right ventricle in congenital heart disease
- To assess results of surgical intervention.[48]

Thallium Imaging

Normal Images

In a normal child, Tl-201 is distributed homogeneously throughout the myocardium. In some children there is a relative diminution of tracer uptake at the apex of the heart on the anterior view because of the normal apical myocardial thinning. Accumulation of Tl-201 per gram of tissue is identical in both ventricles. Because the left ventricular wall is normally thicker, only the left ventricle is demonstrated on the Tl-201 images performed at rest. The right ventricle is not usually visualized in normal children. The right ventricle receives 90% of blood flow per gram as compared to the left ventricle, but is approximately one-third the thickness of the left ventricle. When the free wall of the right ventricle is demonstrated, its mass and/or blood flow per gram must be as great as that of the left ventricle. During tachycardia, exercise, and other conditions that result in an increase in the right ventricular muscle mass or blood flow, the free wall can be detected on the anterior and oblique views.

Imaging Technique

Tl-201 myocardial scintigraphy can be performed in the resting state or after exercise. Most children are examined at rest. Tl-201, in the form of thallous chloride, is injected intravenously at a dosage of approximately 1.15 mCi/m^2, with a minimum dosage of 350 μCi. Children are injected in the fasted state and in the upright position when possible. Anterior, 35° to 45° left anterior oblique, and lateral projections are obtained using a single-crystal, gamma scintillation camera and computer. Imaging is begun 5 to 10 minutes after tracer injection. If an exercise study is performed, both immediate and delayed images should be obtained.

Estimated Radiation Absorbed Dose

One estimate of the radiation absorbed dose for Tl-201 is noted in Table 13-1.

Clinical Applications

One application of Tl-201 imaging in infants and children is the noninvasive assessment of right ventricular overload or hypertrophy, or both (Figs. 13-20 – 13-22). Quantitative assessment of right ventricular overload by other noninvasive methods may be difficult. Tl-201 can be used to detect and, in a semiquantitative manner, assess abnormalities of the right ventricle in both primary cardiac disease and right heart disease due to a pulmonary abnormality. This method has been evaluated clinically and in laboratory animals.

In human patients, several reports have demonstrated that visualization of the right ventricle with Tl-201 is associated with elevated right ventricular systolic pressure and increased pulmonary vascular resistance. However, because the right ventricle may be visualized in acute right ventricular volume or pressure overload before the development of myocardial hypertrophy or with tachycardia, the significance of right ventricular Tl-201 uptake in each patient must be carefully evaluated in context of the clinical situation.

Computer techniques have been developed to quantify the amount of right ventricular uptake relative to the background or the left ventricle. Quantitative Tl-201 imaging for assessing the effects of elevated right ventricular pressures in children with congenital heart disease has been evaluated by Rabinovitch, Fischer, and Treves.[49] They studied 24 patients ranging in age from 7 months to 30 years. Eighteen were evaluated before corrective surgery and six were studied after the procedure. All but three patients had congenital heart defects that had resulted in right ventricular pressure or volume overload, or both. Their data suggest that if Tl-201 counts emanating from the right ventricle were less than 1% of the injected dosage or less than 30% of the left ventricular counts, the right ventricular peak systolic pressures were less than 30 mm Hg. This was found to be true in six of their patients. In the remaining 18 patients, there was good correlation between right ventricular/left ventricular Tl-201 count ratios and right ventricular pressure. All patients with right ventricular peak systolic pressures at half systemic levels or greater had a right ventricular/left ventricular Tl-201 count ratio at or greater than 0.4. They concluded that quantitative analysis of myocardial imaging with Tl-201 can be used in patients with congenital heart defects in assessing the severity of pulmonary stenosis or the presence of pulmonary hypertension (Fig. 13-20).

Similarly, Ohsuzu and co-workers[50] demonstrated the favorable clinical role of Tl-201 myocardial imaging in evaluating right ventricular overload by comparing it with the results of electrocardiographic and hemodynamic variables. They studied 47 patients who had chronic right heart overload, including 28 with pressure overload, 14 with volume overload, and five with combined pressure and volume overload. They noted that with pressure overload only, the degree of right ventricular visualization correlated with elevation of right ventricular systolic pressure. With volume overload only, the right ventricular free wall was always visualized; the degree of right ventricular visualization correlated with an increasing pulmonary to systemic flow ratio. Pressure overload could be differentiated from volume overload by the morphologic characteristics of the myocardial image, but not from the degree or type of uptake. With pressure overload, the interventricular septum seemed to appear straight. However, with volume overload, the right ventricular cavity appeared to be dilated and the interventricular septum was convex to the right. The sensitivity of myocardial imaging for the diagnosis of right ventricular pressure overload was 93.3%, which was higher than that achieved by applying electrocardiographic criteria for right ventricular hypertrophy.

Reduto and colleagues[51] reported the radionuclide evaluation of right and left ventricular function after total correction of tetralogy of Fallot. They noted that Tl-201 images showed substantial and quantifiable right ventricular uptake consistent with residual right ventricular hypertrophy in many patients. To define the magnitude of abnormal right ventricular Tl-201 uptake, the right ventricular/left ventricular count densities were obtained from the images. The average number of Tl-201 counts per pixel was determined for both the right ventricle and left ventricle. Patients with residual right ventricular hypertrophy had an average right ventricular/left ventricular ratio of 0.6 (Fig. 13-21).

Chronic pulmonary disease (Fig. 13-22) may lead to severe right-sided heart disease. The most common chronic lung disease associated with right ventricular hypertrophy in children is cystic fibrosis. Newth and associates[52] evaluated 32 children with cystic fibrosis using Tl-201 imaging. They compared the results with those of other nonimaging techniques including electrocardiography, vectorcardiography, and M-mode echocardiography. The patients studied had a wide spectrum of clinical and pulmonary abnormalities. In all patients, the Tl-201 images, like the vectorcardiograms and M-mode echocardiograms, gave a surprisingly high proportion of positive predictions for right ventricular hypertrophy (44%). However, correlation among the noninvasive studies was extremely poor. Investigators suggested caution when using Tl-201 to determine right ventricular hypertrophy for the prediction of the natural history of cor pulmonale and cystic fibrosis. The Tl-201 scans predicted right ventricular hypertrophy successfully in five of the six patients. One false-negative Tl-201 scan was recorded, which was probably due to myocardial ischemia secondary to severe right ventricular failure.

In addition to its usefulness in evaluating abnormalities of the right ventricle, Tl-201 can be used to evaluate the left ventricle in ischemic heart disease in pediatrics. Tl-201 can be injected not only at rest, but immediately after or during exercise. Tl-201 imaging has become a routine procedure in adults to evaluate myocardial ischemia due to coronary artery disease because it provides both anatomic and physiologic confirmation of myocardial ischemia and an accurate means of assessing the results of surgical intervention. In children there can be many causes of myocardial ischemia and infarction (Fig. 13-23, A and B).

Etiology of Myocardial Ischemia and Infarction in Pediatrics
Anomalous left coronary artery
Atherosclerosis (progeria)
Complication of surgery
Coronary calcinosis
Friedreich's ataxia
Mucocutaneous lymph node syndrome
Muscular dystrophy
Hypertrophic subaortic stenosis
Myocarditis
Polyarteritis nodosa
Pulmonary stenosis
Refsum's syndrome
Rubella
Sickle cell anemia
Supravalvular aortic stenosis
Tetralogy of Fallot
Transient myocardial ischemia
Tumors

Differentiation of anomalous origin of the left coronary artery from congestive cardiomyopathy of other causes can be determined. In infants with anomalous origin of the left coronary artery from the pulmonary artery, it is difficult to rule out congestive cardiomyopathy of other causes, such as idiopathic endocardial fibroelastosis. Both conditions result in cardiomegaly with congestive heart failure with either no murmur or with a murmur of mitral insufficiency.[53,54] Gutgesell, Pinsky, and DePuey[55] used Tl-201 myocardial perfusion in 16 children who ranged from 1 month to 4 years of age to distinguish congestive cardiomyopathy from anomalous left coronary artery. Seven had an anomalous left coronary artery and nine had idiopathic congestive cardiomyopathy. Localized abnormalities of Tl-201 uptake were found in each of the seven patients with anomalous left coronary artery, including two asymptomatic 4-year-old children. Tl-201 distribution was normal in five patients with congestive cardiomyopathy, diffusely irregular in three, and absent in the lateroposterior basal segment of the left ventricle in one.

Similarly, Finley and co-workers[56] performed Tl-201 imaging on six patients with an anomalous left coronary artery (Fig. 13-24) arising from the pulmonary artery. Initial images of three children demonstrated anterolateral perfusion defects in agreement with electrocardiographic localization of a myocardial infarct. Repeat imaging in two patients 2 to 3 months after clinical improvement with medical therapy for congestive cardiac failure demonstrated reduction in the size of the perfusion deficit. In three other patients who had surgical correction of the anomalous left coronary artery, Tl-201 imaging performed during exercise demonstrated normal myocardial perfusion 7 to 15 years after surgery. It was concluded from this study that Tl-201 is a helpful adjunct in monitoring changes in myocardial perfusion before and after medical or surgical treatment of anomalous left coronary artery and may shed light on the pathophysiology of the perfusion deficit. As in adults, the practical limitation of Tl-201 imaging in infants and children makes it difficult to detect subtle endocardial myocardial infarction.

Occasionally in the newborn, transient myocardial dysfunction and respiratory distress can result in severe hypoxemia in the absence of congenital heart disease. There is rapid improvement in myocardial function after correction of hypoxemia, but the electrocardiographic evidence of myocardial ischemia can persist after clinical resolution of the syndrome. Myocardial perfusion imaging in this syndrome has been reported to show extremely poor uptake of Tl-201 in the myocardium relative to the lung, a pattern that is suggestive of global myocardial ischemia. An alternative explanation of this imaging pattern is increased pulmonary accumulation of Tl-201 relative to the heart, suggestive of severe left ventricular dysfunction. The radionuclide pattern in this syndrome differs from that noted in children with myocarditis or other forms of congenital myocardial disease in whom tracer uptake is generally normal or nonhomogeneous in nature[57] (Fig. 13-25).

Kawasaki syndrome consists of several weeks of fever that is unresponsive to antibiotics, conjunctivitis, and reddening of the lips, tongue, oropharyngeal mucosa, palms, and soles. Desquamation of the fingertips, cervical adenopathy, and various cardiac abnormalities occur. Cardiac manifestations may be mild, consisting of sinus tachycardia, or severe, resulting in massive myocardial infarction. The prognosis is usually good, but death occurs in 1% to 2% of cases typically because of the cardiac disease.

Pathologically the involvement of the heart in Kawasaki disease is an acute inflammatory process. The characteristic findings include acute vasculitis and perivasculitis, with inflammation of the adventitia of the arterioles, capillaries, and venules, including the vasa vasorum of the coronary arteries. Acute pericarditis, interstitial myocarditis, and endocarditis have also been observed. Interstitial myocarditis consists of infiltration with both polymorphonuclear (PMN) cells and lymphocytes. Later stages of the disease include severe stenosis, thrombosis, and aneurysm formation of the coronary arteries.

Several imaging techniques are available to evaluate the heart in children with Kawasaki disease. Tl-201 perfusion imaging is useful in the detection of myocardial ischemia or infarction (Fig. 13-26). Gated blood pool studies will detect regional or global abnormalities of the LVEF. Echocardiography is useful in detecting coronary artery aneurysms and assessing ventricular function.[58-64]

Clinical experience suggests that stress Tl-201 myocardial perfusion imaging allows an estimate of the regional blood flow to areas supplied by critically narrowed coronary arteries. The coexistence of idiopathic hypertrophic subaortic stenosis in coronary artery disease has been substantiated in many patients. Rubin and associates[65] evaluated stress perfusion imaging with Tl-201 in ten symptomatic patients with idiopathic hypertrophic subaortic stenosis who had angina pectoris and normal coronary angiograms. The at-rest electrocardiogram demonstrated left ventricular hypertrophy with ST segment abnormalities in seven patients, thereby negating any further increase in ST segment abnormalities that developed in these patients during exercise or in the postexercise state. Of the three patients with normal at-rest electrocardiograms, one had significant exercise-induced ST segment depression. Tl-201 myocardial imaging revealed no significant perfusion deficits in nine of the ten patients. In one patient with severe left ventricular hypertrophy, significant perfusion deficits developed after exercise that were not found on the at-rest study. The scan indicated ischemia of the anterior and anterolateral walls and septum. The angiogram revealed unusual hypertrophy of the anterior wall of the left ventricle that could suggest an imbalance between left ventricular muscle mass and coronary blood flow. All seven patients with electrocardiographic evidence of left ventricular hypertrophy demonstrated further ST segment depression of at least 1 mm during exercise. Five patients developed typical anginal chest pain during stress exercise. Nonetheless, six of the seven patients had normal exercise perfusion scintigrams, suggesting that this technique is superior to stress electrocardiographic testing for assessing the status of the coronary arteries in patients with idiopathic hypertrophic subaortic stenosis who have angina pectoris.

Tc-99m Pyrophosphate Imaging

Tc-99m pyrophosphate (PYP) myocardial imaging has been employed infrequently in pediatric patients because the spectrum of myocardial disease differs considerably from adults (see boxed material). Myocardial scintigraphy is not routinely performed in children because of the low incidence of myocardial infarction in this group.

Estimated Radiation Absorbed Dose

One estimate of the radiation absorbed dose for Tc-99m pyrophosphate is noted in Table 13-1.

Imaging Technique

Tc-99m PYP is administered intravenously at a dosage of 8.3 mCi/1.7m^2 (minimum dosage of 2 mCi). Anterior, 30° and 60° left anterior oblique, and left lateral images are obtained immediately following injection and are repeated at 2 and 4 hours. At least 500,000 counts are obtained per image. The multiple views are necessary to differentiate myocardial uptake from overlying skeletal activity. Patients with uncomplicated acute myocardial infarction show peak Tc-99m uptake between 48 and 72 hours after the onset of symptoms.

Clinical Applications

Myocardial scintigraphy with Tc-99m PYP is useful in children with suspected acute infarction in whom other clinical and laboratory findings are not diagnostic (Fig. 13-23A). A negative scintigram is useful, particularly if it is obtained between 1 and 6 days after the onset of symptoms, since the probability that the child has had an acute infarction is greatly reduced and the likelihood of complications is very low. This technique has been used in evaluating pediatric congenital heart disease after cardiac surgery. The diagnosis of myocardial infarction after cardiac surgery is complicated because of the chest pain, enzyme elevation, and electrocardiographic changes that may result from the operation itself.

As in adults, a positive Tc-99m PYP image is not specific for myocardial infarction. We have noted abnormal uptake of Tc-99m PYP in myocarditis, cardiomyopathy, bacterial endocarditis, and trauma associated with the battered child.

PARTICLE BODY IMAGING

Several investigators have used particles for evaluating right-to-left shunts. In 1971, Gates, Orme, and Dore[66] described a method for calculating right-to-left shunts with microspheres. The examination proved useful in detecting and quantifying right-to-left shunts, determining maldistribution of pulmonary blood flow, and evaluating relative systemic-pulmonary shunting procedures[67] (Fig. 13-27). Although the work relating to toxicity (heart, brain, and kidney microemboli) demonstrated a wide margin of safety for the procedure, meticulous attention must be paid to the number and size of particles because of the systemic embolization that occurs with right-to-left shunting.[68-70]

Imaging Technique

The major assumption in quantifying right-to-left shunts with this technique is that particles are completely extracted from the circulation in one pass through either the pulmonary or systemic capillary beds. After the injection of 200 μCi to 400 μCi of Tc-99m microspheres (less than 50,000 particles) in patients with right-to-left shunts, determination of the ratio of particles that lodge in the pulmonary and systemic capillary beds will equal the ratio of pulmonary blood flow to shunt blood flow. Two or more scintigrams, each of 2 minutes' duration, are taken with the gamma camera to produce a whole-body image. The counts from separate scintigrams are summed to obtain the total body count, and the activity of each lung is measured separately. The right-to-left shunt is calculated from the following equation:

$$\frac{\text{Total body count} - \text{total lung count}}{\text{Total body count}}$$

In addition to the particle imaging technique, a right-to-left shunt can be detected by radionuclide angiocardiography with inert gas. Some investigators claim that the use of inert gases and nondiffusable tracers help to define the nature of complex congenital heart abnormalities. Still others have described the calculation of right-to-left shunting from time–activity curves from the left ventricle. The advantage of these methods is elimination of theoretical complications of systemic emboli that can occur with particle imaging.[71]

REFERENCES

1. Blumgart HL, Weiss S: Studies on the velocity of blood flow. J Clin Invest 4:15, 1927
2. Prinzmetal M, Corday E, Bergman HC et al: Radiocardiography: A new method for studying the blood flow through the chambers of the heart in human beings. Science 108:340, 1948
3. Bender MA, Blau M: The autofluoroscope. Nucleonics 21:52, 1963
4. Rosenthall L: Radionuclide venography using technetium-99m and the gamma ray scintillation camera. Am J Roentgenol 97:874, 1966
5. Starshak RJ, Sty JR: Radionuclide angiocardiography: Use in the detection of myocardial rhabdomyoma. Clin Nucl Med 3:106, 1978
6. Rosenthall L: Nucleographic screening of patients for left-to-right cardiac shunts. Radiology 99:601, 1971
7. Kriss JP, Enright LP, Hayden WG et al: Radioisotopic angiocardiography: Findings in congenital heart disease. J Nucl Med 13:31, 1972
8. Wesselhoeft H, Hurley PJ, Wagner HN Jr: Nuclear angiocardiography in the diagnosis of congenital heart disease in infants. Circulation 45:77, 1972
9. Stocker FP, Kinser J, Weber JW: Pediatric radiocardioangiography. Shunt diagnosis. Circulation 47:819, 1973
10. Alazraki NP, Ashburn WL, Hagan A et al: Detection of left-to-right cardiac shunts with the scintillation camera pulmonary dilution curve. J Nucl Med 13:142, 1972
11. Krishnamurthy GT, Srinivasan NV, Blahd WH: Pulmonary hypertension in acquired valvular cardiac disease: Evaluation by a scintillation camera technique. J Nucl Med 13:604, 1972
12. Bianco JA, Shafer RB: Abnormal images of right heart disorders. Clin Nucl Med 4:368, 1979
13. Alderson PO, Gilday DL, Wagner HN: Atlas of pediatric nuclear medicine, pp 138–140. St Louis, CV Mosby, 1978
14. Kriss JP, Freedman GS, Enright LP et al: Radioisotopic angiocardiography. Preoperative and postoperative evaluation of patients with diseases of the heart and aorta. Radiol Clin North Am 9:369, 1971
15. Matin P, Kriss JP: Radioisotopic angiocardiography: Findings in mitral stenosis and mitral insufficiency. J Nucl Med 11:723, 1970
16. Thompson HK Jr, Starmer CF, Whalen RE: Indicator transit time considered as a gamma variate. Circ Res 14:502, 1964
17. Maltz DL, Treves S: Quantitative radionuclide angiocardiography: Determination of $Q_p : Q_s$ in children. Circulation 47:1049, 1973
18. Alderson PO, Jost RG, Strauss AW et al: Detection and quantification of left-to-right cardiac shunts in children: A clinical comparison of count ratio and area ratio techniques. J Nucl Med 16:511, 1975
19. Askenasi J, Ahnberg DS, Korngold E et al: Quantitative radionuclide angiocardiography: Detection and quantitation of left-to-right shunts. Am J Cardiol 37:382, 1976
20. Wells RG, Litwin SB, Sty JR: Radionuclide cardioangiographic demonstration of a coronary artery fistula. Pediatr Radiol 16:61, 1986
21. Berger HV, Matthay RA, Pylik LM et al: First-pass radionuclide assessment of right and left ventricular performance in patients with cardiac and pulmonary disease. Semin Nucl Med 9:275, 1979
22. Marshall RC, Berger HJ, Costin JC et al: Assessment of cardiac performance with quantitative radionuclide angiocardiography: Sequential left ventricular ejection fraction, normalized left ventricular ejection rate, and regional wall motion. Circulation 56:820, 1977
23. Schad N, Nickel O: Assessment of ventricular function with first-pass angiocardiography. Cardiovasc Radiol 2:149, 1979
24. Tobinick E, Schelbert HR, Henning H et al: Right ventricular ejection fraction in patients with acute anterior and inferior myocardial infarction assessed by radionuclide angiocardiography. Circulation 57:1078, 1978
25. Folland ED, Hamilton GW, Larson SM et al: The radionuclide ejection fraction: A comparison of three radionuclide techniques with contrast angiography. J Nucl Med 18:1159, 1977
26. Ashburn WL, Schelbert HR, Verba JW: Left ventricular ejection fraction—A review of several radionuclide angio-

graphic approaches using the scintillation camera. Prog Cardiovasc Dis 20:267, 1978

27. Kurtz D, Ahnberg DS, Freed M et al: Quantitative radionuclide angiocardiography. Determination of left ventricular ejection fraction in children. Br Heart J 38:966, 1976

28. Berger HJ, Gottschalk A, Zaret BL: First pass radionuclide angiocardiography for evaluation of right and left ventricular performance: Computer applications and technical considerations. In Bacharadi SL, Alpert NM, Shames DM (eds): Nuclear Cardiology: Selected Computer Aspects, pp 29–44. New York, Society of Nuclear Medicine, 1978

29. Pierson RN Jr, Alam S, Kemp HG et al: Radiocardiography in clinical cardiology. Semin Nucl Med 9:85, 1977

30. Anthony CL, Arnon RG, Ritch CW: Pediatric Cardiology. New York, Medical Publishing Co, 1979

31. Hellenbrand WE, Berger HJ, O'Brien RT et al: Left ventricular performance in thalassemia: Combined non-invasive radionuclide and echo cardiographic assessment (abstr). Circulation 56:III-49, 1977

32. Reduto LA, Berger HJ, Johnstone DE et al: Radionuclide assessment of exercise right and left ventricular performance following total surgical correction of tetralogy of Fallot (abstr). Circulation 58:II-145, 1978

33. Jones RH, Sabiston DC Jr, Bates BB et al: Quantitative radionuclide angiocardiography for determination of chamber to chamber cardiac transit times. Am J Cardiol 30:855, 1972

34. Levine FH, Lappas DL, Pohost GH et al: Radionuclide and hemodynamic assessment of pulmonary bypertension and right ventricular function after mitral valve replacement (abstr). Am J Cardiol 43:406, 1979

35. Costellana F: Approaches to modelling radiocardiographic data. In Pierson RN Jr (ed): Quantitative nuclear cardiography, p 217. New York, Wiley, 1975

36. Alexander J, Dainiak N, Berger HJ et al: Serial assessment of doxorubicin cardiotoxicity with quantitative radionuclide angiocardiography. N Engl J Med 300:278, 1979

37. McRae J, Sugar RM, Shipley B et al: Alterations in tissue distribution of 99mTc-pertechnetate in rats given stannous tin. J Nucl Med 15:151, 1974

38. Khentigan A, Garrett M, Lum D et al: Effects of prior administration of Sn(II) complexes on in vivo distribution of 99mTc-pertechnetate. J Nucl Med 17:380, 1976

39. Pavel DG, Zimmer M, Patterson VN: In vivo labeling of red blood cells with 99mTc: A new approach to blood pool visualization. J Nucl Med 10:305, 1977

40. Stokely EM, Parkey RW, Bonte FJ et al: Gated pool imaging following 99mTc stannous pyrophosphate imaging. Radiology 120:433, 1976

41. Hamilton RG, Alderson PO: A comparative evaluation of techniques for rapid and efficient in vivo labeling of red cells with 99mTc pertechnetate. J Nucl Med 18:1010, 1979

42. Winchel HS: Effect of tin-induced enzymes on pertechnetate distribution. J Nucl Med 17:941, 1976

43. Kappos A, Maines MD: Tin: A potent inducer of heme-oxygenase in kidney. Science 192:4,60,234, 1976

44. Chervu LR: Radiopharmaceuticals in cardiovascular nuclear medicine. Semin Nucl Med 9:241, 1979

45. Hamilton RG, Alderson PO: A comparative evaluation of techniques for rapid and efficient in vivo labeling of red cells with [99mTc] pertechnetate. J Nucl Med 18:1010, 1977

46. Liberthson RR, Pohost GM, Dinsmore RE et al: Atrial septal defect: Pre and post operative evaluation by gated cardiac scanning (abstr). J Nucl Med 19:750, 1979

47. Das SK, Brady TJ, Thrall JH et al: Evidence of cardiac dysfunction in asymptomatic patients with prior myocarditis. Circulation 57, 1978

48. Sty JR, Starshak RJ: Thallium-201 myocardial imaging in children. J Am Coll Cardiol 5:128S, 1985

49. Rabinovitch M, Fischer KC, Treves S: Quantitative thallium 201 myocardial imaging in assessing right ventricular pressure in patients with congenital heart defects. Br Heart J 45:198, 1981

50. Ohsuzu F, Handa S, Kondo M et al: Thallium-201 myocardial imaging to evaluate right ventricular overloading. Circulation 61:620, 1980

51. Reduto LA, Berger HJ, Johnstone DE et al: Radionuclide assessment of right and left ventricular exercise reserve after total correction of tetralogy of Fallot. Am J Cardiol 45:1013, 1980

52. Newth CJL, Corey ML, Fowler RS et al: Thallium myocardial perfusion scans for the assessment of right ventricular hypertrophy in patients with cystic fibrosis. Am Rev Respir Dis 124:463, 1981

53. Wesselhoeft H, Fawcett JS, Johnson AL: Anomalous origin of the left coronary artery from the pulmonary trunk. Its clinical spectrum, pathology and pathophysiology based on a review of 140 cases with seven further cases. Circulation 38:403, 1968

54. Askenazi J, Nadas AS: Anomalous left coronary artery originating from the pulmonary artery: Report on 15 cases. Circulation 51:976, 1975

55. Gutgesell HP, Pinsky WW, DePuey EG: Thallium-201 myocardial perfusion imaging in infants and children. Circulation 61:596, 1980

56. Finley JP, Howman-Giles R, Gilday DL et al: Thallium-201 myocardial imaging in anomalous left coronary artery arising from the pulmonary artery. Am J Cardiol 42:675, 1978

57. Finley JP, Howman-Giles RB, Gilday DL et al: Transient myocardial ischemia of the newborn infant demonstrated by thallium myocardial imaging. J Pediatr 94:263, 1978

58. Landing BH, Larson EJ: Are infantile periarteritis nodosa with coronary artery involvement and fatal mucocutaneous lymph node syndrome the same? Comparison of 20 patients from North America with patients from Hawaii and Japan. Pediatrics 59:651, 1977

59. Kato H, Koike S, Yamamoto M et al: Coronary aneurysms in infants and young children with acute febrile mucocutaneous lymph node syndrome. J Pediatr 86:892, 1975

60. Fujiwara H, Hamashima Y: Pathology of the heart in Kawasaki disease. Pediatrics 61:100, 1978

61. Morens DM, Nahmias AJ: Kawasaki disease: A new pediatric enigma. Hosp Pract 13:109, 119, 1978

62. Treves S, Hill TC, VanPraagh R, Holman BL: Computed tomography of the heart using thallium-201 in children. Radiology 133:707, 1979

63. Ueda K, Saito A, Nakano H, Yano M: Thallium 201 scintigraphy in an infant with myocardial infarction following mucocutaneous infarction lymph node syndrome. Pediatr Radiol 9:183, 1980

64. Kato S, Ohta T, Suzuki Y, Matsuyama S: Thallium-201

myocardial imaging in Kawasaki disease. Tokai J Exp Clin Med 7:69, 1982

65. Rubin KA, Morrison J, Padnick MB et al: Idiopathic hypertrophic subaortic stenosis: Evaluation of anginal symptoms with thallium-201 myocardial imaging. Am J Cardiol 44:1040, 1979

66. Gates GF, Orme HW, Dore EK: Measurement of cardiac shunting with technetium labeled albumin aggregates. J Nucl Med 12:746, 1971

67. Gates GF: Radionuclide Scanning in Cyanotic Heart Disease. Springfield, IL, Charles C Thomas, 1974

68. Kennady JC, Taplin GV: Albumin macroaggregates for brain scanning: Experimental basis and safety in primates. J Nucl Med 6:566, 1965

69. Kennady JC, Taplin GV: Safety of measuring regional cerebrocortical blood flow with radioalbumin macroaggregates. J Nucl Med 7:354, 1966

70. Verhas M, Schoutens A, Demol O et al: Use of 99mTc-labeled albumin microspheres in cerebral vascular disease. J Nucl Med 17:170, 1976

71. Parker JA, Treves S: Radionuclide detection, localization, and quantitation of intracardiac shunts and shunts between the great arteries. In Holman BL, Sonnenblick EH, Lesch ED (eds): Principles of Cardiovascular Nuclear Medicine, pp 189–218. New York, Grune & Stratton, 1978

72. MIRD Dose Estimate Report No. 8: Summary of current radiation dose estimates to normal humans from 99m-Tc as sodium pertechnetate. J Nucl Med 17:74, 1976

73. Malmud L: Dosimetry of 99m-Tc-labeled blood pool scanning agents. Clin Nucl Med 3:421, 1978

74. Atkins HL, Budinger TF, Lebowitz E et al: Thallium-201 for medical use. Part 3: Human distribution and physical imaging properties. J Nucl Med 18:133, 1977

75. Graham LS, Krishnamurthy GT, Blahd BT: Dosimetry of skeletal-seeking radiopharmaceuticals. J Nucl Med 15:496, 1974

Atlas Section

Figure 13-1 ***Normal Analog Angiocardiogram (0.5-second images)***

Examination of a 6-year-old patient demonstrates normal transit of the radiopharmaceutical through *(1)* the superior vena cava, *(2)* right atrium, *(3)* right ventricle, *(4)* pulmonary arteries, *(5)* left heart, and *(6)* aorta. (Sty JR: Atlas of pediatric nuclear cardiography. Clin Nuc Med 5:Part I 373-386, Part II 424-438, 1980)

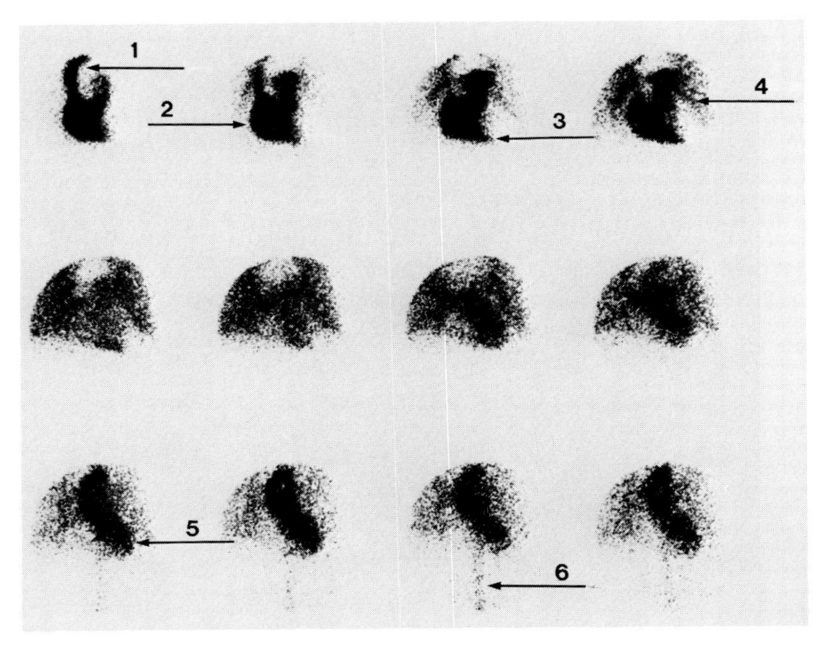

Figure 13-2 ***Azygos Continuation of the Inferior Vena Cava (angiocardiogram, posterior images)***

Radiopharmaceutical, which was injected through a foot vein, was visualized passing through the *(1)* inferior vena cava, *(2)* azygos vein, and *(3)* arch of azygos vein. There is lack of visualization of the inferior vena cava in the upper abdomen. Tracer passes superiorly in the azygos vein and enters the right atrium via the superior vena cava.

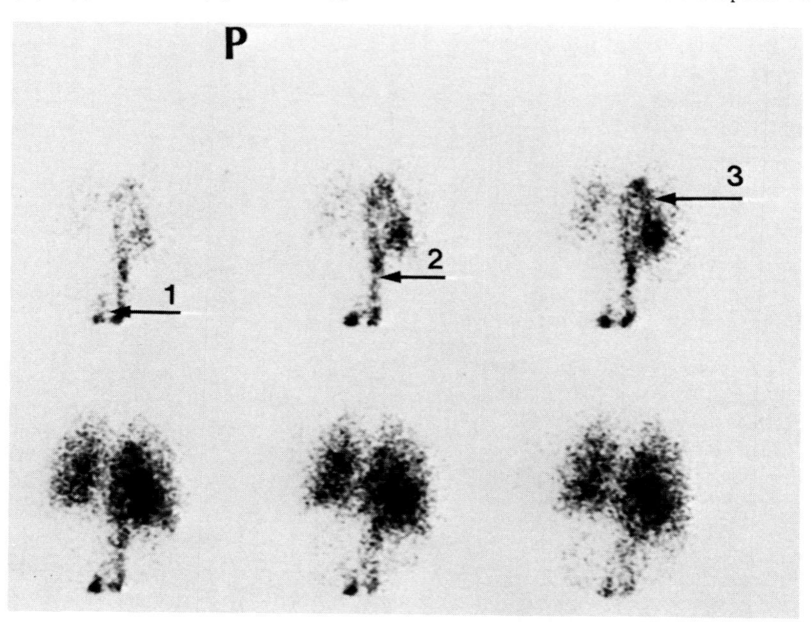

Figure 13-3 ***Tricuspid Atresia***

Analog angiocardiogram, [anterior *(A)* projection (0.25-second images)], of a newborn shows *(1)* tracer in the superior vena cava, *(2)* tracer in the right atrium, *(3)* absence of tracer in the muscle mass between right atrium and left ventricle, and *(4)* tracer in the left ventricle. Note early visualization of the left atrium and left ventricle, without visualization of the right ventricle.

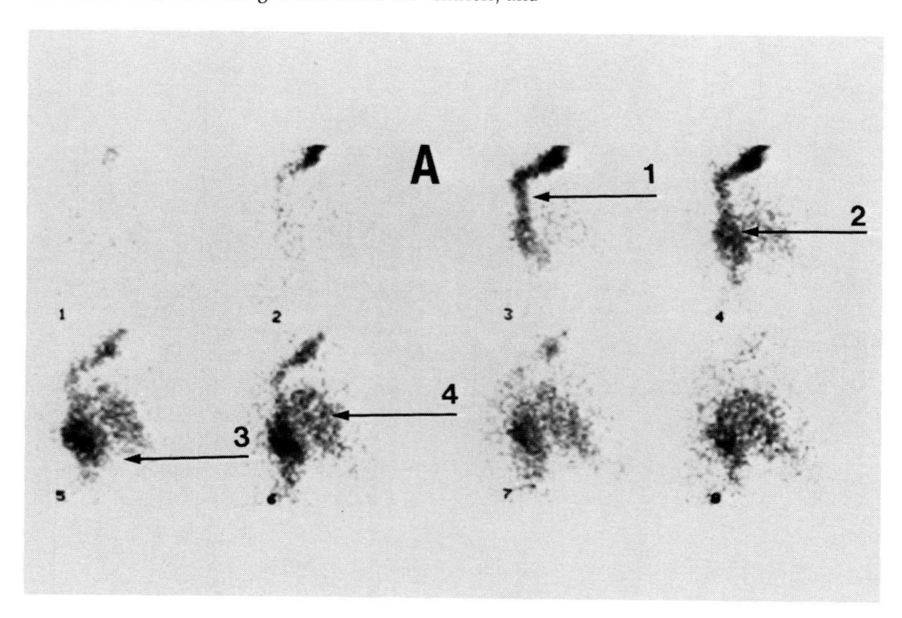

Figure 13-4 ***Total Anomalous Venous Return***

Radionuclide angiocardiogram (anterior images) of a newborn was made after injection of the radiopharmaceutical through the jugular vein. The flow of the radiopharmaceutical from the lung to the portal vein is demonstrated *(arrow)*. The radiopharmaceutical then flows through the liver. This indicates total anomalous pulmonary venous connection to the portal vein. Also note early appearance of tracer in the left heart and aorta due to the obligatory intracardiac right-to-left shunt.

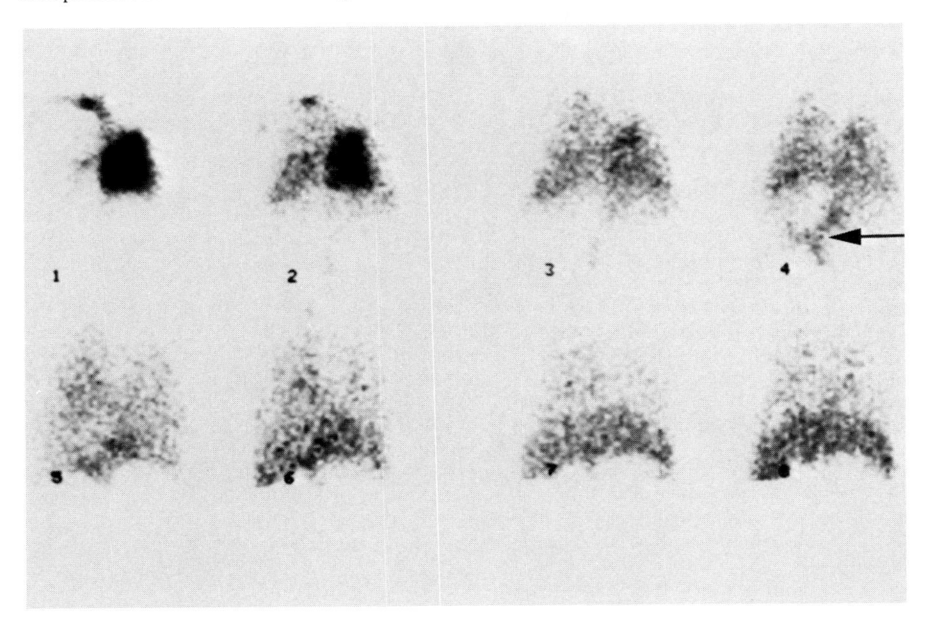

Figure 13-5 ***Evaluation of Surgically Corrected Atresia of the Right Pulmonary Artery***

After postoperative reconstruction of the right pulmonary artery, radionuclide angiocardiogram (1-second anterior images) demonstrates radioactivity in the *(1)* right ventricle, *(2)* perfused left lung, *(3)* perfused right lung, *(4)* left ventricle, and *(5)* bronchial blood flow. The angiocardiogram confirms relatively normal blood flow to the right lung from the right heart, indicating a satisfactory surgical result. (Sty JR: Atlas of pediatric nuclear cardiography. Clin Nuc Med 5:Part I 373-386, Part II 424-438, 1980)

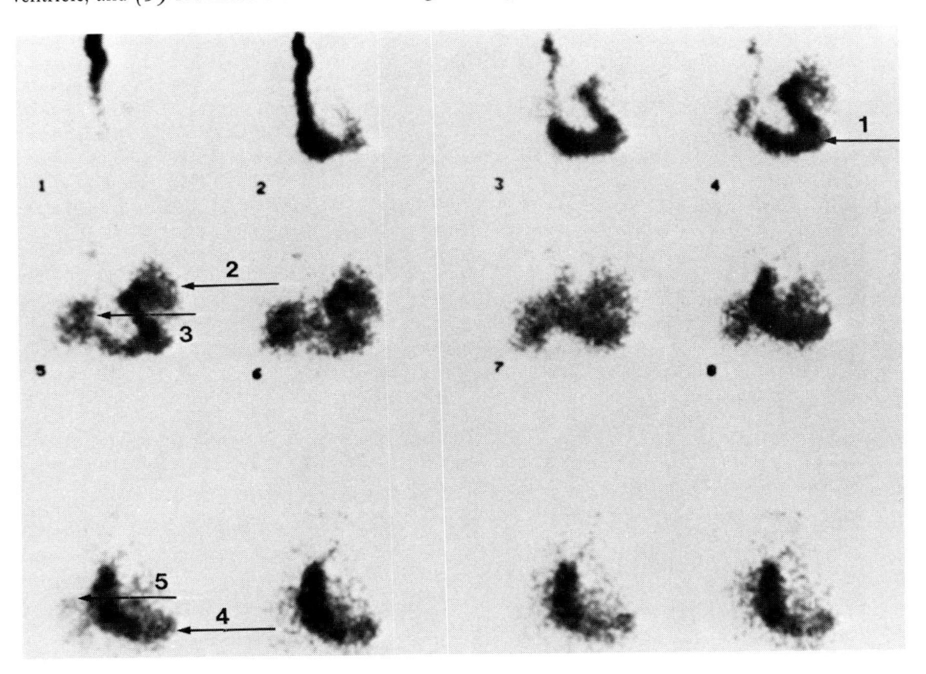

Figure 13-6 ***Transposition of the Great Vessels***

The radionuclide angiocardiogram of a newborn demonstrates early radioactivity in the aorta *(arrow)*. Computer-generated curves from areas of interest over the aorta *(A)* and lung *(L)* show tracer appearing in the aorta prior to the lung. This is indicative of transposition of the great vessels with a right-to-left shunt. (Sty JR: Atlas of pediatric nuclear cardiography. Clin Nuc Med 5:Part I 373 – 386, Part II 424 – 438, 1980)

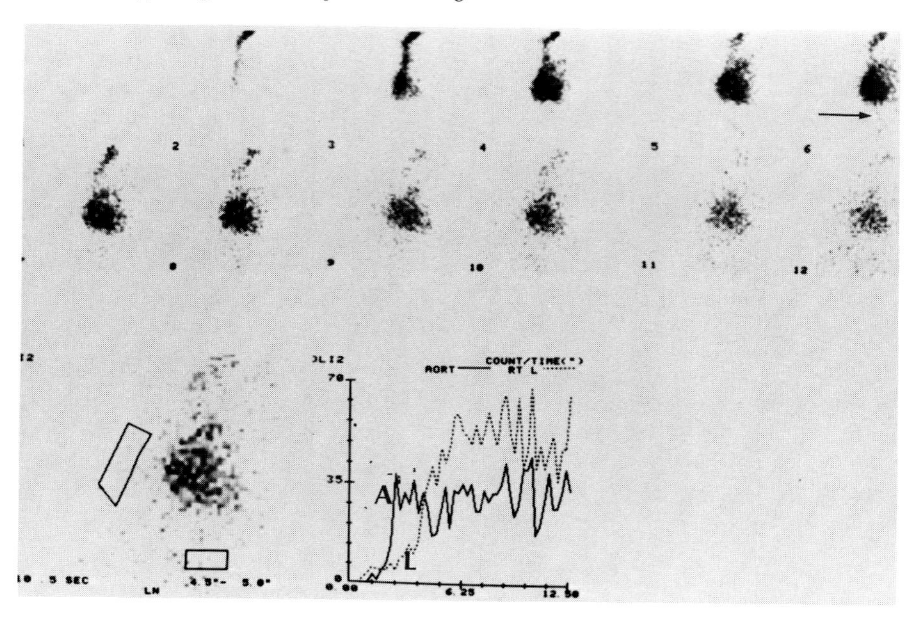

Figure 13-7 ***Evaluation of Surgically Corrected Transposition***

This patient is postoperative with a Mustard procedure for transposition of the great arteries. The radiopharmaceutical was injected into the superior vena cava through the left subclavian vein. The radionuclide angiocardiogram (anterior images) shows flow *(1)* entering right atrium, *(2)* entering anatomic left ventricle, *(3)* filling the lung, and *(4)* proceeding to the left atrium, anatomic right ventricle, and into the aorta. A satisfactory surgical result is therefore confirmed.

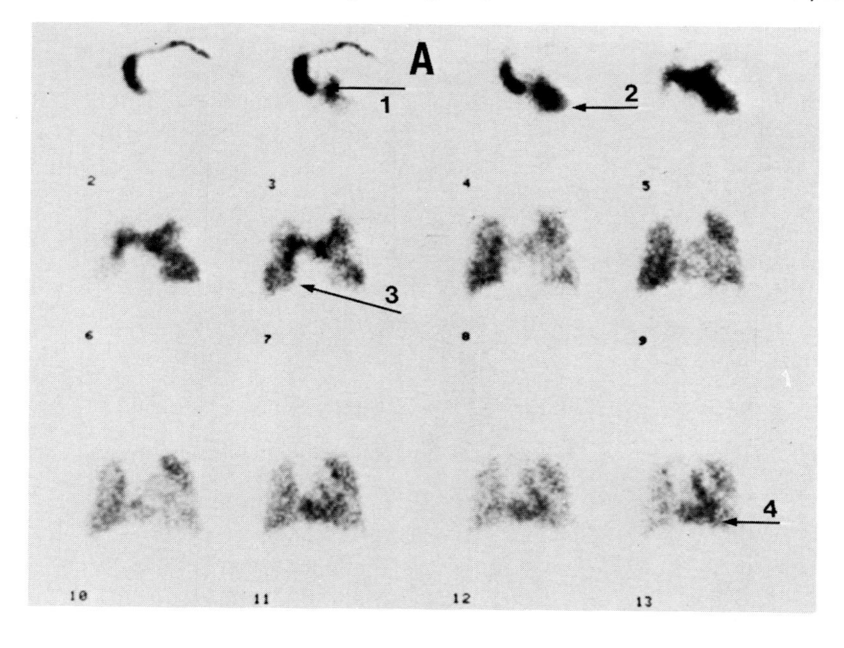

Figure 13-8 **_Palliative Shunt Evaluation_**

This patient is postoperative from a Glenn anastomosis. Radiopharmaceutical was injected through an antecubital vein. The analog angiocardiogram demonstrates the anastomosis between the superior vena cava and the right pulmonary artery, with early visualization of the right lung. This indicates a patent Glenn anastomosis. (*1*, superior vena cava; *2*, right lung; *3*, left ventricle; *4*, position of the descending aorta)

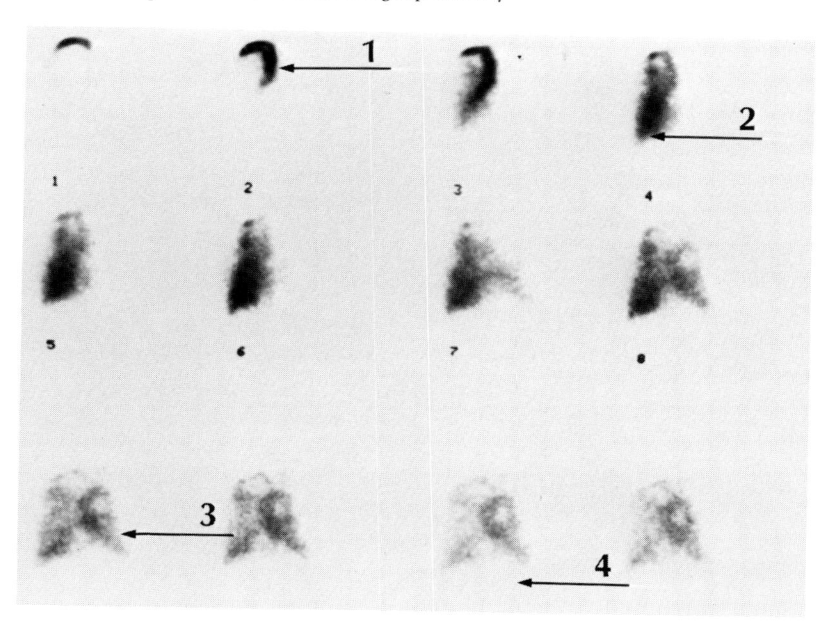

Figure 13-9 **_Left-to-Right Shunt Quantification_**

This is a method for quantifying the ratio of pulmonary to systemic blood flow. *(1)* A pulmonary time-activity histogram is obtained from a region of interest over the right lung. *(2)* A gamma variate that represents the idealized curve of lung blood flow without recirculation is fit to the lung curve. *(3)* The gamma variate is subtracted from the original lung curve *(arrowhead)*, and the resulting curve represents lung recirculation *(arrow)*. *(4)* A gamma variate is fit to the new histogram *(arrowhead)*. *(5)* The areas under the two gamma variate curves are proportional to the pulmonary and systemic blood flow: $Q_p : Q_s$ = area A/(area A — area B). The 1.1 : 1 ratio obtained in this example indicates no significant left-to-right shunting.

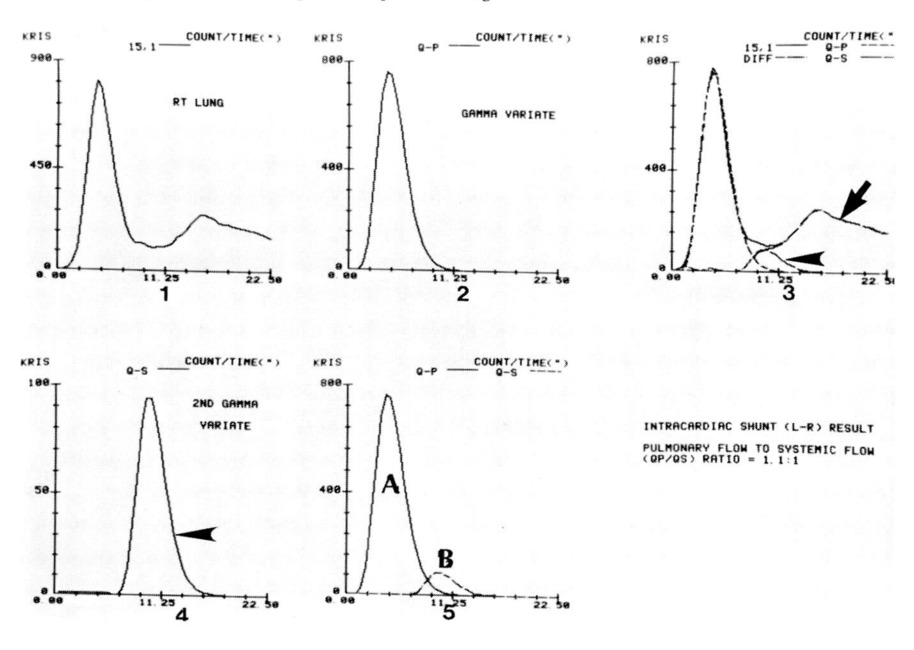

Figure 13-10 *Left-to-Right Shunt Quantification*

These are studies made of a 3-year-old child with atrial and ventricular septal defects. The pulmonary time-activity histogram shows persistent increased activity in the lung following the initial bolus *(arrow)*. Gamma variate analysis demonstrates a prominent recirculation curve *(curved arrows)*. The $Q_p : Q_s$ ratio measures 2.7 : 1, indicating a fairly large left-to-right shunt.

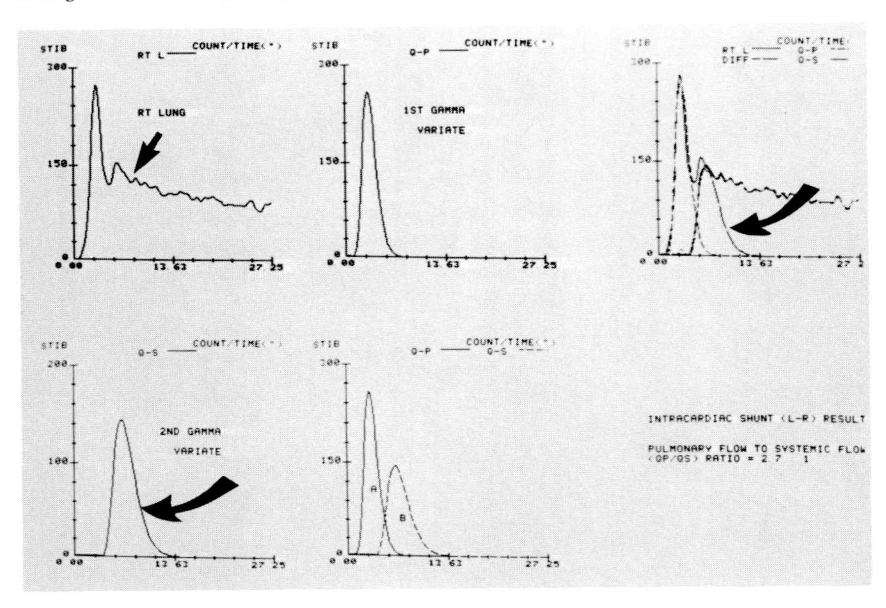

Figure 13-11 *Determination of Level of Shunt*

This patient has a patent ductus arteriosus (PDA). The left image shows areas of interest over the right atrium and right ventricle. Time-activity curves generated from these areas show no evidence of early recirculation; therefore, an intracardiac shunt is excluded. The lung curve (not shown) demonstrated a $Q_p : Q_s$ ratio of 2.2 : 1, owing to extracardiac shunting in the PDA.

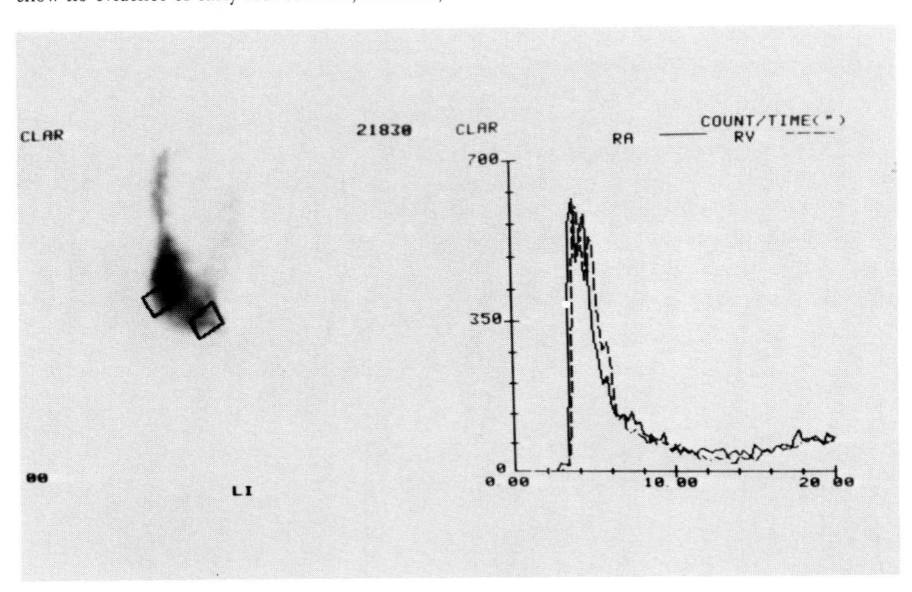

Figure 13-12 ***Determination of Level of Shunt***

(A) Visual inspection of the radionuclide angiocardiogram (anterior images) during the levophase demonstrates the lung, left ventricle, and an anomalous vessel *(arrow)*. Note that the anomalous connection returns blood from the aorta to the right ventricle. *(B)* The composite image during the levophase of the radionuclide angiogram demonstrates this abnormality *(arrow)* to better advantage. At surgery, the presence of a coronary artery fistula to the right ventricle was confirmed.

A

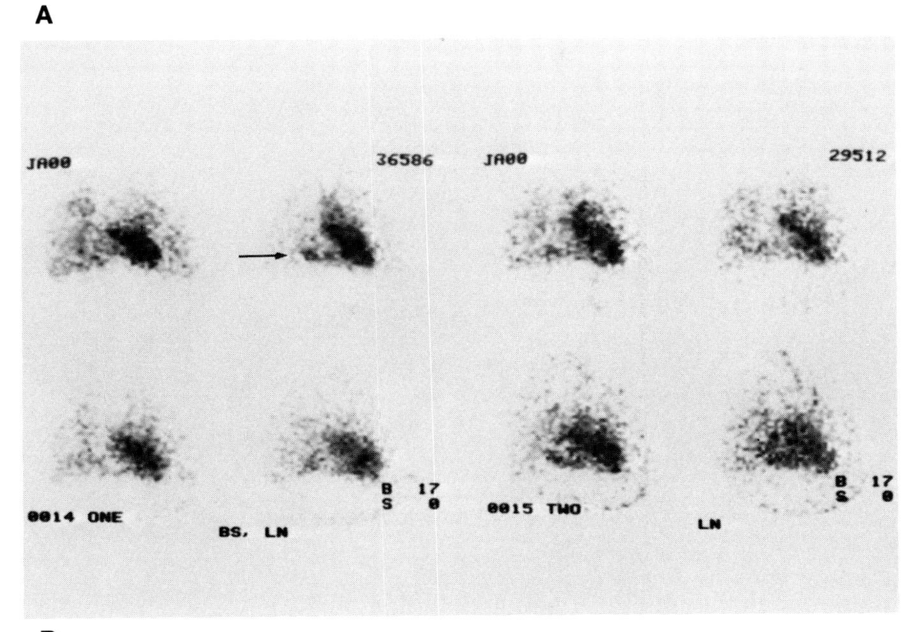

B

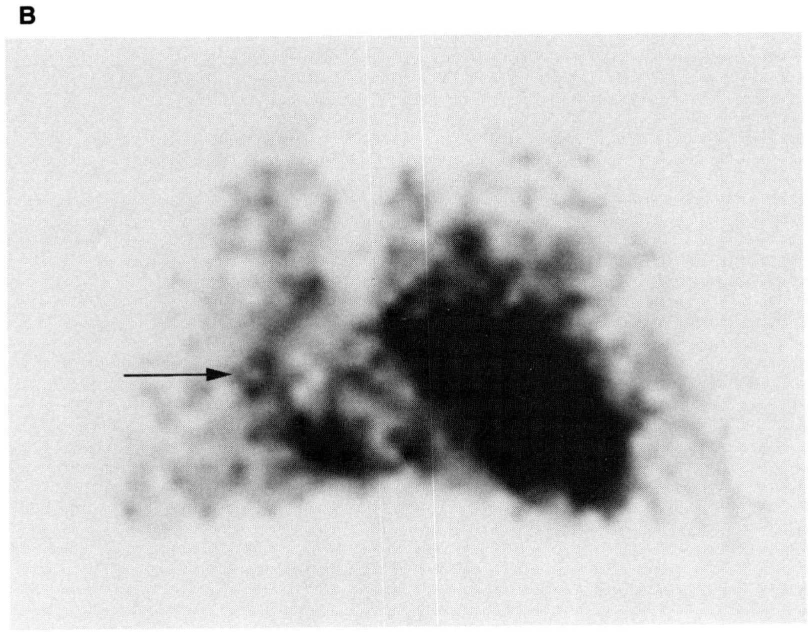

Figure 13-13 ***Left Ventricular Ejection Fraction (LVEF), First-Pass Technique***

First-pass radionuclide angiocardiogram (anterior images) demonstrates *(left)* end-diastolic phase of the cardiac cycle and region of interest and *(right)* end-systolic phase of the cardiac cycle with an area used for background subtraction. Time-activity curves (not shown) were generated from these regions of interest and the LVEF calculated.

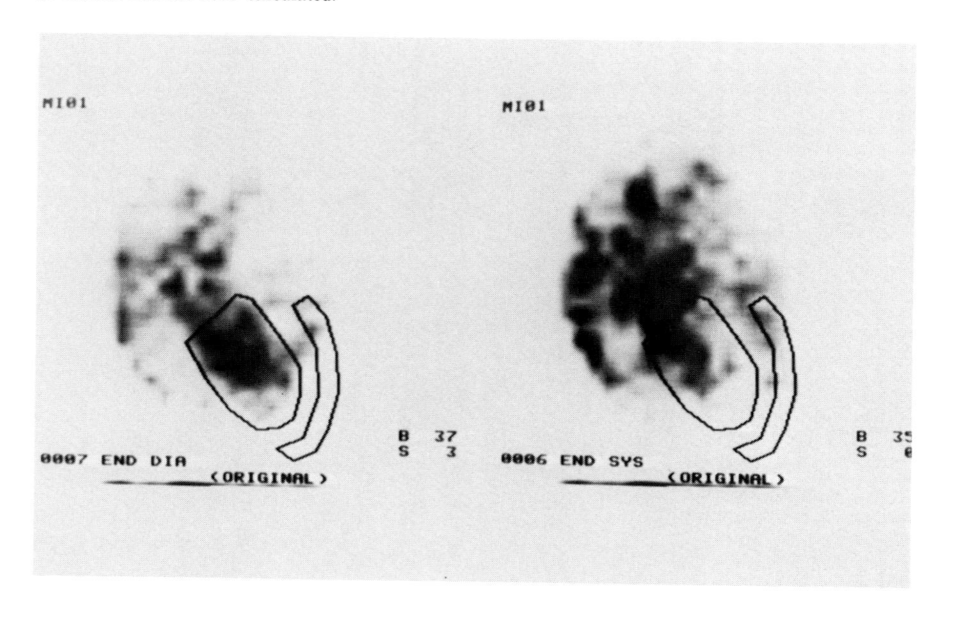

Figure 13-14 ***Right Ventricular Ejection Fraction (RVEF), First-Pass Technique***

First-pass radionuclide angiocardiogram (anterior images) demonstrates *(left)* a region of interest of the right ventricle during end-diastole with a region used for background subtraction and *(right)* a region of interest of the right ventricle during end-systole with a region used for background subtraction. The RVEF is determined by a technique similar to that used for LVEF calculation.

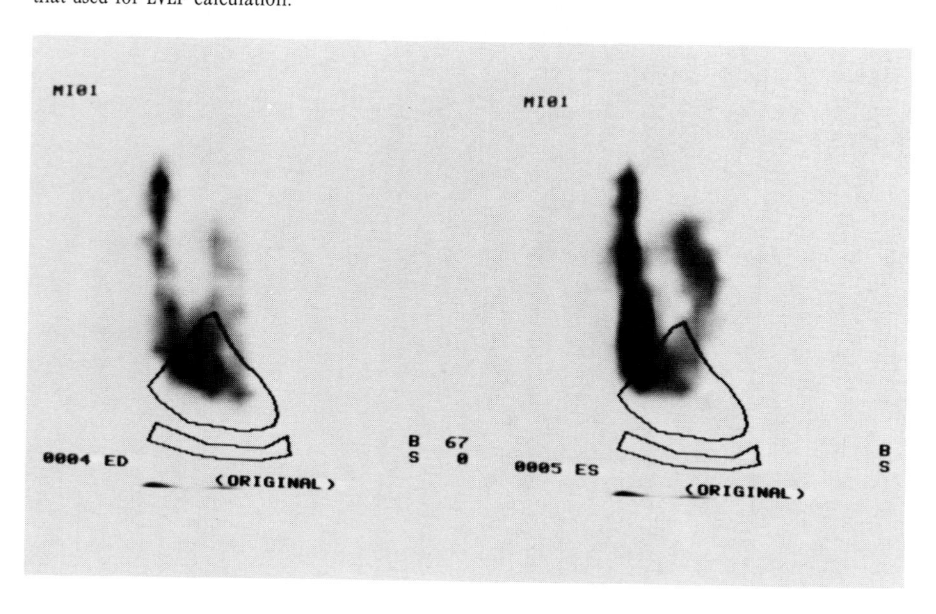

Figure 13-15 ***Normal Gated Blood Pool Examination***

Images of a 9-year-old child were acquired in the left anterior oblique (LAO) position. The four images on the left show different phases of the cardiac cycle. The four images on the right are computer-processed to demonstrate the cardiac chambers and interventricular septum to better advantage.

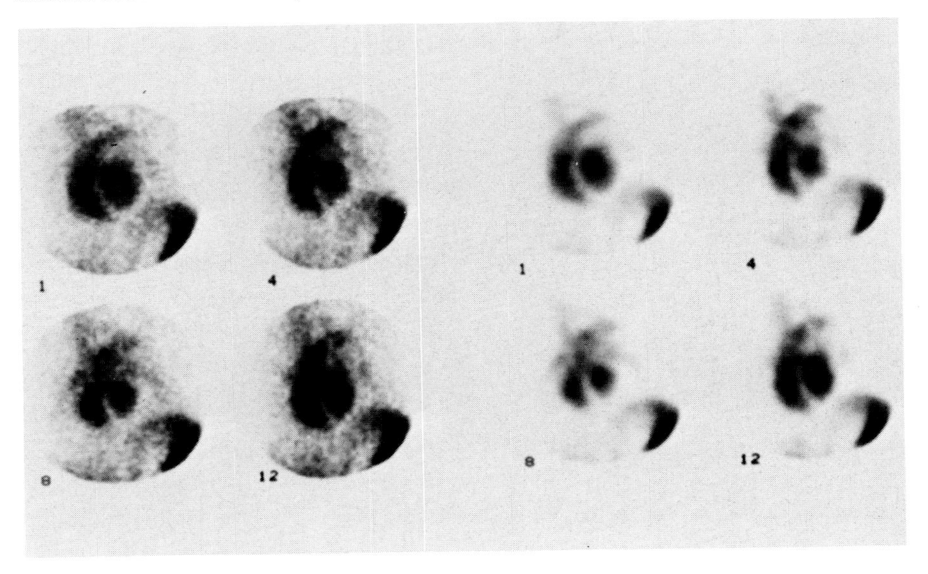

Figure 13-16 ***Idiopathic Hypertrophic Subaortic Stenosis***

In this, the gated blood pool examination, the LAO position is used for the four images on the left, and the anterior position is used for the four images on the right. Images of a 13-year-old patient show *(1)* asymmetric septal hypertrophy, *(2)* left ventricle, and *(3)* hypertrophied left ventricular myocardium during end-systole. (Left ventricular ejection traction = 0.92)

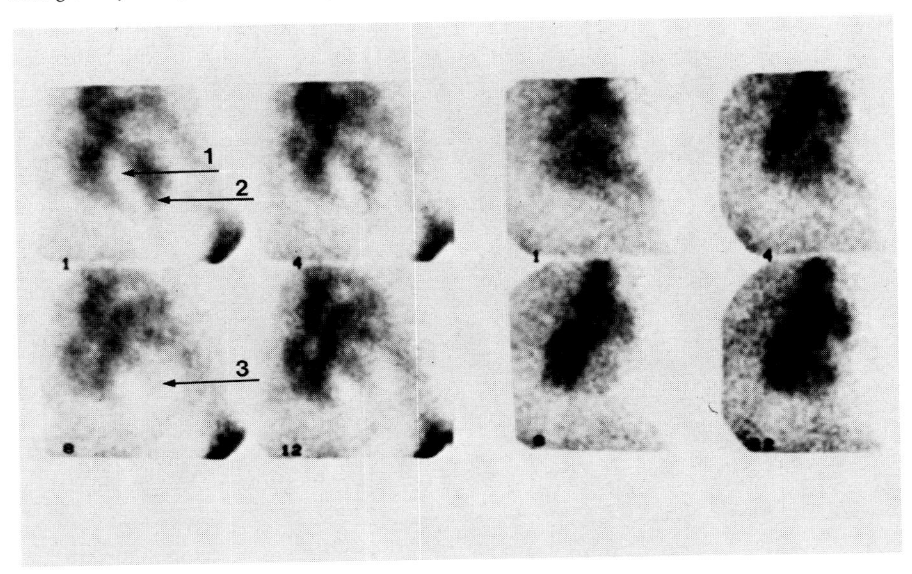

Figure 13-17 *Aortic Valvular Stenosis*

These gated blood pool images are in the left anterior oblique (LAO) *(left)* and the right anterior oblique (RAO) *(right)* positions. Images of the 10-year-old patient show *(1)* normal septum, *(2)* left ventricle, and *(3)* hypertrophied left ventricular myocardium during end-systole. (left ventricular ejection fraction = 0.90) (Sty JR: Atlas of pediatric nuclear cardiology. Clin Nuc Med 5:Part I 373-386, Part II 424-438, 1980)

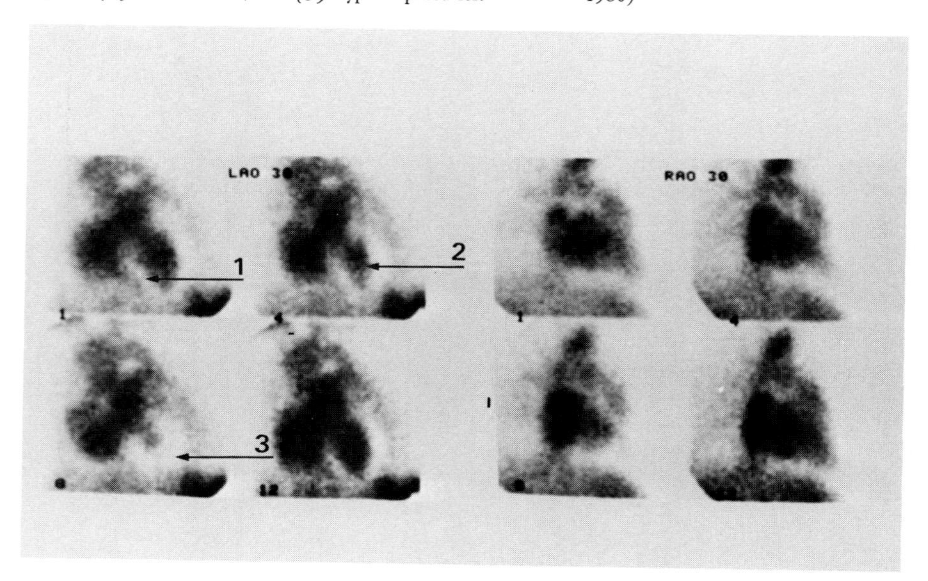

Figure 13-18 *Anomalous Origin of the Left Coronary Artery with Left Ventricular Aneurysm*

This gated blood pool image of an 8-year-old child is in the anterior position. The arrow marks the left ventricular aneurysm during systole. The ventricular aneurysm occurred because of myocardial damage secondary to the anomalous left coronary artery. (Sty JR: Atlas of pediatric nuclear cardiology. Clin Nuc Med 5:Part I 373-386, Part II 424-438, 1980)

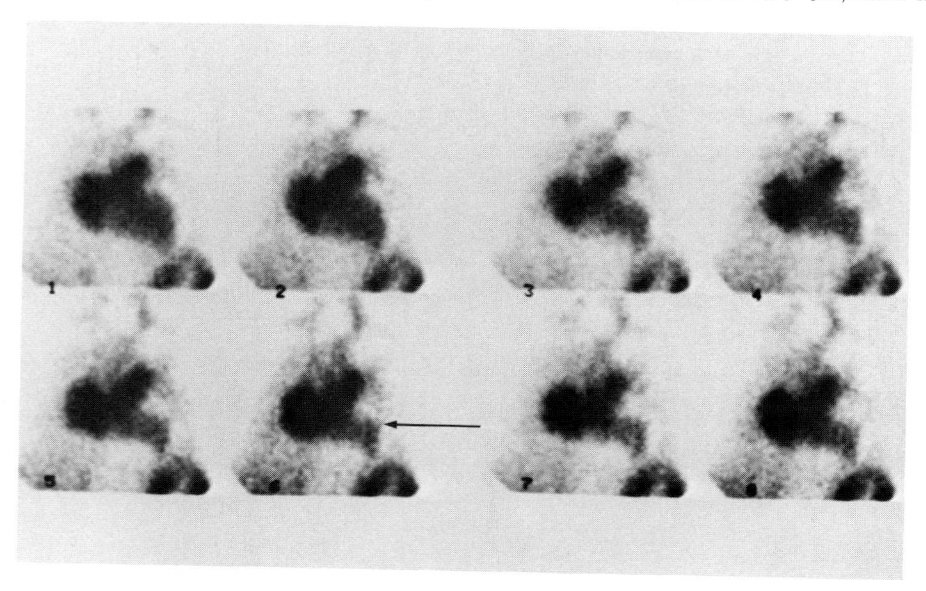

Figure 13-19 ***Myocarditis***

These gated blood pool images are in the anterior *(left)* and LAO *(right)* positions. The patient is 6 years old. Note the large left ventricular cavity with poor ventricular contraction *(arrow)*. (Sty JR: Atlas of pediatric nuclear cardiology. Clin Nuc Med 5:Part I 373-386, Part II 424-438, 1980)

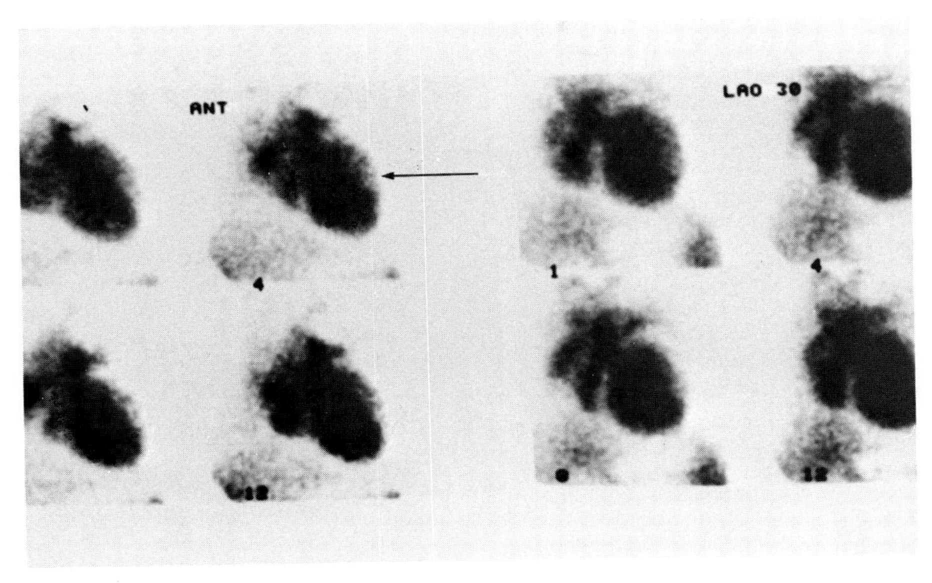

Figure 13-20 ***Right Ventricular Overload***

Thallium-201 myocardial imaging in the anterior *(upper left)*, LAO *(upper right)*, and RAO *(lower)* was performed in a 4-year-old child with a ventricular septal defect and pulmonary hypertension. There is marked accumulation of Tl-201 in the right ventricle *(arrow)*. Both ventricular chambers are dilated. (Sty JR: Atlas of pediatric nuclear cardiology. Clin Nuc Med 5:Part I 373-386, Part II 424-438, 1980)

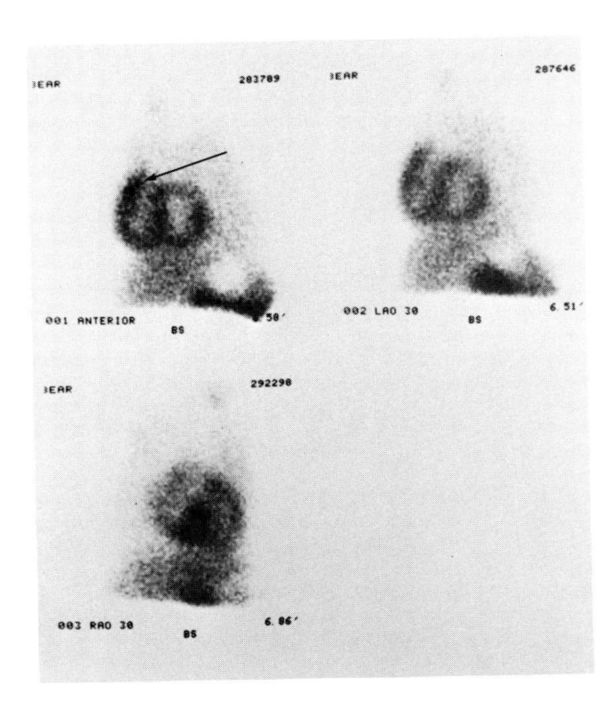

Figure 13-21 ***Right Ventricular Overload***

Tl-201 myocardial imaging in the anterior *(upper left),* 30° LAO *(upper right),* and 60° LAO *(lower)* positions was performed in a 3-year-old child with tetralogy of Fallot and absence of the pulmonary valves. There is abnormal accumulation of radiopharmaceutical in the free wall of the right ventricle *(arrow),* indicating right ventricular hypertrophy. This chamber is significantly dilated. (Sty JR: Atlas of pediatric nuclear cardiology. Clin Nuc Med 5:Part I 373-386, Part II 424-438, 1980)

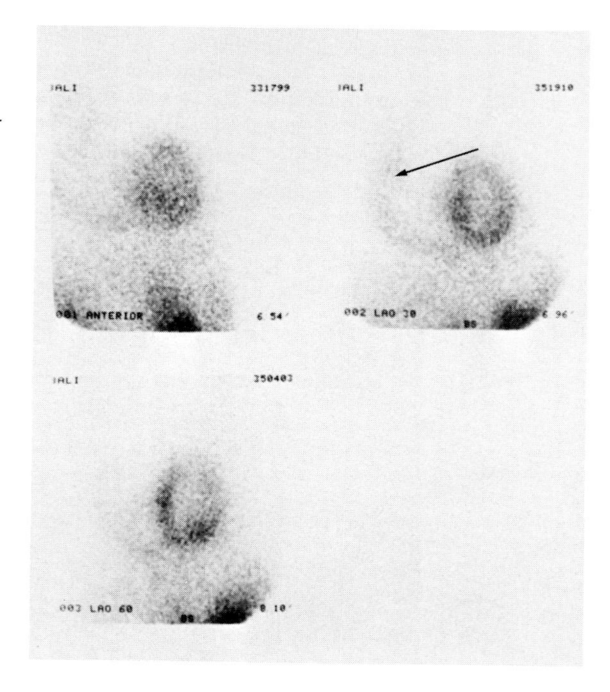

Figure 13-22 ***Right Ventricular Overload***

Tl-201 myocardial imaging in the anterior *(left)* and 30° LAO *(right)* positions was performed in a 17-year-old with cystic fibrosis and cor pulmonale. There is uptake of tracer in the free wall of the ventricle *(arrow).* The right ventricular chamber is significantly dilated. (Sty JR: Atlas of pediatric nuclear cardiology. Clin Nuc Med 5:Part I 373-386, Part II 424-438, 1980)

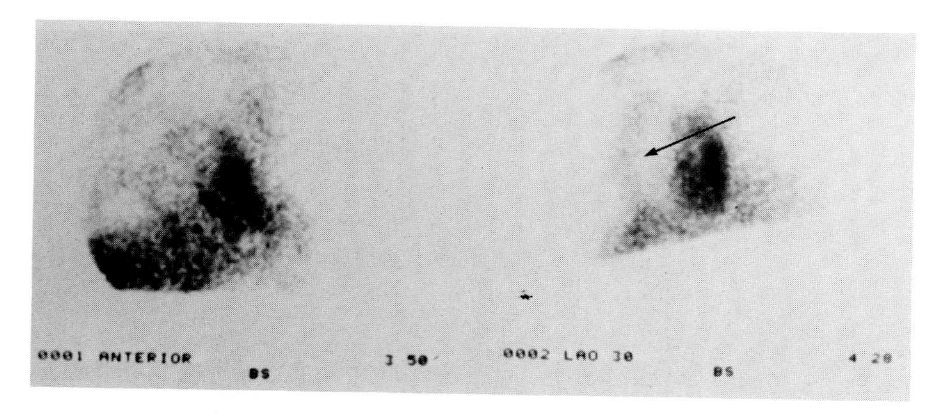

Figure 13-23 ***Myocardial Infarction***

(A) Tc-99m pyrophosphate scan consists of 1-hr RAO *(upper left)* and LAO images *(upper right)*, and 2-hr anterior *(lower left)* and LAO images *(lower right)*. These scans of a 14-month-old infant, who had surgical repair of transposition of the great arteries, demonstrate abnormal uptake of the radiopharmaceutical in the lateral aspect of the left ventricle, indicating a recent myocardial infarction *(arrow)*. ***(B)*** Tl-201 imaging in the same patient in the anterior *(upper left)*, 30° LAO *(upper right)*, and 60° LAO positions demonstrates a perfusion deficit in the posterolateral aspect of the left ventricle *(arrow)* corresponding to the area of the infarction.

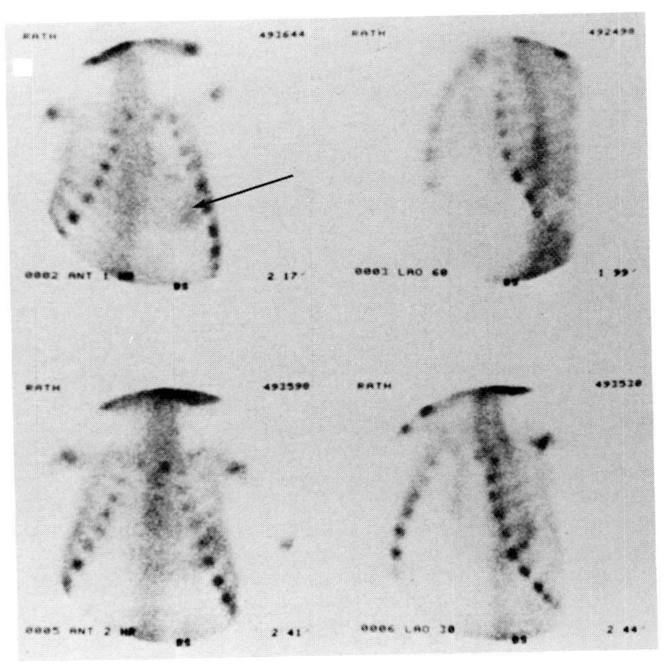

A

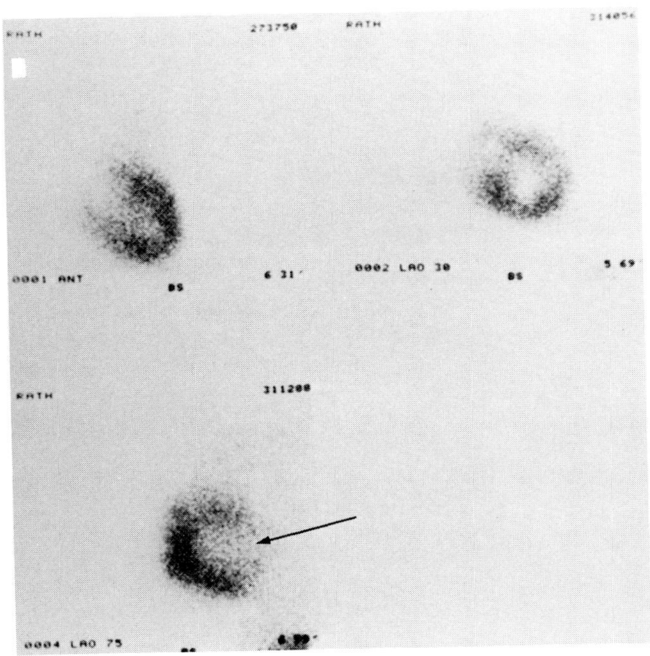

B

Figure 13-24 *Anomalous Origin of Left Coronary Artery*

Tl-201 myocardial imaging in the anterior *(upper left)*, 30° LAO *(upper right)*, 45° LAO *(lower left)*, and 60° LAO *(lower right)* positions was performed in a 6-year-old with anomalous origin of the left coronary artery from the pulmonary artery. A perfusion defect is present in the lateral aspect of the free wall of the left ventricle *(arrow)*. The left ventricular chamber is dilated. (Sty JR: Atlas of pediatric nuclear cardiology. Clin Nuc Med 5:Part I 373–386, Part II 424–438, 1980)

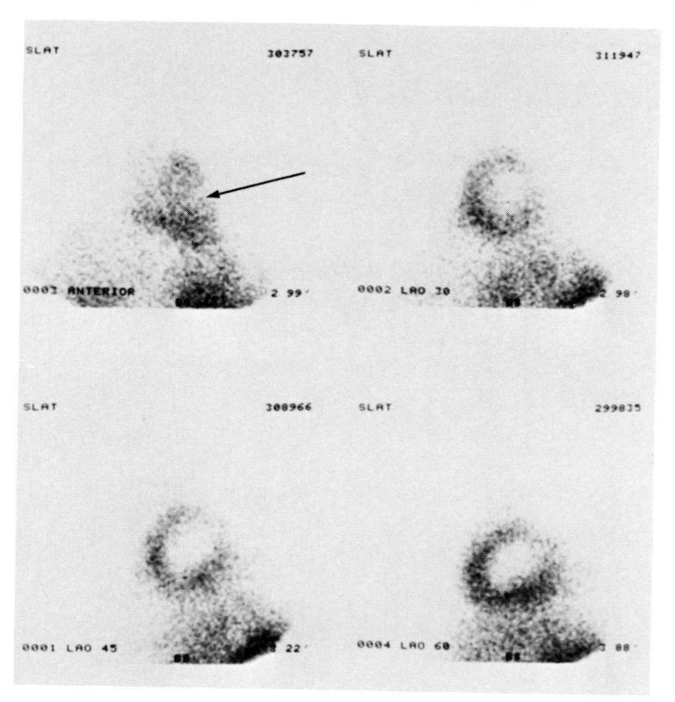

Figure 13-25 *Perinatal Asphyxia*

This anterior Tl-201 myocardial scan is of a newborn infant with perinatal asphyxia. Note the lack of radiopharmaceutical in the myocardium and an excess accumulation in the lung. (Sty JR: Atlas of pediatric nuclear cardiology. Clin Nuc Med 5:Part I 373–386, Part II 424–438, 1980)

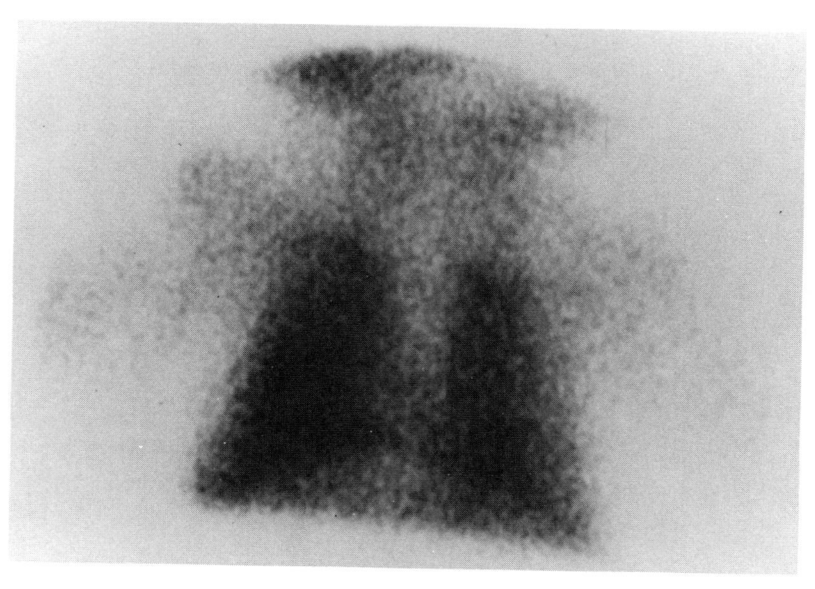

Figure 13-26 ***Kawasaki Syndrome***

Tl-201 myocardial imaging in the anterior *(upper left)*, 30° LAO *(upper right)*, and 45° LAO *(lower)* positions was performed in a 2-year-old child with Kawasaki syndrome. The ventricular chamber is markedly dilated, and there is a large perfusion deficit in the anterolateral aspect of the free wall of the left ventricle *(arrow)*.

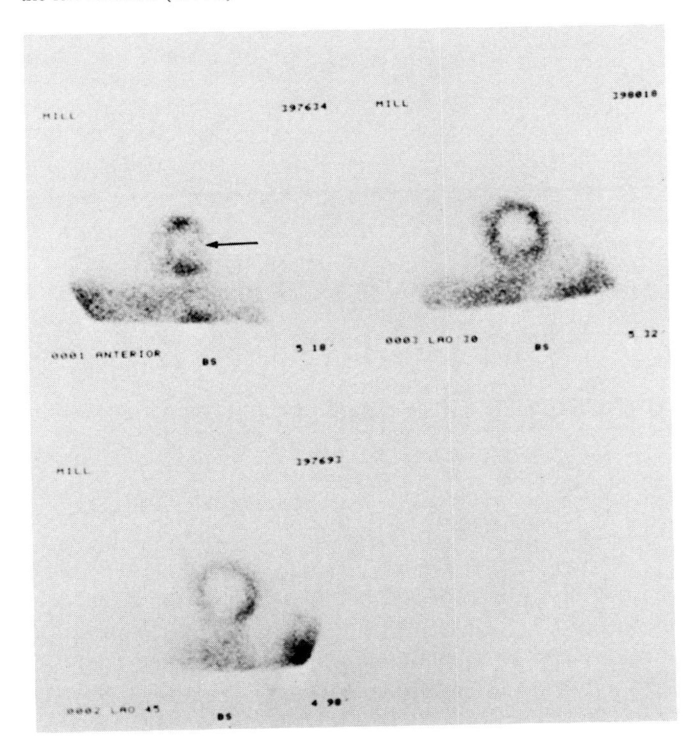

Figure 13-27 ***Right-to-Left Shunt Evaluation by Particle Body Imaging with Tc-99m Macroaggregated Albumin***

Anterior views of the chest *(A)*, abdomen *(B)*, and a lateral view of head *(C)* demonstrate a hyperperfused left upper lobe. Note accumulation of tracer in the brain and kidneys from right-to-left shunting. This child's right-to-left shunt measured 30%.

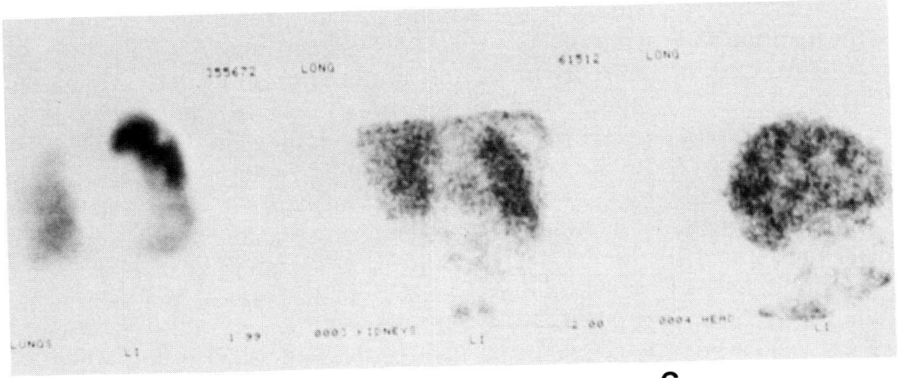

A **B** **C**

B. David Collier
Walter J. Slizofski
Arthur Z. Krasnow

Chapter 14 SPECT BONE IMAGING (Lumbar Spine, Hips, Knees, and Temporomandibular Joint)

At first glance, single photon emission computed tomography (SPECT) might appear to be an unnecessary addition to a nearly ideal nuclear medicine procedure. Planar bone scintigraphy is generally acknowledged to be the most sensitive imaging test for detecting most forms of skeletal pathology. A total body bone scan usually is the first imaging test to be ordered when screening patients for metastatic bone disease. Based on this remarkable efficacy, the bone scan has become the most common nuclear medicine imaging procedure performed in the United States.

Why, then, is SPECT bone imaging necessary? Because, there are still large numbers of patients with benign bone and joint disorders who are neither easily nor routinely examined with planar bone scintigraphy. One such example is the patient with symptomatic temporomandibular joint (TMJ) dysfunction. This disorder may affect to varying degrees as much as a quarter of the adult population.[2] These patients often complain of joint pain, limited opening, joint locking, and/or an audible opening click, all of which may be successfully treated if correctly diagnosed. SPECT rather than planar bone scanning is the scintigraphic method of choice for examining the TMJ.[3–6] The adult patient with low back pain (LBP) also may benefit from SPECT bone imaging. While planar bone scintigraphy has been applied to this patient population, recent evidence indicates that SPECT frequently detects causes of LBP which are not correctly identified and localized by planar bone scintigraphy.[7–9] These two situations, in addition to the other skeletal applications of SPECT cited below, suggest that many patients who formerly would not have been referred for bone scintigraphy can now be expected to benefit from SPECT bone imaging. Thus, the nuclear medicine department that offers SPECT bone imaging will be able to perform more examinations and help more patients.

The diagnostic advantage of SPECT bone imaging becomes significant when examining large and anatomically complex bony structures. To date, SPECT bone imaging has been applied principally to the evaluation of patients with pain and dysfunction in the lumbar spine, hips, knees, and TMJs. In these instances, planar imaging often superimposes substantial underlying and overlying activity on the bone structures being investigated. SPECT, however, will remove such unwanted activity from the tomographic plane of medical interest. In the hips, for example, the acetabulum extends downward behind the femoral head. During planar imaging, activity originating in the underlying acetabulum may obscure the photopenic defect, which is typical of avascular necrosis (AVN) of the femoral head. SPECT bone imaging separates the plane containing the photopenic defect from the underlying and overlying distributions of activity. Thus, SPECT provides both superior positional information and improved image contrast, which in turn facilitate detection of femoral head AVN.[10–12]

As of 1986, many nuclear medicine departments that perform a high volume of bone scan examinations were not yet offering SPECT bone imaging. The commitment to total body imaging for oncology studies and the more recent trend toward flow study, blood pool, and delayed planar imaging for certain benign skeletal diseases have made scintigraphic bone imaging one of the most time-consuming and expensive nuclear medicine procedures. Therefore, the recommendation that SPECT be added to many bone scans will be well received only if the examination can be performed quickly and result in unique diagnostic information. Fortunately, as will be described below, currently available rotating gamma camera devices allow a single technologist to perform as many as five SPECT examinations in a single day and still have time for other planar imaging studies.

TECHNIQUES

Many patients referred for SPECT imaging of the back, hip, knee, or TMJ will also benefit from radionuclide flow study and blood pool imaging. At our institution, the imaging sequence for all such patients is as follows:

1. Flow study (5 seconds/frame)
2. Blood pool image (500,000 counts)
3. Planar images
4. SPECT

For some adults, the diagnostic quality of SPECT bone imaging may be count-limited. Therefore, an adult dose of 25 mCi of Tc-99m methylene diphosphonate (MDP) is used. The mechanisms for bony localization of this radiopharmaceutical and the generally favorable radiation dosimetry have been previously reported and reviewed.[13]

Imaging Procedure

Only gamma cameras which have passed quality control tests should be used for SPECT imaging.[14,15] Table 14-1 gives the guidelines for the type and frequency of special SPECT quality control testing performed at our institution. Routine quality control for planar imaging is not listed. The circular 400-mm field-of-view gamma cameras* used at our institution for over 2000 SPECT bone scans, using a standard protocol (64 × 64 matrix, 64 projections over 360° of circular or eliptical rotation, 20 seconds per projection), has routinely produced SPECT images of good technical quality. While

* General Electric Medical Systems, Milwaukee, Wisconsin

Table 14-1
Special Quality Control for SPECT Bone Imaging

Test	*Frequency*	*Comment*
Uniformity-flood* 3 M-count	Daily	
Uniformity-flood† 30 M-count	Weekly	Also after camera servicing or power outages
Center of rotation	Weekly	Same as above
Phantom for linearity and resolution	Monthly	Same as above

* *Use a flat and uniform cobalt-57 flood source.*
† *Weekly uniformity-flood is acquired for each SPECT collimator.*

high-resolution collimators, fanbeam collimators,[16] angled collimators,[17] or special-purpose "head collimators"[18] may be used, the low-energy, general-purpose collimator provides an acceptable trade-off between resolution and sensitivity.

Positioning for SPECT bone imaging requires that the patient remain supine and motionless on the imaging table for 21 minutes of data acquisition. Patients with back, hip, or knee pain will often want to move slightly during the exam. Patient motion is in fact the most common technical problem encountered with SPECT bone imaging, and any measures which

Table 14-2
Patient Positioning for SPECT Bone Imaging

Bone Structures	*Special Positioning*	*Pitfalls*
Knees	• Place 2- to 3-inch pad between knees. • Secure knees with straps to prevent motion. • Secure feet in neutral position to prevent rotation.	With obese patients, both knees may not fit in field of view.
Hips and pelvis	• Empty bladder immediately before exam. • Position hips symmetrically and secure knees and/or feet to prevent motion.	Bladder filling during exam creates artifacts.
Lumbar spine	• Keep arms out of field of view. • A pillow under the knees may relieve back pain.	Patients with back pain will often move during the exam.
TMJ	• Tape forehead to table with head in hyperextension. • Secure jaw in closed position.	Check lateral view to be sure the chin is in the field of view.

will increase patient comfort and motivation are warranted. Most patients are more comfortable on the scanning table if their arms and legs are loosely strapped in place. An additional pillow under the knees or head will often help. Bilaterally symmetric skeletal positioning is also important for an optimal SPECT examination. For example, having both knees equally extended and in a neutral position will result in SPECT bone images for which any right-to-left asymmetry will be due to skeletal pathology rather than faulty positioning. Special considerations for the positioning of patients during SPECT bone imaging are listed in Table 14-2.

Computer Acquisition

For routine data processing, we first perform a uniformity correction and then a nine-point spatial smooth on each of the 64 projections. For our SPECT system, this is equivalent to a Hanning filter of 0.8 cycles/cm in both the "X" and "Y" directions. After this "preprocessing" (which the latest software will perform "on the fly" during data acquisition), one-pixel-thick transaxial, coronal, and sagittal tomograms are reconstructed using filtered backprojection with a ramp filter. Attenuation correction is not used. For most bone SPECT studies, a linear map is used when photographing SPECT images onto transparencies. However, in certain instances, such as when searching for a photopenic defect in the femoral head, the log map shows relatively low count areas to better advantage.

Our routine protocol for SPECT bone imaging does not place excessive demands on either the technologist's time or the scanning facilities. The addition of SPECT to each routine low back, hip, knee, and TMJ bone scan requires an extra 5 minutes for equipment set-up and patient positioning, 21 minutes for data acquisition, and 15 minutes for processing and filming. Furthermore, with automated scanning and processing software, the technologist can "set up and walk away" so that actually only 20 minutes of his or her time are required for the SPECT portion of the examination.

While this basic protocol for acquiring and processing SPECT bone data has been used for over 95% of the exams performed at our institution, other techniques may provide improved image quality at the cost of longer imaging or computer reconstruction times. Fanbeam collimation or other high-resolution collimation,[16] longer imaging times, higher radiopharmaceutical doses, novel reconstruction techniques (such as distance-weighted filtered backprojection[19,20]), and asymmetric Tc-99m energy windows[21] may produce a visible improvement in image quality. Be aware, however, that making such changes in the routine SPECT bone protocol (or the choice of a 128 × 128 matrix, 128 projections, or a field-of-view either larger or smaller than 400 mm) may also necessitate changes in quality control, acquisition, and processing methods, which may not be obvious to the inexperienced user.

Pitfalls

Artifacts and technical pitfalls may limit the diagnostic value of SPECT bone images. However, with attention to routine gamma camera quality control and the previously described essentials of patient positioning, data acquisition, and data processing, most technical problems can be prevented. Since 1981 our clinical experience with over 2000 SPECT bone examinations has uncovered only four noteworthy technical pitfalls. In order of decreasing frequency these are:

1. Patient motion (Fig. 14-4)
2. Bony structures moving in-and-out of the field of view as the gamma camera rotates around the patient (Fig. 14-5)
3. Bladder filling during SPECT of the hips and pelvis (Fig. 14-6)
4. Low count study (Fig. 14-4)

SPECT bone imaging artifacts due to gamma camera nonuniformity or inappropriate center of rotation have not occurred in our clinical experience with SPECT bone imaging.

RADIATION ABSORBED DOSES

Estimates of radiation absorbed doses for Tc-99m diphosphonates are given in Table 14-3. The critical

Table 14-3
Radiation Absorbed Dose Estimates

Organ	*RAD/mCi**
Bladder (critical organ)	0.2
Bone	0.07
Kidney	0.04
Liver	0.003
Ovaries	0.03
Testes	0.011
Whole body	0.22

** Modified from Harbert JC et al: Absorbed dose estimates. Clin Nuc Med 9:210,1984*

organ is the one that typically receives the largest radiation absorbed dose.

INTERPRETATION OF SPECT BONE EXAMINATIONS

When reviewing normal SPECT bone images (Figs. 14-1–14-3), it is important to note that SPECT increases image contrast by removing activity from in front and from behind the plane of medical interest. In addition, SPECT provides positional information, which allows the observer to distinguish between bony structures that overlap on planar images. However, the spatial resolution of SPECT is inferior to that of planar imaging, and therefore SPECT supplements but does not replace planar bone imaging.

Optimal review and interpretation of scintigraphic bone examinations requires that all images be simultaneously available. All three orthogonal sets of SPECT images, in addition to the planar images, often are needed when attempting to identify and localize a skeletal abnormality. A left-to-right asymmetry on a SPECT examination should be considered abnormal only when identified in all three tomographic planes. When the SPECT technique is first adopted by a department, physicians seeking to assure themselves that images have been optimally displayed will need to review these images at the computer terminal. Later, when technologists become adept at photographing SPECT bone images onto transparencies, no more than 10% of cases will require such computer review. Some physicians, however, prefer to use the more flexible computer display for all image review and interpretation.

Both radiographic and clinical correlation are necessary before the clinical significance of an abnormal SPECT bone imaging study can be fully appreciated. For example, when interpreted without radiographs, SPECT rarely can identify a specific cause of LBP in adult patients with no history of malignancy. Trauma, infection, degenerative disk disease, articular facet osteoarthritis, spondylolysis, and numerous other forms of bone and joint pathology all can produce increased scintigraphic activity in the lumbar spine. Similarly, a history of malignancy with the potential for metastatic bone disease changes the significance of positive scintigraphic findings. In such instances, SPECT and planar bone scintigraphy complement rather than replace radiography. The greatest diagnostic value of these imaging tests is realized when all modalities are simultaneously available for the rendering of a combined interpretation. This diagnostic approach allows one to correlate the superior anatomic detail of radiography with the functional information available from bone scintigraphy.

DISCUSSION

Many bony structures suitable for SPECT imaging have not been studied in detail, and the potential of SPECT bone imaging in oncology has not been thoroughly investigated. However, experience to date has shown a role for SPECT in examining patients with low back, hip, knee, or TMJ pain.

Low Back Pain

For the diagnostic evaluation of patients with LBP due to benign bone and joint pathology, SPECT bone imaging has two key advantages over planar technique: more lesions are detected and the lesions are more accurately localized.

The application of planar bone scintigraphy to the study of LBP has been limited by the inability to distinguish between abnormalities lying in a lumbar vertebral body and those found in the posterior neural arch. For LBP patients, not only detection but also localization of a scintigraphic abnormality may be important. One such example is the altered alignment of the vertebral bodies in high-grade spondylolisthesis, which may subsequently lead to osteoarthritic changes in the vertebral body end-plates. This may be associated with increased scintigraphic activity at both the bilateral pars interarticularis defects and at the sites of vertebral body osteoarthritis. When using planar imaging techniques, such vertebral body and posterior neural arch abnormalities are superimposed on one another in the straight posterior planar views. The abnormalities also are difficult to localize on posterior oblique views. With SPECT, more accurate scintigraphic localization is possible.

To date, the SPECT technique has been of greatest value in imaging painful sites of spondylolysis/spondylolisthesis, pseudarthrosis following failed spinal fusion, articular facet osteoarthritis, and sacral pathology[8,9,22] (Figs. 14-7–14-10). Other promising applications include evaluation of ankylosing spondylitis[7] and detection of occult fractures, infection, and metastases.

Hip Pain

When prospectively compared with conventional radiography or planar bone scintigraphy, SPECT bone imaging has been shown to be superior in the detection of AVN of the femoral head[10–12] (Fig. 14-11). Magnetic resonance imaging (MRI) also is a highly sensitive diagnostic test for this disorder. A recently reported prospective comparison of MRI with planar, but not SPECT, bone scintigraphy found MRI to be the more sensitive examination.[23] Another recent study favored MRI over SPECT and planar bone scintigraphy for the noninvasive diagnosis of femoral head AVN.[12] The authors of this later study emphasized their difficulties in proving the status of the femoral head. Further prospective studies comparing MRI with SPECT including patients with both early asymptomatic disease and late-stage AVN are needed to determine the relative diagnostic value of these two examinations.

SPECT has been prospectively compared with both planar bone scintigraphy and conventional radiography in 21 patients with clinically diagnosed AVN.[11] A final diagnosis of AVN was established for 15 of these symptomatic patients who had a total of 20 involved hips. The scintigraphic studies were considered positive for AVN only if a relatively photopenic zone could be identified within the femoral head. Using SPECT bone imaging, 12 of the 15 symptomatic patients and 17 of the 20 involved hips (sensitivity of 0.85) were correctly identified, whereas with planar imaging only 8 of 15 symptomatic patients and 11 of 20 involved hips (sensitivity of 0.55) were detected. There were no false-positive interpretations of either scintigraphic technique. The radiographic appearance of a subchondral fracture (the most specific radiographic finding of early femoral head AVN) was present in only six instances. It was concluded that by identifying a photopenic defect that is not evident on planar views, SPECT can contribute to the accurate diagnosis of AVN of the femoral head.

Approximately one quarter of the SPECT examinations for femoral head AVN are likely to show artifacts created by rapid bladder filling (Fig. 14-6). Bladder catheterization will prevent this artifact, but the catheter tubing must be run down between the patient's legs with the collection bag over the end of the table. Placing the collection bag in the field-of-view at the patient's side will create an even *more* annoying artifact.

Knee Pain

In patients with significant pain and disability, arthroscopy is commonly used to provide direct visual inspection of the knee. The availability of such a highly specific, invasive, and relatively expensive test creates the need for a noninvasive screening examination. This role can be filled by SPECT bone imaging, which is a highly sensitive, albeit nonspecific, screening examination for degenerative cartilage damage, longstanding meniscus tears, and other internal derangements of the knee[24,25] (Figs. 14-12 and 14-13). In addition, successful use of planar bone scanning to detect acute meniscus tears has been reported.[26] Thus, the conservative orthopaedist can use SPECT bone scanning to help select those patients who require invasive diagnostic and therapeutic procedures. Furthermore, by using the scintigraphic examination as a "roadmap" to guide subsequent arthroscopy, the orthopaedist can concentrate on the sites of potentially significant pathology.

It is generally accepted that degenerative cartilage damage and synovitis are associated with increased scintigraphic activity. While there is little reason to believe that Tc-99m MDP localizes in an acutely torn meniscus *per se*, it is tempting to speculate that altered joint mechanics and degenerative changes secondary to a long-standing meniscus tear will eventually lead to increased uptake of the scanning agent in the subchondral bone adjacent to the torn meniscus. When increased scintigraphic activity is identified over the medial or lateral compartments of a painful knee, it is usually not possible to distinguish a torn meniscus from cartilage damage, synovitis, intra-articular loose bodies, or most other morbid conditions associated with chronic knee pain. However, it is the ability to detect and localize such lesions, rather than the ability to make a specific diagnosis, which creates a role for SPECT bone imaging as a pre-arthroscopic screening examination.

Temporomandibular Joint Pain

SPECT bone imaging has been shown to be a valuable test for confirming internal derangement of the TMJ.[3–6,16] When used to examine 36 patients undergoing preoperative evaluation, the sensitivity of SPECT (0.94) was comparable to arthrography (0.96), and significantly better than planar bone scanning (0.76) or conventional radiography (0.04)[3] (Fig. 14-14). Lower sensitivities for SPECT have been reported in several series which used arthrography rather than surgery to confirm the status of the TMJ.[4,5] One recent report identified patients with clinical symptoms of TMJ dysfunction, positive SPECT examinations, but normal disk position on arthrography.[5] These results suggest that all functionally significant derangements of the TMJ are not routinely detected by arthrography. Both MRI and CT can identify an anteriorly displaced disk within the TMJ and therefore compete with SPECT for use as a noninvasive screening procedure. While a prospective study of CT and SPECT has not yet been published, a prospective comparison of MRI with SPECT found the sensitivity of SPECT to be slightly lower (0.76 *vs.* 0.88); however, SPECT rather than MRI or arthrography correlated best with TMJ symptomatology.[5]

When compared with planar bone imaging, SPECT frequently provides images of greater clarity, which in many instances yield unique diagnostic information. For patients with low back, hip, knee, and TMJ pain, the diagnostic value of SPECT bone imaging has been established. As many as five patients with such complaints can easily be imaged in a single day using currently available rotating gamma camera devices. For these reasons, SPECT bone imaging is frequently performed in many nuclear medicine departments.

Additional clinical applications of SPECT bone imaging are anticipated. The potential of SPECT bone imaging using radiopharmaceuticals other than Tc-99m phosphates and the usefulness of SPECT for detecting bone metastases has not yet been thoroughly investigated. Furthermore, additional improvements in instrumentation and technique, such as fanbeam collimators and distance-weighted filtered backprojection, may soon come into common clinical use. Thus, SPECT bone imaging will play an increasingly important role in the clinical practice of nuclear medicine.

REFERENCES

1. Mettler FA, Williams AG, Christie JH et al: Trends and utilization of nuclear medicine in the United States: 1972–1982. J Nucl Med 26:201, 1985

2. Solberg WK, Woo MW, Houston JB: Prevalence of mandibular dysfunction in young adults. J Am Dent Assoc 98:25, 1979

3. Collier BD, Carrera GF, Messer EM et al: Internal derangement of the temporomandibular joint: Detection by single-photon emission computed tomography. Radiology 149:557, 1983

4. O'Mara RE, Katzberg RW, Weber DA et al: Skeletal imaging and SPECT in temporomandibular joint disease (abstr). Radiology 149:P101, 1983

5. Krasnow AZ, Collier BD, Kneeland JB et al: High-resolution NMR, SPECT, & planar scintigraphic imaging of the temporomandibular joint (abstr). J Nucl Med 27:952, 1986

6. Schroeder G, Reich RH, Kleba C et al: SPECT or planar imaging in temporomandibular joint disease? (abstr) J Nucl Med 27:953, 1986

7. Jacobsson H, Larsson SA, Vesterskold L et al: The application of single photon emission computed tomography to the diagnosis of ankylosing spondylitis of the spine. Br J Radiol 57:133, 1984

8. Collier BD, Johnson RP, Carrera GF et al: Painful spondylolysis or spondylolisthesis studied by radiology and single-photon emission computed tomography. Radiology 154:207, 1985

9. Slizofski WJ, Collier BD, Flatley TJ et al: Painful pseudoarthrosis following lumbar spinal fusion: Detection by SPECT bone scintigraphy. Skeletal Radiol (in press)

10. Stromqvist B, Brismar J, Hansson LI: Emission tomography in femoral neck fracture for evaluation of avascular necrosis. Acta Orthop Scand 54:872, 1983

11. Collier BD, Carrera GF, Johnson RP et al: Detection of femoral head avascular necrosis in adults by SPECT. J Nucl Med 26:979, 1985

12. Feiglin D, Levine M, Stulberg B et al: Comparison of planar (PBS) and SPECT scanning in the diagnosis of avascular necrosis (AVN) of the femoral head (FH) (abstr). J Nucl Med 27:952, 1986

13. Galasko CSB, Weber DA: Radionuclide scintigraphy in orthopaedics, pp 1–16, 34–53. New York, Churchill Livingstone, 1984

14. Croft BY: single-photon emission computed tomography, pp 177–234. Chicago, Yearbook Medical Publishers, 1986

15. Eisner RL: Principles of instrumentation in SPECT. J Nucl Med Tech 13:23, 1985

16. Tsui BMW, Gullberg GT, Edgerton ER et al: Design and clinical utility of a fan beam collimator for SPECT imaging of the head. J Nucl Med 27:810, 1986

17. Esser PD, Alderson PO, Mitnick RJ et al: Angled-collimator SPECT (A-SPECT): An improved approach to cranial single photon emission tomography. J Nucl Med 25:805, 1984

18. Palmer DW, Knobel J, Collier BD et al: Collimator for emission tomography of the head. The Radiological Society of North America 69th Annual Meeting, Chicago, November 1983

19. Hellman RS, Nowak DJ, Collier BD et al: Evaluation of distance-weighted SPECT reconstruction for skeletal scintigraphy. Radiology 159:473, 1986

20. Nowak DJ, Eisner RL, Fajman WA: Distance-weighted backprojection: A SPECT reconstruction technique. Radiology 159:531, 1986

21. Jaszczak RJ, Greer KL, Floyd CE Jr et al: Improved SPECT quantification using compensation for scattered photons. J Nucl Med 24:893, 1984

22. Jarritt PH, Ell PJ: Gamma camera emission tomography: Quality control & clinical applications, pp 158–160. London, Current Medical Literature Ltd, 1984

23. Alavi A, Mitchell M, Kundel H et al: Comparison of RN, MRI, and XCT imaging in diagnosis of avascular necrosis (AVN) of the femoral head (abstr). J Nucl Med 27:952, 1986

24. Collier BD, Johnson RP, Carrera GF et al: Chronic knee pain assessed by SPECT: Comparison with other modalities. Radiology 157:795, 1985

25. Fajman WA, Diehl M, Dunaway E et al: Tomographic and planar radionuclide imaging in patients with suspected meniscal injury: Arthroscopic correlation (abstr). J Nucl Med 26:P77, 1985

26. Marymont JV, Lynch MA, Henning CE: Evaluation of meniscus tears of the knee by radionuclide imaging. Am J Sports Med 11:432, 1983

Atlas Section

Figure 14-1 ***Normal SPECT Exam of the Knees in a 20-Year-Old Male***

Coronal image *(A)* through the femoral condyles and tibial plateaux; sagittal image *(B)* is through midpatella; and transaxial image *(C)* is through the femoral condyles near the joint line. Activity originating in subchondral bone over the medial and lateral compartments is substantially less intense than activity over the growth plates. Activity in the patella has an intensity similar to that of the subchondral bone of the medial and lateral compartments. Intensity equal to or greater than the growth plates would indicate the presence of an abnormality in the patella or the bone adjacent to the weight-bearing surfaces. The orientation for these and all SPECT images that follow is *R*, right; *L*, left; *A*, anterior; *P*, posterior. (Collier BD, Johnson RP, Carrera GF et al: Chronic knee pain assessed by SPECT: Comparison with other modalities. Radiology 157:795, 1985)

A

R L

B

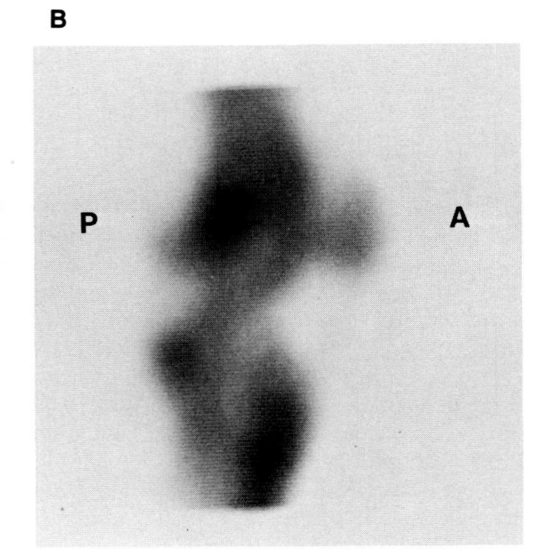

C

A
R L
P

Figure 14-2 ***Normal SPECT Exam of the Adult Lumbar Spine***

Midline sagittal *(A)*, L4 transaxial *(B)*, and Sl transaxial *(C)* images show individual lumbar vertebral bodies *(straight arrows)* separate from the spinous processes *(curved arrows)* and other posterior neural arch elements. Because of the lordotic curve of the lumbar spine, coronal images *(D)* include vertebral bodies of the upper lumbar spine in the same planes as the posterior neural arch elements of the lower lumbar spine.

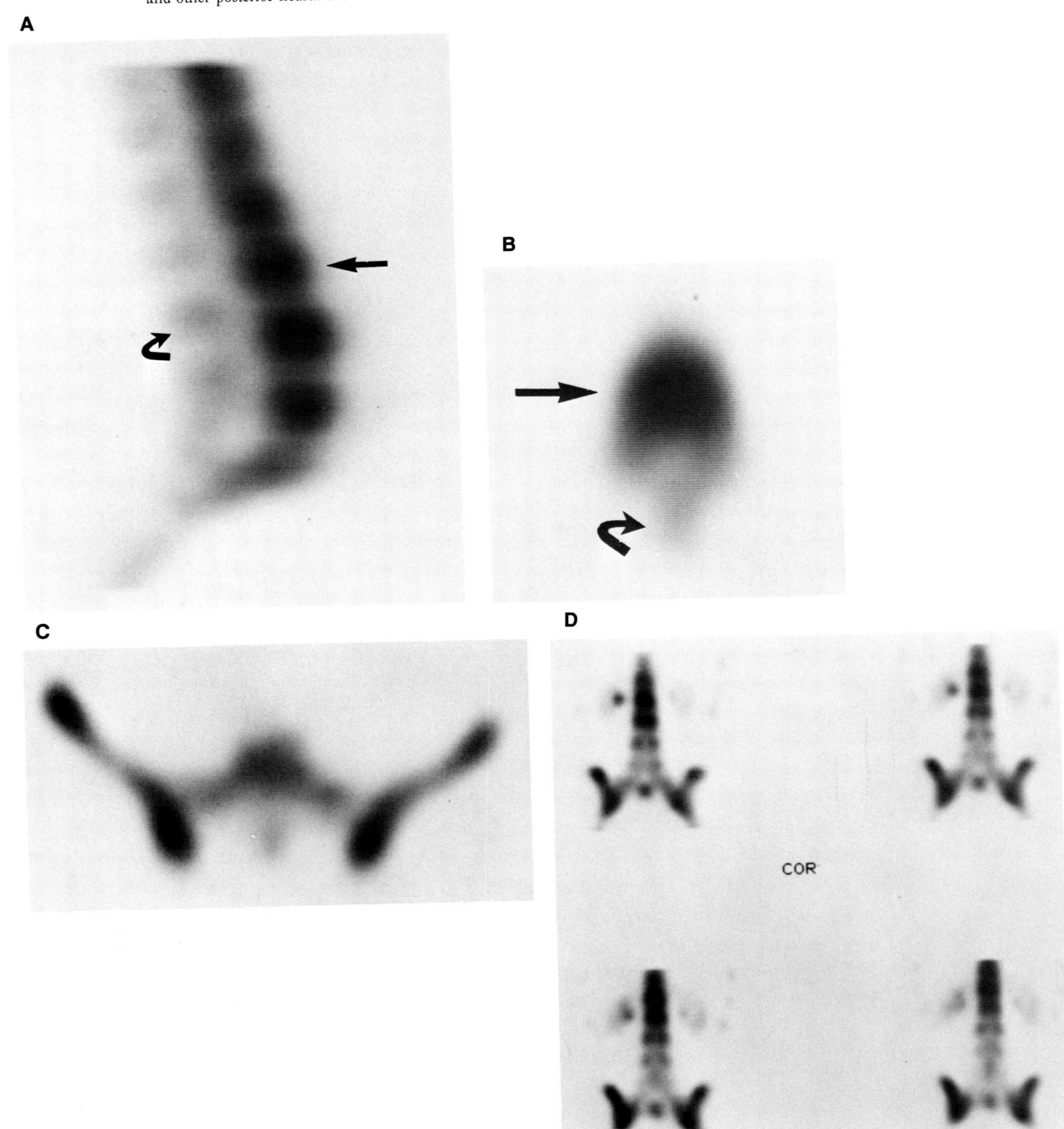

Figure 14-3 ***Normal Transaxial SPECT Image of the Adult Temporomandibular Joint (TMJ)***

When interpreting these studies, the intensity of activity at the TMJ *(straight arrows)* is compared to that of the adjacent calvarium. The normal TMJ usually shows slightly more intense activity than that seen in the adjacent portion of the temporal bone. Incidently noted on this examination is the markedly increased activity in the alveolar ridge of the maxilla due to recent dental extractions *(curved arrow)*; such intense activity in a TMJ would be considered abnormal. (Krasnow AZ, Collier BD, Kneeland JB et al: High-resolution NMR, SPECT, and planar scintigraphic imaging of the temporomandibular joint (abstr). J Nucl Med (in press)

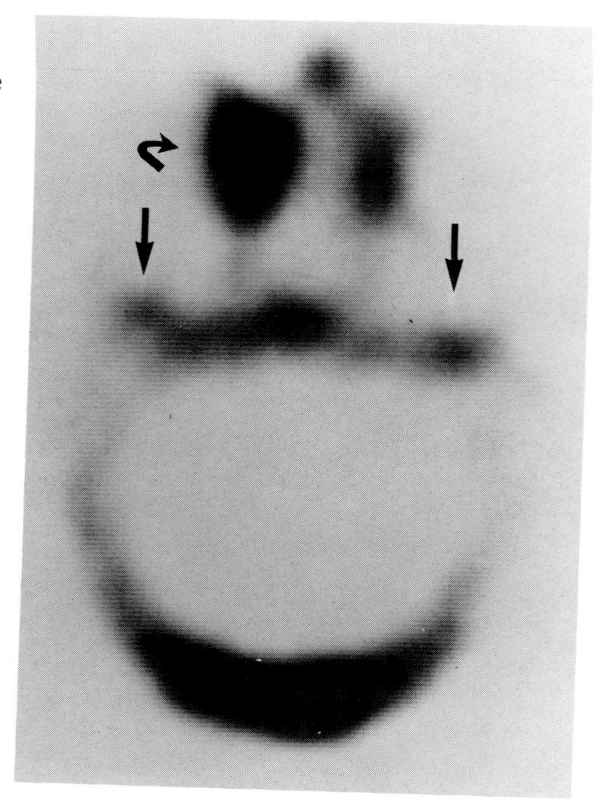

Figure 14-4 ***Artifacts***

(A) Transaxial image of the L5 vertebra in a normal 18-year-old male volunteer. *(B)* Artifacts were then created in a second study by having the patient roll from side to side during data acquisition. *(C)* Low count study was obtained using 2 sec/projection data acquisition instead of the usual 20 sec/projection. (Collier BD, Dellis CJ, Peck DC et al: Bone SPECT. J Nucl Med Tech 13:230, 1985)

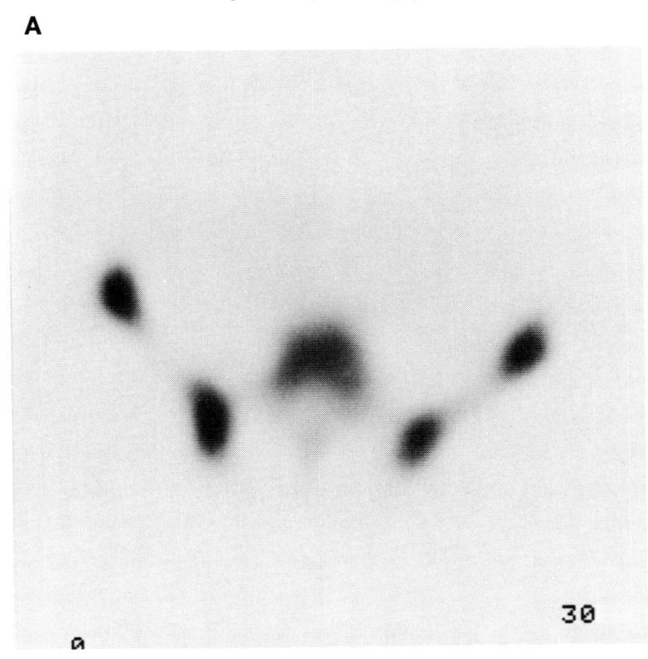

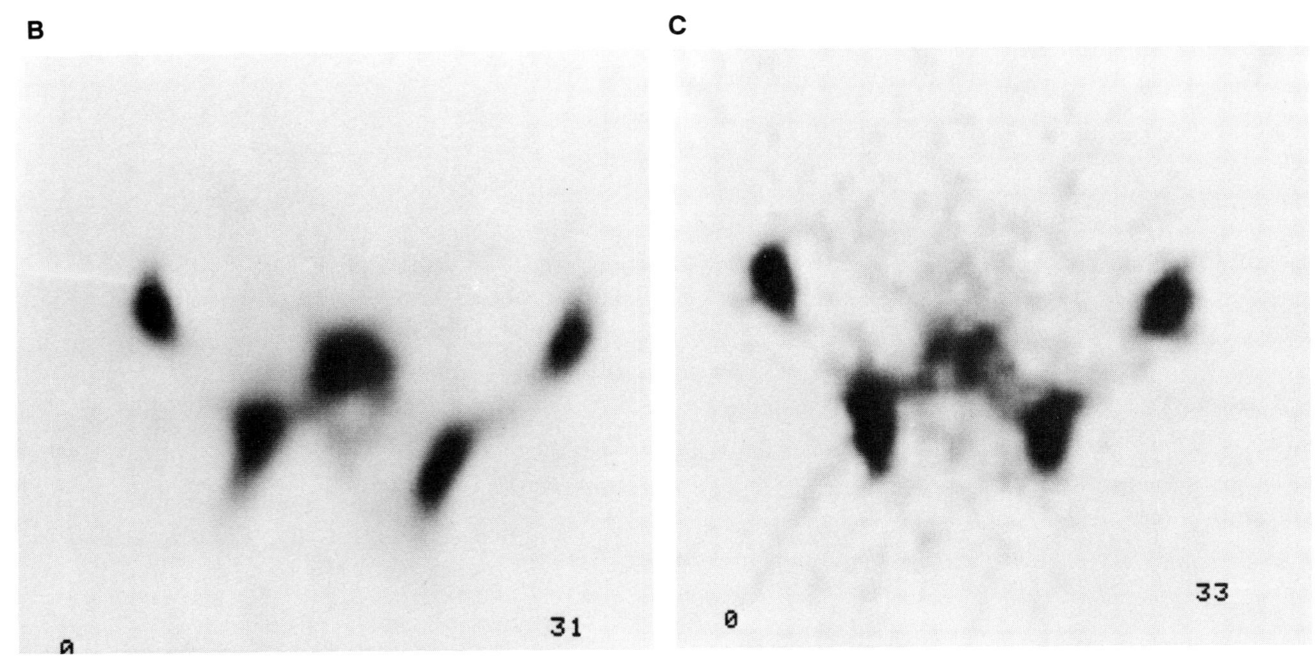

Figure 14-5 *"In and Out of the Field of View" Artifact*

Transaxial image of the knees exhibits an "in and out of the field of view" artifact. The knees were completely included in the field of view when the rotating gamma camera was lateral to the patient. However, as the camera passed, first posterior, and then anterior to the patient, the lateral aspect of the left knee was out of the field of view.

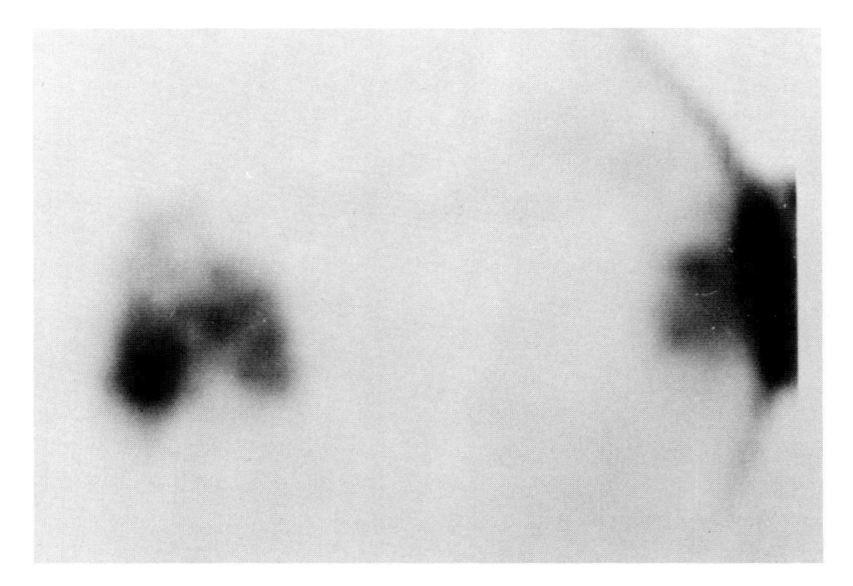

Figure 14-6 ***"Bladder-Filling" Artifact***

Transaxial image of the femoral heads exhibits a "bladder-filling" artifact. Broad rays project beyond the bladder and, in part, obscure the femoral heads. This technical problem and the "in-and-out" problem described in Figure 14-5 are types of incomplete angular sampling artifacts. (Collier BD, Carrera GF, Johnson RP et al: Detection of femoral head avascular necrosis in adults by SPECT. J Nucl Med 26:979, 1985)

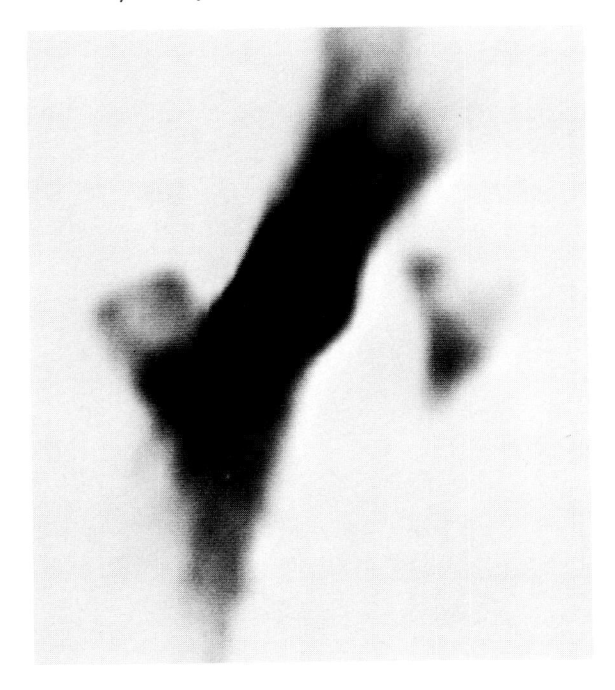

Figure 14-7 ***Painful Spondylolysis***

(A) Left posterior oblique (LPO) radiograph shows spondylolysis of L5 on the left *(arrow)*. ***(B)*** The right posterior oblique (RPO) radiograph is normal. (***C, D,*** and ***E***) Posterior, RPO, and LPO planar bone scintigrams show minimal increased activity over the left side of L5 *(arrows)*. On the oblique views this increased activity appears to be located in the posterior neural arch. Coronal ***(F)*** and transaxial ***(G)*** SPECT images localize the increased activity to the left side of the L5 neural arch *(arrows)*. This indicates that the spondylolysis is causing active bony repair. Spinal fusion relieved the low back pain of this 22-year-old man. (Collier BD, Johnson RP, Carrera GF et al: Painful spondylolysis or spondylolisthesis studied by radiology and single-photon emission computed tomography. Radiology 154:207, 1985)

A

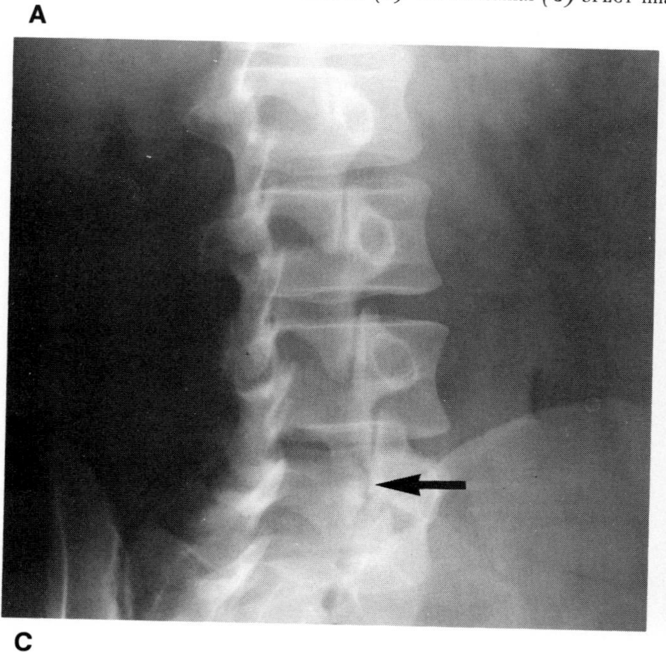

B

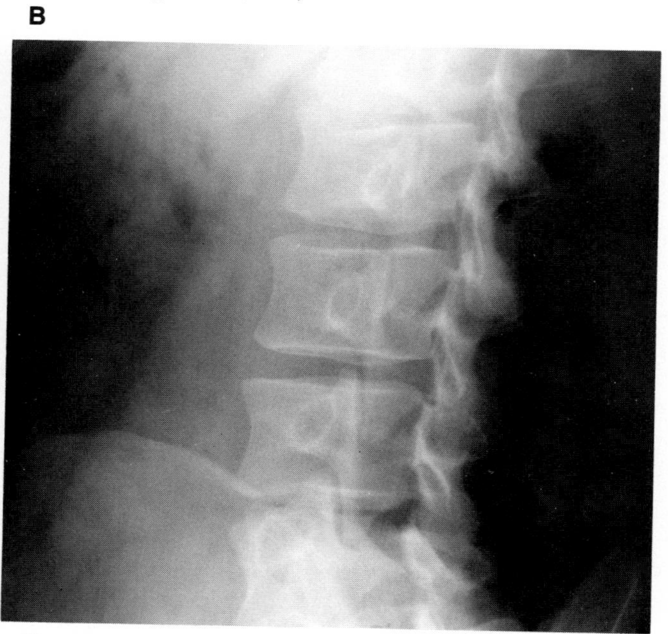

C

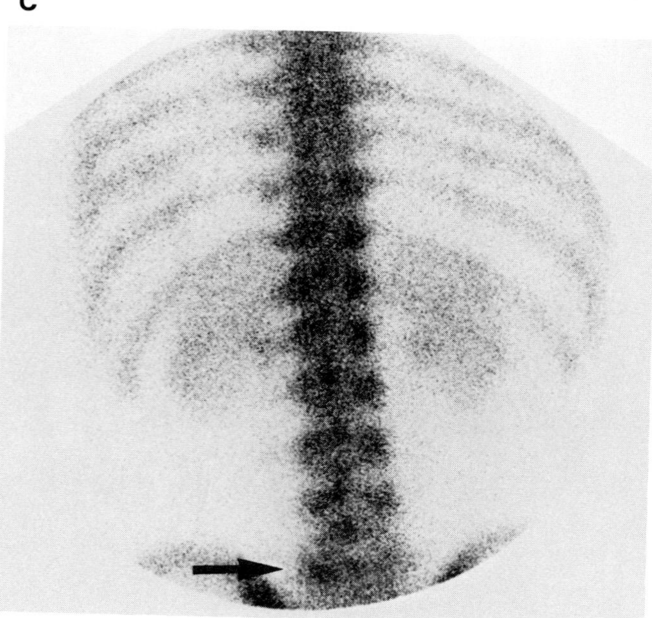

D

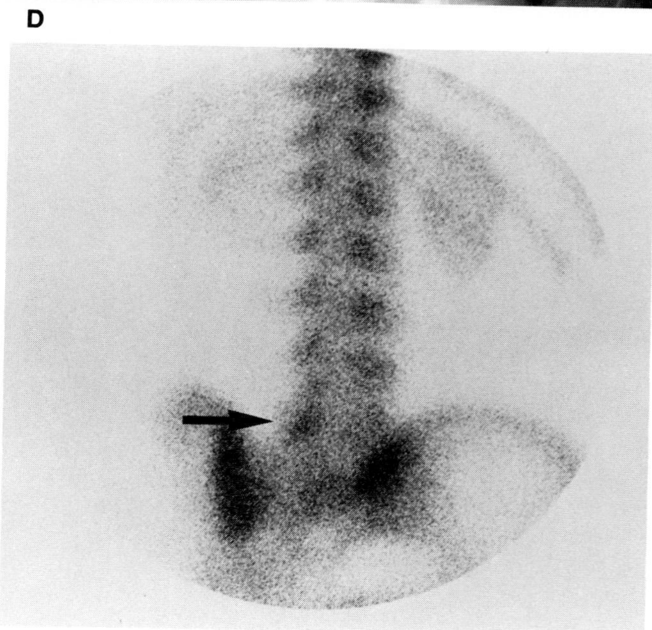

E

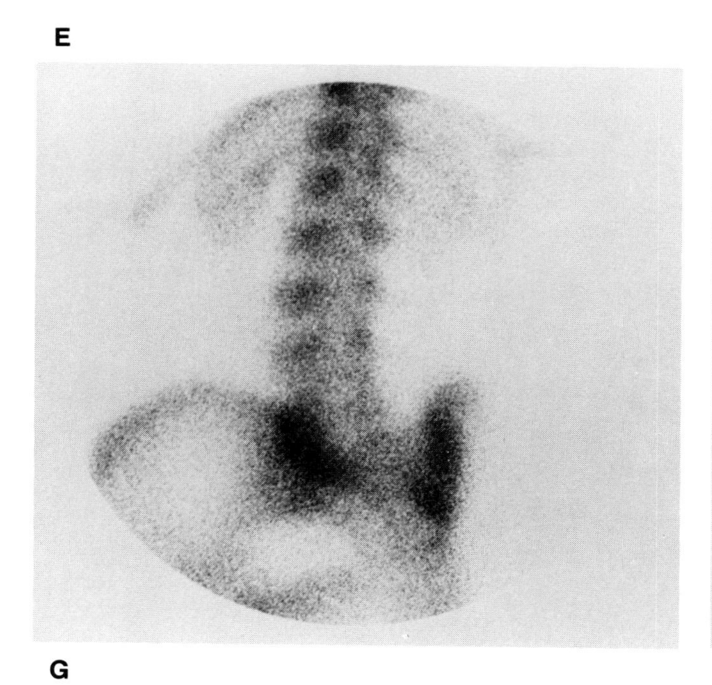

F

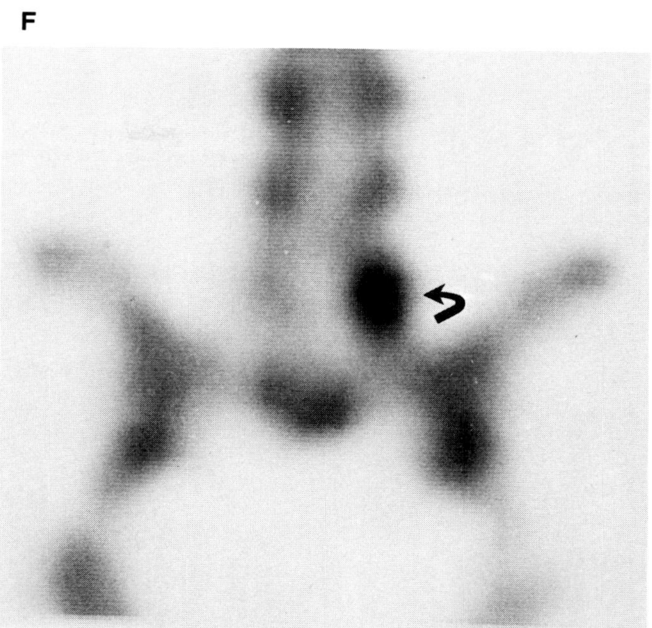

G

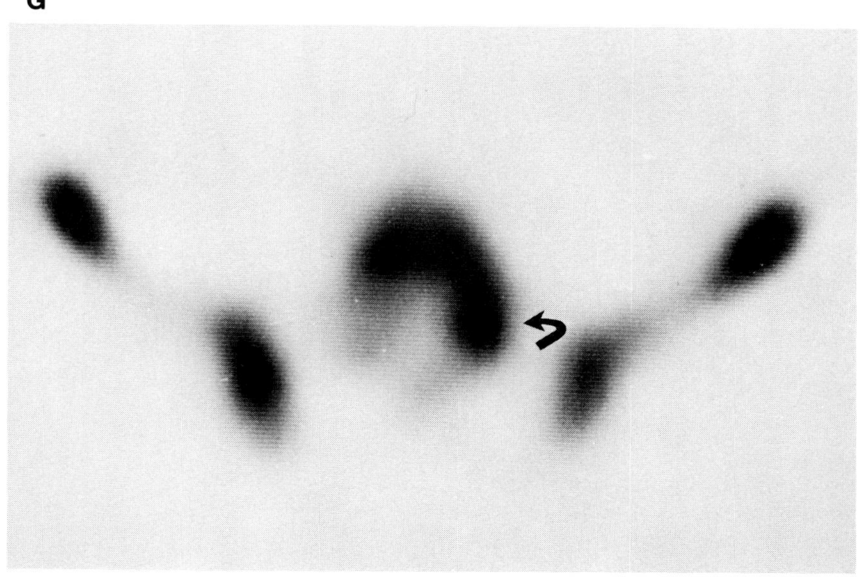

Figure 14-8 ***Painful Spondylolysis***

(***A*** and ***B***) RPO and LPO radiographs of an 18-year-old boy with a 1-year history of severe low back pain show a defect of the pars interarticularis on the right side of L5 *(arrow)*. (***C, D,*** and ***E***) Posterior, LPO, and RPO planar bone scintigrams are normal. ***(F)*** Transaxial L5 SPECT image shows increased activity over the right side of the L5 neural arch at the site of the pars defect *(arrow)*. (Collier BD, Johnson RP, Carrera GF et al: Painful spondylolysis or spondylolisthesis studied by radiology and single-photon emission computed tomography. Radiology 154:207, 1985)

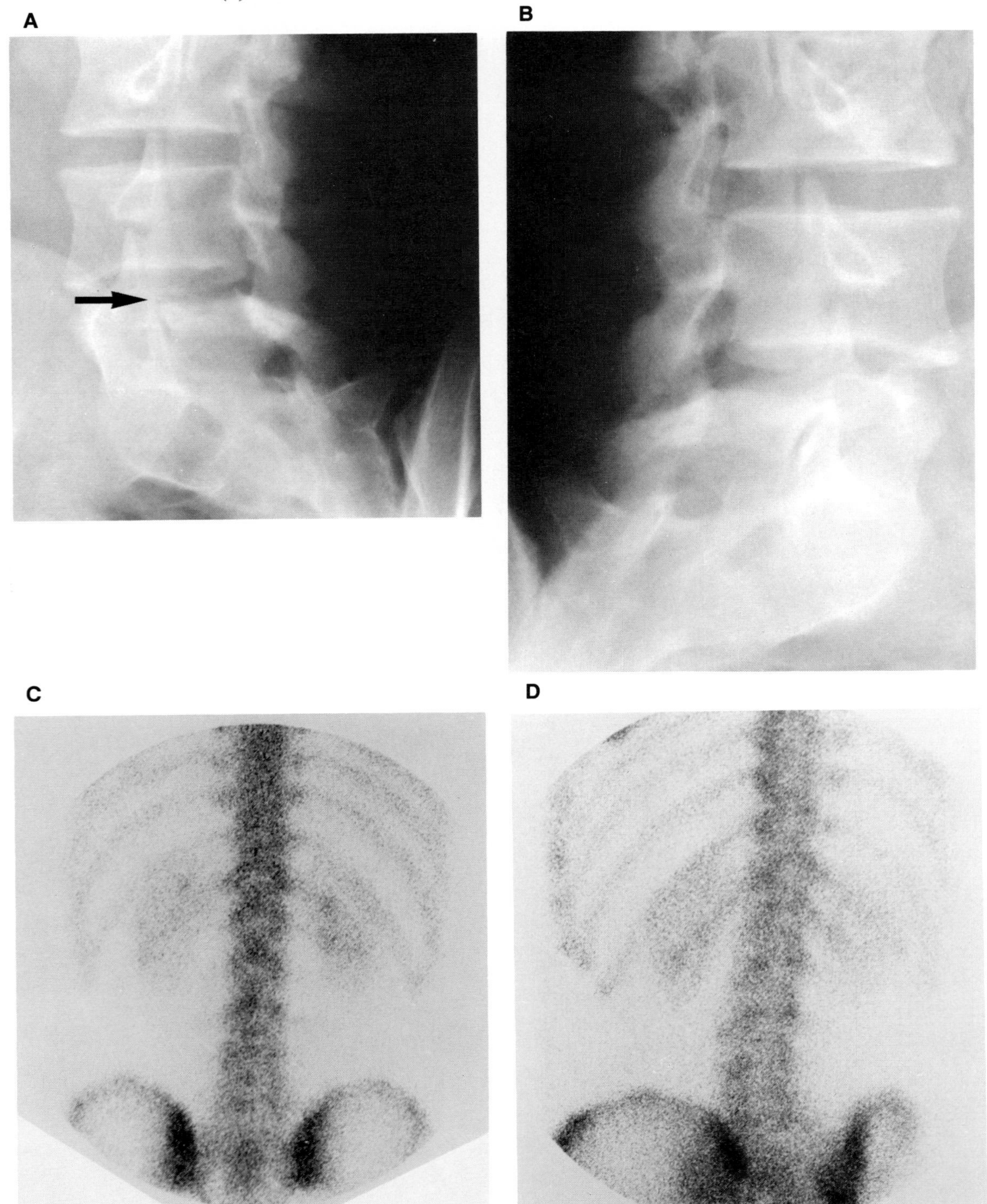

E

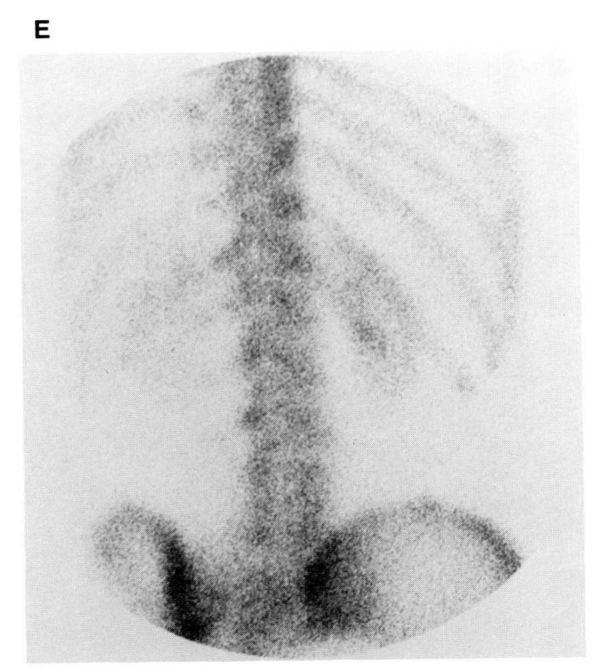

F

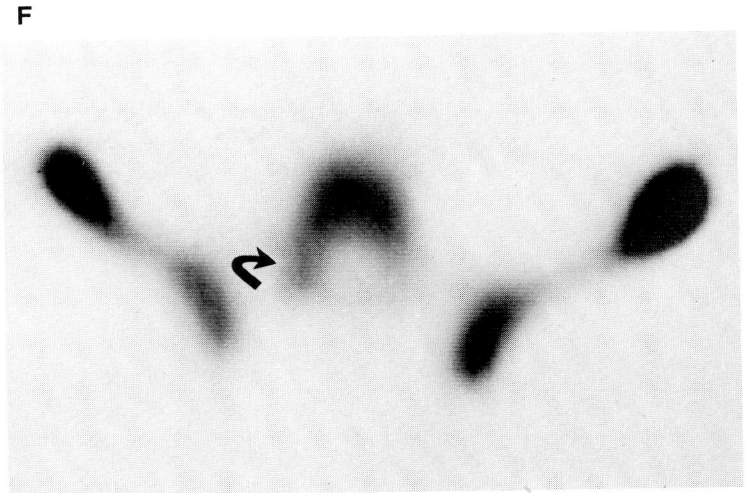

Figure 14-9 ***Painful Pseudarthrosis***

A 66-year-old woman was studied two years after a lateral transverse process spinal fusion from L3 to the sacrum. Increased activity within the fusion due to the pseudarthrosis can be seen on the transaxial ***(A)*** and sagittal ***(B)*** images *(arrows)*, but is not evident on the posterior view planar image ***(C)***. (Slizofski WJ, Collier BD, Flatley TJ et al: Painful pseudarthrosis following lumbar spinal fusion: Detection by SPECT bone scintigraphy. Skeletal Radiol, in press)

A

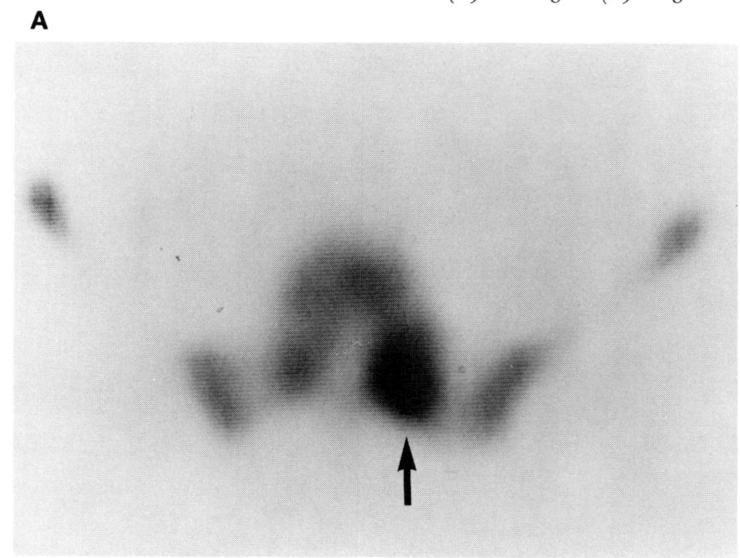

B

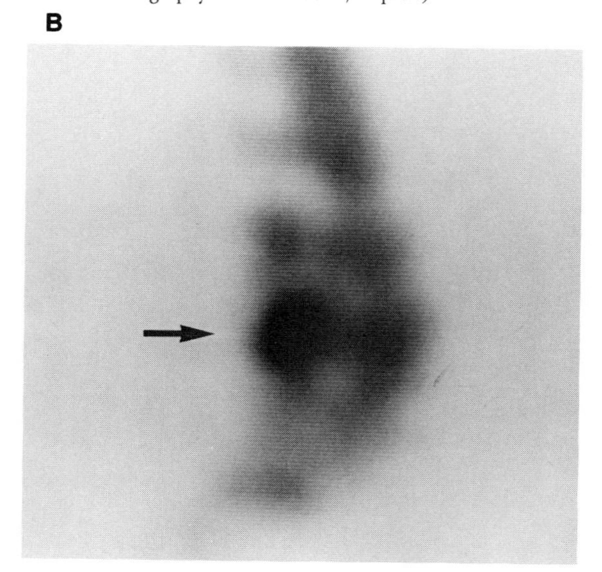

C

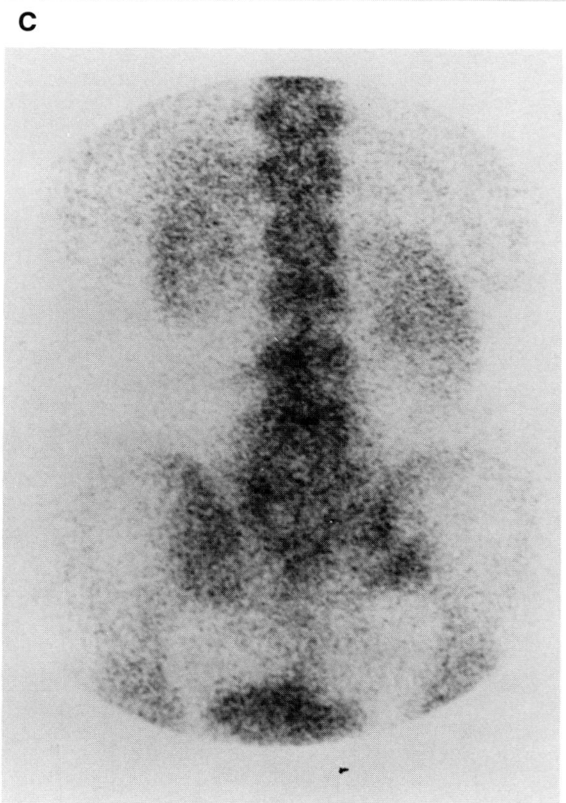

Figure 14-10 ***Sacral Fracture***

In this 27-year-old woman imaged 3 months following injury, increased activity near the left sacroiliac joint on the planar image *(A)* is more easily identified and localized on the transaxial SPECT image *(B)*. Coronal CT image *(C)* confirms the fracture site *(arrow)*.

A B

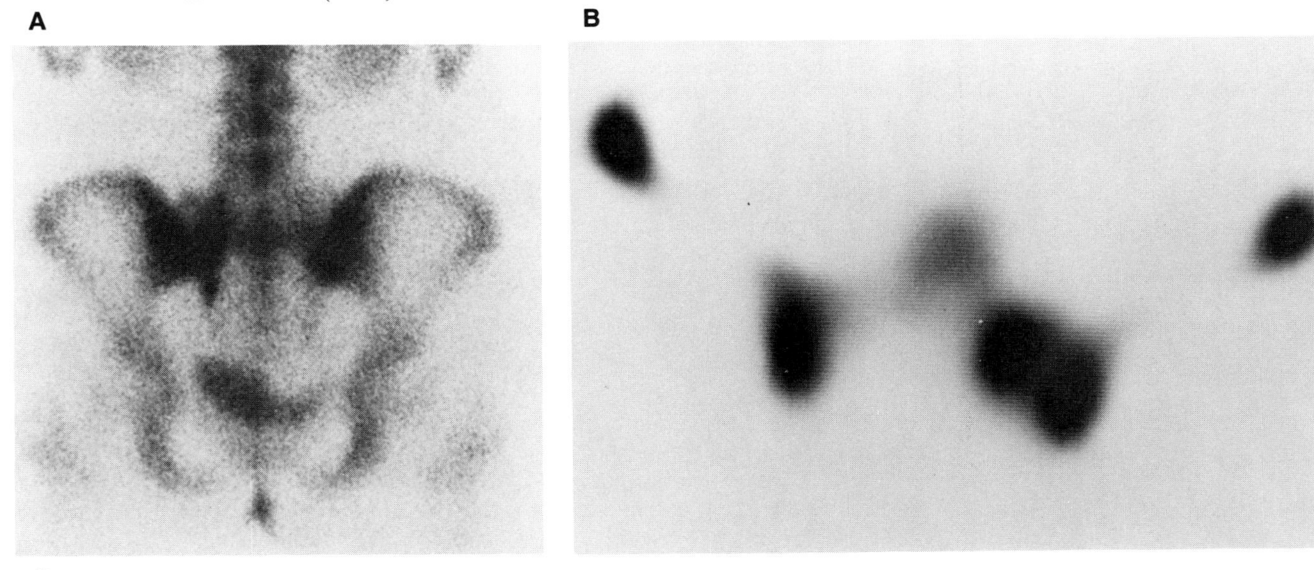

C

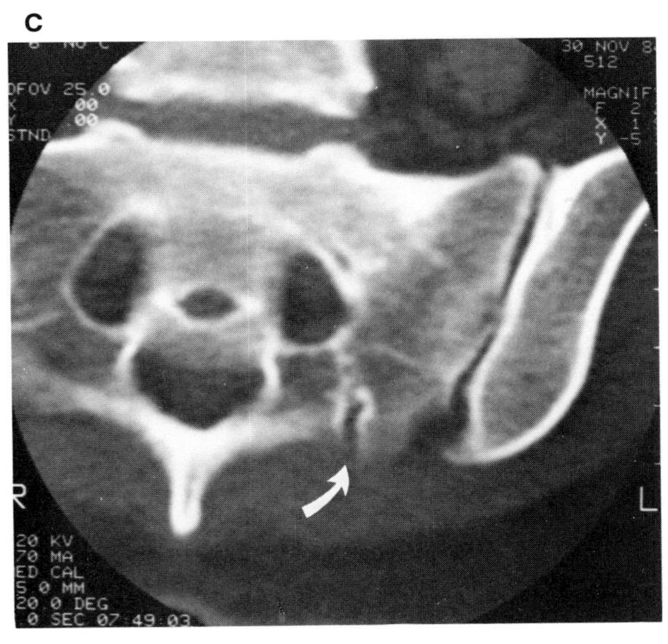

Figure 14-11 Bilateral Femoral Head Avascular Necrosis (AVN)

A 52-year-old woman, who had been treated with prednisone, had experienced 4 months of left hip pain with the right hip asymptomatic at the time of imaging. *(A)* Left hip radiograph shows sclerosis and flattening of the femoral head, in contrast to a normal right hip radiograph *(B)*. Anterior view planar bone scintigram *(C)* shows increased activity over the acetabulum and proximal left femur without convincing evidence of a photon-deficient defect in the left femoral head. Coronal *(D)* and transaxial *(E)* SPECT images through both hips clearly demonstrate a central photon-deficient defect surrounded by increased activity within the left femoral head *(straight arrows)*. In addition, a photon-deficient defect is seen within the asymptomatic right femoral head *(curved arrows)*. Histogram analysis *(F)* confirms the photon-deficient defects within both femoral heads. One month later, the patient developed right hip pain. Subsequent bilateral total hip replacements for relief of pain were performed with histologic evidence of AVN found in the resected femoral heads. (Collier BD, Carrera GF, Johnson RP et al: Detection of femoral head avascular necrosis in adults by SPECT. J Nucl Med 26:979, 1985)

A

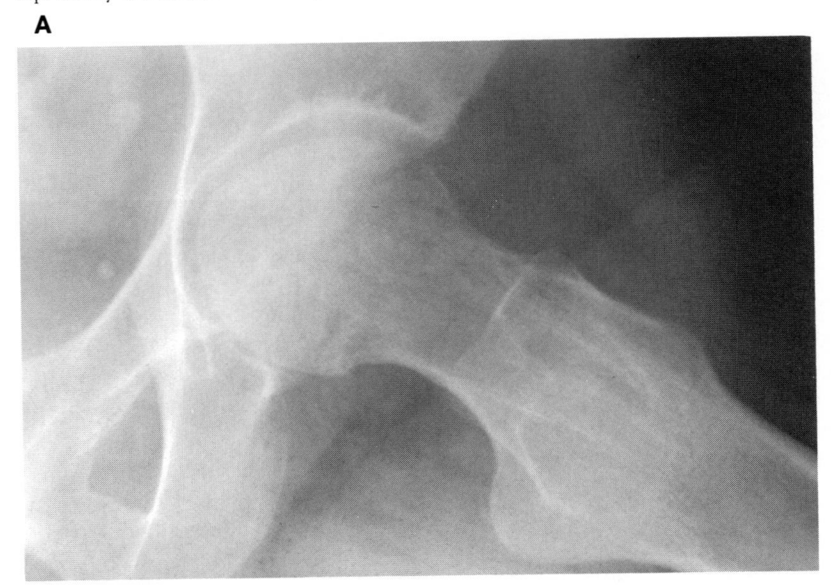

B

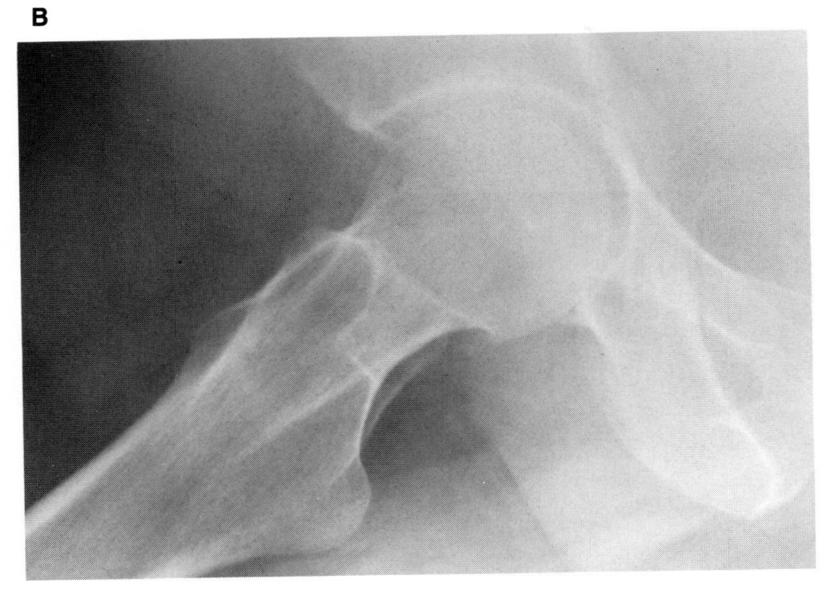

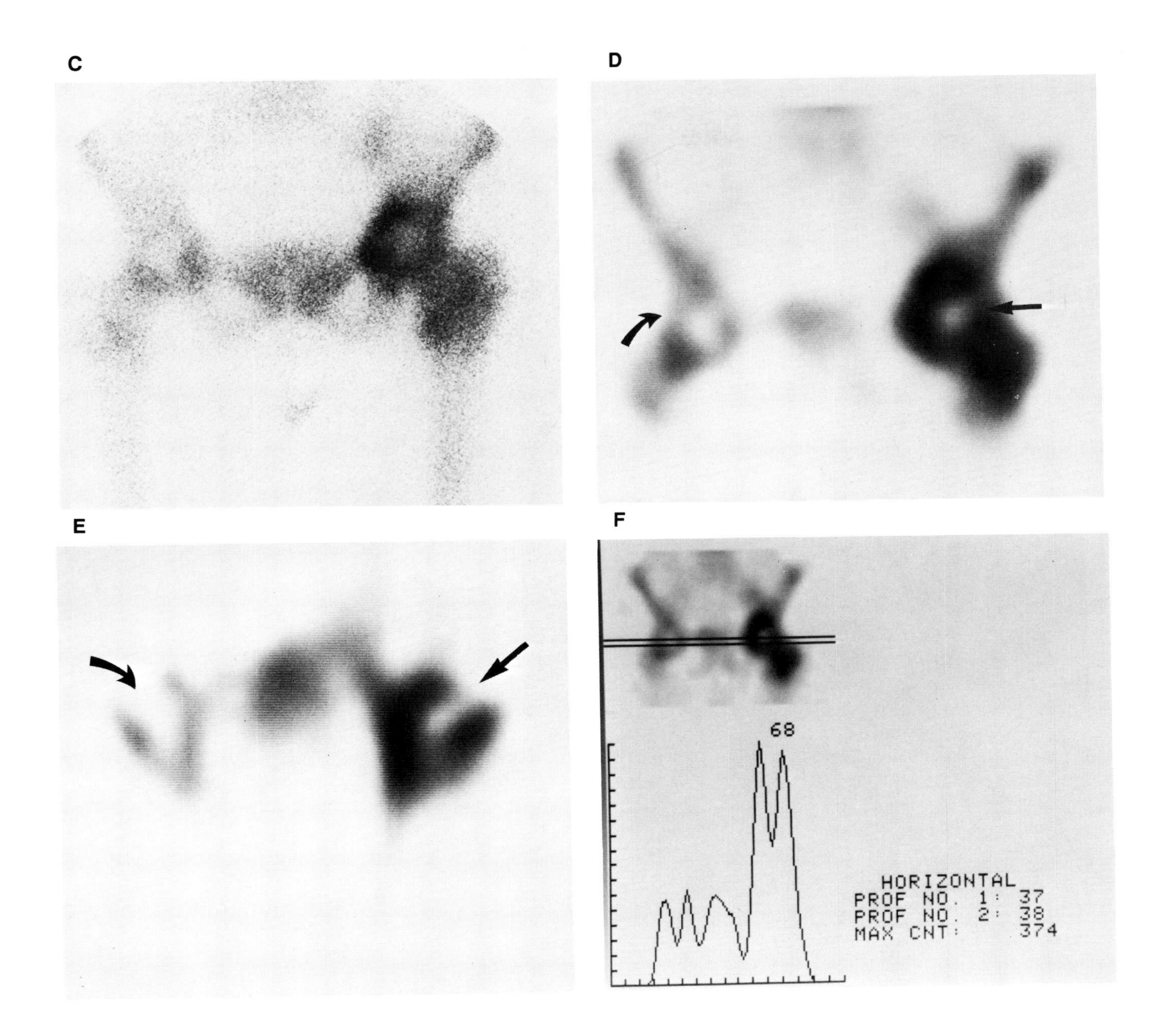
C
D
E
F
68
HORIZONTAL
PROF NO. 1: 37
PROF NO. 2: 38
MAX CNT: 374

Figure 14-12 *Torn Lateral Meniscus*

A 63-year-old man presented with 2 months of right knee pain, swelling, and locking. He had no history of trauma. Lateral joint line tenderness was present. Anteroposterior *(A)* and tangential *(B)* patellar radiographs show only small patellar osteophytes. *(C)* Anterior view planar bone scintigram shows moderately intense, increased activity over both patellae without any major scintigraphic abnormalities in the medial and lateral compartments of the painful right knee. *(D)* Transaxial SPECT image at the level of the joint space shows a crescent-shaped band of increased activity in the lateral compartment of the right knee *(arrow)*. This finding corresponded closely to the location of a longitudinal lateral meniscus tear identified during a subsequent arthroscopic examination. Cartilage damage and synovitis were seen by the arthroscopist along the articular surfaces adjacent to the torn meniscus. (Collier BD, Johnson RP, Carrera GF et al: Chronic knee pain assessed by SPECT: Comparison with other modalities. Radiology 157:795, 1985)

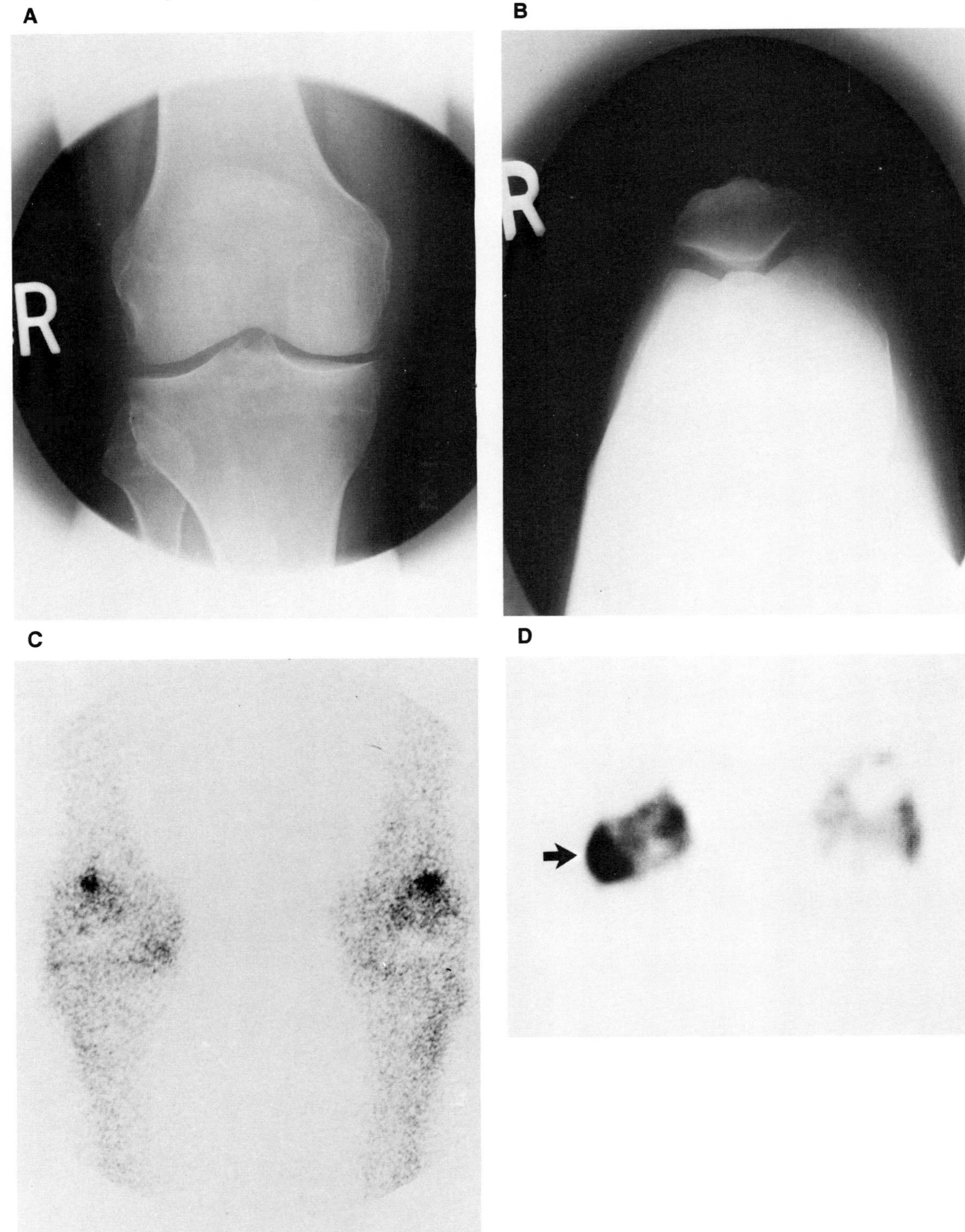

Figure 14-13 ***Triple Compartment Osteoarthritis of the Knee***

Radiographs *(not shown)* of a 30-year-old woman with a 3-year history of left knee pain demonstrated medial compartment narrowing, varus angulation, and osteophytes involving both the medial and lateral compartments. Anterior *(A)* and lateral *(B)* planar bone scintigrams show increased activity over the medial and lateral compartments of the left knee with the patella obscured by these superimposed regions of increased activity. *(C)* Coronal SPECT bone scintigram through the weight-bearing surfaces of the medial and lateral compartments confirms the increase in subchondral bony repair at the level of the joint space in the left knee. Incidental note is made of increased activity in the medial compartment of the right knee. *(D)* Sagittal SPECT image of the left knee shows increased activity in the patella *(straight arrow)*, which is easily separated on this tomographic image from the increased activity along the dorsal surface of the distal femoral condyle *(curved arrow)*. (Collier BD, Johnson RP, Carrera GF et al: Chronic knee pain assessed by SPECT: Comparison with other modalities. Radiology 157:795, 1985)

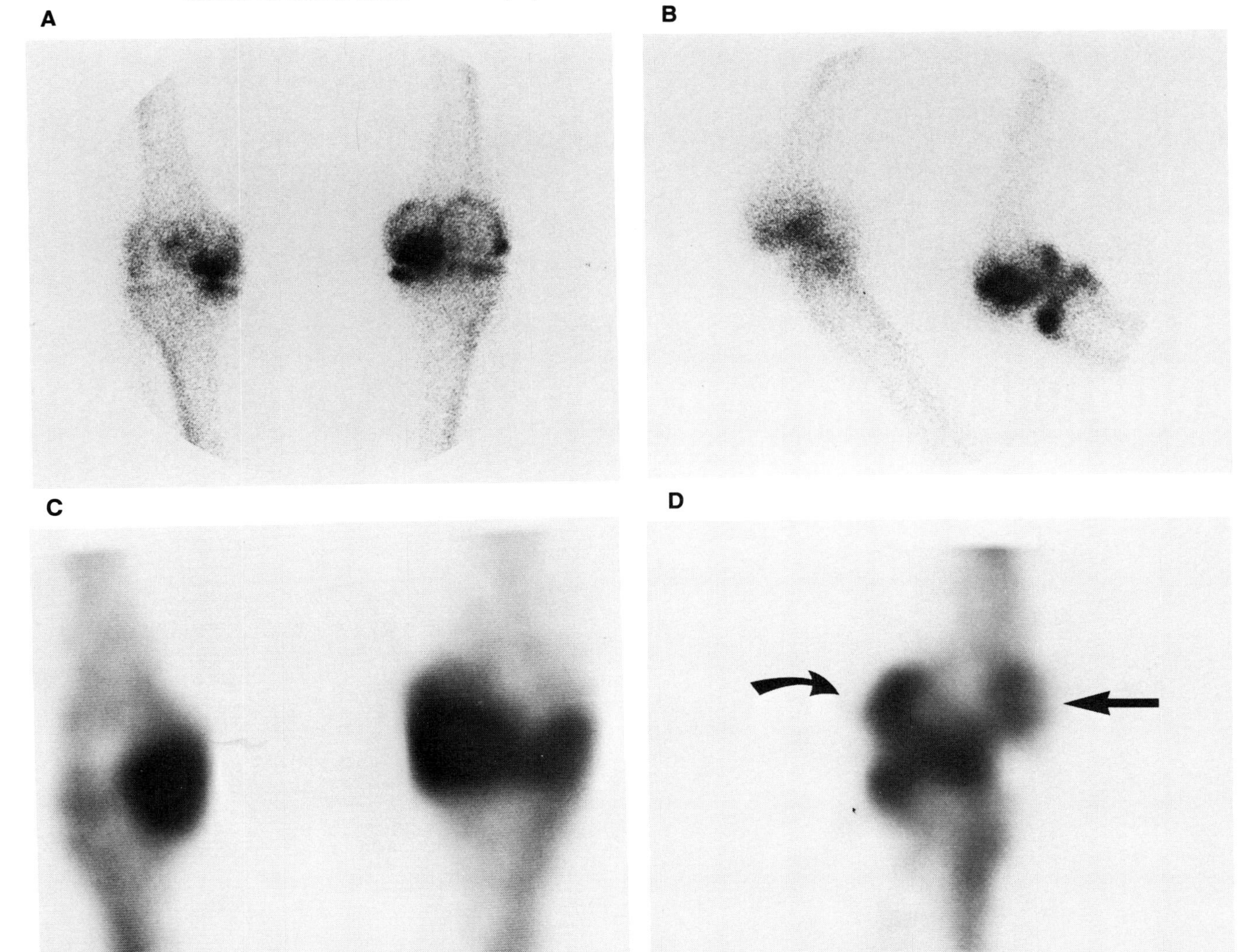

Figure 14-14 ***Right Temporomandibular Joint (TMJ) Internal Derangement***

A 43-year-old woman had pain and limited opening of the jaw. Right *(A)* and left *(B)* lateral planar images show no abnormality. Transaxial *(C)* and coronal *(D)* SPECT images show increased right TMJ activity *(arrows)*. Arthrography subsequently showed an abnormal anteriorly displaced disc. (Collier BD, Carrera GF, Messer EM et al: Internal derangement of the temporomandibular joint: Detection by single-photon emission computed tomography. Radiology 149:557, 1983)

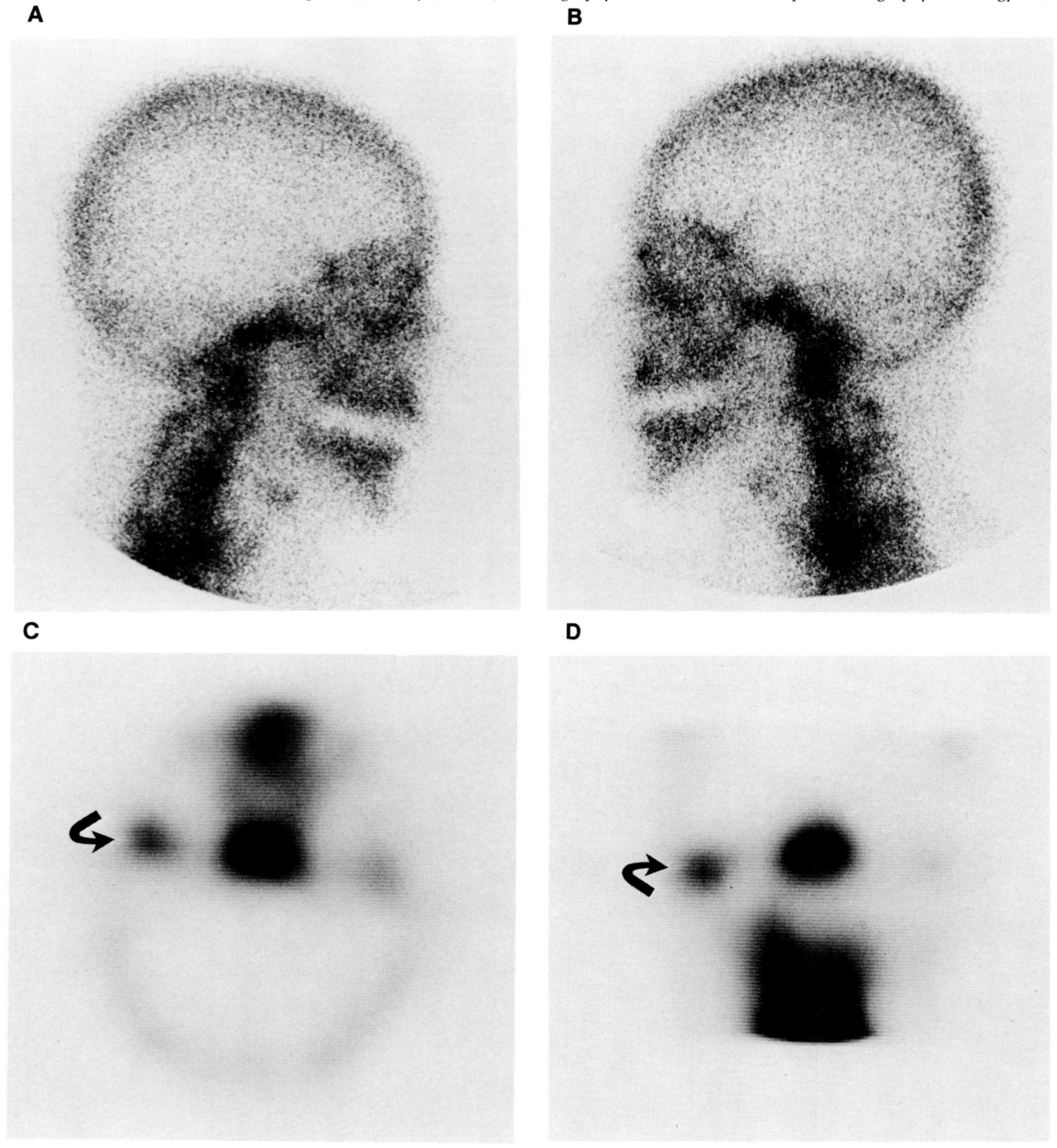

Tapan K. Chaudhuri

Chapter 15 NUCLEAR DACRYOCYSTOGRAPHY

The lacrimal drainage apparatus (LDA) consists of upper and lower canaliculi that drain the puncta at the medial canthus of each eyelid and join to form the common canaliculus, which passes medially into the lacrimal sac (Fig. 15-1). The nasolacrimal duct (NLD) is approximately 2 cm long and drains from the sac into the nasal cavity below the inferior turbinate.

Blockage of the normal pathway of drainage of tears from the eye into the nose and resultant epiphora (tearing) is a fairly common ophthalmologic problem. Due to the small size of the LDA, determination of the exact site of blockage is difficult. Yet, determination of the site of obstruction is important to directing the mode of therapy. The site of obstruction (Fig. 15-2) may be at the level of the individual canaliculus, common canacliculus, sac–duct junction, or at any level of the NLD.

By placing a drop of radioactive material into the conjunctival sac, it is possible to image the flow of tears through LDA using a gamma camera and a micropinhole collimator.[1–31] Throughout this chapter the technique of imaging LDA using radioactive material will be called *nuclear dacryocystography* (DCG) and the image obtained will be called a *nuclear dacryocystogram*. Nuclear DCG is simple, safe, and much more physiologic than conventional radiographic DCG for studying the patency of the LDA.

The test is indicated in any patient with epiphora and suspected obstruction in the LDA. It is also useful in evaluating the efficacy of surgery (plastic repair of the proximal canaliculus or dacryocystorhinostomy) to relieve obstruction.

The causes of obstruction in LDA vary from fractures of the facial bone adjacent to LDA to recurrent infections arising in the nose or sinus, intrinsic fungal infection, neoplasm, and congenital anomalies.

TECHNIQUE

Imaging Procedure

No patient preparation is needed. Technetium-99m pertechnetate is diluted in sterile, pyrogen-free, buffered normal saline to a concentration of 10 mCi/ml. One drop (0.01 ml or 10 μl) of this diluted solution containing 100 μCi Tc-99m O_4 is placed into the conjunctival sac near the lateral canthus with a micropipette (4) or a 23-gauge needle attached to a tuberculin syringe. The number of drops needs to be adjusted depending on the concentration of Tc-99m O_4. The instilled dose should not exceed 100 μCi. Imaging begins immediately. This study requires a gamma camera with a special, high-resolution, micro-pinhole collimator.

Immediately after instillation of the radiopharmaceutical, the patient is positioned (Fig. 15-3) sitting with the eye at variable distances (1 cm–5 cm) from the micro-pinhole collimator attached to the gamma camera. A close-up view produces a greatly magnified high-resolution image. A pinhole insert with an aperture size of approximately 1 mm is required for best resolution.[7] Such inserts are commercially available from a number of sources and are interchangeable with the standard 4 mm to 5 mm aperture inserts found in the conventional gamma camera pinhole collimator.

Sequential images of both eyes together or separately (for further improvement of resolution) are obtained every 2 to 4 seconds for 40 seconds, and then every 1 minute for 5 minutes. More delayed images may be necessary in some patients with slow filling of LDA. A radioactive marker is useful in correlating the nuclear DCG images with the surface anatomy. Even if the patient is asymptomatic in one eye, the study should be performed in both eyes so that the contralateral asymptomatic eye provides a normal control for comparison. A head immobilizer is often used to hold the head in place.

Noninvasive Interventional Maneuvers

Several noninvasive interventional maneuvers (*e.g.*, sac massage, artificial tearing, and eyewash) may be used for sluggish or delayed flow and for confusing collections of radioactivity on the lid or in the canthal regions.

Sac Massage. Sac massage is carried out by gently massaging the lacrimal sac, using a fingertip covered by a sterile gauze sponge. This maneuver helps in the evaluation of sluggish flow in the nasolacrimal duct.

Artificial Tearing. Two or three drops of sterile normal saline may be placed into the lateral canthal region to facilitate the flow, in case of sluggish or delayed flow.

Eyewash. Pooling of the radioactivity in the inner canthus or inadvertent spill onto the lids, when it happens, makes identification of the radioactivity within the lacrimal sac difficult. In this situation, the confusing collections of radioactivity on the lids or in the medial canthal regions can be washed away (ocular lavage) by thoroughly cleansing the area with water and a gauze sponge. The laboratory's emergency eyewash apparatus may be used for this purpose.

Tearing of the radioisotope onto the cheek is not an uncommon problem. It can be prevented by keeping the volume of Tc drops as minimal as possible (2 μl – 10 μl). Tearing (due to too large a drop, baggy eyelid, or inadvertent spill) is recognized by direct visual identification at the time of dropping Tc into the conjunctival sac or by identifying the location (along the cheek) and characteristics (broad, ragged area of activity) of the activity in the image. Once this happens, the study can be salvaged by wiping out the activity spilled over the face as much as possible and re-imaging. Usually the activity along the nasolacrimal duct is stronger and therefore distinguishable from the weak residual activity over the face. Alternatively, the cheek area can be covered by a thin lead sheet.

If the distinction is still difficult, the study needs to be repeated.

PHYSIOLOGIC MECHANISM OF THE RADIOPHARMACEUTICAL

The Tc-99m O_4 instilled in the eye is carried down with tears through the LDA. This normal drainage of radioactive tears helps in the imaging of the LDA.

ESTIMATED RADIATION ABSORBED DOSE

The estimated radiation absorbed dose in the lens is 0.02 rad/100 μCi (unobstructed flow) and 0.50 rad/100 μCi (obstructed flow).

The radiation dose to the lens from a complete contrast DCG procedure is approximately 2 rads.

VISUAL DESCRIPTION AND INTERPRETATION

In the normal individual, radioactivity collects along the palpebral fissures and drains medially into the lacrimal sac within 5 to 10 seconds. Over the next 30 to 50 seconds, good filling of the canaliculi, the lacrimal sac, and the NLD to its outlet in the inferior nasal meatus occurs. The canaliculi, the lacrimal sac, and the NLD are usually well defined on the longer duration, delayed images.

Abnormalities consist of obvious obstruction and lack of filling in any of these structures and/or prolongation of clearance time through LDA beyond 5 minutes.

DISCUSSION

The nuclear DCG has no significant limitation from a radiation standpoint. The amount of radioactivity used is so minimal that even the usual relative contraindication of pregnancy and lactation need not be considered.

The common method of instilling Tc-99m O_4 into the eye for nuclear DCG is from a needle point or a syringe-hub end. The disadvantages of this technique are twofold: statistical variation in drop size and hence the variation in instilled dose is wide; physiological discomfort of the patient is considerable, who must watch a needle over the eye.

A micropipette with calibrated tip (*e.g.*, 10 μl, 20 μl) offers the advantage of being more accurate, patient-acceptable, and having a broad spectrum of dose range with a single unit at economical cost.[4]

Nuclear DCG is a simple, nontraumatic, relatively harmless, and useful method of evaluating LDA using direct instillation of a pertechnetate droplet in the eye(s). It involves little or no patient discomfort and is readily performed by an experienced technologist with minimal or no medical supervision.

Contrast DCG, although providing similar information, requires local conjunctival anesthesia, dilatation of the inferior punctum, and injection of contrast material through a cannula placed in the canaliculus. It also requires the presence of an ophthalmologist. Often the cannulation is technically difficult and sometimes impossible. The contrast DCG is contraindicated in the presence of acute infection or sensitivity to local anesthetic.

While contrast DCG is traumatic, nonphysiologic, and offers detailed anatomy, nuclear DCG is nontraumatic, more physiologic, and provides functional assessment of LDA, but has the serious limitation of not defining pathologic anatomy. Contrast DCG outlines the anatomy well but often misses minor obstructions and functional blocks. Obstructions detected by nuclear DCG may be missed when the ducts are cannulated and injected with contrast media, because the pressure used to inject the contrast media overcomes the resistance of obstruction. Nuclear DCG also has the advantage of assessing the success of dacryocystorhinostomy; contrast DCG, being traumatic, is undesirable in and around fragile, healing, postoperative tissue. Our experience[3,5] indicates that both sensitivity and specificity of nuclear DCG are 100%, provided the disease does not involve the puncta or lids.

The nuclear DCG is primarily a physiologic test that provides information about the way tears flow through the LDA. The contrast DCG is used primarily as an anatomic study and provides a detailed picture of the pathologic anatomy of obstructed LDA. In most instances in which obstruction is suspected, the nuclear DCG is recommended as the initial imaging study because it is simpler to perform, causes less discomfort, and delivers much less radiation to the lens. When information obtained is inadequate or if confirmation is needed, the contrast DCG is recommended. In no instance has a contrast DCG uncovered an obstruction that was not detected by nuclear DCG. Neither test is particularly useful when the primary disease involves the puncta or lids, except as a way to rule out an associated distal abnormality or an unsuspected contralateral abnormality.

REFERENCES

1. Brizel HE, Sheila CW, Brown M: The effects of radiotherapy on the nasolacrimal system as evaluated by dacryoscintigraphy. Radiology 116:373, 1975
2. Carlton WH, Trueblood JH, Rossomondo RM: Clinical evaluation of microscintigraphy of the lacrimal drainage apparatus. J Nucl Med 14:89, 1973
3. Chaudhuri TK, Saparoff GR, Doland KD et al: A comparative study of contrast dacryocystogram (DCG) and nuclear DCG. J Nucl Med 15:482, 1974
4. Chaudhuri TK: A versatile way of instilling Tc-pertechnetate for nuclear dacryocystography. J Nucl Biol Med 20:84, 1976
5. Chaudhuri TK: Clinical evaluation of nuclear dacryocystography. Clinical Nucl Med 1:83, 1976
6. Chaudhuri TK: Nuclear dacryocystography. In Kirchner PT (ed): Nuclear Medicine Review Syllabus, pp 306–310, 321. New York, Society of Nuclear Medicine, 1980
7. Chaudhuri TK, Chaudhuri TK: Minipinhole collimator as an insert to conventional pinhole collimator to obtain super resolution in certain imaging procedures. Proceedings of the Third World Congress of Nuclear Medicine and Biology, pp 2590–2592. Paris, Pergamon Press, 1982
8. Chavis RM, Walham RA, Maisey MN: Quantitative lacrimal scintillography. Arch Ophthalmol 96:2066, 1978
9. Dressler J, v. Denffer H: Erste erfahrungen mit der Funktionsszintigraphie der tränenwege. In Hifer R (ed): Radioaktive isotope in Klinik und Forschung, pp 378. Munich, Urban & Schwarzenberg, 1975
10. Dressler J, v. Denffer H, Pabst HW et al: Vergleichende untersuchungen zum kinitischen verhalten von ^{99m}Tc-pertechnetat, ^{99m}Tc-humanalbumin-mikrospharen und ^{99m}Tc-schwefelkolloid in der tränenflussigkeit. In Pabst HW, Hör G, Schmidt HA (eds): Fortschritte der nuklearmedizin in klinischer und technologischer sicht, pp 822–826. Stuttgart, Schattauer, 1975
11. Gullotta U, Denffer HV: Dacryocystography: An Atlas and Textbook. New York, Thieme-Stratton, 1980
12. Harris CC, Goodrich JK, Chandler AC et al: Nuclear dacryocystography. In Croll MN, Brady LW, Carmichael P et al (eds): Nuclear Ophthalmology, pp 226–234. New York, John Wiley & Sons, 1976
13. Hurwitz JJ, Maisey MN, Welham RAN: Quantitative lacrimal scintillography. I. Method and physiological application. Br J Ophthalmol 59:308, 1975
14. Hurwitz JJ, Maisey MN, Welham RAN: Quantitative lacrimal scintillography. II. Lacrimal pathology. Br J Ophthalmol 59:313, 1975
15. Hurwitz JJ, Welham RA, Maisey MN: Intubation macrodacryocystography and quantitative scintillography: The "complete" lacrimal assessment. Trans Am Acad Ophthalmol Otolaryngol 81:575, 1975
16. Mayer PB, Dausch D: Klinische erfahrungen mit der radionuklid-dakryozystographie. Klin Mbl Augenheilk 167:421, 1975

17. Pink V, Gliem H: Funktionsszintigraphische untersuchungen nach der dakryozystorhinostomie. Klin Mbl Augenheilk 167:830, 1975

18. Robertson JS, Brown ML, Colvard DM: Radiation absorbed dose to the lens in dacryoscintigraphy with $^{99m}TcO_4$. Radiology 113:747, 1979

19. Rossomondo RM, Carlton WH, Trueblood JN et al: A new method of evaluating lacrimal drainage. Arch Ophthalmol 88:523, 1972

20. Saparoff G, Chaudhuri TK, Chaudhuri TK et al: Contrast dacryocystography vs ^{99m}Tc-lacrimal scan. Trans Am Acad Ophthalmol Otolaryngol 79:133, 1975

21. Saparoff GR, Chaudhuri TK, Chaudhuri TK, et al: Nuclear lacrimal scan vs dacryocystography. Trans Am Acad Ophthalmol Otolaryngol 81:566, 1976

22. Sørensen T, Jensen FT: Methodological aspects of tear flow determination by means of a radioactive tracer. Acta Ophthalmol 55:726, 1977

23. Sørensen T, Jensen FT: Lacrimal pathology evaluated by dynamic lacrimal scintigraphy. Acta Ophthal 58:597, 1980

24. Theodossiadis G, Panopoulos M, Chatzoulis D et al: How do tears drain: Technetium studies. Can J Ophthalmol 14:169, 1979

25. Trueblood JH, Rossomondo RM, Carlton WH et al: Corneal contact times of ophthalmic vehicles. Evaluation by microscintigraphy. Arch Ophthalmol 93:127, 1975

26. v. Denffer H, Bofilias I, Michejew P et al: Untersuchungen zur erfassung der dynamik des tränenabtransportes mit der radionukliddakryographie. Zbl ges Ophthalmol 116:551, 1979

27. v. Denffer H, Dressler J: Radionuklid-dacryocystographie in der diagnostik von stenosen der tränenableitenden wege. Albrecht v Graefes Arch Ophthalmol 191:321, 1974

28. v. Denffer H, Dressler J: Resorption von ^{99m}Tc-pertechnetat durch cornea und conjunctiva. In Quality Factors in Nuclear Medicine. Proceedings of XIII Int Meeting Soc Nucl Med, pp 1031–1037. Kobenhaven, Fadl's, 1975

29. v. Denffer H, Dressler J: Radionukliddakryographie in klinik und forschung. Klin Mbl Augenheilk 169:66, 1976

30. v. Denffer H, Dressler J, Gullotta U: Vergleichende nuklearmedizinische und röntgenologische untersuchungen bei sturungen der tränendrainage. Ber de Vers DOG München, Bergmann, 1977:591–597.

31. v. Denffer H, Gullotta U, Dressler J: Dakryozystographie und radionukliddakryographie. In Hanselmayer H (ed): Neue Erkenntnisse uber Erkrankungen der Tränenwege, pp 19–28. Stuttgart, Enke, 1981

Atlas Section

Figure 15-1 ***Normal Anatomy of the Lacrimal Drainage Apparatus (LDA)***

(A) Vertical part of canaliculus is 2 mm in length; ***(B)*** horizontal part of canaliculus, 8 mm; ***(C)*** fundus of sac, 3 mm to 5 mm; ***(D)*** body of sac, 10 mm; ***(E)*** intraosseous part of nasolacrimal duct, 12 mm: ***(F)*** meatal part of nasolacrimal duct, 5 mm; and ***(G)*** inferior meatus, 20 mm. (Modified from Chaudhuri TK: Clinical evaluation of nuclear dacryocystography. Clin Nucl Med 1(July):83, 1976

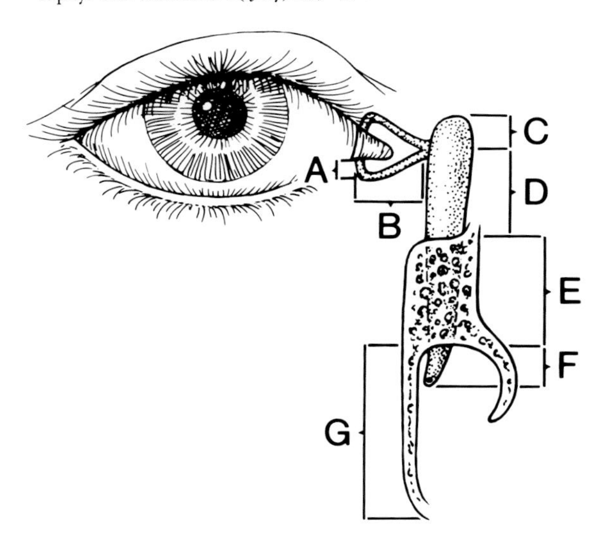

Figure 15-2 ***Sites of Blockage in the Lacrimal Drainage Apparatus***

Blockage can occur at ***(A)*** the level of individual canaliculus, ***(B)*** the level of common canaliculus, ***(C)*** the sac-duct junction, or ***(D)*** the level of the distal duct. (Modified from Chaudhuri TK: Clinical evaluation of nuclear dacryocystography. Clin Nucl Med 1(July):83, 1976)

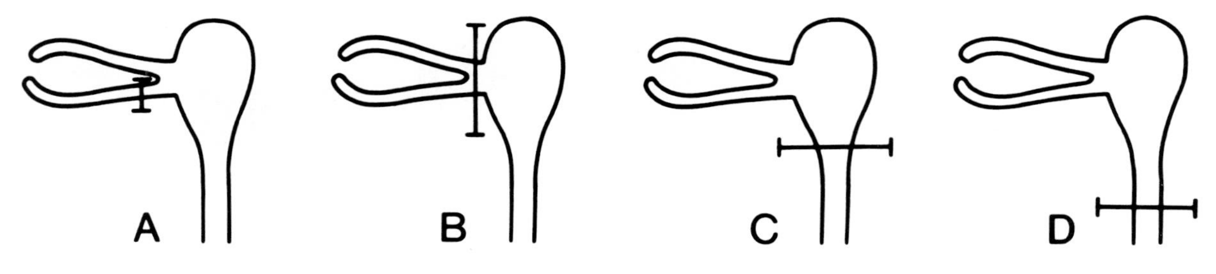

Figure 15-3 ***Patient Position in Front of the Gamma Camera***

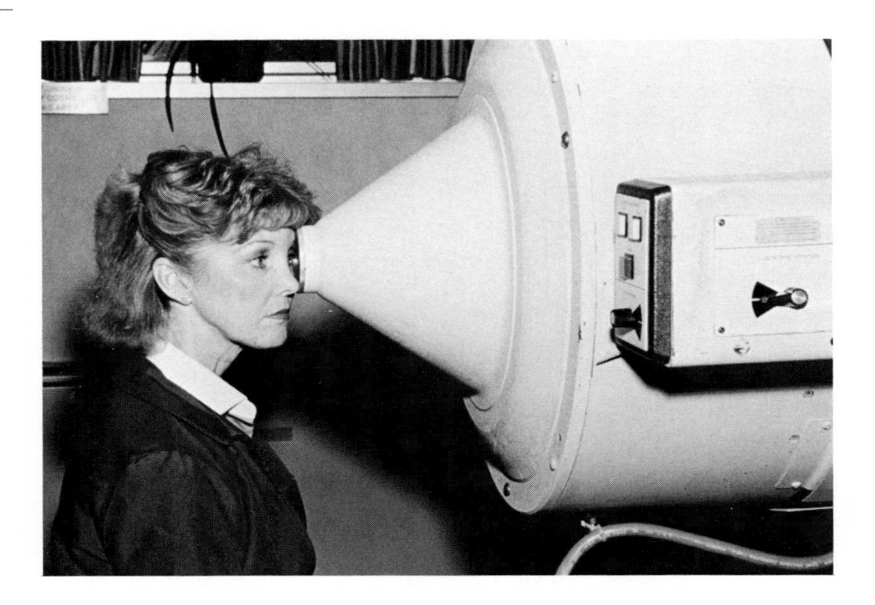

Figure 15-4 ***Normal Nuclear Dacryocystograms (DCG)***

(A) Normal sequential nuclear DCG of the left eye. (*c*, canaliculus; *cc*, common canaliculus; *s*, sac; *d*, duct; *m*, meatus) ***(B)*** Normal contrast DCG (left eye). Arrow points to the dye in the inferior meatus. (Modified from: Chaudhuri TK: Clinical evaluation of nuclear dacryocystography. Clin Nucl Med 1(July):83, 1976)

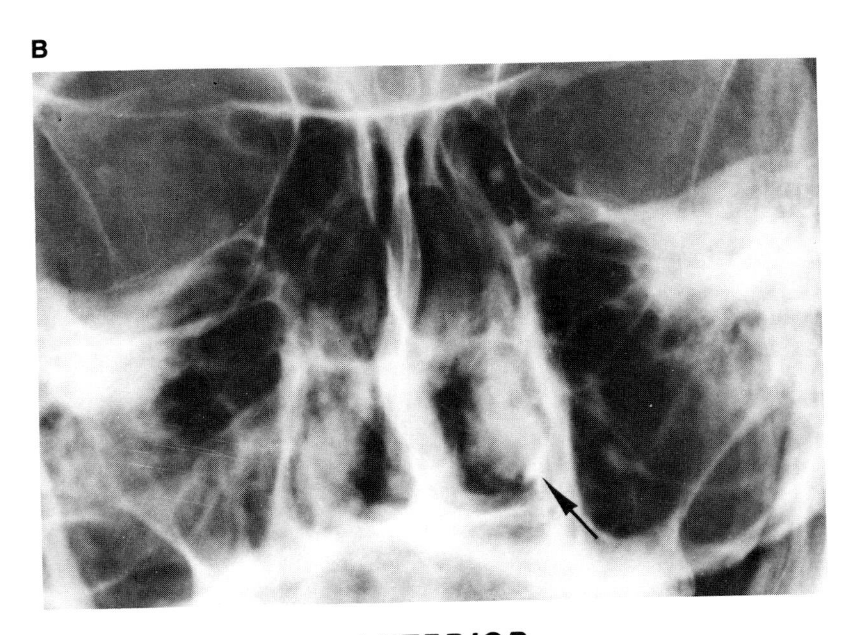

Figure 15-5 ***Normal Nuclear Dacryocystograms (DCG)***

Normal nuclear DCG done bilaterally shows lacrimal sac *(curved arrows),* nasolacrimal duct *(arrowheads),* and inferior meatus *(straight arrows).*

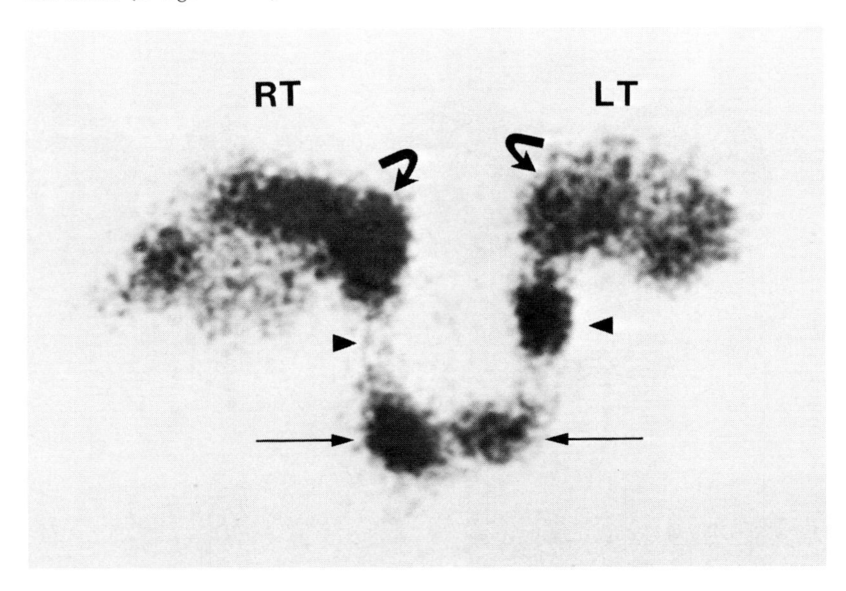

Figure 15-6 ***Normal Nuclear Dacryocystograms (DCG)***

Normal nuclear DCG of both eyes is superimposed on patient's surface anatomy. (Chaudhuri TK: Clinical evaluation of nuclear dacryocystography. Clin Nucl Med 1(July):83, 1976)

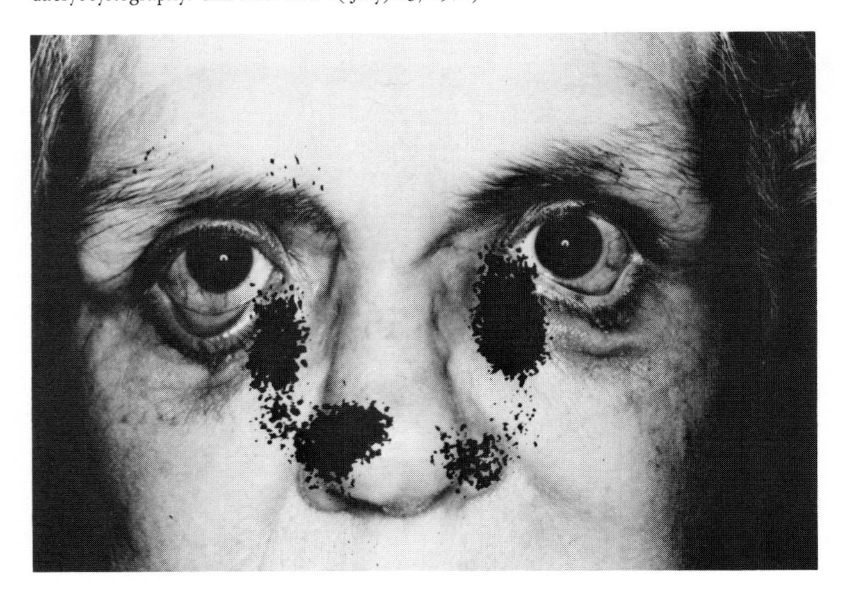

Figure 15-7 ***Bilateral Obstruction: Canaliculus***

Bilateral obstruction at the level of canaliculus *(arrows)*. (Modified from Chaudhuri TK: Clinical evaluation of nuclear dacryocystography. Clin Nuc Med 1(July):83, 1976)

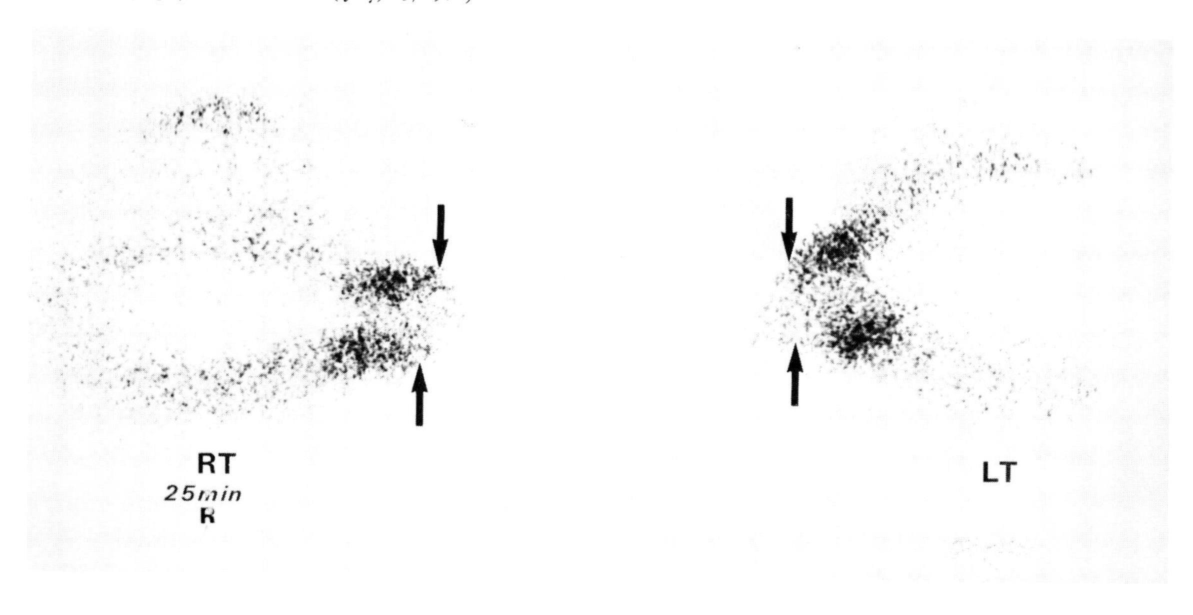

Figure 15-8 ***Discontinuity of Canaliculus***

Anatomical discontinuity of right lower canaliculus (*a*) is secondary to a laceration of the right lower lid caused by trauma. The right lower canaliculus has been completely transected. The nuclear DCG shows ragged distribution of radioactivity along the right lower canaliculus. Note the smooth outline of radioactivity along the normal right upper canaliculus (*b*) and the left upper (*c*) and left lower (*d*) canaliculi. (Modified from Chaudhuri TK: Clinical evaluation of nuclear dacryocystography. Clin Nucl Med 1(July):83, 1976

NUCLEAR DACRYOCYSTOGRAM

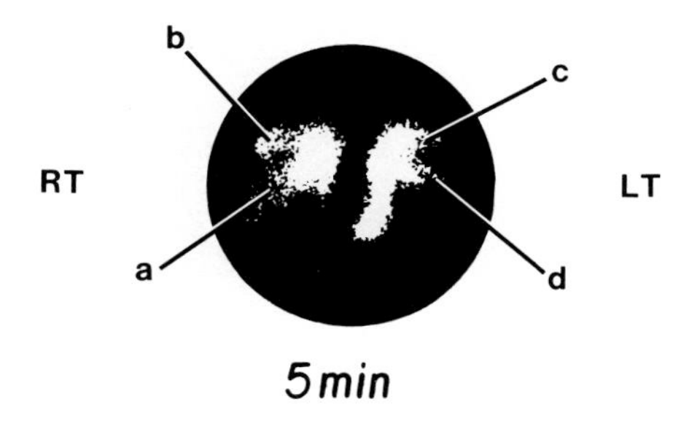

ANTERIOR

Figure 15-9 ***Obstruction: Common Canaliculus—SAC Junction***

Obstruction has occurred at the junction of common canaliculus and lacrimal sac on the left *(curved arrow)*. Also note obstruction in mid-duct on the right *(straight arrow)*.

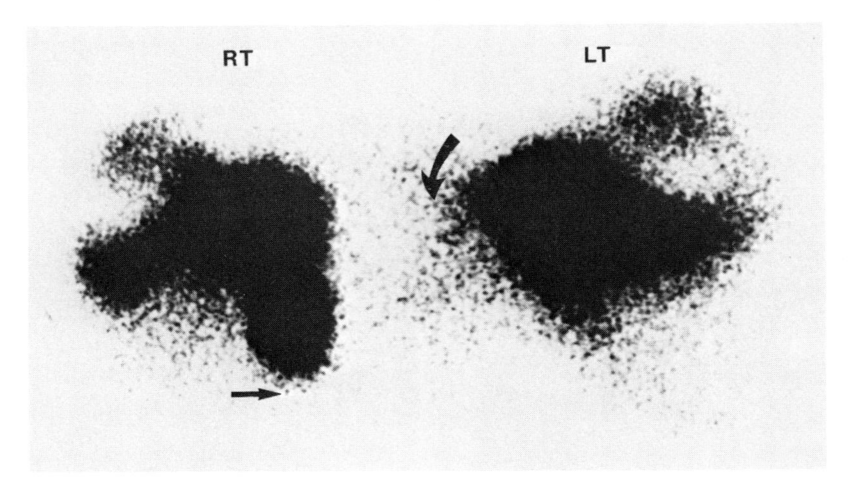

Figure 15-10 ***Obstruction: SAC—Duct Junction***

(A) Obstruction has occurred at the junction of the lacrimal sac and nasolacrimal duct on the right. The left side is normal. The straight arrows mark the region of common canaliculus. The curved arrows point to the lacrimal sac (note clearance of radioactivity from the left lacrimal sac). *(B)* Contrast DCG also shows obstruction *(arrow)* distal to sac on the right. (Figure 10A—Chaudhuri TK: Clinical evaluation of nuclear dacryocystography. Clin Nucl Med 1(July):83, 1976)

A

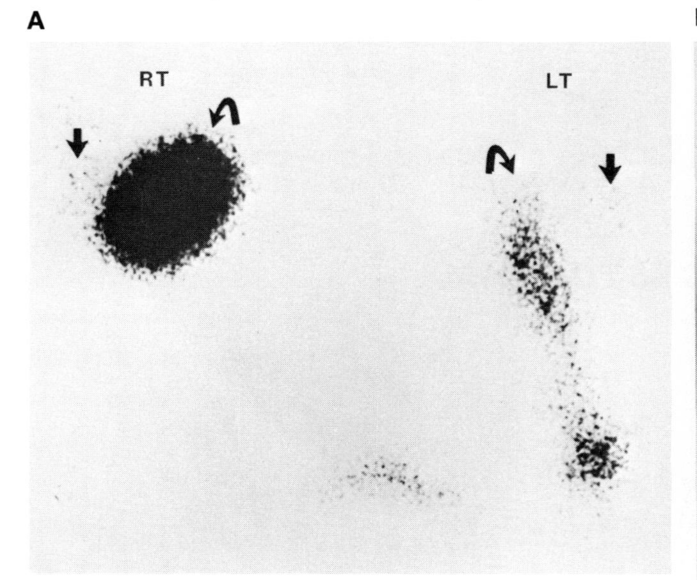

B

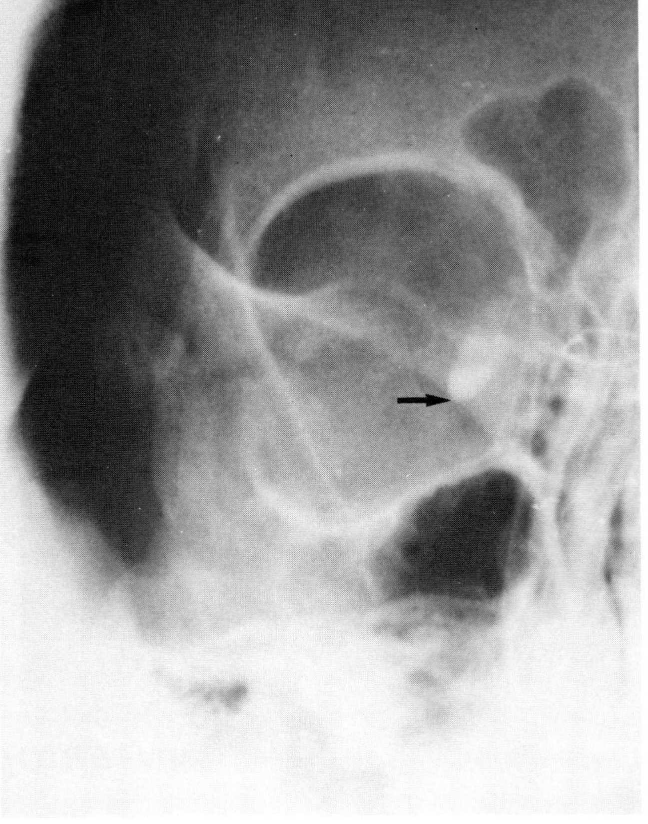

Figure 15-11 ***Obstruction: Lacrimal Duct***

(A) Nuclear DCG shows obstruction *(arrow)* in the lacrimal duct on the left. *(B)* Contrast DCG also shows obstruction *(arrow)* in left nasolacrimal duct.

A

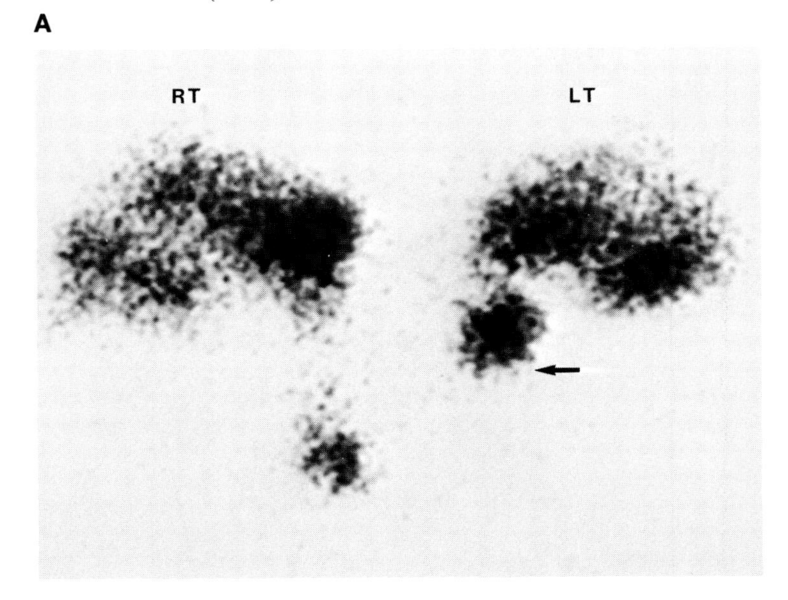

B

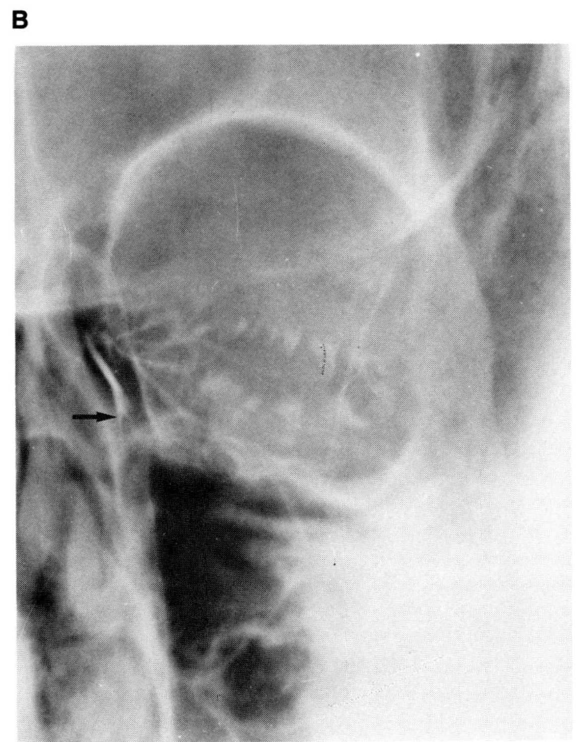

Figure 15-12 ***Functional Block on the Left***

The patient complained of intermittent tearing on the left side. An irrigation study indicated no anatomical block, but the nuclear DCG revealed functional obstruction. Straight arrows mark common canaliculus; curved arrows mark lacrimal sac (note the hang-up of activity in the left sac, while the right sac has emptied into the duct). ***Comment:*** An irrigation study involves injecting sterile normal saline through a cannula placed in the canaliculus. A patent LDA is indicated by the patient's subjective taste of salt solution in the throat. (Chaudhuri TK: Clinical evaluation of nuclear dacryocystography. Clin Nucl Med 1(July):83, 1976)

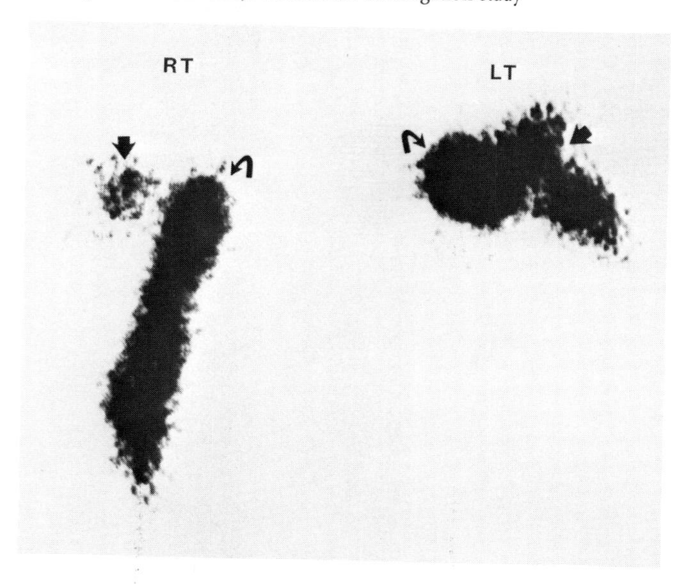

Figure 15-13 ***Successful Dacryocystorhinostomy on the Left***

Preoperative *(left)* image shows no flow distal to the sac *(arrows)* but the postoperative *(right)* image shows normal flow. (Chaudhuri TK: Clinical evaluation of nuclear dacryocystography. Clin Nucl Med 1(July):83, 1976)

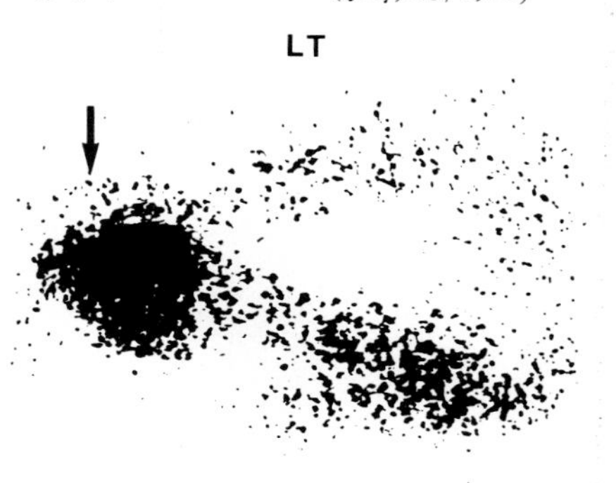

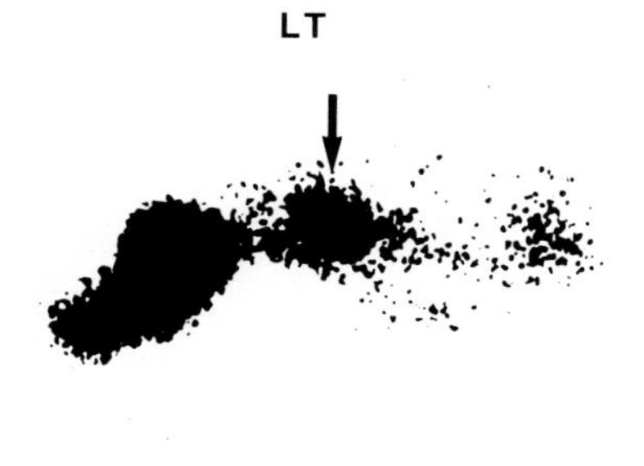

Preop **Postop**

Figure 15-14 ***Unsuccessful Dacryocystorhinostomy on the Left***

Both preoperative *(left)* and postoperative *(right)* nuclear DCG images show absence of flow on the left. The right side is normal. (Chaudhuri TK: Clinical evaluation of nuclear dacryocystography. Clin Nucl Med 1(July):83, 1976)

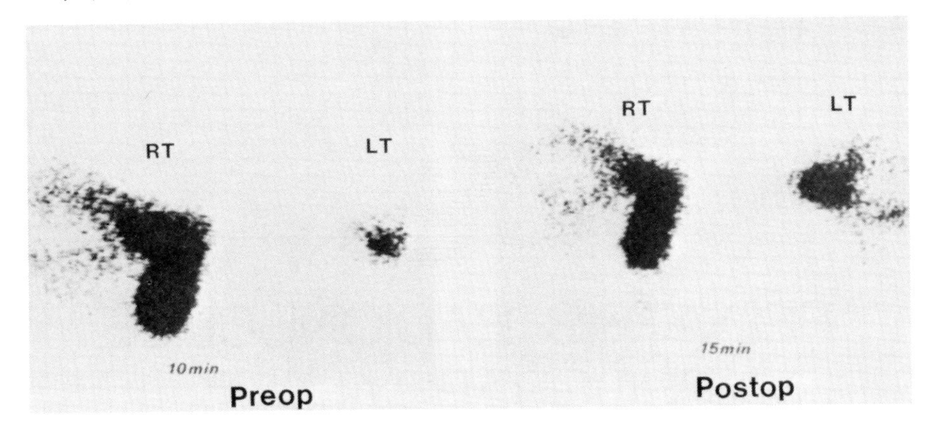

John R. Sty
Robert G. Wells

Chapter 16 PEDIATRIC RADIONUCLIDE LYMPHOGRAPHY

Investigation in the late 1940s demonstrated that Au-198 accumulated in reticuloendothelial cells following parenteral injection and, because of this property, could be used for evaluating the lymphatic system.[1-3] Subsequently, other radiopharmaceuticals including Cr-51 serum albumin, Tc-99m serum albumin, Tc-99m sulfur colloid, Hg-197 sulfur colloid, Tc-99m antimony sulfur colloid, Tc-99m stannous phytate, and In-111 phosphate colloid[4-13] have been employed. Presently, Tc-99m antimony colloid is the most commonly used radiopharmaceutical.

THEORY AND TECHNIQUE

Tc-99m antimony trisulfide colloid is the radiopharmaceutical of choice for interstitial lymphoscintigraphy in children. It has a particle size of 4 to 12 nanometers, a 6-hour physical half-life, a photon energy ideal for imaging with a gamma scintillation camera, and tendency to clear from subcutaneous injection sites rapidly. The injection site is determined by the specific region of the lymphatic system to be evaluated. For inguinal, femoral, pelvic and abdominal lymph nodes, the subcutaneous tissues of the dorsum of the feet are injected. A dosage of 100 μCi to 500 μCi of Tc-99m antimony trisulfide colloid is usually employed. The estimated radiation absorbed dose for radionuclide lymphography is listed in Table 16-1.[14]

Interstitially injected Tc-99m antimony sulfur colloid is removed from the site of injection by lymphatic flow and regional tissue macrophages. It enters the lymph nodes through afferent capsular vessels. During its passage through the lymph node, portions of tracer become incorporated within the sinusoids or are actively phagocytized by the reticuloendothelial cells. The lymph and radiopharmaceutical which exit the lymph node through the efferent lymphatic system become incorporated into the next most proximal node in the chain. The quantity of radiopharmaceutical that is not deposited in the lymph nodes eventually empties into the venous system. From here it is removed by the hepatic and splenic reticuloendothelial cells. The trapped radiopharmaceutical contained within the lymph nodes allows for visualization of the lymph node chain[15-22] (Figs. 16-1 and 16-2).

The appearance time of the radiopharmaceutical in the lymphatic vessels is immediate. Within 5 minutes the lymphatic channels of the lower extremities, pelvis, and retroperitoneum are demonstrated to the level of the cisterna chyli. After this time interval there is progressive extraction of the radiopharmaceutical by the major lymph node groups in the iliofemoral, pelvic, and paraspinal regions. The liver is seen within the first 10 minutes of the examination.

Table 16-1
Estimated Radiation Absorbed Doses for Tc-99m Antimony Colloid Lymphography

Organ	*Rad/mCi*	*Estimated Exposure of 100 μCi – 500 μCi Dose (rad)*
Injection site	30	3 – 15
Lymph nodes	0.3	0.03 – 0.15
Kidney	0.016	0.0016 – 0.008
Liver	0.01	0.001 – 0.005
Ovaries	0.07	0.007 – 0.035
Testes	0.07	0.007 – 0.035
Whole body	0.001 – 0.02	0.0001 – 0.01

Instrumentation for lymphoscintigraphy consists of a single-crystal, gamma scintillation camera with parallel, converging, or pinhole collimation, depending on the size of the child or the particular area of interest. The patients are imaged in the supine position. Views are tailored to the area of interest and the clinical problem under investigation. Images are obtained immediately after interstitial injection. Delayed images at 5 minutes and 20 minutes are also obtained to evaluate the anatomic distribution of deposited radiopharmaceutical. Computer enhancement is sometimes helpful because of the low count density.[23]

The interpretation of a lymphoscintigraphic study is based on the appearance time of radiopharmaceutical and its distribution during the vascular phase and delayed images. During the vascular phase the size, appearance time, and clearing of the radiopharmaceutical constitute the major criteria for interpretation. For example, with obliterated lymphatics, the lymphatic channels are not seen, and dermal backflow may be demonstrated. The contralateral extremity can be used as a comparison for symmetry, provided that there is a unilateral abnormality. Experience and knowledge of the disease that is being investigated is essential. The interpretation of the delayed images is based on comparison of lymph node groups on the contralateral side. This comparison is based on the size, number, and distribution of the nodes. It is not possible with lymphoscintigraphic technique to determine if a node is replaced by a pathologic process or is absent. Small defects in nodes cannot be appreciated with this technique because of the lack of spatial resolution. There is usually normal variation in the number of nodes in the iliofemoral and pelvic region. However, the advantages of the technique outweigh the disadvantages (Table 16-2).

Table 16-2
Lymphoscintigraphy as an Imaging Method

Advantages	*Disadvantages*
Noninvasive	Poor spatial resolution
Physiologic	
Minimal adverse reactions	
Easy to perform	
Good patient tolerance	

DISCUSSION

There are many applications of interstitial lymphoscintigraphy in children.

Clinical Applications in Pediatrics
Lymphedema
Primary
Secondary
Lymphangioma
Lymphatic disruption
Lymphangiectasia
Intestinal
Pulmonary
Chylous ascites
Chylothorax
Metastasis

It can be used to evaluate children and adolescents with lower leg edema. Lymphedema in children has a variety of etiologies. It can occur after surgical procedures, from tumors or infection (parasitic or bacterial), or as a congenital abnormality. The lymphedemas may be divided into aplastic, hypoplastic, and hyperplastic varieties, based on the number and caliber of lymphatics.

Primary Lymphedema

Primary lymphedema is due to a developmental defect in the lymphatic system. Although abnormalities of the lymphatic system are presumably present at birth in primary lymphedema, many cases do not present clinically until later childhood or adulthood. Kinmouth and associates proposed a classification system of primary lymphedema based on the age at presentation: *congenital,* present at birth; *praecox,* prior to age 35; and *tarda,* after age 35.[24]

The reason for the frequent delay in clinical presentation of primary lymphedema is not clear. Some authors have suggested that congenitally abnormal lymphatics may be modified by hormonal or inflammatory effects. There is an increased incidence in females, and the peak age of presentation is at puberty, consistent with the suggestion of hormonal influences. In addition, positive contrast lymphangiography frequently demonstrates abnormal lymphatics in the clinically normal contralateral leg. There is an unexplained predilection for the left lower extremity.

The hypoplastic type accounts for the majority of cases of primary lymphedema. Lymphoscintigraphy may show no detectable flow from the injection site, or there may be marked delay in flow and poor visualization of lymphatics and nodes in the affected extremity (Fig. 16-3). The hyperplastic form of primary lymphedema is less common.[25]

Nonne-Milroy disease (Fig. 16-4) is a form of primary lymphedema that exhibits an autosomal dominant pattern of inheritance. In addition to extremity lymphedema, associated findings can include chylous ascites, protein-losing enteropathy, and chylous pleural effusion. Jackson, Bowen, and Lentle[26] reported lymphoscintigraphic findings in three patients with Nonne-Milroy disease. All demonstrated clinically evident lower extremity edema, which was bilateral in two patients. Scintigraphic abnormality was present in all three patients, ranging from total obstruction of lymphatic flow to lymphatic pooling. The pelvic and periaortic nodes on the affected side, if seen, were small and few in number.

A form of lymphedema has been described in some patients with neurofibromatosis. Positive-contrast lymphangiographic findings have been described in two patients. In one report, a patient with gross enlargement of an upper extremity showed marked dilatation of lymphatic vessels and no visualization of lymphatic valves. In another report, a child with elephantiasis neuromatosis of a lower extremity was studied with positive-contrast lymphangiography. Dilated, irregular lymphatics and enlarged lymph nodes with prominent filling defects were demonstrated.[27–29]

Lymphoscintigraphic findings in a case of neurofibromatosis have also been reported (Fig. 16-5). The affected leg demonstrated dilated lymphatic channels and enlarged iliac lymph nodes. Because of the limited spatial resolution of lymphoscintigraphy, the filling defects in lymph nodes are not usually appreciated.[30]

Lymphatic abnormalities have been described in Noonan's syndrome, in which patients exhibit many of the phenotypic features of Turner syndrome. These patients have short stature, delayed puberty, craniofacial abnormalities, webbed neck, cardiovascular abnormalities, and lymphatic abnormalities. Reported lymphatic abnormalities include lymphedema, pulmonary lymphangiectasia, and intestinal lymphangiectasia.

Secondary Lymphedema

Secondary lymphedema may occur secondary to obstruction at any level of the lymphatic vessels or lymph nodes. Etiologies include surgery (particularly regional node dissection), trauma, neoplasm, radiation, and infection. Venous obstruction complicates many cases of secondary lymphedema.

Lymphoscintigraphy often demonstrates "dermal backflow" in cases of acquired lymphatic obstruction. Dermal lymphatics function as collateral pathways of lymph flow when major deep lymphatics are absent or obstructed. In this pattern, tracer is identified as extending up the superficial tissues of the affected extremity; normal deep lymphatic channels are usually not visible, or are decreased. Lymph node activity is typically decreased as well. The dermal backflow pattern is not specific for a particular disease, but may be observed with any of the various etiologies of acquired lymphatic obstruction in the extremities[31] (Figs. 16-6–16-8).

Lymphangioma

Congenital lymphangioma usually presents as a soft mass in an extremity. This lesion may be difficult to distinguish clinically from other vascular lesions. Differentiation is important, because the therapy is different. Radionuclide lymphography is useful in demonstrating lymphangiomas. The vascular phase typically shows tracer accumulation in the lymphangioma. Delayed images are usually normal, since the radiopharmaceutical is deposited in nodes and has cleared from the vascular compartment (Fig. 16-9).

Lymphatic Disruption

Lymphoscintigraphy is useful in documenting lymphatic leakage, whether congenital or acquired. Common sites of leakage include the thorax and abdomen. Chylous ascites may be primary (fistulous, exudative), or acquired (trauma, neoplastic); see the additional discussion below. A rent in a dilated lymphatic vessel leads to the formation of a fistula which can rarely be demonstrated with positive contrast lymphography. Lymphoscintigraphy will frequently show tracer accumulation in the peritoneal cavity, but it, too, will not identify the specific site of the fistula.

Lymphangietasia

In the exudative type of chylous ascites, the lymphatic system is dysplastic. An example of this malformation is lymphangiectasia. In this entity, no discrete site of lymphatic leakage can be found.

Intestinal

Intestinal lymphangiectasia was first described as a clinical entity in 1961. It presents with hypoproteinia, lymphopenia, and digestive symptoms of variable severity. Pathologic examination demonstrates telangiectatic lymphatic vessels in the submucosa of the intestine. The pathophysiology is not well understood; however, electron microscopy suggests obstruction to the flow of lymph. The obstruction is thought to increase the intraluminal pressure, resulting in dilatation of the lymphatic vessels.

In addition to the congenital variety, secondary lymphangiectasia of the intestine may occur with retroperitoneal fibrosis, tumors, scleroderma, pancreatitis, severe congestive heart failure, radiation enteritis, tuberculosis, and lymphatic malignancy. Lymphoscintigraphy in this condition after pedal injection demonstrates normal inguinal nodes, abdominal nodes, and liver; however, accumulation of radiopharmaceutical can also be identified in the gastrointestinal tract.[32]

Pulmonary

Pulmonary lymphangiectasia is a rare condition which produces respiratory distress and marked cyanosis in the neonate. Survival beyond the neonatal period is rare. Pulmonary lymphangiectasia in such cases usually is an isolated abnormality. However pulmonary lymphangiectasia can be associated with a more generalized lymphangiectasia and in infants with specific congenital heart lesions. This variability of presentation has resulted in a classification of pulmonary lymphangiectasia into three groups. One group consists of those infants with isolated abnormalities; another, those with associated congenital heart disease; and lastly, those infants with generalized lymphangiectasia. The first two groups are usually seen within the neonatal period. The infants with generalized lymphangiectasia show little radiographic pulmonary abnormality. Their major problem is usually the associated intestinal lymphangiectasia and protein-losing enteropathy.

The etiology of lymphangiectasia is believed to lie in an early embryonic arrest of pulmonary lymphatic development and development of large, dilated, obstructed lymphatic channels. In those cases where congenital heart disease and severe pulmonary venous obstruction are present, it is believed that lymphatic dilatation occurs because of abnormal lymphatic drainage secondary to the high venous pressure. Pathologic examination in cases of pulmonary lymphangiectasia shows extremely edematous lungs with numerous overdistended, even cystic, lymphatic channels scattered throughout the pulmonary parenchyma and subpleural spaces. Lymphoscintigraphy in cases of pulmonary lymphangiectasia will show tracer accumulation in regions of dilated lymphatics within the lung[33] (Fig. 16-10).

Chylous Ascites

Acquired chylous ascites may develop secondary to trauma or tumor with complete lymphatic obstruction. Radionuclide lymphography demonstrates no visualization of the abdominal nodes, liver, and spleen. Instead, the radiopharmaceutical accumulates in the peritoneal cavity. Currently, lymphoscintigraphy can be used to detect chylous ascites but cannot distinguish the etiology.

Chylothorax

The etiologies of chylothorax are similar to those of chylous ascites: idiopathic, neoplasm, and trauma. The congenital or idiopathic varieties appear to respond to conservative therapy, such as multiple thoracentesis. When associated with malignancy, chylothorax is usually indicative of widespread disease or mechanical obstruction of the lymphovenous circulation and conservative management is usually warranted. Those cases due to trauma may be secondary to cardiac surgery, subclavian catheterization, or neck surgery, and require clinical judgment in management. An initial nonoperative treatment of either repeated thoracentesis or tube thoracostomy is recommended, with surgery reserved for patients who have either no decrease in chylous drainage after 2 weeks or suffer electrolyte and nutritional complications. Lymphoscintigraphy has two roles in the evaluation of chylothorax. First, it can confirm that the fluid in the chest is chyle, if subcutaneously injected tracer flows into the involved hemithorax. Second, it can be used to quantify the amount of chyle accumulating in the chest and in this way be used to assess the results of conservative management[34] (Fig. 16-11).

Renal Transplantation

Radionuclide lymphoscintigraphy can also be used to evaluate lymphatic abnormality following renal transplantation.[35,36] It is possible to image the surgical lymphatic interruption, quantify flow, and characterize the precise nature and origin of the abnormality.

Metastasis

In lymphatic obstruction due to metastatic tumor, lymph flow is slower than normal and many nodes will not accumulate tracer, secondary to replacement with malignancy (Figs. 16-12–16-15). However, lymphoscintigraphy usually provides clinically less satisfactory results in children with malignant abdominal lymphadenopathy than ultrasonography, computed tomography, and positive contrast lymphography. Lymphoscintigraphy lacks the spatial resolution of these procedures for investigating lymph node involvement and it cannot be substituted for them in evaluating malignant lymph node disease.

Summary

Currently, interstitial lymphoscintigraphy is a valuable, but uncommonly employed technique in children. Its role is primarily restricted to the study of lymphatic leakage (congenital or acquired), and to the etiologic investigation of lymphedema. It is of limited value in evaluating lymph node disease from metastatic abnormalities, due to the poor spatial resolution.[23]

REFERENCES

1. Hahn PF, Sheppard CW: Selective radiation obtained by the intravenous administration of colloidal radioactive isotopes in diseases of the lymphoid system. South Med J 39:558, 1946
2. Sheppard CW, Goodell JPB, Hahn PF: Colloidal gold containing the radioactive isotope Au-198 in selective internal radiation therapy of disease of the lymphoid system. J Lab Clin Med 32:1437, 1947
3. Walker LA: Localization of radioactive colloids in lymph nodes. J Lab Clin Med 36:440, 1950
4. Hauser W, Atkins HL, Richards P: Lymph node scanning with 99mTc sulfur colloid. Radiology 92:1369, 1969
5. Cox PH: Dynamic studies of lymph drainage using labelled human serum albumin. Proceedings of 12th Annual Meeting Jugoslav Soc Nucl Med Ohrid 76, 1972
6. Schenck P, Wenkel K zum, Becker J: Die szintigraphie des parasternalen lymph system. (Scintigraphy of the parasternal lymph system). Nucl Medizen 5:388, 1966
7. Davis MA, Jones AG, Trindade H: A rapid and accurate method for sizing radiocolloids. J Nucl Med 15:923, 1974
8. Szymendera J, Zoltowski T, Radwan M et al: Chemical and electron microscope observations of a safe PVP-stabilized colloid for liver and spleen scanning. J Nucl Med 12:212, 1971
9. Cox PH: The preparation of mercuric sulphide suspension for lymphatic investigations and the influence of particle size on its physiological properties. Radioisotopy 12:997, 1971
10. Warbick A, Ege GN, Henkelman RM et al: An evaluation of radiocolloid sizing techniques. J Nucl Med 18:827, 1977
11. Ege GN, Warbick A: Lymphoscintigraphy: A comparison of 99Tc(m) antimony sulphide and 99Tc(m) stannous phytate. Br J Radiol 52:124, 1979
12. Goodwin DA, Finston RA, Colombetti LG et al: ^{111}In for imaging: Lymph node visualization. Radiology 94:175, 1970
13. Garzon OL, Palcos MC, Radicella R: A technetium-99m labeled colloid. Int J Appl Radiat Isot 16:613, 1965
14. Harbert JC, Pollina R: Absorbed dose estimates from radionuclides. Clin Nucl Med 9(4):210, 1984
15. Sage HH, Sinha BK, Kizilay D et al: Radioactive colloidal gold measurements of lymph flow and functional patterns of lymphatics and lymph nodes in the extremities. J Nucl Med 5:626, 1964
16. Pearlman AW: Abdominal lymph node scintiscanning with radioactive gold (Au-198) for evaluation and treatment of patients with lymphoma. Am J Roentgenol Radium Ther Nucl Med 109:780, 1970
17. Fairbanks VF, Tauxe WN, Kiely JM et al: Scintigraphic visualization of abdominal lymph nodes with 99mTc-pertechnetate labeled sulfur colloid. J Nucl Med 13:185, 1972
18. Hauser W, Atkins HL, Richards P: Lymph node scanning with 99mTc-sulfur colloid. Radiology 92:1369, 1969
19. Kazem I, Antoniades J, Brady LW et al: Clinical evaluation of lymph node scanning utilizing colloidal gold 198. Radiology 90:905, 1968
20. Drewett J: Lymph node scanning and imaging using colloidal 198 Au. J Nucl Med 14:471, 1973
21. Gates GF, Dore EK: Primary congenital lymphedema in infancy evaluated by isotope lymphangiography. J Nucl Med 12:315, 1971
22. Dunson GL, Thrall JH, Stevenson JS et al: 99mTc minicolloid for radionuclide lymphography. Radiology 109:387, 1973
23. Sty JR, Starshak RJ: Atlas of pediatric radionuclide lymphography. Clin Nucl Med 7:428, 1982
24. Kinmonth JB, Taylor GW, Tracy GD, March JD: Primary lymphoedema. Br J Surg 45:1, 1957
25. Chilvers AS, Kinmonth JB: Operations for lymphoedema of the lower limbs. A study of the results in 108 operations utilizing vascularized dermal flaps. J Cardiovasc Surg 16:115, 1975
26. Jackson FI, Bowen P, Lentle BC: Scintilymphangiography with 99mTc-antimony sulfide colloid in hereditary lymphedema (Nonne-Milroy disease). Clin Nucl Med 3:296, 1978
27. Preston JM, Starshak RJ, Oechler HW: Neurofibromatosis: Unusual lymphangiographic findings. AJR 132:474, 1979
28. Casselman ES, Miller WT, Lin SR et al: Von Recklinghausen's disease: Incidence of roentgenographic findings with a clinical review of the literature. CRC Crit Rev Diagn Imaging 9:387, 1977
29. Lanstov VP, Kramorev VA: State of the lymphatic system in neurofibromatosis (according to roentgenolymphographic data). Vestn Rentgenol Radiol 2:99, 1976
30. Sty JR, Starshak RJ, Woods GA: Neurofibromatosis: Lymphoscintigraphic observations. Clin Nucl Med 6:264, 1981
31. Sty JR, Boedecker RA, Scanlon GT et al: Radionuclide "dermal backflow" in lymphatic obstruction. J Nucl Med 20:905, 1979
32. Soucy JP, Eybalin MC, Taillefer R et al: Lymphoscintigraphic demonstration of intestinal lymphangiectasia. Clin Nucl Med 8:535, 1983

33. Sty JR, Thomas JP Jr, Wolff MH, Litwin SB: Lymphoscintigraphy. Pulmonary lymphangiectasia. Clin Nucl Med 9:716, 1984

34. Gates GF, Dore EK, Kanchanapoom V: Thoracic duct leakage in neonatal chylothorax visualized by 198 Au lymphangiography. Radiology 105:619, 1972

35. Jackson FI, Lentle BC, Higgins MR et al: Lymphocutaneous fistula following renal transplantation. Clin Nucl Med 5:19, 1980

36. Sty JR, Thomas JP Jr, Abrahams J: Radionuclide lymphography: A demonstration of surgical lymphatic interruption. Clin Nucl Med 3:412, 1978

Atlas Section

Figure 16-1 **_Normal Lymphoscintigraphy_**

(A) Anterior view of the lower extremity during the vascular phase demonstrates normal lymphatic vessels. Five-minute delayed images of the lower extremities *(B)* and pelvis and abdomen *(C)* demonstrate clearing from the vascular channels, as well as early lymph-node filling.

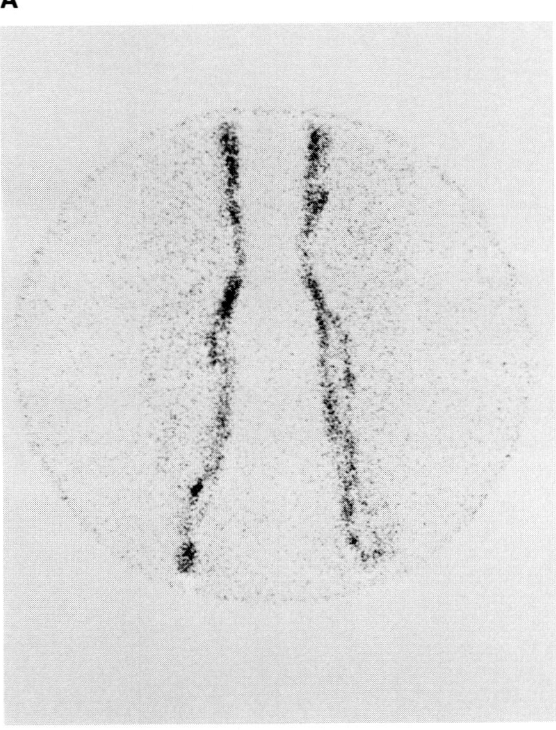

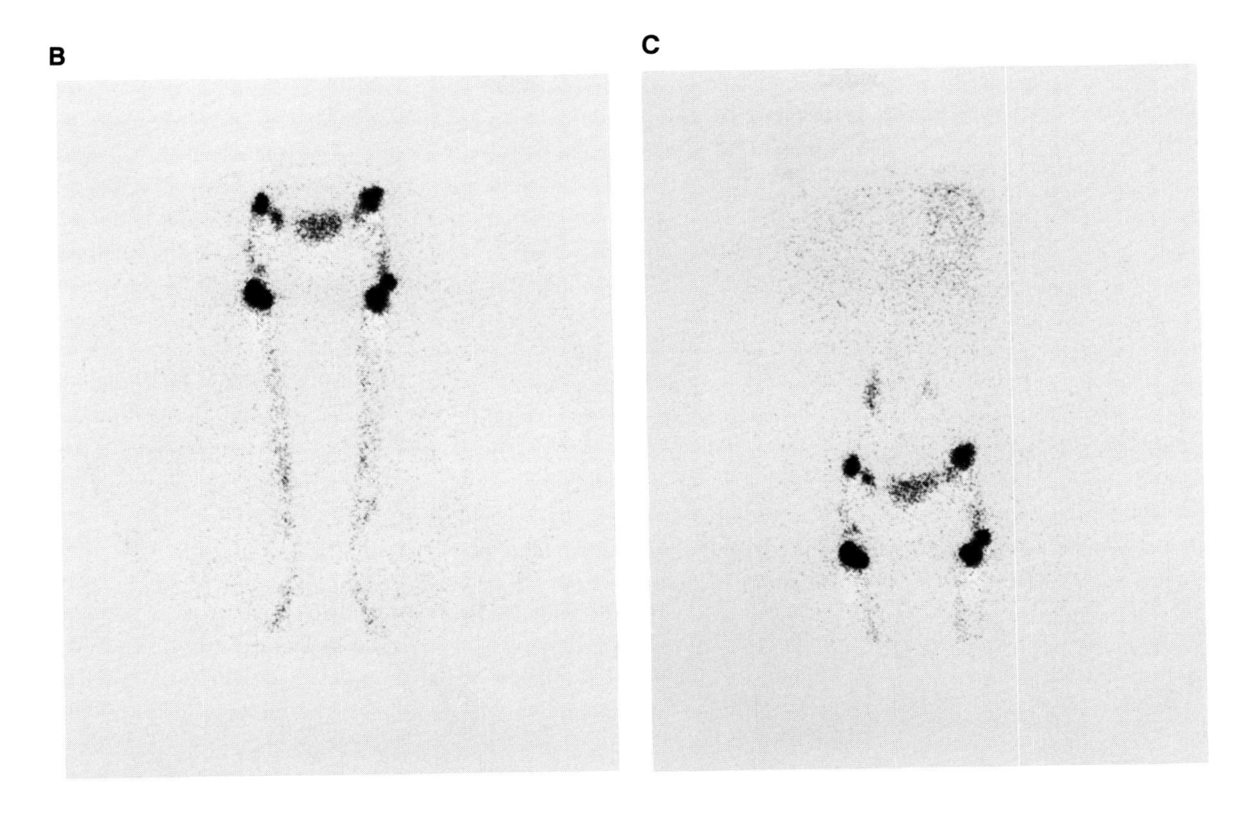

Figure 16-2 ***Normal Lymphoscintigraphy***

Posterior view of the abdomen and pelvis is from a normal lymphoscintigraphy performed in a 2-year-old child. (*S*, spleen; *L*, liver; *P*, posterior view) (Sty JR, Starshak RJ: Atlas of Pediatric Radionuclide Lymphography. Clin Nuc Med 7:428, 1982)

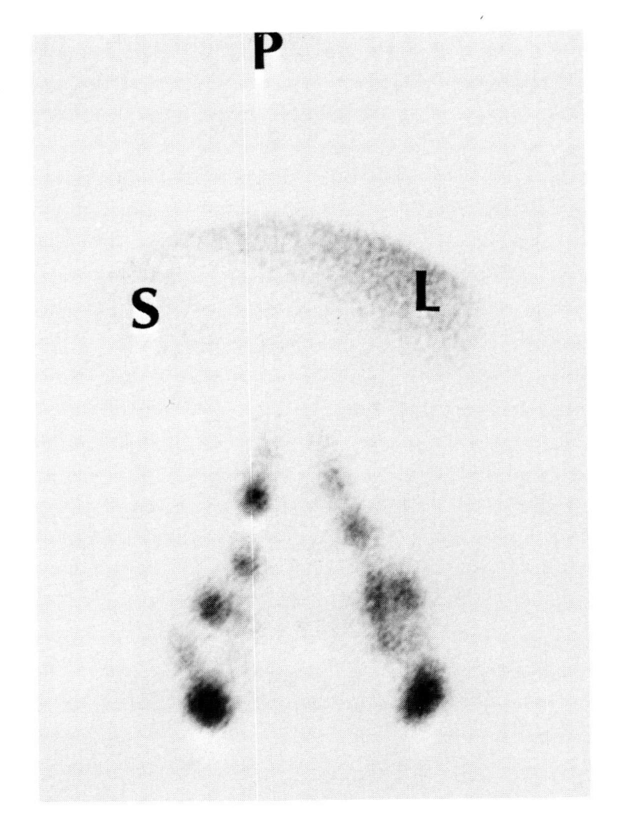

Figure 16-3 ***Congenital Lymphedema***

Anterior views of the lower extremities *(A)* and pelvis *(B)* in a 10-year-old child with hypoplastic lymphedema are notable for the lack of lymphatic flow from the injection site in the left foot and the absence of left pelvic node filling.

A

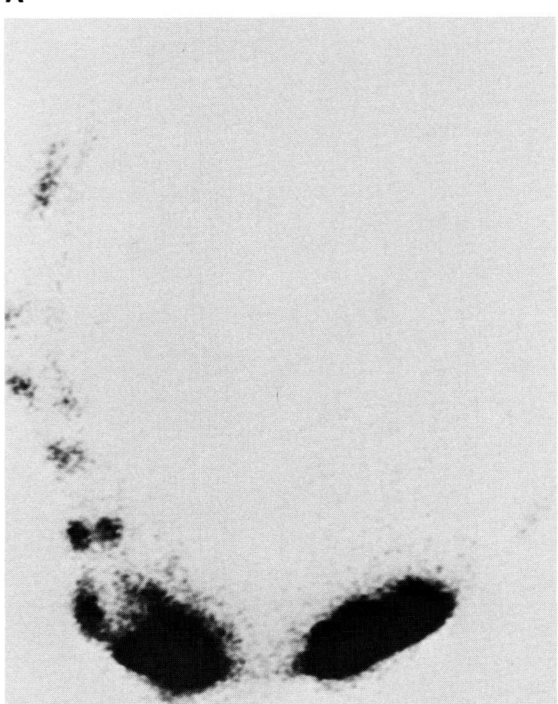

B

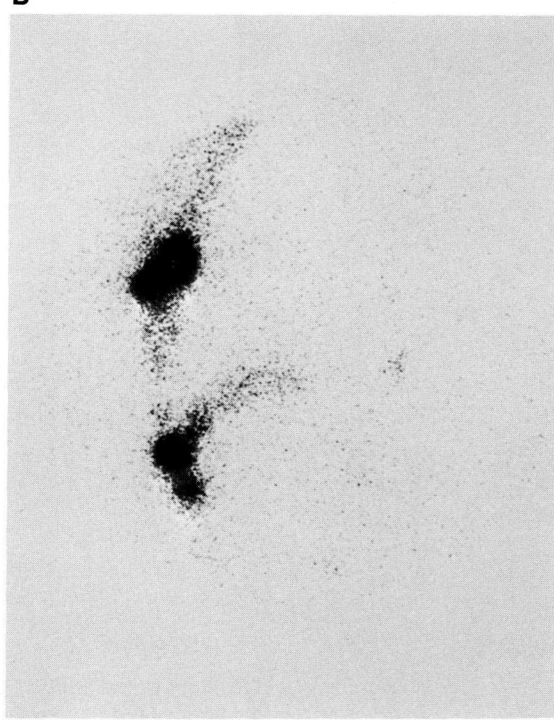

Figure 16-4 ***Nonne-Milroy Disease***

(A) Anterior view of the vascular phase of the examination demonstrates early pooling of the radiopharmaceutical in the lymphatics of the left calf *(arrows)* and a lack of left pelvic node filling. *(B)* The delayed image demonstrates filling of the right iliac and abdominal nodes without appearance in the left pelvic nodes.

A

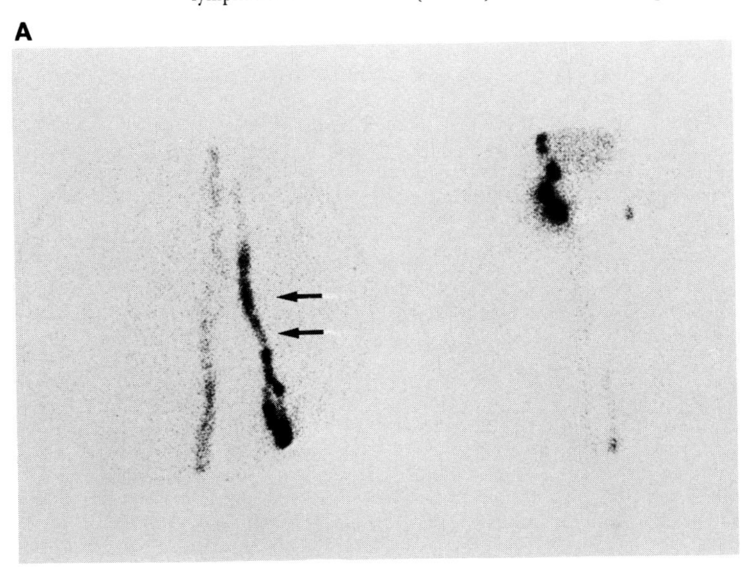

B

Figure 16-5 ***Neurofibromatosis and Elephantiasis***

(A) The immediate anterior view obtained of the right leg of this 8-year-old child demonstrates an enlarged lymphatic channel. ***(B)*** The delayed posterior pelvic view demonstrates enlarged nodes. These abnormalities were secondary to neurofibromatosis and elephantiasis. (Sty JR, Starshak RJ, Woods BS: Neurofibromatosis: Lymphoscintigraphic Observations. Clin Nuc Med 6:264, 1981)

A

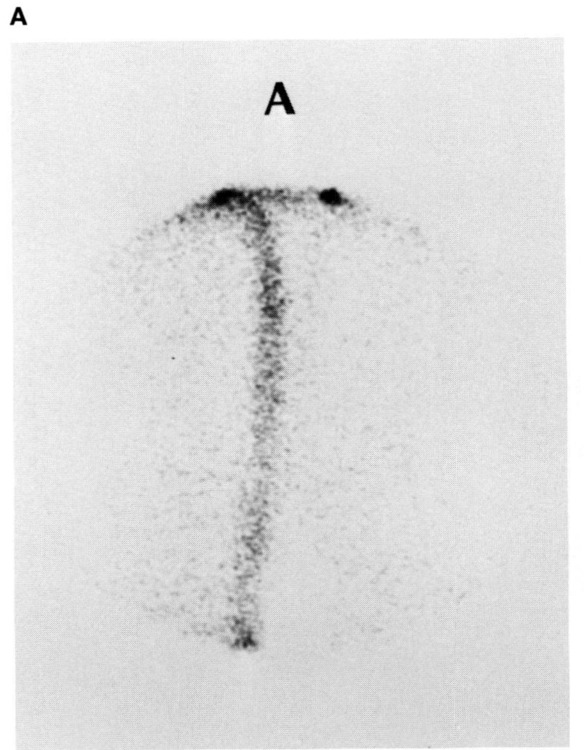

B

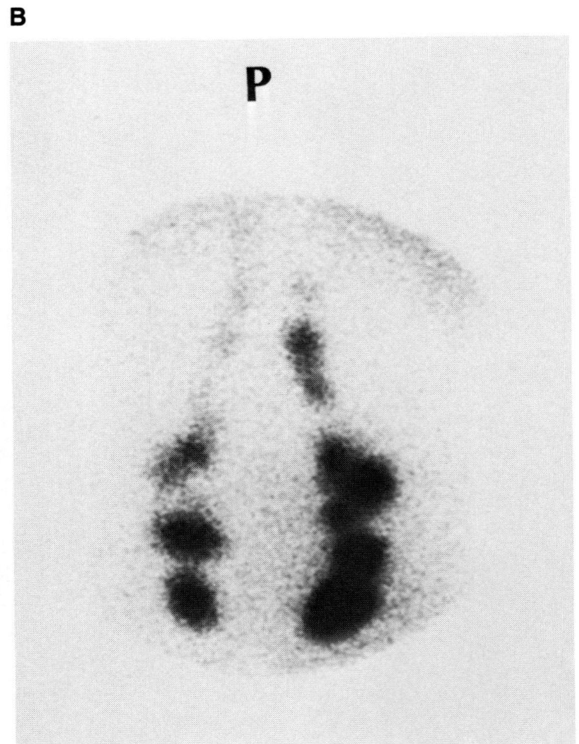

Figure 16-6 ***Acquired Lymphatic Obstruction Secondary to Lymph Node Dissection***

Lymphoscintigraphic anterior views of the calves *(A)* and pelvis *(B)* of a 12-year-old boy with left femoral node dissection demonstrate dermal backflow in the left leg *(A, arrow)* and fewer lymph nodes at the surgical site *(B, arrow; L = left)*. The dermal backflow indicates obstruction. (Sty JR, Starshak RJ: Atlas of Pediatric Radionuclide Lymphography. Clin Nuc Med 7:428, 1982)

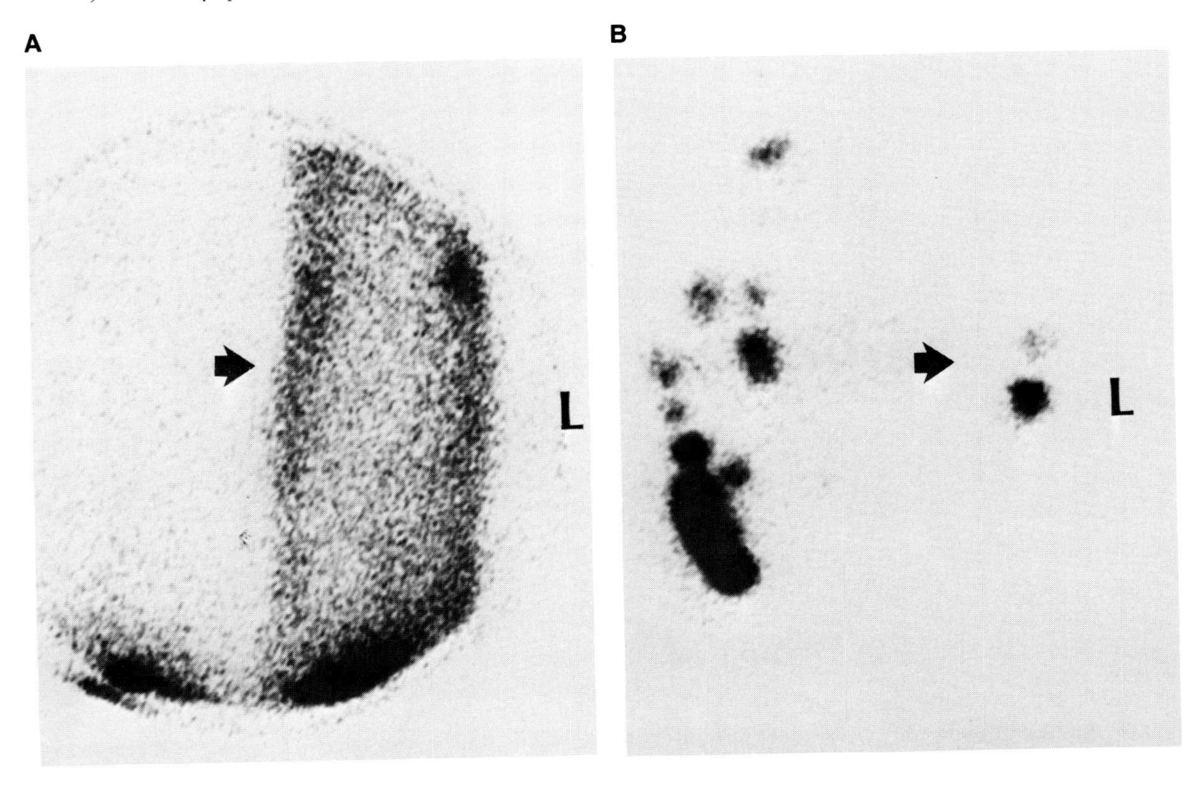

Figure 16-7 ***Acquired Lymphatic Obstruction Secondary to Pelvic Irradiation***

These images are of a 14-year-old child with bilateral lower extremity edema after pelvic irradiation. The anterior views of the calves *(A)* and midthighs *(B)* show engorged lymphatic vessels, which were attributed to changes secondary to pelvic irradiation.

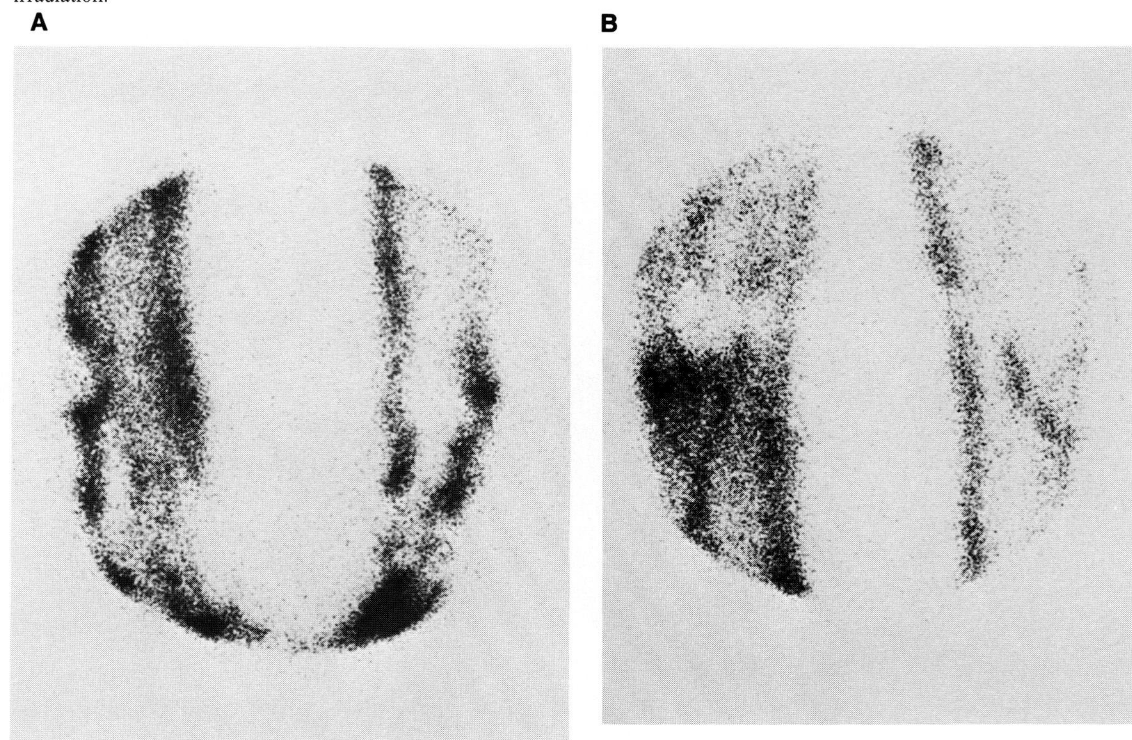

Figure 16-8 ***Acquired Lymphatic Obstruction Secondary to Hidradenitis***

Lymphoscintigraphic images of the right arm *(R)* and left arm *(L)* in an 8-year-old child demonstrate lymphatic engorgement and patchy tracer deposition in the nodes of the left arm *(arrow)*. This was attributed to hidradenitis of the left axilla.

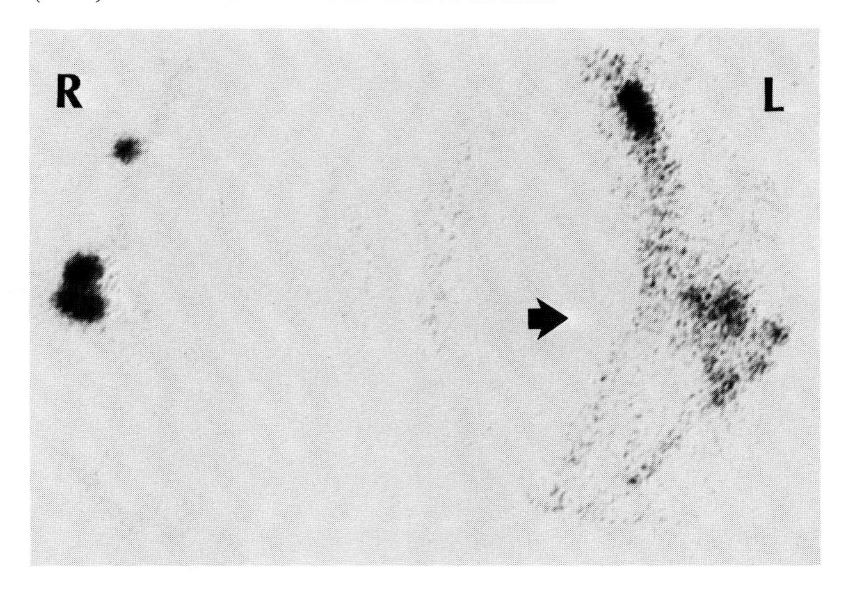

Figure 16-9 ***Lymphangioma***

The immediate images *(upper)*, which were obtained during the vascular phase in an 8-year-old boy, demonstrate radioactivity in the lower aspect of the left lateral thigh *(arrows)*. Five-minute delayed images *(lower)* show clearing of the radioactivity. The abnormal radioactivity was secondary to pooling of radioactivity within a lymphangioma. (Wells RG, Ruskin JA, Sty JR: Lymphoscintigraphy: Lower Extremity Lymphangioma. Clin Nuc Med 11:523, 1986)

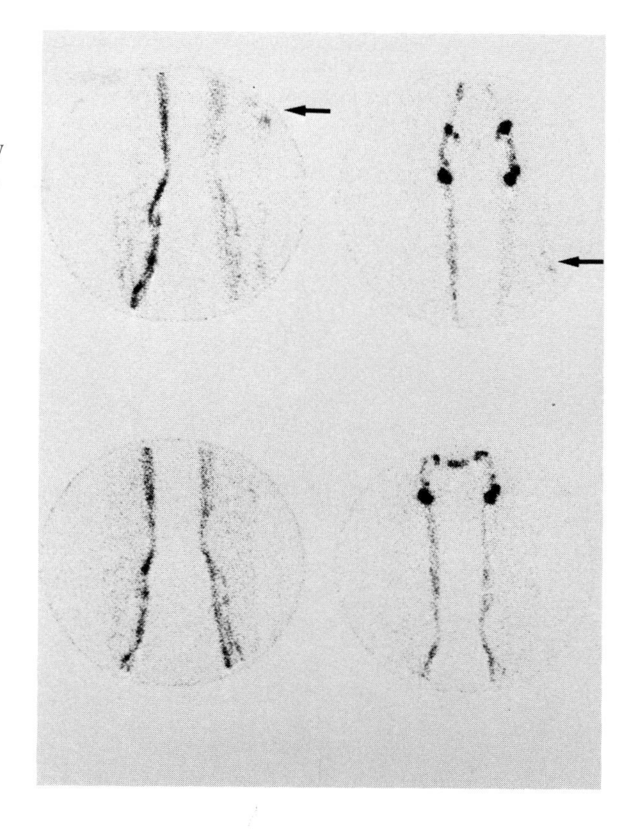

Figure 16-10 ***Lymbangiectasia***

This anterior chest image was obtained after a pedal injection. Of note is the tracer accumulation in the nodes at the base of the neck and in the right hilus *(arrow)*. (Sty JR, Thomas JP Jr, Wolff MH, Litwin SB: Lymphoscintigraphy: Pulmonary lymphangiectasia. Clin Nuc Med 9:716, 1984)

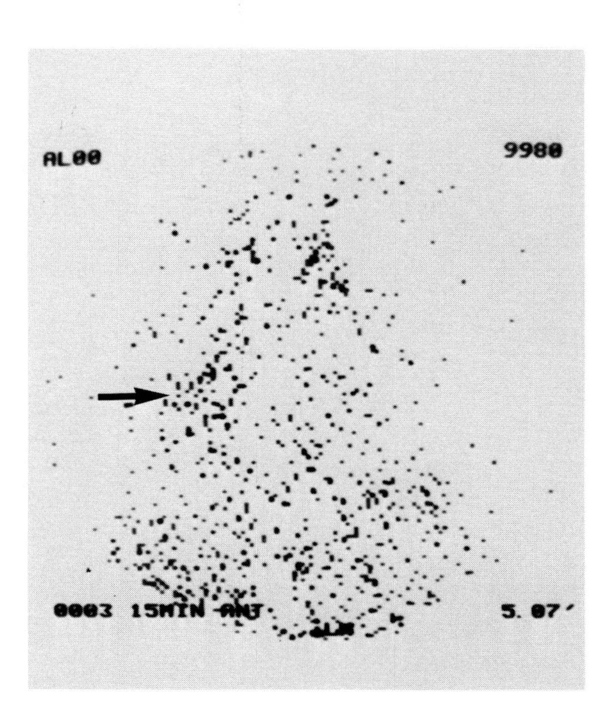

Figure 16-11 ***Congenital Chylothorax***

This posterior *(P)* view demonstrates the radiopharmaceutical in the right thorax *(arrow)*, indicating a chylothorax. (Sty JR, Starshak RJ: Atlas of Pediatric Radionuclide Lymphography. Clin Nuc Med 7:428, 1982)

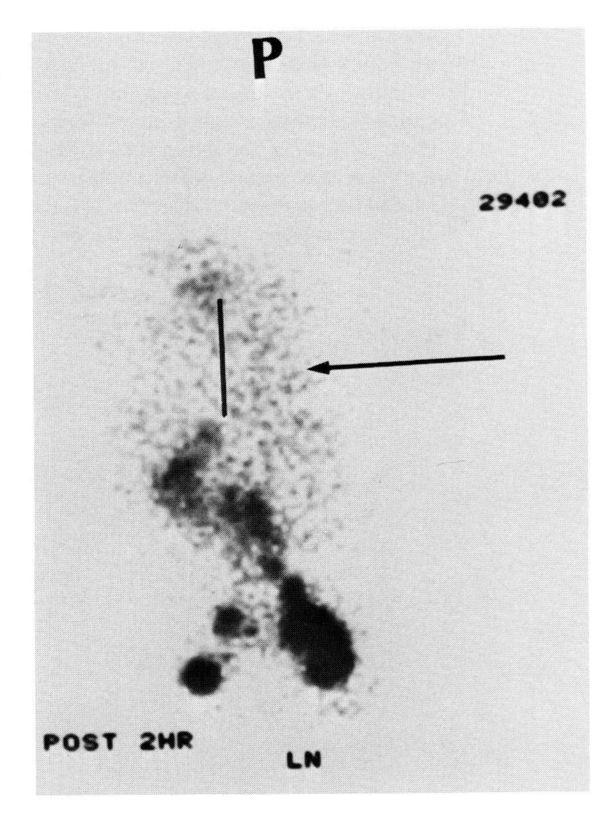

Figure 16-12 ***Nodal Metastasis***

Lymphoscintigraphy was performed in a child with nodal metastatic neuroblastoma. ***(A)*** The anterior view of the pelvis demonstrates lateral displacement of the iliac nodes *(arrows)*, which was caused by an enlarged periaortic node that had been replaced with tumor and did not accumulate the radiotracer. This is an example of the "flare sign." ***(B)*** In the abdomen no radioactive tracer is noted within the abdominal nodes, compatible with replacement by metastasis. ***(C)*** Serendipitously, the image of the liver demonstrated photon deficiencies *(arrow)*, which were secondary to liver metastasis. (Sty JR, Starshak RJ: Atlas of Pediatric Radionuclide Lymphography. Clin Nuc Med 7:428, 1982)

A

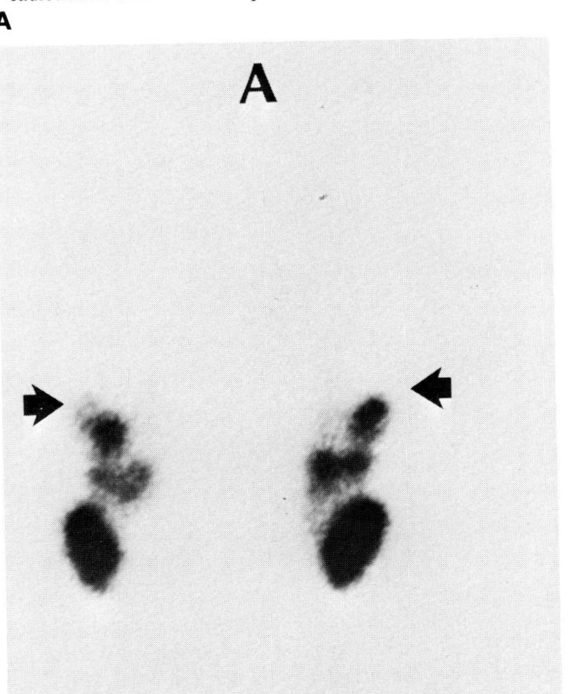

B

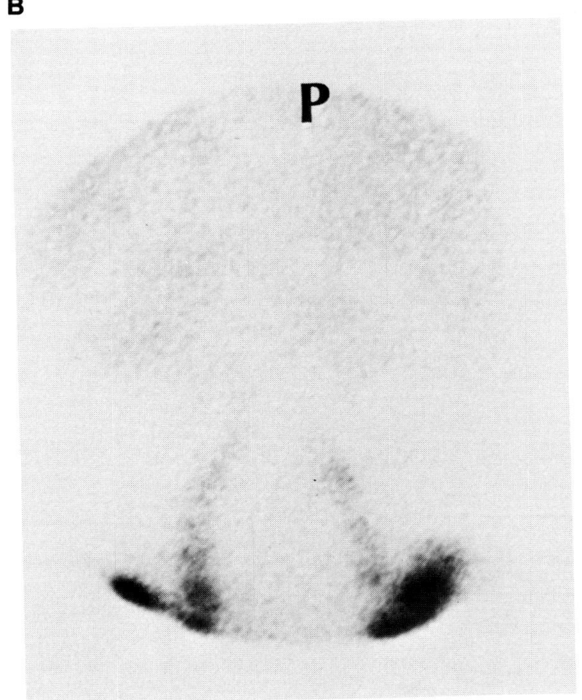

C

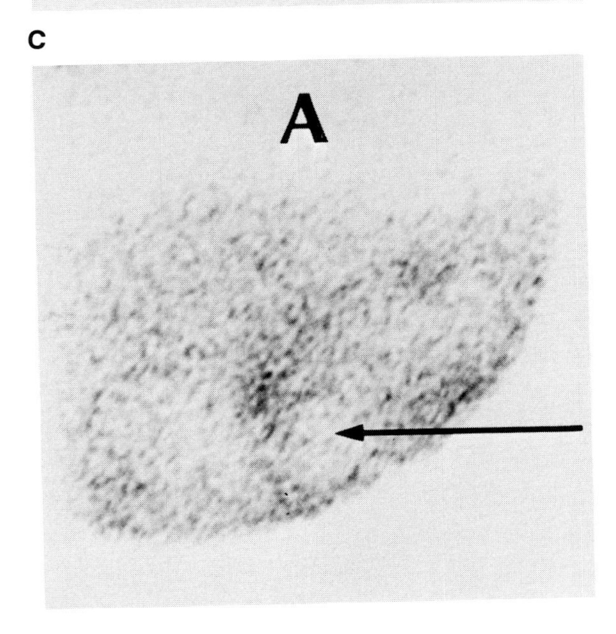

Figure 16-13 *Nodal Metastasis*

These are anterior *(A)* and posterior *(P)* images of the abdomen in a 3-year-old girl with metastatic lymph node disease. The abdominal lymph nodes are large, displaced, and partially replaced by apparent metastasis. (Sty JR, Starshak RJ: Atlas of Pediatric Radionuclide Lymphography. Clin Nuc Med 7:428, 1982)

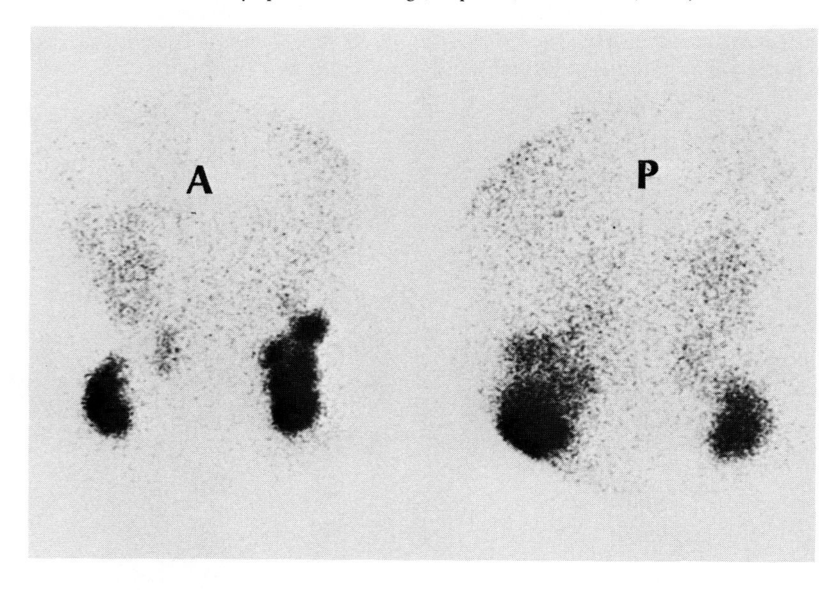

Figure 16-14 ***B-cell (American Burkitt) Lymphoma***

(A) This posterior *(P)* view of the pelvis demonstrates nonvisualization of the left iliac nodes *(arrow)* compatible with involvement by the lymphoma. ***(B)*** The posterior *(P)* view of the abdomen demonstrates enlarged and partially replaced nodes. (Sty JR, Starshak RJ: Atlas of Pediatric Radionuclide Lymphography. Clin Nuc Med 7:428, 1982)

A

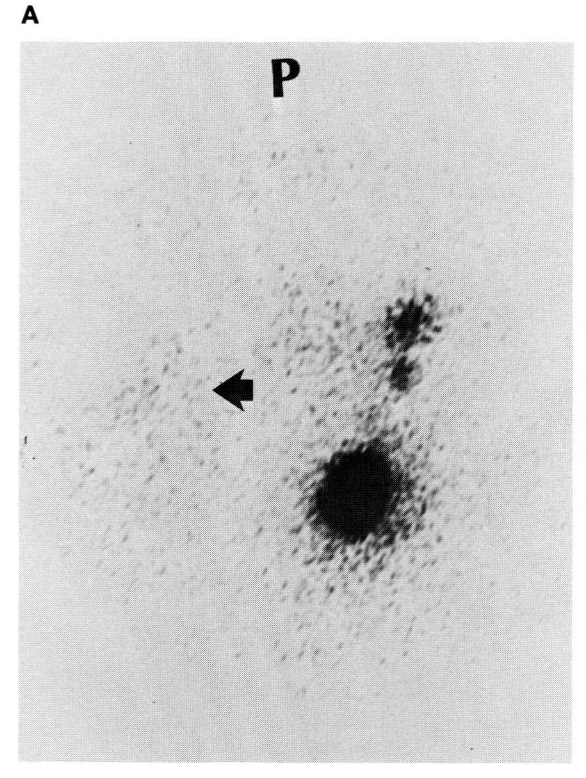

B

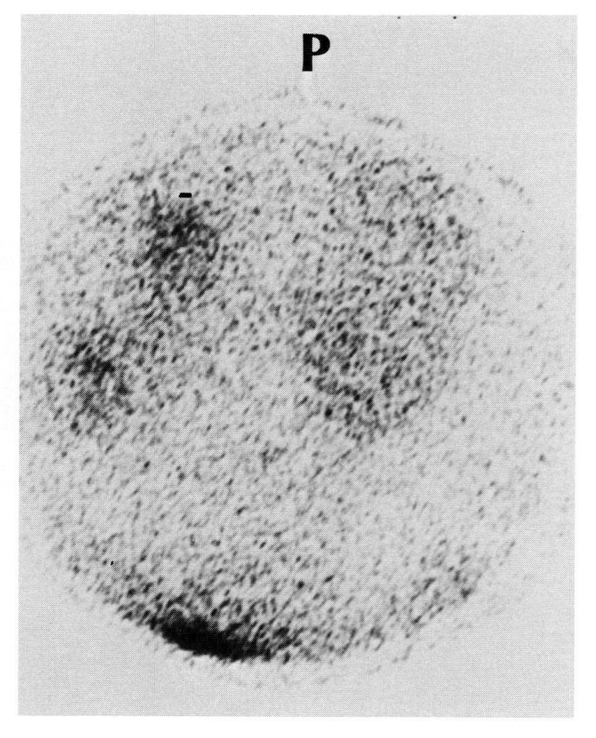

Figure 16-15 ***Hodgkin's Disease***

(A) The anterior pelvic view in a 13-year-old boy with Hodgkin's disease demonstrates minimal filling of a left iliac lymph node *(arrow)*. *(B)* Abnormalities of this type are better demonstrated with contrast lymphangiography *(arrow indicates nodes involved with metastatic lymphoma)*. (Sty JR, Starshak RJ: Atlas of Pediatric Radionuclide Lymphography. Clin Nuc Med 7:428, 1982)

A

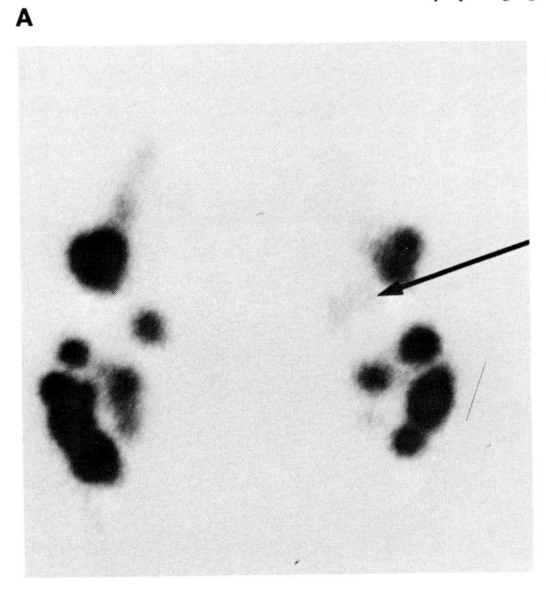

B

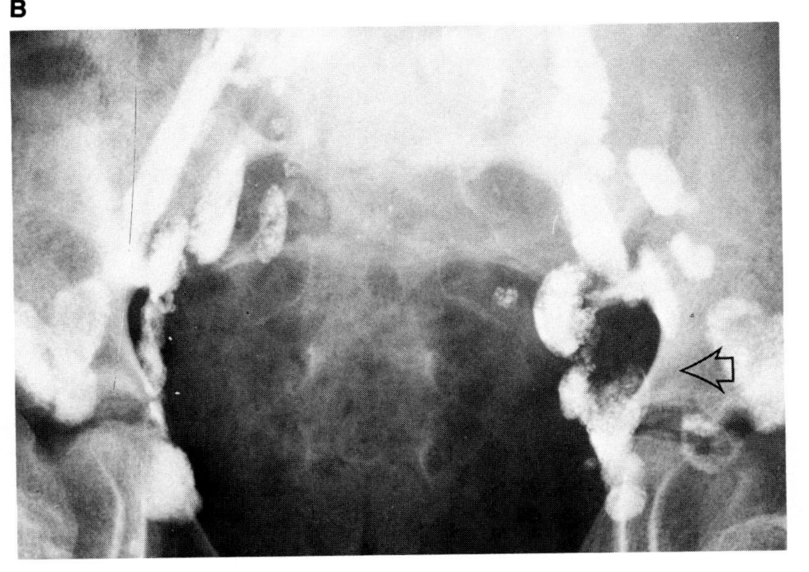

Index

Page numbers in *italics* denote figures.
Page numbers followed by "t" denote tables.

A

Abscesses. *See also* specific locations
- In-111 uptake in, 301
- occult, *310*
- soft tissues, *317*
- WBC imaging, *312*

Achalasia, 1–6, *19, 20, 22*
Acid reflux test, *22*
Acquisition protocol, 2–3, 8, *126*
ACTH. *See* Adrenocorticotrophic hormone (ACTH)
Adenocarcinoma. *See* specific organs
Adenoma. *See also* specific organs
- of adrenal glands, 52–62
- aldosteronism and, *65*
- histological types, 198
- hyperplasia and, 56
- parathyroid, 115

Adrenal gland
- adenoma of, 52–59, *60–71*
- adrenalectomy and, *96–102*
- beta blockade of, 73
- cyst in, *68*
- functional variations, 55t
- hyperfunction of, 57
- medulla scanning, 72–84, *88–114*
- MIGB uptake, 76–77, 84, *88–114*
- ovarian tumors and, 54
- pheochromocytoma, *102*
- radioactivity and, *61*

Adrenocorticotrophic hormone (ACTH)
- adenoma and, *65*
- albumin and, 2, 4
- aldosterone and, 52, 53t, 57, *65*
- Cushing's syndrome, 53t
- LDL and, 55
- levels of, 62
- zona glomerulosa (zG) and, 52

Aldosteronism, 52, 57, *61, 65*
Alkalosis, 53t
Alpha adrenergic agonists, 73
Alpha-methyl paratyrosine, 74
Amenorrhea, in hyperandrogenism, *66*
Americum-241, 177–178
Amine precursors, 74
Amiodarone, 181
Amyloid, in thyroid carcinoma, 199
Analog images
- DPTA and, 202

Analog images (*continued*)
MIGB and, 75
parathyroid scanning, 120
RETS and, 2, 5, *10–11*
Tc-99m DTPA and, 202
Anaplastic carcinoma, 199
Androgens, abnormal levels of, 53–54
Anger scintillation camera, 239, 331
Angiography
anastomosis of SVC and RPA, *349*
angiocardiography and, *352*
aortic valvular stenosis, *354*
azygos vein, *346*
cardiac catheterization and, 333
digital subtraction, *135*
idiopathic hypertrophic subaortic stenosis, *353*
inferior vena cava, *346*
myocardial infarction, *357*
myocarditis, *355*
neuroendocrine tumors, 81
normal angiocardiogram, *345*
parathyroid, 122
right pulmonary artery, *347*
right ventricular overload, *355*
scrotum and, 273
transposition of vessels, *348*
tricuspid atresia, *346*
Angiotensinogen, 52
Anticoagulation therapy, *257*
Antihypertensives, adrenal function and, 55t
Antimetabolites, in Cushing's syndrome, 58
Antireflux procedure, *26*
Antithyroid antibodies, 197–199
Aorta
acute tubular necrosis, *214*
A-V fistula, *213*
K/A ratio and, *203*
normal angiogram of, *345*
perfusion of, 273
renal artery thrombosis, *212*
RTS, *209–210*
stenosis of, 336, 338, *354*
valvular angiocardiography, *354*
Aperistalsis, in PSS, *28*
Appendicitis, In-111 WBC imaging, *320*
APUD. *See* Amine precursors
Arteriovenous fistula, *213*
Arthroscopy, 364
Artifacts, 119, *162, 192,* 362
ASD. *See* Atrial septal defect
Askanazy cells, 197
Aspiration. *See* Fine needle aspiration; specific organs
Atrial septal defect (ASD), 334, 337, *350*
Atrioventricular valve, 337
Autonomic nervous system, 72, 76, 84
Avascular necrosis, of femur, 363
Azygos vein, *346*

B

Baker cyst, 242, *267*
Barium, esophagus and, 4, 6, *33, 62*
Bell-clapper testis, 272, *276*
Bile duct disease, 41–51
atresia, 43t, *49*
obstruction, 42, 43t, *49–50*
Bilirubin level, imaging and, 41–42
Binucleation, in thyroid carcinoma, 199
Blalock-Taussig shunt, 333
Blood. *See* Blood pool; Cardiovascular system
Blood-brain barrier, MIGB uptake in, 76
Blood pool
cardiac (CBP), *47*
GBPI methods, 336–341, *353*
I-131 imaging, 145, *155*
labeling of, 239, 336
SPECT in, 361
of testis, 273
venography of, 240, 242, *247*
WBC imaging, 303
Bolus
movement of, 2, 6–7, *23*
renal artery thrombosis, *212*
Bone. *See also* Bone marrow
In-111 imaging in, 301
metastases in, *168*
pathological LBP and, 363
SPECT method, 360
Bone marrow
biopsy in, 74, 83
criticality of, 146–147
I-111 imaging, 301, *311*
red marrow as, 42t
Bowels, *61,* 78
Brain, metastases in, *170*
Breast cancer, thyroid metastases, 199
Bronchus
aspiration in, *38*
bronchoesophageal fistula in, *35*
bronchogenic carcinoma in, 199
bronchoscopy in, *35*
in OMD, *17*
Burkitt's lymphoma, *412*

C

Calcitonin, 82t, 198–199
Calf
lymphatics of, *405*
muscles of, *245–246*

occlusive thrombosis of, *251*
vein imaging of, 240
Canaliculi, 384, 387, *391*
Carcinoid tumors, 81–84, *110*
Cardia, adenocarcinoma, *37*
Cardiovascular system. *See also* specific organs; techniques
anomalies in, 340
autonomic innervation of, 84
cardiac blood pool and, *47*
cardiac cycle in, 332–336, *353*
catheterization and, 331
GBPI, 336, 337
MIGB uptake in, 112
myopathy in, 338
pheochromocytoma and, *99*
pulmonary circulation in, 334–335
radioactivity in, 145
radio-pharmaceuticals and, 334
RNA in, 332, 334
stroke-volume and, 337
Carney's triad, MIGB in, 79
Carotid body tumors, 82
Castration, spermatic chord and, 274
Catecholamines
assay of, 73
biosynthesis of, 74
epinephrine as, 72
excretion and, 72, 82
MIGB and, 77–78, 84
norepinephrine, 74, 76, 88
pheochromocytoma, 80
Cellulitis, 242, *264*
Cervical lymph node, *163*
Cervical sympathectomy, 76
Chelating agents, for WBC labeling, 300
Chemodectomas, 82t, *110*
Chest. *See also* specific organs
CT scan, *99*
lymphogram of, *409*
pain in, 7, *24*
scintigraphy of, 144–149, *150–175*
Children
cardiology in, 331–342, *343–359*
edema in, 397
lymphography in, 396–401, *402–411*
thyroid examination in, 178
Cholecystitis
acalculous, *47*
acute, 43t, *47*
cholecystectomies for, *47*
chronic, 43t
In-111 WBC imaging, *316*
intrahepatic, 43t
suppurative, *316*
Cholescintigram, 46
Cholestasis, drug-induced, *51*
Cholesterol, 55, 58
Choriocarcinoma, 82t
Chromaffin cells, 72
Chylothorax, 399, *410*
Chylous ascites, 398–399
Cisterna chyli, PRL and, 396
Clonidine, catecholamines and, 73
Cobalt, *156,* 272
Cocaine, MIGB uptake and, 76
Colitis, WBC imaging, *320*
Collecting system activity, *221, 225*
Collimator
in DCG, 384
fanbeam collimation, 362
in HB imaging, 41
for I-123 MIGB, 75
for I-131 imaging, 146, *152*
in parathyroid imaging, 116, *131*
pinhole system, *152,* 397
in RBC venography, 239
for RNI of scrotum, 272
in SPECT, 361
in WBC imaging, *30*
Colloids, thyroid and, *187,* 198–199
Colostomy, In-111 WBC imaging, *323*
Common duct obstruction, 43t, *49–50*
Common iliac veins, 241
Compton diffusion, 177
Computer tomography (CT). *See also* specific techniques
abdominal, *97*
abscesses and, *310*
AG imaging, 54
chest, *99*
digital images, 2–3, 75, 120, 273
GBPI analysis, 336–337
In-111 labeling and, 303
intra-arterial studies, 122
in MIGB, 75, *99*
minicomputers in, 54, 117
neuroblastomas and, 74
neuroendocrine tumors and, 81
in parathyroid scanning, 122
in pheochromocytomas, 73, 82–83, *99*
RETS, 2–3
subtraction methods, 119–120
of TMJ, 364
Congenital heart anomalies, RNA and, 335
Congestive heart failure, 335
Conjunctival sac, 383, 385. *See also* Lacrimal drainage apparatus (LDA)
Coronary arteries, 341, 351
Corpora cavernosa, *280*
Cor pulmonale, 339, *356*
Corticotropin-releasing hormone (CRH), 52

Courvoisier's law, *47*
Creatinine clearance, MIGB and, 76
Cremasteric muscle, 272, *279*
Cricopharynx, 7, *14*
CT. *See* Computer tomography
Cushing's syndrome
  adenoma in, *62*
  AG imaging and, 54
  antimetabolites, *69*
  carcinoma in, *64*
  clinical findings in, 52–53, 57
  hyperplasia in, *61, 63*
  scans in, 56
Cynotic heart disease, 333
Cystic duct, 43t, *47*
Cystic fibrosis, 335–336, 339
Cytotechnologist, 194

D

Dacryocystography (DCG), 383–384, *386–395*
  canaliculus obstruction in, *392*
  dacryocystorhinostomy in, *394–395*
  LDA blockage in, *388*
  patient positioning in, *388*
  radiation absorbed dose in, 384
  SAC junction in, *392*
  tearing and, *394*
  technique in, *388*
DCG. *See* Dacryocystography
Deep vein thrombosis (DVT)
  acute, 242
  bilateral, 241, *255*
  calf, *253, 255–256, 259*
  cellulitis in, 242, *264*
  chronic, 242, *261*
  drainage in, 240
  femoral, *253–256*
  iliac vein, 241, *259*
  occlusive, *251*
  progressive, *257*
  pulmonary embolization, 242, *259*
  resolving, *258–259*
DeQuevervain's subacute thyroiditis, 180
Dexamethasone, aldosteronism and, 53–57, *61*
Dialysis catheters, WBC imaging and, 301–304
Dioxyribonucleic acid (DNA), 197
Disofenin, 41
Diuretics, 55t, 58, *70*
Diverticulitis, WBC imaging and, *319*
DNA. *See* Dioxyribonucleic acid
Dopamine, synthesis of, 74
Dosimetry, 75–77, 83, *118*
Doughnut sign, *101*
Downscatter, imaging and, 119–120
Dual isotope subtraction, 119, *125–126*
Duodenal activity, in HB imaging, 41
DVT. *See* Deep vein thrombosis
Dynamic rotating display, 79
Dysphagia, 2–6
  antireflux procedures and, *26*
  in DES, *33*
  esophageal, 3, *14, 19–20, 25*
  lusoria, *40*

E

Echography, 179, 335
Edema, *205, 227,* 397
Effective renal plasma flow (ERPF), 202, *203–214*
  in chronic renal rejection, *226*
  excretion index, *209*
  in normal RTS, *210, 214*
  in transplant rejection, *221, 225*
Ejection fraction, RN technique and, 335
Electrocardiogram, 336, 341
Electron microscopy, 81, 83
Elephantiasis, *406*
Embolism, pulmonary, 242, 341–342
Emptying time, 3
Empyema, WBC imaging and, *321*
Encapsulated carcinoma, 180
Endoluminal catheter, 6
Endoscopy, in DES, *22*
Enteropathy, 399
Epidermoid carcinoma, RETS in, *38*
Epididymis
  appendix of, *286, 288*
  epididymitis in, *283, 286, 288–291*
  missed torsion and, *283*
  normal anatomy of, *285*
  orchitis in, *283, 288, 290*
  perfusion of, 273
  scrotum and, 272, 274, *276, 288–289*
  varicocele in, *298*
Epiphora, investigation for, 383
Epiphrenic diverticulae, *35*
ERPF. *See* Effective renal plasma flow
Erythrocytes. *See* Red blood cells
Esophagus
  achalasia, *19–21*
  aperistalsis in, *20*
  Barrett's, 4, *36, 159*
  cancer of, 5, *38*
  DES, 4t, *22–23*
  diverticulae, 5, 7, *33*
  esophagectomy and, *38, 40*
  esophagitis and, 4, *38, 40*
  esophagogram and, 5–6
  I-131 in, *158–159*

motility disorders, 5–6, *25, 30*
nutcracker type, *24*
pharyngeal regurgitation in, 5
segments of, 3, *22, 28*
sphincter, lower, 6–7
stenosis, diagnosis of, *36*
External beam radiation, 74, 83
Extra-adrenal lesions, 72, 79
Eyes, sequential images of, 384

F

Fallot's tetralogy, 333, 335–336, 339, *356*
False-negative scans, 79, 82. *See also* specific images
Felty's syndrome, *322*
Femoral vein
deep, 242, *263*
normal, 240–241
occult duplication of, *269*
RBC venogram of, *263, 268*
thrombosis of, *255–256, 263*
valves in, *254*
Femur
AVN of, 360, 363–364, *378*
blood supply to, 272, *279*
I-131 imaging in, *168*
Fertility, testicular tortion and, *284*
Fine needle aspiration
advantages of, 200
false negatives in, 199
false positives in, 199
limitations of, 196
for lymphocele, *235*
preparation for, 194
radiation therapy and, 200
sampling procedure of, 194, *195*
techniques in, 194, *195*
thyroid nodules and, 193–200
Fistulae. *See* specific types
Flare sign, in lymph nodes, *411*
Fluorescent thyroid scanning (FTS), 176–183, *184–192*
adenomas, *187–188*
artifacts in, *192*
cold nodules and, 179
colloid cyst, *187*
devices used, *185*
goiter and, *190*
hypothyroidism and, 180–181, *191*
iodine flooding and, 179
in normal patients, *186*
pitfalls in, 182
photon flux, 177
thyroiditis and, 180
Flush aortography, 73
Follicular adenoma, 198
Fractures, pathological, *103*
FTS. *See* Fluorescent thyroid scanning
Full width half maximum (FWHM), 177
Fungal infection, *174*

G

Gallbladder
ACI radioactivity, 56
carcinoma, metastatic, *48*
HB imaging in, 43t
hydrops of, *47*
imaging, 41, *70*
isotope transit time, 42t
normal cholescintigram in, *46*
Tc-99 radiation in, 42t
Gallium markers, 301, 303, *310*
Gamma ray camera
in AG imaging, 54
in DCG, 383–384
fluorescent imaging, 176
as imaging device, 145
for parathyroid scanning, 117
PRL studies, 397
radiation measurement, 147
RNA and, 333
in SPECT, 75, 361
TI-201 methods, 338
variation methods, 334
in WBC imaging, 300
Ganglioneuroblastomas, 74
Gastrointestinal (GI) tract
achalasia and, *21*
adenocarcinomas and, *172*
DES and, *22*
diagnosis of, *37*
emptying time, 3, 7, *11–12, 30*
esophageal motility disorder, *25*
evaluation of, 1, *38*
fundus, 2–3, 13
in HB imaging, 43t
hiatus hernia and, *30*
study of, 3, 7, *37*
tracer in, 1, 42t
Gated blood pool imaging (GBPI)
abnormalities and, 337
angiocardiography of, *353*
computer analysis of, 336–337
exercise and, 338
in Kawasaki syndrome, 341
normal, 336, *353*
PNC studies, 336–338
GBPI. *See* Gated blood pool imaging
Geiger-Muller counter, 331
Geriatric patients, RETS and, *35*

Giant cells, in thyroid nodule, 197
Glenn shunt, 333, *349*
Glomerular filteration function, *203*
Goiter. *See also* Thyroid
  cytology of, 198
  fluorescent scanning, 181, 190
  Hürthle's cells and, 198
  types of, 178, 198
Grave's disease, 197
Groin shield, *279*
Groin trauma, *288*

H
Hanning filter, 362
Hashimoto's thyroiditis, 180, 197–199
Heart. *See also* Cardiovascular system; specific disorders
  attenuation of, 2
  innervation of, 77
  MIGB uptake in, 76, 79, 84, *112*
  normal scintigram of, 42, *88*
  radiation and, 77
  RNA and, 332
Heartburn, *30*
Heller's myotomy, *20–21*
Hematocele, *205, 212, 237, 293*
Hemiscrotum, 274
  epididymitis and, *289*
  inguinal hernia and, *294*
  RNI of, *282, 284*
  in spermatic cord torsion, *280*
  in testicular abscesses, *292*
  in varicocele, *298*
Hemithyroidectomy, *137*
Hemocytometer, 196
Hemorrhagic cyst, *330*
Hepatic abscess, *312*
Hepatic-to-cardiac ratio, *51*
Hepatic uptake curves, 42
Hepatitis, neonatal, *49*
Hepato-biliary imaging, 41–44, *45–51*. *See also* Gallbladder; Liver
  hepatic duct in, 43t
  radioisotopes in, 42
Hepatocytes, 42, *51*
Hernia hiatus, 3, 5, 7, *30–31, 294*
High pressure liquid chromatography (HPLC), 72
Hip(s), SPECT in, 360–363
Hippuran, RTS, 202, *208, 226*
Histograms, 81
  pulmonary, *349*
  in quantitative RNA, *334*
  time-activity, 3, *12*
Hodgkin's disease, 199, *413*
Hormonogenesis, 181
Horner's syndrome, 77, *111*
Hot nodules, scanning of, 180
Hürthle cells, 197–198
Hydroceles, in epididymitis, *290–291, 295*
Hydrops, *47*
Hyperandrogenism, 53–54
  adrenal adenoma and, *66*
  ovary and, *67*
  types of, 57t
Hypercalcemia, thyroid and, 115
Hyperchloresterolemia, 55t, 58, *69*
Hypercortisolism, 52–53
Hypertension, *65,* 72
Hyperthyroidism. *See also* Thyroid
  evaluation of, 178
  familial, 115–116
  fluorescent scan, 181, *189, 191*
  MEN-1 syndrome, 116
  primary, 115–116
  radioiodine therapy, 178
  secondary, 116
  in subacute thyroiditis, 116
  subtotal parathyroidectomy, 116
Hypoglycemia, MIGB uptake and, 76
Hypokalemia, 52, 53t, *65*
Hypopharynx. *See also* Pharynx
  emptying of, 2–3, *16–17*
  radionuclide in, *14, 17, 27*
  normal radioactivity values, 4t
  time-activity curves and, 3, 5
Hypothyroidism, scanning and, 180, *189*

I
Idiopathicsubaortic stenosis, *353*
Iliac arteries, *291*
Iliac lymph nodes, 398
Iliac veins
  compression of, *265*
  DVT of, 241, *259*
  external, 240, *265*
  RBC venography, *265–266*
  scrotum and, 272, *279*
  thrombosis and, *247*
  variation in, 241
Immunohistochemicals, 83
Incubation studies, 76
Indium-111 imaging, 299–300. *See also* specific disorders, methods
  abscesses, 301, *312, 317*
  algorithm for, *310*
  catheter injection in, *329*
  gallbladder in, *316*
  GI tract in, 300, *314, 319–320, 323, 326*
  hemorrhage and, *328, 330*

injection sites, *327*
kidney in, *318*
liver in, *312*
lung in, *321*
lymph node uptake, *323*
oxine in, 299–300, *305*
pancreas and, *313, 318–319*
patient preparation, 302t
potential problems, 303
tube sites, *327*
utility of, 301
vascular grafts, *322*
veins in, *325, 328, 330*
wound abscesses, *315*
Inferior vena cava, 240, *346*
Inguinal hernia, *294*
Intra-cardiac baffle, 333
Intravenous urography, 73, 79
Iodine imaging. *See also* specific disorders, methods
beta emission, 123, 146
diet and, 144
dose of, *153*
etiologies of, 148
flooding and, 179
gamma emissions, 146
goiters and, 181
hippuran images, 201–202, *207, 230*
iliac crests, *169*
indirect measurement of, 147
metastatic uptake and, *165*
neck and chest of, 147
normal I-111 scans, *311*
oral administration, 147
patient preparation in, 144
radiation dose of, 146–147
scintigraphy and, 144–149, *150–175*
sodium iodide liquid, 119, 145
thyroid uptake and, 74, 179–180
WBC of, *324*
Ischemic bowel disease, *326*
Islet cell carcinoma, 82t

K

K/A ratio. *See* Kidney-aortic (KA) ratio
Kidney. *See also* Kidney-aortic (KA) ratio
in acute tubular necrosis, *214*
autotransplantation of, *95*
photopenic defect in, *204*
RAD in, 77, *207*
simultaneous imaging with MIGB, 75
Kidney-aortic (KA) ratio
in acute tubular necrosis, *214, 217, 219*
Kawasaki syndrome, 340–341, *359*
in kidney activity, *203*
in RTS, 202, *203, 210*
in transplant rejection, *221–225*
Kinetic analysis, of radiation, 179
Kinevac, in gallbladder imaging, 41
Knees
image of artifacts, *367*
normal, *369*
osteoarthritis, *381*
SPECT in, 360
K rays, 177–179
Krypton, in RETS, 4

L

Lacrimal drainage apparatus (LDA), 383–385, *386–395*
anatomy of, *387*
blockade in, *388*
clearance time, 384
DCT in, *393*
duct obstruction of, 383, *393*
I-123 MIGB uptake in, 78
in normal DCG, 384
radioactive marker in, 384
Lactation, thyroid examination and, 178
Laxatives, 55, *61*
LDA. *See* Lacrimal drainage apparatus
Leading-edge transit time, 42
Left coronary artery, *354, 358*
Left ventricle
angiocardiogram in, *252*
aneurysm in, *354*
aortic stenosis in, 336
dysfunctions of, 335–336, 340
LVEF, 335
myocardial perfusion, 338
normal angiogram, *345*
pressure in, 339
time-activity curve, 335
Legs. *See also* Deep vein thrombosis (DVT); Femur; Knees
lymphatics in, *398*
normal venous system of, 240
PRL of, *406*
visualization of deep veins, 240
Leucocytes. *See also* Indium-111 imaging; specific types
buttons, 299, *305*
donor cells and, 302
labeling, 299–300, *305*
leucopenia, 300
localization of, 301
migration of, 301, 303
Line spread functions (LSF), 177
Lipoproteins, 55, 58
Liver
infection in, *312*
In-111 imaging of, 301, *311*

Liver (*continued*)
lymphography of, 396
metastases in, *103, 157*
MIGB radiation in, 77
normal images of, *88*
Tc-99 radiation in, 42t
uptake of I-131, *157*
Low back pain (LBP), 360, 363
Low-density lipoproteins (LDL), 55
Lower dysphagia, RETS in, *32*
Lower esophageal sphincter pressure, 7
Luer-Lok syringes, 194
Lugol's solution, 54, 74, 179
Lumbar spine, SPECT in, 360, 363
Lungs. *See also* Pulmonary system
adenocarcinoma in, *171*
DVT scan, 243
metastases in, *103*
miliary nodules in, *166*
RNA in, 332, 334
Lymphatic system. *See also* Lymphoscintigraphy
abnormalities in, *12,* 398
channels in, 399
dissection and, *407*
lymphangiectasia, 398–399, *409*
lymphangioma, 398, *409*
lymphedema in, 397–398, *404*
lymphocele in, *234–236*
lymphocytes, 197, 300, 303
lymphomas, 199, *256*
metastases in, *104, 161,* 197, 399–400, *412*
obstruction in, 398, *407–408*
parathyroid and, 122
PRL and, 396, *409*
symptography of, 396
thyroid and, 197
Lymphedema, 397–398, *404*
Lymphoceles, *234–236*
Lymphoscintigraphy
in chylothorax, 399
of intestinal lymphangiectasia, 399
iodine flooding in, *186*
lymphatic disruption in, 398
normal, 397, *402*
in pediatrics, 396–401, *402–413*
positive contact, 400
radiopharmaceuticals in, 396
of secondary lymphedema, 398, *404*

M

Macrophages, 197
Magnetic resonance imaging (MRI), 122, 363–364
Manometry, esophageal, 6, *15, 25, 28*
Marinelli formula, 147
Markers. *See* specific types
Marrow. *See* Bone marrow
May-Grunwald-Giemsa (MGG) stain, 196
Meckel cell carcinoma, 82
Mediastinum
I-131 in, *158, 160, 163*
lymph nodes, *163*
parathyroid adenoma and, 122
Medical data systems (MDS), 117
Medical internal radiation dose (MIRD), 77, 147
Medullary carcioma, thyroid, 198
Melanomas, MIGB and, 82
Meniscus, of knee, 364, *380*
Menstruation, RTS in, *213*
MEN syndromes, 78–79, *106,* 115, *137*
Metabolic alkalosis, 52
Metabolic inhibitors, 55t
Meta-iodobenzylguanidine (MIGB) methods
abnormalities in, 79
adrenal medulla and, *91*
analysis in, 75
cardiac uptake in, *112*
cervical paraganglioma, *100*
dose in, 74, 78
imaging procedures in, 74
metabolism and, 76
metastases and, *92*
nervous system and, 84
neuroblastoma and, 81t, *108*
normal appearance in, 77–78, *88*
pheochromocytoma, 73, 75, 79–80, *93*
radiation dose in, 74, 77t, 78, 79
radiolabels for, 75
SPECT and, *107*
techniques in, 74, 84
Metastases. *See also* specific types
blood pool and, *155*
carcinoid tumor in, *110*
gallbladder and, *48*
hepatic, *48, 103*
I-131 uptake in, *163*
lymphatic, 399–400
MIGB methods, *92*
neuroblastoma, *108*
nodal, *411*
pheochromocytoma, 75, *102*
PRL and, *411*
pulmonary, *103, 165–166*
thyroid and, *48, 171,* 176t
MGG stain. *See* May-Grunwald-Geimsa (MGG) stain
MIGB. *See* Meta-iodobenzylguanidine (MIGB) methods
MIRD. *See* Medical internal radiation dose
Mitral stenosis, 336
M-mode echocardiography, 339
Monocytes, in WBC imaging, 303

Motility disorders. *See* specific disorders; site
Mouth, 2, *17*
MRI. *See* Magnetic resonance imaging
Müllerian duct, of testis, 271
Multinodular goiter, *142*
Multiphasic ingestion, 3
Multiple endocrine neoplasia, *90*
Muscles
  dystrophy of, 7
  hyperemia of, 283
  RBC scintigraphy of, 240
  thrombosis in, 242
Mustard procedure, 333, *348*
Myasthenia gravis, *27*
Mycotic aneurysm, *325*
Myocardium. *See also* Cardiovascular system
  abscesses in, *313*
  in children, 338–341
  congenital defects in, 340
  infarction, 340
  ischemia in, 340
  MIGB imaging and, 75
  myocarditis in, 338, *355*
  myotonia in, *18*
  neonate functions, 340
  thallium imaging in, 338, 340
Myxedema, idiopathic, 197

N

Nasopharynx. *See also* Pharynx
  I-131 uptake, *154*
  MIGB and, 77, *93*
Neck, scintigraphy
  false positives in, *33*
  interpretation of, 147
  I-131 radioactivity in, *158*
  RADs in, 147
Needle aspiration. *See* Fine needle aspiration
Nephrolithiasis, 115
Nephrotomography, 73
Neuroblastomas
  chemotherapy for, 83
  metastases in, *108*
  MIGB and, 72, 74, 81t, *109*
  radiation for, 83
  therapy in, 84, *108*
Neuroectodermal syndromes, *97*
Neuroendocrine tumors, 78–82, 84
Neurofibromatosis
  lymphedema and, 398
  MIGB and, 79, 80t
  pheochromocytomas and, *96*
  PRL and, *406*
Neurosecretory granules, 81
Neutron activation analysis, 181
Neutrophil labeling, 300
Nitroprusside, 74
Non-Hodgkin's lymphomas, 199
Nonne-Milroy disease, 398, *405*
Noonan's syndrome, 398
Norepinephrine, 74, 76, *88*
NP-59 methods, 56t, *62–63*
Nubbin sign, 273, *280*
Nuclear magnetic resonance (NMR), 73
Nutcracker esophagus, *24*

O

Oat cell carcinoma, 82t
Obesity, in Cushing's syndrome, 53t
Oculopharyngeal muscular dystrophy (OMD), *15–17*
Oddi's sphincter, 46, *50*
Odynophagia, 3
Oliguria, DTPA study, *237*
Opthalmology, *385*
Oral cavity, 3, 4t, *13*
Oral contraceptives, 55t
Orchitis, RNI in, *291*
Oropharynx, 1, *27, 154, 158*
Orthopedists, SPECT and, 360–365, *366–382*
Osteoarthritis, 363, *381*
Osteomyelitis, WBC imaging, *317*
Ovary
  androgens and, 54
  MIGB radiation in, 77
  RAD in, *307*
  Tc-99 radiation in, 42t
  tumors in, 57, *67*

P

Pancreas
  carcinoma in, *47*
  WBC imaging, *313, 318–319*
Papillae
  carcinoma thyroid and, 198
  meningioma and, *174*
  metastases from, *163*
Paracolic abscesses, *316*
Paradidymis, *286*
Paraesophagus. *See also* Esophagus
  hernia in, 7, *32*
  linear focus of activity in, *38*
  paraganglioma, 72, 81, 82t, *100*
Parathyroid, 115–123, *124–143*
  adenoma of, 115, 119, *138*
  carcinoma of, 115
  cyst of, 196–197
  ectopics, 120, *134, 136, 140*
  hormone in, 115, 197

Parotid gland, MIGB visualization of, 77, *88, 93, 111*
Particle body imaging, 341–342
Patella, *369*
Patent ductus arteriosus (PDA), 334
Pediatric nuclear cardiology (PNC), 331–342, *343–359*
Pediatric radionuclide lymphography (PRL), 396–401, *402–411*
  in Burkitt's lymphoma, *412*
  in chylothorax, *410*
  in elephantiasis, *406*
  in Hodgkin's disease, *413*
  lymphatics in, *404, 407–409*
  metastases in, *411*
  for neurofibromatosis, *406*
  in Nonne-Milroy disease, *405*
  normal, *403*
  radiation absorbed dose in, 397
Pelvis
  abscesses in, *315–316*
  bone and, *169*
  irritation, *408*
  lymphatic flow in, 398, *405*
  struma ovarii and, *174*
  venography and, *256, 266*
Penis, *279–280*
Peptic esophagitis, 5, *19, 25, 30, 38*
Peptic hormones, tumors and, 81
Perirenal hyperemia, *212, 237*
Peristalsis, *15, 19, 27*
Peritoneum, WBC imaging, *314*
Perivascular lesions, 242
Pharynx, 2. *See also* Hypopharynx
  diverticulum and, *29–30*
  I-131 uptake, *154*
  MIGB and, 77, *93*
  nasal regurgitations and, 7, *17*
Phenoxybenzamine, 73
Phenylpropanolamine, 76, 78
Pheochromocytoma, 72–84, *85–105*
  embryology of, 72
  intra-adrenal, 79, *93–95*
  malignant, 72, 83, *101–103*
  metastases in, 74–75, *103*
  in neurofibromatosis, *96*
  therapy of, 73–74
Phlebitis, 242, *250*
Pituitary adenoma, *62,* 115
Planar bone imaging, SPECT for, 360, 363
Plasma, *63,* 72, *95,* 299, *305*
Platelets, in WBC imaging, 84, 303
Pleural effusions, WBC imaging of, *321*
Pneumatic dilation, 7, *21*
Pneumonia, WBC imaging of, *321*
Polychromatic beams, 177
Polycystic ovary syndrome, 54
Polymorphonuclear leukocytes, 300
Popliteal vein, 240–241, *264, 267*
Potassium perchlorate, 74, 332
Pregnancy
  I-131 use in, 145
  MI and, 339
  thyroid examination in, 178
Primary aldosteronism, 52–57
PRL. *See* Pediatric radionuclide lymphography
Profunda femoris, 240
Progressive systemic sclerosis, *28*
Prostate, carcinoma of, *67*
Protein, iodination of, 177
Provodine-iodine, 194
Psammoma bodies, 198
Pseudoarthritis, SPECT in, 363
Pudendal vessels, 272–273, *278–279*
Pulmonary system, 301
  arteries of, 332–334, *345*
  aspirations, *17*
  embolization in, 242, *270*
  fibrosis, *165–166*
  hypertension, 335–339
  venous obstruction in, 399
Pyriform sinus, *14*

R

RAD. *See* Radiation absorbed dose
Radiation absorbed dose (RAD), 84
  in AG imaging, 54
  clearance curves, *153*
  estimation of, 55, 178
  GBPI, 337
  for I-123 MIGB, 75
  for I-131, 146–147, 337
  MIGB and, 75
  in parathyroid scanning, 118
  PNC and, 332t
  in RTS, *207*
  in SPECT, 362
  for WBC labeling, 300
Radiation pneumonitis, *165–166*
Radiocolloid statis, in OMD, *16*
Radioenzymatic assay, 81, 115
Radionuclide angiocardiography (RNA), 271–275, *276–298*
Radionuclide dilution curve, 333
Radionuclide esophageal transit study (RETS), 1–8, *9–40*
Radiopharmaceuticals. *See* specific types
Raynaud's disease, RETS in, *18*
RBC. *See* Red blood cells
Rectilinear scanner, 145, *151*
Red blood cells (RBC), 75, 239, 242, 303, *305*
Red marrow, radiation in, 42t

Reed-Sternberg cells, 199
Reflux esophagitis, *28*
Region of interest (ROI)
  in parathyroid esophagitis, 117, *127*
  in RETS, *11*
  of thyroid, 118
  time-activity curves and, *129*
Regitine, as adrenal blocker, 74
Regurgitation, 7, *25, 29–30, 33*
Renal excretion, *51*
Renal transplant scintigraphy (RTS), 201–202, *203–238*
  abscess in, *206*
  arteriovenous fistula in, *204, 213*
  collecting system in, *204–206, 228*
  DTPA in, *204–206*
  excretory index of, 202
  flow in, *204–206*
  glomerulonephritis in, *206*
  hematoma in, *206*
  hippuran in, 202, *204–206*
  I-131 images in, *210*
  implications of, *204–205*
  infarcts, *204, 206, 237*
  lymphocele, *206, 234*
  pyelonephritis, *206*
  RAD in, *207*
  rejection in, *206, 221–226*
  renal vessels and, *204, 237*
  thrombosis, *206*
  tubular necrosis, *206, 214–220*
  urinoma, *204, 206, 232*
  vascular blush in, *238*
  wound dehiscence in, *204*
Renin, release of, 52
Renograms. *See* Renal transplant scintigraphy (RTS)
Reserpine alkaloids, 76
Reticuloendothelial cells, 396
Retroperitoneal metastases, *104*
RETS. *See* Radionuclide esophageal transit study
Rheumatic carditis, 335
Right subclavian vein hematoma, *328*
Right ventricle
  in cystic fibrosis, 335
  dysfunction of, 335–336, *345, 356*
  ejection fraction, *352*
  hypertrophy, 335
  normal angiogram of, *345*
  pressure in, 339
  RVEF in, *352*
  TI-201 in, 339, *356*
  ventricular ratio in, 339
ROI. *See* Region of interest
R-to-R interval, 337
RTS. *See* Renal transplant scintigraphy (RTS)
R-wave, of ECG, 336

S
Sacrum, 363, *377*
Salivary glands. *See also* specific type
  anti-depressants and, *111*
  I-131 uptake and, *154, 174*
  MIGB in, 76–78, 84
  in parathyroid imaging, 120
  ROI in, *127*
Saphenous veins, in DVT, 240
Schwannoma, MIGB and, 82t
Scintigraphy. *See* specific organs, methods
Scleroderma, 1, 4, *18*
Scrotum, 271–275, *276–298*
  epididymitis and, *290*
  hematocele of, *293*
  hyperemia, *283–284*
  perfusion of, 273
  rugae destruction, *293*
  ultrasound use in, 271
Secondary sex characteristics, 53
Segmental esophagus, 5, *12*
Shunt procedures, 332–342, *343–359*
  blood flow in, 334
  evaluation of, *349, 359*
  left-to-right, 333, *349–350*
  level of, 333–335, *350*
  palliative, *349*
  quantification of, 333–334, *349–350*
  right-to-left, 332–333, 341–342, *359*
  RNA and, 342
Sickle cell crisis, 335
Sinacalide, 41, *50*
Single photon emission computed tomography (SPECT). *See also* specific organs; pathologies
  arthroscopy and, 364, *369, 380*
  artifacts in, 362, *366, 369, 371, 377*
  AVN methods, 360–361
  bladder-filling in, *371*
  blood pool imaging in, 361
  computer acquisition for, 362
  for hips, 360
  histogram analysis of, *378*
  I-123 uptake in, 78, *98*
  imaging procedures, 361
  in meniscus tears, 364
  MIGB and, 78, 98, *107*
  normal lumbar spine, *367*
  in oncology, 361
  in osteoarthritis of knee, 363, *381*
  for parathyroid, 122
  of patella, *370*
  in pheochromocytoma, 79
  pseudoarthritis and, 363, *375*
  radiation in, 362–365
  sacroiliac joint and, *377*

Single photon emission computed tomography (SPECT) (*continued*)
sacrum in, 363, *377*
spinal fusion and, *376*
in spondylolisthesis, 363
spondylosis and, *372, 374*
TMJ and, 364, *368, 382*
Sinusitis, imaging for, *329*
Skeletal systems. *See* specific locations; methods
Skull, imaging and, *93*
Small intestine, Tc-99 in, 42t
Soleus, 240, *246*
Sonography. *See* Ultrasonography
SPECT. *See* Single photon emission computed tomography
Spectrophotometry, 177, 181
Spermatic chord
blood supply of, 272, *278*
in epididymitis, *290*
missed torsion and, *283*
RNI of, 273, *279*
testicular abscess and, *292*
torsion of, 271, 274, *280*
Spinal fusion, SPECT and, *376*
Spironolactone, 58
Spleen
In-111 absorbtion by, 301, 303, *311, 328*
MIGB in, 76–78, 84
normal scintigram of, *88*
simultaneous imaging, 75
Spondylolisthesis, 363
Spondylosis, *372–375*
SSKI. *See* Super-saturated potassium iodide
Star pattern, radioactivity and, 146, *151*
Static imaging, 2–3, *13, 33*
Steinert's disease, *18*
Stress perfusion imaging, 341
Struma ovarii, *174*
Subcellular tissue fractionation, 76
Subclavian vein infection, *40, 325*
Submandibular glands, 77, 83
Sulfur, 2, 4
Superior vena cava (SVC)
angiography of, *345*
nerve phlebitis, *264*
normal veins, 241, *264*
obstruction of, 332
radioactivity in, 334
thrombosis of, 241–242, *264*
Super-saturated potassium iodide (SSKI), 54
Swallowing. *See* Dysphagia
Sympathectomy, 77, *111*
Sympathetic nervous system, MIGB and, 76, 78–79, 84
Sympathoadrenal imaging, 72–84, *85–114*
Sympathoblasts, primitive, 74
Systematic-pulmonary shunting, 341
Systemic lupus erythematosus, *18*

T

Technitium-99m (Tc-99m) scintigraphy. *See also* Venography
acute tubular necrosis, *214*
antimony trisulphide colloid in, 396–397
collecting system obstruction, *228*
colloid scan, *103*
computer acquisition in, 202
DCG in, 383–385, *386–395*
DMSA, 79
HSA labeling, 336
imaging study in, *95,* 202
iminodiacetic acid and, 41–42, 43t
indium and, 303
iodine and, 273
kidney images, 79
lymphocele and, *236*
macro-aggregated albumin and, *359*
neuroblastomas, 74
particle body imaging, 342
RADS in, *207,* 240
RBC and, 242, 336
renal artery thrombosis, *212*
scrotum imaging, 271–275, *276–298*
SPECT and, 361
thyroid images with, *159*
transplant rejection, *221–225*
urinary obstruction and, *227*
urinoma in, *229*
Temporomandibular joint (TMJ), 360, 364, *382*
Testes
abnormalities of, 272–273
abscess in, *292*
anatomy of, *276–285*
androgens and, 54
appendages of, *285–286, 288, 290*
atrophy of, 274
bell-clapper deformity, *276*
blood supply to, 272
carcinoma of, *296*
development of, 271–272
epididymitis and, *289–290*
groin shield and, 272
hydrocele of, *290*
infarction of, *284*
ischemia, *280*
migration of, 271–272
normal visualization of, *279*
pain in, 271–275, *276–298*
RADs in, *207*
RNI of, *279–292*

Tc-99 in, 42t, 271, 273
testis redux in, 272, *279–280*
torsion of, *268, 270,* 272, *280–282*
Tetralogy of Fallot, 333, 335–336, 339, *356*
Thallium imaging, 116, 179, 335–391
Thermography, for cold nodules, 179
Thrombophlebitis
in femoral DVT, *253*
In-111 WBC imaging, *330*
in pulmonary embolization, 243
Tc-99m RBC in, 241
Thrombosis. *See* Deep vein thrombosis (DVT)
Thymus, I-131 imaging of, *160*
Thyroglossal duct, remnant of, *161*
Thyroid. *See also* Goiter
adenoma, *139*
bed of, *155–156, 167*
benign nodules of, 179, 196
biopsy technique for, 193
carcinoma of, 82, 84, *105,* 141, *164, 172–173,* 198–199
colloid cyst, 180
cystic lesions, 196–197
cytology of, 193–200
fluorescent scanning of, 176–183, *184–192*
hyperthyroidism, *189*
I-131 images of, *160*
intra-thyroid content, 177, 180, 191
metastases, 146, *169–170, 179,* 180
MIGB in, 77–78, 93
nodules of, 179, *187*
RAD in, 178
radioiodine uptake and, 54, 56, 82–84, 179
releasing hormone (TRH) in, 144
scrotum and, 272
smear of, 196–197
stimulating hormone (TSH) in, 74, 145, *155,* 176
thyroiditis, 178, 180, *189,* 197
thyroxine, 74, 144, 200
thyrotoxicosis, 178, 181
Time-activity curves
in achalasia, *21*
of hypopharynx, *14*
pulmonary, 334
in quantitative RNA, 333
RETS and, 1–12, 201, *210*
thyroid and, 129
Tomography. *See* Computer tomography
Trachea
aspiration in, 5, *17*
fistula in, 6, *35*
RETS in, 6
tracheostomy, *162, 329*
Tracheostomy tube, as artifact, *162*
Tricyclic compounds, 76, 78, *111*
Tubules, renal, *203, 214, 220*
Tunica albuginea, *293*
Tunica vaginalis, 271–272, *276*
Turner's syndrome, 398

U

Ultrasonography
abscessses and, *310*
in adrenal glands, 57, 68
In-111 labeling and, 303
in lymph node malignancy, 400
for lymphocele, *235*
methods of, 121–122
in MIGB, 75, 79
in neuroblastomas, 74
for neuroendocrine tumors, 81
in pheochromocytoma, 73
RNA and, 333
in testis, *292–296*
Ureter
collecting system obstruction, *228*
partial obstruction of, *227*
reflux of urine into, *234*
RTS and, *205*
urinoma, *229–231*
Urinary bladder
imaging of, 78, *226, 368*
in lymphocele, *234*
MIGB rejection in, *226*
normal scintigram of, *88*
RADs in, 77, *207*
in RNI scrotum, *279*
SPECT in, *368*
Tc-99 radiation in, 42t
in testicular torsion, *282*
Urine
abnormalities in, *233*
aldosterone excretion in, *65*
corticosteroids, *63*
in Cushing's syndrome, 53t
iodine in, 179
menustruation and, *213*
MIGB excretion and, 77–78
obstruction in, *203, 227*
in renal transplant, *229*
steroids in, *62*
urinoma and, *205, 229–231*

V

Valleculae sinus, *14*
Valsalva-type maneuvers, 333
Varicocele, *297–298*

Vascular graft, WBC imaging, *322*
Vas deferens, *276, 285–286*
Vasectomy, RNI after, *279*
Veins, thrombosis of, 239–243, *244–270. See also* specific types
Vena cava, *97, 213*
Venography, 73, 83, 239–243, *244–270. See also* Technitium-99m
  abnormal, 241
  artifacts in, 241, *245*
  in Baker's cyst, *267*
  calf and, *251–252*
  in cellulitis, *264*
  contrast venography and, *264*
  DVT and, 242, *261*
  extrinsic venous compression and, *256*
  false-negative, *262*
  femoral vein in, *248, 268–269*
  method of, 239
  in obesity, *246*
  pelvic hematoma, *266*
  peronial vein in, *249*
  popliteal vein, *267*
  in pulmonary embolization, 243
  RAD in, 240
Ventricles. *See* Left ventricle; Right ventricle
Vim-Silverman needles, 193
Virilization, in hyperandogenism, *66*
Visual grading scale, 76
Volume overload, MI and, 339
Von Hippel-Lindau disease, 79–80

W

Warthin's tumor, *174*
Washout, kidney and, *221–225*
Waterston shunt, 333
White blood cells (WBC). *See* Indium-111 imaging; Leucocytes
Wilson cloud chamber, 331
Wolffian duct, of testis, 271
Wolft-Chaikoff effect, 179
Wounds, *232,* 301, *315*
Wright's stain, 196

X

Xylocaine, in thyroid, 194
Xyphoid, visualization of, *159*

Z

Zenker's diverticulum, *29, 159*
Zollinger Ellison syndrome, 115

ISBN 0-397-50789-5

Arboviruses

Molecular Biology, Evolution and Control

Edited by

Nikos Vasilakis

Department of Pathology
Center for Biodefense and Emerging Infectious Diseases
Center of Tropical Diseases and Institute of Human Infections and Immunity
University of Texas Medical Branch
Galveston, TX
USA

and

Duane J. Gubler

Program on Emerging Infectious Diseases
Duke-NUS Graduate Medical School Singapore
Singapore

Caister Academic Press

Caister Academic Press
Norfolk, UK

www.caister.com

British Library Cataloguing-in-Publication Data
A catalogue record for this book is available from the British Library

ISBN: 978-1-910190-21-0 (paperback)
ISBN: 978-1-910190-22-7 (ebook)

Cover design adapted from an image designed by Shawna Machado.

Contents

	Contributors	v
	Foreword	ix
	Preface	xi
1	The Arboviruses: Quo Vadis? Duane J. Gubler and Nikos Vasilakis	1
Part I	Molecular Biology	7
2	The Taxonomy of Arboviruses Nicole C. Arrigo, Scott C. Weaver and Charles H. Calisher	9
3	Genomic Organization of Arboviral Families Nikos Vasilakis, Amy Lambert, N. James MacLachlan and Aaron C. Brault	31
4	Host Metabolism and its Contribution in Flavivirus Biogenesis Rushika Perera and Richard J. Kuhn	45
5	Vector-borne Bunyavirus Pathogenesis and Innate Immune Evasion Brian Friedrich, Birte Kalveram and Shannan L. Rossi	61
6	Vector-borne Rhabdoviruses Ivan V. Kuzmin and Peter J. Walker	71
7	Alphavirus–Host Interactions Kate D. Ryman and William B. Klimstra	89
8	Molecular Interactions Between Arboviruses and Insect Vectors: Insects' Immune Responses to Virus Infection Natapong Jupatanakul and George Dimopoulos	107
Part II	Viral Diversity and Evolution	119
9	Genetic Diversity of Arboviruses Kenneth A. Stapleford, Gonzalo Moratorio and Marco Vignuzzi	121
10	Ecological and Epidemiological Factors Influencing Arbovirus Diversity, Evolution and Spread Roy A. Hall, Sonja Hall-Mendelin, Jody Hobson-Peters, Natalie A. Prow and John S. Mackenzie	135
11	Role of Inter- and Intra-host Genetics in Arbovirus Evolution Alexander T. Ciota and Gregory D. Ebel	167
12	Arbovirus Genomics and Metagenomics Adam Fitch, Matthew B. Rogers, Lijia Cui and Elodie Ghedin	175
13	Role of Vertical Transmission in Mosquito-borne Arbovirus Maintenance and Evolution Robert B. Tesh, Bethany G. Bolling and Barry J. Beaty	191

14	The Boundaries of Arboviruses: Complexities Revealed in Their Host Ranges, Virus–Host Interactions and Evolutionary Relationships Goro Kuno	219
Part III	Arbovirus Diagnosis and Control	269
15	Laboratory Diagnosis of Arboviruses Amy J. Lambert and Robert S. Lanciotti	271
16	Conventional Vector Control: Evidence it Controls Arboviruses Scott Ritchie and Gregor Devine	281
17	Biological Control of Arbovirus Vectors Thomas Walker and Steven P. Sinkins	291
18	RNA Interference: a Pathway to Arbovirus Control Kathryn A. Hanley and Christy C. Andrade	303
19	Genetically Modified Vectors for Control of Arboviruses Ken E. Olson and Alexander W.E. Franz	315
20	Arbovirus Vaccines Scott B. Halstead	337
21	Small Molecule Drug Development for Dengue Virus Qing-Yin Wang and Pei-Yong Shi	373
Part IV	Future Trends	383
22	Arbovirology: Back to the Future Robert B. Tesh and Charles H. Calisher	385
	Index	391